BAYESIAN INFERENCE
IN STATISTICAL ANALYSIS

BAYESIAN INFERENCE IN STATISTICAL ANALYSIS

GEORGE E. P. BOX and GEORGE C. TIAO
Department of Statistics
University of Wisconsin

ADDISON-WESLEY PUBLISHING COMPANY
Reading, Massachusetts · Menlo Park, California · London · Don Mills, Ontario

This book is in the

ADDISON-WESLEY SERIES IN BEHAVIORAL SCIENCE: QUANTITATIVE METHODS

Consulting Editor

FREDERICK MOSTELLER

 Printed in the United States of America. Published simultaneously in Canada. Library of Congress Catalog Card No. 78-172804.

To BARBARA, HELEN, and HARRY

PREFACE

The object of this book is to explore the use and relevance of Bayes' theorem to problems such as arise in scientific investigation in which inferences must be made concerning parameter values about which little is known *a priori*.

In Chapter 1 we discuss some important general aspects of the Bayesian approach, including: the role of Bayesian inference in scientific investigation, the choice of prior distributions (and, in particular, of noninformative prior distributions), the problem of nuisance parameters, and the role and relevance of sufficient statistics.

In Chapter 2, as a preliminary to what follows, a number of standard problems concerned with the comparison of location and scale parameters are discussed. Bayesian methods, for the most part well known, are derived there which closely parallel the inferential techniques of sampling theory associated with t-tests, F-tests, Bartlett's test, the analysis of variance, and with regression analysis. These techniques have long proved of value to the practicing statistician and it stands to the credit of sampling theory that it has produced them. It is also encouraging to know that parallel procedures may, with at least equal facility, be derived using Bayes' theorem. Now, practical employment of such techniques has uncovered further inferential problems, and attempts to solve these, using sampling theory, have had only partial success. One of the main objectives of this book, pursued from Chapter 3 onwards, is to study some of these problems from a Bayesian viewpoint. In this we have in mind that the value of Bayesian analysis may perhaps be judged by considering to what extent it supplies insight and sensible solutions for what are known to be awkward problems.

The following are examples of the further problems considered:

1. How can inferences be made in small samples about parameters for which no parsimonious set of sufficient statistics exists?
2. To what extent are inferences about means and variances sensitive to departures from assumptions such as error Normality, and how can such sensitivity be reduced?
3. How should inferences be made about variance components?
4. How and in what circumstances should mean squares be pooled in the analysis of variance?

5. How can information be pooled from several sources when its precision is not exactly known, but can be estimated, as, for example, in the "recovery of interblock information" in the analysis of incomplete block designs?
6. How should data be transformed to produce parsimonious parametrization of the model as well as to increase sensitivity of the analysis?

The main body of the text is an investigation of these and similar questions with appropriate analysis of the mathematical results illustrated with numerical examples. We believe that this (1) provides evidence of the value of the Bayesian approach, (2) offers useful methods for dealing with the important problems specifically considered and (3) equips the reader with techniques which he can apply in the solution of new problems.

There is a continuing commentary throughout concerning the relation of the Bayes results to corresponding sampling theory results. We make no apology for this arrangement. In any scientific discussion alternative views ought to be given proper consideration and appropriate comparisons made. Furthermore, many readers will already be familiar with sampling theory results and perhaps with the resulting problems which have motivated our study.

This book is principally a bringing together of research conducted over the years at Wisconsin and elsewhere in cooperation with other colleagues, in particular David Cox, Norman Draper, David Lund, Wai-Yuan Tan, and Arnold Zellner. A list of the consequent source references employed in each chapter is given at the end of this volume.

An elementary knowledge of probability theory and of standard sampling theory analysis is assumed, and from a mathematical viewpoint, a knowledge of calculus and of matrix algebra. The material forms the basis of a two-semester graduate course in Bayesian inference; we have successfully used earlier drafts for this purpose. Except for perhaps Chapters 8 and 9, much of the material can be taught in an advanced undergraduate course.

We are particularly indebted to Fred Mosteller and James Dickey, who patiently read our manuscript and made many valuable suggestions for its improvement, and to Mukhtar Ali, Irwin Guttman, Bob Miller, and Steve Stigler for helpful comments. We also wish to record our thanks to Biyi Afonja, Yu-Chi Chang, William Cleveland, Larry Haugh, Hiro Kanemasu, David Pack, and John MacGregor for help in checking the final manuscript, to Mary Esser for her patience and care in typing it, and to Greta Ljung and Johannes Ledolter for careful proofreading.

The work has involved a great deal of research which has been supported by the Air Force Office of Scientific Research under Grants AF-AFOSR-1158-66, AF-49(638) 1608 and AF-AFOSR 69-1803, the Office of Naval Research under Contract ONR-N-00014-67-A-0128-0017, the Army Office of Ordnance Research under Contract DA-ARO-D-31-124-G917, the National Science Foundation under Grant GS-2602, and the British Science Research Council.

The manuscript was begun while the authors were visitors at the Graduate School of Business, Harvard University, and we gratefully acknowledge support from the Ford Foundation while we were at that institution. We must also express our gratitude for the hospitality extended to us by the University of Essex in England when the book was nearing completion.

We are grateful to Professor E. S. Pearson and the Biometrika Trustees, to the editors of *Journal of the American Statistical Association* and *Journal of the Royal Statistical Society Series B*, and to our coauthors David Cox, Norman Draper, David Lund, Wai-Yuan Tan, and Arnold Zellner for permission to reprint condensed and adapted forms of various tables and figures from articles listed in the principal source references and general references. We are also grateful to Professor O. L. Davies and to G. Wilkinson of the Imperial Chemical Industries, Ltd., for permission to reproduce adapted forms of Tables 4.2 and 6.3 in *Statistical Methods in Research and Production*, 3rd edition revised, edited by O. L. Davies.

We acknowledge especial indebtedness for support throughout the whole project by the Wisconsin Alumni Research Foundation, and particularly for their making available through the University Research Committee the resources of the Wisconsin Computer Center.

Madison, Wisconsin
August 1972

G.E.P.B.
G.C.T.

CONTENTS

CHAPTER 1

NATURE OF BAYESIAN INFERENCE

1.1 INTRODUCTION AND SUMMARY

Opinion as to the value of Bayes' theorem as a basis for statistical inference has swung between acceptance and rejection since its publication in 1763. During periods when it was thought that alternative arguments supplied a satisfactory foundation for statistical inference Bayesian results were viewed, sometimes condescendingly, as an interesting but mistaken attempt to solve an important problem. When subsequently it was found that initially unsuspected difficulties accompanied the alternatives, interest was rekindled. Bayes' mode of reasoning, finally buried on so many occasions, has recently risen again with astonishing vigor.

In addition to the present growing awareness of possible deficiencies in the alternatives, three further factors account for the revival. First, the work of a number of authors, notably Fisher, Jeffreys, Barnard, Ramsey, De Finetti, Savage, Lindley, Anscombe and Stein, has, although not always directed to that end, helped to clarify and overcome some of the philosophical and practical difficulties.

Second, while other inferential theories had yielded nice solutions in cases where rather special assumptions such as Normality and independence of errors could be made, in other cases, and particularly where no sufficient statistics existed, the solutions were often unsatisfactory and messy. Although it is true that these special assumptions covered a number of situations of scientific interest, it would be idle to pretend that the set of statistical problems whose solution has been or will be needed by the scientific investigator coincides with the set of problems thus amenable to convenient treatment. Data gathering is frequently expensive compared with data analysis. It is sensible then that hard-won data be inspected from many different viewpoints. In the selection of viewpoints, Bayesian methods allow greater emphasis to be given to scientific interest and less to mathematical convenience.

Third, the nice solutions based on the rather special assumptions have been popular for another reason—they were easy to compute. This consideration has much less force now that the desk calculator is no longer the most powerful instrument for executing statistical analysis. Suppose, using a desk calculator, it takes five hours to perform a data analysis appropriate to the assumption that errors are Normal and independent, then the five hundred hours it might take to explore less restrictive assumptions could be prohibitive. By contrast, the use of an electronic computer can so reduce the time base that, with general programs available, the wider analysis can be almost as immediate and economic as the more restricted one.

Scientific investigation uses statistical methods in an iteration in which controlled data gathering and data analysis alternate. Data analysis is a subiteration in which inference from a tentatively entertained model alternates with criticism of the conditional inference by inspection of residuals and other means. Statistical inference is thus only one of the responsibilities of the statistician. It is however an important one. Bayesian inference alone seems to offer the possibility of sufficient flexibility to allow reaction to scientific complexity free from impediment from purely technical limitation.

A prior distribution, which is supposed to represent what is known about unknown parameters before the data is available, plays an important role in Bayesian analysis. Such a distribution can be used to represent prior knowledge or relative ignorance. In problems of scientific inference we would usually, were it possible, like the data "to speak for themselves." Consequently, it is usually appropriate to conduct the analysis as if a state of relative ignorance existed *a priori*. In this book, therefore, extensive use is made of "noninformative" prior distributions and very little of informative priors. The aim is to obtain an inference which would be appropriate for an unprejudiced observer. The understandable uneasiness felt by some statisticians about the use of prior distributions is often associated with the fear that the prior may dominate and distort "what the data are trying to say." We hope to show by the examples in this book that, by careful choice of model structure and appropriate noninformative priors, Bayesian analysis can produce the reverse of what is feared. It can permit the data to comment on dubious aspects of a model in a manner not otherwise possible.

The usefulness of a theory is customarily assessed by tentatively adopting it, and then considering whether its consequences agree with common sense, and whether they provide insight where common sense fails. It was in this spirit that some years ago the authors with others began research in applications of the Bayesian theory of inference. A series of problems were selected in the solution of which difficulties or inconsistencies had been encountered with other approaches. Because Bayesian analysis of these problems has seemed consistently helpful and interesting, we believe it is now appropriate to bring this and other related work together, and to consider its wider aspects.

The objective of this book, therefore, is to explore Bayesian inference in statistical analysis. The book consists of ten chapters. Chapter 1 discusses the role of statistical inference in scientific investigation. In the light of that discussion the nature of Bayesian inference, including the choice of noninformative prior distributions, is considered. The chapter ends with an account of the role and relevance of sufficient statistics, and discusses the problem of nuisance parameters.

In Chapter 2 a number of standard Normal theory inference problems concerning location and scale parameters are considered. Bayes' solutions are given which closely parallel sampling theory techniques† associated with t-tests, F-tests, the analysis of variance and regression analysis. While these procedures have long proved valuable to practising statisticians, efforts to extend them in important directions using non-Bayesian theories have met serious difficulties. An advantage of the Bayes approach is that it can be used to explore the consequences of any type of probability model, without restriction to those having special mathematical forms. Thus, in Chapter 3 the problem of making inferences about location parameters is considered for a wider class of parent probability models of which the Normal distribution is a member. In this framework, we show how it is possible to assess to what extent inferences about location parameters are sensitive to departures from Normality. Further, it is shown how we can use the evidence from the data to make inferences about the form of the parent distributions of the observations. The analysis is extended in Chapter 4 to the problem of comparing variances.

Chapters 5 and 6 discuss various random effect and mixed models associated with hierarchical and cross classification designs. With sampling theory, one experiences a number of difficulties in estimating means and variance components in these models. Notably one encounters problems of negative variance estimates, of eliminating nuisance parameters, of constructing confidence intervals, and of pooling variance estimates. Analysis, from a Bayesian standpoint, is much more tractable, and in particular provides an interesting and sensible solution to the pooling dilemma.

Chapter 7 deals with two further important problems in the analysis of variance. The first concerns the estimation of means in the one-way classification. When it is sensible to regard such means as themselves a sample from a population, the appropriate Bayesian analysis shows that there are then *two* sources of information about the means and appropriately combines them. The chapter ends with a discussion of the recovery of interblock information in the balanced incomplete block design model. This is again a problem in which two sources of information need to be appropriately combined and for which the sampling theory solution is unsatisfactory.

In Chapters 8 and 9 a general treatment of linear and nonlinear Normal multivariate models is given. While Bayesian results associated with standard linear models are discussed, particular attention is given to the problem of estimating common location parameters from several equations. The latter problem is of considerable practical importance, but is difficult to tackle by sampling theory methods, and has not previously received much attention.

† We shall assume in this book that the reader has some familiarity with standard ideas of the sampling theory approach explained for example in Mood and Graybill (1963) and Hogg and Craig (1970).

Finally, in Chapter 10, we consider the important problem of data transformation from a Bayesian viewpoint. The problem is to select a transformation which, so far as possible, achieves Normality, homogeniety of variance, and simplicity of the expectation function in the transformed variate.

A bald statement of a mathematical expression, however correct, frequently fails to produce understanding. Many Bayesian results are of particular interest because they seem to provide a kind of higher intuition. Mathematical results which at first seemed puzzling have later been seen to provide a maturer kind of common sense. For this reason, throughout this book, individual mathematical formulae are carefully analyzed and illustrated with examples and diagrams. Also, appropriate approximations are developed when they provide deeper understanding of a situation, or where they simplify calculation. For the convenience of the reader a number of short summaries of formulas and calculations are given in appropriate places.

1.1.1 The Role of Statistical Methods in Scientific Investigation

Statistical methods are tools of scientific investigation. Scientific investigation is a controlled learning process in which various aspects of a problem are illuminated as the study proceeds. It can be thought of as a major iteration within which secondary iterations occur. The major iteration is that in which a tentative conjecture suggests an experiment, appropriate analysis of the data so generated leads to a modified conjecture, and this in turn leads to a new experiment, and so on. An idealization of this process is seen in Fig. 1.1.1, involving an alternation between *conjecture* and *experiment* carried out via experimental *design* and data *analysis*.† As indicated by the zig-zag line at the bottom of the figure, most investigations involve not one but a number of alternations of this kind.

An efficient investigation is one where convergence to the objective occurs as quickly and unambiguously as possible. A basic determinant of efficiency, which we must suppose is outside the control of the statistician, is the originality, imagination, and subject matter knowledge of the investigator. Apart from this vital determining factor, however, efficiency is decided by the appropriateness and force of the methods of design and analysis employed. In moving from conjecture to experimental data, (D), experiments must be designed which make best use of the experimenter's current state of knowledge and which best illuminate his conjecture. In moving from data to modified conjecture, (A), data must be analyzed so as to accurately present information in a manner which is readily understood by the experimenter.

† The words design and experiment are broadly interpreted here to refer to any data gathering process. In an economic study, a conjecture might lead the investigator to study the functional relationship between money supply and interest rate. The difficult decision as to what types of money supply and interest rate data to use, here constitutes the design. In social studies a particular sample survey might be the experiment.

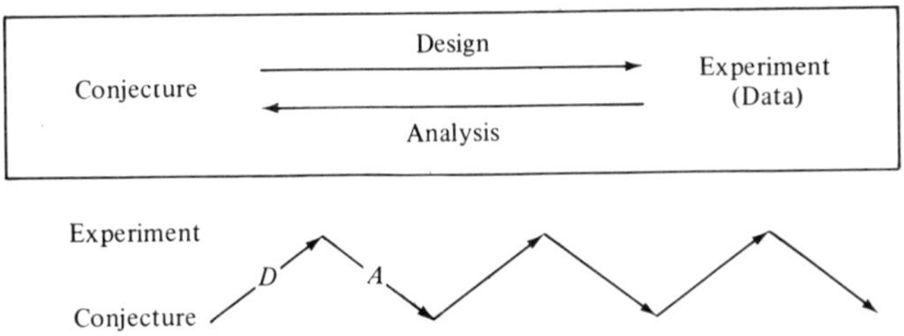

Fig. 1.1.1 Iterative process of scientific investigation (the alternation between conjecture and experiment).

A full treatise on the use of statistical methods in scientific investigation therefore would necessarily include consideration of statistical design as well as statistical analysis. The aims of this book are, however, much more limited. We shall not discuss experimental design, and will be concerned only with one aspect of statistical analysis, namely, *statistical inference*.

1.1.2 Statistical Inference as one Part of Statistical Analysis

For illustration, suppose we were studying the useful life of batteries produced by a particular machine. It might be appropriate to assume tentatively that the observed lives of batteries coming from the machine were distributed independently and Normally about some mean θ with variance σ^2. The probability distribution of a projected sample of n observations $\mathbf{y}' = (y_1, \ldots, y_n)$ would then be

$$p(\mathbf{y} \mid \theta, \sigma^2) \propto \sigma^{-n} \exp\left[-\frac{1}{2\sigma^2}\sum_{t=1}^{n}(y_t - \theta)^2\right], \qquad -\infty < y_t < \infty. \quad (1.1.1)$$

Given the value of the parameters θ and σ^2, this expression permits the calculation of the probability density $p(\mathbf{y} \mid \theta, \sigma^2)$ associated with any *hypothetical* data set $\mathbf{y}$ *before* any data is taken. For statistical analysis this is, in most cases, the converse of what is needed. The analyst already has the data but he does not know θ and σ^2. He can, however, use $p(\mathbf{y} \mid \theta, \sigma^2)$ indirectly to make *inferences* about the values of θ and σ^2, given the n data values.

Two of the methods by which this may be attempted employ

a. Sampling Theory,

b. Bayes' Theorem.

We now give a brief description of each of these approaches using the Normal probability model (1.1.1) for illustration.

Sampling Theory Approach

In this approach inferences are made by directing attention to a reference set of hypothetical data vectors $\mathbf{y}_1, \mathbf{y}_2, \ldots \mathbf{y}_j, \ldots$ which could have been generated by the probability model $p(\mathbf{y} \mid \theta_0, \sigma_0^2)$ of (1.1.1), where θ_0 and σ_0^2 are the hypothetical true values of θ and σ^2. Estimators $\hat{\theta}(\mathbf{y})$ and $\hat{\sigma}^2(\mathbf{y})$, which are functions of the data vector $\mathbf{y}$, are selected. By imagining values $\hat{\theta}(\mathbf{y}_j)$ and $\hat{\sigma}^2(\mathbf{y}_j)$ to be calculated for each hypothetical data vector $\mathbf{y}_j$, reference sets are generated for $\hat{\theta}(\mathbf{y})$ and $\hat{\sigma}^2(\mathbf{y})$. Inferences are then made by comparing the values of $\hat{\theta}(\mathbf{y})$ and $\hat{\sigma}_2(\mathbf{y})$ actually observed with their "sampling distributions" generated by the reference sets.

The functions $\hat{\theta}(\mathbf{y})$ and $\hat{\sigma}^2(\mathbf{y})$ are usually chosen so that the sampling distributions of the extimators $\hat{\theta}(\mathbf{y}_j)$ and $\hat{\sigma}^2(\mathbf{y}_j)$ are, in some sense, concentrated as closely as possible about the true values θ_0 and σ_0. To provide some idea of how far away from the true values the calculated quantities $\hat{\theta}(\mathbf{y})$ and $\hat{\sigma}^2(\mathbf{y})$ might be, *confidence intervals* are calculated. For example, the $1 - \alpha$ confidence interval for θ would be of the form

$$\bar{\theta}_1(\mathbf{y}) < \theta < \bar{\theta}_2(\mathbf{y}),$$

where $\bar{\theta}_1(\mathbf{y})$ and $\bar{\theta}_2(\mathbf{y})$ would be functions of $\mathbf{y}$, chosen so that in repeated sampling the computed confidence intervals included the value θ_0, a proportion $1-\alpha$ of the time.

Bayesian Approach

In a Bayesian approach, a different line is taken. As part of the model a *prior* distribution $p(\theta, \sigma^2)$ is introduced. This is supposed to express a state of knowledge or ignorance about θ and σ^2 before the data are obtained. Given the prior distribution, the probability model $p(\mathbf{y} \mid \theta, \sigma^2)$ and the data $\mathbf{y}$, it is now possible to calculate the probability distribution $p(\theta, \sigma^2 \mid \mathbf{y})$ of θ and σ^2, given the data $\mathbf{y}$. This is called the *posterior* distribution of θ and σ^2. From this distribution inferences about the parameters are made.

1.1.3 The Question of Adequacy of Assumptions

Consider the battery-life example and suppose $n = 20$ observations are available. Then, whichever method of inference is used, *conditional on the assumptions* we can summarize all the information in the 20 data values in terms of inferences about just *two* parameters, θ and σ^2.

The inferences are, in particular, conditional on the adequacy of the probability model in (1.1.1). It is not difficult, however, to imagine situations in which this model, and therefore the associated inferences, could be inadequate. It might happen, for example, that during the period of observation, a quality characteristic x of a chemical additive, used in making the batteries, could vary and could cause, via an approximate linear relationship, a corresponding change in the mean life time of the batteries. In this case, a more appropriate model might be

$$p(\mathbf{y} \mid \mathbf{x}, \sigma^2, \theta_1, \theta_2) \propto \sigma^{-20} \exp\left[-\frac{1}{2\sigma^2} \sum_{t=1}^{20} (y_t - \theta_1 - \theta_2 x_t)^2\right],$$

$$-\infty < y_t < \infty. \qquad (1.1.2)$$

Alternatively, it might be suspected that the first battery of a production run was always faulty, in which case a more adequate model could be

$$p(\mathbf{y} \mid \sigma_1^2, \sigma^2, \theta_1, \theta) \propto \sigma_1^{-1}\sigma^{-19} \exp\left[-\frac{1}{2\sigma_1^2}(y_1 - \theta_1)^2 - \frac{1}{2\sigma^2}\sum_{t=2}^{20}(y_t - \theta)^2\right],$$

$$-\infty < y_t < \infty. \qquad (1.1.3)$$

Again it could happen that successive observations were not distributed independently but followed some time series, or it might be that their distribution was highly non-Normal. The reader will have no difficulty in inventing many other situations that might arise and the probability models that could describe them.

Clearly the inferences which can be made will depend upon which model is selected. Whence it is seen that a basic dilemma exists in all statistical analysis. Such analysis implies the summarizing of information contained in a body of data via a probability model containing a minimum of parameters. We need such a summary to see clearly, and so to make progress, but if the model were inappropriate the summary could distort and exclude relevant information.

1.1.4 An Iterative Process of Model Building in Statistical Analysis

Because we can *never* be sure that a postulated model is entirely appropriate, we must proceed in such a manner that inadequacies can be taken account of and their implications considered as we go along. To do this we must regard statistical analysis, which is a step in the major iteration of Fig. 1.1.1, as itself an iteration. To be on firm ground we must do more than merely postulate a model; we must build and test a tentative model at each stage of the investigation.

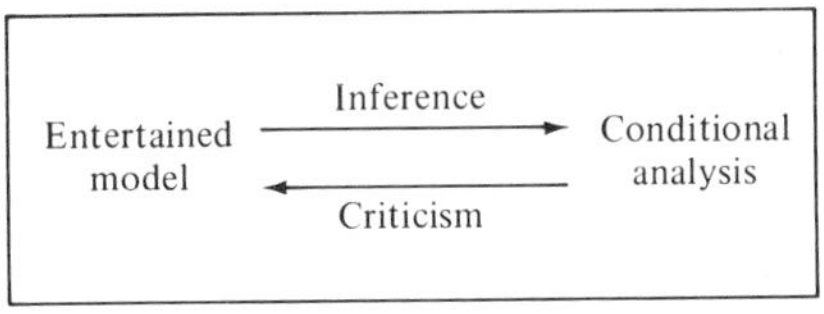

Box & Tiao
1.1-2

Fig. 1.1.2 Statistical analysis of data as an iterative process of model building.

Only when the analyst and the investigator are satisfied that no important fact has been overlooked and that the model is adequate to the purpose, should it be used to further the major iteration. The iterative model building process† taking place *within a statistical analysis* is depicted in Fig. 1.1.2.

The process usually begins by the postulating of a model worthy to be tentatively entertained. The data analyst will have arrived at this tentative model

† A fuller discussion is found, for example, in Box and Jenkins (1970), where in the context of time series analysis the steps in this iteration are discussed in terms of model identification, model fitting, and model diagnostic checking.

in cooperation with the scientific investigator. They will choose it so that, in the light of the then available knowledge, it best takes account of relevant phenomena in the simplest way possible. It will usually contain unknown parameters. Given the data the analyst can now make statistical inferences about the parameters conditional on the correctness of this first tentative model. These inferences form part of the conditional analysis. *If the model is correct*, they provide all there is to know about the problem under study, given the data.

Up to now the analyst has proceeded as if he believed the model absolutely. He now changes his role from tentative sponsor to tentative critic and broadens his analysis with computations throwing light on the question: "Is the model adequate?" Residual quantities are calculated which, while they would contain no information if the tentative model were true, could suggest appropriate modifications if it were false. Resulting speculation as to the appropriateness of the initially postulated model and possible need for modification, again conducted in cooperation with the investigator, may be called model *criticism*.†

For example, suppose the Normal probability model of (1.1.1) was initially postulated, and a sample of 20 successive observations were taken from a production run. These would provide the data from which, conditional on the model, inferences could be made about θ and σ^2.

An effective way of criticizing the adequacy of the assumed model (1.1.1) employs what is called "an analysis of residuals." Suppose for the moment that θ and σ^2 were known; then if the model (1.1.1) were adequate, the quantities $u_1 = (y_1 - \theta)/\sigma, \ldots, u_t = (y_t - \theta)/\sigma, \ldots$ would be a random sample from a Normal distribution with zero mean and unit variance. Such a sequence would by itself be informationless, and is sometimes referred to as white noise. Thus, a check on model adequacy would be provided by inspection of the quantities $y_t - \theta = u_t\sigma,\ t = 1, 2, \ldots$. Any suggestion that these quantities were nonrandom, or that they were related to some other known variable, could provide a hint that the entertained model (1.1.1) should be modified.

In practice, θ would be unknown but we could proceed by substituting the sample mean $\bar{y}$. The resulting quantities $r_t = y_t - \bar{y},\ t = 1, 2, \ldots,$ would for this example be the residuals. If, for example, they seemed to be correlated with the amount of additive x_t, this would suggest that a model like (1.1.2) might be more appropriate. This new model might then be entertained, and the iterative process of Fig. 1.1.2 repeated.

Useful devices for model criticism have been proposed, in particular by Anscombe (1961), Anscombe and Tukey (1963), and Daniel (1959). Many of these involve plotting residuals in various ways. However, these techniques are not part of statistical inference as we choose to consider it, but of model criticism which is an essential adjunct to inference in the adaptive process of data analysis depicted in Fig. 1.1.2.

† This apt term is due to Cuthbert Daniel.

1.1.5 The Role of Bayesian Analysis

The applications of Bayes' theorem which we discuss, therefore, are examples of statistical inference. While inference is only a part of statistical analysis, which is in turn only a part of design and analysis, used in the investigatory iteration, nevertheless it is an important part.

Among different systems of statistical inference, that derived from Bayes' theorem will, we believe, be seen to have properties which make it particularly appropriate to its role in scientific investigation. In particular:

1. Precise assumption introduced on the left in Fig. 1.1.2 leads, via a *leak proof* route, to consequent inference on the right.
2. It follows that, given the model, Bayesian analysis automatically makes use of all the information from the data.
3. It further follows that inferences that are unacceptable *must* come from inappropriate assumption and not from inadequacies of the inferential system. Thus all parts of the model, including the prior distribution, are exposed to appropriate criticism.
4. Because this system of inference may be readily applied to any probability model, much less attention need be given to the mathmetical convenience of the models considered and more to their scientific merit.
5. Awkward problems encountered in sampling theory, concerning choice of estimators and of confidence intervals, do not arise.
6. Bayesian inference provides a satisfactory way of explicitly introducing and keeping track of assumptions about prior knowledge or ignorance. It should be recognized that some prior knowledge is employed in all inferential systems. For example, a sampling theory analysis using (1.1.1) is made, as is a Bayesian analysis, as if it were believed *a priori* that the probability distribution of the data was *exactly* Normal, and that each observation had exactly the *same* variance, and was distributed *exactly* independently of every other observation. But after a study of residuals had suggested model inadequacy, it might be desirable to reanalyse the data in relation to a less restrictive model into which the initial model was embeded. If non-Normality was suspected, for example, it might be sensible to postulate that the sample came from a wider class of parent distributions of which the Normal was a member. The consequential analysis could be difficult via sampling theory but is readily accomplished in a Bayesian framework (see Chapters 3 and 4). Such an analysis allows evidence *from the data* to be taken into account about the form of the parent distribution besides making it possible to assess to what extent the prior assumption of exact Normality is justified.

The above introductory survey suggests that Bayes' theorem provides a system of statistical inference suited to iterative model building, which is in turn an essential part of scientific investigation. On the other hand, we have pointed out that statistical inference (Bayesian or otherwise) is only a part of statistical method. It is, we believe, equally unhelpful for enthusiasts to seem to claim that Bayesian analysis can do everything, as it is for its detractors to seem to assert that it can do nothing.

1.2 NATURE OF BAYESIAN INFERENCE

1.2.1 Bayes' Theorem

Suppose that $\mathbf{y}'=(y_1, \ldots, y_n)$ is a vector of n observations whose probability distribution $p(\mathbf{y} \mid \boldsymbol{\theta})$ depends on the values of k parameters $\boldsymbol{\theta}' = (\theta_1, \ldots, \theta_k)$. Suppose also that $\boldsymbol{\theta}$ itself has a probility distribution $p(\boldsymbol{\theta})$. Then,

$$p(\mathbf{y} \mid \boldsymbol{\theta})p(\boldsymbol{\theta}) = p(\mathbf{y}, \boldsymbol{\theta}) = p(\boldsymbol{\theta} \mid \mathbf{y})p(\mathbf{y}). \tag{1.2.1}$$

Given the observed data $\mathbf{y}$, the conditional distribution of $\boldsymbol{\theta}$ is

$$p(\boldsymbol{\theta} \mid \mathbf{y}) = \frac{p(\mathbf{y} \mid \boldsymbol{\theta})p(\boldsymbol{\theta})}{p(\mathbf{y})}. \tag{1.2.2}$$

Also, we can write

$$p(\mathbf{y}) = E\, p(\mathbf{y} \mid \boldsymbol{\theta}) = c^{-1} = \begin{cases} \int p(\mathbf{y} \mid \boldsymbol{\theta})p(\boldsymbol{\theta})\, d\boldsymbol{\theta} & \boldsymbol{\theta} \text{ continuous} \\ \Sigma\, p(\mathbf{y} \mid \boldsymbol{\theta})p(\boldsymbol{\theta}) & \boldsymbol{\theta} \text{ discrete} \end{cases} \tag{1.2.3}$$

where the sum or the integral is taken over the admissible range of $\boldsymbol{\theta}$, and where $E[f(\boldsymbol{\theta})]$ is the mathematical expectation of $f(\boldsymbol{\theta})$ with respect to the distribution $p(\boldsymbol{\theta})$. Thus we may write (1.2.2) alternatively as

$$p(\boldsymbol{\theta} \mid \mathbf{y}) = cp(\mathbf{y} \mid \boldsymbol{\theta})p(\boldsymbol{\theta}). \tag{1.2.4}$$

The statement of (1.2.2), or its equivalent (1.2.4), is usually referred to as *Bayes' theorem*. In this expression, $p(\boldsymbol{\theta})$, which tells us what is known about $\boldsymbol{\theta}$ without knowledge of the data, is called the *prior* distribution of $\boldsymbol{\theta}$, or the distribution of $\boldsymbol{\theta}$ *a priori*. Correspondingly, $p(\boldsymbol{\theta} \mid \mathbf{y})$, which tells us what is known about $\boldsymbol{\theta}$ given knowledge of the data, is called the *posterior* distribution of $\boldsymbol{\theta}$ given $\mathbf{y}$, or the distribution of $\boldsymbol{\theta}$ *a posteriori*. The quantity c is merely a "normalizing" constant necessary to ensure that the posterior distribution $p(\boldsymbol{\theta} \mid \mathbf{y})$ integrates or sums to one.

In what follows we sometimes refer to the prior distribution and the posterior distribution simply as the "prior" and the "posterior", respectively.

Bayes' Theorem and the Likelihood Function

Now given the data $\mathbf{y}$, $p(\mathbf{y} \mid \boldsymbol{\theta})$ in (1.2.4) may be regarded as a function not of $\mathbf{y}$ but of $\boldsymbol{\theta}$. When so regarded, following Fisher (1922), it is called the *likelihood function* of $\boldsymbol{\theta}$ for given $\mathbf{y}$ and can be written $l(\boldsymbol{\theta} \mid \mathbf{y})$. We can thus write Bayes' formula as

$$p(\boldsymbol{\theta} \mid \mathbf{y}) = l(\boldsymbol{\theta} \mid \mathbf{y})p(\boldsymbol{\theta}). \tag{1.2.5}$$

In other words, then, Bayes' theorem tells us that the probability distribution for $\boldsymbol{\theta}$ posterior to the data $\mathbf{y}$ is proportional to the product of the distribution for $\boldsymbol{\theta}$ prior to the data and the likelihood for $\boldsymbol{\theta}$ given $\mathbf{y}$. That is,

$$\text{posterior distribution} \propto \text{likelihood} \times \text{prior distribution}.$$

The likelihood function $l(\boldsymbol{\theta} \mid \mathbf{y})$ plays a very important role in Bayes' formula. It is *the* function through which the data $\mathbf{y}$ modifies prior knowledge of $\boldsymbol{\theta}$; it can therefore be regarded as representing the information about $\boldsymbol{\theta}$ coming from the data.

The likelihood function is defined up to a multiplicative constant, that is, multiplication by a constant leaves the likelihood unchanged. This is in accord with the role it plays in Bayes' formula, since multiplying the likelihood function by an arbitrary constant will have no effect on the posterior distribution of $\boldsymbol{\theta}$. The constant will cancel upon normalizing the product on the right hand side of (1.2.5). It is only the relative value of the likelihood which is of importance.

The Standardized Likelihood

When the integral $\int l(\boldsymbol{\theta} \mid \mathbf{y})\, d\boldsymbol{\theta}$, taken over the admissible range of $\boldsymbol{\theta}$, is finite, then occasionally it will be convenient to refer to the quantity

$$\frac{l(\boldsymbol{\theta} \mid \mathbf{y})}{\int l(\boldsymbol{\theta} \mid \mathbf{y})\, d\boldsymbol{\theta}}. \tag{1.2.6}$$

We shall call this the *standardized likelihood*, that is, the likelihood scaled so that the area, volume, or hypervolume under the curve, surface, or hypersurface, is one.

Sequential Nature of Bayes' Theorem

The theorem in (1.2.5) is appealing because it provides a mathematical formulation of how previous knowledge may be combined with new knowledge. Indeed, the theorem allows us to continually update information about a set of parameters $\boldsymbol{\theta}$ as more observations are taken.

Thus, suppose we have an initial sample of observations $\mathbf{y}_1$, then Bayes' formula gives

$$p(\boldsymbol{\theta} \mid \mathbf{y}_1) \propto p(\boldsymbol{\theta})\, l(\boldsymbol{\theta} \mid \mathbf{y}_1). \tag{1.2.7}$$

Now, suppose we have a second sample of observations $\mathbf{y}_2$ distributed independently of the first sample, then

$$\begin{aligned} p(\boldsymbol{\theta} \mid \mathbf{y}_2, \mathbf{y}_1) &\propto p(\boldsymbol{\theta})\, l(\boldsymbol{\theta} \mid \mathbf{y}_1)\, l(\boldsymbol{\theta} \mid \mathbf{y}_2) \\ &\propto p(\boldsymbol{\theta} \mid \mathbf{y}_1)\, l(\boldsymbol{\theta} \mid \mathbf{y}_2). \end{aligned} \tag{1.2.8}$$

The expression (1.2.8) is precisely of the same form as (1.2.7) except that $p(\boldsymbol{\theta} \mid \mathbf{y}_1)$, the posterior distribution for $\boldsymbol{\theta}$ given $\mathbf{y}_1$, plays the role of the prior distribution for the second sample. Obviously this process can be repeated any number of times. In particular, if we have n independent observations, the posterior distribution can, if desired, be recalculated after each new observation, so that at the mth stage the likelihood associated with the mth observation is combined with the posterior distribution of $\boldsymbol{\theta}$ after $m-1$ observations to give the new posterior distribution

$$p(\boldsymbol{\theta} \mid y_1, \ldots, y_m) \propto p(\boldsymbol{\theta} \mid y_1, \ldots, y_{m-1}) l(\boldsymbol{\theta} \mid y_m), \qquad m = 2, \ldots, n \tag{1.2.9}$$

where

$$p(\boldsymbol{\theta} \mid y_1) \propto p(\boldsymbol{\theta}) l(\boldsymbol{\theta} \mid y_1).$$

Thus, Bayes' theorem describes, in a fundamental way, the process of learning from experience, and shows how knowledge about the state of nature represented by $\boldsymbol{\theta}$ is continually modified as new data becomes available.

1.2.2 Application of Bayes' Theorem with Probability Interpreted as Frequencies

Mathematically, Bayes' formula is merely a statement of conditional probability, and as such its validity is not in question. What *has* been questioned is its applicability to general problems of scientific inference. The difficulties concern

a. the meaning of probability, and

b. the choice of, and necessity for, the prior distribution.

Specific examples can be found of applications of Bayes' theorem where the probabilities involved may be directly interpreted in terms of frequencies and may therefore be said to be objective, and where the prior probabilities can be supposed exactly known. The validity of applications of this sort has not been in serious dispute. An example of this situation is described by Fisher (1959, p.19). In this example, there are mice of two colors, black and brown. The black mice are of two genetic kinds, homozygotes (*BB*) and heterozygotes (*Bb*), and the brown mice are of one kind (*bb*). It is known from established genetic theory that the probabilities associated with offspring from various matings are as follows:

Table 1.2.1

Probabilities for genetic character of mice offspring

Mice	*BB* (black)	*Bb* (black)	*bb* (brown)
BB mated with *bb*	0	1	0
Bb mated with *bb*	0	$\frac{1}{2}$	$\frac{1}{2}$
Bb mated with *Bb*	$\frac{1}{4}$	$\frac{1}{2}$	$\frac{1}{4}$

Suppose we have a "test" mouse which is black and has been produced by a mating between two (*Bb*) mice. Using the information in the last line of the table, it is seen that, in this case, the prior probabilities of the test mouse being homozygous (BB) and heterozygous (Bb) are precisely known, and are $\frac{1}{3}$ and $\frac{2}{3}$ respectively. Given this prior information, Fisher supposed that the test mouse was now mated with a brown mouse and produced (by way of data) seven black offspring. One can then calculate, as Fisher did, the probabilities, posterior to the data, of the test mouse being homozygous (BB) and heterozygous (Bb) using Bayes' theorem.

Specifically, if we use θ to denote the test mouse being (BB) or (Bb),

$$\theta = \begin{cases} 0 & (BB) \\ 1 & (Bb) \end{cases}$$

then the prior knowledge is represented by the distribution

$$p(\theta = 0) = \Pr(BB) = \tfrac{1}{3}, \qquad p(\theta = 1) = \Pr(Bb) = \tfrac{2}{3}.$$

Further, letting $\mathbf{y}$ denote the offspring, we have the likelihood

$$l(\theta = 0 \mid \mathbf{y} = 7 \text{ black}) \propto \Pr(7 \text{ black} \mid BB) = 1,$$
$$l(\theta = 1 \mid \mathbf{y} = 7 \text{ black}) \propto \Pr(7 \text{ black} \mid Bb) = (\tfrac{1}{2})^7.$$

It follows from (1.2.5) that

$$p(\theta = 0 \mid \mathbf{y} = 7 \text{ black}) \propto \tfrac{1}{3}, \qquad p(\theta = 1 \mid y = 7 \text{ black}) \propto (\tfrac{2}{3})(\tfrac{1}{2})^7.$$

Upon normalizing the posterior probabilities are then

$$p(\theta = 0 \mid \mathbf{y} = 7 \text{ black}) = \Pr(BB \mid 7 \text{ black}) = \tfrac{64}{65},$$
$$p(\theta = 1 \mid \mathbf{y} = 7 \text{ black}) = 1 - \Pr(BB \mid 7 \text{ black}) = \tfrac{1}{64}.$$

which represent the posterior knowledge of the test mouse being (BB) or (Bb). We see that, given the genetic characteristics of the offspring, the mating results of 7 black offspring changes our knowledge considerably about the test mouse being (BB) or (Bb), from a prior probability ratio of 2:1 in favor of (Bb) to a posterior ratio of 64:1 against it.

As an illustration of the sequential nature of Bayes' theorem, suppose the 7 black offspring are viewed as a sequence of seven independent observations; then, if we let $\mathbf{y}' = (y_1, \ldots, y_7)$, the likelihood can be written

$$l(\theta \mid \mathbf{y} = 7 \text{ black}) = l(\theta \mid y_1 = \text{black}) \cdots l(\theta \mid y_7 = \text{black})$$

where

$$l(\theta \mid y_m = \text{black}) \propto \begin{cases} 1 & \theta = 0 \\ \frac{1}{2} & \theta = 1 \end{cases}, \qquad m = 1, \ldots, 7.$$

Applying (1.2.9), the changes in the probabilities of the test mouse being (BB) or (Bb) after the mth observation, $m = 1, \ldots, 7$, are given in Table 1.2.2.

Table 1.2.2

Probabilities for the test mouse being homozygous and heterozygous

Mice	Probabilities	
	$\theta = 0\ (BB)$	$\theta = 1\ (Bb)$
Initial	$\frac{1}{3}$	$\frac{2}{3}$
1st black	$\frac{1}{2}$	$\frac{1}{2}$
2nd black	$\frac{2}{3}$	$\frac{1}{3}$
3rd black	$\frac{4}{5}$	$\frac{1}{5}$
4th black	$\frac{8}{9}$	$\frac{1}{9}$
5th black	$\frac{16}{17}$	$\frac{1}{17}$
6th black	$\frac{32}{33}$	$\frac{1}{33}$
7th black	$\frac{64}{65}$	$\frac{1}{65}$

This shows the increasing certainty of the test mouse being (BB) as more and more black offspring are observed.

Other applications of this sort are to be found in the theory of design of sampling inspection schemes. See, for example, Barnard (1954). In these examples, all the probabilities, both prior and posterior, are *objective* in the sense that they may be given a direct limiting frequency interpretation and are, in principle, subject to experimental confirmation.

In most scientific applications, however, exactly known objective prior distributions are rarely available.

1.2.3 Application of Bayes' Theorem with Subjective Probabilities

Following Ramsay (1931), De Finetti (1937), and Savage (1954, 1961a, b, 1962), we shall in this book regard probability as a mathematical expression of our degree of belief with respect to a certain proposition. In this context the concept of verification of probabilities by repeated experimental trials is regarded merely as a means of calibrating a subjective attitude. Thus, to say that one feels the probability is one half that Miss *A* and Mr. *B* will get married means that we have the same belief in the proposition "Mr. *B* will marry Miss *A*" as we would in the proposition "a toss of a fair coin will produce a head." We do not need to imagine an infinite series of situations in half of which *A* and *B* are wedded, and in half of which they are not.

The actual elucidation of what is believed by a particular person can be attempted in terms of betting odds. If, for example, the value of a continuous parameter θ is in question, we may, in suitable circumstances, infer an experimenter's prior distribution by asking at what value θ_0 he would be prepared to bet at particular odds that $\theta > \theta_0$. Given that a subjective probability distribution of this kind represents *a priori* what a person believes, then the posterior distribution obtained by combining this prior with the likelihood function shows how the prior beliefs are modified by information coming from the data.

Estimation of a Physical Constant

To consolidate ideas, we consider the example illustrated in Fig. 1.2.1. Suppose two physicists, A and B, are concerned with obtaining more accurate estimates of some physical constant θ, previously known only approximately. Suppose physicist A, being very familiar with this area of study, can make a moderately good guess of what the answer will be, and that his prior opinion about θ can be approximately represented by a Normal distribution centered at 900, with a standard deviation of 20. Thus

$$p_A(\theta) = \frac{1}{\sqrt{2\pi}\,20} \exp\left[-\frac{1}{2}\left(\frac{\theta - 900}{20}\right)^2\right]. \tag{1.2.10a}$$

According to A, *a priori* $\theta \sim N(900, 20^2)$ where the notation means that θ is distributed Normally with "mean 900 and variance 20^2." This would imply, in particular, that to A the chance that the value of θ could differ from 900 by more than 40 was only about one in twenty. By contrast, we suppose that B has had little previous experience in this area ,and his rather vague prior beliefs are represented by the Normal distribution

$$p_B(\theta) = \frac{1}{\sqrt{2\pi}\,80} \exp\left[-\frac{1}{2}\left(\frac{\theta - 800}{80}\right)^2\right]. \tag{1.2.10b}$$

Thus, according to B, $\theta \sim N(800, 80^2)$. He centers his prior at 800 and is considerably less certain about θ than A is. To B, a value anywhere between 700 and 900 would certainly be plausible. The curves in Fig. 1.2.1(a) labelled $p_A(\theta)$ and $p_B(\theta)$ show these prior distributions for A and B.

Suppose now that an unbiased method of experimental measurement is available and that an observation y made by this method, to a sufficient approximation, follows a Normal distribution with mean θ and standard deviation 40, that is $y \sim N(\theta, 40^2)$. If now a single observation y is made, the standardized likelihood function is represented by a Normal curve† centered at y with standard deviation 40. Then we can apply Bayes' theorem to show how each man's opinion regarding θ is modified by the information coming from that piece of data.

If *a priori* $\theta \sim N(\theta_0, \sigma_0^2)$, and the standardized likelihood function is represented by a Normal curve centered at y with standard deviation σ, then it is

---

† We refer to the function

$$f(x) = (2\pi\sigma^2)^{-1/2} \exp\left[-\frac{1}{2}\left(\frac{x - \mu}{\sigma}\right)^2\right]$$

as the Normal function, and the corresponding curve as the Normal curve. When the Normal function is employed to represent a probability distribution, it becomes the Normal distribution. The standardized likelihood function in this example is a Normal function, but it is not a probability distribution.

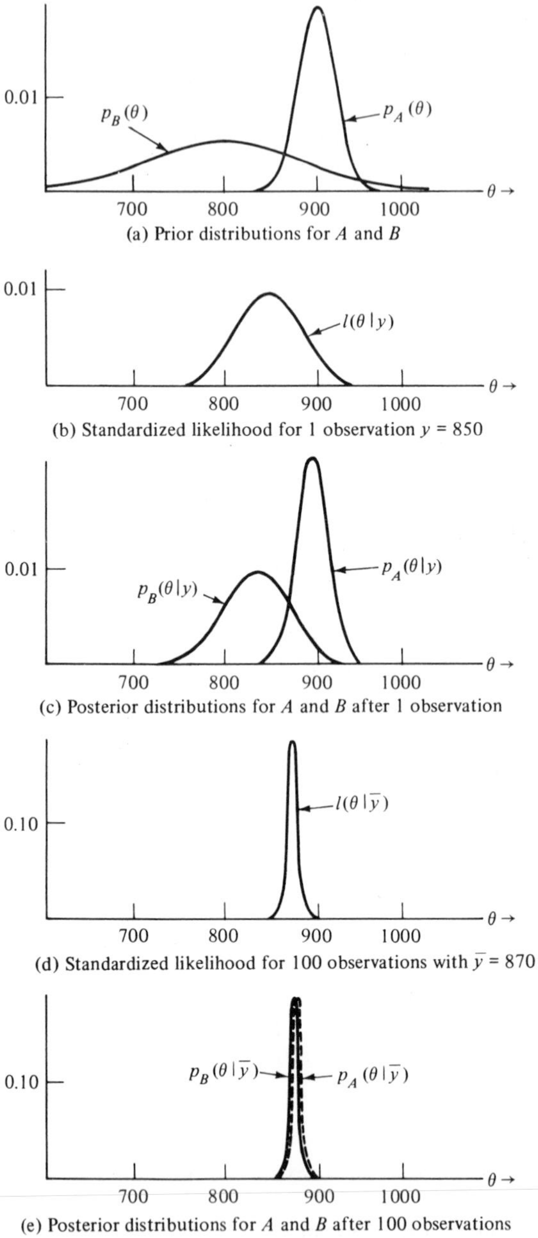

Fig. 1.2.1 Prior and posterior distributions for physicists A and B.

shown in Appendix A1.1 that the posterior distribution of θ given $y, p(\theta \mid y)$, is the Normal distribution $N(\bar{\theta}, \bar{\sigma}^2)$ where

$$\bar{\theta} = \frac{1}{w_0 + w_1}(w_0\theta_0 + w_1 y), \qquad \frac{1}{\bar{\sigma}^2} = w_0 + w_1$$

with

$$w_0 = \frac{1}{\sigma_0^2} \quad \text{and} \quad w_1 = \frac{1}{\sigma^2}. \tag{1.2.11}$$

The posterior mean $\bar{\theta}$ is a weighted average of the prior mean θ_0 and the observation y, the weights being proportional to w_0 and w_1 which are, respectively, the reciprocal of the variance of the prior distribution of θ and that of the observation. This is an appealing result, since the reciprocal of the variance is a measure of information which determines the weight to be attached to a given value. The variance of the posterior distribution is the reciprocal of the sum of the two measures of information w_0 and w_1, reflecting the fact that the two sources of information are pooled together.

Suppose the result of the single observation is $y = 850$; then the likelihood function is shown in Fig. 1.2.1(b). Physicist A's posterior opinion now is represented by the Normal distribution $p_A(\theta \mid y)$ with mean 890 and standard deviation 17.9, while that for B is represented by the Normal distribution $p_B(\theta \mid y)$ with mean 840 and standard deviation 35.78. These posterior distributions are shown in Fig. 1.2.1(c). The complete inferential process is sketched in Table 1.2.3.

Table 1.2.3

Prior and posterior distributions of θ for physicists A and B.

Prior distribution	Likelihood from data	Posterior distribution
A		A
$\theta \sim N(900, 20^2)$	$N(850, 40^2)$	$\theta \sim N(890, 17.9^2)$
B		B
$\theta \sim N(800, 80^2)$		$\theta \sim N(840, 35.70^2)$

We see that after this single observation the ideas of A and B about θ, as represented by the posterior distributions, are much closer than before, although they still differ considerably. We see that A, relatively speaking, did not learn much from the experiment, while B learned a great deal. The reason, of course, is that to A, the uncertainty in the measurement, as reflected by $\sigma = 40$, was larger than the uncertainty in his prior ($\sigma_0 = 20$). On the other hand, the uncertainty in the measurement was considerably smaller than that in B's prior ($\sigma_0 = 80$).

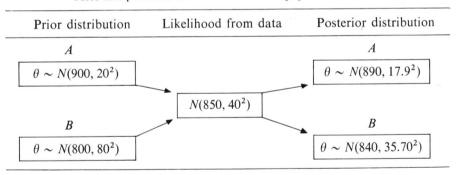

For A, the prior has a stronger influence on the posterior distribution than has the likelihood, while for B the likelihood has a stronger influence than the prior.

Suppose 99 further independent measurements are made and the sample mean $\bar{y} = \frac{1}{100} \Sigma y_i$ of the entire 100 observations is 870. In general, the likelihood function of θ *given* n independent observations from the Normal population $N(\theta, \sigma^2)$, is

$$l(\theta \mid \mathbf{y}) \propto \left(\frac{1}{\sqrt{2\pi}\,\sigma}\right)^n \exp\left[-\frac{1}{2\sigma^2} \Sigma (y_i - \theta)^2\right]. \tag{1.2.12}$$

Also since

$$\Sigma (y_i - \theta)^2 = \Sigma (y_i - \bar{y})^2 + n(\theta - \bar{y})^2, \tag{1.2.13}$$

and, given the data, $\Sigma (y_i - \bar{y})^2$ is a fixed constant, the likelihood is

$$l(\theta \mid \mathbf{y}) \propto \exp\left[-\frac{1}{2}\left(\frac{\theta - \bar{y}}{\sigma/\sqrt{n}}\right)^2\right], \tag{1.2.14}$$

which is a Normal function centred about $\bar{y}$ with standard deviation $\sigma/\sqrt{n}$.

In the present example, therefore, the likelihood is the Normal function centered at $\bar{y} = 870$ with standard deviation $\sigma/\sqrt{n} = \frac{40}{10} = 4$ shown in Fig. 1.2.1(d). We can thus apply the result in (1.2.11) as if $\bar{y}$ were a single observation with variance σ^2/n, that is, with weight n/σ^2. The posterior distribution of θ obtained by combining the likelihood function (1.2.14) with a Normal prior $N(\theta_0, \sigma_0^2)$ is the Normal distribution $N(\bar{\theta}_n, \bar{\sigma}_n^2)$, where

$$\bar{\theta}_n = \frac{1}{w_0 + w_n}(w_0\theta_0 + w_n\bar{y}), \qquad \frac{1}{\bar{\sigma}_n^2} = w_0 + w_n, \tag{1.2.15}$$

with

$$w_0 = \frac{1}{\sigma_0^2} \quad \text{and} \quad w_n = \frac{n}{\sigma^2}.$$

Thus the posterior distributions of A and B are $N(871.2, 3.9^2)$ and $N(869.8, 3.995^2)$, respectively. These two distributions, shown in Fig. 1.2.1e), are, for all practical purposes, the same, and are closely approximated by the Normal distribution $N(870, 4^2)$, which is the standardized form of the likelihood function in (1.2.14). Thus, after 100 observations, A and B would be in almost complete agreement. This is because the information coming from the data almost completely overrides prior differences.

Influence of the Prior Distribution on the Posterior Distribution

In the above example, we were concerned with the value of a *location* parameter θ, namely, the mean of a Normal distribution. In general, we shall say that a parameter η is a location parameter if addition of a constant c to all the observations changes η to $\eta + c$.

In this example, the contribution of the prior in helping to determine the posterior distribution of the location parameter θ was seen to depend on its sharpness or flatness *in relation* to the sharpness or flatness of the likelihood with which it was to be combined (see again Fig. 1.2.1). After a single observation, the likelihood was not sharply peaked relative to either of the prior distributions $p_A(\theta)$ or $p_B(\theta)$. These priors were therefore influential in deciding the posterior distribution. Because of this, the two different priors, when combined with the same likelihood, produced different posterior distributions. On the other hand, after 100 observations, both the priors $p_A(\theta)$ and $p_B(\theta)$ were rather flat *compared with* the likelihood function $l(\theta \mid \mathbf{y}) = l(\theta \mid \bar{y})$. These priors were therefore not very influential in deciding the corresponding posterior distributions of the location parameter θ. We can say that, after 100 observations, the priors were *dominated* by the likelihood.

1.2.4 Bayesian Decision Problems

The problems which we treat in this book are nearly all concerned with the situation common in scientific inference where the prior distribution is dominated by the likelihood. However, we must at least mention the important topic of Bayesian decision analysis [Schlaifer (1959), Raiffa and Schlaifer (1961), and DeGroot (1970)], where it is often not true that the prior is dominated by the likelihood. In Bayesian decision analysis, it is supposed that a choice has to be made from a set of available actions $(a_1, \ldots, a_r)$, where the payoff or utility of a given action depends on a state of nature, say θ, which is unknown. The decision maker's knowledge of θ is represented by a posterior distribution which combines prior knowledge of θ with the information provided by an experiment, and he is then supposed to choose that action which maximizes the *expected* payoff over the posterior distribution. An important application of such analysis is to business decision problems, such as whether or not to introduce a new industrial product. In such problems, a subjective prior distribution based, for example, on the opinion of an executive concerning the potential size θ of a market may be influential in determining the posterior distribution.

The fact that in such situations different decisions can result from different choices of prior distribution has worried some statisticians. We feel, however, that making explicit the dependence of the decision on the choice of what is believed to be true is an advantage of Bayesian analysis rather than the reverse. Suppose four different executives, after careful consideration, produce four different prior distributions for the size of a potential market and separate analyses are made for each. Then either (1) the decision (e.g. whether to market the product) will be the same in spite of differences in the priors, or (2) the decision will be different. In either case the Bayesian decision analysis will be valuable. In the first case, the ultimate arbiter would be reassured that such differences in opinion did not logically lead to differences on what the appropriate *action* should be. In the second case, it would be clear to him that *on present evidence* a real conflict existed. He would, in this case, either have to take the responsibility of ignoring the judgement of one or more of his executives, or of arranging that further data be obtained to resolve the conflict. Far from nullifying the value of Bayesian analysis, the fact that such analysis shows to what extent different decisions may or may not be appropriate when different prior opinions are held, seems to enhance it. For problems of this kind any procedure which took no account of such opinion would seem *necessarily* ill conceived.

1.2.5 Application of Bayesian Analysis to Scientific Inference

Important as the topic is, in this book, our concern will not be with statistical decision problems but with statistical inference problems such as occur in scientific investigation. By statistical inference we mean inference about the state of nature made in terms of probability, and a statistical inference problem is regarded as solved as soon as we can make an appropriate probability statement about the state of nature in question. Usually the state of nature is described by the value of one or more parameters. Such a parameter θ could, for example, be the velocity of light or the thermal conductivity of a certain alloy. Thus, a solution to the inference problem is supplied by a posterior distribution $p(\boldsymbol{\theta} \mid \mathbf{y})$ which shows what can be inferred about the parameters $\boldsymbol{\theta}$ from the data $\mathbf{y}$ given a relevant prior state of knowledge represented by $p(\boldsymbol{\theta})$.

Dominance of the Likelihood in the Normal Theory Example

Let us return again to the example of Section 1.2.3 concerning the estimation of the location parameter θ of a Normal distribution. In general, if the prior distribution is Normal $N(\theta_0, \sigma_0^2)$ and n independent observations with average $\bar{y}$ are taken from the distribution $N(\theta, \sigma^2)$, then from (1.2.15) the posterior distribution of θ is

$$\theta \sim N(\bar{\theta}_n, \bar{\sigma}_n^2), \quad \text{with} \quad \bar{\theta}_n = \frac{1}{w_0 + w_n}(w_0\theta_0 + w_n\bar{y}) \quad \text{and} \quad \bar{\sigma}_n^{-2} = w_0 + w_n,$$

where $w_0 = \sigma_0^{-2}$ is the weight associated with the prior distribution and $w_n = n/\sigma^2$ is the weight associated with the likelihood. In this expression, if w_0 is small *compared* with w_n, then *approximately* the posterior distribution is numerically equal to the standardized likelihood, and is

$$N\left(\bar{y}, \frac{\sigma^2}{n}\right). \tag{1.2.16}$$

Strictly speaking, this result is attained only when the prior variance σ_0^2 becomes infinite so that w_0 is zero. Such a limiting prior distribution would, however, by itself make little theoretical or practical sense. For, when $\sigma_0^2 \to \infty$, in the limit the prior density becomes uniform over the entire line from $-\infty$ to ∞, and is therefore not a proper density function. Furthermore, it represents a situation where all values of θ from $-\infty$ to ∞ are equally acceptable *a priori*. But it is difficult, if not impossible, to imagine a practical situation where sufficiently extreme values could not be virtually ruled out. The practical situation is represented *not* by the limiting case where $w_0 = 0$, but by the case where w_0 is small compared with w_n, that is, where the prior is locally flat so that the likelihood dominates the prior.

It is, therefore, important to note that the use of the limiting posterior in (1.2.16) corresponding to $w_0 = 0$ to supply a numerical approximation to the practical situation is not the same thing as *assuming* w_0 is actually zero. Limiting cases of this kind are frequently used in this book, but it must be remembered this is for the purpose of supplying a numerical approximation and for this purpose only.

"*Proper*" *and* "*Improper*" *Prior Distributions*

A basic property of a probability density function $f(x)$ is that it integrates or sums over its admissible range to 1, that is,

$$\left.\begin{matrix}\int f(x)\,dx \\ \Sigma f(x)\end{matrix}\right\} = 1 \qquad \begin{cases}(x \text{ continuous}),\\ (x \text{ discrete}).\end{cases}$$

Now, if $f(x)$ is uniform over the entire line from $-\infty$ to ∞,

$$f(x) = \kappa, \qquad -\infty < x < \infty, \quad \kappa > 0, \tag{1.2.17}$$

then it is not a proper density since the integral

$$\int_{-\infty}^{\infty} f(x)\,dx = \kappa \int_{-\infty}^{\infty} dx$$

does not exist no matter how small κ is. Density functions of this kind are sometimes called *improper* distributions. As another example, the function

$$f(x) = \kappa x^{-1}, \qquad 0 < x < \infty, \quad \kappa > 0 \tag{1.2.18}$$

is also improper. In this book, density functions of the types in (1.2.17) and (1.2.18) are frequently employed to represent the *local* behavior of the prior distribution in the region where the likelihood is appreciable, but *not* over its entire admissible range. By supposing that to a sufficient approximation the prior follows the form (1.2.17) or (1.2.18) only over the range of appreciable likelihood and that it suitably tails to zero outside that range we ensure that the priors actually used are proper. Thus, by employing the distributions in a way that makes practical sense we are relieved of a theoretical difficulty.

The Role of the Dominant Likelihood in the Analysis of Scientific Experiments

It is often appropriate to analyze data from scientific investigations on the assumption that the likelihood dominates the prior. Two reasons for this are:

1. A scientific investigation is not usually undertaken unless information supplied by the investigation is likely to be considerably more precise than information already available. For instance, suppose a physical constant θ had been estimated at 0.85 ± 0.05; then usually there would be no justification for making

a new determination whose accuracy was $\pm$ 0.25,† but there might be considerable justification for making one whose accuracy was $\pm$ 0.01. In brief, a scientific investigation is not usually undertaken unless it is likely to increase knowledge by a substantial amount. Therefore, as is illustrated in Figs. 1.2.2 and 1.2.3, analysis with priors which are dominated by the likelihood often realistically represents the true inferential situation. Situations of this kind have been referred to by Savage (1962) and Edwards, Lindman, and Savage (1963) as those where the principle of "precise measurement" or "stable estimation" applies.

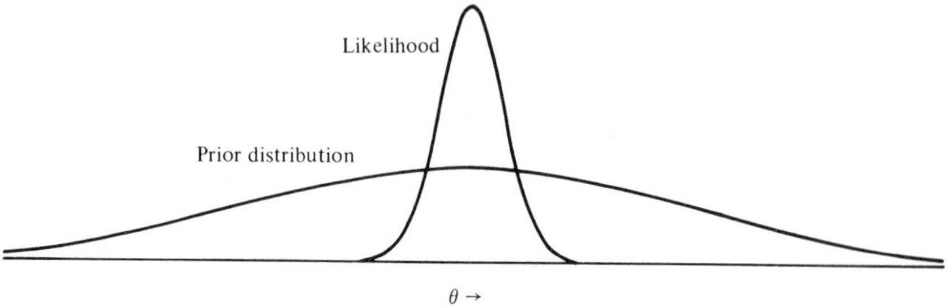

Fig. 1.2.2 Dominant likelihood (often appropriate to the analysis of scientific data).

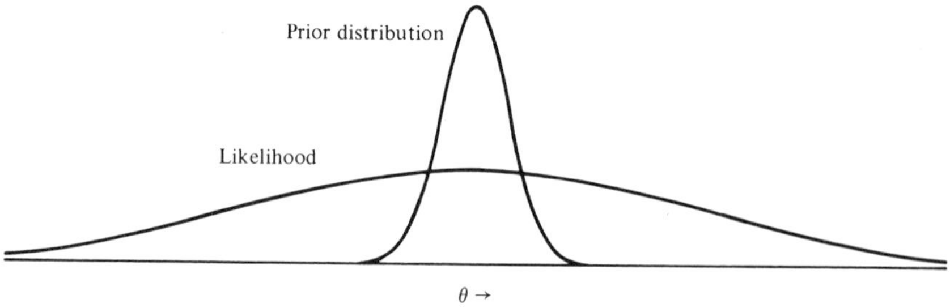

Fig. 1.2.3 Dominant prior (rarely appropriate to the analysis of scientific data).

2. Even when a scientist holds strong prior beliefs about the value of a parameter θ, nevertheless, in reporting his results it would usually be appropriate and most convincing to his colleagues if he analyzed the data against a *reference* prior which is dominated by the likelihood. He could then say that, irrespective of what he or anyone else believed to begin with, the posterior distribution represented what someone who *a priori* knew very little about θ should believe in the light of the data.‡

---

† Special circumstances could, of course, occur when the new determination *was* justified; for example, if it were suspected that the original method of determination might be subject to a major bias.

‡ As a *separate issue* his colleagues might also like to know what his prior opinion was and how this would affect the conclusions.

In judging the data in relation to a "neutral" reference prior, the scientist employs what may be called the "jury principle." Cases are tried in a law court before a jury which is carefully screened so that it has no possible connection with the principals and the events of the case. The intention is clearly to ensure that information gleaned from "data" or testimony may be assumed to dominate prior ideas that members of the jury may have concerning the possible guilt of the defendant.

The Reference Prior

In the above we have used the word *reference* prior. In general we mean by this a prior which it is convenient to use as a standard. In principle, a reference prior might or might not be dominated by the likelihood, but in this book reference priors which are dominated by the likelihood are often employed.

Dominant Likelihood and Locally Uniform Priors

The argument so far has been illustrated by the single example concerning the location parameter θ of a Normal distribution with a Normal prior. In particular, we have used this example to illustrate the important situation where the likelihood dominates the prior. We now consider the dominant likelihood idea more generally.

In general, a prior which is dominated by the likelihood is one which does not change *very much* over the region in which the likelihood is appreciable and does not assume large values outside that range (see Fig. 1.2.2). We shall refer to a prior distribution which has these properties as a *locally uniform* prior. For such a prior distribution we can approximate the result from Bayes' formula by substituting a constant for the prior distribution so that

$$p(\theta \mid \mathbf{y}) = \frac{l(\theta \mid \mathbf{y})\, p(\theta)}{\int l(\theta \mid \mathbf{y})\, p(\theta)\, d\theta} \doteq \frac{l(\theta \mid \mathbf{y})}{\int l(\theta \mid \mathbf{y})\, d\theta}. \tag{1.2.19}$$

Thus, for a locally uniform prior, the posterior distribution is approximately numerically equal to the standardized likelihood as we have previously found in (1.2.16) for the very special case of a Normal prior dominated by a Normal likelihood.

Difficulties Associated with Locally Uniform Priors

Historically, the choice of a prior to characterize a situation where "nothing (or, more realistically, little) is known *a priori*" has long been, and still is, a matter of dispute. Bayes tentatively suggested that where such knowledge was lacking concerning the nature of the prior distribution, it might be regarded as uniform. This suggestion is usually referred to as Bayes' postulate. He seemed, however, to have been himself so doubtful as to the validity of this postulate that he did not publish it, and his work was presented (Bayes, 1763) to the Royal Society posthumously by his friend Richard Price. This was accompanied by Price's own commentary which might not have reflected Bayes' final view. Fisher (1959) pointed out that although Bayes considered this postulate in his essay, in his actual mathematics he avoided its use as open to dispute and showed by example how the prior distribution could be determined by an auxilliary experiment. The postulate was accepted without question by later writers such as Laplace, but its reckless application led unfortunately to the falling into disrepute of the theorem itself.

We now examine some objections which have been made to Bayes' postulate, and then discuss ways which have been proposed to overcome these objections and extend the concept. In refutation of Bayes' postulate, it has been argued that, if the distribution of a continuous parameter θ were taken locally uniform, then the distribution of log θ, θ^{-1}, or some other transformation of θ (which might provide equally sensible bases for parametrizing the problem) would not be locally uniform. Thus, application of Bayes' postulate to different transformations of θ would lead to posterior distributions from the same data which were inconsistent.

This argument is of course correct, but the arbitrariness of the choice of parametrization does not by itself mean that we should not employ Bayes' postulate in practice. Arbitrariness exists to some extent in the specification of any statistical model. The only realistic expectation from a statistical analysis is that the conclusions will provide a good enough *approximation* to the truth. In applied (as opposed to pure) mathematics, arbitrariness is inadmissible only in so far as it produces results outside acceptable limits of approximation. In particular:

a) If, as would often be the case, the range of uncertainty for θ was not large compared with its mean value, then *over this range*, transformations such as the logarithmic and the reciprocal would be nearly linear, in which case approximate uniformity for θ would *imply* approximate uniformity for the transformed θ.

b) Although the argument (a) would fail for an extreme transformation such as θ^{10}, it is equally true that a rational experimenter would not agree to employ a uniform distribution after such a transformation. Thus, suppose that an investigator was concerned with measuring the specific gravity θ of a sample of ore; he expected that θ would be about 5 and felt happy with the idea that the probability that θ lay between 4 and 5 was about the same as the probability that θ lay between 5 and 6. A uniform distribution on θ^{10} would imply that the probability that it lay between 5 and 6 was almost six times as great as the probability that it lay between 4 and 5. Once he understood the implication of taking a constant prior distribution for this extreme transformation, he would be unwilling to accept it.

c) For large or even moderate-sized samples, fairly drastic modification of the prior distribution may only lead to minor modification of the posterior density. Thus, for independent observations $y_1, \ldots, y_n$, the posterior distribution can be written

$$p(\theta \mid y_1, \ldots, y_n) \propto p(\theta) \prod_{i=1}^{n} p(y_i \mid \theta). \tag{1.2.20}$$

and, for sufficiently large n, the n terms introduced by the likelihood will tend to overwhelm the single term contributed by the prior [see Savage, (1954)]. An illuminating illustration of the robustness of inference, under sensible modification of the prior, is provided by the study of Mosteller and Wallace (1964) on disputed authorship.

The above arguments indicate only that arbitrariness in the choice of the transformation in terms of which the prior is supposed locally uniform is often not catastrophic and that effects on the posterior distribution are likely to be of order n^{-1} and not of order 1 in relation to the data. For instance, we shall discuss in Chapter 2 a Bayesian derivation of Student's t distribution, and in so doing we must choose a prior distribution for the dispersion of the supposed Normal distribution of the observations. In various contexts, the dispersion of a Normal distribution can with some justification be measured in terms of $\sigma^2, \sigma, \log \sigma$, σ^{-1}, or σ^{-2}. Depending on which of these metrics are regarded as locally uniform, a t distribution is obtained having $n - 3$, $n - 2$, $n - 1$, n, or $n + 1$ degrees of freedom, respectively. What we have in this case is an uncertainty in the degrees of freedom (which in turn implies an uncertainty in the variance of the posterior distribution) of order n^{-1}. This degree of arbitrariness would not matter very much for large samples but it would have an appreciable effect for small samples. We are thus led to ask whether there is some way of eliminating, or at least reducing it so that the situation where "little is known *a priori*" can be more closely and meaningfully approximated.

1.3 NONINFORMATIVE PRIOR DISTRIBUTIONS

In this section we present an argument for choosing a particular metric in terms of which a locally uniform prior can be regarded as noninformative about the parameters. It is important to bear in mind that one can never be in a state of *complete* ignorance; further, the statement "knowing little *a priori*" can only have meaning *relative* to the information provided by an experiment. For instance, in Fig. 1.2.1, physicist A's prior knowledge is substantial compared with the information from a single observation but it is noninformative relative to that from a hundred observations. Now, a prior distribution is supposed to represent knowledge about parameters before the outcome of a projected experiment is known. Thus, the main issue is how to select a prior which provides little information relative to what is expected to be provided by the intended experiment. We consider first the case of a single parameter.

1.3.1 The Normal Mean θ (σ^2 Known)

Suppose $\mathbf{y}' = (y_1, \ldots, y_n)$ is a random sample from a Normal distribution $N(\theta, \sigma^2)$, where σ is a supposed known. Then, from (1.2.14), the likelihood function of θ is

$$l(\theta \mid \sigma, \mathbf{y}) \propto \exp\left[-\frac{n}{2\sigma^2}(\theta - \bar{y})^2\right] \tag{1.3.1}$$

where, as before, $\bar{y}$ is the average of the observations. The standardized likelihood function of θ is graphically represented by a Normal curve located by $\bar{y}$, with standard deviation $\sigma/\sqrt{n}$. Figure 1.3.1(a) shows a set of standardized likelihood curves which could result from an experiment in which $n = 10$ and $\sigma = 1$. Three different situations are illustrated with data giving averages of $\bar{y} = 6$, $\bar{y} = 9$, and $\bar{y} = 12$. Now it could happen that the quantity of immediate scientific interest was not θ itself but the reciprocal $\kappa = \theta^{-1}$. In that case the likelihood is

$$l(\kappa \mid \sigma, \mathbf{y}) \propto \exp\left[-\frac{n}{2\sigma}(\kappa^{-1} - \bar{y})^2\right], \tag{1.3.2}$$

and the standardized likelihood curves would have the appearance shown in Fig. 1.3.1(b).

In our previous discussion of the Normal mean, the prior was taken to be locally uniform in θ, which implies of course that it is *not* uniform in κ. We now consider whether this choice can be justified, and whether the principle can be extended to a wider context.

Data Translated Likelihood and Non-informative Prior

Our problem is to express the idea that little is known *a priori* relative to what the data has to tell us about a parameter θ. What the data has to tell us about θ is expressed by the likelihood function, and in the case of the Normal mean with n and σ^2 known, the data enter the likelihood only via the sample average $\bar{y}$. Figure 1.3.1(a) illustrates how, when the likelihood is expressed in terms of θ, the sample average $\bar{y}$ affects only the *location* of the likelihood curve. Different sets of data *translate* the likelihood curve on the θ axis but leave it otherwise unchanged. On the other hand, Fig. 1.3.1(b) illustrates how, when the likelihood is expressed in terms of $\kappa = \theta^{-1}$, both the location and the spread of the likelihood curve are changed when the data (and hence $\bar{y}$) are changed.

Now, in general, suppose it is possible to express the unknown parameter θ in terms of a metric $\phi(\theta)$, so that the corresponding likelihood is *data translated.* This means that the likelihood curve for $\phi(\theta)$ is completely determined *a priori except for its location* which depends on the data yet to be observed. Then to say that we know little *a priori* relative to what the data is going to tell us, may be expressed by saying that we are almost equally willing to accept one value of $\phi(\theta)$ as another. This state of indifference may be expressed by taking $\phi(\theta)$ to be locally

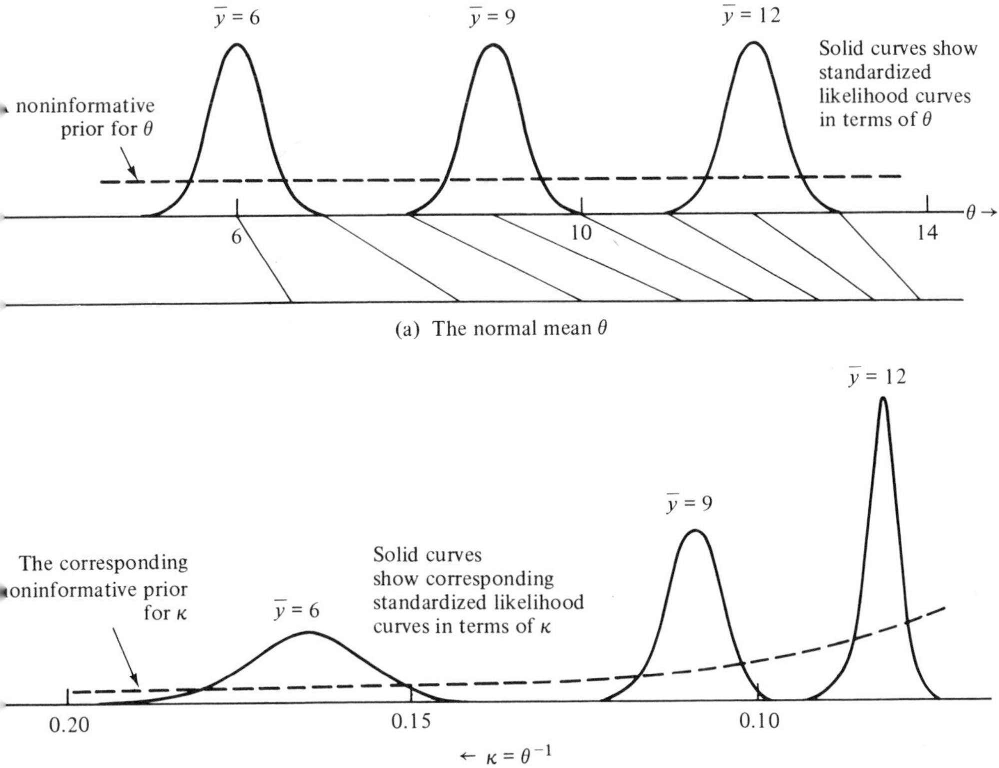

(b) Reciprocal of the normal mean $\kappa = \theta^{-1}$

Fig. 1.3.1 Noninformative prior distributions and standardized likelihood curves: (a) for the Normal mean θ, and (b) for $\kappa = \theta^{-1}$.

uniform, and the resulting prior distribution is called *noninformative* for $\phi(\theta)$ with respect to the data.

In the particular case of the Normal mean, the likelihood of θ is a Normal curve completely known *a priori* except for location which is determined by $\bar{y}$. That is, the likelihood is data translated in the original metric θ. Therefore, in this case, $\phi(\theta) = \theta$ and a noninformative prior is locally uniform in θ itself. That is, locally

$$p(\theta \mid \sigma) \propto c. \tag{1.3.3}$$

This noninformative prior distribution is shown in Fig. 1.3.1(a) by the dotted line. Since

$$p(\kappa \mid \sigma) \doteq p(\theta \mid \sigma) \left|\frac{d\theta}{d\kappa}\right| = p(\theta \mid \sigma)\theta^2 \propto \kappa^{-2}, \tag{1.3.4}$$

the corresponding noninformative prior for κ is not uniform but is locally proportional to θ^2, that is, to κ^{-2}. In general, if the noninformative prior is locally uniform in $\phi(\theta)$, then the corresponding noninformative prior for θ is locally proportional to $|d\phi/d\theta|$, assuming the transformation is one to one.

It is to be noted that we regard this argument only as indicating in what metric (transformation) the *local* behaviour of the prior should be uniform. Figure 1.3.2 illustrates what might be the situation over a wider range of the parameter. Here $p(\theta \mid \sigma)$ is a proper distribution which is merely flat over the region of interest. Similarly, $p(\kappa \mid \sigma)$ is a proper distribution obtained by transformation which is proportional to κ^{-2} over the region of interest. This point is important, because it would be inappropriate mathematically and meaningless practically to suppose, for example, that $p(\theta \mid \sigma)$ was uniform over an infinite range, or that $p(\kappa \mid \sigma)$ was proportional to κ^{-2} over an infinite range. We do not assume this nor do we need to.

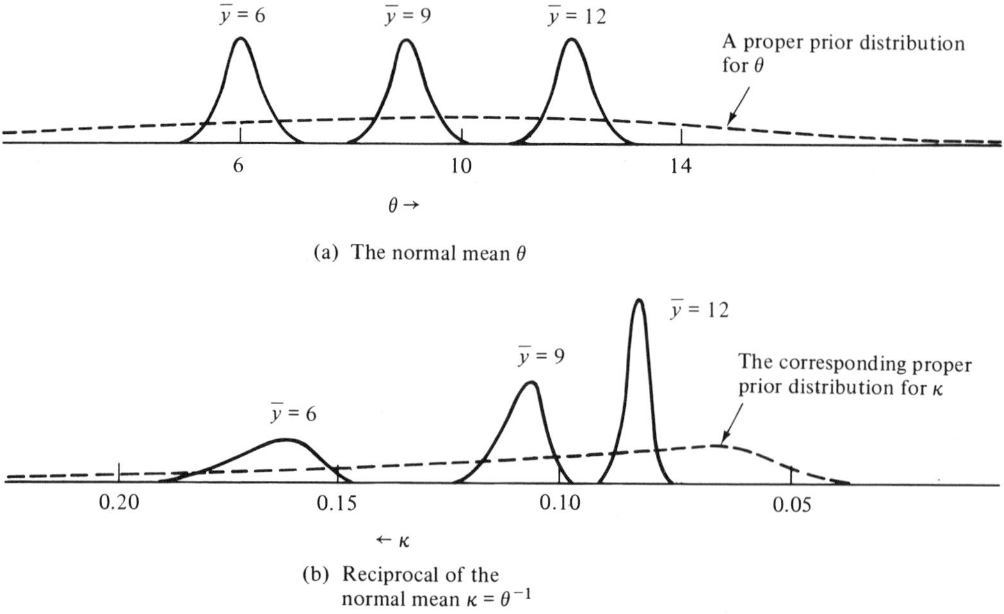

Fig. 1.3.2 Noninformative prior distributions and standardized likelihood curves: (a) for the Normal mean θ, and (b) for $\kappa = \theta^{-1}$ seen over a wider range of parameter values.

Posterior Distribution of the Normal Mean θ

On multiplying the likelihood in (1.3.1) by the locally uniform noninformative prior in (1.3.3), and introducing the appropriate normalizing constant, we have

$$p(\theta \mid \sigma, \mathbf{y}) \doteq \left(\frac{2\pi\sigma^2}{n}\right)^{-1/2} \exp\left[-\frac{n}{2\sigma^2}(\theta - \bar{y})^2\right], \qquad -\infty < \theta < \infty. \qquad (1.3.5)$$

That is, when it is desired to assume little prior knowledge about θ relative to that which would be supplied from the data, and given a sample of n observations from a Normal distribution with known variance σ^2, then *a posteriori* θ is approximately Normally distributed with mean $\bar{y}$ and variance σ^2/n.

As an example, Fig. 1.3.3 shows the posterior distribution calculated from (1.3.5) when a sample of 16 observations has been taken whose average value is $\bar{y} = 10$, it being known that $\sigma = 8$. The figure shows θ distributed about $\bar{y} = 10$ with standard deviation $\sigma/\sqrt{n} = 2$. It is perhaps appropriate to emphasize the meaning which attaches to this distribution. To someone who, before the data was collected, was indifferent to the choice of θ in the relevant range, the posterior distribution represents what, given the data, his attitude should now be. He could, for example, state that the probability that θ was less than 8 was 15.9%, this being the size of the shaded area shown in the figure. Relative to the same state of prior indifference he could, moreover, employ the same posterior distribution of Fig. 1.3.3 to obtain, by transformation, the posterior distribution for any function $\kappa(\theta)$ which was of interest. For example he could state that the probability that κ was greater than 1/8 was 15.9%. Other probabilities are readily obtained by using a table of the Normal probability integral, such as Table I at the end of the book.

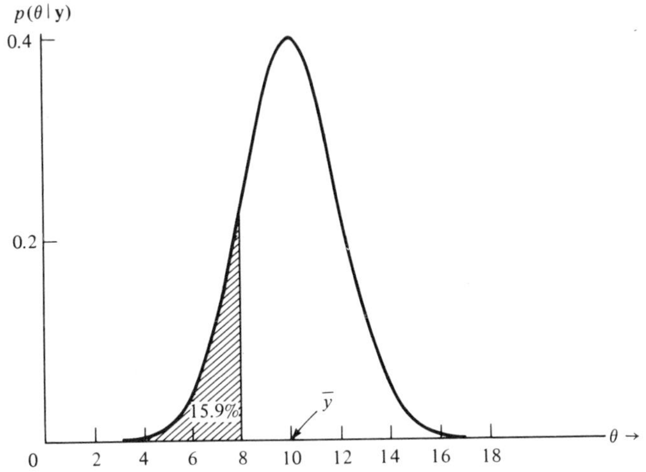

Fig. 1.3.3 Posterior distribution of the Normal mean θ (noninformative prior), when $\bar{y} = 10$, $\sigma = 8$, $n = 16$.

1.3.2 The Normal Standard Deviation σ (θ known)

As a second example, consider the choice of a noninformative prior distribution for σ, the standard deviation of a Normal distribution for which the mean θ is supposed known. In this case, the likelihood is

$$l(\sigma \mid \theta, \mathbf{y}) \propto \sigma^{-n} \exp\left(-\frac{ns^2}{2\sigma^2}\right), \qquad (1.3.6)$$

where

$$s^2 = \Sigma\,(y_u - \theta)^2/n.$$

For illustration, suppose there are $n = 10$ observations, then Fig. 1.3.4(a) shows the standardized likelihood curves for σ with $s = 5$, $s = 10$, and $s = 20$. Clearly, in the original metric σ, the likelihood curves are not data translated. According to the principle stated in the preceding section therefore a noninformative prior should *not* be taken to be locally uniform in σ.

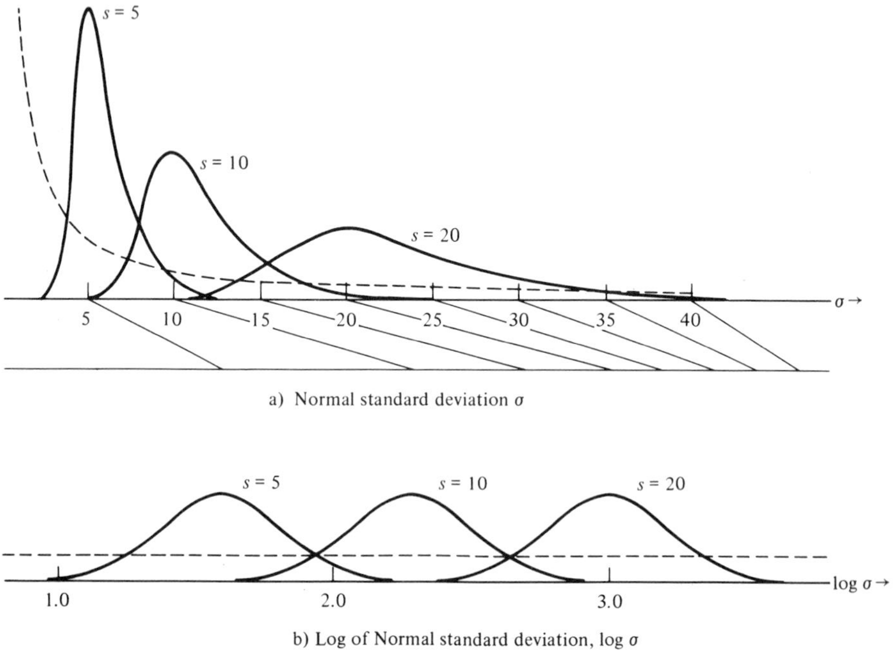

Fig. 1.3.4 Noninformative prior distributions and standardized likelihood curves: (a) for the Normal standard deviation σ, and (b) for $\log \sigma$ (broken curves are noninformative priors and solid curves are the standard likelihoods).

Figure 1.3.4(b) shows, however, that the corresponding likelihood curves in terms of $\log \sigma$ are exactly data translated. To see this mathematically, note that multiplication by the constant s^n leaves the likelihood unchanged. Therefore we can express the likelihood of $\log \sigma$ as

$$l(\log \sigma \mid \theta, \mathbf{y}) \propto \exp\left\{-n(\log \sigma - \log s) - \frac{n}{2} \exp\left[-2(\log \sigma - \log s)\right]\right\}. \qquad (1.3.7)$$

Thus, in this logarithmic metric the data acting through s serve only to relocate the likelihood. A noninformative prior should therefore be locally uniform in $\log \sigma$. When expressed in the metric σ, the noninformative prior is thus locally proportional to σ^{-1},

$$p(\sigma \mid \theta) \propto \left| \frac{d \log \sigma}{d\sigma} \right| = \sigma^{-1}. \tag{1.3.8}$$

If we use this prior distribution, then the posterior distribution of σ is

$$p(\sigma \mid \theta, \mathbf{y}) \dot{\propto} \sigma^{-(n+1)} \exp\left(-\frac{ns^2}{2\sigma^2}\right). \tag{1.3.9}$$

It will be seen in Section 2.3, where the implication of this distribution is discussed in greater detail, that the normalizing constant required to make the distribution integrate to unity is

$$k = \frac{(ns^2)^{n/2}}{2^{(n/2)-1}\,\Gamma(n/2)}. \tag{1.3.10}$$

Thus, given a sample $\mathbf{y}$ of n observations from a Normal distribution $N(\theta, \sigma^2)$, with θ known and little prior information about σ relative to that supplied by the data, the posterior distribution of σ is approximately

$$p(\sigma \mid \theta, \mathbf{y}) \doteq \frac{(ns^2)^{n/2}}{2^{(n/2)-1}\,\Gamma(n/2)} \sigma^{-(n+1)} \exp\left(-\frac{ns^2}{2\sigma^2}\right), \qquad \sigma > 0. \tag{1.3.11}$$

and the corresponding posterior distribution of any function of σ may be found by an appropriate transformation of (1.3.11).

Figure 1.3.5 illustrates the situation where the sample standard deviation calculated from $n = 10$ observations is

$$s = \left[\frac{\Sigma\,(y_u - \theta)^2}{10}\right]^{1/2} = 1.0.$$

The distribution shows what, given the assumptions and the data, can be said about σ. Tail area probabilities are readily found using the fact that (1.3.11) implies that ns^2/σ^2 has the "chi-square" (χ^2) distribution with n degrees of freedom,

$$p(\chi^2) = \frac{1}{\Gamma(n/2)2^{n/2}} (\chi^2)^{(n/2)-1} \exp\left(-\tfrac{1}{2}\chi^2\right), \qquad \chi^2 > 0. \tag{1.3.12}$$

For instance, suppose we wish to find the probability that σ is greater than $\sigma_0 = 1.5$. We have

$$\frac{ns^2}{\sigma_0^2} = \frac{10}{1.5^2} = 4.4,$$

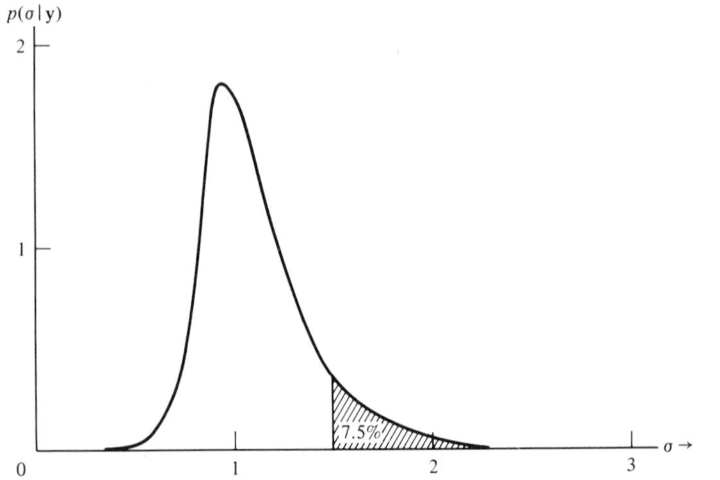

Fig. 1.3.5 Posterior distribution of the Normal standard deviation σ (noninformative prior), when $s = 1$ and $n = 10$.

so that, with χ^2_ν referring to a chi-square variate with ν degrees of freedom, the required probability corresponding to the shaded area in the diagram can be obtained from a table of χ^2 integral and is found to be

$$\Pr\{\chi^2_{10} < 4.4\} = 7.5\%.$$

1.3.3 Exact Data Translated Likelihoods and Noninformative Priors

We can summarize the above discussion of the choice of prior for a single parameter as follows.

If $\phi(\theta)$ is a one-to-one transformation of θ, we shall say that a prior distribution of θ which is locally proportional to $|d\phi/d\theta|$ is *noninformative* for the parameter θ if, in terms of ϕ, the likelihood curve is *data translated*, that is, the data only serve to change the location of the likelihood $l(\phi \mid \mathbf{y})$. Mathematically, a data translated likelihood must be expressible in the form

$$l(\theta \mid \mathbf{y}) = g\,[\phi(\theta) - f(\mathbf{y})], \tag{1.3.13}$$

where $g(x)$ is a known function independent of the data $\mathbf{y}$ and $f(\mathbf{y})$ is a function of $\mathbf{y}$.

The examples we have so far considered are both special cases of the above principle. For the Normal mean, $\phi(\theta) = \theta, f(\mathbf{y}) = \bar{y}$, and for the Normal standard deviation, $\phi(\sigma) = \log \sigma,\ f(\mathbf{y}) = \log s$.

In particular, we see that any likelihood of the form

$$l(\sigma \mid \mathbf{y}) \propto l\left[\frac{s(\mathbf{y})}{\sigma}\right] \tag{1.3.14}$$

can be bought into the form

$$l(\sigma \mid \mathbf{y}) = g[\log \sigma - \log s(\mathbf{y})] \tag{1.3.15}$$

so that it is data translated in terms of the logarithmic transformation $\phi(\sigma) = \log \sigma$.

The choice of a prior which is locally uniform in the metric ϕ for which the likelihood is data translated, can be viewed in another way. Let

$$l(\phi \mid \mathbf{y}) = g[\phi - f(\mathbf{y})], \tag{1.3.16}$$

and assume that the function g is continuous and has a unique maximum $\hat{g}$. Let α be an *arbitrary* positive constant such that $0 < \alpha < \hat{g}$. Then, for any given α, there exist two constants c_1 and c_2 $(c_1 < c_2)$, independent of $\mathbf{y}$ such that $g[\phi - f(\mathbf{y})]$ is greater than α for ϕ in the interval

$$f(\mathbf{y}) + c_1 < \phi < f(\mathbf{y}) + c_2. \tag{1.3.17}$$

This interval may be called the α highest likelihood interval. Now suppose the transformation from ϕ to λ is monotone. Then the corresponding α highest likelihood interval for λ is

$$\phi^{-1}[f(\mathbf{y}) + c_1] < \lambda < \phi^{-1}[f(\mathbf{y}) + c_2]. \tag{1.3.18}$$

We see that, in terms of ϕ, the length of the interval in (1.3.17) is $(c_2 - c_1)$ independent of the data $\mathbf{y}$, while for the metric λ the corresponding length

$$\phi^{-1}[f(\mathbf{y}) + c_2] - \phi^{-1}[f(\mathbf{y}) + c_1]$$

will in general depend upon $\mathbf{y}$ (except when the transformation is linear). For example, in the case of the Normal mean, $\phi(\theta) = \theta$, $f(\mathbf{y}) = \bar{y}$, and for $n = 10$, $\sigma = 1$,

$$g(x) = \exp\left(-\frac{n}{2\sigma^2}x^2\right) = \exp(-5x^2)$$

so that $\hat{g} = 1$. Suppose we take $\alpha = 0.05$; then

$$c_1 = -0.77, \qquad c_2 = 0.77.$$

For the three cases $\bar{y} = 6$, $\bar{y} = 9$, and $\bar{y} = 12$ considered earlier, the corresponding 0.05 highest likelihood intervals for θ are

$$\begin{array}{ccc} 6 \pm 0.77 & 9 \pm 0.77 & 12 \pm 0.77 \\ (5.23,\ 6.77), & (8.23,\ 9.77), & (11.23,\ 12.77), \end{array} \tag{1.3.19}$$

having the same length $c_2 - c_1 = 1.54$. However, in terms of the metric $\lambda = -\kappa = -1/\theta$, which is a monotone increasing function of θ, the 0.05 highest likelihood interval is

$$-(\bar{y} - 0.77)^{-1} < \lambda < -(\bar{y} + 0.77)^{-1},$$

so that for the three values of $\bar{y}$ considered we have

$\bar{y}$	6	9	12	
Interval	$(-0.191, -0.148)$	$(-0.122, -0.102)$	$(-0.089, -0.078)$	(1.3.20)
Length	0.043	0.020	0.011	

If we say we have little *a priori* knowledge about a parameter θ relative to the information expected to be supplied by the data, then we should be equally willing to accept the information from one experimental outcome as that from another. Since the information from the data is contained in the likelihood, this is saying that, over a relevant region of θ, we would have no *a priori* preference for one likelihood curve over another. This state of local indifference can then be represented by assigning approximately *equal* probabilities to all α-highest likelihood intervals. Now, in terms of ϕ for which the likelihood is data translated, the intervals all have the same length, so that the prior density must be locally uniform.

In the above example we would assign equal prior probabilities to the three intervals in (1.3.19), and the corresponding one in (1.3.20). It then follows that the noninformative prior distribution is locally uniform in θ but is locally proportional to $|d\theta/d\lambda| = \lambda^{-2}$ in terms of λ.

1.3.4 Approximate Data Translated Likelihood

As might be expected, a transformation which allows the likelihood to be expressed *exactly* in the form (1.3.13) is not generally available. However, for moderate sized samples, because of the insensitivity of the posterior distribution to minor changes in the prior, all that it would seem necessary to require is a transformation $\phi(\theta)$ in terms of which the likelihood is approximately data translated. That is to say, the likelihood for ϕ is nearly independent of the data $\mathbf{y}$ except for its location.

The Binomial Mean π

To illustrate the possibilities we consider the case of n independent trials, in each of which the probability of success is π. The probability of y successes in n trials is given by the binomial distribution

$$p(y \mid \pi) = \frac{n!}{y!\,(n-y)!}\,\pi^y\,(1-\pi)^{n-y}, \qquad y = 0, \ldots, n, \tag{1.3.21}$$

so that the likelihood is

$$l(\pi \mid y) \propto \pi^y(1-\pi)^{n-y}. \tag{1.3.22}$$

Suppose for illustration there are $n = 24$ trials. Then Fig.1.3.6(a) shows the standardized likelihood for $y = 3, y = 12$, and $y = 21$ successes. Figure 1.3.6(b) is the corresponding diagram obtained by plotting in the transformed metric

$$\phi(\pi) = \sin^{-1}\sqrt{\pi}. \tag{1.3.23}$$

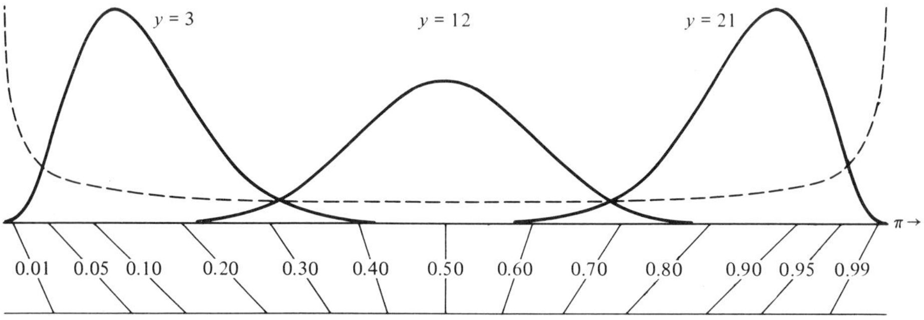

a) The binomial mean π

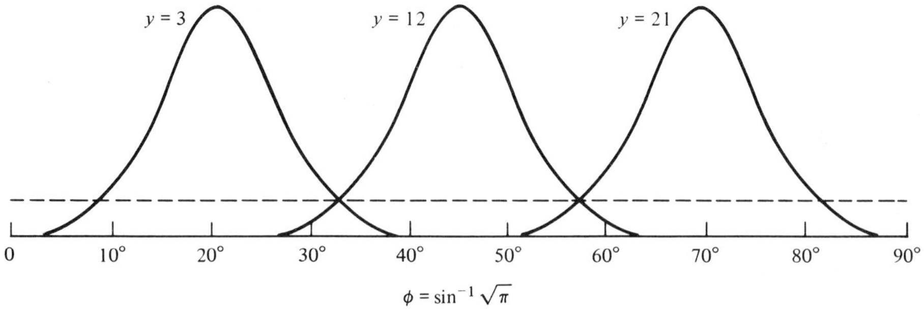

b) The transformed mean $\phi = \sin^{-1}\sqrt{\pi}$

Fig. 1.3.6 Noninformative prior distributions and standardized likelihood curves: (a) for the binomial mean π, and (b) for the transformed mean $\phi = \sin^{-1}\sqrt{\pi}$ (broken curves are the noninformative priors and solid curves are standardized likelihoods).

Although in terms of ϕ the likelihood curves are not exactly identical in shape and spread, they are nearly so. In this metric the likelihood curve is very nearly data translated and a locally uniform prior distribution is nearly noninformative. This in turn implies that the corresponding nearly noninformative prior for π is proportional to

$$p(\pi) \propto \left|\frac{d\phi}{d\pi}\right| = [\pi(1-\pi)]^{-\frac{1}{2}}. \qquad (1.3.24)$$

If we employ this approximately noninformative prior, indicated in Fig. 1.3.6(a) and (b) by the dotted lines, then as was noted by Fisher (1922),

$$p(\pi \mid y) \propto \pi^{y-\frac{1}{2}}(1-\pi)^{n-y-\frac{1}{2}}, \qquad 0 < \pi < 1. \qquad (1.3.25)$$

After substitution of the appropriate normalizing constant, we find that the corresponding posterior distribution for π is the beta distribution

$$p(\pi \mid y) = \frac{\Gamma(n+1)}{\Gamma(y+\frac{1}{2})\,\Gamma(n-y+\frac{1}{2})}\,\pi^{y-\frac{1}{2}}(1-\pi)^{n-y-\frac{1}{2}} \qquad 0 < \pi < 1. \qquad (1.3.26)$$

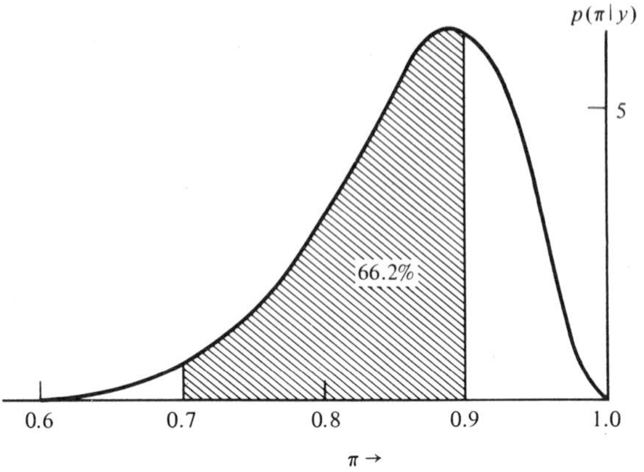

Fig. 1.3.7 Posterior distribution of the binomial mean π (noninformative prior) for 21 successes out of 24 trials.

Figure 1.3.7 shows the posterior distribution of π given that 21 out of 24 binomial trials (a proportion of 0.875) are successes. For illustration, tail area probabilities can be obtained by consulting the incomplete beta function tables.

The shaded area shown in the diagram is the probability that the parameter π lies between 0.7 and 0.9, and this is given by

$$\int_{0.7}^{0.9} \frac{\Gamma(25)}{\Gamma(21.5)\Gamma(3.5)}\,\pi^{20.5}(1-\pi)^{2.5}\,d\pi = 66.2\%.$$

We note in passing that, for this example where we have a moderately sized sample of $n = 24$ observations, the posterior density is not very sensitive to the precise choice of a prior. For instance, while for 21 successes the noninformative prior (1.3.24) yielded a posterior density proportional to $\pi^{20.5}(1-\pi)^{2.5}$, for a uniform prior in the original metric π the posterior density would have been proportional to $\pi^{21}(1-\pi)^3$. The use of the noninformative prior for π, rather than the uniform prior, is in general merely equivalent to reducing the number of successes and the number of "not successes" by 0.5.

Derivation of Transformations Yielding Approximate Data Translated Likelihoods

We now consider methods for obtaining parameter transformations in terms of which the likelihood is approximately data translated as in the binomial case. Again, let $\mathbf{y}' = (y_1, \ldots, y_n)$ be a random sample from a distribution $p(y \mid \theta)$. When the distribution obeys certain regularity conditions, Johnson (1967, 1970), then for sufficiently large n, the likelihood function of θ is approximately Normal, and remains approximately Normal under mild one-to-one transformations of θ. In such a case, the logarithm of the likelihood is approximately quadratic, so that

$$L(\theta \mid \mathbf{y}) = \log l(\theta \mid \mathbf{y}) = \log \prod_{u=1}^{n} p(y_u \mid \theta)$$

$$\doteq L(\hat{\theta} \mid \mathbf{y}) - \frac{n}{2}(\theta - \hat{\theta})^2 \left(-\frac{1}{n} \frac{\partial^2 L}{\partial \theta^2} \right)_{\hat{\theta}}. \tag{1.3.27}$$

where $\hat{\theta}$ is the maximum likelihood estimate of θ. In general, the quantity

$$\left(-\frac{1}{n} \frac{\partial^2 L}{\partial \theta^2} \right)_{\hat{\theta}}$$

is a positive function of $\mathbf{y}$. For the moment we shall discuss the situation in which it can be expressed as *a function of $\hat{\theta}$ only*, and write

$$J(\hat{\theta}) = \left(-\frac{1}{n} \frac{\partial^2 L}{\partial \theta^2} \right)_{\hat{\theta}}. \tag{1.3.28}$$

Now the logarithm of a Normal function $p(x)$ is of the form

$$\log p(x) = \text{const} - \tfrac{1}{2}(x - \mu)^2/\sigma^2 \tag{1.3.29}$$

and, given the location parameter μ, is completely determined by its standard deviation σ. Comparison of (1.3.27) and (1.3.29) shows that the standard deviation of the likelihood curve is approximately equal to $n^{-\frac{1}{2}} J^{-\frac{1}{2}}(\hat{\theta})$. Now suppose $\phi(\theta)$ is a one-to-one transformation; then,

$$J(\hat{\phi}) = \left(-\frac{1}{n} \frac{\partial^2 L}{\partial \phi^2} \right)_{\hat{\phi}} = \left(-\frac{1}{n} \frac{\partial^2 L}{\partial \theta^2} \right)_{\hat{\theta}} \left(\frac{d\theta}{d\phi} \right)^2_{\hat{\theta}} = J(\hat{\theta}) \left(\frac{d\theta}{d\phi} \right)^2_{\hat{\theta}}. \tag{1.3.30}$$

It follows that if $\phi(\hat{\theta})$ is chosen such that

$$\left| \frac{d\theta}{d\phi} \right|_{\hat{\theta}} \propto J^{-1/2}(\hat{\theta}), \tag{1.3.31}$$

then $J(\hat{\phi})$ will be a constant independent of $\hat{\phi}$, and the likelihood will be approximately data translated in terms of ϕ. Thus, the metric for which a locally uniform prior is approximately noninformative can be obtained from the relationship

$$\frac{d\phi}{d\theta} \propto J^{1/2}(\theta) \qquad \text{or} \qquad \phi \propto \int^{\theta} J^{1/2}(t)\, dt. \tag{1.3.32}$$

This, in turn, implies that the corresponding noninformative prior for θ is

$$p(\theta) \propto \left| \frac{d\phi}{d\theta} \right| \propto J^{1/2}(\theta). \tag{1.3.33}$$

As an example, consider again the binomial mean π. The log likelihood is

$$L(\pi \mid y) = \log l(\pi \mid y) = \text{const} + y \log \pi + (n - y) \log (1 - \pi). \tag{1.3.34}$$

Thus

$$\frac{\partial L}{\partial \pi} = \frac{y}{\pi} - \frac{n - y}{1 - \pi}, \qquad \frac{\partial^2 L}{\partial \pi^2} = -\frac{y}{\pi^2} - \frac{n - y}{(1 - \pi)^2}. \tag{1.3.35}$$

For $y \neq 0$ and $y \neq n$, by setting $\partial L / \partial \pi = 0$, one obtains the maximum likelihood estimates as $\hat{\pi} = y/n$, so that

$$J(\hat{\pi}) = \left(-\frac{1}{n} \frac{\partial^2 L}{\partial \pi^2} \right)_{\hat{\pi}} = \left(\frac{1}{\hat{\pi}} + \frac{1}{1 - \hat{\pi}} \right) = \frac{1}{\hat{\pi}(1 - \hat{\pi})}, \tag{1.3.36}$$

which is a function of $\hat{\pi}$ only, whence the noninformative prior for π is proportional to

$$J^{1/2}(\pi) \propto \pi^{-1/2} (1 - \pi)^{-1/2}, \tag{1.3.37a}$$

which is the prior used in (1.3.24). Also, the transformation

$$\phi = \int^{\pi} t^{-1/2}(1 - t)^{-1/2} \, dt \propto \sin^{-1}\sqrt{\pi} \tag{1.3.37b}$$

is precisely the metric employed in plotting the nearly data translated likelihood curves in Fig. 1.3.6. We recognize the $\sin^{-1}\sqrt{\pi}$ transformation as the well-known asymptotic variance stabilizing transformation for the binomial, originally proposed by Fisher. [See, for example, Bartlett (1937) and Anscombe (1948a)].

In the above we have specifically supposed that the quantity

$$\left(-\frac{1}{n} \frac{\partial^2 L}{\partial \theta^2} \right)_{\hat{\theta}}$$

is a function of $\hat{\theta}$ only. It can be shown that this will be true whenever the observations $\mathbf{y}$ are drawn from a distribution $p(y \mid \theta)$ of the form

$$p(y \mid \theta) = h(y) w(\theta) \exp [c(\theta) u(y)], \tag{1.3.38}$$

where the range of y does not depend upon θ. For the cases of the Normal mean θ with σ^2 known, the Normal standard deviation σ with θ known and the binomial mean π, the distributions are of this form. In fact, this is the form for which a single sufficient statistic for θ exists, a concept which will be discussed later in Section 1.4.

The Poisson Mean λ

As a further example, consider the Poisson distribution with mean λ,

$$p(y \mid \lambda) = \frac{\lambda^y}{y!} \exp(-\lambda), \qquad y = 0, \ldots, \infty, \tag{1.3.39}$$

which is of the form in (1.3.38). Suppose $\mathbf{y}' = (y_1, \ldots, y_n)$ is a set of n independent frequencies each distributed as (1.3.39). Then, given $\mathbf{y}$, the likelihood is

$$l(\lambda \mid \mathbf{y}) \propto \lambda^{n\bar{y}} \exp(-n\lambda), \qquad \bar{y} = \frac{1}{n} \Sigma\, y_u. \tag{1.3.40}$$

Thus,

$$L(\lambda \mid \mathbf{y}) = \text{const} + n\bar{y} \log \lambda - n\lambda \tag{1.3.41}$$

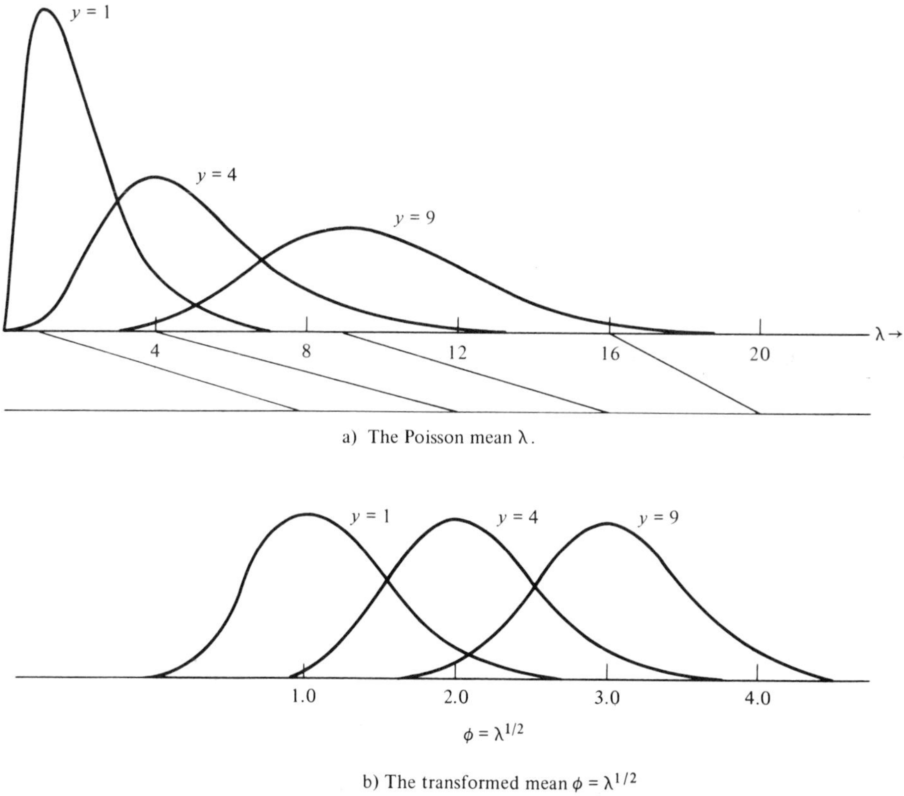

Fig. 1.3.8 Standardized likelihood curves: (a) for the Poisson mean λ, and (b) for the transformed mean $\phi = \lambda^{1/2}$.

and

$$\frac{\partial L}{\partial \lambda} = \frac{n\bar{y}}{\lambda} - n, \qquad \frac{\partial^2 L}{\partial \lambda^2} = \frac{-n\bar{y}}{\lambda^2}.$$

For $\bar{y} \neq 0$, the maximum likelihood estimate of λ obtained from $\partial L/\partial\lambda = 0$ is $\hat{\lambda} = \bar{y}$ so that

$$J(\hat{\lambda}) = \left(-\frac{1}{n}\frac{\partial^2 L}{\partial \lambda^2}\right)_{\hat{\lambda}} = \frac{1}{\hat{\lambda}}. \tag{1.3.42}$$

According to (1.3.33), a noninformative prior for λ is

$$p(\lambda) \propto J^{1/2}(\lambda) \propto \lambda^{-1/2}, \tag{1.3.43}$$

and $\phi = \lambda^{1/2}$ is the metric for which the approximate noninformative prior is locally uniform. The effectiveness of the transformation in achieving data translated curves is illustrated in Fig. 1.3.8(a) and (b), with $n = 1$ and $\bar{y} = y = 1$, $y = 4$, and $y = 9$.

Using the noninformative prior (1.3.43), the posterior distribution of λ is

$$p(\lambda \mid \mathbf{y}) = c\lambda^{n\bar{y}-\frac{1}{2}} \exp(-n\lambda), \qquad \lambda > 0, \tag{1.3.44}$$

where, on integration, the Normalizing constant is found to be

$$c = n^{-(n\bar{y}+\frac{1}{2})}\left[\Gamma(n\bar{y} + \tfrac{1}{2})\right]^{-1}.$$

Equivalently, we have that $n\lambda$ is distributed as $\frac{1}{2}\chi^2$ with $2n\bar{y} + 1$ degrees of freedom.

Figure 1.3.9 shows the posterior distribution of λ, given that $n = 1$ and a frequency of $y = 2$ has been observed, where little is known about λ *a priori*. The shaded area

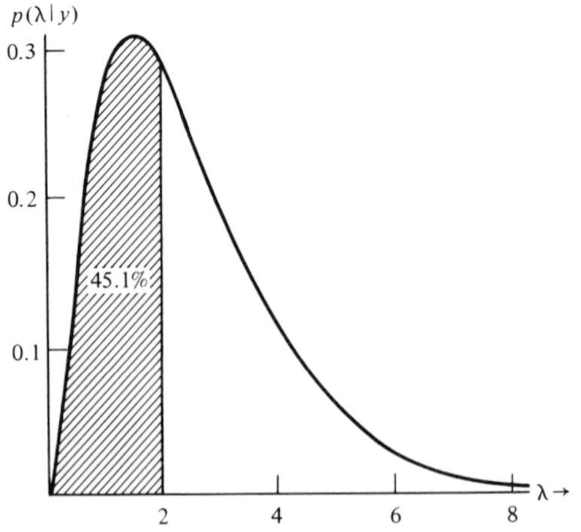

Fig. 1.3.9 Posterior distribution of the Poisson mean λ (noninformative prior) for an observed frequency $y = 2$.

corresponds to the probability that $\lambda < 2$ which is

$$\Pr\{\tfrac{1}{2}\chi_5^2 < 2\} = \Pr\{\chi_5^2 < 4\} = 45.1\%.$$

1.3.5 Jeffreys' Rule, Information Measure, and Noninformative Priors

In general, the distribution $p(y \mid \theta)$ need not belong to the family defined by (1.3.38), and the quantity

$$\left(-\frac{1}{n}\frac{\partial^2 L}{\partial \theta^2}\right)_{\hat{\theta}},$$

in (1.3.27) is a function of all the data $\mathbf{y}$. The argument leading to the approximate noninformative prior in (1.3.33) can then be modified as follows.

It is to be noted that, for given θ,

$$-\frac{1}{n}\frac{\partial^2 L}{\partial \theta^2} = -\frac{1}{n}\sum_{u=1}^{n}\frac{\partial^2 \log p(y_u \mid \theta)}{\partial \theta^2} \tag{1.3.45}$$

is the average of n identical functions of $(y_1, \ldots, y_n)$, respectively. Now suppose θ_0 is the true value of θ so that $\mathbf{y}$ are drawn from the distribution $p(y \mid \theta_0)$. It then follows that, for large n, the average converges in probability to the expectation of the function, that is, to

$$\mathop{E}_{y\mid\theta_0}\left[-\frac{\partial^2 \log p(y \mid \theta)}{\partial \theta^2}\right] = -\int \frac{\partial^2 \log p(y \mid \theta)}{\partial \theta^2}\, p(y \mid \theta_0)\, dy = a(\theta, \theta_0),$$

assuming that the expectation exists. Also, for large n, the maximum likelihood estimate $\hat{\theta}$ converges in probability to θ_0. Thus, we can write, approximately,

$$\left(-\frac{1}{n}\frac{\partial^2 L}{\partial \theta^2}\right)_{\hat{\theta}} \doteq a(\hat{\theta}, \theta_0) \doteq a(\hat{\theta}, \hat{\theta}) = \mathscr{I}(\hat{\theta}), \tag{1.3.46}$$

where $\mathscr{I}(\theta) = a(\theta, \theta)$ is the function

$$\mathscr{I}(\theta) = -\mathop{E}_{y\mid\theta}\left[\frac{\partial^2 \log p(y \mid \theta)}{\partial \theta^2}\right] = \mathop{E}_{y\mid\theta}\left[\frac{\partial \log p(y \mid \theta)}{\partial \theta}\right]^2. \tag{1.3.47}$$

Consequently, if we use $\mathscr{I}(\hat{\theta})$, which depends on $\hat{\theta}$ only, to approximate

$$\left(-\frac{1}{n}\frac{\partial^2 L}{\partial \theta^2}\right)_{\hat{\theta}}$$

in (1.3.27), then, arguing exactly as before, we find that the metric $\phi(\theta)$ for which a locally uniform prior is approximately noninformative is such that

$$\frac{d\phi}{d\theta} \propto \mathscr{I}^{1/2}(\theta) \qquad \text{or} \qquad \phi \propto \int^{\theta} \mathscr{I}^{1/2}(t)\, dt. \tag{1.3.48}$$

Equivalently, the noninformative prior for θ should be chosen so that, locally,

$$p(\theta) \propto \mathscr{I}^{1/2}(\theta). \tag{1.3.49}$$

It is readily confirmed that, when the distribution $p(y \mid \theta)$ is of the form (1.3.38), $J(\hat{\theta}) \equiv \mathscr{I}(\hat{\theta})$. Thus, the prior in (1.3.33), when applicable, is in fact identical to the prior of (1.3.49) and the latter form can be used generally.

For illustration, consider again the binomial mean π. The likelihood is equivalent to the likelihood from a sample of n independent point binomials identically distributed as

$$p(x \mid \pi) = \pi^x (1 - \pi)^{1-x}, \qquad x = 0, 1. \tag{1.3.50}$$

Thus

$$\frac{\partial^2 \log p}{\partial \pi^2} = -\frac{x}{\pi^2} - \frac{1-x}{(1-\pi)^2}. \tag{1.3.51}$$

Since $\underset{x\mid\pi}{E}(x) = \pi$, it follows that

$$\mathscr{I}(\pi) = \underset{x\mid\pi}{E}\left[-\frac{\partial^2 \log p}{\partial \pi^2}\right] = \pi^{-1}(1-\pi)^{-1}, \tag{1.3.52}$$

whence, according to (1.3.49), the noninformative prior of π is locally proportional to $\pi^{-1/2}(1-\pi)^{-1/2}$ as obtained earlier in (1.3.37a). Also, we see from (1.3.36) and (1.3.52) that $J(\hat{\pi})$ and $\mathscr{I}(\hat{\pi})$ are identical.

Now, the quantity $\mathscr{I}(\theta)$ obtained in (1.3.47) above is Fisher's measure of information about θ in a single observation y. More generally, Fisher's measure (1922, 1925) of information about θ in a sample $\mathbf{y}' = (y_1, \ldots, y_n)$ is defined as

$$\mathscr{I}_n(\theta) = \underset{\mathbf{y}\mid\theta}{E}\left(-\frac{\partial^2 L}{\partial \theta^2}\right) = \underset{\mathbf{y}\mid\theta}{E}\left(\frac{\partial L}{\partial \theta}\right)^2, \tag{1.3.53}$$

where, as before, L is the log likelihood and the expectation is taken with respect to the distribution $p(\mathbf{y} \mid \theta)$. When $\mathbf{y}$ is a random sample, $\mathscr{I}_n(\theta) = n\mathscr{I}(\theta)$. Thus, (1.3.49) can be expressed by the following rule.

Jeffreys' rule: The prior distribution for a single parameter θ is approximately noninformative if it is taken proportional to the square root of Fisher's information measure.

This rule for the choice of a noninformative prior distribution was first given by Sir Harold Jeffreys (1961), who justified it on the grounds of its invariance under parameter transformations. For, suppose $\phi = \phi(\theta)$ is a one-to-one transformation of θ; then it is readily seen that

$$\mathscr{I}(\phi) = \mathscr{I}(\theta)\left(\frac{d\theta}{d\phi}\right)^2. \tag{1.3.54}$$

Now, if some principle of choice led to $p(\theta)$ as a noninformative prior for θ, the same principle should lead to

$$p(\phi) = p(\theta)\left|\frac{d\theta}{d\phi}\right| \tag{1.3.55}$$

as a noninformative prior for ϕ. The principle of choice in (1.3.49) precisely satisfies this requirement: if we use it, then the prior of ϕ is

$$p(\phi) \propto \mathscr{I}^{1/2}(\phi) = \mathscr{I}^{1/2}(\theta)\left|\frac{d\theta}{d\phi}\right| \propto p(\theta)\left|\frac{d\theta}{d\phi}\right|. \tag{1.3.56}$$

The Location Scale Family

For illustration we show how results in (1.3.3) and (1.3.8) concerning the parameters (θ, σ) for the Normal distribution may, as an approximation, be extended to cover the general location-scale family of distributions

$$p(y \mid \theta, \sigma) \propto \sigma^{-1} h\left(\frac{y-\theta}{\sigma}\right). \tag{1.3.57}$$

where the range of y does not involve (θ, σ) and h satisfies certain regularity conditions. Suppose we have a sample of n independent observations $\mathbf{y}' = (y_1, \ldots, y_n)$ from this distribution.

a) θ unknown, σ known. The likelihood is

$$l(\theta \mid \sigma, \mathbf{y}) \propto \prod_{u=1}^{n} h\left(\frac{y_u-\theta}{\sigma}\right). \tag{1.3.58}$$

Now

$$\frac{\partial \log p(y \mid \theta, \sigma)}{\partial \theta} = -\frac{1}{\sigma}\frac{h'(x)}{h(x)}, \quad \text{where} \quad x = \left(\frac{y-\theta}{\sigma}\right). \tag{1.3.59}$$

Thus, from (1.3.47)

$$\mathscr{I}(\theta) = \underset{y|\theta}{E}\left[\frac{\partial \log p(y \mid \theta, \sigma)}{\partial \theta}\right]^2 = \frac{1}{\sigma^2}\,\underset{x}{E}\left[\frac{h'(x)}{h(x)}\right]^2, \tag{1.3.60}$$

where the expectation on the extreme right is taken over the distribution $p(x) = h(x)$. Since this expectation does not involve θ and σ^2 is known, $\mathscr{I}(\theta)$ = constant. So we are led to take θ locally uniform *a priori*,

$$p(\theta \mid \sigma) \propto \mathscr{I}^{1/2}(\theta) = \text{constant}, \tag{1.3.61}$$

and the corresponding posterior distribution of θ is approximately

$$p(\theta \mid \sigma, \mathbf{y}) \propto \prod_{u=1}^{n} h\left(\frac{y_u-\theta}{\sigma}\right), \qquad -\infty < \theta < \infty. \tag{1.3.62}$$

b) *σ unknown, θ known.* Here the likelihood is

$$l(\sigma \mid \theta, \mathbf{y}) \propto \sigma^{-n} \prod_{u=1}^{n} h\left(\frac{y_u - \theta}{\sigma}\right). \tag{1.3.63}$$

Since

$$\frac{\partial \log p(y \mid \theta, \sigma)}{\partial \sigma} = -\frac{1}{\sigma}\left[1 + \frac{xh'(x)}{h(x)}\right],$$

it follows that

$$\mathscr{I}(\sigma) = \frac{1}{\sigma^2} \underset{x}{E}\left[1 + \frac{xh'(x)}{h(x)}\right]^2 \propto \frac{1}{\sigma^2}. \tag{1.3.64}$$

Consequently, Jeffreys' rule leads to taking the prior

$$p(\sigma \mid \theta) \propto \frac{1}{\sigma} \qquad \text{or} \qquad p\,(\log \sigma) \propto \text{const.} \tag{1.3.65}$$

The corresponding posterior distribution of σ is then

$$p(\sigma \mid \theta, \mathbf{y}) \propto \sigma^{-(n+1)} \prod_{u=1}^{n} h\left(\frac{y_u - \theta}{\sigma}\right), \qquad \sigma > 0. \tag{1.3.66}$$

Caution in the Application of Jeffreys' Rule

Jeffreys' rule given by (1.3.49), like most rules, should not be mechanically applied. It is obviously inapplicable for example when $\mathscr{I}(\theta)$ does not exist. Furthermore, as Jeffreys pointed out, the rule has to be modified in certain circumstances which we will discuss later in Section 1.3.6. We believe that it is best, to treat as basic the idea of seeking a transformation for which the likelihood is approximately data translated, to treat each problem individually, and to regard equation (1.3.49) as a means whereby *in appropriate circumstances* this transformation can be determined.

Dependence of the Noninformative Prior Distribution on the Probability Model

In the development followed above the form of the noninformative prior distribution depends upon the probability model of the observations. It might be argued that this dependence is objectionable because the representation of *total* ignorance about a parameter ought to be the same, whatever the nature of a projected experiment. On the contrary, the position adopted above is that we seek to represent not total ignorance but an amount of prior information which is small *relative* to what the particular projected experiment can be expected to provide. The form of the prior *must* then depend on the expected likelihood.

As a specific example, suppose for example π is the proportion of successes in a Bernoulli population. Now π can be estimated (1) by counting the number of successes

y in a *fixed number of trials* n and using the fact that y has the binomial distribution in (1.3.21), or (2) by counting the number of trials z until a *fixed number of successes* r is obtained and supposing that z has the Pascal distribution

$$\binom{z-1}{r-1}\pi^r(1-\pi)^{z-r}, \qquad z = r, r+1, \ldots . \tag{1.3.67a}$$

These two experiments lead, on the argument given above, to slightly different noninformative priors. Specifically, for the binomial case, the information measure is, from (1.3.52),

$$E\left(-\frac{\partial^2 L}{\partial \pi^2}\right) = n\pi^{-1}(1-\pi)^{-1},$$

whence

$$p(\pi) \propto \pi^{-\frac{1}{2}}(1-\pi)^{-\frac{1}{2}}$$

and the corresponding noninformative prior is locally uniform in $\phi = \sin^{-1}\sqrt{\pi}$ as in (1.3.37a, b). On the other hand, it is easily shown that the information measure for the Pascal experiment is

$$E\left(-\frac{\partial^2 L}{\partial \pi^2}\right) = r\pi^{-2}(1-\pi)^{-1}, \tag{1.3.67b}$$

whence

$$p(\pi) \propto \pi^{-1}(1-\pi)^{-\frac{1}{2}} \tag{1.3.67c}$$

and the corresponding noninformative prior is locally uniform in

$$\phi = \log\frac{1-\sqrt{1-\pi}}{1+\sqrt{1-\pi}}.$$

Now, these two kinds of experiments would lead to exactly the same likelihood when $n = z$ and $y = r$; and when this is so it has been argued that inference about π from both experiments ought to be identical. The use of the above two noninformative priors however will not yield this result. For illustration let us suppose there were 24 trials with 21 successes. If to arrive at this result sampling had been continued till the number of *trials* was 24 the posterior distribution obtained with the appropriate noninformative prior would have been

$$p(\pi \mid y = 21) \propto \pi^{20.5}(1-\pi)^{2.5}.$$

However if sampling had been continued till the number of *successes* was 21 then the posterior distribution obtained with the appropriate noninformative prior in (1.3.67c) would have been

$$p(\pi \mid z = 24) \propto \pi^{20}(1-\pi)^{2.5}.$$

This says that when we sample till the number of successes reaches a certain value some downward adjustment of probability is needed relative to sampling with fixed n. We find this result much less surprising than the claim that they ought to agree.

In general we feel that it is sensible to choose a noninformative prior which expresses ignorance *relative* to information which can be supplied by a particular experiment. If the experiment is changed, then the expression of relative ignorance can be expected to change correspondingly.

1.3.6 Noninformative Priors for Multiple Parameters

We now extend previous ideas to include multiparameter models. We begin by considering the Normal linear model with σ assumed known.

The Parameters θ in a Normal Linear Model, σ Assumed Known

Suppose $\mathbf{y}' = (y_1, \ldots, y_n)$ is a set of Normally and independently distributed random variables having common variance σ^2, and the expected value of y_u is a linear function of k parameters $\boldsymbol{\theta}' = (\theta_1, \ldots, \theta_k)$ such that

$$E(y_u) = \theta_1 x_{u1} + \theta_2 x_{u2} + \cdots + \theta_k x_{uk}, \qquad u = 1, 2, \ldots, n, \tag{1.3.68}$$

where the x's are known constants.

This Normal linear model is of basic importance. In particular, for suitable choice of the x's, it provides the structure for general Analysis of Variance and for Regression (Least Squares) Analysis. Special cases include models already considered. For example, the model (1.1.1) for a Normal sample is obtained by setting $k = 1, \theta_1 = \theta$, and $x_{u1} = 1$ $(u = 1, 2, \ldots, n)$.

In general, if $\mathbf{X}$ is the $n \times k$ matrix $\{x_{uj}\}$ of known constants, then the n equations may be written concisely as

$$E(\mathbf{y}) = \mathbf{X}\boldsymbol{\theta}.$$

If σ is known, then the likelihood is

$$l(\boldsymbol{\theta} \mid \sigma, \mathbf{y}) \propto \exp\left[-\frac{1}{2\sigma^2}(\mathbf{y} - \mathbf{X}\boldsymbol{\theta})'(\mathbf{y} - \mathbf{X}\boldsymbol{\theta})\right]. \tag{1.3.69}$$

We shall suppose that the rank of the matrix $\mathbf{X}$ is k. The quadratic form in (1.3.69) may then be written

$$(\mathbf{y} - \mathbf{X}\boldsymbol{\theta})'(\mathbf{y} - \mathbf{X}\boldsymbol{\theta}) = (\mathbf{y} - \hat{\mathbf{y}})'(\mathbf{y} - \hat{\mathbf{y}}) + (\boldsymbol{\theta} - \hat{\boldsymbol{\theta}})'\mathbf{X}'\mathbf{X}(\boldsymbol{\theta} - \hat{\boldsymbol{\theta}}), \tag{1.3.70}$$

where

$$\hat{\boldsymbol{\theta}} = (\mathbf{X}'\mathbf{X})^{-1}\mathbf{X}'\mathbf{y}$$

is the vector of least squares estimate of $\boldsymbol{\theta}$, and $\hat{\mathbf{y}} = \mathbf{X}\hat{\boldsymbol{\theta}}$ is the vector of fitted values so that $(\mathbf{y} - \hat{\mathbf{y}})'(\mathbf{y} - \hat{\mathbf{y}})$ is a function of data not involving $\boldsymbol{\theta}$. The

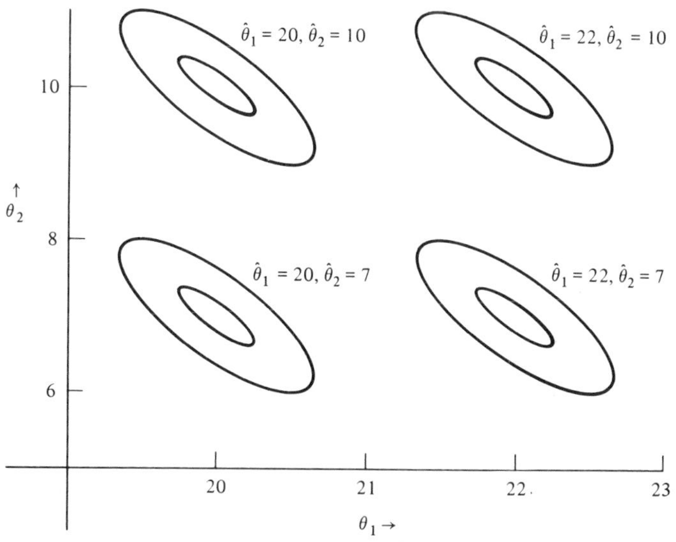

Fig. 1.3.10 Normal linear model: contours of likelihood function for different sets of data.

likelihood can therefore be expressed as

$$l(\boldsymbol{\theta} \mid \sigma, \mathbf{y}) \propto \exp\left[-\frac{1}{2\sigma^2}(\boldsymbol{\theta} - \hat{\boldsymbol{\theta}})'\mathbf{X}'\mathbf{X}(\boldsymbol{\theta} - \hat{\boldsymbol{\theta}})\right], \tag{1.3.71}$$

which is in the form of a multivariate Normal function† centered at $\boldsymbol{\theta}$ and having covariance matrix $\sigma^2(\mathbf{X}'\mathbf{X})^{-1}$. The likelihood contours in the parameter space of $\boldsymbol{\theta}$ are thus ellipses ($k = 2$), ellipsoids ($k = 3$), or hyperellipsoids ($k > 3$) defined by

$$(\boldsymbol{\theta} - \hat{\boldsymbol{\theta}})'\mathbf{X}'\mathbf{X}(\boldsymbol{\theta} - \hat{\boldsymbol{\theta}}) = \text{const.} \tag{1.3.72}$$

Figure 1.3.10 illustrates the case $k = 2$, where likelihood contours are shown for different sets of data yielding different values of $\hat{\theta}_1$ and $\hat{\theta}_2$. The likelihood is data translated. Specifically, from (1.3.71) it is seen that, as soon as an experimental

---

† We refer to the function

$$f(\mathbf{x}) = \frac{|\mathbf{V}|^{-1/2}}{(2\pi)^{p/2}} \exp\left[-\tfrac{1}{2}(\mathbf{x} - \boldsymbol{\mu})'\, \mathbf{V}^{-1}\, (\mathbf{x} - \boldsymbol{\mu})\right],$$

where $\mathbf{x}' = (x_1, \ldots, x_p)$ and $\boldsymbol{\mu}' = (\mu_1, \ldots, \mu_p)$ are $p \times 1$ vectors, and $\mathbf{V}$ is a $p \times p$ positive definite symmetric matrix, as the multivariate Normal function. When $\mathbf{x}$ are random variables, the function becomes the multivariate Normal distribution with mean vector $\boldsymbol{\mu}$ and covariance matrix $\mathbf{V}$.

design has been decided and hence $\mathbf{X}$ is known, all features of the likelihood except its location are known prior to the data being taken. By contrast, location of the likelihood is determined through $\hat{\boldsymbol{\theta}}$ *solely* by the data. The idea that little is known *a priori* relative to what the data will tell us is therefore expressed by a prior distribution such that, locally,

$$p(\boldsymbol{\theta} \mid \sigma) \propto c. \tag{1.3.73}$$

On multiplying the likelihood in (1.3.71) by a nearly constant prior, and introducing the appropriate normalizing constant so that the posterior distribution integrates to one, we have approximately

$$p(\boldsymbol{\theta} \mid \sigma, \mathbf{y}) \doteq \frac{|\mathbf{X}'\mathbf{X}|^{1/2}}{(2\pi\sigma^2)^{k/2}} \exp\left[-\frac{1}{2\sigma^2}(\boldsymbol{\theta} - \hat{\boldsymbol{\theta}})' \mathbf{X}'\mathbf{X}(\boldsymbol{\theta} - \hat{\boldsymbol{\theta}})\right],$$

$$-\infty < \theta_i < \infty, \quad i = 1, \ldots, k, \tag{1.3.74}$$

a multivariate Normal distribution which we denote by $N_k[\hat{\boldsymbol{\theta}}, \sigma^2(\mathbf{X}'\mathbf{X})^{-1}]$. This distribution will be discussed in detail in Section 2.7.

For certain common situations discussed more fully in Chapter 7 this formulation of the multivariate prior distribution may be inappropriate. For example, it may be known that the θ's themselves are a sample from some distribution. When such information is available and particularly when k is large, the locally uniform prior of (1.3.73) may supply an inadequate approximation.

Multiparameter Data Translated Likelihoods

In general, suppose the distribution of the data $\mathbf{y}$ involves k parameters $\boldsymbol{\theta}' = (\theta_1, \ldots, \theta_k)$. A data translated likelihood must be of the form

$$l(\boldsymbol{\theta} \mid \mathbf{y}) = g[\boldsymbol{\phi} - \mathbf{f}(\mathbf{y})], \tag{1.3.75}$$

where $g(\mathbf{x})$ is a known function independent of $\mathbf{y}$, $\boldsymbol{\phi}' = (\phi_1, \ldots, \phi_k)$ is a one-to-one transformation of $\boldsymbol{\theta}$, and the elements $f_1(\mathbf{y}), \ldots, f_k(\mathbf{y})$ of the $k \times 1$ vector $\mathbf{f}(\mathbf{y})$ are k functions of $\mathbf{y}$. Extending the single parameter case, a noninformative prior is supposed locally uniform in $\boldsymbol{\phi}$. The corresponding noninformative prior in $\boldsymbol{\theta}$ is then

$$p(\boldsymbol{\theta}) \propto |J|, \tag{1.3.76}$$

where

$$|J| = \left|\frac{\partial(\phi_1, \ldots, \phi_k)}{\partial(\theta_1, \ldots, \theta_k)}\right|_+ = \begin{vmatrix} \dfrac{\partial\phi_1}{\partial\theta_1} & \cdots & \dfrac{\partial\phi_1}{\partial\theta_k} \\ \vdots & & \vdots \\ \dfrac{\partial\phi_k}{\partial\theta_1} & \cdots & \dfrac{\partial\phi_k}{\partial\theta_k} \end{vmatrix}_+$$

is the absolute value of the Jacobian of the transformation. For the Normal linear model (1.3.68), $\boldsymbol{\theta} = \boldsymbol{\phi}$ so that a noninformative prior is locally uniform in $\boldsymbol{\theta}$ as given earlier in (1.3.73).

Multiparameter Problems Involving Location and Scale Parameters

Special care must be exercised in choosing noninformative priors for location and scale parameters simultaneously. As mentioned earlier, we regard a parameter η as a *location* parameter of a distribution $p(y)$ if addition of a constant c to y changes η to $\eta + c$. Thus, the Normal mean and, more generally, the parameter θ in the family of distributions (1.3.57) are location parameters. The elements of $\boldsymbol{\theta}$ in the Normal linear model (1.3.68) are also location parameters in a general sense, since it can be shown that they are location parameters of the distribution of the least squares estimates $\hat{\boldsymbol{\theta}}$. On the other hand, a *scale* parameter λ of a distribution $p(y)$ is such that multiplication of y by a constant c changes λ to $|c|\,\lambda$. Examples of scale parameters are the Normal standard deviation and, more generally, the parameter σ in (1.3.57) and in the linear model (1.3.68).

Normal Mean θ and Standard Deviation σ

We first consider the choice of prior in relation to a sample from a Normal distribution $N(\theta, \sigma^2)$, where θ and σ are both unknown. The likelihood of (θ, σ) is

$$l(\theta, \sigma \mid \mathbf{y}) \propto \sigma^{-n} \exp\left[-\frac{1}{2\sigma^2} \Sigma\, (y_u - \theta)^2\right]. \tag{1.3.77a}$$

Now

$$\begin{aligned} \Sigma\,(y_u - \theta)^2 &= \Sigma\,(y_u - \bar{y})^2 + n(\bar{y} - \theta)^2 \\ &= (n-1)s^2 + n(\bar{y} - \theta)^2, \end{aligned}$$

where $s^2 = \Sigma\,(y_u - \bar{y})^2/(n-1)$. Thus,

$$l(\theta, \sigma \mid \mathbf{y}) \propto \sigma^{-n} \exp\left[-\frac{n(\theta - \bar{y})^2}{2\sigma^2} - \frac{(n-1)s^2}{2\sigma^2}\right]. \tag{1.3.77b}$$

Also, since multiplication by the constant s^n leaves the likelihood unchanged,

$$l(\theta, \sigma \mid \mathbf{y}) \propto \left(\frac{s}{\sigma}\right)^n \exp\left[-\frac{n(\theta - \bar{y})^2}{2s^2}\left(\frac{s^2}{\sigma^2}\right) - \frac{(n-1)s^2}{2\sigma^2}\right], \tag{1.3.77c}$$

which can be written,

$$l(\theta, \sigma \mid \mathbf{y}) \propto \exp\left\{-\frac{n}{2}\left(\frac{\theta - \bar{y}}{s}\right)^2 \exp\left[-2(\log\sigma - \log s)\right]\right\}$$

$$\times \exp\left\{-n(\log\sigma - \log s) - \left(\frac{n-1}{2}\right) \exp\left[-2(\log\sigma - \log s)\right]\right\} \tag{1.3.77d}$$

and is therefore of the form

$$l(\theta, \sigma \mid \mathbf{y}) \propto F\left(\frac{\theta - \bar{y}}{s}, \log \sigma - \log s\right) G(\log \sigma - \log s). \tag{1.3.77e}$$

To aid understanding of this expression, Fig. 1.3.11 shows likelihood contours for θ and $\log \sigma$ given by four different samples of $n = 10$ observations. For fixed s a change in $\bar{y}$ relocates the likelihood surface on the θ axis. For fixed $\bar{y}$ a magnification in s appropriately relocates the likelihood surface on the $\log \sigma$ axis, while at the same time its spread along the θ axis is correspondingly magnified. This magnification reflects the greater uncertainty in the likelihood about θ which occurs when a larger σ is implied by an increase in s. It would usually be the case, however, that prior opinions about θ bear little relationship to those about σ, so that such magnifications would be irrelevant to the choice of transformations of θ. Thus, we are led to seek a transformation which, *apart from this inherent magnification along the θ axis*, is such that the data serves only to relocate the likelihood function. In this case the appropriate transformation is clearly obtained in terms of θ and $\log \sigma$. A noninformative prior is therefore taken to be one for which approximately $\log \sigma$ and θ are locally uniform,

$$p(\theta, \log \sigma) \propto c. \tag{1.3.78}$$

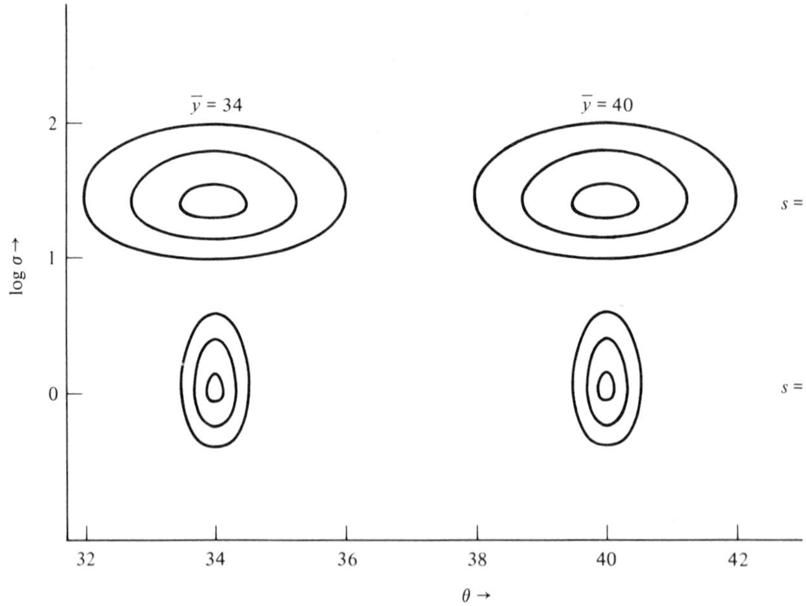

Fig. 1.3.11 The Normal mean θ and standard deviation σ: contours of likelihood function of $(\theta, \log \sigma)$ for different sets of data.

Equivalently

$$p(\theta, \sigma) \propto \sigma^{-1}. \tag{1.3.79}$$

Employing the prior in (1.3.79) with the likelihood (1.3.77b), we found that the posterior distribution of (θ, σ) is

$$p(\theta, \sigma \mid \mathbf{y}) \propto \sigma^{-(n+1)} \exp\left[-\frac{n(\theta - \bar{y})^2}{2\sigma^2} - \frac{(n-1)s^2}{2\sigma^2}\right], \qquad -\infty < \theta < \infty, \quad \sigma > 0, \tag{1.3.80}$$

which will be discussed in detail in Section 2.4.

Normal Linear Model, σ Unknown

We now turn to the case of the general Normal linear model in (1.3.68), and suppose that the standard deviation σ as well as the parameters $\boldsymbol{\theta}$ are unknown. The likelihood can be written

$$l(\boldsymbol{\theta}, \sigma \mid \mathbf{y}) \propto \left(\frac{s}{\sigma}\right)^n \exp\left[-\frac{(n-k)s^2}{2\sigma^2} - \frac{s^2}{2\sigma^2}\,\frac{(\boldsymbol{\theta} - \hat{\boldsymbol{\theta}})'\,\mathbf{X}'\mathbf{X}(\boldsymbol{\theta} - \hat{\boldsymbol{\theta}})}{s^2}\right], \tag{1.3.81}$$

where

$$s^2 = \frac{1}{(n-k)}(\mathbf{y} - \hat{\mathbf{y}})'\,(\mathbf{y} - \hat{\mathbf{y}}).$$

By comparing (1.3.81) with (1.3.77b–e), it is evident that in terms of $\boldsymbol{\theta}$ and $\log \sigma$, for any fixed s, a change in $\hat{\boldsymbol{\theta}}$ merely relocates the likelihood. On the other hand, for fixed $\hat{\boldsymbol{\theta}}$, the volume enclosed in a given contour is proportional to s^k. Arguing as in the case of (θ, σ), we seek a transformation which, apart from this inherent magnification in the space of $\boldsymbol{\theta}$, is such that the data serves only to relocate the likelihood surface. Clearly, the appropriate transformation is obtained in terms of $\boldsymbol{\theta}$ and $\log \sigma$. A noninformative prior in this case is then one for which approximately $\log \sigma$ and $(\theta_1, \ldots, \theta_k)$ are locally uniform. Specifically, it is assumed that, locally,

$$p(\boldsymbol{\theta}, \log \sigma) \propto c \qquad \text{or} \qquad p(\boldsymbol{\theta}, \sigma) \propto 1/\sigma. \tag{1.3.82}$$

The corresponding posterior distribution is then

$$p(\boldsymbol{\theta}, \sigma \mid \mathbf{y}) \propto \sigma^{-(n+1)} \exp\left[-\frac{(n-k)s^2}{2\sigma^2} - \frac{(\boldsymbol{\theta} - \hat{\boldsymbol{\theta}})'\mathbf{X}'\mathbf{X}(\boldsymbol{\theta} - \hat{\boldsymbol{\theta}})}{2\sigma^2}\right]$$

$$-\infty < \theta_i < \infty, \quad i = 1, \ldots, k; \quad \sigma > 0, \tag{1.3.83}$$

which will be further discussed in Section 2.7.

Prior Independence Between Parameters

In some examples, certain parameters or sets of parameters may be judged *a priori* to be distributed independently of certain other parameters or sets of parameters. When this is so, the choice of prior distribution is sometimes simplified because the independent sets of parameters may be separately considered.

In particular, it is usually appropriate to take location parameters to be distributed independently of scale parameters. This is because any prior idea one might have about the location θ of a distribution would usually not be much influenced by one's idea about the value of its scale parameter σ. Thus $p(\theta \mid \sigma) \doteq p(\theta)$. Consider the case of the Normal distribution. We have seen that for the case where σ is known, a noninformative prior for θ is obtained by taking $p(\theta \mid \sigma)$ locally uniform. With the additional independence assumption this implies that $p(\theta)$ should be uniform. A similar argument leads to our taking $p(\sigma) \propto 1/\sigma$. Thus

$$p(\theta, \sigma) \doteq p(\theta)p(\sigma) \propto 1/\sigma \tag{1.3.84}$$

as in (1.3.79).

In certain circumstances, such an assumption of independence between θ and σ could appear inappropriate. If, for example, we know that grains of sand, with mean weight one milligram, were to be weighed, we should expect that σ would be less than a milligram, and so it has been argued that if θ is small, then σ is likely to be small also, while if θ is large, σ is likely to be large.

To this we reply, (1) that dependence of this kind is most often associated with a natural constraint on the data (for example, that all observations must be non-negative); (2) that this kind of dependence is usually removed when a more appropriate metric is adopted in terms of which the data are not so constrained; (3) that usually when there is no such constraint one does not expect this dependence. We illustrate with the following examples.

An example of prior dependence removed by appropriate transformation. Suppose we knew that the mean income θ of a certain community was \$5000 and the standard deviation was \$1000. Then given another community where the mean income was known to be \$50,000, we might guess that the standard deviation was closer to \$10,000 than to \$1000. In other words, prior beliefs about θ and σ would be dependent. However, this guess is clearly based on the supposition that, in examples of this kind, it is the coefficient of variation σ/θ and not σ itself which is more likely to be approximately constant over different values of θ. But if this supposition is correct, then log income is the appropriate quantity to consider. For, if we denote income by y and suppose that ϕ and λ are the mean and standard deviation of log y, then

$$\phi = E(\log y) \doteq \log \theta, \qquad \lambda = \sqrt{\operatorname{var}(\log y)} \doteq \sigma/\theta.$$

Measured in the logarithmic metric, the standard deviation λ can realistically be assumed independent of the mean ϕ *a priori*. We notice that, in this example, if y is an observed income, $0 < y < \infty$ but $-\infty < \log y < \infty$.

An example where measurements may be negative. Except in examples like the above, where the measurement scale has a natural constraint such as a truncation at zero, values of θ which are small in magnitude need not be associated with small values of σ. For example, suppose we were checking the declination of a compass from magnetic north, using an instrument which could detect a declination from $-180°$ to $180°$. For a properly constructed compass, we would expect the declination θ to be close to zero but this would not ordinarily affect our ideas about σ.

In this book, we shall usually assume that location and scale parameters are approximately independent *a priori*. In particular, arguing as before, for the parameters $(\boldsymbol{\theta}, \sigma)$ in the linear Normal model, we suppose that $p(\boldsymbol{\theta}, \sigma) \doteq p(\boldsymbol{\theta})\,p(\sigma)$ so that $p(\boldsymbol{\theta}) \doteq p(\boldsymbol{\theta} \mid \sigma)$ and $p(\sigma) \doteq p(\sigma \mid \boldsymbol{\theta})$. It then follows that $p(\boldsymbol{\theta}, \sigma) \propto \sigma^{-1}$ as given in (1.3.82).

Extension of Jeffreys' Rule to Multiparameter Models

In the multiparameter examples discussed above, transformations were available which had the property that, apart from the inherent magnifying effect of the scale factor—in the location parameter space, the likelihood was data translated. Although transformations of this kind are available for many of the applications discussed in later chapters, they are not available in general. In some cases, then, to obtain noninformative prior distributions, we must rely on a somewhat less satisfactory argument leading to the multiparameter version of Jeffreys' rule.

If the distribution of $\mathbf{y}$, depending on k parameters $\boldsymbol{\theta}$, obeys certain regularity conditions, then, for sufficiently large samples, the likelihood function for $\boldsymbol{\theta}$ and for mild transformations of $\boldsymbol{\theta}$ approaches a multivariate Normal distribution. The log likelihood is thus approximately quadratic,

$$L(\boldsymbol{\theta} \mid \mathbf{y}) = \log l(\boldsymbol{\theta} \mid \mathbf{y}) \doteq L(\hat{\boldsymbol{\theta}} \mid \mathbf{y}) - \frac{n}{2}(\boldsymbol{\theta} - \hat{\boldsymbol{\theta}})'\mathbf{D}_{\hat{\boldsymbol{\theta}}}(\boldsymbol{\theta} - \hat{\boldsymbol{\theta}}), \tag{1.3.85}$$

where $\hat{\boldsymbol{\theta}}$ is the vector of maximum likelihood estimates of $\boldsymbol{\theta}$ and $-n\mathbf{D}_{\hat{\boldsymbol{\theta}}}$ is the $k \times k$ matrix of second derivatives evaluated at $\hat{\boldsymbol{\theta}}$, that is,

$$\mathbf{D}_{\hat{\boldsymbol{\theta}}} = \left\{-\frac{1}{n}\frac{\partial^2 L}{\partial\theta_i\,\partial\theta_j}\right\}_{\hat{\boldsymbol{\theta}}}, \qquad i, j = 1, \ldots, k.$$

In general, $\mathbf{D}_{\hat{\boldsymbol{\theta}}}$ will depend upon all the data $\mathbf{y}$. But for large n, it can be closely approximated by

$$\mathbf{D}_{\hat{\boldsymbol{\theta}}} \doteq \frac{1}{n}\,\mathscr{I}_n(\hat{\boldsymbol{\theta}}), \tag{1.3.86}$$

which is a function of $\hat{\boldsymbol{\theta}}$ only. Specifically, $\mathscr{I}_n(\boldsymbol{\theta})$ is the matrix function

$$\mathscr{I}_n(\boldsymbol{\theta}) = E\left\{-\frac{\partial^2 L}{\partial\theta_i\,\partial\theta_j}\right\}, \tag{1.3.87}$$

where the expectation is taken with respect to the data distribution $p(\mathbf{y} \mid \boldsymbol{\theta})$. In other words, $\mathscr{I}_n(\boldsymbol{\theta})$ is the information matrix associated with the sample $\mathbf{y}$.

Now, ideally one would seek a transformation $\boldsymbol{\phi}(\hat{\boldsymbol{\theta}})$ such that $\mathscr{I}_n(\hat{\boldsymbol{\phi}})$ would be a constant matrix independent of $\hat{\boldsymbol{\phi}}$ so that the likelihood would be approximately data translated. Because this is not possible in general, we may seek a transformation $\boldsymbol{\phi}$ which ensures that the *content* of the approximate likelihood region of $\boldsymbol{\phi}$,

$$(\boldsymbol{\phi} - \hat{\boldsymbol{\phi}})' \mathscr{I}_n(\hat{\boldsymbol{\phi}}) (\boldsymbol{\phi} - \hat{\boldsymbol{\phi}}) < \text{const.}, \tag{1.3.88}$$

remains constant for different $\hat{\boldsymbol{\phi}}$. Since the square root of the determinant, $|\mathscr{I}_n(\hat{\boldsymbol{\phi}})|^{1/2}$, measures the volume of the likelihood region, the above requirement is equivalent to asking for a transformation for which the $|\mathscr{I}_n(\hat{\boldsymbol{\phi}})|$ is independent of $\hat{\boldsymbol{\phi}}$. To find such a transformation, note that

$$\mathscr{I}_n(\boldsymbol{\phi}) = \mathbf{A}\mathscr{I}_n(\boldsymbol{\theta})\mathbf{A}', \tag{1.3.89}$$

where $\mathbf{A}$ is the $k \times k$ matrix of partial derivatives,

$$\mathbf{A} = \left[\frac{\partial(\theta_1, \ldots, \theta_k)}{\partial(\phi_1, \ldots, \phi_k)} \right].$$

Thus,

$$|\mathscr{I}_n(\boldsymbol{\phi})| = |\mathbf{A}|^2 \, |\mathscr{I}_n(\boldsymbol{\theta})|, \tag{1.3.90}$$

whence the above requirement will be satisfied if $\boldsymbol{\phi}$ is such that

$$|\mathbf{A}| = \left| \frac{\partial(\theta_1, \ldots, \theta_k)}{\partial(\phi_1, \ldots, \phi_k)} \right| \propto |\mathscr{I}_n(\boldsymbol{\theta})|^{-1/2}, \tag{1.3.91}$$

and an approximate noninformative prior is one which is locally uniform in $\boldsymbol{\phi}$. The corresponding noninformative in $\boldsymbol{\theta}$ is then

$$p(\boldsymbol{\theta}) = p(\boldsymbol{\phi}) \left| \frac{\partial(\phi_1, \ldots, \phi_k)}{\partial(\theta_1, \ldots, \theta_k)} \right|_+ ;$$

that is,

$$p(\boldsymbol{\theta}) \propto |\mathscr{I}_n(\boldsymbol{\theta})|^{1/2}. \tag{1.3.92}$$

From the above we obtain the following rule:

Jeffreys' rule for multiparameter problems: The prior distribution for a set of parameters is taken to be proportional to the square root of the determinant of the information matrix.

As in the case of a single parameter, Jeffreys derived this general rule by *requiring* invariance under parameter transformation. He himself pointed out that this multiparameter rule must be applied with caution, especially where scale and location parameters occur simultaneously. We first consider an application where no difficulty occurs.

Multinomial Distribution

Consider the derivation of an appropriate prior for the parameters of the multinomial distribution. Suppose the result of a trial must be to produce one of m different outcomes, the probabilities for which are $\pi_1, \pi_2, \ldots, \pi_m$. Thus, the trial might consist of the random drawing with replacement of a ball from a bag. The probabilities could then refer to the proportions $\boldsymbol{\pi}' = (\pi_1, \ldots, \pi_m)$ of balls of m different colors where $\pi_m = 1 - \sum_{i=1}^{m-1} \pi_i$. Suppose n independent trials were made, resulting in a sample of $\mathbf{y}' = (y_1, \ldots, y_m)$ balls of the various types, where $y_m = n - \sum_{i=1}^{m-1} y_i$. Then

$$p(\mathbf{y} \mid \boldsymbol{\pi}) = \frac{n!}{(y_1!)(y_2!) \ldots (y_m!)} (\pi_1^{y_1})(\pi_2^{y_2}) \cdots (\pi_m^{y_m}) \tag{1.3.93a}$$

so that

$$L = \log l(\boldsymbol{\pi} \mid \mathbf{y}) = \sum_{j=1}^{m} y_j \log \pi_j. \tag{1.3.93b}$$

On differentiating, we obtain

$$b_{jj} = \frac{\partial^2 L}{(\partial \pi_j)^2} = -\frac{y_j}{\pi^2} + \frac{y_m}{\pi_m^2} \tag{1.3.94}$$

and

$$b_{ij} = \frac{\partial^2 L}{\partial \pi_i \, \partial \pi_j} = -\frac{y_m}{\pi_m^2}, \qquad i, j = 1, \ldots, m-1.$$

Taking expectations over the distribution $p(\mathbf{y} \mid \boldsymbol{\pi})$, we have

$$-E(b_{jj}) = \frac{n}{\pi_j} + \frac{n}{\pi_m}, \qquad -E(b_{ij}) = \frac{n}{\pi_m}. \tag{1.3.95}$$

After some algebraic reduction we find that

$$|\mathscr{I}_n(\boldsymbol{\pi})| = -|E\{b_{ij}\}| = n(\pi_1 \cdot \pi_2 \cdots \pi_m)^{-1}. \tag{1.3.96}$$

Thus, Jeffreys' rule says that we should take for a noninformative prior

$$p(\boldsymbol{\pi}) \propto (\pi_1 \cdot \pi_2 \cdots \pi_m)^{-1/2}. \tag{1.3.97}$$

For the case where little is known *a priori* about the probabilities, this leads to the posterior density

$$p(\boldsymbol{\pi} \mid \mathbf{y}) \propto \pi_1^{y_1 - \frac{1}{2}} \pi_2^{y_2 - \frac{1}{2}} \cdots \pi_m^{y_m - \frac{1}{2}}. \tag{1.3.98}$$

which is proportional to the likelihood for $\boldsymbol{\pi}$, with each cell frequency reduced by one half. In the particular case $k = 2$, we obtain the binomial result (1.3.26) considered earlier.

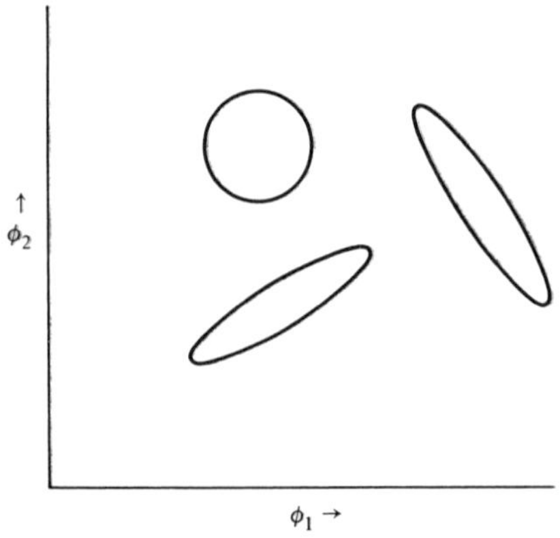

Fig. 1.3.12 Likelihood regions of different shape having the same size.

Some Comments on the Application of the Multiparameter Jeffreys' Rule

The multiparameter version of Jeffreys' rule leading to $p(\boldsymbol{\theta}) \propto |\mathscr{I}_n(\boldsymbol{\theta})|^{\frac{1}{2}}$ corresponds with a less stringent and less convincing transformation requirement on the likelihood than data translation. Specifically, if approximate Normality of the likelihood function is assumed, the rule seeks a transformation which ensures, irrespective of the data, that corresponding likelihood regions for $\boldsymbol{\phi}$ are of the same size. As illustrated in Fig. 1.3.12 for the case of $k = 2$ parameters, regions can of course be of the same size and yet be very different.

A further difficulty associated with the blanket application of Jeffreys' rule arises when parameters of different kinds are considered simultaneously. We have already seen, for example, that when the mean θ and standard deviation σ of a Normal distribution are being considered simultaneously—see Fig. 1.3.11, it is not usually appropriate to seek a transformation which produces likelihood regions of the same size. To further appreciate the difficulty, we discuss again the choice of prior for the location and scale parameters of the Normal distribution $N(\theta, \sigma^2)$.

Location and Scale Parameters

Jeffreys argued that in cases where θ and σ are known to be independent *a priori*, priors for the two parameters should be considered separately, leading to $p(\theta, \sigma) \doteq p(\theta)\, p(\sigma) \propto \sigma^{-1}$ as in (1.3.84).

Now, the information matrix of (θ, σ) is

$$\mathscr{I}_n(\theta, \sigma) = n \begin{bmatrix} \sigma^{-2} & 0 \\ 0 & 2\sigma^{-2} \end{bmatrix} = n \begin{bmatrix} \mathscr{I}(\theta) & 0 \\ 0 & \mathscr{I}(\sigma) \end{bmatrix}. \tag{1.3.99}$$

If the prior independence of (θ, σ) were ignored, then application of the generalized rule would lead to a prior locally proportional to

$$|\mathscr{I}_n(\theta, \sigma)|^{1/2} \propto \sigma^{-2}, \tag{1.3.100}$$

which differs from the prior in (1.3.84) by an additional factor σ^{-1}. Some light is thrown on this factor if we consider the problem from a transformation point of view.

Suppose $\kappa(\theta)$ and $\phi(\sigma)$ are, respectively, one-to-one transformations of θ and σ. Then the information matrix of (κ, ϕ) is

$$\mathscr{I}_n[\kappa(\theta), \phi(\sigma)] = n\begin{bmatrix} \sigma^{-2}\left|\dfrac{d\theta}{d\kappa}\right|^2 & 0 \\ 0 & 2\sigma^{-2}\left|\dfrac{d\sigma}{d\phi}\right|^2 \end{bmatrix} = n\begin{bmatrix} \mathscr{I}(\kappa) & 0 \\ 0 & \mathscr{I}(\phi) \end{bmatrix}, \qquad (1.3.101)$$

and by setting $\phi(\sigma) = \log\sigma$, the lower right hand element in the matrix, representing information about $\phi(\sigma)$, is made independent of σ so that

$$\mathscr{I}_n[\kappa(\theta), \log\sigma] = n\begin{bmatrix} \sigma^{-2}\left|\dfrac{d\theta}{d\kappa}\right|^2 & 0 \\ 0 & 2 \end{bmatrix}. \qquad (1.3.102)$$

This reflects the fact that whatever transformation $\kappa(\theta)$ is made on θ, information about κ will be inversely proportional to σ^{-2}—data having a small σ lead to a more accurate determination of the location parameter θ or of any transformation $\kappa(\theta)$ of it. On the other hand, the metric for which the information is independent of θ is clearly θ itself.

The situation can be related to the likelihood regions illustrated in Fig. 1.3.11. First, the zero off-diagonal elements of the matrix (1.3.102) reflects the fact that the "axes" of the likelihood contours lie parallel to the θ and $\log\sigma$ axes. Second, independence of $\mathscr{I}(\log\sigma)$ and $\mathscr{I}(\theta)$ with respect to θ corresponds to the fact that, for a fixed s, the likelihood is merely relocated when $\bar{y}$ is changed. Third, constancy of $\mathscr{I}(\log\sigma)$ has led to a relocation of the likelihood along the $\log\sigma$ axis when s is changed. Finally, the same change in s magnifies the spread of the likelihood along the θ axis because $\mathscr{I}(\theta) \propto \sigma^{-2}$.

Thus, when the assumption of prior independence of θ and σ is incorporated, the fact that the information of θ, or any function of it, is proportional to σ^{-2} is clearly irrelevant to the choice of a noninformative prior for θ. The additional factor σ^{-1} in (1.3.100) arises only from a misapplication which ignores prior independence.

A similar situation occurs for the linear Normal model in (1.3.68). The information matrix for $\boldsymbol{\theta}$ and log is σ

$$\mathscr{I}_n(\boldsymbol{\theta}, \log\sigma) = \begin{bmatrix} \mathscr{I}_n(\boldsymbol{\theta}) & 0 \\ 0 & 2n \end{bmatrix}, \quad \text{where} \quad \mathscr{I}_n(\boldsymbol{\theta}) = \sigma^{-2}(\mathbf{X}'\mathbf{X})^{-1}. \qquad (1.3.103)$$

Thus, $$|\mathscr{I}_n(\boldsymbol{\theta}, \log\sigma)|^{1/2} \propto \sigma^{-k}.$$

Arguing exactly as above, the factor σ^{-k} is irrelevant and the appropriate prior distribution is $p(\boldsymbol{\theta}, \log\sigma) \propto c$ or $p(\boldsymbol{\theta}, \sigma) \propto \sigma^{-1}$ as in (1.3.82).

Finally, for the general location-scale family of (1.3.57),

$$p(y \mid \theta, \sigma) \propto \sigma^{-1} h\left(\frac{y-\theta}{\sigma}\right),$$

the information matrix of $(\theta, \log \sigma)$ is

$$\mathscr{I}_n(\theta, \log \sigma) = n \begin{bmatrix} b_1 \sigma^{-2} & 0 \\ 0 & b_2 \end{bmatrix}, \qquad (1.3.104)$$

where b_1 and b_2 are two positive constants independent of (θ, σ). If θ and σ are considered independent *a priori*, then Jeffreys' rule applied separately yields

$$p(\theta, \log \sigma) \propto c \qquad \text{or} \qquad p(\theta, \sigma) \propto \frac{1}{\sigma}, \qquad (1.3.105)$$

whence the corresponding posterior distribution of (θ, σ) is

$$p(\theta, \sigma \mid \mathbf{y}) \propto \sigma^{-(n+1)} \prod_{u=1}^{n} h\left(\frac{y_u - \theta}{\sigma}\right), \qquad -\infty < \theta < \infty, \quad \sigma > 0. \qquad (1.3.106)$$

Necessity for Individual Consideration of Multiparameter Prior Distributions

Choice of noninformative prior distributions in multiparameter problems requires careful consideration in each particular instance. Although devices such as Jeffreys' rule can be suggestive, it is necessary to investigate transformation implications for each example in the light of any appropriate knowledge of prior independence.

1.3.7 Noninformative Prior Distributions: A Summary

In the previous sections, methods have been developed for selecting prior distributions to represent the situation where little is known *a priori* in relation to the information from the data. The concept of a "data translated likelihood" leads to a class of what we call "noninformative" priors. In the case of a single parameter, the resulting prior distributions correspond exactly to those proposed by Jeffreys on grounds of invariance.

A noninformative prior does not necessarily represent the investigator's prior state of mind about the parameters in question. It ought, however, to represent an "unprejudiced" state of mind. In this book, noninformative priors are frequently employed as a point of *reference* against which to judge the kind of unprejudiced inference that can be drawn from the data.

The phrase "knowing little" can only have meaning relative to a specific experiment. The form of a noninformative prior thus depends upon the experiment to be performed, and for two different experiments, each of which can throw light on the same parameter, the choice of "noninformative" prior can be different.

Table 1.3.1

A summary of noninformative prior and corresponding posterior distributions

Parameter(s)	Noninformative Prior	Posterior
Binomial mean π	$\pi^{-\frac{1}{2}}(1-\pi)^{-\frac{1}{2}}$	$\pi^{y-\frac{1}{2}}(1-\pi)^{n-y-\frac{1}{2}}$
Multinomial means $\pi_1, \ldots, \pi_m$	$\pi_1^{-\frac{1}{2}} \ldots \pi_m^{-\frac{1}{2}}$	$\pi_1^{y_1-\frac{1}{2}} \ldots \pi_m^{y_m-\frac{1}{2}}$
Poisson mean λ	$\lambda^{-\frac{1}{2}}$	$\lambda^{n\bar{y}-\frac{1}{2}} \exp(-n\lambda)$
Normal mean θ (σ known)	c	$\exp\left[-\frac{n(\theta-\bar{y})^2}{2\sigma^2}\right]$
Normal standard deviation σ (θ known)	σ^{-1}	$\sigma^{-(n+1)} \exp\left[-\frac{\Sigma(y_u-\theta)^2}{2\sigma^2}\right]$
Normal θ and σ	σ^{-1}	$\sigma^{-(n+1)} \exp\left[-\frac{n(\theta-\bar{y})^2}{2\sigma^2} - \frac{\Sigma(y_u-\bar{y})^2}{2\sigma^2}\right]$
Normal linear model $\boldsymbol{\theta}$ (σ known)	c	$\exp\left[-\frac{(\boldsymbol{\theta}-\hat{\boldsymbol{\theta}})'\mathbf{X}'\mathbf{X}(\boldsymbol{\theta}-\hat{\boldsymbol{\theta}})}{2\sigma^2}\right]$
Normal linear model $\boldsymbol{\theta}$ and σ	σ^{-1}	$\sigma^{-(n+1)} \exp\left[-\frac{(\mathbf{y}-\hat{\mathbf{y}})'(\mathbf{y}-\hat{\mathbf{y}}) + (\boldsymbol{\theta}-\hat{\boldsymbol{\theta}})'\mathbf{X}'\mathbf{X}(\boldsymbol{\theta}-\hat{\boldsymbol{\theta}})}{2\sigma^2}\right]$
Location-scale family θ (σ known)	c	$\prod_{u=1}^{n} h\left(\frac{y_u-\theta}{\sigma}\right)$
Location-scale family σ (θ known)	σ^{-1}	$\sigma^{-(n+1)} \prod_{u=1}^{n} h\left(\frac{y_u-\theta}{\sigma}\right)$
Location-scale family θ and σ	σ^{-1}	$\sigma^{-(n+1)} \prod_{u=1}^{n} h\left(\frac{y_u-\theta}{\sigma}\right)$

When more than one parameter is involved, the problem of choosing noninformative priors can be complex. Each problem has to be considered on its merits. Careful consideration must be given to transformation implications and to knowledge of prior independence.

Considerable literature exists concerning the choice of prior distribution to characterize a state of "ignorance". Jeffreys himself has discussed a number of criteria for choosing prior distributions, invariance being the most important among them. Other contributions in this area include those of Huzurbazar (1955), Perks (1947), Welch and Peers (1963), Novick and Hall (1965), Novick (1969), Hartigan (1964, 1965) and Jaynes (1968).

Table 1.3.1 summarizes the results obtained in Sections 1.3.1–6 of the noninformative priors, and the corresponding posteriors for the parameters of the various models considered. The distributions are given in unnormalized form and we suppose that n observations are available.

1.4 SUFFICIENT STATISTICS

When discussing the example in which the mean θ of a Normal distribution was supposed unknown but σ was known, we found in (1.3.1) that the only function of the observations appearing in the likelihood, apart from the sample size n, was the sample average $\bar{y}$. Thus, for example, if σ was known to be equal to unity, the likelihood would be

$$l(\theta \mid \sigma, \mathbf{y}) = \exp\left[-\frac{n}{2}(\bar{y} - \theta)^2\right].$$

Since the data enter Bayes' formula only through the likelihood, it follows that all other aspects of the data, with the exception of $\bar{y}$, are irrelevant in deciding the posterior distribution of θ and hence in making inferences about θ. In these circumstances, following Fisher (1922, 1925), $\bar{y}$ is said to be *sufficient* for θ, and is called a *sufficient statistic* for θ. Similarly when, for a Normal sample, θ is assumed known but σ is unknown, the likelihood corresponds to (1.3.6) and the posterior distribution employs the data only through n and $s^2 = \Sigma(y - \theta)^2/n$. In this case, s^2 is said to be sufficient for σ (or for σ^2). Further, when both θ and σ are unknown, the joint likelihood function for (θ, σ), given a random sample from the Normal distribution $N(\theta, \sigma^2)$, is

$$l(\theta, \sigma \mid \mathbf{y}) = \sigma^{-n} \exp\left[-\frac{1}{2\sigma^2} \sum_{u=1}^{n} (y_u - \theta)^2\right], \tag{1.4.1}$$

which, as in (1.3.77a), by setting $s^2 = \Sigma (y_u - \bar{y})^2/(n - 1)$ can be written

$$l(\theta, \sigma \mid \mathbf{y}) = \sigma^{-n} \exp\left\{-\frac{1}{2\sigma^2}\left[(n - 1)s^2 + n(\bar{y} - \theta)^2\right]\right\}. \tag{1.4.2}$$

Apart from n, the only functions of the observations involved are $\bar{y}$ and s^2, which are said to be *jointly sufficient* for θ and σ. If n is given, $\bar{y}$ and s^2 can be constructed from knowledge of Σy_u and $\Sigma (y_u - \bar{y})^2$ or from Σy_u and Σy_u^2, so that any one of these pairs of quantities are jointly sufficient for θ and σ.

Of course, it would also be true that the *three* quantities Σy_u, $\Sigma (y_u - \bar{y})^2$, Σy_u^2 were sufficient for θ and σ. However, since

$$\Sigma (y_u - \bar{y})^2 = \Sigma y_u^2 - n^{-1} (\Sigma y_u)^2$$

there is an obvious redundancy. The notion is therefore used of a "minimally sufficient set" of statistics, which contains the smallest number of independent functions of the data needed to write down the likelihood. In this example, assuming n given, a minimally sufficient set contains two functions which could be chosen in any way which allows the likelihood to be written down. In particular, they could be any one of the three choices $(\bar{y}, s^2)$, $(\Sigma y_u, \Sigma(y_u - \bar{y})^2)$, or $(\Sigma y_u, \Sigma y_u^2)$. Since n is also needed to write down the likelihood, this quantity is sometimes treated as a statistic and added to the sufficient set.

We see that for Normal samples sufficient statistics are available which conveniently "match" the parameters. For example, $\bar{y}$ and s^2 are sample quantities which one would expect to supply information about θ and σ.

Convenient parsimonious sets of sufficient statistics do exist for some other distributions. In general, however, if we have k parameters in the likelihood, the minimal sufficient set of $q \geqslant k$ functions of the data which appear in the likelihood function will not be such that $q = k$. Consider, for example, the distribution

$$p(y \mid \theta, \sigma) \propto \sigma^{-1} \exp\left[-\left(\frac{y - \theta}{\sigma}\right)^4\right], \quad -\infty < y < \infty. \tag{1.4.3}$$

The likelihood function based upon n independent observations is

$$l(\theta, \sigma \mid \mathbf{y}) \propto \sigma^{-n} \exp\left[-\sum_{u=1}^{n} \left(\frac{y_u - \theta}{\sigma}\right)^4\right] \tag{1.4.4}$$

$$= \sigma^{-n} \exp\left[\frac{1}{\sigma^4}\left(-S_4 + 4\theta S_3 - 6\theta^2 S_2 + 4\theta^3 S_1 - \theta^4\right)\right] \tag{1.4.5}$$

where $S_p = \sum_{u=1}^{n} y_u^p$. In this case, then, a minimal sufficient set of statistics for the $k = 2$ parameters θ and σ consists of the $q = 4$ sums S_1, S_2, S_3, and S_4. Note, however, that if the power in the exponent in (1.4.3) had not been an integer (suppose, for example, it had been 4.1 instead of 4.0) then no finite expansion of the form of (1.4.5) would have been possible, and the $q = n$ observations themselves would have been a minimally sufficient set of statistics.

In general, we may define sufficient statistics as follows.

Definition (1.4.1) Let $\mathbf{y}' = (y_1, \ldots, y_n)$ be a vector of observations whose distribution depends upon the k parameters $\boldsymbol{\theta}' = (\theta_1, \ldots, \theta_k)$. Let $\mathbf{t}' = (t_1, \ldots, t_q)$

be q functions of $\mathbf{y}$. Then the set of statistics $\mathbf{t}$ is said to be jointly sufficient for $\boldsymbol{\theta}$ if the likelihood function $l(\boldsymbol{\theta} \mid \mathbf{y})$ can be expressed in the form

$$l(\boldsymbol{\theta} \mid \mathbf{y}) \propto g(\boldsymbol{\theta} \mid \mathbf{t}), \tag{1.4.6}$$

and provided the ranges of $\boldsymbol{\theta}$, if dependent on the observations, can also be expressed as functions of $\mathbf{t}$.

Thus, considering distributions which have been used as examples in this chapter, sufficient statistics exist for the parameters (θ, σ) in the Normal distribution, for $(\boldsymbol{\theta}, \sigma)$ of the Normal linear model in (1.3.68), for the Poisson mean λ in (1.3.39), for the binomial mean π in (1.3.21), for the multinomial means $(\pi_1, \ldots, \pi_m)$ in (1.3.93a), and for (θ, σ) in the "fourth power" distribution in (1.4.3). In the case of a single parameter θ, if $\mathbf{y}$ is a random sample from the distribution $p(y \mid \theta)$ and the range of y does not depend on θ, then it was shown by Pitman (1936) that a *single* sufficient statistic exists for θ if and only if $p(y \mid \theta)$ is a member of the exponential family previously referred to in (1.3.38).

When the ranges of the observations $\mathbf{y}$ are independent of $\boldsymbol{\theta}$, a useful property of sufficient statistics is given by the following lemma.

Lemma 1.4.1 Let $\mathbf{t}$ be jointly sufficient for $\boldsymbol{\theta}$, having joint distribution $p(\mathbf{t} \mid \boldsymbol{\theta})$. Then,

$$l(\boldsymbol{\theta} \mid \mathbf{y}) \propto l_1(\boldsymbol{\theta} \mid \mathbf{t}) \qquad \text{where} \qquad l_1(\boldsymbol{\theta} \mid \mathbf{t}) \propto p(\mathbf{t} \mid \boldsymbol{\theta}). \tag{1.4.7}$$

In other words, the likelihood function obtained from the distribution of $\mathbf{t}$ is the same as that obtained from the distribution of $\mathbf{y}$.

Proofs of the lemma can be found in standard text books such as Kendall and Stuart (1961) and Wilks (1962). Two examples follow.

1. *Normal mean, variance known.* We have seen in (1.3.1) that if $\mathbf{y}' = (y_1, \ldots, y_n)$ is a random sample from $N(\theta, \sigma^2)$, where σ^2 is assumed known, then the likelihood function is

$$l(\theta \mid \sigma, \mathbf{y}) \propto \exp\left[-\frac{n}{2\sigma^2}(\theta - \bar{y})^2\right]. \tag{1.4.8}$$

Now, the sample mean $\bar{y}$ is distributed as

$$p(\bar{y} \mid \theta, \sigma^2) = \frac{\sqrt{n}}{\sqrt{2\pi}\,\sigma} \exp\left[-\frac{n}{2\sigma^2}(\bar{y} - \theta)^2\right], \qquad -\infty < \bar{y} < \infty, \tag{1.4.9}$$

so that, given $\bar{y}$,

$$l_1(\theta \mid \sigma, \bar{y}) \propto \exp\left[-\frac{n}{2\sigma^2}(\theta - \bar{y})^2\right], \tag{1.4.10}$$

which is the same as in (1.4.8).

2. *Normal distribution, both mean and variance unknown.* In this case, the likelihood function $l(\theta, \sigma \mid \mathbf{y})$ based upon the n independent observations $\mathbf{y}$ is that given in (1.4.2). Now, it is well known that

a) $\bar{y}$ is distributed as $N(\theta, \sigma^2/n)$,

b) $(n-1)s^2 = \Sigma_u(y_u - \bar{y})^2$ is distributed as $\sigma^2\chi^2_{n-1}$, and

c) $\bar{y}$ and s^2 are statistically independent.

Thus,

$$p(\bar{y}, s^2 \mid \theta, \sigma) = p(\bar{y} \mid \theta, \sigma^2)\, p(s^2 \mid \sigma^2), \tag{1.4.11}$$

where $p(\bar{y} \mid \theta, \sigma^2)$ is that in (1.4.9) and

$$p(s^2 \mid \sigma^2) = \frac{1}{\Gamma[\frac{1}{2}(n-1)]}\left(\frac{n-1}{2\sigma^2}\right)^{\frac{1}{2}(n-1)} (s^2)^{\frac{1}{2}(n-1)-1} \times \exp\left[-\frac{(n-1)s^2}{2\sigma^2}\right], \qquad s^2 > 0. \tag{1.4.12}$$

It follows that, given $(\bar{y}, s^2)$,

$$l(\theta, \sigma \mid \bar{y}, s^2) \propto \sigma^{-n} \exp\left\{-\frac{1}{2\sigma^2}\left[(n-1)\,s^2 + n(\theta - \bar{y})^2\right]\right\}, \tag{1.4.13}$$

which is identical with (1.4.2).

1.4.1 Relevance of Sufficient Statistics in Bayesian Inference

Sufficient statistics play a vital role in sampling theory. For, if inferences about fixed parameters are to be made using the distributional properties of statistics which are functions of the data, then, to avoid inefficiency due to leakage of information, it is essential that a small minimally sufficient set of statistics be available containing all the information about the parameters.

By happy mathematical accident such sets of sufficient statistics do exist for a number of important distributions and, in particular, for the Normal distribution. However, serious difficulties can accompany the exploration of less restricted models which may be motivated by scientific interest, but for which no convenient set of sufficient statistics happens to be available.

Because Bayesian analysis is concerned with the distribution of *parameters*, given known (fixed) data, it does not suffer from this artificial constraint. It does not matter whether or not the distribution of interest happens to have the special form which yields sufficient statistics. For example, the likelihood, and hence the posterior density, can be calculated with almost the same ease from (1.4.4), which expresses the likelihood in terms of the data, as it can from (1.4.5), which expresses it in terms of the sufficient statistics alone. Furthermore, very little more effort would be needed to compute (1.4.4) if the power in the exponent were 4.1 or some other noninteger value.

1.4.2 An Example Using the Cauchy Distribution

To illustrate these points further, we consider the problem of making inferences about the location parameter θ of the Cauchy distribution

$$p(y \mid \theta) = \pi^{-1}[1 + (y - \theta)^2]^{-1}, \qquad -\infty < y < \infty, \tag{1.4.14}$$

where from the form of the distribution it is apparent that no summarizing statistics exist and that the observations themselves are a minimum sufficient set. This fact does not embarrass the Bayesian approach. The Cauchy distribution is a special case of the location scale family (1.3.57) and, using the argument leading to (1.3.61), a noninformative prior is locally uniform in θ, whence the posterior density function is immediate. To provide numerical illustration, a sample of $n = 5$ observations (11.4, 7.3, 9.8, 13.7, 10.6) were randomly drawn from the Cauchy distribution shown in Fig. 1.4.1. Thus, assuming little were known about θ *a priori*, the posterior distribution is approximately

$$p(\theta \mid \mathbf{y}) \doteq cH(\theta), \qquad -\infty < \theta < \infty, \tag{1.4.15}$$

where

$$H(\theta) = 10^5[1 + (7.3 - \theta)^2]^{-1}[1 + (9.8 - \theta)^2]^{-1} \dots [1 + (13.7 - \theta)^2]^{-1},$$

the factor 10^5 is merely a convenient multiplier, and c is the normalizing constant. The posterior distribution obtained by evaluating this expression for a suitable series of values of θ is shown in Fig. 1.4.2. Thus, in spite of the fact that we do not have a sufficient statistic for θ, the posterior distribution, from which inferences can be made, is easily determined.

Calculation of $p(\theta \mid \mathbf{y})$

To obtain the density explicitly, we need to determine the normalizing factor c. That is, we have to evaluate the integral

$$c^{-1} = \int_{-\infty}^{+\infty} H(\theta)\, d\theta. \tag{1.4.16}$$

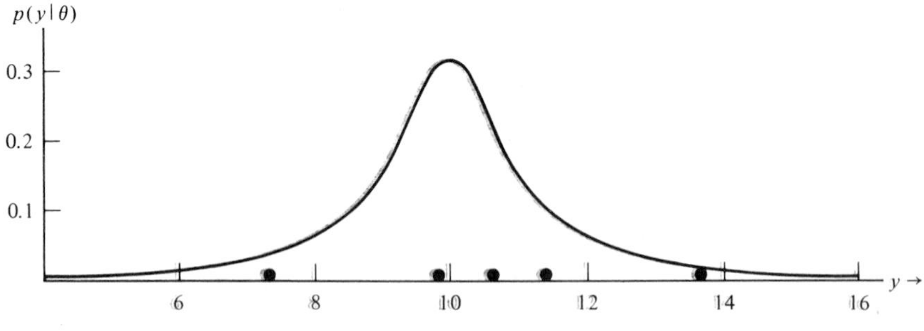

Fig. 1.4.1 Density curve of a Cauchy distribution (dots show 5 observations drawn from the distribution).

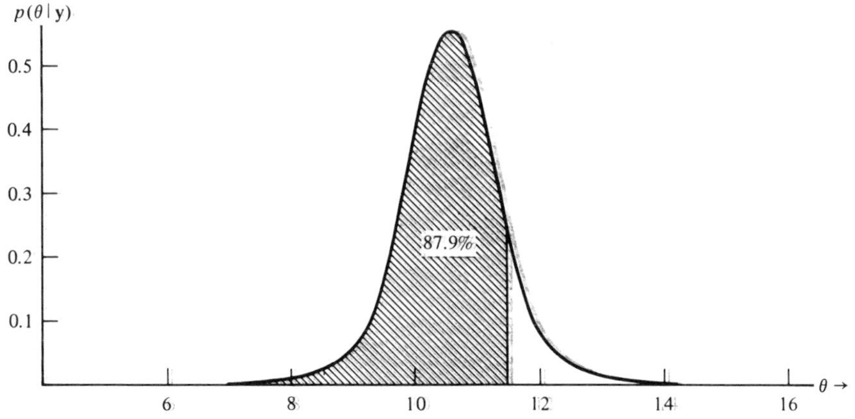

Fig. 1.4.2 Posterior distribution for the location parameter θ of a Cauchy distribution (noninformative prior), given the data shown in Fig. 1.4.1.

Also, to obtain the probability that θ is less than some value θ_0, we need to evaluate the ratio

$$\left[\int_{-\infty}^{\theta_0} H(\theta)\,d\theta\right] \Big/ \left[\int_{-\infty}^{+\infty} H(\theta)\,d\theta\right]. \tag{1.4.17}$$

Frequently the integrals which occur in applications of Bayes' formula will not possess convenient solutions in closed form, or solutions which have been tabulated. However, this is of little practical importance.

In the first place, the best way to convey to the experimenter what the data tell him about θ is to show him a picture of the posterior distribution. For this purpose, the scale of the vertical axis is superfluous and it is sufficient to plot $H(\theta)$. In the second place, just as the question of the existence of convenient sufficient statistics ought to be irrelevant, so should the question of whether or not an integral happens to be one which can be expressed in terms of tabled functions. As the Bayesian approach sets us free from the yoke of sufficiency, so numerical integration and the availability of computers set us free from the need to worry about the "integrability in closed form" of the function.

For univariate distributions, even summing ordinates or drawing the distribution on graph paper and counting squares could approximate the integral with sufficient accuracy for most practical purposes. Since small differences in probability cannot be appreciated by the human mind, there seems little point in being excessively precise about uncertainty.

For illustration, a specimen calculation using Simpson's rule is shown in Table 1.4.1. Suppose for the Cauchy example that for some reason $p(\theta \mid \mathbf{y})$ itself is needed and not merely $H(\theta)$, and that the probability $\Pr\{\theta < 11.5\}$ is required. The first column in the table shows θ at intervals of 0.5 over the range of interest. The second column shows the corresponding value of $H(\theta)$ to the nearest whole number. The third column shows the approximate integral given by Simpson's rule. Thus, for example,

$$\int_{-\infty}^{8.5} H(\theta)\,d\theta \doteq \frac{0.5}{3}[(1\times 0)+(4\times 1)+(2\times 2)+(4\times 5)+(1\times 11)] = 6.5. \quad (1.4.18)$$

Proceeding in this way we find that

$$c^{-1} = \int_{-\infty}^{+\infty} H(\theta)\,d\theta \doteq 535.0 \quad (1.4.19)$$

so that $c = 0.001869$, whence $p(\theta \mid \mathbf{y})$ is calculated and entered in the fourth column. The values for the cumulative probability

$$\int_{-\infty}^{\theta_0} p(\theta \mid \mathbf{y})\,d\theta = c\int_{-\infty}^{\theta_0} H(\theta)\,d\theta \quad (1.4.20)$$

are given in the fifth column. In particular (see Fig. 1.4.2), we find that

Table 1.4.1

Calculation of the posterior density function and cumulative distribution function for the location parameter θ of a Cauchy distribution

θ or θ_0	$H(\theta)$	$\int_{-\infty}^{\theta_0} H(\theta)\,d\theta$	$p(\theta\|\mathbf{y}) = cH(\theta)$	$\int_{-\infty}^{\theta_0} p(\theta\|\mathbf{y})\,d\theta = c\int_{-\infty}^{\theta_0} H(\theta)\,d\theta$
6.5	0	0	0.000	0·000
7.0	1		0.002	
7.5	2	1.0	0.004	0.002
8.0	5		0.009	
8.5	11	6.5	0.021	0.012
9.0	28		0.052	
9.5	83	40.8	0.155	0.076
10.0	196		0.336	
10.5	291	233.8	0.544	0.437
11.0	250		0.467	
11.5	129	470.5	0.241	0.879
12.0	47		0.088	
12.5	17	526.2	0.032	0.983
13.0	7		0.013	
13.5	2	534.0	0.004	0.998
14.0	1		0.002	
14.5	0	535.0	0.000	1.000

$$c^{-1} = \int_{-\infty}^{+\infty} H(\theta)\,d\theta = 535.0, \qquad c = 0.001869.$$

$\Pr\{\theta < 11.5\} = 0.879$. Values of the cumulative probability for intermediate values can be found by interpolation in column five. Greater accuracy is obtainable by using a finer interval in θ.

1.5 CONSTRAINTS ON PARAMETERS

Examples occur later in this book where, as part of the model, certain constraints must be imposed on the values which the parameters $\boldsymbol{\theta}$ can take. Such problems can usually be dealt with by choosing the prior distribution so as to include the constraint. Alternatively, it is sometimes more convenient to solve a fictitious unconstrained problem, and then modify the solution to take account of the constraint.

In general, let Ω be the unconstrained parameter space of $\boldsymbol{\theta}$ and let C be a constraint, such that

$$C: \boldsymbol{\theta} \in \Omega_C, \tag{1.5.1}$$

where Ω_C is a subspace of Ω. Let $\boldsymbol{\theta}_S$ be a subset of $\boldsymbol{\theta}$ (which could be $\boldsymbol{\theta}$ itself), and $R_S \subset \Omega_C$ be a region in the parameter space of $\boldsymbol{\theta}_S$.

Then, by definition of conditional probability, the posterior probability that $\boldsymbol{\theta}_S \in R_S$, given the constraint C, is

$$\Pr\{\boldsymbol{\theta}_S \in R_S \mid C, \mathbf{y}\} = \Pr\{\boldsymbol{\theta}_S \in R_S \mid \mathbf{y}\} \frac{\Pr\{C \mid \boldsymbol{\theta}_S \in R_S, \mathbf{y}\}}{\Pr\{C \mid \mathbf{y}\}}. \tag{1.5.2}$$

It follows that the posterior distribution of $\boldsymbol{\theta}_S$ given the constraint C can be written

$$p(\boldsymbol{\theta}_S \mid C, \mathbf{y}) = p(\boldsymbol{\theta}_S \mid \mathbf{y}) \frac{\Pr\{C \mid \boldsymbol{\theta}_S, \mathbf{y}\}}{\Pr\{C \mid \mathbf{y}\}}. \tag{1.5.3}$$

It sometimes happens that C takes the form

$$C: \mathbf{f}(\boldsymbol{\theta}) = \mathbf{d} \tag{1.5.4}$$

where $\mathbf{f}$ is a vector of q functions of $\boldsymbol{\theta}$, and $\mathbf{d}$ a vector of q constants. In this case,

$$p(\boldsymbol{\theta}_S \mid C, \mathbf{y}) = p(\boldsymbol{\theta}_S \mid \mathbf{y}) \frac{p(C \mid \boldsymbol{\theta}_S, \mathbf{y})}{p(C \mid \mathbf{y})}. \tag{1.5.5}$$

Note that $\Pr(C \mid \mathbf{y})$ and $p(C \mid \mathbf{y})$, are constants independent of $\boldsymbol{\theta}_S$. Thus, the posterior distribution of $\boldsymbol{\theta}_S$, given the constraint C, is equal to the posterior distribution for $\boldsymbol{\theta}_S$ which would have been obtained if no such constraint were applied, multiplied by a modifying factor. This modifying factor is proportional to the conditional probability, or the conditional density, of the constraint C given $\boldsymbol{\theta}_S$.

As an example, suppose we wished to make inferences about the percentage conversion of a certain chemical obtained in a particular experiment. Suppose the percentage conversion was determined by a biological assay method which was unbiased but was subject to fairly large approximately Normally distributed errors having known standard deviation $\sigma = 4$. Suppose finally that the results of four analytical determinations were $y_1 = 93$, $y_2 = 101$, $y_3 = 100$ and $y_4 = 98$, yielding a sample average of $\bar{y} = 98$. We need to consider what inferences could be made about θ, bearing in mind that values of θ greater than 100 are impossible.

If it were reasonable to suppose that in the relevant neighborhood the prior of θ is locally uniform for $\theta < 100$, then the problem could be solved by the straightforward application of Bayes' theorem. We have

$$p(\theta \mid \mathbf{y}) \propto l(\theta \mid \mathbf{y})p(\theta) \tag{1.5.6}$$

where in the *relevant* region

$$p(\theta) \begin{cases} \doteq \text{const} & \theta < 100, \\ = 0 & \theta > 100. \end{cases} \tag{1.5.7}$$

Thus

$$p(\theta \mid \mathbf{y}) = c\frac{\sqrt{n}}{\sqrt{2\pi}\,\sigma}\exp\left[-\frac{n}{2\sigma^2}(\theta - \bar{y})^2\right], \qquad \theta < 100, \tag{1.5.8}$$

where

$$c^{-1} = \frac{\sqrt{n}}{\sqrt{2\pi}\,\sigma}\int_{-\infty}^{100}\exp\left[-\frac{n}{2\sigma^2}(\theta - \bar{y})^2\right]d\theta, \qquad n = 4, \qquad \bar{y} = 98.$$

As illustrated in Fig. 1.5.1, the posterior distribution of θ is proportional to the unconstrained likelihood for $\theta < 100$ and is zero for $\theta > 100$. The posterior distribution for θ is thus a Normal distribution $N(\bar{y}, \sigma^2/n)$, with $\bar{y} = 98$ and $\sigma/\sqrt{n} = 2$, truncated from above at $\theta = 100$. The normalizing constant in (1.5.8) is $c = 1.189$. It is chosen so that its reciprocal $c^{-1} = 0.8413$ is the area under unit Normal curve truncated at one standard deviation above the mean.

To illustrate that results of this kind can be obtained equally well by an application of (1.5.3), consider first the (fictitious) unconstrained problem in which θ is not limited in any way and is supposed to have a locally uniform distribution over the whole region in which the likelihood is appreciable. With this setup the *unconstrained* posterior distribution of θ would be an untruncated Normal distribution having a standard deviation of 2 and centered at $\bar{y} = 98$.

For this problem, the constraint is $C: \theta < 100$ so that the modifying factor of (1.5.3) is

$$\frac{\Pr\{C \mid \theta, \mathbf{y}\}}{\Pr\{C \mid \mathbf{y}\}} = \frac{\Pr\{\theta < 100 \mid \theta, \mathbf{y}\}}{\Pr\{\theta < 100 \mid \mathbf{y}\}}. \tag{1.5.3}$$

Now the numerator of this expression is one if $\theta < 100$, and zero otherwise. Furthermore, the denominator is the unconstrained posterior probability that $\theta < 100$ given the data. This is the probability that an unrestricted Normally distributed random variable with mean 98 and standard deviation 2 would not exceed 100 and is precisely the same as c^{-1} in (1.5.8). The effect, then, of the modifying factor is to multiply the Normal density by 1.189 if $\theta < 100$, and by zero if $\theta > 100$, leading to the same result as before.

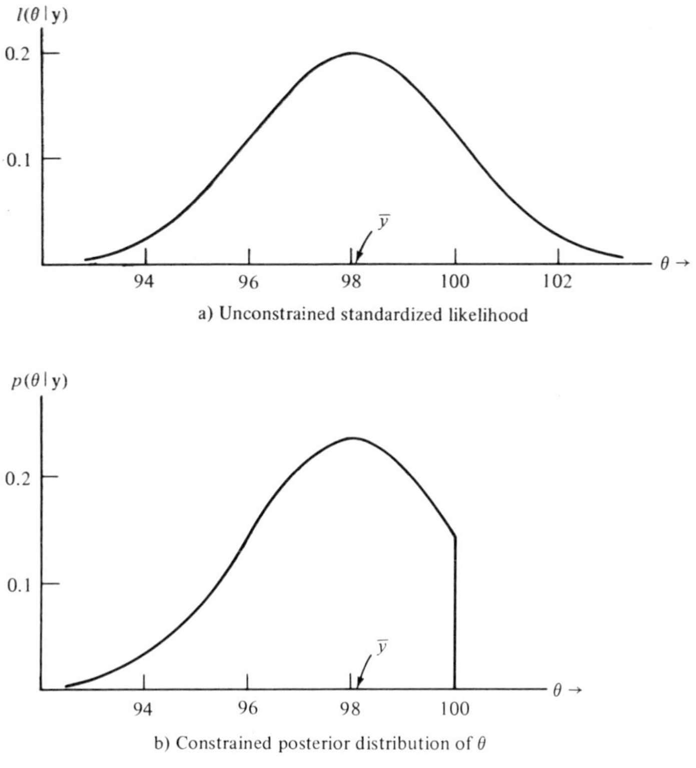

Fig. 1.5.1 Posterior distribution of the Normal mean θ subject to the constraint $\theta < 100$.

The device of first solving a fictitious unconstrained problem and then applying constraints to the solution is particularly useful for variance component problems discussed in Chapters 5 and 6. In deriving solutions and in understanding them, it is helpful to begin by solving the unconstrained problem, which allows negative components of variance, and then constraining the solution.

1.6 NUISANCE PARAMETERS

Frequently the distribution of observations $\mathbf{y}$ depends not only upon a set of r parameters $\boldsymbol{\theta}_1 = (\theta_1, \ldots, \theta_r)$ of interest, but also on a set of, say $t - r$ further nuisance parameters $\boldsymbol{\theta}_2 = (\theta_{r+1}, \ldots, \theta_t)$. Thus we may wish to make inferences about the mean $\theta = \theta_1$ of a Normal population with unknown variance $\sigma^2 = \theta_2$. Here the parameter of interest, or inference parameter, is the mean $\theta = \theta_1$, while the nuisance, or incidental, parameter is $\sigma^2 = \theta_2$. Again, we may wish to make inferences about a single parameter θ_i in the Normal theory linear model of (1.3.68) involving $k + 1$ parameters $\theta_1, \theta_2, \ldots, \theta_k$ and σ^2. In this case, $\theta_1, \ldots, \theta_{i-1}, \theta_{i+1}, \ldots, \theta_k$ and σ^2 all occur as nuisance parameters. In the above examples, where sufficient statistics exist for all the parameters, no particular difficulty is encountered with the sampling theory approach. But when this is not so, difficulties arise in dealing with nuisance parameters by non-Bayesian methods. Furthermore, even when sufficient statistics are available, examples can occur in sampling theory where there is difficulty in eliminating nuisance parameters. One such example is the Behrens–Fisher problem which will be discussed in Section 2.5.

In the Bayesian approach, "overall" inferences about $\boldsymbol{\theta}_1$ are completely determined by the posterior distribution of $\boldsymbol{\theta}_1$, obtained by "integrating out" the nuisance parameter $\boldsymbol{\theta}_2$ from the joint posterior distribution of $\boldsymbol{\theta}_1$ and $\boldsymbol{\theta}_2$. Thus,

$$\int_{R_2} p(\boldsymbol{\theta}_1, \boldsymbol{\theta}_2 \mid \mathbf{y})\, d\boldsymbol{\theta}_2 = p(\boldsymbol{\theta}_1 \mid \mathbf{y}), \tag{1.6.1}$$

where R_2 denotes the appropriate region of $\boldsymbol{\theta}_2$.

Now we can write the joint posterior distribution as the product of the conditional (posterior) distribution of $\boldsymbol{\theta}_1$ given $\boldsymbol{\theta}_2$ and the marginal (posterior) distribution of $\boldsymbol{\theta}_2$,

$$p(\boldsymbol{\theta}_1, \boldsymbol{\theta}_2 \mid \mathbf{y}) = p(\boldsymbol{\theta}_1 \mid \boldsymbol{\theta}_2, \mathbf{y})p(\boldsymbol{\theta}_2 \mid \mathbf{y}). \tag{1.6.2}$$

The posterior distribution of $\boldsymbol{\theta}_1$ can be written

$$p(\boldsymbol{\theta}_1 \mid \mathbf{y}) = \int_{R_2} p(\boldsymbol{\theta}_1 \mid \boldsymbol{\theta}_2, \mathbf{y})p(\boldsymbol{\theta}_2 \mid \mathbf{y})\, d\boldsymbol{\theta}_2, \tag{1.6.3}$$

in which the marginal (posterior) distribution $p(\boldsymbol{\theta}_2 \mid \mathbf{y})$ of the nuisance parameters acts as a weight function multiplying the conditional distribution $p(\boldsymbol{\theta}_1 \mid \boldsymbol{\theta}_2, \mathbf{y})$ of the parameters of interest.

1.6.1 Application to Robustness Studies

It is often helpful in understanding a problem and the nature of the conclusions which can safely be drawn, to consider not only $p(\boldsymbol{\theta}_1 \mid \mathbf{y})$, but also the components of the integrand on the right-hand side of (1.6.3). One is thus led to consider the conditional distributions of $\boldsymbol{\theta}_1$ for particular values of the nuisance parameters $\boldsymbol{\theta}_2$ in relation to the distribution of the postulated values of the nuisance parameters.

In particular, in judging the robustness of the inference relative to characteristics such as non-Normality and lack of independence between errors, the nuisance parameters $\boldsymbol{\theta}_2$ can be measures of departure from Normality and independence. The distribution of the parameters of interest $\boldsymbol{\theta}_1$, conditional on some specific choice $\boldsymbol{\theta}_2 = \boldsymbol{\theta}_{20}$, will indicate the nature of the inference which we could draw if the corresponding set of assumptions (for example, the assumptions of Normality with uncorrelated errors) are made, while the marginal posterior density $p(\boldsymbol{\theta}_2 = \boldsymbol{\theta}_{20} \mid \mathbf{y})$ reflects the plausibility of such assumptions being correct. The marginal distribution $p(\boldsymbol{\theta}_1 \mid \mathbf{y})$, obtained by integrating out $\boldsymbol{\theta}_2$, indicates the overall inference which can be made when proper weight is given to the various possible assumptions in the light of the data and their initial plausibility.

Examples using such an approach will be considered in detail in Chapters 3 and 4.

1.6.2 Caution in Integrating Out Nuisance Parameters

As has been emphasized by Barnard†, cation should be exercised in integrating out nuisance parameters. In particular, if the conditional distribution $p(\boldsymbol{\theta}_1 \mid \boldsymbol{\theta}_2, \mathbf{y})$ in (1.6.3) changes drastically as $\boldsymbol{\theta}_2$ is changed, we would wish to be made aware of this. It is true that whatever the situation, $p(\boldsymbol{\theta}_1 \mid \mathbf{y})$ will theoretically yield the distribution of $\boldsymbol{\theta}_1$. However, in cases where $p(\boldsymbol{\theta}_1 \mid \boldsymbol{\theta}_2, \mathbf{y})$ changes rapidly as $\boldsymbol{\theta}_2$ is changed, great reliance is being placed on the precise applicability of the weight function $p(\boldsymbol{\theta}_2 \mid \mathbf{y})$, and it would be wise to supplement $p(\boldsymbol{\theta}_1 \mid \mathbf{y})$ with auxiliary information.

Thus, if we wished to make inferences about $\boldsymbol{\theta}_1$ alone by integrating out $\boldsymbol{\theta}_2$ and it was found that $p(\boldsymbol{\theta}_1 \mid \boldsymbol{\theta}_2, \mathbf{y})$ was very sensitive to changes in $\boldsymbol{\theta}_2$, it would be important to examine carefully the distribution $p(\boldsymbol{\theta}_2 \mid \mathbf{y})$, which summarizes the information about $\boldsymbol{\theta}_2$ in the light of the data and prior knowledge. Two situations might occur:

1. If $p(\boldsymbol{\theta}_2 \mid \mathbf{y})$ were sharp, with most of its probability mass concentrated over a small region about its mode $\hat{\boldsymbol{\theta}}_2$, then integrating out $\boldsymbol{\theta}_2$ would be nearly equivalent to assigning the modal value to $\boldsymbol{\theta}_2$ in the conditional distribution $p(\boldsymbol{\theta}_1 \mid \boldsymbol{\theta}_2, \mathbf{y})$, so that

$$p(\boldsymbol{\theta}_1 \mid \mathbf{y}) \doteq p(\boldsymbol{\theta}_1 \mid \hat{\boldsymbol{\theta}}_2, \mathbf{y}). \tag{1.6.4}$$

   Thus, even though inferences on $\boldsymbol{\theta}_1$ would have been sensitive to $\boldsymbol{\theta}_2$ over a wider range, the posterior distribution $p(\boldsymbol{\theta}_2 \mid \mathbf{y})$ would contain so much information about $\boldsymbol{\theta}_2$ as essentially to rule out values of $\boldsymbol{\theta}_2$ not close to $\hat{\boldsymbol{\theta}}_2$.

2. On the other hand, if $p(\boldsymbol{\theta}_2 \mid \mathbf{y})$ were rather flat, indicating there was little information about $\boldsymbol{\theta}_2$ from prior knowledge and from the sample, this sensitivity would warn that

† Personal communication.

we should if possible obtain more information about $\boldsymbol{\theta}_2$ so that the inferences about $\boldsymbol{\theta}_1$ could be sharpened. If this were not possible, then as well as reporting the marginal distribution $p(\boldsymbol{\theta}_1 \mid \mathbf{y})$ obtained by integration, it would be wise to add information showing how $p(\boldsymbol{\theta}_1 \mid \boldsymbol{\theta}_2, \mathbf{y})$ changed over the range in which $p(\boldsymbol{\theta}_2 \mid \mathbf{y})$ was appreciable.

1.7 SYSTEMS OF INFERENCE

It is not our intention here to study exhaustively and to compare the various systems of statistical inference which have from time to time been proposed. We assume familarity with the sampling theory approach to statistical inference, and, in particular, with significance tests, confidence intervals, and the Neyman-Pearson theory of hypothesis testing. The main differences between sampling theory inference and Bayesian inference are outlined below.

In sampling theory we are concerned with making inferences about unknown parameters in terms of the sampling distributions of statistics, which are functions of the observations.

1. The probabilities we calculate refer to the frequency with which different values of statistics (arising from sets of data *other* than those which have actually happened) could occur for some *fixed but unknown values of the parameters.* The theory does not employ a prior distribution for the parameters, and the relevance of the probabilities generated in such manner to inferences about the parameters has been questioned (see, for example, Jeffreys, 1961). Furthermore, once we become involved in discussing the probabilities of sets of data which have not actually occurred, we have to decide which "reference set" of groups of data which have not actually occurred we are going to contemplate, and this can lead to further difficulties (see, for example, Barnard 1947).

2. If we accept the relevance of sampling theory inference, then for finite samples we can claim to know *all* that the data have to tell us about the parameters *only* if the problem happens to be one for which all aspects of the data which provide information about the parameter values are expressible in terms of a convenient set of sufficient statistics.

3. Usually when making inferences about a set of parameters of primary interest, we must also take account of nuisance parameters necessary to the specification of the problem. Except where suitable sufficient statistics exist, it is difficult to do this with sampling theory.

4. Using sampling theory it is difficult to take account of constraints which occur in the specification of the parameter space.

By contrast, in Bayesian analysis, inferences are based on probabilities associated with *different* values of parameters which could have given rise to the *fixed* set of data which has actually occurred. In calculating such probabilities we must make assumptions about prior distributions, but we are not dependent upon the existence of sufficient statistics, and no difficulty occurs in taking account of parameter constraints.

1.7.1 Fiducial Inference and Likelihood Inference

Apart from the sampling and the Bayesian approaches, two other modes of inference, proposed originally by Fisher (1922, 1930, 1959), have also attracted the attention of statisticians. These are fiducial inference and likelihood inference. Both are in the spirit of Bayesian theory rather than sampling theory, in that they consider inferences that can be made about *variable* parameters given a set of data *regarded as fixed.* Indeed, fiducial inference has been described by Savage (1961b) as "an attempt to make the Bayesian omelette without breaking the Bayesian eggs." It has been further developed by Fraser (1968), and Fraser and Haq (1969), using what they refer to as structural probability.

Although this approach does not employ prior probability, Fisher made it clear that fiducial inference was intended to cover the situation where nothing was known about the parameters *a priori*, and the solutions which are accessible to this method closely parallel those obtained from Bayes theorem with non-informative prior distributions. For example, one early application of fiducial inference was to the so-called Behrens–Fisher problem of comparing the means of two Normally distributed populations with unknown variances not assumed to be equal. The fiducial distribution of the difference between the population means is identical with the posterior distribution of the same quantity, first obtained by Jeffreys (1961), when noninformative prior distributions are taken for the means and variances. By contrast, Welch's sampling theory solution (1938, 1947), does not parallel this result. We discuss the Bayesian solution in Section 2.5 in more detail.

While Fisher's employment of maximum likelihood estimates was followed up with enthusiasm, by sampling theorists, few workers took account of his suggestion for considering not only the maximum but the whole likelihood function. Notable exceptions are to be found in Barnard (1949), Barnard, Jenkins and Winsten (1962) and Birnbaum (1962). Barnard has frequently stated his opinion that inferences ought to be drawn by studying the likelihood function. As we have seen earlier in Section 1.3, if the likelihood function can be plotted in an appropriate metric, it is identical with the posterior distribution using a non-informative prior. However, the likelihood approach also suffers from fundamental difficulties. It is meaningless, for example, to integrate the likelihood in an attempt to obtain "marginal likelihoods." Yet if the whole likelihood function supplies information about the parameters jointly, one feels it should be able to say something about them individually. Thus, while fiducial inference and likelihood inference each can lead to an analysis similar to a Bayesian analysis with a noninformative prior, each is frustrated in its own particular way from possessing generality.

APPENDIX A1.1

COMBINATION OF A NORMAL PRIOR AND A NORMAL LIKELIHOOD

Suppose *a priori* a parameter θ is distributed as

$$p(\theta) = \frac{1}{\sqrt{2\pi}\,\sigma_0} \exp\left[-\frac{1}{2}\left(\frac{\theta - \theta_0}{\sigma_0}\right)^2\right], \qquad -\infty < \theta < \infty, \tag{A1.1.1}$$

and the likelihood function of θ is proportional to a Normal function

$$l(\theta \mid \mathbf{y}) \propto \exp\left[-\frac{1}{2}\left(\frac{\theta - x}{\sigma_1}\right)^2\right], \tag{A1.1.2}$$

where x is some function of the observations $\mathbf{y}$. Then the posterior distribution of θ given the data $\mathbf{y}$ is

$$p(\theta \mid \mathbf{y}) = \frac{p(\theta)l(\theta \mid \mathbf{y})}{\int_{-\infty}^{\infty} p(\theta)l(\theta \mid \mathbf{y})\,d\theta}$$

$$= \frac{f(\theta \mid \mathbf{y})}{\int_{-\infty}^{\infty} f(\theta \mid \mathbf{y})\,d\theta}, \qquad -\infty < \theta < \infty, \tag{A1.1.3}$$

where

$$f(\theta \mid \mathbf{y}) = \exp\left\{-\frac{1}{2}\left[\left(\frac{\theta - \theta_0}{\sigma_0}\right)^2 + \left(\frac{x - \theta}{\sigma_1}\right)^2\right]\right\}. \tag{A1.1.4}$$

Using the identity

$$A(z - a)^2 + B(z - b)^2 = (A + B)(z - c)^2 + \frac{AB}{A + B}(a - b)^2 \tag{A1.1.5}$$

with

$$c = \frac{1}{A + B}(Aa + Bb),$$

we may write

$$\left(\frac{\theta - \theta_0}{\sigma_0}\right)^2 + \left(\frac{x - \theta}{\sigma_1}\right)^2 = (\sigma_0^{-2} + \sigma_1^{-2})(\theta - \bar{\theta})^2 + d,$$

where

$$\bar{\theta} = \frac{1}{\sigma_0^{-2} + \sigma_1^{-2}}(\sigma_0^{-2}\theta_0 + \sigma_1^{-2}x),$$

and d is a constant independent of θ. Thus,

$$f(\theta \mid \mathbf{y}) = \exp\left(-\frac{d}{2}\right)\exp\left[-\tfrac{1}{2}(\sigma_0^{-2} + \sigma_1^{-2})(\theta - \bar{\theta})^2\right], \tag{A1.1.6}$$

so that

$$\int_{-\infty}^{\infty} f(\theta \mid \mathbf{y})\, d\theta = \exp\left(-\frac{d}{2}\right) \int_{-\infty}^{\infty} \exp\left[-\tfrac{1}{2}(\sigma_0^{-2} + \sigma_1^{-2})(\theta - \bar{\theta})^2\right] d\theta$$

$$= \sqrt{2\pi}\,(\sigma_0^{-2} + \sigma_1^{-2})^{-1/2} \exp(-d/2). \qquad \text{(A1.1.7)}$$

It follows that

$$p(\theta \mid \mathbf{y}) = \frac{(\sigma_0^{-2} + \sigma_1^{-2})^{1/2}}{\sqrt{2\pi}} \exp\left[-\tfrac{1}{2}(\sigma_0^{-2} + \sigma_1^{-2})(\theta - \bar{\theta})^2\right],$$

$$-\infty < \theta < \infty, \qquad \text{(A1.1.8)}$$

which is the Normal distribution

$$N[\bar{\theta}, (\sigma_0^{-2} + \sigma_1^{-2})^{-1}].$$

CHAPTER 2

STANDARD NORMAL THEORY INFERENCE PROBLEMS

2.1 INTRODUCTION

Variation in experimental observations may be associated with known or unknown causes. Suppose observations are made many times under conditions in which suspected causes of variation are held constant. Then we can usefully think of the unknown causes generating a hypothetical *distribution* of observations which are said to differ because of "experimental error." Furthermore, we can sometimes regard a particular observation y as a *random* drawing from this hypothetical underlying distribution $p(y)$. Given a sample of observations, statistical inferences can then be made about the distribution from which the observations are supposed to have been drawn.

Scientific investigation often focuses attention upon certain aspects of distributions which are of special importance. In particular, attention may be focussed on the *location* and *spread* of distributions.

Measures of Location

The investigator often desires to know to what extent some treatment he has applied has caused a shift in the *location* of a distribution. He might, for example, want to discover by how much the yield of a particular product in a chemical reaction is changed by modifying the method of preparation. Suppose he makes n runs using a standard method A, and a further n runs using a modified method B, and he gets a scatter of yield values for method A and a second scatter for method B. He is concerned not with this particular set of $2n$ values *per se*, but rather with what they might allow him to infer about the overall superiority (or otherwise) of the modification. Thus, he might wish to draw inferences about the relative locations of the two underlying distributions from which the observations are presumed to be samples.

Many different measures of location might be considered. Examples are the mean

$$\mu = E(y) = \int_{-\infty}^{\infty} yp(y)\,dy, \tag{2.1.1}$$

and the median m defined by

$$\int_{-\infty}^{m} p(y)\,dy = \int_{m}^{\infty} p(y)\,dy. \tag{2.1.2}$$

In general, we shall say that a parameter θ is a measure of location if addition of a constant c to all the observations (changing y to $y + c$) changes θ to $\theta + c$.

Measures of Spread, Scale Parameters

Another feature of distribution which is of special interest is its *spread*. Thus, a chemist who is developing a new analytical method may wish to compare the spreads of the distributions of results obtained by two different techniques. There are many different parameters that can be used to measure the spread of a distribution, for example, the variance σ^2, the standard deviation σ, the precision constant σ^{-2}, and the mean deviation $E|y - E(y)|$. In this book, a parameter θ ($\theta > 0$) qualifies as a spread parameter if a linear recoding of the observations in which y is changed to $(a + by)$ changes the parameter θ to $|b|^q\theta$. This implies that the logarithm, $\log\theta$, is changed by a fixed constant. Further, if γ and θ are two spread parameters of a distribution then they are related by $\gamma = c\theta^\alpha$.† In terms of the logarithm, $\log\gamma$ and $\log\theta$ are linear functions of each other.

A particular class of spread parameters are *scale* parameters. A spread parameter is a scale parameter if a linear recoding of the data changes θ to $|b|\theta$. Thus, a scale parameter has the same units of measurement as the observations. In particular σ, σ^2, and $\log\sigma$ are all spread parameters, but among this group only σ is a scale parameter.

2.1.1 The Normal Distribution

Since location and spread are characteristics of a distribution which are of particular importance, and since statistical models should include the smallest number of parameters needed to realistically describe the phenomenon under study, it is natural to seek, for representational purposes, some convenient two-parameter family of probability distributions with one parameter measuring location and one measuring spread. A distribution of this kind, which lends itself to ready manipulation and simple analysis, and has been widely used to represent the distributions of observational data, is the Normal or Gaussian distribution

$$p(y) = (2\pi\sigma^2)^{-\frac{1}{2}} \exp - \left[\frac{(y-\theta)^2}{2\sigma^2}\right], \qquad -\infty < y < \infty \qquad (2.1.3)$$

which we denote as $N(\theta, \sigma^2)$.

For this distribution, the mean θ, which is also the mode and the median, is a location parameter. The standard deviation σ is a spread parameter and, since it has the same units as y, is also a scale parameter.

The Role of the Normal Distribution

While the assumption of Normality can lead to greatly simplified analysis it would be cynical and unrealistic to argue that this was the only reason for its use. When

---

† To see this, let $\gamma = f(\theta)$. Then for some (q, q'), $|b|^{q'}\gamma = f(\theta\,|b|^q)$, that is, $|b|^{q'}f(\theta) = f(\theta\,|b|^q)$; thus, we obtain the functional equation $t_1 f(\theta) = f(t_2\theta)$, whose general solution is $f(\theta) = c\theta^\alpha$.

extensive sets of data have been examined it has often been found that they were roughly Normally distributed either in the original metric or in some simple transformation such as the logarithm. Furthermore, under "central limit conditions," theory suggests that data distributions should "cluster" about the Normal and not about some other distribution. These central limit conditions arise when the errors in the observations behave like linear aggregates of independent component errors which are of comparable importance.

For example, suppose that, from an experiment on a small scale reactor, an engineer obtained a certain number of grams of chemical product a sample of which was sent for chemical analysis. From the results, he might calculate a quantity y which measured the "yield" of a particular chemical entity of interest. This calculated yield, y, might typically be a function of z_1, the weight of crude material; of z_2, the weight of sample taken; of z_3, z_4, z_5, various analytical results; and of $z_6, z_7, z_8, z_9, \ldots, z_p$, experimental conditions such as temperature, pressure, concentration and flow rate.

Thus, the final calculated yield might be written as

$$y = f(z_1, \ldots, z_p), \tag{2.1.4}$$

where p would be moderately large, and the variables $z_1, \ldots, z_p$ would all be subject to error. To study the effect of the errors, suppose $z_i = \xi_i + \varepsilon_i$ $(i = 1, \ldots, p)$, where the ε_i is a random variable distributed with zero mean and variance σ_i^2. If the ε_i's could be assumed to be small compared to the corresponding ξ_i's—we could, perhaps, expect that no z would be measured with more than say a 20 per cent error—then, after expanding f about the point $\boldsymbol{\xi} = (\xi_1, \ldots, \xi_p)$

$$y \doteq f(\xi_1, \ldots, \xi_p) + \varepsilon, \tag{2.1.5}$$

where

$$\varepsilon = \omega_1 \varepsilon_1 + \cdots + \omega_p \varepsilon_p$$

and

$$\omega_i = \left. \frac{\partial f}{\partial z_i} \right|_{\boldsymbol{\xi}}.$$

If the ε_i's were independent, the variance of y would now be approximately

$$\operatorname{Var}(y) \doteq \sum_{i=1}^{p} (\omega_i \sigma_i)^2, \tag{2.1.6}$$

so that the contribution of any particular source of error would be large or small depending upon whether $(\omega_i \sigma_i)^2$ was large or small. Usually some of the contributions would be negligible compared with the others but a number would be comparable. With suitable regularity conditions, it is possible to show that the distribution of y will approach the Normal as the number of these comparable contributions increases, and we shall call such a tendency a *central limit effect* and the conditions in which it operates central limit conditions.

The above argument does not, of course, imply that all distributions met in practice can safely be assumed Normal. It does suggest that, in situations where there are a number of sources of variation none of which dominates the others, distributions which are roughly Normal will occur. We discuss in more detail questions of non-Normality in Chapter 3.

2.1.2 Common Normal-theory Problems

Statisticians have used Normal distributions to represent the variation of observational data in many different applications. In particular, they have considered the comparison of the means of two or more such distributions, both when the variances of the distributions are known and when they are unknown, and when these variances can, and cannot, be assumed equal. They have also considered problems of comparing variances of two or more Normal distributions. As an extension of the problem of comparing means, they have investigated inferences which can be made about the parameters of the general *linear model* (1.3.68). The associated analysis is often referred to as *regression analysis* or *least squares* analysis.

Sampling Theory Approaches—The Confidence Distribution

The most common approach to these problems has used sampling theory. In this theory an appropriate criterion, which is a function of the observations, is first selected. For example, in the problem of comparing the means θ_2 and θ_1 of two Normal distributions with known variances σ_2^2 and σ_1^2, the observed difference $\bar{y}_2 - \bar{y}_1$ of the sample means is the criterion considered. Inferences about the possible values that $\theta_2 - \theta_1$ might take, in the light of the data, are made by considering the distribution of $\bar{y}_2 - \bar{y}_1$ in repeated sampling from the same populations.

Sampling theory inferences are usually made in terms of significance tests and confidence intervals. Although we assume that the reader is familiar with these ideas, it is useful to recall certain features of confidence intervals. With samples of n_1 and n_2 observations from distributions $N(\theta_1, \sigma_1^2)$ and $N(\theta_2, \sigma_2^2)$, the criterion $\bar{y}_2 - \bar{y}_1$ is distributed as $N(\theta_2 - \theta_1, \sigma^2)$, where $\sigma = (\sigma_1^2/n_1 + \sigma_2^2/n_2)^{1/2}$. The $1 - \alpha$ confidence interval for $\theta_2 - \theta_1$ may take the form $(\bar{y}_2 - \bar{y}_1) \pm \sigma u_{\alpha/2}$, where $u_{\alpha/2}$ is the unit Normal deviate which cuts off an upper tail area of $\alpha/2$ of the $N(0, 1)$ distribution. The meaning to be associated with the confidence interval is that, in repeated sampling from the same populations, the calculated intervals will cover the true location difference $\theta_2 - \theta_1$, which is *supposed fixed*, a proportion $1 - \alpha$ of the time. In practice, a variety of intervals could be calculated for different values of $1 - \alpha$. A device for simultaneously showing all such intervals is the *confidence distribution*.

Suppose that having computed $\bar{y}_2 - \bar{y}_1$ from a particular set of data, we drew a Normal distribution centered at $\bar{y}_2 - \bar{y}_1$ with standard deviation σ. Then this diagram would immediately allow the intervals for every value of $1 - \alpha$ to be obtained. Thus, if in Fig. 2.1.1 any interval were marked off, then, the total area under the distribution curve enclosed within two verticals drawn at the ends of the interval, would supply the value of $1 - \alpha$. It should be carefully noted that the

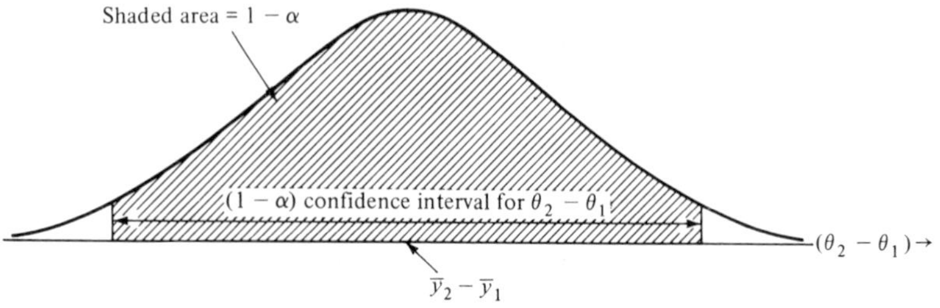

Fig. 2.1.1 Confidence distribution for the difference of two Normal means $\theta_2 - \theta_1$.

confidence distribution is simply a convenient device for associating the value of $1 - \alpha$ with an interval. The confidence distribution is *not* a probability distribution of $\theta_2 - \theta_1$; it will turn out in some common cases, however, that it *is* numerically identical to the posterior distribution obtained with a noninformative prior distribution.

Bayesian Approach

Procedures for comparing Normal means and Normal variances are often given special names. Some of these are Student's "t" test, the Behrens–Fisher test, the analysis of variance, Normal theory least squares, and Bartlett's test for the comparison of variances. In this chapter, we discuss a number of these Normal theory problems from a Bayesian viewpoint. In most cases, the analysis will be given on the basis of noninformative reference priors. The results will therefore be appropriate to situations in which either (a) little is known *a priori* about the parameters, or (b) it is desired to know what the inference would be *if* little were known *a priori*.

Most of the results presented in this chapter have been given earlier by Jeffreys (1961), Savage (1961a), Lindley (1965) and others. This review, however, serves to bring the results together here as a starting point for the consideration in later chapters of a wider class of problems.

2.1.3 Distributional Assumptions

Normal theory problems are those in which the data are treated as a sample of n observations $(y_1, \ldots, y_n)$, where y_u, $u = 1, \ldots, n$, is Normally distributed with mean η_u and variance σ_u^2. That is,

$$p(y_u) = (2\pi\sigma_u^2)^{-1/2} \exp\left[-\frac{1}{2}\left(\frac{y_u - \eta_u}{\sigma_u}\right)^2\right], \qquad -\infty < y_u < \infty. \qquad (2.1.7)$$

As before, we shall say y_u is distributed as $N(\eta_u, \sigma_u^2)$. The corresponding notation $N_q(\boldsymbol{\eta}, \boldsymbol{\Sigma})$ is used to denote the q dimensional multivariate Normal distribution

$$p(\mathbf{y}) = [(2\pi)^q\,|\boldsymbol{\Sigma}|]^{-1/2} \exp\left[-\tfrac{1}{2}(\mathbf{y} - \boldsymbol{\eta})'\,\boldsymbol{\Sigma}^{-1}(\mathbf{y} - \boldsymbol{\eta})\right], \qquad (2.1.8)$$

where $\mathbf{y}' = (y_1, \ldots, y_q)$ $(-\infty < y_u < \infty)$, and the quantities

$$\boldsymbol{\eta}' = (\eta_1, \ldots, \eta_q),$$

$$\boldsymbol{\Sigma} = \{\sigma_{ij}\} \qquad i = 1, \ldots, q, \quad j = 1, \ldots, q,$$

are respectively the $q \times 1$ vector of means and the $q \times q$ covariance matrix.

We make the important assumption throughout this chapter that the n observations $\mathbf{y}' = (y_1, \ldots, y_n)$ of the sample are *independently* distributed. Thus, the n-variate vector $\mathbf{y}$ is distributed as $N_n(\boldsymbol{\eta}, \boldsymbol{\Sigma})$ with

$$\boldsymbol{\eta}' = (\eta_1, \ldots, \eta_n) \qquad \text{and} \qquad \boldsymbol{\Sigma} = \begin{bmatrix} \sigma_1^2 & & & & \\ & \cdot & & & \\ & & \cdot & & \\ & & & \cdot & \\ & & & & \sigma_n^2 \end{bmatrix}. \tag{2.1.9}$$

That observations are independently distributed is perhaps the most sensitive of the assumptions made in the theory that follows. Violation of this assumption can cause dramatic differences in the inferences which may be legitimately drawn from a set of observations, Box (1954), Zellner and Tiao (1964). In those situations where planned experiments can be independently conducted the independence assumption will often be plausible; however, for data, such as economic data, over whose generation the investigator has no control, the independence assumption is frequently quite untenable and, if legitimate inferences are to be made, models which can take direct account of data dependence must be employed; see, for example, Box and Jenkins (1970).

In most of the problems discussed in this chapter, it is also assumed that each observation has the same variance, so that $\boldsymbol{\Sigma} = \mathbf{I}\sigma^2$ where $\mathbf{I}$ is an identity matrix of appropriate size. This assumption, which amounts to saying that observations have equal weight, is often a plausible approximation; however, situations do occur when it is not. For example, in some investigations it is the percentage error rather than the absolute error that one would expect to have constant variance. In such a case logarithms of the observations would have constant variance, and an analysis in terms of the logarithms might be more appropriate. We take up this general problem of data transformation to achieve closer correspondence with assumptions in Chapter 10.

When we refer to a set of random variables which are Normally and *independently* distributed with the *same variance*, we shall sometimes say that they are *spherically Normal*. This terminology is used because the density contours of the distribution $N_n(\boldsymbol{\eta}, \mathbf{I}\sigma^2)$ in the space of the random variables are hyperspheres. It is convenient to use $\mathbf{1}_n$ to denote an $n \times 1$ vector of ones. If, for example, n spherically Normal observations have the same mean so that $\eta_1 = \eta_2 = \cdots = \eta_n = \theta$, then $\boldsymbol{\eta} = \theta\mathbf{1}_n$, and the distribution is $N_n(\theta\mathbf{1}_n, \mathbf{I}\sigma^2)$.

2.2 INFERENCES CONCERNING A SINGLE MEAN FROM OBSERVATIONS ASSUMING COMMON KNOWN VARIANCE

Suppose that the n observations $\mathbf{y}' = (y_1, \ldots, y_n)$ can be regarded as a random sample from a Normal population $N(\theta, \sigma^2)$ with σ^2 known. Thus, it is assumed that the n observations are spherically Normally distributed as

$$N_n(\theta \mathbf{1}_n, \mathbf{I}\sigma^2).$$

We consider what inferences can be made about the unknown mean θ.

Given the assumptions, the sample mean, $\bar{y} = (1/n)\,\Sigma\, y_i$, is a sufficient statistic for θ, and the sampling distribution of $\bar{y}$ is $N(\theta, \sigma^2/n)$. As in (1.4.10), the likelihood function is therefore

$$l(\theta \mid \mathbf{y}) \propto p(\bar{y} \mid \theta, \sigma^2) \qquad (2.2.1)$$
$$\propto \exp\left[-\frac{n}{2\sigma^2}(\theta - \bar{y})^2\right],$$

so that the posterior distribution for θ is

$$p(\theta \mid \mathbf{y}) \propto p(\theta) l(\theta \mid \mathbf{y}) \qquad (2.2.2)$$
$$\propto p(\theta) p(\bar{y} \mid \theta, \sigma^2),$$

where $p(\theta)$ is the prior distribution.

As in (1.3.3), a locally uniform distribution in θ itself provides the appropriate noninformative reference prior, and inferences can be made by treating $p(\theta)$ as constant in (2.2.2). Thus, approximately,

$$p(\theta \mid \mathbf{y}) \propto \exp\left[-\frac{n}{2\sigma^2}(\theta - \bar{y})^2\right]. \qquad (2.2.3)$$

The normalizing constant k which makes the right-hand side of (2.2.3) integrate to one is

$$k^{-1} = \int_{-\infty}^{\infty} \exp\left[-\frac{n}{2\sigma^2}(\theta - \bar{y})^2\right] d\theta = \left(\frac{2\pi\sigma^2}{n}\right)^{1/2},$$

so that finally

$$p(\theta \mid \mathbf{y}) = \left(2\pi\frac{\sigma^2}{n}\right)^{-1/2} \exp\left[-\frac{n}{2\sigma^2}(\theta - \bar{y})^2\right], \qquad -\infty < \theta < \infty. \qquad (2.2.4)$$

We may summarize the above into the following:

Theorem 2.2.1 Let the sample quantity $\bar{y}$ be distributed as $N(\theta, \sigma^2/n)$ in which θ is unknown and σ^2 is known. If the prior distribution of θ is locally uniform, then given $\bar{y}$, the posterior distribution of θ is approximated by $N(\bar{y}, \sigma^2/n)$.

For illustrations consider the following example

2.2.1 An Example

Table 2.2.1 gives the observed breaking strength (in grams) for 20 samples of yarn taken randomly from spinning machines in a certain production area. Past experience indicates that observations of this kind are Normally distributed about a mean value θ with standard deviation $\sigma = (20)^{1/2} = 4.472$ (grams). What conclusion can be drawn about θ?

Table 2.2.1

Observed breaking strength of 20 samples of yarn (in grams)

46	58	40	47	47
53	43	48	50	55
49	50	52	56	49
54	51	50	52	50
		$\bar{y} = 50$		

On the basis of prior local indifference and Normality, and given the information that $\bar{y} = 50$ and $\sigma^2 = 20$, θ will have the approximate posterior distribution $N(50, 1)$.

The posterior density function of θ is shown in Fig. 2.2.1. This distribution implies that for an investigator who knew relatively little about θ *a priori* the probability *a posteriori* that, for example, θ is greater than 52 is given to a close approximation by the tail area to the right of $\theta = 52$ grams.

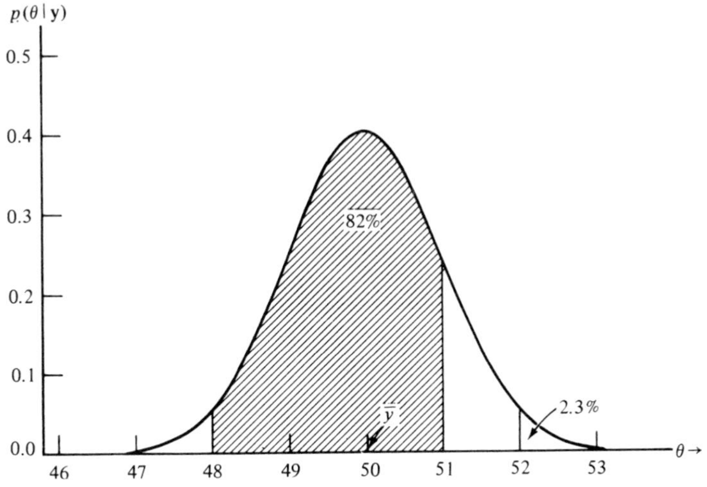

Fig. 2.2.1 Breaking strength data: posterior distribution of θ with σ assumed known and equal to 4.472.

To obtain such probabilities we use a table of the probability integral of the Normal curve, such as Table I (at the end of the book). In this table u is used to indicate a *unit Normal deviate* having the distribution $N(0, 1)$. The table gives values of u_α such that

$$\Pr\{u > u_\alpha\} = \Pr\{u < -u_\alpha\} = \alpha.$$

In the present example, the quantity $\sqrt{n}(\theta - \bar{y})/\sigma$ is distributed as u so that we can evaluate $\Pr\{\theta > \theta_0\}$ by finding an α in Table I such that $u_\alpha = \sqrt{n}(\theta_0 - \bar{y})/\sigma$ In particular, if $\theta_0 = 52$, $u_\alpha = (52 - 50)/1 = 2$, and $\alpha \doteq 2{\cdot}3\%$.

As a further example, suppose the investigator requires the probability that $48 < \theta < 51$, corresponding to the shaded area in Fig. 2.2.1. The required probability is

$$\begin{aligned}\Pr\{48 < \theta < 51 &= 1 - \Pr\{\theta > 51\} - \Pr\{\theta < 48\}\\ &= 1 - \Pr\left\{u > \frac{51 - 50}{1}\right\} - \Pr\left\{u < \frac{48 - 50}{1}\right\}\\ &= 1 - \Pr\{u > 1\} - \Pr\{u > 2\}.\end{aligned}$$

Using the table, this probability is approximately 82%.

2.2.2 Bayesian Intervals

All the information coming from the data on the basis of prior local indifference and the Normal assumption is contained in the posterior distribution sketched in Fig. 2.2.1. To convey to the experimenter what, on this basis, he is entitled to believe about the mean breaking strength θ, it would be best to show him this sketch. With slightly increased convenience to the statistician and lessened convenience to the experimenter, the same information could be conveyed by telling him that the posterior distribution of θ is $N(50, 1)$ and by supplying him with a table of Normal probabilities. Another way to summarize partially the information contained in the posterior distribution is to quote one or more intervals which contain stated amounts of probability.† Sometimes the problem itself will dictate certain limits which are of special interest. In the breaking strength example, 48 and 51 grams might have been of interest because they were specified limits between which the true mean strength should lie. The 82 percent probability will then be the chance that the mean breaking strength lies within specification.

A rather different situation occurs when there are no limits of special interest but an interval is needed to show a range within which "most of the distribution lies." It seems sensible that such an interval should have the property that the density for every point inside the interval is greater than that for every point

† Posterior intervals based on noninformative priors are called "credible intervals" by Edwards, Lindman, and Savage (1963), and "Bayesian confidence intervals" by Lindley (1965).

outside the interval. A second desirable property is that for a given probability content the interval should be as short as possible. A moment's reflection will reveal that these two requirements are equivalent. For a posterior distribution with two tails like the Normal, such an interval is obtained by arranging its extremes to have equal density.

Since for this type of interval every point included has higher probability density than every point excluded, we shall call it a *highest posterior density* interval, or an H.P.D. interval for short.

Intervals derived from sampling theory are customarily associated with 90, 95, or 99 percent of the total probability, with 95 percent the most popular choice. When several intervals are given, it is wise to choose them with their probability content differing enough so as to provide a fairly detailed picture of the posterior distribution. For example, 50, 75, 90, 95, and 99 percent intervals might be quoted together with the mode. More fundamentally, it is desirable to quote intervals adequate to outline important features of the posterior distribution. Thus, while a single 95 percent interval might give sufficient information about the nature of a Normal posterior distribution, several intervals might be needed to indicate less familiar distributions. For example, several intervals in the upper range might be needed to show the nature of a distribution which had a very long tail. As the number of intervals quoted increases, so we come closer to specifically defining the entire posterior distribution itself, and as we have said, presentation of the entire posterior distribution is always desirable if this can be conveniently accomplished. For the breaking strength data, the statement that the mean θ is Normally distributed about $\bar{y} = 50$ grams, with a 95 percent H.P.D. interval extending roughly from 48 to 52 grams, would be sufficient for persons having knowledge of the Normal distribution.

We shall return to the discussion of H.P.D. intervals and regions in Section 2.8 where a more formal treatment is given.

2.2.3 Parallel Results from Sampling Theory

In making inferences about the Normal mean θ with σ assumed known, Bayesian results, on the basis of a noninformative reference prior, parallel sampling theory results, in the sense that the posterior distribution of θ is numerically identical to the sampling theory confidence distribution. This is because the confidence intervals are based upon the sufficient statistic $\bar{y}$, and in both theories the quantity

$$u = \sqrt{n}(\theta - \bar{y})/\sigma$$

is distributed as $N(0, 1)$. Thus in Fig. 2.2.1, the interval (48, 51) which contains 82 percent of the posterior probability, is also an 82 percent confidence interval for the mean breaking load θ. Also, for this problem, the $(1 - \alpha)$ H.P.D. interval is numerically equivalent to the $(1 - \alpha)$ *shortest* confidence interval. Thus, for the breaking strength example, the interval (48, 52) which is the 95 percent H.P.D. interval is also the 95 percent shortest confidence interval for θ.

While Bayesian and confidence intervals for θ are numerically identical, it is important to remember that their interpretations are quite different under the two theories. In the Bayesian formulation, θ is a random variable and a $(1-\alpha)$ interval is one which on the basis of a given $\bar{y}$, computed from the observed data, includes values of θ whose probability mass is a proportion $(1-\alpha)$ of the total posterior probability. In sampling theory, however, θ is an unknown but *fixed* constant and a $(1-\alpha)$ confidence interval is a realization of a random interval, for example, $\bar{y} \leqslant (\sigma/\sqrt{n})u_{\alpha/2}$, which in repeated sampling from the same population will cover θ a proportion $1-\alpha$ of the time.

2.3 INFERENCES CONCERNING THE SPREAD OF A NORMAL DISTRIBUTION FROM OBSERVATIONS HAVING COMMON KNOWN MEAN

Suppose that a random sample $\mathbf{y}' = (y_1, \ldots, y_n)$ is drawn from a Normal population $N(\theta, \sigma^2)$, in which θ is known, and we wish to make inferences about the spread of this distribution.

The likelihood function of σ^2 is

$$l(\sigma^2 \mid \mathbf{y}) \propto p(\mathbf{y} \mid \sigma^2) \propto \sigma^{-n} \exp\left[-\frac{1}{2\sigma^2} \Sigma(y_u - \theta)^2\right]. \tag{2.3.1}$$

The posterior distribution of σ^2 is then

$$p(\sigma^2 \mid \mathbf{y}) \propto p(\sigma^2) l(\sigma^2 \mid \mathbf{y})$$

$$\propto p(\sigma^2)\sigma^{-n} \exp\left(-\frac{ns^2}{2\sigma^2}\right), \tag{2.3.2}$$

where $s^2 = (1/n)\, \Sigma\, (y_u - \theta)^2$ is the sample variance and $p(\sigma^2)$ is the prior distribution.

Alternatively, since $l(\sigma^2 \mid \mathbf{y})$ involves only the sample quantity s^2 which is sufficient for σ^2, we can employ Lemma 1.4.1 (on page 62) to derive (2.3.2) from the distribution of s^2. It will be recalled that if $\mathbf{y}' = (y_1, \ldots, y_n)$ are spherically Normal with zero means and common variance σ^2, then the distribution of s^2 is $(\sigma^2/n)\, \chi^2_n$, where χ^2 is a chi-square variable with n degrees of freedom. That is,

$$p(s^2 \mid \sigma^2) = \left[\Gamma\left(\frac{n}{2}\right)\right]^{-1} \left(\frac{n}{2}\right)^{n/2} \sigma^{-n} (s^2)^{(n/2)-1} \exp\left(-\frac{ns^2}{2\sigma^2}\right), \qquad s^2 > 0. \tag{2.3.3}$$

Since for given s^2 the quantity $[\Gamma(n/2)]^{-1}\, (n/2)^{n/2}\, (s^2)^{(n/2)-1}$ is a fixed constant, the posterior distribution of σ^2, given s^2, is

$$p(\sigma^2 \mid s^2) \propto p(\sigma^2) p(s^2 \mid \sigma^2) \propto p(\sigma^2)\sigma^{-n} \exp\left(-\frac{ns^2}{2\sigma^2}\right) \tag{2.3.4}$$

which is identical with (2.3.2).

Following our discussion in Section 1.3.2, we adopt as our noninformative reference prior a distribution locally uniform in $\log \sigma$. This implies that

$$p(\sigma^2) \propto \sigma^{-2}. \tag{2.3.5}$$

Using (2.3.5) and applying the integral formula (A2.1.2) in Appendix A2.1 to find the appropriate normalizing constant, we obtain

$$p(\sigma^2 \mid \mathbf{y}) = k(\sigma^2)^{-[(n/2)+1]} \exp\left(-\frac{ns^2}{2\sigma^2}\right), \qquad \sigma^2 > 0, \qquad (2.3.6)$$

where

$$k = \left[\Gamma\left(\frac{n}{2}\right)\right]^{-1}\left(\frac{ns^2}{2}\right)^{n/2}.$$

From (2.3.6) the posterior distribution of the standard deviation σ is then

$$p(\sigma \mid \mathbf{y}) = k'\sigma^{-(n+1)} \exp\left(-\frac{ns^2}{2\sigma^2}\right), \qquad \sigma > 0, \qquad (2.3.7)$$

where

$$k' = \left[\tfrac{1}{2}\Gamma\left(\frac{n}{2}\right)\right]^{-1}\left(\frac{ns^2}{2}\right)^{n/2}$$

as we have seen earlier in (1.3.11). In addition, we can also deduce from (2.3.6) the posterior distribution of $\log \sigma^2 = 2 \log \sigma$ as

$$p(\log \sigma^2 \mid \mathbf{y}) = \left[\Gamma\left(\frac{n}{2}\right) 2^{n/2}\right]^{-1}\left(\frac{ns^2}{\sigma^2}\right)^{n/2} \exp\left[-\frac{1}{2}\left(\frac{ns^2}{\sigma^2}\right)\right]$$

$$= \left[\Gamma\left(\frac{n}{2}\right) 2^{n/2}\right] \exp\left[\frac{n}{2}(\log n + \log s^2 - \log \sigma^2)\right.$$

$$\left. - \tfrac{1}{2}\exp(\log n + \log s^2 - \log \sigma^2)\right],$$

$$-\infty < \log \sigma^2 < \infty. \qquad (2.3.8)$$

2.3.1 The Inverted χ^2, Inverted χ, and the Log χ Distributions

Before showing how this theory can be used, we consider some properties of the distributions of σ^2, σ, and $\log \sigma^2$. We first consider some sampling results. Using the assumptions and notation of the previous section, the sampling distribution of the quantity

$$\chi_n^2 = \sum_{u=1}^{n} \frac{(y_u - \theta)^2}{\sigma^2} = \frac{ns^2}{\sigma^2}$$

is

$$p(\chi_n^2) = \left[\Gamma\left(\frac{n}{2}\right) 2^{n/2}\right]^{-1} (\chi^2)^{(n/2)-1} \exp(-\tfrac{1}{2}\chi^2), \qquad \chi^2 > 0. \qquad (2.3.9)$$

which is the χ^2 distribution with n degrees of freedom given earlier in (1.3.12).

Further, the sampling distribution of $\chi_n = \sqrt{n}\, s/\sigma$ (the positive square root of χ_n^2) is

$$p(\chi_n) = \left[\Gamma\left(\frac{n}{2}\right)2^{(n/2)-1}\right]^{-1} \chi^{n-1} \exp\left(-\tfrac{1}{2}\chi^2\right), \qquad \chi > 0. \qquad (2.3.10)$$

This is known as the χ distribution with n degrees of freedom. Finally, the sampling distribution of

$$\log \chi_n^2 = (\log n + \log s^2 - \log \sigma^2)$$

is

$$p(\log \chi_n^2) = \left[\Gamma\left(\frac{n}{2}\right)2^{n/2}\right]^{-1} (\chi^2)^{n/2} \exp\left(-\tfrac{1}{2}\chi^2\right)$$

$$= \left[\Gamma\left(\frac{n}{2}\right)2^{n/2}\right]^{-1} \exp\left[\frac{n}{2}\log \chi^2 - \tfrac{1}{2}\exp(\log \chi^2)\right],$$

$$-\infty < \log \chi^2 < \infty. \qquad (2.3.11)$$

In the corresponding Bayesian analysis, it is the reciprocals χ_n^{-2} and χ_n^{-1} which naturally appear. The "inverted" χ^2 and the "inverted" χ distributions having n degrees of freedom are derived from (2.3.9) and (2.3.10) by making the transformations

$$\chi_n^{-2} = \frac{1}{\chi_n^2}, \qquad \chi_n^{-1} = \frac{1}{\chi_n}$$

to yield

$$p(\chi_n^{-2}) = \left[\Gamma\left(\frac{n}{2}\right)2^{n/2}\right]^{-1} (\chi^{-2})^{-[(n/2)+1]} \exp\left(-\frac{1}{2\chi^{-2}}\right), \qquad \chi^{-2} > 0, \qquad (2.3.12)$$

and

$$p(\chi_n^{-1}) = \left[\Gamma\left(\frac{n}{2}\right)2^{(n/2)-1}\right]^{-1} (\chi^{-1})^{-(n+1)} \exp\left(-\frac{1}{2(\chi^{-1})^2}\right), \qquad \chi^{-1} > 0. \qquad (2.3.13)$$

Now comparing the posterior distributions in (2.3.6) and (2.3.7) with (2.3.12) and (2.3.13), we see that *a posteriori*, the quantities σ^2/ns^2 and $\sigma/\sqrt{n}\, s$ are distributed respectively as χ_n^{-2} and χ_n^{-1}. Further, from (2.3.8) and (2.3.11) the posterior distribution of $(\log n + \log s^2 - \log \sigma^2)$ is that of $\log \chi_n^2$. In dealing with the posterior distributions of such a quantity as σ^2/ns^2, it must be remembered that σ^2 is the random variable and s^2 is a fixed quantity computed from the observed data.

To summarize:

Theorem 2.3.1 Let the sample quantity s^2 be distributed as $(\sigma^2/\nu)\chi_\nu^2$. If the prior distribution of $\log \sigma$ is locally uniform, then, given s^2, σ^2 is distributed *a posteriori* as $(\nu s^2)\chi_\nu^{-2}$, σ is distributed as $(\sqrt{\nu}\, s)\chi_\nu^{-1}$, and finally $\log \sigma^2$ is distributed as $\log \nu s^2 - \log \chi_\nu^2$ (or $\log \nu s^2 + \log \chi_\nu^{-2}$).

In what follows we shall always refer to the constant ν as the number of *degrees of freedom* of the distribution.

2.3.2 Inferences About the Spread of a Normal Distribution

The χ^{-2}, χ^{-1}, and $\log \chi^2$ distributions provide posterior distributions for σ^2, σ, and $\log \sigma^2$. To partially summarize the information contained in these distributions H.P.D. intervals may be quoted. For example, if we are specifically interested in making inferences about the variance σ^2, then the end points of a $(1 - \alpha)$ H.P.D. interval in σ^2 are given by two values (σ_0^2, σ_1^2), such that

a) $p(\sigma_0^2 \mid \mathbf{y}) = p(\sigma_1^2 \mid \mathbf{y})$,

b) $\Pr\{\sigma_0^2 < \sigma^2 < \sigma_1^2 \mid \mathbf{y}\} = 1 - \alpha$.

While the probability contained in a given interval of σ^2 will, of course, be the same as that in the corresponding intervals of σ and $\log \sigma^2$, it is not true that limits of an H.P.D. interval in σ^2 will correspond to the limits of H.P.D. intervals in σ and $\log \sigma^2$. This is easily seen by noting that the posterior distributions $p(\sigma^2 \mid \mathbf{y})$, $p(\sigma \mid \mathbf{y})$, and $p(\log \sigma^2 \mid \mathbf{y})$ in (2.3.6)–(2.3.8) are not proportional to one another. Consequently, the pair of values (σ_0^2, σ_1^2) which satisfy

$$p(\sigma_0^2 \mid \mathbf{y}) = p(\sigma_1^2 \mid \mathbf{y})$$

will not make

$$p(\sigma_0 \mid \mathbf{y}) = p(\sigma_1 \mid \mathbf{y}) \quad \text{or} \quad p(\log \sigma_0^2 \mid \mathbf{y}) = p(\log \sigma_1^2 \mid \mathbf{y}).$$

In other words, H.P.D. intervals are not invariant under transformation of the parameter unless the transformation is linear. This raises the question of parameterization in quoting H.P.D. intervals.

Standardized H.P.D. Intervals

Inferential statements about the spread of a Normal distribution could be made in terms of the variance σ^2, the standard deviation σ, or the precision constant σ^{-2} The H.P.D. intervals associated with a given probability would be slightly different depending on which of these metrics was used, although each interval would exactly include the stated probability. In this book when a noninformative reference prior is assumed for any parameter, we shall for definiteness present *standardized* H.P.D. intervals. These will be H.P.D. intervals calculated in that metric in which the noninformative prior is locally uniform. Thus, for the spread of the Normal distribution, we shall quote intervals which are H.P.D. intervals for $\log \sigma$. Since

H.P.D. intervals are equivalent under linear transformation of $\log \sigma$, we may employ intervals for any member of the class of transformations

$$c + q \log \sigma \tag{2.3.14}$$

where c and q are two arbitrary constants. In particular, we may work with

$$\log \chi_n^2 = \log \frac{ns^2}{\sigma^2} = \log ns^2 - 2 \log \sigma. \tag{2.3.15}$$

Limits of H.P.D. intervals for $\log \chi_n^2$ may be obtained from Table II (at the end of this book). To avoid calculation of logarithms, the limits are given in terms of $\chi_n^2 = ns^2/\sigma^2$. Specifically, for various combinations of (α, n), Table II provides the lower limit $\underline{\chi}^2(n, \alpha)$ and the upper limit $\bar{\chi}^2(n, \alpha)$ for which

a) $$p(\log \underline{\chi}^2) = p(\log \bar{\chi}^2), \tag{2.3.16}$$

that is,

$$(\underline{\chi}^2)^{n/2} \exp(-\tfrac{1}{2}\underline{\chi}^2) = (\bar{\chi}^2)^{n/2} \exp(-\tfrac{1}{2}\bar{\chi}^2)$$

and

b) $$\Pr\{\underline{\chi}^2 < \chi^2 < \bar{\chi}^2\} = 1 - \alpha. \tag{2.3.17}$$

The use of the table is illustrated by the following example.

2.3.3 An Example

Practical situations where θ is known but σ^2 is unknown are extremely rare, but for illustration we consider again the breaking strength data in Table (2.2.1). Suppose it were *known* that $\theta = 50$ and it was desired to consider the inferential situation concerning σ^2 in the light of the noninformative reference prior distribution $p(\log \sigma^2) \propto$ constant. Then, $\log \sigma^2$ would be distributed *a posteriori* as $\log ns^2 - \log \chi_n^2$. For this example,

$$n = 20,$$

$$s^2 = \sum \frac{(y_u - 50)^2}{20} = \frac{348}{20} = 17.4,$$

$$\log n = \log 20 = 2.9957,$$

$$\log s^2 = 2.8565,$$

$$\log ns^2 = 5.8522.$$

The distribution of $\log \sigma^2 = 5.8522 - \log \chi^2_{20}$ is shown in Fig. 2.3.1. It is almost symmetrical about $\log s^2$. Also shown in the same figure is the appropriate 95 percent H.P.D. interval. The abscissas of the distribution in the figure are labelled in terms of σ^2, but plotted on a logarithmic scale to emphasize that it is *for this scaling* that the ordinates at the extremes of the H.P.D. interval are equal. The limits of the 95 percent interval are obtained from Table II as follows:

Corresponding to $\alpha = 0.05$ and $n = 20$, we find

$$\underline{\chi}^2 = 9.9579, \qquad \bar{\chi}^2 = 35.227.$$

Thus, the lower and upper limits are

$$\sigma_0^2 = \frac{ns^2}{\bar{\chi}^2} = \frac{348}{35.227} = 9.8788$$

and

$$\sigma_1^2 = \frac{ns^2}{\underline{\chi}^2} = \frac{348}{9.9579} = 34.9471.$$

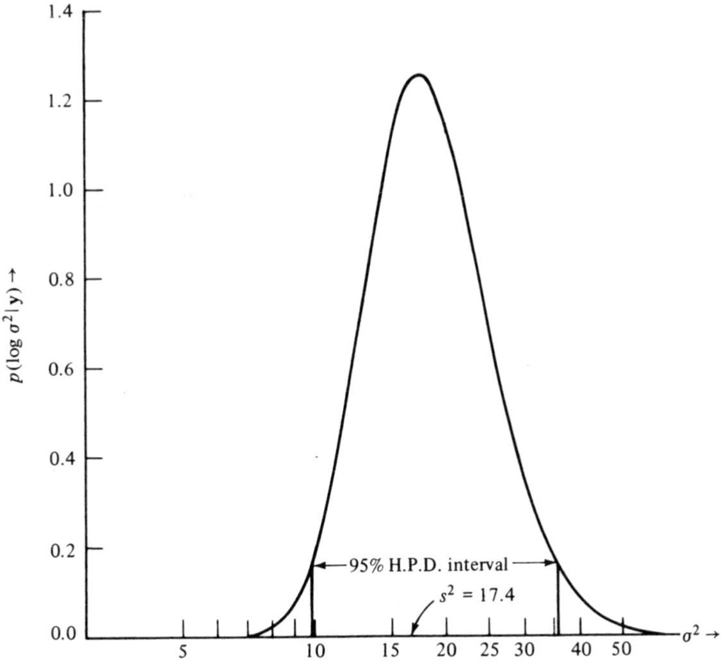

Fig. 2.3.1 Breaking strength data: posterior distribution of $\log \sigma^2$ (scaled in σ^2).

2.3.4 Relationship to Sampling Theory Results

As was the case for the location parameter, parallel results in Bayesian and sampling theories exist for making inferences about the spread of the Normal distribution. The posterior distributions of σ^2, σ, and $\log\sigma$ in (2.3.6)–(2.3.8) are numerically identical to the corresponding confidence distributions. This is because the confidence intervals are based upon the sufficient statistic s^2, and in both theories the quantity ns^2/σ^2 is distributed as χ^2_n. Shortest confidence intervals, like H.P.D. intervals, are not invariant under nonlinear transformation. For example, the $(1-\alpha)$ shortest interval for σ does not correspond to the $(1-\alpha)$ shortest confidence interval for $\log\sigma$ and one faces the same problem of deciding on the choice of parameterization.

2.4 INFERENCES WHEN BOTH MEAN AND STANDARD DEVIATION ARE UNKNOWN

Suppose a random sample of n independent observations is drawn from a Normal population $N(\theta, \sigma^2)$ but both θ and σ are unknown. As we have seen in Section 1.4, the sample mean $\bar{y}$ and the sample variance $s^2 = \nu^{-1}\sum(y_u - \bar{y})^2$, $\nu = n - 1$, are jointly sufficient for (θ, σ), and are distributed independently as $N(\theta, \sigma^2/n)$ and $(\sigma_\nu^2/\nu)\chi_\nu^2$ respectively. From (1.4.13) the likelihood function is

$$l(\theta, \sigma \mid \mathbf{y}) \propto p(\bar{y} \mid \theta, \sigma^2)\, p(s^2 \mid \sigma^2)$$

$$\propto \sigma^{-n} \exp\left\{-\frac{1}{2\sigma^2}[\nu s^2 + n(\theta - \bar{y})^2]\right\}, \tag{2.4.1}$$

where $p(\bar{y} \mid \theta, \sigma^2)$ and $p(s^2 \mid \sigma^2)$ are given in (1.4.9) and (1.4.12), respectively. Thus, given the data $\mathbf{y}$, the joint posterior distribution of (θ, σ) is

$$p(\theta, \sigma \mid \mathbf{y}) \propto p(\theta, \sigma) p(\bar{y} \mid \theta, \sigma^2) p(s^2 \mid \sigma^2), \tag{2.4.2}$$

where $p(\theta, \sigma)$ is the prior distribution.

Following the discussion in Section 1.3.6. leading to (1.3.84), we assume that *a priori* θ and σ are approximately independent, so that

$$p(\theta, \sigma) \doteq p(\theta)\, p(\sigma), \tag{2.4.3}$$

and we adopt as our noninformative reference priors for θ and σ

$$p(\theta) \propto c \tag{2.4.4}$$

$$p(\log\sigma) \propto c \qquad \text{or} \qquad p(\sigma) \propto \sigma^{-1}. \tag{2.4.5}$$

From (2.4.1), (2.4.4) and (2.4.5), the corresponding posterior distribution is

$$p(\theta, \sigma \mid \mathbf{y}) = k\sigma^{-(n+1)} \exp\left\{-\frac{1}{2\sigma^2}[\nu s^2 + n(\theta - \bar{y})^2]\right\}, \qquad -\infty < \theta < \infty, \quad \sigma > 0, \tag{2.4.6}$$

where $\nu = n - 1$ and k is the appropriate Normalizing constant. Using the integral formulae (A2.1.4) and (A2.1.9) in Appendix A2.1, we find

$$k = \sqrt{\frac{n}{2\pi}} \left[\tfrac{1}{2}\Gamma\left(\frac{\nu}{2}\right)\right]^{-1} \left(\frac{\nu s^2}{2}\right)^{\nu/2}. \tag{2.4.7}$$

Contours of $p(\theta, \sigma \mid \mathbf{y})$

On the basis of locally uniform prior distributions for θ and $\log \sigma$, inferences about (θ, σ) for a particular sample $\mathbf{y}$ should be based upon the posterior distribution in (2.4.6). Since $p(\theta, \sigma \mid \mathbf{y})$ is a function of two variables, a plot of the entire distribution would involve constructing a three-dimensional figure; however, a sufficient understanding of the nature of the distribution is obtained by plotting probability density contours in the (θ, σ) plane. Each contour is a curve in the (θ, σ) plane,

$$p(\theta, \sigma \mid \mathbf{y}) = c, \tag{2.4.8}$$

where $c > 0$ is a suitable constant. On taking logarithms in (2.4.8), a density contour is defined by

$$-(n+1)\log \sigma - \frac{1}{2\sigma^2}[\nu s^2 + n(\theta - \bar{y})^2] = d, \tag{2.4.9}$$

where d is a function of c. By differentiating the left-hand side of (2.4.9), it is easily shown that the mode of $p(\theta, \sigma \mid \mathbf{y})$ is

$$\hat{\theta} = \bar{y}, \qquad \hat{\sigma} = \left(\frac{\nu s^2}{n+1}\right)^{1/2}. \tag{2.4.10}$$

Thus, $-\infty < d < d_0$ where

$$\begin{aligned} d_0 &= -(n+1)\log \hat{\sigma} - \frac{1}{2\hat{\sigma}^2}[\nu s^2 + n(\hat{\theta} - \bar{y})^2] \\ &= -(n+1)(\log \hat{\sigma} + \tfrac{1}{2}). \end{aligned} \tag{2.4.11}$$

Three or four suitably chosen contours would usually allow the investigator to have a good appreciation of the main features of the distribution.

Probability Content of a Contour

For a given contour, it will certainly be useful to know the posterior probability content of its interior region. The required probability is the double integral

$$\int_R p(\theta, \sigma \mid \mathbf{y})\, d\theta\, d\sigma, \tag{2.4.12}$$

where the region

$$R: -(n+1)\log\sigma - \frac{1}{2\sigma^2}[\nu s^2 + n(\theta - \bar{y})^2] < d,$$

which, in principle, can be evaluated by numerical methods. As an approximation, we may use the fact that, for large samples the joint distribution tends to Normality, [see Jeffreys (1961)]. Therefore,

$$-2\log\frac{p(\theta, \sigma \mid \mathbf{y})}{p(\hat{\theta}, \hat{\sigma} \mid \mathbf{y})} \mathrel{\dot{\sim}} \chi_2^2, \tag{2.4.13}$$

where the symbol "$\dot{\sim}$" means "approximately distributed as". It follows that the contour defined by

$$\log p(\theta, \sigma \mid \mathbf{y}) = \log p(\hat{\theta}, \hat{\sigma} \mid \mathbf{y}) - \tfrac{1}{2}\chi^2(2, \alpha), \tag{2.4.14}$$

where $\chi^2(2, \alpha)$ is the upper 100α per cent point of a χ^2 distribution with 2 degrees of freedom, encloses a region whose probability content is approximately $(1-\alpha)$. That is, it contains approximately a proportion $(1-\alpha)$ of the posterior probability. Equivalently, the contour is given by

$$-(n+1)\log\sigma - \frac{1}{2\sigma^2}[\nu s^2 + n(\theta - \bar{y})^2] = d_0 - \tfrac{1}{2}\chi^2(2, \alpha). \tag{2.4.15}$$

In practice, a good idea of the distribution can be obtained by plotting the approximate 50, 75, and 95 percent contours together with the mode.

2.4.1 An Example

Using the breaking strength data of Table 2.2.1, suppose now that both σ and θ are unknown. The necessary sample quantities are $\bar{y} = 50$, $\nu s^2 = \Sigma(y_u - \bar{y})^2 = 348$, $\nu = 19$ so that $s^2 = 18.32$ and $s = 4.29$, whence the joint posterior distribution for σ and θ, on the basis of the noninformative reference prior distributions (2.4.4) and (2.4.5), is

$$p(\theta, \sigma \mid \mathbf{y}) = k\sigma^{-21}\exp\left\{-\frac{1}{2\sigma^2}[348 + 20(\theta - 50)^2]\right\}.$$

From (2.4.10), the mode is

$$\hat{\theta} = \bar{y} = 50, \qquad \hat{\sigma} = \left(\frac{348}{21}\right)^{1/2} = 4.07,$$

so that, according to (2.4.11),

$$d_0 = -21(\log 4.07 + 0.5) = -39.98.$$

The contour which contains approximately a proportion $(1 - \alpha)$ of the posterior probability is therefore given by

$$(-21)\log\sigma - \frac{1}{2\sigma^2}[348 + 20(\theta - 50)^2] = d,$$

where

$$d = -39.98 - 0.5\chi^2(2, \alpha).$$

Values of $\chi^2(n, \alpha)$, for various values of n and α, are given in Table III (at the end of this book). For the 50, 75, and 96 percent contours, we find

$$\chi^2(2, 0.50) = 1.39, \qquad \chi^2(2, 0.25) = 2.77, \qquad \text{and} \qquad \chi^2(2, 0.05) = 5.99,$$

whence the corresponding values for $-d$ are (40.68, 41.37, 42.98), respectively. These three contours, together with the mode, are shown in Fig. 2.4.1.

2.4.2 Component Distributions of $p(\theta, \sigma \mid \mathbf{y})$

The joint posterior distribution of (θ, σ) in (2.4.6) can be written as the product

$$p(\theta, \sigma \mid \mathbf{y}) = p(\theta \mid \sigma, \mathbf{y})p(\sigma \mid \mathbf{y}), \tag{2.4.16}$$

where the first factor is the conditional posterior distribution of θ, given σ, and the second factor is the marginal posterior distribution of σ.

The Conditional Distribution of θ, Given σ

Treating σ as a known constant in (2.4.6), the conditional posterior distribution of θ, given σ, is

$$p(\theta \mid \sigma, \mathbf{y}) = \frac{\sqrt{n}}{\sqrt{2\pi}\,\sigma}\exp\left[-\frac{n}{2\sigma^2}(\theta - \bar{y})^2\right], \qquad -\infty < \theta < \infty. \tag{2.4.17}$$

That is, given σ, θ is distributed *a posteriori* as $N(\bar{y}, \sigma^2/n)$.

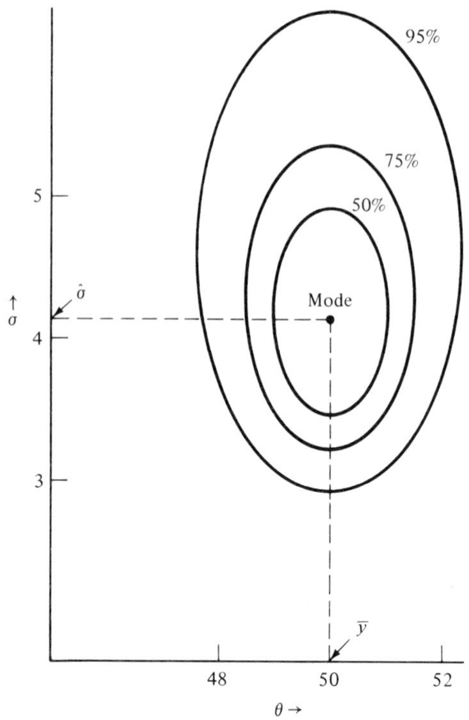

Fig. 2.4.1 Breaking strength data: contours of $p(\theta, \sigma \mid \mathbf{y})$.

The Marginal Distribution of σ

The marginal posterior distribution of σ is

$$p(\sigma \mid \mathbf{y}) = \frac{p(\theta, \sigma \mid \mathbf{y})}{p(\theta \mid \sigma, \mathbf{y})}. \tag{2.4.18}$$

From (2.4.6) and (2.4.17), we find that

$$p(\sigma \mid \mathbf{y}) = k'\sigma^{-(\nu+1)} \exp\left(-\frac{\nu s^2}{2\sigma^2}\right), \qquad \sigma > 0, \tag{2.4.19}$$

with

$$k' = \left[\tfrac{1}{2}\Gamma\left(\frac{\nu}{2}\right)\right]^{-1}\left(\frac{\nu s^2}{2}\right)^{\nu/2}.$$

Thus, σ is distributed *a posteriori* as $\sqrt{\nu}\, s\chi_\nu^{-1}$. This distribution is of exactly the same form as that in (2.3.7). The only difference is that whereas with θ known the distribution of $\sigma/(\sqrt{n}s)$, with $ns^2 = \Sigma(y_u - \theta)^2$, had n degrees of freedom; with θ unknown the distribution of $\sigma/(\sqrt{\nu}s)$, with $\nu s^2 = \Sigma(y_u - \bar{y})^2$, has $\nu = n - 1$

degrees of freedom. Inferences about σ, with θ unknown, can thus be made following the analysis in Section 2.3.

The Marginal Posterior Distribution of θ

The marginal posterior distribution of θ can be obtained by integrating out σ from the joint posterior distribution of θ and σ,

$$p(\theta \mid \mathbf{y}) = \int_0^\infty p(\theta, \sigma \mid \mathbf{y}) d\sigma \tag{2.4.20}$$

$$= \int_0^\infty p(\theta \mid \sigma, \mathbf{y}) p(\sigma \mid \mathbf{y}) d\sigma.$$

Making use of (A2.1.4) in Appendix A2.1, we find from (2.4.6) that

$$p(\theta \mid \mathbf{y}) = \frac{(s/\sqrt{n})^{-1}}{B(\frac{1}{2}\nu, \frac{1}{2})\sqrt{\nu}} \left[1 + \frac{n(\theta - \bar{y})^2}{\nu s^2}\right]^{-\frac{1}{2}(\nu+1)}, \qquad -\infty < \theta < \infty, \tag{2.4.21}$$

where $B(p, q)$ is the complete beta function $B(p, q) = \Gamma(p)\Gamma(q)/\Gamma(p+q)$.

Setting

$$t = \frac{\theta - \bar{y}}{s/\sqrt{n}},$$

we then have

$$p\left(t = \frac{\theta - \bar{y}}{s/\sqrt{n}} \,\middle|\, \mathbf{y}\right) = \frac{1}{B(\frac{1}{2}\nu, \frac{1}{2})\sqrt{\nu}} \left(1 + \frac{t^2}{\nu}\right)^{-\frac{1}{2}(\nu+1)}, \qquad -\infty < t < \infty, \tag{2.4.22}$$

which will be recognized as Student's t distribution with $(\nu = n - 1)$ degrees of freedom. It should be remembered that θ is the random variable in this expression and $(\bar{y}, s^2)$ are known sample quantities. As was first shown by Gossett (1908) (2.4.22) provides the sampling distribution of $(\theta - \bar{y})/(s/\sqrt{n})$, in which θ is regarded as fixed and $\bar{y}$ and s are random variables. The distribution of θ in (2.4.21) will be denoted by $t(\bar{y}, s^2/n, \nu)$.

From the above discussions, we have the following general results.

Theorem 2.4.1 Let the sample quantities $\bar{y}$ and s^2 be independently distributed as $N(\theta, \sigma^2/n)$ and $(\sigma^2/\nu)\chi_\nu^2$, respectively. Suppose *a priori* that θ and $\log \sigma$ are approximately independent and locally uniform. Then, given $(\bar{y}, s^2)$, (a) σ is distributed as $\sqrt{\nu} s \chi_\nu^{-1}$, (b) conditional on σ, θ is distributed as $N(\bar{y}, \sigma^2/n)$, and (c) unconditionally, θ is distributed as $t(\bar{y}, s^2/n, \nu)$.

2.4.3 Posterior Intervals for θ

The posterior distribution of θ in (2.4.21) is a symmetric distribution centered at $\bar{y}$ with scaling factor $s/\sqrt{n}$. To obtain H.P.D. intervals for θ we use the fact that the $t(0, 1, \nu)$ distribution of $t = (\theta - \bar{y})/(s/\sqrt{n})$ has been tabulated.

If we denote $t_{(\alpha/2)}(\nu)$ as the value for which

$$\Pr\{t > t_{(\alpha/2)}(\nu)\} = \alpha/2,$$

then from symmetry of the t distribution, $\Pr\{t < -t_{(\alpha/2)}(\nu)\} = \alpha/2$ so that $\Pr\{|t| < t_{(\alpha/2)}(\nu)\} = 1 - \alpha$. It follows that the limits of the $(1 - \alpha)$ H.P.D. intervals of $t(\bar{y}, s^2/n, \nu)$ are given by

$$\bar{y} \pm \left(\frac{s}{\sqrt{n}}\right) t_{(\alpha/2)}(\nu). \tag{2.4.23}$$

Table IV, at the end of this book, gives values of $t_{(\alpha/2)}(\nu)$ for various values of α and ν. The use of the table is illustrated by the following example.

An Example

Consider again the yarn breaking strength data of Table 2.2.1. Suppose little is known about θ and σ *a priori*, and we wish to find a 95 percent H.P.D. interval for θ on this basis, then, $\bar{y} = 50$, $s/\sqrt{n} = 0.96$, $\nu = 19$, and from Table IV, $t_{0.025}(19) = 2.093$, so that the 95 per cent limits are $50 \pm 0.96(2.093)$ or (47.99, 52.01).

2.4.4 Geometric Interpretation of the Derivation of $p(\theta \mid \mathbf{y})$

The implications of the integration in expression (2.4.20) are illustrated in Fig. 2.4.2. The conditional distribution $p(\theta|\sigma, \mathbf{y})$ of θ for any fixed σ is a Normal distribution $N(\bar{y}, \sigma^2/n)$. The joint distribution $p(\theta, \sigma|\mathbf{y})$ can be regarded as an aggregation of such conditional Normal distributions weighted by the marginal distribution $p(\sigma|\mathbf{y})$. When σ is unknown, $p(\theta|\mathbf{y})$ is obtained by averaging these conditional Normal distributions $p(\theta|\sigma, \mathbf{y})$ using as a weight function the marginal distribution of σ. The resulting t distribution for θ has the characteristics we would expect of a distribution obtained this way. In particular, for large ν, the distribution of θ will be very nearly Normal because the conditional Normal distributions $p(\theta|\sigma, \mathbf{y})$ will be averaged only over a narrow "weight" distribution $p(\sigma|\mathbf{y})$, with σ close to s. But, for small ν, the distribution of θ will be leptokurtic with much heavier tails because the conditional Normal distributions $p(\theta|\sigma, \mathbf{y})$ will be averaged over a more disperse distribution $p(\sigma|\mathbf{y})$, in which values of σ widely different from s are given considerable weight.

Figure 2.4.2 is drawn using the breaking strength data for which $\bar{y} = 50$, $s = 4.29$ and $n = 20$. For *each given* σ, θ has a Normal distribution with mean 50 and standard deviation $\sigma/\sqrt{20}$. Also σ is marginally distributed as $\sqrt{19}(4.29)\chi_{19}^{-1}$. Finally, θ has the t distribution with 19 degrees of freedom, centered at the value $\bar{y} = 50$ and scaled by the quantity $s/\sqrt{20} \doteq 0.96$. The three conditional distributions of θ which are shown are for the arbitrarily chosen values $\sigma = 5.6$, 4.47 and 3.2.

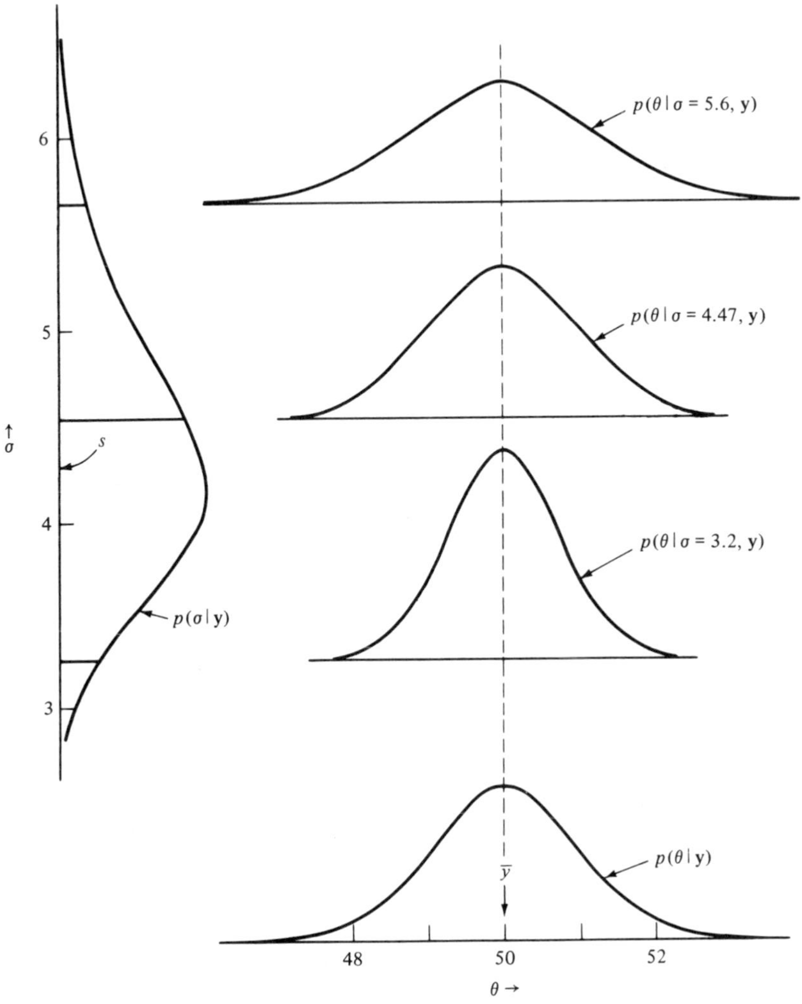

Fig. 2.4.2 Breaking strength data: component distributions of $p(\theta, \sigma \mid \mathbf{y})$.

2.4.5 Informative Prior Distribution of σ

We discuss in this section how prior knowledge or ignorance about σ affects inferences about θ. When *a priori* θ is locally uniform and independent of σ, (2.4.2) and (2.4.3) imply that the joint distribution of (θ, σ) can be expressed in the general form

$$p(\theta, \sigma \mid \mathbf{y}) = p(\theta \mid \sigma, \mathbf{y})\, p(\sigma \mid \mathbf{y}) \tag{2.4.24}$$

where

$$p(\theta \mid \sigma, \mathbf{y}) \propto p(\bar{y} \mid \theta, \sigma^2), \tag{2.4.25}$$

and

$$p(\sigma \mid \mathbf{y}) \propto p(\sigma)\, p(s^2 \mid \sigma^2). \tag{2.4.26}$$

Thus the conditional posterior distribution of θ is

$$p(\theta \mid \sigma, \mathbf{y}) = \frac{\sqrt{n}}{\sqrt{2\pi}\sigma} \exp\left[-\frac{n}{2\sigma^2}(\theta - \bar{y})^2\right], \qquad -\infty < \theta < \infty. \tag{2.4.27}$$

Also, the marginal posterior distribution of θ, obtained by integrating out σ from (2.4.24), is

$$p(\theta \mid \mathbf{y}) = \int_0^\infty p(\theta \mid \sigma, \mathbf{y})\, p(\sigma \mid \mathbf{y})\, d\sigma. \tag{2.4.28}$$

This integration makes it possible to see how various states of knowledge about σ are taken account of in the Bayesian analysis. First, consider the situation where σ is supposed exactly *known*. For definiteness suppose, as we did earlier, that σ was known to be equal to 4.47. This would correspond to the supposition that $p(\sigma)$ was a spike with all its probability concentrated at the point $\sigma = 4.47$ and with zero density elsewhere. The posterior distribution $p(\sigma \mid \mathbf{y})$ is proportional to $p(\sigma)\, p(s^2 \mid \sigma^2)$ and so would also have all its probability concentrated at $\sigma = 4.47$. From (2.4.28), the posterior distribution of θ would then degenerate to the conditional Normal distribution with $\sigma = 4.47$, with all other conditional Normal distributions having zero weight.

This may be contrasted with the situation already considered where very little is known about σ *a priori*, so that essentially all the information about σ has to come from the sample itself. This information is represented by the marginal posterior distribution of σ in (2.4.19) and supplies appropriate weights for the conditional Normal distributions. Integration over these weights gives the marginal t distribution for θ as has already been illustrated in Fig. 2.4.2.

We have discussed two extreme cases: in the first a great deal was known about σ and in the second little was known about σ. Sometimes σ is not precisely known but there may be preliminary information which it is desired to incorporate. Such knowledge may be expressed by employing an appropriate informative prior distribution for σ in (2.4.26).

For example, suppose a preliminary sample is available from which an estimate s_1 having ν_1 degrees of freedom is obtained. Then, on the assumption that initially $\log \sigma$ is locally uniform, the posterior distribution of $\sigma/\sqrt{\nu_1} s_1$, *given the preliminary sample*, is that of $\chi_{\nu_1}^{-1}$. This posterior distribution now becomes the appropriate prior distribution for the new sample and we have

$$p(\sigma) = k\sigma^{-(\nu_1+1)} \exp\left(-\frac{\nu_1 s_1^2}{2\sigma^2}\right), \qquad \sigma > 0, \tag{2.4.29}$$

with

$$k = \left[\tfrac{1}{2}\Gamma\left(\frac{\nu_1}{2}\right)\right]^{-1} \left(\frac{\nu_1 s_1^2}{2}\right)^{\nu_1/2}.$$

Combining this distribution with $p(s^2 \mid \sigma^2)$ in (2.4.26) and applying the integral formula (A2.1.4) in Appendix A2.1, we obtain the posterior distribution of σ,

$$p(\sigma \mid \mathbf{y}) = k'\sigma^{-(\nu'+1)} \exp\left(-\frac{\nu' s'^2}{2\sigma^2}\right), \qquad \sigma > 0, \tag{2.4.30}$$

where

$$k' = \left[\tfrac{1}{2}\Gamma\left(\frac{\nu'}{2}\right)\right]^{-1}\left(\frac{\nu' s'^2}{2}\right)^{\nu'/2},$$

$$\nu' = \nu + \nu_1 \qquad \text{and} \qquad \nu' s'^2 = \nu_1 s_1^2 + \nu s^2.$$

That is, σ distributed as $\sqrt{\nu'}s'\chi_{\nu'}^{-1}$.

Substituting (2.4.30) into (2.4.28), we get

$$p(\theta \mid \mathbf{y}) = \frac{(s'/\sqrt{n})^{-1}}{B(\frac{1}{2}\nu', \frac{1}{2})\sqrt{\nu'}}\left[1 + \frac{n(\theta - \bar{y})^2}{\nu' s'^2}\right]^{-\frac{1}{2}(\nu'+1)}, \qquad -\infty < \theta < \infty, \tag{2.4.31}$$

so that $t = (\theta - \bar{y})/(s'/\sqrt{n})$ has the distribution

$$p(t) = \frac{1}{B(\frac{1}{2}\nu', \frac{1}{2})\sqrt{\nu'}}(1 + t^2/\nu')^{-\frac{1}{2}(\nu'+1)}, \qquad -\infty < t < \infty, \tag{2.4.32}$$

which is a t distribution with $\nu' = \nu + \nu_1$ degrees of freedom. That is, θ is distributed *a posteriori* as $t(\mathbf{y}, s'^2/n, \nu')$. The effect on the posterior distribution of using (2.4.29) instead of (2.4.5) to represent $p(\sigma)$ is to change the degrees of freedom from ν to $\nu' = \nu + \nu_1$ and the scaling factor from $s/\sqrt{n}$ to $s'/\sqrt{n}$, with $\nu' s'^2 = \nu_1 s_1^2 + \nu s^2$ and $\nu' = \nu_1 + \nu$.

Often, even though some previous information is available about σ, its relevance to the particular experiment might be questionable and one might prefer to consider, either as an alternative or in addition, what inferences were possible from the new sample alone in the light of a noninformative reference prior for σ.

2.4.6 Effect of Changing the Metric of σ for Locally Uniform Prior

It was argued in Section 1.3 that when we wished to analyze data as if little were known about σ *a priori*, we should take $\log \sigma$ as locally uniform. By further adopting a locally uniform and independent prior distribution for θ, the information supplied by spherically Normally distributed observations $\mathbf{y}' = (y_1, \ldots, y_n)$ then results in a $t(\bar{y}, s^2/n, n-1)$ posterior distribution for θ. To study the effect of changing the metric of σ, suppose it is assumed either that $\log \sigma$ or that some power of σ is locally uniform. Equivalently

$$p(\sigma) \propto \begin{cases} \sigma^{q-1} & \text{if } \sigma^q \text{ assumed locally uniform,} \\ \sigma^{-1} & \text{if } \log\sigma \text{ assumed locally uniform } (q = 0). \end{cases} \tag{2.4.33}$$

From (2.4.26) the posterior distribution of σ is then

$$p(\sigma \mid \mathbf{y}) \propto p(\sigma)\, p(s^2 \mid \sigma^2)$$
$$\propto \sigma^{-(n-q+1)} \exp\left(-\nu s^2/2\sigma^2\right), \qquad \sigma > 0. \tag{2.4.34}$$

The appropriate normalizing constant can be obtained by using (A2.1.4) in Appendix A2.1. Substituting (2.4.34) into (2.4.28), we obtain

$$p(\theta \mid \mathbf{y}) = \frac{(s_{\nu-q}/\sqrt{n})^{-1}}{B[\frac{1}{2}(\nu - q), \frac{1}{2}]\sqrt{\nu - q}} \left[1 + \frac{n(\theta - \bar{y})^2}{(\nu - q)\, s_{\nu-q}^2}\right]^{-\frac{1}{2}(\nu-q+1)},$$
$$-\infty < \theta < \infty, \tag{2.4.35}$$

where $\nu = n - 1$, and

$$s_{\nu-q}^s = \Sigma\,(y_i - \bar{y})^2/(\nu - q).$$

Thus, if σ^q were assumed locally uniform *a priori*, the posterior distribution of θ would be $t(\bar{y}, s_{\nu-q}^2/n, \nu - q)$.

This derivation serves to emphasize the comparative insensitivity of the posterior distribution of θ to the choice of prior on σ, at least for moderate sized samples. A change of q, even by one unit, can produce a major change in the prior distribution of σ. Yet such a change has only a minor effect on $p(\theta \mid \mathbf{y})$ provided n is not too small.

2.4.7 Elimination of the Nuisance Parameter σ in Bayesian and Sampling Theories

We have seen that with appropriate assumptions and interpretations, the quantity

$$\frac{\bar{y} - \theta}{s/\sqrt{n}} \tag{2.4.36}$$

has the $t(0, 1, n - 1)$ distribution, both from a sampling theory viewpoint and from a Bayesian viewpoint. The corresponding confidence distribution and the posterior distribution for θ are therefore numerically identical. While both the Bayesian and the sampling results are independent of the unknown nuisance parameter σ, it is important to distinguish the manner in which σ is eliminated under the two theories. In the Bayesian framework σ is eliminated by integration. The conditional posterior distributions of θ for various values of σ are averaged over a weight function which is the marginal posterior distribution of σ. This method of eliminating σ has the advantage that it may be used for the elimination of any set of nuisance parameters. However, using sampling theory, the confidence distribution can be obtained only because it so happens the sampling distribution of the "pivotal" quantity $\sqrt{n}(\bar{y} - \theta)/s$ is independent of σ. Unhappily, this sampling theory procedure cannot be generally applied because quantities which are functions of the observations and the parameters of interest, but whose sampling distributions do not involve the nuisance parameters, do not usually exist.

2.5 INFERENCES CONCERNING THE DIFFERENCE BETWEEN TWO MEANS

An investigator is often interested in the *difference* between mean values. Suppose for instance it was decided to investigate a possible mean difference in the breaking strength of yarn made by standard and modified spinning machines, and that data were available from a random sample of n_1 standard machines and from a corresponding sample of n_2 modified machines.

To produce a mathematical formulation of the situation it may be supposed that the n_1 observations $(y_{11}, \ldots, y_{1n_1})$ are as if drawn independently from a Normal population $N(\theta_1, \sigma_1^2)$ and the n_2 observations $(y_{21}, \ldots, y_{2n_2})$ are as if drawn independently from another Normal population $N(\theta_2, \sigma_2^2)$ with all four parameters θ_1, θ_2, σ_1^2, and σ_2^2 unknown. Consider the problem of obtaining the posterior distribution of $\gamma = (\theta_2 - \theta_1)$.

2.5.1 Distribution of $\theta_2 - \theta_1$ when $\sigma_1 = \sigma_2$

We first discuss the situation where the variances σ_1^2 and σ_2^2, although unknown, can be assumed equal, so that $\sigma_1^2 = \sigma_2^2 = \sigma^2$. The quantities $(\bar{y}_1, \bar{y}_2, s^2)$, where

$$s^2 = \nu^{-1}(\nu_1 s_1^2 + \nu_2 s_2^2) = [(n_1 - 1) + (n_2 - 1)]^{-1}\,[\Sigma(y_{1i} - \bar{y}_1)^2 + \Sigma(y_{2i} - \bar{y}_2)^2]$$

and

$$\nu = n_1 + n_2 - 2,$$

are then jointly sufficient for $(\theta_1, \theta_2, \sigma^2)$, and are independently distributed as $N(\theta_1, \sigma^2/n_1)$, $N(\theta_2, \sigma^2/n_2)$, and $(\sigma^2/\nu)\chi_\nu^2$, respectively. Assuming that *a priori* θ_1, θ_2 and $\log \sigma$ are approximately independent and locally uniformly distributed, it can be readily verified by using Theorems 2.2.1 (on page 82) and 2.3.1 (on page 88) that the joint posterior distribution of θ_1, θ_2 and σ^2 is

$$p(\theta_1, \theta_2, \sigma^2 \mid \mathbf{y}) = p(\sigma^2 \mid s^2)p(\theta_1 \mid \sigma^2, \bar{y}_1)p(\theta_2 \mid \sigma^2, \bar{y}_2), \tag{2.5.1}$$

where σ^2 has the $(\nu s^2)\chi_\nu^{-2}$ distribution and, for given σ^2, the parameters θ_1 and θ_2 are independent, having Normal distributions $N(\bar{y}_1, \sigma^2/n_1)$ and $N(\bar{y}_2, \sigma^2/n_2)$, respectively. Thus, the joint distribution of $\gamma = \theta_2 - \theta_1$ and σ^2 is

$$p(\gamma, \sigma^2 \mid \mathbf{y}) = p(\sigma^2 \mid s^2)p(\gamma \mid \sigma^2, \bar{y}_2 - \bar{y}_1), \tag{2.5.2}$$

where, for given σ^2, γ has the Normal distribution $N[\bar{y}_2 - \bar{y}_1, \sigma^2(1/n_1 + 1/n_2)]$. Applying formula (A2.1.2) in Appendix A2.1 to integrate out σ^2, we deduce the posterior distribution of γ as

$$p(\gamma \mid \mathbf{y}) = \frac{[s^2(1/n_1 + 1/n_2)]^{-1/2}}{B(\tfrac{1}{2}, \tfrac{1}{2}\nu)\sqrt{\nu}} \left\{1 + \frac{[\gamma_2 - \gamma_1 - (\bar{y}_2 - \bar{y}_1)]^2}{\nu s^2(1/n_1 + 1/n_2)}\right\}^{-\frac{1}{2}(\nu+1)},$$

$$-\infty < \gamma_1 - \gamma_2 < \infty. \tag{2.5.3}$$

Thus γ is distributed as $t[\bar{y}_2 - \bar{y}_1, s^2(1/n_1 + 1/n_2), n_1 + n_2 - 2]$, or, equivalently,

$$t = \frac{\gamma - (\bar{y}_2 - \bar{y}_1)}{s(1/n_1 + 1/n_2)^{1/2}} \tag{2.5.4}$$

has the $t(0, 1, n_1 + n_2 - 2)$ distribution. Again, this exactly parallels the corresponding sampling theory result, and a confidence distribution for γ exactly matches the posterior distribution.

As before, inferences about the difference γ can best be made by studying the complete posterior distribution in (2.5.3), but H.P.D. intervals having any desired probability content may be constructed using Table III (at the end of this book), and will partially represent the information contained in the distribution.

2.5.2 Distribution of $\theta_2 - \theta_1$ when σ_1^2 and σ_2^2 are not Assumed Equal

In some experimental situations, the assumption that $\sigma_1^2 = \sigma_2^2$ is inappropriate. For instance, in the textile example quoted above, the two kinds of spinning machines might differ substantially not only in their means but in their variances as well. Suppose we assume that the two sets of observations are random samples independently drawn from the Normal populations $N(\theta_1, \sigma_1^2)$ and $N(\theta_2, \sigma_2^2)$. Then the quantities $(\bar{y}_1, s_1^2)$ are jointly sufficient for (θ_1, σ_1^2) having independent sampling distributions $N(\theta_1, \sigma_1^2/n_1)$ and $(\sigma_1^2/\nu_1)\chi_{\nu_1}^2$, respectively. Similarly, $(\bar{y}_2, s_2^2)$ are jointly sufficient for (θ_2, σ_2^2) having independent sampling distributions $N(\theta_2, \sigma_2^2/n_2)$ and $(\sigma_2^2/\nu_2)\chi_{\nu_2}^2$.

Assuming for our noninformative reference prior that $\theta_1, \theta_2, \log\sigma_1$ and $\log\sigma_2$ are approximately independent and locally uniform, it can be verified by using Theorem 2.4.1 (on page 97) that the joint posterior distribution of θ_1 and θ_2 is

$$p(\theta_1, \theta_2 \mid \mathbf{y}) = p(\theta_1 \mid \bar{y}_1, s_1^2)p(\theta_2 \mid \bar{y}_2, s_2^2), \tag{2.5.5}$$

where

$$p(\theta_1 \mid \bar{y}_1, s_1^2) = \frac{(s_1/\sqrt{n_1})^{-1}}{B(\frac{1}{2}, \frac{1}{2}\nu_1)\sqrt{\nu_1}}\left[1 + \frac{n_1(\theta_1 - \bar{y}_1)^2}{\nu_1 s_1^2}\right]^{-\frac{1}{2}(\nu_1+1)}, \qquad -\infty < \theta_1 < \infty,$$

$$p(\theta_2 \mid \bar{y}_2, s_2^2) = \frac{(s_2/\sqrt{n_2})^{-1}}{B(\frac{1}{2}, \frac{1}{2}\nu_2)\sqrt{\nu_2}}\left[1 + \frac{n_2(\theta_2 - \bar{y}_2)^2}{\nu_2 s_2^2}\right]^{-\frac{1}{2}(\nu_2+1)}, \qquad -\infty < \theta_2 < \infty,$$

that is, *a posteriori* θ_1 and θ_2 are independent, and distributed as $t(\bar{y}_1, s_1^2/n_1, \nu_1)$ and $t(\bar{y}_2, s_2^2/n_2, \nu_2)$, respectively.

The posterior distribution of $\gamma = \theta_2 - \theta_1$ can be obtained, for example, by making the transformation

$$\gamma = \theta_2 - \theta_1, \qquad \gamma_1 = \theta_1 \tag{2.5.6}$$

and integrating out γ_1 from the joint distribution of (γ_1, γ).

The distribution of $\gamma = \theta_2 - \theta_1$ cannot be expressed in terms of simple functions but may be computed by numerical integration to any desired accuracy. For the purpose of tabulation, following Fisher (1935), it is convenient to define the quantity

$$\tan\phi = \frac{s_1/\sqrt{n_1}}{s_2/\sqrt{n_2}} \tag{2.5.7}$$

and write

$$\tau = \frac{\gamma - (\bar{y}_2 - \bar{y}_1)}{(s_1^2/n_1 + s_2^2/n_2)^{1/2}} = \frac{\theta_2 - \bar{y}_2}{s_2/\sqrt{n}} \cos\phi - \frac{\theta_1 - \bar{y}_1}{s_1/\sqrt{n_1}} \sin\phi$$

$$= t_2 \cos\phi - t_1 \sin\phi, \tag{2.5.8}$$

where t_1 and t_2 are independently distributed as $t(0, 1, \nu_1)$ and $t(0, 1, \nu_2)$, respectively. The posterior distribution of $\theta_2 - \theta_1$ can then be determined from the distribution of τ. The density function of τ may now be obtained from the joint distribution of t_1 and t_2 by performing the integration

$$p(\tau \mid \nu_1, \nu_2, \phi) = \iint_{t_2 \cos\phi \,-\, t_1 \sin\phi \,=\, \tau} p(t_1)p(t_2)\, dt_1\, dt_2, \qquad -\infty < \tau < \infty, \tag{2.5.9}$$

where

$$p(t_i) = \frac{1}{B(\tfrac{1}{2}, \tfrac{1}{2}\nu_i)\sqrt{\nu_i}} \left(1 + \frac{t_i^2}{\nu_i}\right)^{-\frac{1}{2}(\nu_i + 1)}, \qquad i = 1, 2.$$

This integration process can be viewed geometrically in Fig. 2.5.1. In the figure, the joint distribution of t_1 and t_2, which is the product $p(t_1)p(t_2)$, is illustrated by the three contours. Also shown is the desired distribution of τ. For a given value of τ, say $\tau = \tau_0$, the equation

$$\tau_0 = t_2 \cos\phi - t_1 \sin\phi \tag{2.5.10}$$

determines a line in the (t_1, t_2) plane. By varying the value of τ in its permissible range, we get a set of parallel lines as illustrated in the figure. The density function of τ at τ_0, $p(\tau = \tau_0 | \nu_1, \nu_2, \phi)$, is obtained simply by "aggregating" all the joint densities of (t_1, t_2) on the line (2.5.10). Repeating this process for all values of τ, we obtain the entire distribution of τ as shown.

To perform the integration in (2.5.9), we may make the transformation

$$\begin{cases} \tau = t_2 \cos\phi - t_1 \sin\phi, \\ z = t_2 \sin\phi + t_1 \cos\phi. \end{cases} \tag{2.5.11}$$

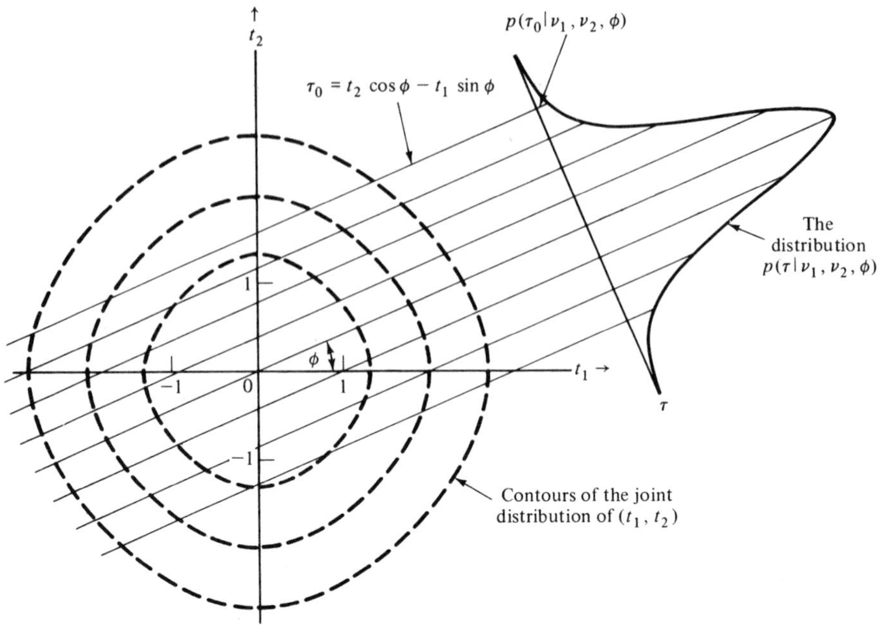

Fig. 2.5.1 Geometric interpretation of the integration process in producing the distribution of τ.

Since the transformation has unit Jacobian, it follows that

$$p(\tau \mid \nu_1, \nu_2, \phi) = k \int_{-\infty}^{\infty} \left[1 + \frac{(z \cos\phi - \tau \sin\phi)^2}{\nu_1}\right]^{-\frac{1}{2}(\nu_1+1)}$$

$$\times \left[1 + \frac{(z \sin\phi + \tau \cos\phi)^2}{\nu_2}\right]^{-\frac{1}{2}(\nu_2+1)} dz, \qquad -\infty < \tau < \infty, \tag{2.5.12}$$

where

$$k^{-1} = B(\tfrac{1}{2}\nu_1, \tfrac{1}{2}) B(\tfrac{1}{2}\nu_2, \tfrac{1}{2}) (\nu_1 \nu_2)^{1/2}.$$

Note that the distribution of τ depends only on three parameters (ν_1, ν_2, ϕ).

A result which is identical to (2.5.12) was justified by Fisher (1935, 1939) using fiducial theory. Fisher refers to earlier work by Behrens (1929), and the distribution (2.5.12) has come to be called the Behrens–Fisher distribution. A table of percentage points of the distribution was tabulated by Sukhatmé (1938). The Behrens–Fisher distribution is a symmetric distribution and is like the t-distribution in its general appearance. Since Bayesian H.P.D. intervals are numerically (but not, of course, logically) identical with Fisher's fiducial intervals, they can be obtained using Sukhatmé's table. The Bayesian derivation was first given by Jeffreys (1961).

2.5.3 Approximations to the Behrens–Fisher Distribution

As we emphasized, where possible it is desirable to show the whole posterior distribution rather than particular intervals. Although the distribution is not exactly expressible in terms of tabled functions, a number of attempts have been made to approximate it in terms of tabled functions. We employ an approximation proposed by Patil (1964), who fits a scaled t distribution to the distribution of τ by equating the second and the fourth moments. It is shown by Patil that τ is approximately distributed as $t(0, a^2, b)$, where

$$
\begin{aligned}
a^2 &= \left(\frac{b-2}{b}\right) f_1, \\
b &= 4 + \frac{f_1^2}{f_2}, \\
f_1 &= \left(\frac{\nu_2}{\nu_2 - 2}\right)\cos^2\phi + \left(\frac{\nu_1}{\nu_1 - 2}\right)\sin^2\phi, \\
f_2 &= \frac{\nu_2^2}{(\nu_2-2)^2(\nu_2-4)}\cos^4\phi + \frac{\nu_1^2}{(\nu_1-2)^2(\nu_1-4)}\sin^4\phi,
\end{aligned}
\tag{2.5.13}
$$

and, as before,

$$
\cos^2\phi = \frac{s_2^2/n_2}{(s_1^2/n_1 + s_2^2/n_2)} \qquad \text{and} \qquad \sin^2\phi = 1 - \cos^2\phi.
$$

To this degree of approximation, the difference of the mean values $\gamma = (\theta_2 - \theta_1)$ is distributed *a posteriori* as $t[\bar{y}_2 - \bar{y}_1, a^2(s_1^2/n_1 + s_2^2/n_2), b]$. Having calculated a and b, tables of the density of Student's t distribution may be used to obtain approximate ordinates from which the posterior distribution can be sketched, and the probability integral of t may be used to obtain Bayesian intervals for probability levels not tabulated by Sukhatmé.

2.5.4 An Example

Suppose, in the spinning modification experiment the following results are obtained:

$$
\begin{aligned}
\bar{y}_1 &= 50, \qquad n_1 = 20, \qquad s_1^2 = 12, \\
\bar{y}_2 &= 55, \qquad n_2 = 12, \qquad s_2^2 = 40.
\end{aligned}
$$

Using the approximation in (2.5.13), we find

$$
\cos^2\phi = \frac{\frac{40}{12}}{\frac{12}{20} + \frac{40}{12}} = \frac{50}{59}, \qquad \sin^2\phi = 1 - \cos^2\phi = \frac{9}{59},
$$

$$
\nu_1 = n_1 - 1 = 19, \qquad \nu_2 = n_2 - 1 = 11,
$$

so that

$$f_1 = \left(\frac{11}{9}\right)\left(\frac{50}{59}\right) + \left(\frac{19}{17}\right)\left(\frac{9}{59}\right) = 1.2063,$$

$$f_2 = \left(\frac{11}{9}\right)^2\left(\frac{1}{7}\right)\left(\frac{50}{59}\right)^2 + \left(\frac{19}{17}\right)^2\left(\frac{1}{15}\right)\left(\frac{9}{59}\right)^2 = 0.1552,$$

from which

$$b = 4 + \frac{f_1^2}{f_2} = 13.376,$$

$$a^2 = \left(\frac{b-2}{b}\right)f_1 = 1.026.$$

Thus, on the basis of a locally uniform prior for $(\theta_1, \theta_2, \log\sigma_1, \log\sigma_2)$, the posterior distribution of the difference in means $\theta_2 - \theta_1$ of standard and modified spinning machines would be closely approximated by a $t(5, 4.035, 13.38)$ distribution. That is, by a scaled t-distribution having $b \doteq 13.4$ degrees of freedom centered at $\bar{y}_2 - \bar{y}_1 = 5$, with scale factor

$$a\sqrt{(s_1^2/n_1 + s_2^2/n_2)} = (4.035)^{1/2} = 2.009.$$

This approximate distribution is shown by the broken curve in Fig. 2.5.2. Also shown by the solid curve, in the same figure, is the exact distribution of

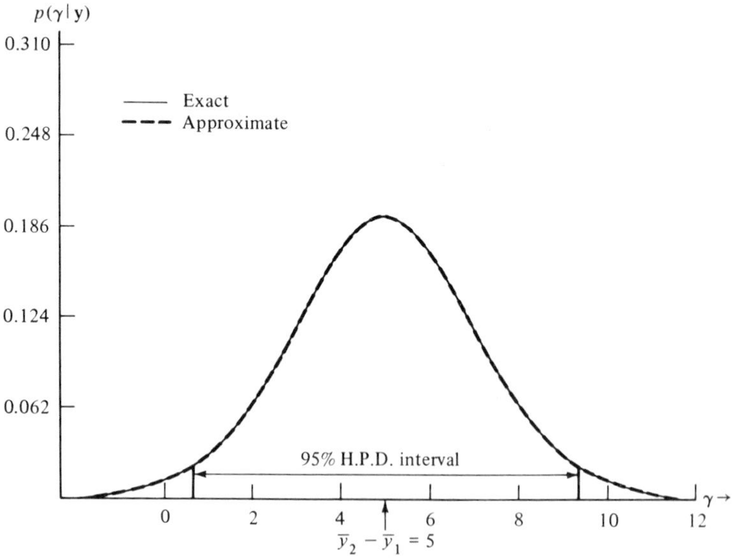

Fig. 2.5.2 Comparison of spinning machines: posterior distribution of $\gamma = \theta_2 - \theta_1$.

$\gamma = (\theta_2 - \theta_1)$. Table 2.5.1 gives a specimen of the exact and approximate density of γ. The agreement is exceedingly close. Limits of the approximate 95 percent H.P.D. interval are given by

$$\bar{y}_2 - \bar{y}_1 \pm a\sqrt{(s_1^2/n_1 + s_2^2/n_2)}\, t_{0.025}(b).$$

From Table IV (at the end of this book), $t_{0.025}(13.38)$ is nearly 2.15 so that the extreme points are (0.673, 9.327). This interval is also shown in Fig. 2.5.2.

No results paralleling the Bayesian (Behrens–Fisher) solution are available from sampling theory. Welch (1938, 1947) has considered the problem from the sampling viewpoint but the solution he produced does not correspond with the Bayesian solution.

Table 2.5.1

Exact and approximate posterior density of γ

γ	Exact density	Approximate density
5.0	0.19411	0.19490
5.2	0.19310	0.19387
5.4	0.19010	0.19080
5.6	0.18523	0.18581
5.8	0.17863	0.17907
6.0	0.17054	0.17081
6.2	0.16120	0.16130
6.4	0.15089	0.15083
6.6	0.13990	0.13970
6.8	0.12851	0.12820
7.0	0.11700	0.11660
7.2	0.10561	0.10515
7.4	0.09453	0.09405
7.6	0.08395	0.08347
7.8	0.07398	0.07354
8.0	0.06474	0.06434
8.4	0.04859	0.04832
8.8	0.03562	0.03547
9.2	0.02558	0.02554
9.6	0.01805	0.01809
10.0	0.01257	0.01264

2.6 INFERENCES CONCERNING A VARIANCE RATIO

Consider again the example just discussed in Section 2.5, where the investigator was interested in studying the effect of a spinning modification on the breaking strength of yarn. Previously we supposed that his interest centered on a possible difference in the means. However, variability of the strength of yarn is also of great practical importance, and we now consider how inferences might be made about a possible change in the variability of yarn between the standard machines and the modified machines.

We suppose, as before, that n_1 independent observations $\mathbf{y}_1' = (y_{11}, \ldots, y_{1n_1})$ are as if drawn from a Normal population $N(\theta_1, \sigma_1^2)$, and n_2 observations $\mathbf{y}_2' = (y_{21}, \ldots, y_{2n_2})$ are as if drawn from $N(\theta_2, \sigma_2^2)$. We first derive the posterior distribution of the variance ratio σ_2^2/σ_1^2. Again assuming for our noninformative reference prior that $\theta_1, \theta_2, \log \sigma_1$, and $\log \sigma_2$ are approximately independent and locally uniform, by using Theorem 2.4.1 (on page 97) we find that *a posteriori* $\sigma_1^2/v_1 s_1^2$ and $\sigma_2^2/v_2 s_2^2$ are independent and distributed as $\chi_{v_1}^{-2}$ and $\chi_{v_2}^{-2}$, respectively. Consequently, the posterior distribution of the ratio

$$F = \frac{\sigma_2^2/s_2^2}{\sigma_1^2/s_1^2} = \frac{\sigma_2^2/\sigma_1^2}{s_2^2/s_1^2} = \frac{s_1^2/s_2^2}{\sigma_1^2/\sigma_2^2} \tag{2.6.1}$$

has the usual F distribution with (v_1, v_2) degrees of freedom, that is,

$$p(F) = \frac{(v_1/v_2)^{v_1/2}}{B(v_1/2, v_2/2)} F^{\frac{1}{2}v_1 - 1} \left(1 + \frac{v_1}{v_2} F\right)^{-\frac{1}{2}(v_1 + v_2)}, \qquad F > 0. \tag{2.6.2}$$

Again, with appropriate interpretation this parallels the sampling theory result. In repeated samples from Normal populations the quantity $(s_1^2/s_2^2)/(\sigma_1^2/\sigma_2^2)$, in which s_1^2/s_2^2 is the random variable and the ratio σ_1^2/σ_2^2 is an unknown fixed constant, is also distributed as an F variable with (v_1, v_2) degrees of freedom. It follows that the posterior distribution of σ_2^2/σ_1^2, on the basis of a locally uniform prior for $(\theta_1, \theta_2, \log \sigma_1, \log \sigma_2)$, is numerically equivalent to the corresponding confidence distribution.

2.6.1 H.P.D. Intervals

In comparing the spread of two Normal distributions, there are many measures that could be used besides σ_2^2/σ_1^2, for example, the reciprocal ratio σ_1^2/σ_2^2, the ratio of standard deviations σ_2/σ_1, and $\log \sigma_2 - \log \sigma_1$. Since $\log \sigma_1$ and $\log \sigma_2$ are assumed locally uniform *a priori*, the standardized H.P.D. intervals will be the H.P.D. intervals of $\log \sigma_2 - \log \sigma_1$. Noting that H.P.D. intervals are equivalent under linear transformation of $\log \sigma_2 - \log \sigma_1$,

$$c + q(\log \sigma_2 - \log \sigma_1) \tag{2.6.3}$$

where c and q are two arbitrary constants, we may consider simply the logarithm of F,

$$\log F = (\log \sigma_2^2 - \log \sigma_1^2) - (\log s_2^2 - \log s_1^2). \tag{2.6.4}$$

The distribution of $\log F$ is

$$p(\log F) = \frac{(\nu_1/\nu_2)^{\nu_1/2}}{B(\frac{1}{2}\nu_1, \frac{1}{2}\nu_2)} F^{\nu_1/2}\left(1 + \frac{\nu_1}{\nu_2}F\right)^{-\frac{1}{2}(\nu_1+\nu_2)}, \qquad -\infty < \log F < \infty. \tag{2.6.5}$$

which is the form first derived by Fisher (1924).

In terms of F, the mode of the distribution in (2.6.5) is $F = 1$ which means that the mode of distribution of $\log \sigma_2^2/\sigma_1^2$ is

$$\frac{\sigma_2^2}{\sigma_1^2} = \frac{s_2^2}{s_1^2}. \tag{2.6.6}$$

Limits of H.P.D. intervals of $\log F$ are given in Table V (at the end of this book) for a combination of values (α, ν_1, ν_2). To avoid computing logarithms the limits are given in terms of F. Thus, the values $\underline{F}(\nu_1, \nu_2, \alpha)$ and $\overline{F}(\nu_1, \nu_2, \alpha)$ in the table are such that

a) $$\underline{F}^{\nu_1/2}\left(1 + \frac{\nu_1}{\nu_2}\underline{F}\right)^{-\frac{1}{2}(\nu_1+\nu_2)} = \overline{F}^{\nu_1/2}\left(1 + \frac{\nu_1}{\nu_2}\overline{F}\right)^{-\frac{1}{2}(\nu_1+\nu_2)}$$

and

b) $$\Pr\{F < \underline{F}\} + \Pr\{F > \overline{F}\} = \alpha. \tag{2.6.7}$$

Further, the table only gives $\underline{F}$ and $\overline{F}$ for $\nu_1 \geqslant \nu_2$. For $\nu_2 > \nu_1$, the corresponding limits may be obtained by using the relationships

$$\underline{F}(\nu_2, \nu_1, \alpha) = \frac{1}{\overline{F}(\nu_1, \nu_2, \alpha)} \quad \text{and} \quad \overline{F}(\nu_2, \nu_1, \alpha) = \frac{1}{\underline{F}(\nu_1, \nu_2, \alpha)}. \tag{2.6.8}$$

The use of the table is illustrated by the following example.

2.6.2 An Example

Consider again the spinning modification example for which $n_2 = 12$, $s_2^2 = 40$, $n_1 = 20$ and $s_1^2 = 12$. Figure 2.6.1 shows the posterior distribution of $\log(\sigma_2^2/\sigma_1^2)$. The abcissas are plotted in terms of the variance ratio σ_2^2/σ_1^2. To obtain a 95 per cent H.P.D. interval we have from Table V,

$$\underline{F}(19, 11, 0.05) = 0.352 \qquad \text{and} \qquad \overline{F}(19, 11, 0.05) = 3.150,$$

so that the lower and upper limits (V_0, V_1) are respectively

$$V_0 = \left(\frac{s_2^2}{s_1^2}\right)\underline{F}(19, 11, 0.05) = (3.33)(0.352) = 1.1722$$

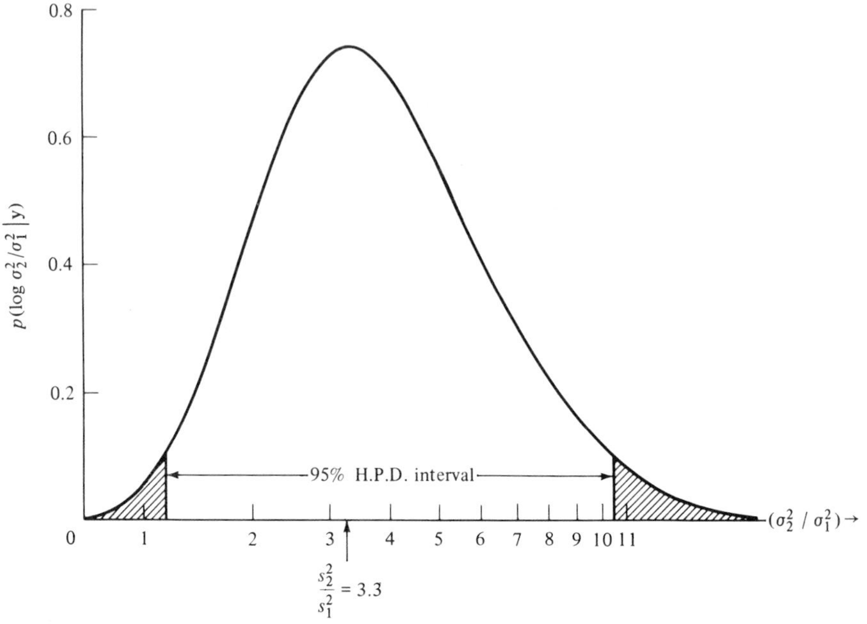

Fig. 2.6.1 Comparison of spinning machines: posterior distribution of log (σ_2^2/σ_1^2) (scaled in σ_2^2/σ_1^2).

and

$$V_1 = \left(\frac{s_2^2}{s_1^2}\right)\bar{F}(19, 11, 0.05) = (3.33)(3.150) = 10.4895,$$

which are also shown in the figure.

A reasonably accurate impression of the distribution $p[\log(\sigma_2^2/\sigma_1^2) \mid \mathbf{y}]$ is gained by stating the mode and a suitable set of H.P.D. intervals. For the present example, the mode of log (σ_2^2/σ_1^2) is at $\sigma_2^2/\sigma_1^2 = s_2^2/s_1^2 = 3.33$ and a suitable set of intervals is given in the following table.

Table 2.6.1

Probability $(1 - \alpha)$ that σ_2^2/σ_1^2 lies within limits (V_0, V_1)

$(1 - \alpha)$	V_0	V_1
0.75	1.8015	6.3836
0.90	1.3853	8.6047
0.95	1.1722	10.4895
0.99	0.8458	15.7842

2.7 ANALYSIS OF THE LINEAR MODEL

We consider in this section the linear model

$$\mathbf{y} = \mathbf{X}\boldsymbol{\theta} + \boldsymbol{\varepsilon}, \tag{2.7.1}$$

where $\mathbf{y}$ is a $n \times 1$ vector of observations, $\mathbf{X}$ a $n \times k$ matrix of known constants, $\boldsymbol{\theta}$ a $k \times 1$ vector of unknown regression coefficients, and $\boldsymbol{\varepsilon}$ a $n \times 1$ vector of random variables having zero means (often referred to as *errors*). Explicitly, the model is

$$\begin{bmatrix} y_1 \\ \vdots \\ y_u \\ \vdots \\ y_n \end{bmatrix} = \begin{bmatrix} x_{11} \cdots x_{1k} \\ \vdots \quad\; \vdots \\ x_{u1} \cdots x_{uk} \\ \vdots \quad\; \vdots \\ x_{n1} \cdots x_{nk} \end{bmatrix} \begin{bmatrix} \theta_1 \\ \vdots \\ \\ \vdots \\ \theta_k \end{bmatrix} + \begin{bmatrix} \varepsilon_1 \\ \vdots \\ \varepsilon_u \\ \vdots \\ \varepsilon_n \end{bmatrix} \tag{2.7.2}$$

and, in particular, for the uth observation

$$y_u = \theta_1 x_{u1} + \cdots + \theta_k x_{uk} + \varepsilon_u. \tag{2.7.3}$$

In this model, y may be called the *dependent* variable, the *response*, or the *output* variable, while the x's are referred to as the *independent* variables or *input* variables. The relationship between the dependent and independent variables, regarded as vectors, is brought out by writing the model in the form

$$\mathbf{y} = \mathbf{x}_1\theta_1 + \cdots + \mathbf{x}_k\theta_k + \boldsymbol{\varepsilon}, \tag{2.7.4}$$

where

$$\mathbf{x}_i = (x_{1i}, \ldots, x_{ui}, \ldots, x_{ni})', \qquad i = 1, \ldots, k,$$

are the columns of the matrix $\mathbf{X}$ which is called the *matrix of independent variables* or the *derivative matrix*.

The model has very wide applicability.

1. It is frequently employed in regression problems where $(x_{u1}, \ldots, x_{uk})$ may be the levels, or functions of levels, of quantities such as temperature and pressure, which have been set or observed in a series of experiments.
2. The model is equally applicable where some, or all, of the $\mathbf{x}_i$'s are vectors of "indicator" variables. An indicator x_{ui}, taking only the values one or zero, may be used to signify the presence, or absence, of a contribution θ_i to the observation y_u.

As a simple example, the model

$$y_u = \theta + \varepsilon_u, \qquad u = 1, \ldots, n \tag{2.7.5a}$$

can be written equally well as

$$y_u = x_{1u}\theta + \varepsilon_u, \qquad x_{1u} = 1, \qquad u = 1, \ldots, n$$

or as

$$\mathbf{y} = \mathbf{x}_1\theta + \boldsymbol{\varepsilon}, \tag{2.7.5b}$$

where $\mathbf{x}_1$ is a $n \times 1$ column of ones. This model is now a special case of (2.7.4). If we assume that ε_u's are independently distributed as $N(0, \sigma^2)$, the model (2.7.5b) is identical to the one discussed in Sections 2.2 and 2.4.

As a second example, suppose the response is expected to be a quadratic function of a variable ξ, so that the model takes the form

$$y_u = \theta_1 + \theta_2\xi_u + \theta_3\xi_u^2 + \varepsilon_u. \tag{2.7.6a}$$

Then writing

$$x_{u1} = 1, \qquad x_{u2} = \xi_u, \qquad x_{u3} = \xi_u^2, \qquad u = 1, \ldots, n,$$

we have

$$\mathbf{y} = \mathbf{x}_1\theta_1 + \mathbf{x}_2\theta_2 + \mathbf{x}_3\theta_3 + \boldsymbol{\varepsilon}. \tag{2.7.6b}$$

Again this model is a special case of (2.7.4).

As a final example, in the one-way classification analysis of variance models, we are typically concerned with the comparison of k treatments for which the mean responses are $\theta_1, \cdots, \theta_k$. Suppose then we have $n = \sum_{i=1}^k n_i$ observations, the first n_1 of these being made with treatment one, the second n_2 with treatment two, etc. The appropriate model written in the form of (2.7.2) would be

$$\begin{bmatrix} y_1 \\ \vdots \\ y_{n_1} \\ \hdashline y_{n_1+1} \\ \vdots \\ y_{n_1+n_2} \\ \hdashline \vdots \\ \hdashline y_{n-n_k+1} \\ \vdots \\ y_n \end{bmatrix} = \begin{bmatrix} 1 & & \\ \vdots & & \\ 1 & & \\ \hdashline & 1 & \\ & \vdots & \\ & 1 & \\ \hdashline & & \\ \hdashline & & 1 \\ & & \vdots \\ & & 1 \end{bmatrix} \begin{bmatrix} \theta_1 \\ \vdots \\ \theta_k \end{bmatrix} + \begin{bmatrix} \varepsilon_1 \\ \vdots \\ \vdots \\ \varepsilon_n \end{bmatrix}. \tag{2.7.7}$$

Thus, models associated with polynomial regression, multiple regression, analysis of variance, randomized block designs, incomplete block designs, and factorial designs are all subsumed in the general linear model.

In this section, we assume that $\boldsymbol{\varepsilon}$ has the multivariate spherically Normal distribution $N_n(\mathbf{0}, \sigma^2\mathbf{I})$ and that the rank of the matrix $\mathbf{X}$ is k, so that

$$p(\mathbf{y} \mid \boldsymbol{\theta}, \sigma^2) = \left(\frac{1}{\sqrt{2\pi}}\right)^n \sigma^{-n} \exp\left[-\frac{1}{2\sigma^2}(\mathbf{y} - \mathbf{X}\boldsymbol{\theta})'(\mathbf{y} - \mathbf{X}\boldsymbol{\theta})\right]$$

$$= \left(\frac{1}{\sqrt{2\pi}}\right)^n \sigma^{-n} \exp\left\{-\frac{1}{2\sigma^2}[\nu s^2 + (\boldsymbol{\theta} - \hat{\boldsymbol{\theta}})'\mathbf{X}'\mathbf{X}(\boldsymbol{\theta} - \hat{\boldsymbol{\theta}})]\right\} \tag{2.7.8}$$

where

$$\hat{\boldsymbol{\theta}} = (\mathbf{X}'\mathbf{X})^{-1}\mathbf{X}'\mathbf{y}, \qquad \nu = n - k,$$

$$s^2 = (1/\nu)(\mathbf{y} - \hat{\mathbf{y}})'(\mathbf{y} - \hat{\mathbf{y}}) \qquad \text{and} \qquad \hat{\mathbf{y}} = \mathbf{X}\hat{\boldsymbol{\theta}}.$$

It then follows that:

a) $\hat{\boldsymbol{\theta}}$ is a vector of statistics jointly sufficient for $\boldsymbol{\theta}$ if σ^2 is known;
b) $\hat{\boldsymbol{\theta}}$ and s^2 are jointly sufficient for $(\boldsymbol{\theta}, \sigma^2)$;
c) $\hat{\boldsymbol{\theta}}$ has the multivariate Normal distribution $N_k[\boldsymbol{\theta}, \sigma^2(\mathbf{X}'\mathbf{X})^{-1}]$, and
d) νs^2 is distributed independently of $\hat{\boldsymbol{\theta}}$ as $\sigma^2\chi^2_\nu$.

We shall distinguish between two situations: the first in which the variance σ^2 is supposed known, and the second where σ^2 is supposed unknown.

2.7.1 Variance σ^2 Assumed Known

If σ^2 is known, then from Lemma 1.4.1 (on page 62) the posterior distribution of $\boldsymbol{\theta}$ is

$$p(\boldsymbol{\theta} \mid \mathbf{y}) \propto p(\boldsymbol{\theta})p(\hat{\boldsymbol{\theta}} \mid \boldsymbol{\theta}, \sigma^2), \tag{2.7.9}$$

where $p(\boldsymbol{\theta})$ is the prior distribution and $p(\hat{\boldsymbol{\theta}} \mid \boldsymbol{\theta}, \sigma^2)$ is the density of a multivariate Normal distribution $N_k[\boldsymbol{\theta}, \sigma^2(\mathbf{X}'\mathbf{X})^{-1}]$. As in (1.3.73), if k is not large, we may employ $p(\boldsymbol{\theta}) \propto$ const. as a noninformative reference prior for $\boldsymbol{\theta}$. Then, the posterior distribution for $\boldsymbol{\theta}$ is approximately

$$p(\boldsymbol{\theta} \mid \mathbf{y}) \propto \exp\left[-\frac{1}{2\sigma^2}(\boldsymbol{\theta} - \hat{\boldsymbol{\theta}})'\mathbf{X}'\mathbf{X}(\boldsymbol{\theta} - \hat{\boldsymbol{\theta}})\right]. \tag{2.7.10}$$

Since

$$\int_R \exp\left[-\frac{1}{2\sigma^2}(\boldsymbol{\theta} - \hat{\boldsymbol{\theta}})'\mathbf{X}'\mathbf{X}(\boldsymbol{\theta} - \hat{\boldsymbol{\theta}})\right] d\boldsymbol{\theta} = \frac{\sigma^k(\sqrt{2\pi})^k}{|\mathbf{X}'\mathbf{X}|^{1/2}},$$

where R is the region $-\infty < \theta_i < \infty$, $i = 1, \ldots, k$, we have

$$p(\boldsymbol{\theta} \mid \mathbf{y}) \doteq \frac{|\mathbf{X}'\mathbf{X}|^{1/2}}{(\sqrt{2\pi}\,\sigma)^k} \exp\left[-\frac{1}{2\sigma^2}(\boldsymbol{\theta} - \hat{\boldsymbol{\theta}})'\mathbf{X}'\mathbf{X}(\boldsymbol{\theta} - \hat{\boldsymbol{\theta}})\right],$$

$$-\infty < \theta_i < \infty, \qquad i = 1, \ldots, k. \tag{2.7.11}$$

All relevant inferences about $\boldsymbol{\theta}$ can now be made from the knowledge that the posterior distribution of $\boldsymbol{\theta}$ is the multivariate Normal distribution $N_k[\hat{\boldsymbol{\theta}}, \sigma^2(\mathbf{X}'\mathbf{X})^{-1}]$.

Some Properties of the Distribution of $\boldsymbol{\theta}$

The density function $p(\boldsymbol{\theta} \mid \mathbf{y})$ is a monotonically decreasing function of the positive definite quadratic form

$$Q(\boldsymbol{\theta}) = (\boldsymbol{\theta} - \hat{\boldsymbol{\theta}})'\mathbf{X}'\mathbf{X}(\boldsymbol{\theta} - \hat{\boldsymbol{\theta}}), \tag{2.7.12}$$

so that

$$Q(\boldsymbol{\theta}) = c, \qquad c > 0 \tag{2.7.13}$$

defines a contour of the distribution in the k-dimensional space of $\boldsymbol{\theta}$. For $k = 2$, (2.7.13) is the equation of an ellipse; for $k = 3$, it is that of an ellipsoid, and for $k > 3$, that of a hyper-ellipsoid. From properties of the multivariate Normal distribution, Wilks (1962), the quadratic form $Q(\boldsymbol{\theta})$ is distributed, *a posteriori*, as $\sigma^2\chi_k^2$. Therefore the inequality

$$Q(\boldsymbol{\theta}) \leqslant \chi^2(k, \alpha) \tag{2.7.14}$$

defines an ellipsoidal region in the parameter space of $\boldsymbol{\theta}$, and $(1 - \alpha)$ is the posterior probability contained in this region. The interpretation of such a region is discussed later.

Denoting

$$\boldsymbol{\theta} = \begin{bmatrix} \boldsymbol{\theta}_1 \\ \boldsymbol{\theta}_2 \end{bmatrix} \begin{matrix} r \\ k-r \end{matrix}, \qquad \hat{\boldsymbol{\theta}} = \begin{bmatrix} \hat{\boldsymbol{\theta}}_1 \\ \hat{\boldsymbol{\theta}}_2 \end{bmatrix} \begin{matrix} r \\ k-r \end{matrix}, \qquad \mathbf{X}'\mathbf{X} = \overset{\begin{matrix} r & k-r \end{matrix}}{\begin{bmatrix} \mathbf{X}_1'\mathbf{X}_1 & \mathbf{X}_1'\mathbf{X}_2 \\ \mathbf{X}_2'\mathbf{X}_1 & \mathbf{X}_2'\mathbf{X}_2 \end{bmatrix}} \begin{matrix} r \\ k-r \end{matrix},$$

$$(\mathbf{X}'\mathbf{X})^{-1} = \overset{\begin{matrix} r & k-r \end{matrix}}{\begin{bmatrix} \mathbf{C}_{11} & \mathbf{C}_{12} \\ \mathbf{C}_{21} & \mathbf{C}_{22} \end{bmatrix}} \begin{matrix} r \\ k-r \end{matrix}, \tag{2.7.15}$$

where $1 \leqslant r < k$, it follows from properties of the multivariate Normal distribution that:

a) The marginal posterior distribution of a subset of $\boldsymbol{\theta}$, say $\boldsymbol{\theta}_1$, is the multivariate Normal distribution $N_r(\hat{\boldsymbol{\theta}}_1, \sigma^2\mathbf{C}_{11})$ and, in particular, the marginal distribution of θ_i is $N(\hat{\theta}_i, \sigma^2 c_{ii})$, where c_{ii} is the ith diagonal element of $(\mathbf{X}'\mathbf{X})^{-1}$.

b) The conditional posterior distribution of $\boldsymbol{\theta}_2$, given $\boldsymbol{\theta}_1$, is the multivariate Normal distribution $N_{k-r}[\hat{\boldsymbol{\theta}}_{2.1}, \sigma^2(\mathbf{X}_2'\mathbf{X}_2)^{-1}]$, where

$$\hat{\boldsymbol{\theta}}_{2.1} = \hat{\boldsymbol{\theta}}_2 - (\mathbf{X}_2'\mathbf{X}_2)^{-1}\mathbf{X}_2'\mathbf{X}_1(\boldsymbol{\theta}_1 - \hat{\boldsymbol{\theta}}_1). \tag{2.7.16}$$

2.7.2 Variance σ^2 Unknown

In many cases the variance σ^2 will not be known, and information about σ^2 coming from the sample is used. From Lemma 1.4.1 (on page 62) the posterior distribution of $(\boldsymbol{\theta}, \sigma^2)$ is then

$$p(\boldsymbol{\theta}, \sigma^2 \mid \mathbf{y}) \propto p(\boldsymbol{\theta}, \sigma^2)p(s^2 \mid \sigma^2)p(\hat{\boldsymbol{\theta}} \mid \boldsymbol{\theta}, \sigma^2). \tag{2.7.17}$$

From Section (1.3.6) we obtain a noninformative prior with $\boldsymbol{\theta}$ and $\log \sigma$ approximately independent and locally uniform, so that

$$p(\boldsymbol{\theta}, \sigma^2) = p(\boldsymbol{\theta})p(\sigma^2) \propto \sigma^{-2}. \tag{2.7.18}$$

Substituting (2.7.18) into (2.7.17), and following an argument similar to that used in Section 2.4 the joint posterior distribution of $(\boldsymbol{\theta}, \sigma^2)$ may be factored such that

$$p(\boldsymbol{\theta}, \sigma^2 \mid \mathbf{y}) = p(\sigma^2 \mid s^2)p(\boldsymbol{\theta} \mid \hat{\boldsymbol{\theta}}, \sigma^2), \tag{2.7.19}$$

where the marginal posterior distribution of σ^2 is $\nu s^2 \chi_\nu^{-2}$, $\nu = n - k$, and the conditional posterior distribution of $\boldsymbol{\theta}$, given σ^2, is the multivariate Normal distribution given in (2.7.11). Applying formula (A2.1.2) in Appendix A2.1 to integrate out σ^2 from (2.7.19), the (marginal) posterior distribution of $\boldsymbol{\theta}$ is

$$p(\boldsymbol{\theta} \mid \mathbf{y}) = \frac{\Gamma[\frac{1}{2}(\nu + k)]|\mathbf{X}'\mathbf{X}|^{1/2}s^{-k}}{[\Gamma(\frac{1}{2})]^k\Gamma(\frac{1}{2}\nu)(\sqrt{\nu})^k}\left[1 + \frac{(\boldsymbol{\theta} - \hat{\boldsymbol{\theta}})'\mathbf{X}'\mathbf{X}(\boldsymbol{\theta} - \hat{\boldsymbol{\theta}})}{\nu s^2}\right]^{-\frac{1}{2}(\nu+k)},$$

$$-\infty < \theta_i < \infty, \qquad i = 1, \ldots, k, \tag{2.7.20}$$

which is the multivariate t distribution discovered independently by Cornish (1954) and by Dunnet and Sobel (1954). We shall denote the distribution in (2.7.20) as $t_k[\hat{\boldsymbol{\theta}}, s^2(\mathbf{X}'\mathbf{X})^{-1}, \nu]$.

Some Properties of the Distribution of $\boldsymbol{\theta}$

The multivariate t distribution has the following properties:

1. The density function in (2.7.20) is a monotonically decreasing function of the quadratic form $Q(\boldsymbol{\theta}) = (\boldsymbol{\theta} - \hat{\boldsymbol{\theta}})'\mathbf{X}'\mathbf{X}(\boldsymbol{\theta} - \hat{\boldsymbol{\theta}})$, so that, like the multivariate Normal posterior distribution of $\boldsymbol{\theta}$ discussed in Section 2.7.1, contours of $p(\boldsymbol{\theta} \mid \mathbf{y})$ are also ellipsoidal in the parameter space of $\boldsymbol{\theta}$.

2. The quantity $Q(\boldsymbol{\theta})/ks^2$ is distributed *a posteriori* as F with (k, ν) degrees of freedom. The ellipsoidal contour of $p(\boldsymbol{\theta} \mid \mathbf{y})$ defined by

$$\frac{Q(\boldsymbol{\theta})}{ks^2} = F(k, \nu, \alpha), \tag{2.7.21}$$

where $F(k, \nu, \alpha)$ is the upper 100α percentage point of an F distribution with (k, ν) degrees of freedom, will thus delineate a region containing a proportion $(1 - \alpha)$ of the posterior probability. Table VI (at the end of this book) gives values of $F(\nu_2, \nu_1, \alpha)$ for various values of (ν_2, ν_1, α) with $\nu_2 < \nu_1$.

3. The marginal distribution of a r-dimensional subset, $\boldsymbol{\theta}_1$, has the multivariate t distribution $t_r(\hat{\boldsymbol{\theta}}_1, s^2\mathbf{C}_{11}, \nu)$, that is,

$$p(\boldsymbol{\theta}_1 \mid \mathbf{y}) = \frac{\Gamma[\frac{1}{2}(\nu + r)]|\mathbf{C}_{11}^{-1}|^{1/2}s^{-r}}{[\Gamma(\frac{1}{2})]^r\Gamma(\frac{1}{2}\nu)(\sqrt{\nu})^r}\left[1 + \frac{(\boldsymbol{\theta}_1 - \hat{\boldsymbol{\theta}}_1)'\mathbf{C}_{11}^{-1}(\boldsymbol{\theta}_1 - \hat{\boldsymbol{\theta}}_1)}{\nu s^2}\right]^{-\frac{1}{2}(\nu+r)},$$

$$-\infty < \theta_i < \infty, \qquad i = 1, \ldots, r, \tag{2.7.22}$$

where the notation is that employed in (2.7.15). In particular θ_i has the distribution $t(\hat{\theta}_i, s^2 c_{ii}, \nu)$; that is,

$$t = \frac{\theta_i - \hat{\theta}_i}{s\sqrt{c_{ii}}} \tag{2.7.23}$$

has the Student's t distribution with $\nu = n - k$ degrees of freedom.

4. The conditional posterior distribution of a $(k - r)$ dimensional subset of parameters $\boldsymbol{\theta}_2$, given $\boldsymbol{\theta}_1$, is also a multivariate t distribution. Specifically, $\boldsymbol{\theta}_2$ has the distribution

$$t_{k-r}[\hat{\boldsymbol{\theta}}_{2.1}, s_{2.1}^2(\mathbf{X}_2'\mathbf{X}_2)^{-1}, \nu + r], \tag{2.7.24}$$

where

$$\hat{\boldsymbol{\theta}}_{2.1} = \hat{\boldsymbol{\theta}}_2 - (\mathbf{X}_2'\mathbf{X}_2)^{-1}\mathbf{X}_2'\mathbf{X}_1(\boldsymbol{\theta}_1 - \hat{\boldsymbol{\theta}}_1)$$

and

$$s_{2.1}^2 = (\nu + r)^{-1}[\nu s^2 + (\boldsymbol{\theta}_1 - \hat{\boldsymbol{\theta}}_1)'\mathbf{C}_{11}^{-1}(\boldsymbol{\theta}_1 - \hat{\boldsymbol{\theta}}_1)].$$

5. Suppose $\boldsymbol{\phi} = (\phi_1, \cdots, \phi_m)'$, $m \leqslant k$, is a set of parameters such that

$$\boldsymbol{\phi} = \mathbf{D}\boldsymbol{\theta}, \tag{2.7.25}$$

where $\mathbf{D}$ is an $m \times k$ matrix of rank m. Then *a posteriori* $\boldsymbol{\phi}$ is distributed as

$$\boldsymbol{\phi} \sim t_m[\mathbf{D}\hat{\boldsymbol{\theta}}, s^2\mathbf{D}(\mathbf{X}'\mathbf{X})^{-1}\mathbf{D}', \nu]. \tag{2.7.26}$$

All the above results, of course, parallel sampling theory results. In particular, the posterior distribution (2.7.20) is identical to the confidence distribution of $\boldsymbol{\theta}$, and the ellipsoidal posterior region enclosed by the contour given in (2.7.21) is numerically equivalent to a $(1 - \alpha)$ confidence region for $\boldsymbol{\theta}$.

2.7.3 An Example

A micro method for weighing extremely light objects is believed to have an error which is approximately Normal. It is also believed that over the relevant range of weights the error has expectation zero and constant variance. The following data (Table 2.7.1) are available from observations made in 18 weighings of two speci-

Table 2.7.1

Observed weights of light objects in micrograms

A Only	*B* Only		*A* and *B* Together	
$109^{(8)}$	$114^{(1)}$	$129^{(5)}$	$217^{(2)}$	$233^{(3)}$
$85^{(4)}$	$121^{(6)}$	$98^{(11)}$	$203^{(7)}$	$221^{(9)}$
	$140^{(12)}$	$134^{(14)}$	$243^{(10)}$	$221^{(13)}$
	$122^{(15)}$	$133^{(16)}$	$229^{(17)}$	
	$125^{(18)}$			

mens, A and B. Two weighings were available for specimen A alone, nine for specimen B alone, and seven for specimens A and B together. The numbers in round brackets associated with each weight refer to the order in which the weighings were made. In the present analysis this is assumed to be irrelevant. As explained in Section 1.1, in a full statistical treatment the analysis about to be described would be regarded as tentative and would be followed by criticism which would certainly include plotting residuals against time order and checks on additivity and other assumptions. This could in turn lead to modification of the model form. We will not follow through with this process here but will use the data to illustrate only the initial inferential analysis.

We postulate the standard linear model

$$\mathbf{y} = \mathbf{X}\boldsymbol{\theta} + \boldsymbol{\varepsilon}.$$

The 18×2 matrix $\mathbf{X}$ and the vector of observations $\mathbf{y}$ are shown below. The elements of the first column of $\mathbf{X}$ take values one or zero, depending upon whether specimen A was, or was not, present. Similarly, the elements of the second column indicate the presence, or absence, of specimen B.

$$\mathbf{y} = \begin{bmatrix} 109 \\ 85 \\ 114 \\ 129 \\ 121 \\ 98 \\ 140 \\ 134 \\ 122 \\ 133 \\ 125 \\ 217 \\ 233 \\ 203 \\ 221 \\ 243 \\ 221 \\ 229 \end{bmatrix}, \qquad \mathbf{X} = \begin{bmatrix} 1 & 0 \\ 1 & 0 \\ 0 & 1 \\ 0 & 1 \\ 0 & 1 \\ 0 & 1 \\ 0 & 1 \\ 0 & 1 \\ 0 & 1 \\ 0 & 1 \\ 0 & 1 \\ 1 & 1 \\ 1 & 1 \\ 1 & 1 \\ 1 & 1 \\ 1 & 1 \\ 1 & 1 \\ 1 & 1 \end{bmatrix}.$$

For the analysis, we require

$$\mathbf{X}'\mathbf{X} = \begin{bmatrix} 9 & 7 \\ 7 & 16 \end{bmatrix}, \qquad (\mathbf{X}'\mathbf{X})^{-1} = \begin{bmatrix} 0.1684 & -0.0737 \\ -0.0737 & 0.0947 \end{bmatrix},$$

$$\mathbf{X}'\mathbf{y} = \begin{bmatrix} 1761 \\ 2683 \end{bmatrix}, \qquad \hat{\boldsymbol{\theta}} = \begin{bmatrix} \hat{\theta}_1 \\ \hat{\theta}_2 \end{bmatrix} = \begin{bmatrix} 98.895 \\ 124.421 \end{bmatrix},$$

$$\sum_{i=1}^{18} y_i^2 = 510{,}501, \qquad \hat{\boldsymbol{\theta}}'\mathbf{X}'\mathbf{y} = 507{,}976,$$

$$\sum_{i=1}^{18} (y_i - \hat{y}_i)^2 = \sum_{i=1}^{18} y_i^2 - \hat{\boldsymbol{\theta}}'\mathbf{X}'\mathbf{y} = 2{,}525,$$

$$s^2 = 157.8, \qquad s = 12.56,$$

$$Q(\boldsymbol{\theta}) = (\boldsymbol{\theta} - \hat{\boldsymbol{\theta}})'\mathbf{X}'\mathbf{X}(\boldsymbol{\theta} - \hat{\boldsymbol{\theta}})$$

$$= 9(\theta_1 - 98.9)^2 + 14(\theta_1 - 98.9)(\theta_2 - 124.4) + 16(\theta_2 - 124.4)^2.$$

Given the assumptions and the data, all that can be said about the vector of parameters $\boldsymbol{\theta}' = (\theta_1, \theta_2)$, in the light of the noninformative reference prior $p(\boldsymbol{\theta}, \sigma^2) \propto \sigma^{-2}$, is contained in the joint posterior distribution of (θ_1, θ_2), which from (2.7.20) is the bivariate t distribution

$$t_2\left\{\begin{bmatrix} 98.9 \\ 124.4 \end{bmatrix}, 157.8\begin{bmatrix} 0.1684 & -0.0737 \\ -0.0737 & 0.0947 \end{bmatrix}, 16\right\}.$$

Elliptical contours of constant density which include 75, 90, and 95 percent of the probability are shown in Fig. 2.7.1. From (2.7.21) the equations of these contours are readily computed by setting $Q(\boldsymbol{\theta})/2s^2$ equal in turn to 1.51, 2.67 and 3.63 which are, respectively, the upper 25, 10 and 5 percent points of the F distribution with 2 and 16 degrees of freedom.

The marginal probability distributions for θ_1 and θ_2 are also shown. From (2.7.23) the distribution for θ_1 is the scaled t distribution

$$t(98.9, 157.8 \times 0.1684, 16),$$

and for θ_2 is the scaled t distribution

$$t(124.4, 157.8 \times 0.0947, 16).$$

On each of these distributions, pairs of ordinates of equal heights have been drawn to mark off "contour points" which include respectively, 75, 90 and 95 percent of the marginal probability.

The problem of making inferences about $\boldsymbol{\theta}$ can be summarized as follows:

Given the assumptions, all that can be known about $\boldsymbol{\theta} = (\theta_1, \theta_2)$ is contained in the joint posterior bivariate t distribution. Mathematically speaking, therefore, the problem is solved as soon as this distribution is written down. To assist in the mental appreciation of the joint posterior distribution however, various relevant features may be described. These features are all *derived* from the original joint distribution and of course add no new mathematical information, but

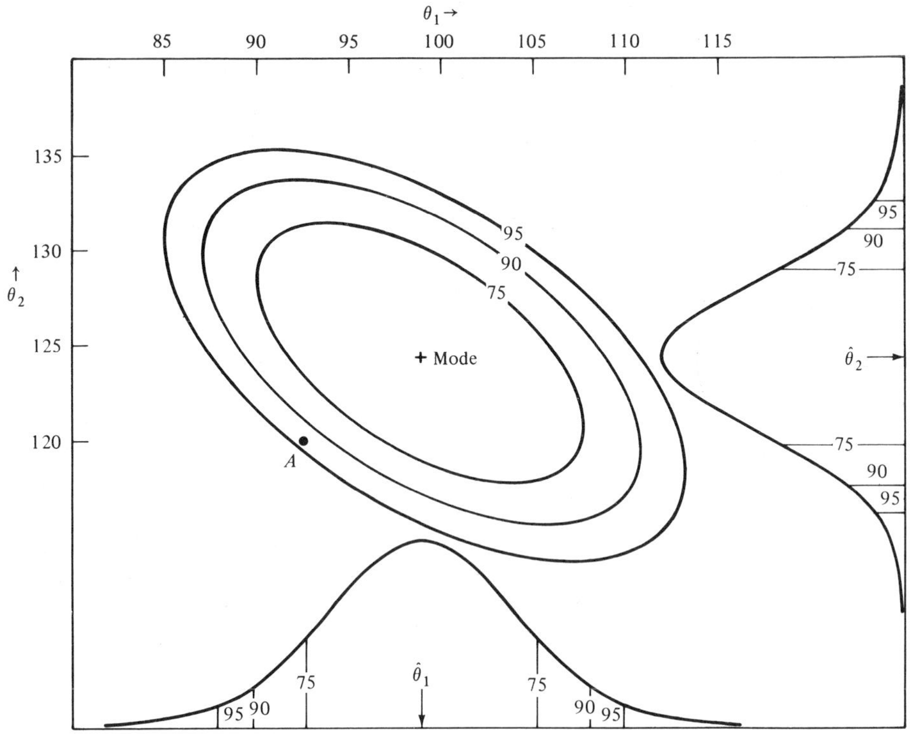

Fig. 2.7.1 The weighing example: contours of joint posterior distribution and marginal distributions of θ_1 and θ_2.

consideration of these features can lead to a better mental comprehension of the implications of the posterior distribution, and so to better inferences. Features of particular importance are (a) the H.P.D. regions, and (b) the marginal distributions.

H.P.D. Regions for θ_1 and θ_2

As we have mentioned, the elliptical contours of the joint posterior distribution shown in Fig. 2.7.1 include, respectively, 75, 90 and 95 percent of the probability. This uses the fact that *a posteriori* the quantity $Q(\boldsymbol{\theta})/2s^2$ is distributed as F with (2, 16) degrees of freedom. Further, since from (2.7.20) the joint posterior distribution of $p(\theta_1, \theta_2 \mid \mathbf{y})$ is a monotonically decreasing function of $Q(\boldsymbol{\theta})/2s^2$, every point inside a given contour has a higher posterior density than every point outside. The regions bounded by the three elliptical contours are thus the 75, 90, and 95 percent highest posterior density (H.P.D.) regions in (θ_1, θ_2), respectively.

The pairs of vertical lines shown in the marginal distributions similarly provide the limits of the 75, 90, and 95 percent H.P.D. intervals for the marginal distributions.

Marginal Distributions—Joint Inferences versus Marginal Inferences

If we wish to make probability statements about θ_1 or θ_2 individually, this can be done by using the marginal posterior distribution $p(\theta_1 \mid \mathbf{y})$ or $p(\theta_2 \mid \mathbf{y})$. It is important, of course, not to make the elementary mistake of supposing that *joint* probability statements can generally be made using the marginal distributions. For example, it is easily seen from Fig. 2.7.1 that values of θ_1 and θ_2 can readily be found which are each individually plausible, although they are not jointly so.

Consider, for example, the point $A = (92.5, 120)$ which is well inside the 90 percent H.P.D. intervals of the marginal distributions, but is excluded by the 90 percent H.P.D. region of the joint distribution. The common sense of the situation is easy to see. When we are considering the marginal distribution of, say θ_1, we are asking what values can θ_1 take irrespective of θ_2. It is implicit here that θ_2 is allowed to vary over all its admissable values with the appropriate probability density. But when we are considering the joint distribution, we are interested in the plausibility of the pair (θ_1, θ_2). Now it is certainly true in the example that for some values of θ_2, the value $\theta_1 = 92.5$ is a very plausible one. It does not happen to be very plausible, however, if $\theta_2 = 120$. The distinction between marginal intervals and joint regions becomes more important as the correlation of θ_1 and θ_2 becomes greater. The reader will have no difficulty in imagining a case where diagonal elliptical contours occur which are more attenuated. The marginal intervals would then be even more misleading if they were incorrectly used as indicating plausible combinations of (θ_1, θ_2).

2.8 A GENERAL DISCUSSION OF HIGHEST POSTERIOR DENSITY REGIONS

The above example serves to illustrate the importance of considering jointly the distribution of parameters, and of partially summarizing information in the joint posterior distribution by means of multidimensional regions of highest posterior density. We consider now the general properites and usefulness of such regions.

Suppose in a particular investigation we have computed an appropriate posterior distribution $p(\boldsymbol{\theta} \mid \mathbf{y})$, where $\boldsymbol{\theta}$ is a k-dimensional vector of parameters of interest and $\mathbf{y}$ is a n-dimensional vector of observations. Then, from the Bayesian point of view, all inferential problems concerning $\boldsymbol{\theta}$ may be answered in terms of $p(\boldsymbol{\theta}|\mathbf{y})$. In practice, inference involves a communication with the mind, and usually it is difficult to comprehend a function in k-dimensions. However, there are often specific individual features of $p(\boldsymbol{\theta} \mid \mathbf{y})$ of interest which can be appreciated by one, two, or three dimensional thought. For example, marginal distributions may be of interest. Or we may inspect conditional distributions of a small subset of the parameters for specific values of the other parameters. With high-speed electronic computers available, print-outs of two dimensional sections of such distributions can be readily obtained. The value of such appraisals of the estimation situation is very great, as has been repeatedly pointed out by Barnard.

In searching for ways to summarize the information in the posterior distribution $p(\boldsymbol{\theta} \mid \mathbf{y})$, it is to be noted that although the region over which the posterior density is nonzero may extend over infinite ranges in the parameter space, nevertheless over a substantial part of the parameter space the density may be negligible. It may, therefore, be possible to delineate a comparatively small region which contains most of the probability, or more generally, to delineate a number of regions which contain various stated proportions of the total probability. Obviously there are an infinite number of ways in which such regions can be chosen. In some cases, a specific region (or regions) of interest may be dictated by the nature of the problem. When it is not, we must decide what properties we would like the region to have. As with H.P.D. intervals, either of the following principles of choice seems intuitively sensible:

1. The region should be such that the probability density of every point inside it is at least as large as that of any point outside it.
2. The region should be such that for a given probability content, it occupies the smallest possible volume in the parameter space.

It is easy to show that if either principle is adopted the other follows as a natural consequence. We will in general call the region so produced a region of highest posterior density, or an H.P.D. region. The first principle will be employed to give a formal definition.

Definition. Let $p(\boldsymbol{\theta} \mid \mathbf{y})$ be a posterior density function. A region R in the parameter space of $\boldsymbol{\theta}$ is called an H.P.D. region of content $(1 - \alpha)$ if

a) $\Pr\{\boldsymbol{\theta} \in R \mid \mathbf{y}\} = 1 - \alpha$,

b) for $\boldsymbol{\theta}_1 \in R$ and $\boldsymbol{\theta}_2 \notin R$,

$$p(\boldsymbol{\theta}_1 \mid \mathbf{y}) \geqslant p(\boldsymbol{\theta}_2 \mid \mathbf{y}). \tag{2.8.1}$$

2.8.1 Some Properties of the H.P.D. Region

1. It follows from the definition that for a given probability content $(1 - \alpha)$, the H.P.D. region has the smallest possible volume in the parameter space of $\boldsymbol{\theta}$.

2. If we make the assumption that $p(\boldsymbol{\theta} \mid \mathbf{y})$ is nonuniform over every region in the space of $\boldsymbol{\theta}$, then the H.P.D. region of content $(1 - \alpha)$ is unique. Further, if $\boldsymbol{\theta}_1$ and $\boldsymbol{\theta}_2$ are two points such that $p(\boldsymbol{\theta}_1 \mid \mathbf{y}) = p(\boldsymbol{\theta}_2 \mid \mathbf{y})$, then these two points are simultaneously included in (or excluded by) a $(1 - \alpha)$ H.P.D. region. The converse is also true. That is, if $p(\boldsymbol{\theta}_1 \mid \mathbf{y}) \neq p(\boldsymbol{\theta}_2 \mid \mathbf{y})$, then there exists a $(1 - \alpha)$ H.P.D. region which includes one point but not the other.

Effect of Transformation: *Standardized regions.* Let $\boldsymbol{\phi} = \mathbf{f}(\boldsymbol{\theta})$ define a one-to-one transformation of the parameters from $\boldsymbol{\theta}$ to $\boldsymbol{\phi}$. Any region of content $(1 - \alpha)$ in the space of $\boldsymbol{\theta}$ transforms into a region of the same content in the space of $\boldsymbol{\phi}$, but it is clear from the definition that an H.P.D. region for $\boldsymbol{\theta}$ will not, in general, transform into an H.P.D. region for $\boldsymbol{\phi}$, unless the transformation is linear. As in the univariate case, when a noninformative prior is used, which is equivalent to assuming that some transformed set of parameters $\boldsymbol{\phi} = \mathbf{f}(\boldsymbol{\theta})$ are locally uniformly distributed, then *standardized* H.P.D. regions calculated in terms of $\boldsymbol{\phi}$ are available.

Smallest confidence regions of sampling theory are similarly not invariant under general transformation. On the implied assumption that such lack of invariance is bad, it has been suggested, for example, that the region should be based on the likelihood itself. That is, the boundary of the region should follow a likelihood contour. In particular, Hildreth (1963) has proposed that a $100(1-\alpha)$ percent region R be based on

$$\frac{\int_R l(\boldsymbol{\theta} \mid \mathbf{y})\, d\boldsymbol{\theta}}{\int_\Omega l(\boldsymbol{\theta} \mid \mathbf{y})\, d\boldsymbol{\theta}} = (1-\alpha), \tag{2.8.2}$$

where Ω is the parameter space of $\boldsymbol{\theta}$, with the property for $\boldsymbol{\theta}_1 \in R$ and $\boldsymbol{\theta}_2 \notin R$,

$$l(\boldsymbol{\theta}_1 \mid \mathbf{y}) \geqslant l(\boldsymbol{\theta}_2 \mid \mathbf{y}). \tag{2.8.3}$$

It will be observed that although the inequality (2.8.3) is preserved under general transformation, the equality (2.8.2) will not be preserved. A posterior region based upon the likelihood, which is, in a sense, invariant under general transformation, can be obtained as follows. For a fixed prior distribution $p_0(\boldsymbol{\theta})$, choose a region R such that

a)
$$\frac{\int_R l(\boldsymbol{\theta} \mid \mathbf{y})\, p_0(\boldsymbol{\theta})\, d\boldsymbol{\theta}}{\int_\Omega l(\boldsymbol{\theta} \mid \mathbf{y})\, p_0(\boldsymbol{\theta})\, d\boldsymbol{\theta}} = (1-\alpha), \tag{2.8.4}$$

b) for $\boldsymbol{\theta}_1 \in R$ and $\boldsymbol{\theta}_2 \notin R$,

$$l(\boldsymbol{\theta}_1 \mid \mathbf{y}) \geqslant l(\boldsymbol{\theta}_2 \mid \mathbf{y}). \tag{2.8.5}$$

Both (2.8.4) and (2.8.5) are invariant under general transformations. This region, which tries to make the best of both worlds is, however, a somewhat artificial construction. If we believe in the appropriateness of the prior distribution $p_0(\boldsymbol{\theta})$, then we should surely not adopt a region for which the posterior density for points outside can be greater than that for points inside. The authors feel that in general nonlinear transformation *ought* to change the relative credibility of any two parameter points and that invariance under nonlinear transformation is therefore not to be expected. Insistence on invariance for problems which ought not to be invariant serves only to guarantee inappropriate solutions.

2.8.2 Graphical Representation

Clearly when there are only two parameters a diagram, showing the point of maximum posterior density and, say, a 95 per cent H.P.D. region, would advise the investigator of a great deal of what the data had to tell him. A more informative plot would be one showing simultaneously the boundaries of, say, the 50, 75, 90, 95, and 99 percent H.P.D. regions. In such a case we would be back to the plot of posterior density contours labeled according to their interior probability content. This graphical approach could be extended to three or four parameters by exhibiting a "grid" of two-dimensional (θ_1, θ_2) diagrams for various combinations of θ_3 and θ_4. Plotters can produce such diagrams automatically from digital computer output. Such plotting, which should be part of the normal stock in trade of the modern practicing statistician, is invaluable for appreciating pecularities† in an estimation situation and is, therefore, of particular importance in exploring new problems.

2.8.3 Is $\boldsymbol{\theta}_0$ Inside or Outside a Given H.P.D. Region?

While it is true that all the information about k parameters $\boldsymbol{\theta}$ is contained in the posterior distribution $p(\boldsymbol{\theta} \mid \mathbf{y})$, because of our three-dimensional human limitations it may not be easy to comprehend what a distribution of higher dimensionality implies. Inspection of appropriate conditional and marginal distributions, or at least of H.P.D. regions, will greatly assist understanding. We now discuss one further device which, when used in conjunction with those mentioned above, can further illuminate the situation.

This is a way of answering the question whether or not a particular parameter point of interest $\boldsymbol{\theta}_0$ lies inside or outside a H.P.D. region of content $(1 - \alpha)$. From properties of H.P.D. regions, we see that if R_α is an H.P.D. region of content $(1 - \alpha)$, then the event $\boldsymbol{\theta} \in R_\alpha$ is equivalent to the event that $p(\boldsymbol{\theta} \mid \mathbf{y}) > c$, where c is a suitably chosen positive constant. It follows that the parameter point $\boldsymbol{\theta}_0$ is covered by the H.P.D. region of content $(1 - \alpha)$ if and only if

$$\Pr\{p(\boldsymbol{\theta} \mid \mathbf{y}) > p(\boldsymbol{\theta}_0 \mid \mathbf{y}) \mid \mathbf{y}\} \leqslant 1 - \alpha. \tag{2.8.6}$$

In this expression, the density function $p(\boldsymbol{\theta} \mid \mathbf{y})$ is treated as a random variable. Thus, once the posterior distribution of the quantity $p(\boldsymbol{\theta} \mid \mathbf{y})$ or some monotonic function of it can be determined, this question can be answered. In what follows we consider the specific nature of the region $p(\boldsymbol{\theta} \mid \mathbf{y}) > c$ for a number of examples of interest.

2.9 H.P.D. REGIONS FOR THE LINEAR MODEL: A BAYESIAN JUSTIFICATION OF ANALYSIS OF VARIANCE

As a first example, we return to the linear model discussed in Section 2.7. We have seen in (2.7.20) that in relation to a noninformative prior in $\boldsymbol{\theta}$ and σ the posterior distribution of $\boldsymbol{\theta}$ is in the form of a multivariate t distribution. Further, in (2.7.21) the quantity $Q(\boldsymbol{\theta})/ks^2 = (\boldsymbol{\theta} - \hat{\boldsymbol{\theta}})'\mathbf{X}'\mathbf{X}(\boldsymbol{\theta} - \hat{\boldsymbol{\theta}})/ks^2$ is distributed *a posteriori* as F with (k, ν) degrees of freedom. Suppose we are now interested in the question: Is the parameter point $\boldsymbol{\theta}_0 = (\theta_{10}, \ldots, \theta_{k0})'$ included in the H.P.D. region of content $(1 - \alpha)$? According to the above argument, we then need to calculate the probability of the event $p(\boldsymbol{\theta} \mid \mathbf{y}) > p(\boldsymbol{\theta}_0 \mid \mathbf{y})$.

† Visual display devices, which for example allow the investigator to see the changing two dimensional contours as he moves through a higher dimensional parameter space, can also be very helpful.

Now $p(\boldsymbol{\theta} \mid \mathbf{y})$ is a monotonic decreasing function of the quantity

$$(\boldsymbol{\theta} - \hat{\boldsymbol{\theta}})'\mathbf{X}'\mathbf{X}(\boldsymbol{\theta} - \hat{\boldsymbol{\theta}})/ks^2.$$

The particular point $\boldsymbol{\theta}_0$ is then included in the H.P.D. region of content $(1 - \alpha)$ if and only if

$$(\boldsymbol{\theta}_0 - \hat{\boldsymbol{\theta}})'\mathbf{X}'\mathbf{X}(\boldsymbol{\theta}_0 - \hat{\boldsymbol{\theta}}) < ks^2F(k, \nu, \alpha). \tag{2.9.1}$$

Equivalently, the quantity

$$\Pr\left\{F_{(k,\nu)} < \frac{(\boldsymbol{\theta}_0 - \hat{\boldsymbol{\theta}})'\mathbf{X}'\mathbf{X}(\boldsymbol{\theta}_0 - \hat{\boldsymbol{\theta}})}{ks^2}\right\}, \tag{2.9.2}$$

where $F_{(k,\nu)}$ is an F variable with (k, ν) degrees of freedom, gives the content of the H.P.D. region which just includes the point $\boldsymbol{\theta}_0$.

The results in (2.9.1) and (2.9.2) supply a Bayesian justification for the Analysis of Variance. The $(1 - \alpha)$ H.P.D. region is numerically identical with the $(1 - \alpha)$ smallest confidence region, and the inequality in (2.9.1) is also appropriate to decide if a given point $\boldsymbol{\theta}_0$ lies inside or outside the corresponding confidence region. Further, the complementary probability of (2.9.2) gives the significance level associated with the null hypothesis $\boldsymbol{\theta} = \boldsymbol{\theta}_0$ against the alternative $\boldsymbol{\theta} \neq \boldsymbol{\theta}_0$. Generalization to the corresponding multivariate linear model is discussed in Chapter 8.

2.9.1 The Weighing Example

The weighing example of Section 2.7 may be used for illustration. Suppose that particular interest was associated with the parameter values $\theta_{10} = \theta_{20} = 120$. Such interest could arise because of some theory that both specimens should weigh 120 micrograms. A question of interest would then be whether the parameter point (120, 120) was included or excluded from, say, the 95 percent H.P.D. region. Inspection of Fig. 2.7.1 immediately shows that this point is excluded. But the question can be answered without recourse to the diagram by considering whether $Q(\boldsymbol{\theta}_0)/2s^2$ is greater or less than $F(k, \nu, \alpha)$. We find

$$\frac{Q(\boldsymbol{\theta}_0)}{2s^2} = \frac{9\,(120 - 98.9)^2 + 14\,(120 - 98.9)\,(120 - 120.4) + 16\,(120 - 124.4)^2}{2 \times 157.8}$$

$$= \frac{3015}{315.6} = 9.56,$$

which is greater than $F(2, 16, 0.05) = 3.63$ so that the point (120, 120) is excluded. Again, we emphasize that in the above calculation what we are interested in is the plausibility that θ_1 and θ_2 are simultaneously equal to 120. It is obvious that the plausibility of θ_2 alone being equal to 120 would be a completely different question.

This question would be answered by considering not the joint posterior distribution of (θ_1, θ_2), but the marginal posterior distribution of θ_2.

For purpose of convenient calculation and checking, it is useful to set out the quantities needed in the form of an analysis of variance table as shown in Table 2.9.1. While such a table is customarily interpreted in terms of sampling theory, it has an equally useful function in a Bayesian framework. In particular, the question of whether $\boldsymbol{\theta}_0$ is, or is not, included in a given H.P.D. region is determined by computing the ratio of the mean squares in the last column of the table, and referring the ratio to the $F(k, \nu, \alpha)$ values in Table VI (at the end of this book).

Table 2.9.1

Analysis of variance table to determine whether the point $(\theta_{10} = 120, \theta_{20} = 120)$ is included within a given H.P.D. region

Sources	Sum of squares	Degrees of freedom	Mean square	Mean square ratio
Parameter discrepancy	$(\boldsymbol{\theta}_0 - \hat{\boldsymbol{\theta}})' \mathbf{X}'\mathbf{X}(\boldsymbol{\theta}_0 - \hat{\boldsymbol{\theta}}) = 3{,}015$	2	1,507.5	9.56
Residual	$(\mathbf{y} - \hat{\mathbf{y}})' (\mathbf{y} - \hat{\mathbf{y}}) = 2{,}525$	16	157.8	
Total	$(\mathbf{y} - \mathbf{X}\boldsymbol{\theta}_0)' (\mathbf{y} - \mathbf{X}\boldsymbol{\theta}_0) = 5{,}540$	18		

2.10 COMPARISON OF PARAMETERS

In the previous sections we have discussed problems of deciding whether a particular parameter point $\boldsymbol{\theta}_0$ is, or is not, included in the $(1 - \alpha)$ H.P.D. region. Frequently, the investigator is concerned with the comparative values of parameters rather than with their absolute values. Suppose we have k parameters $\boldsymbol{\theta} = (\theta_1, \cdots, \theta_k)'$. We may define $(k - 1)$ nonredundant comparisons as $(k - 1)$ independent functions

$$\phi_i = f_i(\boldsymbol{\theta}), \qquad i = 1, \ldots, (k - 1), \tag{2.10.1}$$

which are all equal to zero if and only if $\theta_1 = \cdots = \theta_k$. There is clearly a very wide range of choices for such functions. Some thought must therefore be given as to how such comparisons should be parametrized. This question is now considered for two important problems in statistics: (a) the comparison of the means of k Normal distributions, and (b) the comparison of the variances of k Normal distributions.

2.11 COMPARISON OF THE MEANS OF k NORMAL POPULATIONS

Returning to the linear model discussed in Section 2.7, consider the special case where the elements of the $n \times 1$ vector of observations $\mathbf{y}$ are independent samples $\mathbf{y}'_1 = (y_{11}, \ldots, y_{1n_1}), \ldots, \mathbf{y}'_k = (y_{k1}, \ldots, y_{kn_k})$, of size $n_1, \ldots, n_k$ $(\Sigma n_s = n)$ from k Normal populations with means $(\theta_1, \ldots, \theta_k)$, respectively, and common variance σ^2. The posterior distribution of $\boldsymbol{\theta}$ in (2.7.20) reduces to

$$p(\boldsymbol{\theta} \mid \mathbf{y}) \propto \left[1 + \frac{\sum_{i=1}^k n_i(\theta_i - \bar{y}_i)^2}{\nu s^2}\right]^{-\frac{1}{2}(\nu+k)}, \qquad -\infty < \theta_i < \infty, \quad i = 1, \cdots, k, \tag{2.11.1}$$

with

$$\bar{y}_i = \frac{1}{n_i}\Sigma y_{ij}, \qquad \nu = n - k, \qquad \text{and} \qquad s^2 = \frac{1}{n-k}\sum_i\sum_j (y_{ij} - \bar{y}_i)^2.$$

This distribution allows us to decide whether a *particular* set of values of the means $\boldsymbol{\theta}_0 = (\theta_{10}, \ldots, \theta_{k0})'$ is included in the k-dimensional $(1 - \alpha)$ H.P.D. region. In practice, we are most often concerned with such questions as "How different are the effects of treatments 1 and 2?" Such questions must be answered in terms of *comparisons* among the θ_i's.

Statements about the relative rather than the absolute values of the θ_i's may be made in terms of $k - 1$ parameters $\phi_1, \phi_2, \ldots, \phi_{k-1}$, measuring independent *contrasts* between the θ_i's. In particular, the possibility that all the θ_i's are equal (to some unknown θ) corresponds to the possibility that all the ϕ's are zero. For example, suppose $k = 3$. Then we might consider the contrasts

$$\begin{aligned} \phi_1 &= \theta_1 - \theta_2, \\ \phi_2 &= \theta_1 - \theta_3. \end{aligned} \tag{2.11.2}$$

The two equalities $(\phi_1 = 0, \phi_2 = 0)$ together imply that

$$\theta_1 = \theta_2 = \theta_3 = \theta,$$

irrespective of the actual value of the common parameter θ. The possibility that $\theta_1 = \theta_2 = \theta_3$ can therefore be examined by studying the joint distribution of (ϕ_1, ϕ_2), and checking whether the point $(\phi_1 = 0, \phi_2 = 0)$ lies inside or outside a particular H.P.D. region.

2.11.1 Choice of Linear Contrasts

We shall say that a set of $k - 1$ parameters $\boldsymbol{\phi} = (\phi_1, \ldots, \phi_{k-1})'$ measure contrasts among the parameters if there are linear functions

$$\phi_j = \sum_{i=1}^k a_{ji}\theta_i = \mathbf{a}_j'\boldsymbol{\theta}, \qquad j = 1, \ldots, k - 1, \tag{2.11.3}$$

such that $(\phi_1 = 0, \ldots, \phi_{k-1} = 0)$ necessarily implies $(\theta_1 = \cdots = \theta_k = \theta)$ for some unknown θ.

For this property to hold, it is sufficient and necessary that

a) the vectors $\mathbf{a}_1, \ldots, \mathbf{a}_{k-1}$ are linearly independent, and

b) $\mathbf{a}_j\mathbf{1}_k = 0$, that is, $\sum_{i=1}^k a_{ji} = 0$, $j = 1, \ldots, k - 1$.

To see this, write

$$\mathbf{A}\boldsymbol{\theta} = \boldsymbol{\phi}, \tag{2.11.4}$$

where $\mathbf{A}$ is the $(k-1)\times k$ matrix $\mathbf{A} = [\mathbf{a}_1. \ldots, \mathbf{a}_{k-1}]'$. For $\boldsymbol{\phi} = \mathbf{0}$, we have

$$\mathbf{A}\boldsymbol{\theta} = \mathbf{0}, \tag{2.11.5}$$

which defines a system of $(k-1)$ equations in k unknowns. To show sufficiency, suppose the conditions (a) and (b) hold, then since rank $(\mathbf{A}) = k-1$ and $\mathbf{A}\mathbf{1}_k = \mathbf{0}$, it follows that all the solutions of the equations in (2.11.5) must be of the form $\boldsymbol{\theta} = \theta\mathbf{1}_k$, that is, $\theta_1 = \cdots = \theta_k = \theta$. Now to show necessity, suppose $\boldsymbol{\theta} = \theta\mathbf{1}_k$. This means that $\theta\mathbf{1}_k$ satisfies the equations (2.11.5) so that $\mathbf{a}_j'\mathbf{1}_k = 0$, $j = 1, \ldots, k$. Further, the vectors $\mathbf{a}_j$ must be linearly independent. For if not, then there exists a choice of $\boldsymbol{\theta}$ other than $\boldsymbol{\theta} = \theta\mathbf{1}_k$ which also satisfies (2.11.5). But this contradicts the supposition that $\boldsymbol{\theta} = \theta\mathbf{1}_k$.

There are obviously an infinity of ways in which a set of $k-1$ contrasts among the k parameters $\boldsymbol{\theta}$ might be chosen. For example, instead of using the contrasts ϕ_1 and ϕ_2 of (2.11.2) we could use

$$\begin{aligned} \phi_1^* &= \theta_1 - 2\theta_2 + \theta_3 = 2\phi_1 - \phi_2. \\ \phi_2^* &= \theta_1 + \theta_2 - 2\theta_3 = -\phi_1 + 2\phi_2. \end{aligned} \tag{2.11.6}$$

These new contrasts can be expressed in terms of the old contrasts by

$$\begin{bmatrix} \phi_1^* \\ \phi_2^* \end{bmatrix} = \begin{bmatrix} 2 & -1 \\ -1 & 2 \end{bmatrix} \begin{bmatrix} \phi_1 \\ \phi_2 \end{bmatrix}, \tag{2.11.7}$$

so that (ϕ_1^*, ϕ_2^*) is a nonsingular linear transformation of (ϕ_1, ϕ_2).

In general, the vectors $\mathbf{a}_j'$ $(j = 1, 2, \ldots, k-1)$ span a $(k-1)$-dimensional space, and there always exists a nonsingular linear transformation

$$\boldsymbol{\phi}^* = \mathbf{T}\boldsymbol{\phi}, \tag{2.11.8}$$

expressing one parameterization in terms of another.

Since the H.P.D. region is invariant under linear transformation, the question of whether or not a particular point in the space of the contrasts is, or is not, included in a H.P.D. region will be unaffected by the manner in which the contrasts are parametrized. This is illustrated in Fig. 2.11.1, in terms of the parameterizations expressed by the equations in (2.11.2) and (2.11.7). The figure shows a contrast space with rectangular coordinates for ϕ_1 and ϕ_2 and the corresponding oblique coordinates for ϕ_1^* and ϕ_2^* defined by (2.11.7). Also shown is an elliptical H.P.D. region. The question of whether any point P in the contrast space is, or is not, included in this region is, of course, independent of which coordinate system is used to define P. In particular, the point O is the origin for every system of contrasts, so that $\boldsymbol{\phi} = \mathbf{0}$ implies $\theta_1 = \cdots = \theta_k$, however we choose the nonsingular matrix $\mathbf{T}$ in $\boldsymbol{\phi}^* = \mathbf{T}\boldsymbol{\phi}$.

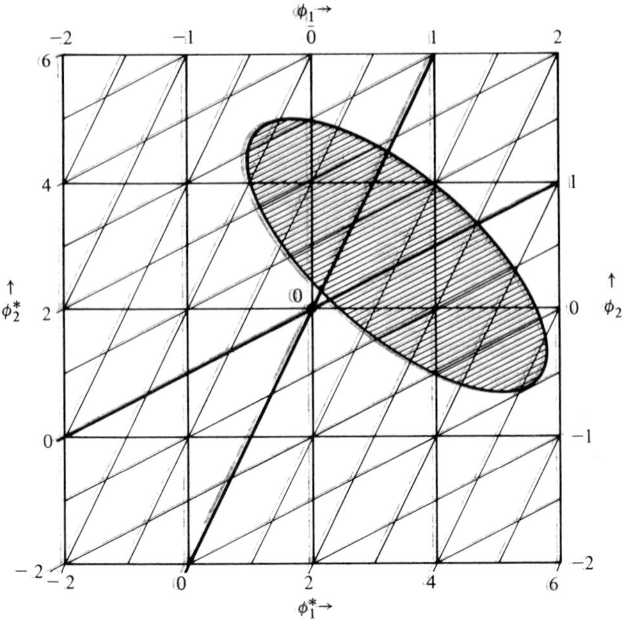

Fig. 2.11.1 Invariance of H.P.D. region under linear transformation from (ϕ_1, ϕ_2) to (ϕ_1^*, ϕ_2^*).

It is convenient here to consider the particular set of contrasts

$$\phi_i = \theta_i - \bar{\theta} \; (i = 1, \cdots, k - 1) \qquad \text{where} \qquad \bar{\theta} = \frac{1}{n} \Sigma n_i \theta_i. \tag{2.11.9}$$

From (2.7.26), we find that the posterior distribution of $\boldsymbol{\phi} = (\phi_1, \ldots, \phi_{k-1})$ is

$$p(\boldsymbol{\phi} \mid \mathbf{y}) \propto \left\{1 + \frac{Q(\boldsymbol{\phi})}{\nu s^2}\right\}^{-\frac{1}{2}[\nu + (k-1)]}, \qquad -\infty < \phi_i < \infty, \quad i = 1, \ldots, k - 1, \tag{2.11.10}$$

where

$$Q(\boldsymbol{\phi}) = \sum_{i=1}^{k} n_i [\phi_i - (\bar{y}_i - \bar{y})]^2,$$

$$\bar{y} = \frac{1}{n} \Sigma n_i \bar{y}_i, \qquad \phi_k = \theta_k - \bar{\theta}, \qquad \text{and} \qquad \sum_{i=1}^{k} n_i \phi_i = 0.$$

Thus, a particular point $\boldsymbol{\phi}_0$ is included in the $(1 - \alpha)$ H.P.D. region if and only if

$$\frac{Q(\boldsymbol{\phi}_0)}{(k-1)s^2} < F(k - 1, \nu, \alpha). \tag{2.11.11}$$

In particular, the point $\boldsymbol{\phi}_0 = \mathbf{0}$ corresponding to $\theta_1 = \cdots = \theta_k$ is included in the $(1 - \alpha)$ region if and only if $Q(\mathbf{0})/(k - 1)s^2 < F(k - 1, \nu, \alpha)$, that is,

$$\frac{\Sigma\, n_i(\bar{y}_i - \bar{y})^2}{(k - 1)s^2} < F(k - 1, \nu, \alpha), \tag{2.11.12}$$

which parallels the well known significance test. Again, the calculations are conveniently set out in the familiar analysis of variance of Table 2.11.1.

Table 2.11.1

Analysis of variance table to determine whether the point $\boldsymbol{\phi} = \mathbf{0}$ $(\theta_1 = \cdots = \theta_k)$ is included in a given H.P.D. region

Sources	Sum of squares	D.F.	Mean square	Mean square ratio.
Inequality of means	$Q(\mathbf{0}) = \Sigma n_i(\bar{y}_i - \bar{y})^2$	$k-1$	$Q(\mathbf{0})/(k-1)$	
Residual	$\Sigma\Sigma(y_{ij} - \bar{y}_i)^2$	$n - k$	s^2	$\dfrac{Q(\mathbf{0})}{(k-1)s^2}$

2.11.2 Choice of Linear Contrasts to Compare Location Parameters

We have tacitly assumed that it was appropriate to compare the θ_i's in terms of *linear* contrasts. There are other ways in which the parameters $\boldsymbol{\theta}$ might have been compared. For example, we might have considered ratios such as θ_2/θ_1 and θ_3/θ_1. Special problems occasionally occur where such ratios are of interest and, when this happens, their distributions may of course be obtained directly from the joint distribution of $\boldsymbol{\theta}$. Linear contrasts are appropriate in the commonly considered situation where interest centers on the *relative* location of k distributions which are otherwise similar. To see this, notice that addition of any constant c to each observation should leave measures of relative location unchanged. Now such an addition will change θ_i to $\theta_i + c$, and if we are to have $k - 1$ independent comparison functions $f_1, \ldots, f_{k-1}$, such that

$$f_j(\boldsymbol{\theta} + c\mathbf{1}_k) = f_j(\boldsymbol{\theta}), \qquad j = 1, 2, \ldots, k - 1, \tag{2.11.13}$$

then the f_j's must be functions of the linear contrasts

$$\phi_j = \sum_{i=1}^{k} a_{ji}\theta_i \qquad \text{where} \qquad \sum_{i=1}^{k} a_{ji} = 0, \qquad j = 1, 2, \ldots, k - 1.$$

We show this for the case $k = 2$. A linear contrast is then proportional to the difference $\theta_2 - \theta_1$. Write

$$f(\theta_1, \theta_2) = g(\phi\ \theta_1) \qquad \text{where} \qquad \phi = \theta_2 - \theta_1. \tag{2.11.14}$$

Now,

$$f(\theta_1, \theta_2) = f(\theta_1 + c,\ \theta_2 + c) \tag{2.11.15}$$

implies that

$$g(\phi, \theta_1) = g(\phi, \ \theta_1 + c). \qquad (2.11.16)$$

Since c is arbitrary, it follows that $g(\phi, \theta_1)$ is independent of θ_1, that is, $f(\theta_1, \theta_2)$ is a function of $\theta_2 - \theta_1$ only.

The simplest functions of linear contrasts, are the contrasts themselves, and consequently the problem of comparing locations has been expressed in these terms.

2.12 COMPARISON OF THE SPREAD OF k DISTRIBUTIONS

Consider now comparison of the spread of k distributions, which may differ only in their location and scale parameters but are otherwise the same. It is reasonable to require that comparisons of k scale parameters $(\theta_1, \ldots, \theta_k)$ should be unaffected by any linear recoding of the data. This would ensure, for example, that the measures comparing scale parameters would be the same whether the observations were in feet or inches. Now a scale parameter θ is such that a linear recoding of the data which changes an observation y to $a + by$ transforms θ to $|b|\theta$. If we are to have $k - 1$ independent comparison functions g_j such that

$$g_j(b\boldsymbol{\theta}) = g_j(\boldsymbol{\theta}), \qquad j = 1, \cdots, k - 1, \qquad (2.12.1)$$

then the g_j's must be functions of linear contrasts ϕ_j among the *logarithms* of the θ's,

$$\phi_j = \sum_{i=1}^{k} a_{ji} \log \theta_i, \qquad \sum_{i=1}^{k} a_{ji} = 0, \qquad j = 1, \ldots, k - 1. \qquad (2.12.2)$$

The simplest such functions are the ϕ_j's themselves. For Normal distributions, where the standard deviations $\sigma_1, \sigma_2, \ldots, \sigma_k$ are scale parameters, suitable comparison functions are

$$\phi_j = \sum_{i=1}^{k} a_{ji} \log \sigma_i = \frac{1}{q} \sum_{i=1}^{k} a_{ji} \log \sigma_i^q. \qquad (2.12.3)$$

The second equality on the right shows that the comparison function will be of the same form for any σ_i^q, and in particular, for the variance σ_i^2.

2.12.1 Comparison of the Spread of k Normal Populations

If we suppose that k samples $(y_{11}, \ldots, y_{1n_1}), \ldots, (y_{k1}, \ldots, y_{kn_k})$, of $n_1, \ldots, n_k$ independent observations, are drawn from Normal populations $N(\theta_i, \sigma_i^2)$ $(i = 1, \ldots, k)$, where both the means θ_i and the variances σ_i^2 are unknown, and that *a priori* θ_i and $\log \sigma_i$ are all approximately independent and locally uniform, it then follows from Theorem 2.4.1. (on page 97) that the joint posterior distribution of $\sigma_1^2, \ldots, \sigma_k^2$ is the product,

$$p(\sigma_1^2, \ldots, \sigma_k^2 \mid \mathbf{y}) = \prod_{i=1}^{k} \omega_i \, (\sigma_i^2)^{-[\frac{1}{2}(\nu_i)+1]} \exp\left(-\frac{\nu_i s_i^2}{2\sigma_i^2}\right), \qquad \sigma_i^2 > 0, \quad i = 1, \ldots, k,$$

(2.12.4)

where

$$\omega_i = \left[\Gamma\left(\frac{\nu_i}{2}\right)\right]^{-1} \left(\frac{\nu_i s_i^2}{2}\right)^{\nu_i/2},$$

$$\nu_i = n_i - 1,$$

$$s_i^2 = \frac{1}{\nu_i} \Sigma\,(y_{ij} - \bar{y}_i)^2,$$

$$\bar{y}_i = \frac{1}{n_i} \sum_{j=1}^{n_i} y_{ij}.$$

To compare the spread of the k Normal populations, we may choose for our comparison functions any $(k-1)$ linearly independent contrasts in $\log \sigma_i$. For this development we set

$$\phi_i = \log \sigma_i^2 - \log \sigma_k^2 = 2(\log \sigma_i - \log \sigma_k), \qquad i = 1, \ldots, (k-1). \qquad (2.12.5)$$

It is straightforward to verify from (2.12.4) that the posterior distribution of $\boldsymbol{\phi}' = (\phi_1, \ldots, \phi_{k-1})$ is

$$p(\boldsymbol{\phi} \mid \mathbf{y}) = \frac{\Gamma(\nu/2)}{\prod_{i=1}^{k} \Gamma(\nu_i/2)} T_1^{\nu_1/2} \cdots T_{k-1}^{(\nu_{k-1})/2} (1 + T_1 + \cdots + T_{k-1})^{-\nu/2},$$

$$-\infty < \phi_i < \infty, \quad i = 1, \ldots, k-1, \qquad (2.12.6)$$

where

$$T_i = \frac{\nu_i s_i^2}{\nu_k s_k^2} e^{-\phi_i}, \qquad i = 1, \ldots, k-1 \quad \text{and} \quad \nu = \nu_1 + \cdots + \nu_k.$$

Upon differentiation, it is readily shown that this distribution has a unique mode at

$$\hat{\phi}_i = \log s_i^2 - \log s_k^2, \qquad i = 1, \ldots, k-1, \qquad (2.12.7)$$

so that T_i can be alternatively written as

$$T_i = \frac{\nu_i}{\nu_k} \exp\left[-(\phi_i - \hat{\phi}_i)\right]. \qquad (2.12.8)$$

A particular point $\boldsymbol{\phi}_0$ is included in the H.P.D. region of content $(1-\alpha)$ if and only if

$$\Pr\{p(\boldsymbol{\phi} \mid \mathbf{y}) > p(\boldsymbol{\phi}_0 \mid \mathbf{y}) \mid \mathbf{y}\} < 1 - \alpha. \qquad (2.12.9)$$

Now, the density function $p(\boldsymbol{\phi} \mid \mathbf{y})$ is a monotonic decreasing function of

$$M = -2 \log W, \qquad (2.12.10)$$

where

$$W = \left(\frac{\nu_k}{\nu}\right)^{-\nu/2} \left[1 + \sum_{i=1}^{k-1} \frac{\nu_i}{\nu_k} \exp\left[-(\phi_i - \hat{\phi}_i)\right]\right]^{-\nu/2} \prod_{i=1}^{k-1} \exp\left[-\tfrac{1}{2}\nu_i(\phi_i - \hat{\phi}_i)\right]. \qquad (2.12.11)$$

In terms of σ_i^2 and s_i^2,

$$W^{2/\nu} = \frac{\prod_{i=1}^{k} (s_i^2/\sigma_i^2)^{\nu_i/\nu}}{\sum_{i=1}^{k} (\nu_i/\nu) s_i^2/\sigma_i^2}, \qquad (2.12.12)$$

which is the ratio of a weighted geometric mean over a weighted arithmetic mean of the quantities s_i^2/σ_i^2, with weights proportional to ν_i. From (2.12.10), the event $p(\boldsymbol{\phi} \mid \mathbf{y}) \geqslant p(\boldsymbol{\phi}_0 \mid \mathbf{y})$ is therefore equivalent to the event $M < -2 \log W_0$, where W_0 is obtained by substituting ϕ_{i0} for ϕ_i in (2.12.11). In particular, we may be interested in the point $\boldsymbol{\phi}_0 = \mathbf{0}$ which corresponds to the situation $\sigma_1^2 = \cdots = \sigma_k^2$. From (2.12.12), on setting $\bar{s}^2 = (1/\nu) \sum_{i=1}^{k} \nu_i s_i^2$, we see that

$$-2 \log W_0 = -\sum_{i=1}^{k} \nu_i (\log s_i^2 - \log \bar{s}^2). \qquad (2.12.13)$$

The expression on the right is identical with Bartlett's criterion for testing homogeneity of variances developed from sampling theory [see Bartlett (1937)].

2.12.2 Asymptotic Distribution of M

We now discuss some asymptotic results from which the distribution of M can be closely approximated. Making use of the Dirichlet integral (A2.1.8) in Appendix A2.1, we obtain the moment generating function of M as

$$E(e^{tM}) = \frac{\Gamma(\nu/2)}{\Gamma[(\nu/2)(1 - 2t)]} \prod_{s=1}^{k} (\nu_s/\nu)^{t\nu_s} \frac{\Gamma[(\nu_s/2)(1 - 2t)]}{\Gamma(\nu_s/2)}. \qquad (2.12.14)$$

Taking logarithms and employing Stirling's series for the logarithm of the Gamma function [see (A2.2.11) in Appendix A2.2], we find the cumulant generating function of M to be

$$\kappa_M(t) = a - \frac{k-1}{2} \log(1 - 2t) + \sum_{r=1}^{\infty} \alpha_r (1 - 2t)^{-(2r-1)}, \qquad (2.12.15)$$

where

$$a = -\sum_{r=1}^{\infty} \alpha_r$$

is a constant independent of t,

$$\alpha_r = \frac{B_{2r}}{2r(2r-1)} 2^{2r-1} \left[\sum_{i=1}^{k} \nu_i^{-(2r-1)} - \nu^{-(2r-1)} \right]$$

and B_{2r} are the Bernoulli numbers.

Now the quantity α_r is of order $\max_i \nu^{-(2r-1)}$. Thus, as the ν_i's become large, α_r tends to zero and in the limit

$$\lim_{\substack{\nu_i \to \infty \\ i=1,\ldots,k}} \kappa_M(t) = -\frac{k-1}{2} \log(1-2t), \tag{2.12.16}$$

which is the cumulant generating function of a χ^2_{k-1} variable. Consequently, as $\nu_i \to \infty$ $(i = 1, \ldots, k)$, M is asymptotically distributed as χ^2_{k-1}, so that

$$\lim_{\nu_i \to \infty} \Pr\{p(\boldsymbol{\phi} \mid \mathbf{y}) > p(\boldsymbol{\phi}_0 \mid \mathbf{y}) \mid \mathbf{y}\} = \lim_{\nu_i \to \infty} \Pr\{M < -2 \log W_0\}$$

$$= \Pr\{\chi^2_{k-1} < -2 \log W_0\}, \tag{2.12.17}$$

For large samples, then, the point $\boldsymbol{\phi}_0$ is included in the $(1-\alpha)$ H.P.D. region if $-2 \log W_0 < \chi^2(k-1, \alpha)$. For moderate sample sizes, a modification, due to Bartlett (1937), can be used to approximate the distribution of M.

Bartlett's Approximation

From (2.12.15) the rth cumulant of M is, to order $\max_i \nu_i^{-1}$

$$\kappa_r(M) = 2^{r-1}(r-1)!\,(k-1)(1+Ar), \tag{2.12.18}$$

where

$$A = \frac{1}{3(k-1)} \left(\sum_{i=1}^{k} \nu_i^{-1} - \nu^{-1} \right).$$

If we take

$$M \mathrel{\dot\sim} (1+A)\chi^2_{k-1} \tag{2.12.19}$$

then

$$\kappa_1(M) = (k-1)(1+A),$$

$$\kappa_r(M) = 2^{r-1}(r-1)!\,(k-1)(1+A)^r,$$

$$= 2^{r-1}(r-1)!\,(k-1)\left[1 + rA + \binom{r}{2}A^2 + \cdots\right]. \tag{2.12.20}$$

Thus, to order $\max_i \nu_i^{-1}$, the cumulants in (2.12.18) and (2.12.20) are identical. It follows that, to this degree of approximation,

$$\Pr\{p(\boldsymbol{\phi} \mid \mathbf{y}) > p(\boldsymbol{\phi}_0 \mid \mathbf{y}) \mid \mathbf{y}\} \doteq \Pr\left\{\chi^2_{k-1} < \frac{-2 \log W_0}{1+A}\right\}. \tag{2.12.21}$$

This differs from (2.12.17) by the factor $(1 + A)$ which can be thought of as the first order adjustment. The distribution may be approximated to greater accuracy [see for example Hartley (1940), Brookner and Wald (1941), and Box (1949)]. We shall not, however, discuss these methods here because Bartlett's approximation, which is easy to apply, appears adequate even for rather small sample sizes.

2.12.3 Bayesian Parallel to Bartlett's Test

We have seen that the sample quantity employed in deciding whether the point $\boldsymbol{\phi} = \mathbf{0}$, that is, $(\sigma_1^2 = \cdots = \sigma_k^2)$, lies inside or outside a H.P.D. region is the same as Bartlett's criterion for testing homogeneity of variances developed in sampling theory. The distribution theory is also parallel, and the procedure for deciding whether or not $\boldsymbol{\phi}_0 = \mathbf{0}$ is included in the $(1 - \alpha)$ H.P.D. region for $\boldsymbol{\phi}$, is identical with Bartlett's test at level α. This correspondence arises because the sampling distribution of s_i^2/σ_i^2, with s_i^2 a random variable and σ_i^2 fixed, is identical to the posterior distribution of s_i^2/σ_i^2, with s_i^2 fixed and σ_i^2 the random variable, under the assumption that θ_i and $\log \sigma_i$ are uniform *a priori*.

A serious difficulty in the practical use of Bartlett's criterion in the context of sampling theory is its extreme sensitivity to non-Normality [Box (1953a, b)]. For samples from non-Normal populations with kurtosis $\gamma_2 = \kappa_4/\kappa_2^2$, for instance, M_0 is asymptotically distributed as $(1 + \gamma_2/2)\chi_{k-1}^2$ and not as χ_{k-1}^2. In this chapter Bayesian procedures are derived under the same assumptions as are customary in the sampling approach. Later in Chapter 4 we turn to the problem of comparing the spread of distributions when Normality cannot be assumed.

2.12.4 An Example

As an example of the Normal theory procedure, suppose an investigation conducted to compare the spread of three distributions yields

$$s_1^2 = 52.785, \qquad s_2^2 = 34.457, \qquad s_3^2 = 66.030,$$

$$\nu_1 = \nu_2 = \nu_3 = 30.$$

From which

$$\bar{s}^2 = 51.091, \qquad \text{and} \qquad \nu = 90.$$

Under the assumptions of Normality and noninformative priors, the joint distribution of the two contrasts

$$\phi_1 = \log \sigma_1^2 - \log \sigma_3^2 \qquad \text{and} \qquad \phi_2 = \log \sigma_2^2 - \log \sigma_3^2$$

is, from (2.12.6),

$$p(\phi_1, \phi_2 \mid \mathbf{y}) = \frac{\Gamma(45)}{[\Gamma(15)]^3} T_1^{15} T_2^{15} (1 + T_1 + T_2)^{-45},$$

$$-\infty < \phi_1 < \infty, \quad -\infty < \phi_2 < \infty,$$

where

$$T_1 = \frac{\nu_1 s_1^2}{\nu_3 s_3^2} e^{-\phi_1} = 0.79941\, e^{-\phi_1},$$

$$T_2 = \frac{\nu_2 s_2^2}{\nu_3 s_3^2} e^{-\phi_2} = 0.52184\, e^{-\phi_2}.$$

Using (2.12.7), the mode of the distribution is

$$\hat{\phi}_1 = \log s_1^2 - \log s_3^2 = -0.2239,$$

$$\hat{\phi}_2 = \log s_2^2 - \log s_3^2 = -0.6504.$$

Figure 2.12.1 shows the mode and contours of the approximate 75, 90, and 95 per cent H.P.D. regions. Making use of Bartlett's approximation, the contours were drawn so that

$$M = -2 \log W = (1 + A)\chi^2(2, \alpha)$$

for $\alpha = (0.25, 0.10, 0.05)$, respectively. For this example, we have from (2.12.10) and (2.12.12)

$$M = -90 \log 3 + 90 \log\{1 + \exp[-(\phi_1 + 0.2239)] + \exp[-(\phi_2 + 0.6504)]\}$$
$$+ 30[(\phi_1 + 0.2239) + (\phi_2 + 0.6504)],$$

$$A = (\tfrac{1}{6})(\tfrac{3}{30} - \tfrac{1}{90}) = 0.0148,$$

and from Table III (at the end of this book), $\chi^2(2, 0.25) = 2.77$, $\chi^2(2, 0.10) = 4.61$, and $\chi^2(2, 0.05) = 5.99$. The contours are nearly elliptical as might be expected because, for samples of this size, ϕ_1 and ϕ_2 will be roughly jointly Normally distributed.

Also shown in the figure is the point $\boldsymbol{\phi} = \mathbf{0}$ corresponding to $\sigma_1^2 = \sigma_2^2 = \sigma_3^2$. For this example, $\boldsymbol{\phi} = \mathbf{0}$ is included in the 90 per cent region but excluded from the 75 per cent region. Of course, if we were merely interested in deciding whether $\boldsymbol{\phi} = \mathbf{0}$ lies inside a particular H.P.D. region, we could employ (2.12.13) directly and obtain

$$-2 \log W_0 = -\sum_{i=1}^{3} \nu_i (\log s_i^2 - \log \bar{s}^2) = 3.1434,$$

so that

$$\frac{-2 \log W_0}{1 + A} = \frac{3.1434}{1.0148} = 3.098,$$

which falls between $\chi^2(2, 0.25)$ and $\chi^2(2, 0.10)$. Thus, we would know straight away that $\boldsymbol{\phi} = \mathbf{0}$ is included in the 90 per cent region but excluded from the 75 per cent region without recourse to the diagram.

By consulting a table of χ^2 probabilities, we can also determine the approximate content of the H.P.D. region which just covers a particular parameter point

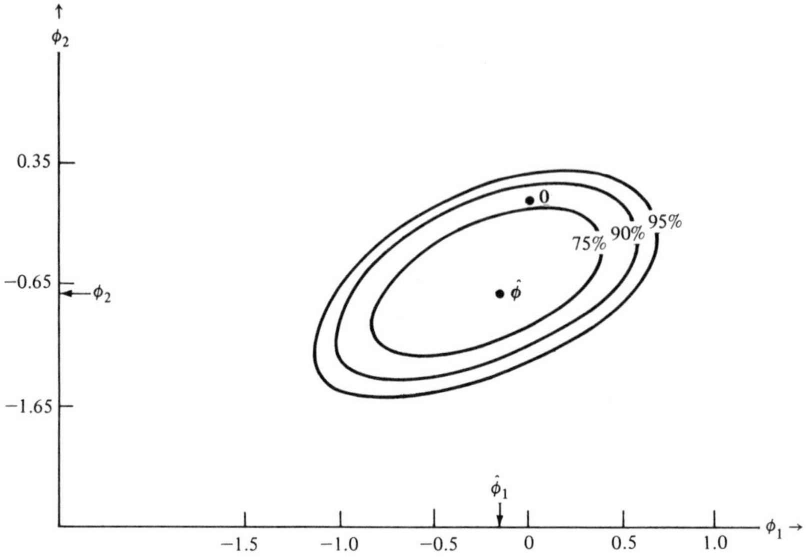

Fig. 2.12.1 Contours of posterior distribution of (ϕ_1, ϕ_2).

$\boldsymbol{\phi} = \boldsymbol{\phi}_0$. For instance, for the point $\boldsymbol{\phi} = \mathbf{0}$, using the Bartlett approximation, we obtain

$$\Pr\{p(\boldsymbol{\phi} \mid \mathbf{y}) > p(\mathbf{0} \mid \mathbf{y}) \mid \mathbf{y}\} = \Pr\{\chi_2^2 < 3.098\} = 0.79.$$

2.13 SUMMARIZED CALCULATIONS OF VARIOUS POSTERIOR DISTRIBUTIONS

Table 2.13.1 provides a summary of the Normal theory posterior distributions, based on noninformative priors, for making inferences about various location and scale parameters. The examples in most cases are those discussed in this chapter.

Table 2.13.1

Summarized calculations of various Normal theory posterior distributions

1. *Normal mean θ, variance σ^2 known*

   From (2.2.4),

$$\theta \sim N\left(\bar{y}, \frac{\sigma^2}{n}\right), \qquad \bar{y} = \frac{1}{n} \Sigma y_j.$$

Table 2.13.1 *Continued*

The $(1 - \alpha)$ H.P.D. interval of θ is

$$\bar{y} - u_{\alpha/2}\frac{\sigma}{\sqrt{n}} < \theta < \bar{y} + u_{\alpha/2}\frac{\sigma}{\sqrt{n}},$$

where $u_{\alpha/2}$ can be obtained from Table I (at the end of this book). Thus, for $(\bar{y} = 50,\ \sigma^2 = 20,\ n = 20)$, then $\theta \sim N(50,1)$. For $\alpha = 0.05$, $u_{0.025} = 1.96 \doteq 2$ and the 95% H.P.D. interval is $(48 < \theta < 52)$.

2. *Normal variance σ^2, θ known.*

From (2.3.6),

$$\sigma^2 \sim (ns^2)\chi_n^{-2}, \qquad s^2 = \frac{1}{n}\Sigma\,(y_j - \theta)^2.$$

The standardized $(1 - \alpha)$ H.P.D. interval is obtained in terms of the metric $\log\sigma$, and the corresponding interval in σ^2 is

$$\frac{ns^2}{\bar{\chi}^2(n,\alpha)} < \sigma^2 < \frac{ns^2}{\underline{\chi}^2(n,\alpha)},$$

where $\underline{\chi}^2(n,\alpha)$ and $\bar{\chi}^2(n,\alpha)$ are given in Table II (at the end of this book). From this suppose $(n = 20,\ ns^2 = 348,\ s^2 = 17.4)$, then $\sigma^2 \sim 348\chi_{20}^{-2}$. For $\alpha = 0.05$, $(\underline{\chi}^2 = 9.96,\ \bar{\chi}^2 = 35.23)$ and the 95% H.P.D. interval is $(9.88 < \sigma^2 < 34.94)$.

3. *Normal variance σ^2, θ unknown.*

From (2.4.19),

$$\sigma^2 \sim (\nu s^2)\chi_\nu^{-2}, \qquad \nu = n - 1, \qquad \text{and} \qquad s^2 = \nu^{-1}\Sigma\,(y_j - \bar{y})^2.$$

As in the case θ is known, the standardized $(1 - \alpha)$ H.P.D. interval is obtained in terms of $\log\sigma$, and the corresponding interval in σ^2 is

$$\frac{\nu s^2}{\bar{\chi}^2(\nu,\alpha)} < \sigma^2 < \frac{\nu s^2}{\underline{\chi}^2(\nu,\alpha)}.$$

4. *Normal mean θ, σ^2 unknown.*

From (2.4.21),

$$\theta \sim t\left(\bar{y}, \frac{s^2}{n}, \nu\right).$$

The $(1 - \alpha)$ H.P.D. interval of θ is

$$\bar{y} - t_{\alpha/2}(\nu)\frac{s}{\sqrt{n}} < \theta < \bar{y} + t_{\alpha/2}(\nu)\frac{s}{\sqrt{n}},$$

Table 2.13.1 *Continued*

where $t_{\alpha/2}(\nu)$ is given in Table IV (at the end of this book). Thus, for $(\bar{y} = 50, \; s/\sqrt{n} = 0.96, \; \nu = 19)$, then $\theta \sim t(50, 0.92, 19)$. For $\alpha = 0.05$, $t_{0.025}(19) = 2.09$ so that the 95% H.P.D. interval is $(47.99 < \theta < 52.01)$.

5. *Difference of two Normal means $\gamma = \theta_2 - \theta_1$, when the variances σ_1^2 and σ_2^2 are unknown but equal.*

From (2.5.3),

$$\gamma \sim t\left[\bar{y}_2 - \bar{y}_1,\; s^2\left(\frac{1}{n_1} + \frac{1}{n_2}\right),\; \nu_1 + \nu_2\right] \qquad (\nu_1 = n_1 - 1,\; \nu_2 = n_2 - 1),$$

$$s^2 = \frac{\nu_1 s_1^2 + \nu_2 s_2^2}{\nu_1 + \nu_2}, \qquad s_1^2 = \nu_1^{-1}\Sigma\,(y_{1j} - \bar{y}_1)^2, \qquad \text{and} \qquad s_2^2 = \nu_2^{-1}\Sigma\,(y_{2j} - \bar{y}_2)^2.$$

The $(1 - \alpha)$ H.P.D. interval of γ is

$$\bar{y}_2 - \bar{y}_1 - t_{\alpha/2}(n_1 + n_2 - 2)\, s\left(\frac{1}{n_1} + \frac{1}{n_2}\right)^{1/2} < \gamma$$

$$< \bar{y}_2 - \bar{y}_1 + t_{\alpha/2}(n_1 + n_2 - 2)\, s\left(\frac{1}{n_1} + \frac{1}{n_2}\right)^{1/2}.$$

For example, $\bar{y}_2 - \bar{y}_1 = 5$, $n_1 = n_2 = 10$, $s^2 = 4$; then, $\gamma \sim t[5, \frac{4}{5}, 18]$. For $\alpha = 0.05$, $t_{0.025}(18) = 2.10$ and the 95% H.P.D. interval is $(3.12 < \gamma < 6.88)$.

6. *Difference of two Normal means $\gamma = \theta_2 - \theta_1$, when σ_1^2 and σ_2^2 are unequal and unknown*

From (2.5.13),

$$\gamma \,\dot{\sim}\, t\left[\bar{y}_2 - \bar{y}_1,\; a^2\left(\frac{s_1^2}{n_1} + \frac{s_2^2}{n_2}\right),\; b\right],$$

where

$$b = 4 + \frac{f_1^2}{f_2}, \qquad a^2 = \left(\frac{b-2}{b}\right) f_1, \qquad f_1 = \left(\frac{\nu_2}{\nu_2 - 2}\right)\cos^2\phi + \left(\frac{\nu_1}{\nu_1 - 2}\right)\sin^2\phi,$$

$$f_2 = \frac{\nu_2^2\cos^4\phi}{(\nu_2 - 2)^2(\nu_2 - 4)} + \frac{\nu_1^2\sin^4\phi}{(\nu_1 - 2)^2(\nu_1 - 4)}, \qquad \text{and} \qquad \cos^2\phi = \frac{s_2^2/n_2}{s_1^2/n_1 + s_2^2/n_2}.$$

The $(1 - \alpha)$ H.P.D. interval is, approximately,

$$\bar{y}_2 - \bar{y}_1 - t_{\alpha/2}(b)\, a\left(\frac{s_1^2}{n_1} + \frac{s_2^2}{n_2}\right)^{1/2} < \gamma < \bar{y}_2 - \bar{y}_1 + t_{\alpha/2}(b)\, a\left(\frac{s_1^2}{n_1} + \frac{s_2^2}{n_2}\right)^{1/2}.$$

Table 2.13.1 *Continued*

Thus, for $\bar{y}_2 - \bar{y}_1 = 5$, $n_1 = 20$, $n_2 = 12$, $s_1^2 = 12$, and $s_2^2 = 40$, we find

$$\cos^2\phi = \tfrac{50}{59}, \qquad \sin^2\phi = \tfrac{9}{59}, \qquad f_1 = 1.2063, \qquad f_2 = 0.1552,$$

$$b = 13.376, \qquad \text{and} \qquad a^2 = 1.026,$$

so that

$$\gamma \mathrel{\dot{\sim}} t(5, 4.035, 13.38).$$

For $\alpha = 0.05$, $t_{0.025}(13.38) \doteq 2.15$ and the 95% H.P.D. interval is approximately $(0.67 < \gamma < 9.33)$.

7. *The ratio of two Normal variances* σ_2^2/σ_1^2, *when* θ_1 *and* θ_2 *are unknown.*

From (2.6.1),

$$\frac{\sigma_2^2}{\sigma_1^2} \sim \frac{s_2^2}{s_1^2} F_{(v_1, v_2)}.$$

The standardized $(1-\alpha)$ H.P.D. interval is expressed in terms of $\log\sigma_2 - \log\sigma_1$, and the corresponding interval in σ_2^2/σ_1^2 is

$$\frac{s_2^2}{s_1^2}\underline{F}(v_1, v_2, \alpha) < \frac{\sigma_2^2}{\sigma_1^2} < \frac{s_2^2}{s_1^2}\bar{F}(v_1, v_2, \alpha),$$

where $\underline{F}(v_1, v_2, \alpha)$ and $\bar{F}(v_1, v_2, \alpha)$ are given in Table V (at the end of this book). Thus, for $s_1^2 = 12$, $s_2^2 = 40$, $n_1 = 20$, $n_2 = 12$, then $\sigma_2^2/\sigma_1^2 \sim (3.33)\ F_{(19,11)}$. For $\alpha = 0.05$, $\underline{F}(19, 11, 0.05) = 0.352$, $\bar{F}(19, 11, 0.05) = 3.15$, and the 95% H.P.D. interval is $(1.172 < \sigma_2^2/\sigma_1^2 < 10.490)$.

8. *The regression coefficients* $\boldsymbol{\theta}$ *of the Normal linear model, with* σ^2 *unknown.*

First, from (2.7.1) and (2.7.8) compute $\mathbf{X}'\mathbf{X}$, $\mathbf{C} = (\mathbf{X}'\mathbf{X})^{-1}$

$$\hat{\boldsymbol{\theta}} = (\mathbf{X}'\mathbf{X})^{-1}\mathbf{X}'\mathbf{y}, \qquad s^2 = v^{-1}(\mathbf{y} - \hat{\mathbf{y}})'(\mathbf{y} - \hat{\mathbf{y}}),$$

where

$$\hat{\mathbf{y}} = \mathbf{X}\hat{\boldsymbol{\theta}} \qquad \text{and} \qquad v = n - k.$$

From (2.7.20),

$$\boldsymbol{\theta} \sim t_k(\hat{\boldsymbol{\theta}}, s^2\mathbf{C}, v).$$

The $(1-\alpha)$ H.P.D. region of $\boldsymbol{\theta}$ is given by

$$\frac{Q(\boldsymbol{\theta})}{ks^2} = \frac{(\boldsymbol{\theta} - \hat{\boldsymbol{\theta}})'(\mathbf{X}'\mathbf{X})(\boldsymbol{\theta} - \hat{\boldsymbol{\theta}})}{ks^2} < F(k, v, \alpha),$$

Table 2.13.1 *Continued*

where $F(k, \nu, \alpha)$ is igven in Table VI (at the end of this book). For inferences about a subset of $\boldsymbol{\theta}$, let

$$\boldsymbol{\theta} = \begin{bmatrix} \boldsymbol{\theta}_1 \\ \boldsymbol{\theta}_2 \end{bmatrix} \begin{matrix} k_1 \\ k_2 \end{matrix}, \qquad \hat{\boldsymbol{\theta}} = \begin{bmatrix} \hat{\boldsymbol{\theta}}_1 \\ \hat{\boldsymbol{\theta}}_2 \end{bmatrix}, \qquad \mathbf{C} = \begin{bmatrix} \mathbf{C}_{11} & \mathbf{C}_{12} \\ \mathbf{C}_{21} & \mathbf{C}_{22} \end{bmatrix}.$$

Then from (2.7.22),

$$\boldsymbol{\theta}_1 \sim t_{k_1}(\hat{\boldsymbol{\theta}}_1, s^2 \mathbf{C}_{11}, \nu).$$

For inferences concerning the ith individual element of $\boldsymbol{\theta}$, $i = 1, ..., k$,

$$\theta_i \sim t(\hat{\theta}_i, s^2 c_{ii}, \nu),$$

where c_{ii} is the ith diagonal element of $\mathbf{C}$. The $(1 - \alpha)$ H.P.D. interval of θ_i is

$$\hat{\theta}_i - t_{\alpha/2}(\nu)\, s\sqrt{c_{ii}} < \theta_i < \hat{\theta}_i + t_{\alpha/2}(\nu)\, s\sqrt{c_{ii}}.$$

Thus, for $k = 2$, $n = 18$,

$$\mathbf{X}'\mathbf{X} = \begin{bmatrix} 9 & 7 \\ 7 & 16 \end{bmatrix}, \qquad \mathbf{C} = \begin{bmatrix} 0.168 & -0.074 \\ -0.074 & 0.095 \end{bmatrix}, \qquad \hat{\boldsymbol{\theta}} = \begin{bmatrix} 98.9 \\ 124.4 \end{bmatrix},$$

and for $\nu = 16$ and $s^2 = 157.8$,

$$\boldsymbol{\theta} = \begin{bmatrix} \theta_1 \\ \theta_2 \end{bmatrix} \sim t_2\left\{ \begin{bmatrix} 98.9 \\ 124.4 \end{bmatrix}, 157.8 \begin{bmatrix} 0.168 & -0.074 \\ -0.074 & 0.095 \end{bmatrix}, 16 \right\}.$$

For $\alpha = 0.05$, $F(2, 16, 0.05) = 3.63$ and the 95% H.P.D. region of $\boldsymbol{\theta}$ is given by

$$\frac{Q(\boldsymbol{\theta})}{2s^2} = \frac{9(\theta_1 - 98.9)^2 + 14(\theta_1 - 98.9)(\theta_2 - 124.4) + 16(\theta_2 - 124.4)^2}{2 \times 157.8} < 3.63.$$

As an example, for the point $\boldsymbol{\theta}_0' = (120, 120)$,

$$\frac{Q(\boldsymbol{\theta}_0)}{2s^2} = \frac{3015}{315.6} = 9.56 > 3.63,$$

and therefore $\boldsymbol{\theta}_0$ is excluded from the 95% region. For individual inferences

$$\theta_1 \sim t(98.9,\ 157.8 \times 0.168,\ 16),$$

$$\theta_2 \sim t(124.4,\ 157.8 \times 0.095,\ 16),$$

and for $\alpha = 0.05$, $t_{0.025}(16) = 2.12$ so that the individual 95% intervals are $(88.25 < \theta_1 < 109.55)$ and $(116.41 < \theta_2 < 132.39)$.

Table 2.13.1 *Continued*

9. *Comparison of k Normal means $\theta_1, \ldots, \theta_k$, with common unknown variance σ^2.*

From (2.11.10), the $(1-\alpha)$ H.P.D. region of the $(k-1)$ linearly independent contrasts

$$\phi_i = \theta_i - \bar{\theta}, \quad i = 1, \ldots, k-1, \qquad \bar{\theta} = \frac{1}{n}\sum^k n_j\theta_j, \qquad n = \sum^k n_j,$$

is given by

$$\frac{\sum^k n_i[\phi_i - (\bar{y}_i - \bar{y})]^2}{(k-1)\,s^2} < F(k-1, \nu, \alpha),$$

where

$$\phi_k = \theta_k - \bar{\theta}, \quad \bar{y} = \frac{1}{n}\sum n_j\bar{y}_j, \qquad s^2 = \nu^{-1}\sum_i\sum_j (y_{ij} - \bar{y}_i)^2, \qquad \text{and} \qquad \nu = n - k.$$

In particular the point $\phi_1 = \cdots = \phi_{k-1} = 0$ (i.e., $\theta_1 = \cdots = \theta_k$) is included in the $(1-\alpha)$ region if and only if

$$\frac{\sum^k n_i(\bar{y}_i - \bar{y})^2}{(k-1)\,s^2} < F(k-1, \nu, \alpha).$$

For example, suppose $k = 3$, $n_1 = 10$, $n_2 = 10$, $n_3 = 12$, $\bar{y}_1 = 8$, $\bar{y}_2 = 10$, $\bar{y}_3 = 7$, and $s^2 = 9$; then $\bar{y} = 8.25$ and $\sum n_i(\bar{y}_i - \bar{y})^2 = 50$. For $\alpha = 0.05$, $F(2, 29, 0.05) = 3.33$ and $\frac{50}{18} \doteq 2.8$ so that the point $\phi_1 = \phi_2 = 0$ (i.e., $\theta_1 = \theta_2 = \theta_3$) is included in the 95% H.P.D. region.

10. *Comparison of k Normal variances $\sigma_1^2, \ldots, \sigma_k^2$, with unknown means $\theta_1, \ldots, \theta_k$.*

From (2.12.11) to (2.12.21), for the $(k-1)$ contrasts

$$\phi_i = \log\sigma_i^2 - \log\sigma_k^2, \qquad i = 1, \ldots, k-1,$$

the H.P.D. region of content approximately $(1-\alpha)$ is given by

$$\frac{-2\log W}{1 + A} < \chi^2(k-1, \alpha),$$

where

$$-2\log W = \nu\log\left(\frac{\nu_k}{\nu}\right) + \nu\log\left[1 + \sum_{i=1}^{k-1}\frac{\nu_i}{\nu_k}\exp\left[-(\phi_i - \hat{\phi}_i)\right]\right] + \sum_{i=1}^{k-1}\nu_i(\phi_i - \hat{\phi}_i),$$

$$A = \frac{1}{3(k-1)}\left(\sum_{i=1}^{k}\nu_i^{-1} - \nu^{-1}\right), \qquad \nu_i = n_i - 1, \qquad \nu = \sum_{i=1}^{k}\nu_i,$$

$$\hat{\phi}_i = \log s_i^2 - \log s_k^2, \qquad s_i^2 = \nu_i^{-1}\sum(y_{ij} - \bar{y}_i)^2,$$

Table 2.13.1 *Continued*

and $\chi^2(k-1, \alpha)$ is given in Table III (at the end of this book). In particular, for $\phi_i = 0$ $(i = 1, ..., k-1)$ (i.e., $\sigma_1^2 = \cdots = \sigma_k^2$),

$$-2\log W = -\sum_{i=1}^{k} \nu_i(\log s_i^2 - \log \bar{s}^2),$$

where

$$\bar{s}^2 = \nu^{-1}\sum_{i=1}^{k} \nu_i s_i^2.$$

Thus, for $k = 3$, $\nu_1 = \nu_2 = \nu_3 = 30$, $s_1^2 = 52.79$, $s_2^2 = 34.46$, $s_3^2 = 66.03$, we have $\nu = 90$, $\bar{s}^2 = 51.09$, $A = 0.015$, $\hat{\phi}_1 = -0.22$, $\hat{\phi}_2 = -0.65$, and

$$\frac{-2\log W}{1+A} = \Big\{-90\log 3 + 90\left[1 + \exp\left[-(\phi_1 + 0.22)\right] + \exp\left[-(\phi_2 + 0.65)\right]\right] + 30\left[(\phi_1 + 0.22) + (\phi_2 + 0.65)\right]\Big\}\Big/ 1.015.$$

For $\alpha = 0.05$, $\chi^2(2, 0.05) = 5.99$ so that the 95% H.P.D. region is given by $-2\log W/(1 + A) < 5.99$. In particular, for $\phi_1 = \phi_2 = 0$ $(\sigma_1^2 = \sigma_2^2 = \sigma_3^2)$

$$\frac{-2\log W}{1+A} = \frac{-\sum_{i=1}^{3} \nu_i(\log s_i^2 - \log \bar{s}^2)}{1+A} = \frac{3.143}{1.015} = 3.098,$$

so that the point $\phi_1 = \phi_2 = 0$ is included in the 95% region.

APPENDIX A2.1

Some Useful Integrals

We here give several integral formulae which are useful in the derivation of a number of distributions discussed in this book.

The Gamma, Inverted Gamma, and Related Integrals

For $a > 0$, $p > 0$,

$$\int_0^\infty x^{p-1}e^{-ax}\,dx = a^{-p}\Gamma(p), \tag{A2.1.1}$$

$$\int_0^\infty x^{-(p+1)}e^{-ax^{-1}}\,dx = a^{-p}\Gamma(p), \tag{A2.1.2}$$

$$\int_0^\infty x^{p-1}e^{-ax^2}\,dx = \tfrac{1}{2}a^{-p/2}\Gamma(p/2), \tag{A2.1.3}$$

and

$$\int_0^\infty x^{-(p+1)}e^{-ax^{-2}}\,dx = \tfrac{1}{2}a^{-p/2}\Gamma(p/2). \tag{A2.1.4}$$

More generally, for $a > 0$, $p > 0$, and $\alpha > 0$,

$$\int_0^\infty x^{p-1}e^{-ax^\alpha}\,dx = \frac{1}{\alpha}\,a^{-p/\alpha}\Gamma\left(\frac{p}{\alpha}\right), \tag{A2.1.5}$$

and

$$\int_0^\infty x^{-(p+1)}e^{-ax^{-\alpha}}\,dx = \frac{1}{\alpha}\,a^{-p/\alpha}\Gamma\left(\frac{p}{\alpha}\right). \tag{A2.1.6}$$

The Dirichlet Integral

For $p_s > 0$, $s = 1, \ldots, n+1$, and $n = 1, 2, \ldots$ the integral

$$\int_R x_1^{p_1-1} \ldots x_n^{p_n-1}(1 - x_1 - \cdots - x_n)^{p_{n+1}-1}dx_1 \ldots dx_n = \frac{\prod_{s=1}^{n+1}\Gamma(p_s)}{\Gamma[\sum_{s=1}^{n+1} p_s]}, \tag{A2.1.7}$$

where $R\colon (x_s > 0, \sum_{s=1}^{n} x_s < 1)$, is known as the Dirichlet integral. Alternatively, it can be expressed in the inverted form

$$\int_0^\infty \cdots \int_0^\infty x_1^{p_1-1} \ldots x_n^{p_n-1}(1 + x_1 + \cdots + x_n)^{-\sum_{s=1}^{n+1} p_s}\,dx_1 \ldots dx_n = \frac{\sum_{s=1}^{n+1}\Gamma(p_s)}{\Gamma[\sum_{s=1}^{n+1} p_s]}. \tag{A2.1.8}$$

The Normal Integrals

For $-\infty < \eta < \infty$, $c > 0$,

$$\int_{-\infty}^\infty \exp\left[-\frac{1}{2}\left(\frac{x-\eta}{c}\right)^2\right]dx = \sqrt{2\pi}\,c. \tag{A2.1.9}$$

Let $\boldsymbol{\eta}$ be a $n \times 1$ vector of constants and $\mathbf{C}$ a $n \times n$ positive definite symmetric matrix. Then

$$\int_{-\infty}^\infty \cdots \int_{-\infty}^\infty \exp\left[-\tfrac{1}{2}(\mathbf{x} - \boldsymbol{\eta})'\mathbf{C}^{-1}(\mathbf{x} - \boldsymbol{\eta})\right]dx_1 \ldots dx_n = (\sqrt{2\pi})^n|\mathbf{C}|^{1/2}, \tag{A2.1.10}$$

where $\mathbf{x} = (x_1, \ldots, x_n)'$.

The t Integrals

For $\nu > 0$,

$$\int_{-\infty}^\infty \left[1 + \left(\frac{x-\eta}{c}\right)^2\Big/\nu\right]^{-\frac{1}{2}(\nu+1)} dx = \frac{\Gamma(\frac{1}{2})\Gamma(\frac{1}{2}\nu)}{\Gamma[\frac{1}{2}(\nu+1)]}\sqrt{\nu c}, \tag{A2.1.11}$$

where η and c are defined in (A2.1.9). For $\nu > 0$,

$$\int_{-\infty}^{\infty} \cdots \int_{-\infty}^{\infty} \left[1 + \frac{(\mathbf{x} - \boldsymbol{\eta})' \mathbf{C}^{-1} (\mathbf{x} - \boldsymbol{\eta})}{\nu}\right]^{-\frac{1}{2}(\nu + n)} dx_1 \ldots dx_n$$

$$= \frac{[\Gamma(\frac{1}{2})]^n \Gamma(\frac{1}{2}\nu)}{\Gamma[\frac{1}{2}(\nu + n)]} (\sqrt{\nu})^n |\mathbf{C}|^{1/2}, \qquad \text{(A2.1.12)}$$

where $\boldsymbol{\eta}$, $\mathbf{C}$, and $\mathbf{x}$ are defined in (A2.1.10).

APPENDIX A2.2

Stirling's Series

To approximate the Gamma function, we often employ an asymptotic series discovered by Stirling, and later investigated by Bayes [see Milne-Thomson (1960)] The expansion involves an interesting class of polynomials discovered by D. Bernoulli.

Bernoulli Polynomials

The Bernoulli polynomial of degree r, $B_r(x)$, is generated by the following function,

$$\frac{te^{xt}}{e^t - 1} = \sum_{r=0}^{\infty} \frac{t^r}{r!} B_r(x). \qquad \text{(A2.2.1)}$$

Upon equating coefficients, we find

$$B_0(x) = 1$$

$$B_1(x) = (x - \tfrac{1}{2})$$

$$B_2(x) = x^2 - x + \tfrac{1}{6}$$

$$B_3(x) = x(x - 1)(x - \tfrac{1}{2}) \qquad \text{(A2.2.2)}$$

$$\cdots\cdots\cdots\cdots\cdots\cdots$$

The Bernoulli numbers B_r are generated by setting $x = 0$ in (A2.2.1),

$$\frac{t}{e^t - 1} = \sum_{r=0}^{\infty} \frac{t_r}{r!} B_r. \qquad \text{(A2.2.3)}$$

In particular $B_1 = -\frac{1}{2}$. Adding $t/2$ to both sides of (A2.2.3), we have

$$\frac{t}{2} \frac{(e^t + 1)}{(e^t - 1)} = \frac{t}{2} + \sum_{r=0}^{\infty} \frac{t^r}{r!} B_r. \qquad \text{(A2.2.4)}$$

Since the left of (A2.2.4) is an even function of t, it follows that

$$B_{2p+1} = 0, \qquad p = 1, 2, \ldots . \tag{A2.2.5}$$

It can also be shown that

$$(-1)^{p-1} B_{2p} > 0 \qquad \text{(that is, } B_{2p} \text{ alternates in sign)} \tag{A2.2.6}$$

and $|B_{2p}|$ tends to infinity as $p \to \infty$. The first few Bernoulli numbers are

$$B_0 = 1, \quad B_1 = -\tfrac{1}{2}, \quad B_2 = \tfrac{1}{6}, \quad B_3 = 0, \quad B_4 = -\tfrac{1}{30}, \quad B_5 = 0, \quad B_6 = \tfrac{1}{42} \cdots . \tag{A2.2.7}$$

Stirling's Series

Stirling's series provides an asymptotic expansion of the logarithm of the Gamma function $\Gamma(p + h)$ which is asymptotic in p for bounded h. We have

$$\log \Gamma(p + h) = \tfrac{1}{2} \log (2\pi) + p \log p - p + B_1(h) \log p - \sum_{r=1}^{n} (-1)^r \frac{B_{r+1}(h)}{r(r+1)p^r} + R_n(p), \tag{A2.2.8}$$

where $B_r(h)$ is the Bernoulli polynomial of degree r given in (A2.2.1). The remainder term $R_n(p)$ is such that

$$|R_n(p)| = \frac{c_n}{|p|^{n+1}}, \tag{A2.2.9}$$

where c_n is some positive constant independent of p. The series in (A2.2.8) is an asymptotic series in the sense that

$$\lim_{|p| \to \infty} |R_n(p)p^n| = 0 \qquad (n \text{ fixed})$$

and

$$\lim_{n \to \infty} |R_n(p)p^n| = \infty \qquad (p \text{ fixed}). \tag{A2.2.10}$$

Thus, even though the series in (A2.2.8) diverges for fixed p, it can be used to approximate $\log \Gamma(p + h)$ when p is large. Setting $h = 0$, we obtain

$$\log \Gamma(p) \doteq \tfrac{1}{2} \log (2\pi) + (p - \tfrac{1}{2}) \log p - p + \sum_{r=1}^{n} \frac{B_{2r}}{2r(2r-1)p^{2r-1}} + R_n'(p), \tag{A2.2.11}$$

where use is made of the fact that $B_{2r+1} = 0$. It can be shown that the remainder term $R_n'(p)$ satisfies the inequality

$$|R_n'(p)| \leqslant \frac{1}{2(n+1)(2n+1)} \left| \frac{B_{2(n+1)}}{p^{2n+1}} \right| . \tag{A2.2.12}$$

This, in conjunction with the fact that $B_{2(n+1)}$ alternates in sign, implies that, for positive p, the value of $\log \Gamma(p)$ always lies between the sum of n terms and the sum of $n + 1$ terms of the series, and that the absolute error of the series is less than the first term neglected. Even though (A2.2.11) is an asymptotic series, surprisingly close approximations can be obtained with it, even for small values of p. Taking the exponential of the log series, we obtain

$$\Gamma(p) = (2\pi)^{1/2} p^{p-(1/2)} e^{-p} \left(1 + \frac{1}{12p} + \frac{1}{288p^2} - \frac{139}{51840p^3} - \frac{571}{248320p^4} \cdots\right), \tag{A2.2.13}$$

which is known as Stirling's series for the Gamma function.

CHAPTER 3

BAYESIAN ASSESSMENT OF ASSUMPTIONS 1. EFFECT OF NON-NORMALITY ON INFERENCES ABOUT A POPULATION MEAN WITH GENERALIZATIONS

3.1 INTRODUCTION

In Chapter 2 a number of conventional problems in statistical inference were considered. Although interpretation was different, in most cases the Bayesian results closely paralleled those from sampling theory.† Whichever mode of inferential reasoning was adopted, certain assumptions were necessary in deriving these results. Specifically, it was assumed that observations could be regarded as normally and independently distributed. Although at first sight this assumption seems restrictive, abundant evidence has accumulated over the years that the basic results are of great usefulness in the actual conduct of scientific investigation. Nevertheless, exploration in new directions is desirable but has tended to be confined by the technical limitations of sampling theory. In particular, development along this route becomes unwieldy unless a set of minimal sufficient statistics happens to exist for the parameters considered.

The Bayesian approach is not restricted in this way, and a principal object of our book is to explore some of the ways in which this flexibility may be put to use.‡ In particular, the consequences of relaxing conventional assumptions may be studied.

In the present chapter the assumption of Normality is relaxed. Specifically inferential problems about means and regression coefficients are considered for a broader class of distributions which includes the Normal as a special case.

3.1.1 Measures of Distributional Shape, Describing Certain Types of non-Normality

Following Fisher, let κ_j be the jth cumulant of the distribution under study. Then $\kappa_1 = \mu$ is the mean, and $\kappa_2 = \sigma^2$ is the variance. Since for the Normal distribution the κ_j are zero for all $j > 2$, each scale-free quantity

$$\gamma_{j-2} = \frac{\kappa_j}{\sigma^j} = \frac{\kappa_j}{\kappa_2^{j/2}}, \qquad j = 3, 4, \ldots, \tag{3.1.1}$$

measures some aspect of non-Normality. The measures γ_1 and γ_2, first considered by Karl Pearson, are of special interest.

† The Behrens–Fisher problem was one instance in which the Bayesian result had no sampling parallel.

‡ Mosteller & Wallace (1964) provide other interesting illustrations of the use to which this freedom can be put.

Skewness

The standardized third cumulant,

$$\gamma_1 = \frac{\kappa_3}{\kappa_2^{3/2}} = \frac{E(y-\mu)^3}{\sigma^3}, \tag{3.1.2}$$

provides a measure of *skewness* of the distribution. Thus in Fig. 3.1.1, γ_1 is zero for the symmetric distribution seen at the center of the diagram; it is negative for the distribution to the left, which is said to be negatively skewed; it is positive for the distribution to the right, which is said to be positively skewed.

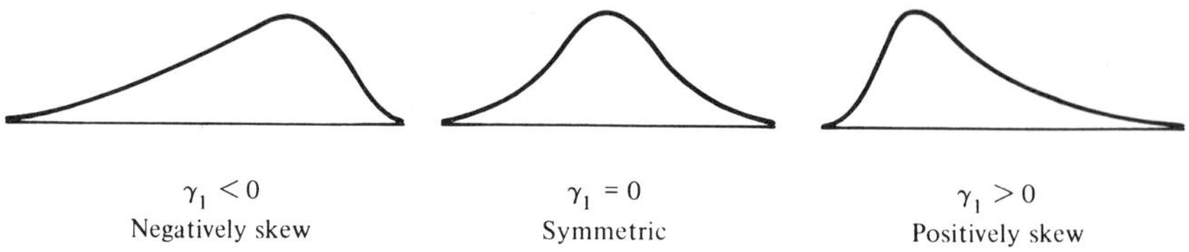

Fig. 3.1.1 Symmetric and skewed distributions.

Kurtosis

The standardized fourth cumulant,

$$\gamma_2 = \frac{\kappa_4}{\kappa_2^2} = \frac{E(y-\mu)^4}{\sigma^4} - 3, \tag{3.1.3}$$

measures a characteristic of the distribution called *kurtosis*.

For the Normal distribution $\gamma_2 = 0$. If $\gamma_2 > 0$, the distribution is said to be *leptokurtic*. If $\gamma_2 < 0$, it is said to be *platykurtic*. The Normal distribution is contrasted with a symmetric leptokurtic and a symmetric platykurtic distribution in Fig 3.1.2. Typically a leptokurtic distribution has less pronounced "shoulders" and heavier tails than the Normal. For instance, Student's t distribution is markedly leptokurtic when the number of degrees of freedom is small. On the other hand, a platykurtic distribution typically has squarer shoulders and lighter tails. As an example, the rectangular distribution is highly platykurtic.

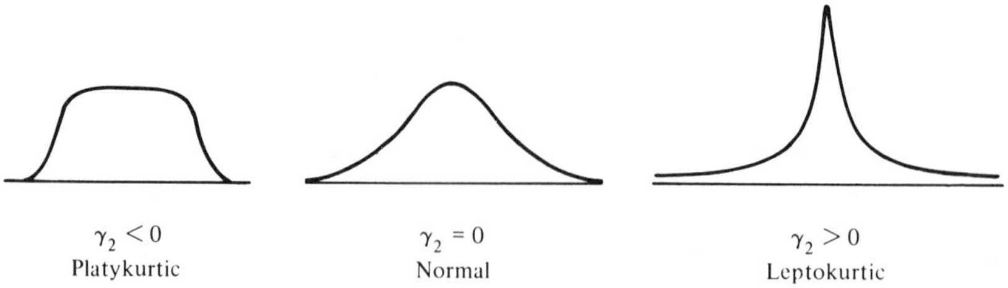

Fig. 3.1.2 The Normal distribution contrasted with a symmetric platykurtic distribution and a symmetric leptokurtic distribution.

Of course, distributions can exhibit kurtosis and skewness at the same time. Thus the χ^2 distribution is leptokurtic and positively skewed (especially when the number of degrees of freedom is small).

3.1.2 Situations where Normality would not be Expected

As we have said in Section 2.1, one can expect the distributions of the observations to be approximately Normal when the experimental conditions are such as to produce a central limit tendency, that is to say, when the errors arise from a variety of independent sources, none of which are dominant.

Nevertheless we would expect that certain kinds of measurements would not be Normally distributed. An example is yarn breaking strength. If the yarn is thought of as being made up of a number of links (like a chain), with the break occurring at the weakest link, and if the distribution of the strength of an *individual link* was Normal, the breaking strength will be distributed like the distribution of the smallest observation from a Normal sample. This extreme value distribution is skewed and highly leptokurtic.

However, this does not mean that the Normal assumption is unrealistic for all experiments where the data are breaking strengths. Consider again the experiment described in Section 2.5 in which two different types of spinning machines were compared The non-Normal error contributed by breaking strength measurement might be a minor contributor to the overall error; for this would usually be dominated by machine differences and sampling errors. The many components associated with these dominant sources would be likely to produce a central limit tendency and approximate Normality might be expected in this example, even though the data analyzed were breaking strengths.

Platykurtic distributions for individual errors can also occur. For example, suppose that in successive runs of a chemical apparatus, temperature varied about a set point. The experimenter might accept minor variations, but when a larger deviation occurred the run might be abandoned and a new run substituted. While the resulting "truncation" could lead to a platykurtic distribution, again, when other sources of error are present, this component could be swamped by the effect of the other errors.

The investigator will rarely be in the position where he can be certain of the precise form of the overall error distribution. Rather, his opinion may be described by a distribution of distributions centered about some central distribution. His state of mind will depend on the experimental setup and situations will sometimes occur where he expects a dominant non-Normal source of variation to determine the overall error distribution. More frequently, the setup will be one where he expects a finite number of sources each to have an important role. While in these circumstances the central limit theorem, which describes an asymptotic property, does not allow him to assume exact Normality, it does provide a basis for thinking of his distribution of distributions as *concentrated about* the

Normal. Furthermore, he is sometimes concerned only with differences between observations drawn from distributions supposed similar except in location. It may then be assumed that the parent distribution of these *differences* is a member of a class of *symmetric* distributions clustered about the Normal.

In this chapter we explore the use of a class of symmetric distributions with kurtosis measured by a non-Normality parameter β which takes the value zero for the Normal distribution. The state of uncertainty about the parent distribution can then be expressed by giving to β an informative unimodal prior probability distribution centered at zero. The sharpness of this prior distribution can be varied so that the effect of varying degrees of uncertainty about Normality can be represented. In particular, when the prior distribution becomes a δ function at $\beta = 0$, exact Normality is assumed. This is an illustration of the manner in which the Bayesian approach can be used to assess the effect of uncertainty in assumptions.

3.2 CRITERION ROBUSTNESS AND INFERENCE ROBUSTNESS ILLUSTRATED USING DARWIN'S DATA

On sampling theory, once the data-generating model is assumed, *criteria* appropriate to that assumption can be derived for inferential purposes. For example, suppose the observations $y_1, \ldots, y_n$ are regarded as a random sample from a Normal population $N(\theta, \sigma^2)$ and it is desired to make inferences about the mean θ. Then, if σ is unknown, the usual criterion is the t statistic

$$t = \frac{\bar{y} - \theta}{s/\sqrt{n}}.$$

Apart from the sample size n, this criterion involves the data only via the sample mean $\bar{y}$ and the sample standard deviation $s = [(n-1)^{-1} \Sigma (y - \bar{y})^2]^{\frac{1}{2}}$, which are jointly sufficient for (θ, σ). Inferences about θ are then based on the sampling distribution of t, assuming a Normal parent distribution. It is customary to justify the use of such a Normal theory criterion in the practical circumstance in which Normality cannot be guaranteed, by arguing that the distribution of *this t criterion* is but little affected by moderate non-Normality in the parent distribution—that is, it is *robust* under non-Normality. We shall refer to this type of insensitivity as *criterion robustness* under non-Normality.

This argument, however, does not take into account the fact that if the parent distribution really differed from the Normal, the appropriate criterion would no longer be the Normal theory statistic. For instance, suppose it was known that the parent distribution was rectangular (uniform); then the same sampling theory arguments previously leading to the t criterion would show that inferences were best made using the criterion

$$W = \frac{\frac{1}{2}(y_{(n)} + y_{(1)}) - \theta}{\frac{1}{2}(y_{(n)} - y_{(1)})/(n-1)},$$

which now involves the data only via $y_{(n)}$ and $y_{(1)}$, the largest and the smallest observations in the sample. Thus, on the assumption that the sample comes from a rectangular distribution, inferences about θ ought to be based not on the distribution of t but on the distribution of W.

The example which follows shows that although the distribution of the Normal theory t criterion is not changed very much by assuming the parent to have some distribution other than the Normal, the *inference* to be drawn from a particular sample can be markedly different when we employ a criterion appropriate to this other distribution. To distinguish it from criterion robustness, the property of insensitivity of inferences to departures from assumptions we shall call *inference robustness*.

Darwin's Data

Consider the analysis of Darwin's data on the difference in heights of self- and cross-fertilized plants quoted by Fisher (1960, p. 37). The data consists of measurements on 15 pairs of plants. Each pair contained a self-fertilized and a cross-fertilized plant grown in the same pot and from the same seed. Following Fisher, we shall treat as our observations the 15 differences y_i $(i = 1, \ldots, 15)$, which are set out in Table 3.2.1 and plotted below the horizontal axis in Fig. 3.2.1.

Table 3.2.1

Darwin's data: differences (in eighths of an inch) of heights of 15 pairs of self- and cross-fertilized plants

49	23	24
−67	28	75
8	41	60
16	14	−48
6	56	29

On sampling theory, given that the differences are a random sample from a Normal parent population $N(\theta, \sigma^2)$, one should interpret these data using the paired t test. In particular, for a significance test appropriate to the hypothesis that $\theta = 0$ against the alternative $\theta > 0$, the associated significance level is 2.485%. The curve on the right in Fig. 3.2.1 is an appropriately scaled t distribution with 14 degrees of freedom centered about $\bar{y} = 20.933$, with scale factor $s/\sqrt{n} = 9.746$, where $s^2 = \Sigma(y_i - \bar{y})^2/(n-1) = 1{,}424.6$. For definiteness, we shall call this distribution the *reference* distribution for θ. This reference distribution may be variously interpreted. It was regarded by Fisher as the fiducial distribution of θ. It is also a "confidence distribution" simultaneously allowing every confidence interval to be associated with its appropriate confidence coefficient. Finally, if as in Section 2.4, we make the assumption that θ and $\log \sigma$ are approximately independent and locally uniform *a priori*, it is the posterior distribution of θ.

Now suppose that instead of assuming Normality for the parent distribution, we supposed it to be uniform over some unknown range $\theta - \sqrt{3}\,\sigma$ to $\theta + \sqrt{3}\,\sigma$.

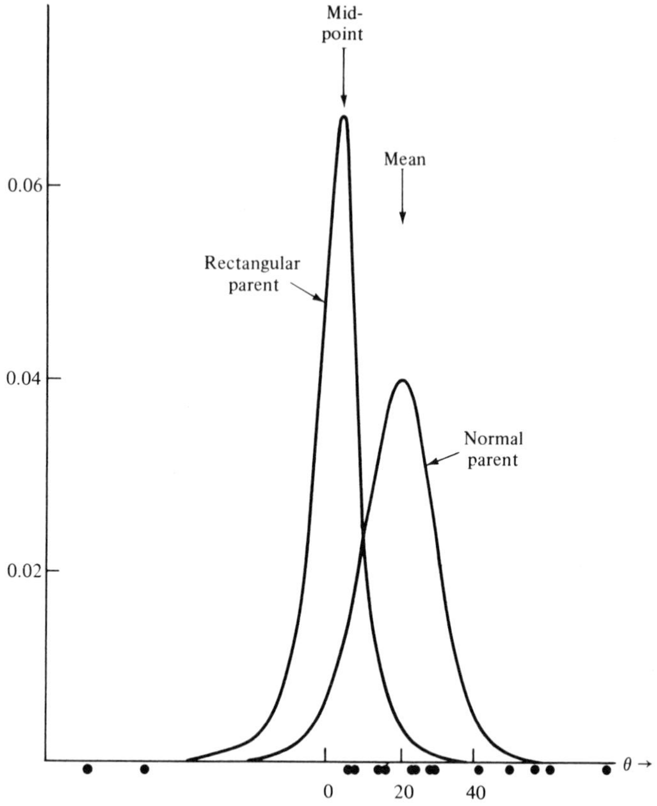

Fig. 3.2.1 Distributions of θ for Darwin's data: Normal and rectangular parent distributions (dots under the horizontal axis are the 15 differences in height recorded by Darwin).

Such a supposition would, of course, be quite ridiculous in the present example. First, we know that many contributing errors arising from genetic differences, soil differences, and so forth, tend to produce a central limit effect, so that we can expect with good reason that the heights themselves and, even more, their differences will be closely Normally distributed. Second, the evidence from the sample itself does not support the uniform assumption.

However, for illustration, let us make the assumption of a rectangular instead of a Normal parent and let us consider the effect of this extreme degree of non-Normality on the distribution of the t statistic. This can be approximately calculated using, for example, the work of Geary (1936), Gayen (1949, 1950), or of Box and Andersen (1955). Following these latter authors, it can be shown that, when the parent is non-Normal, the null distribution of t^2 is approximated by an F distribution with δ and $\delta(n-1)$ degrees of freedom, where

$$\delta = 1 + \frac{E(b-3)}{n},$$

with

$$b = \frac{(n+2)\sum y_i^4}{(\sum y_i^2)^2}, \tag{3.2.1}$$

$$E(b-3) \doteq \gamma_2 - n^{-1}(2\gamma_4 - 3\gamma_2^2 + 11\gamma_2) + n^{-2}(3\gamma_6 - 16\gamma_4\gamma_2 + 15\gamma_2^3 + 38\gamma_4 - 3\gamma_2^2 + 86\gamma_2),$$

$$\gamma_{r-2} = \frac{\kappa_r}{\kappa_2^{r/2}}, \qquad r = 3, 4, \ldots,$$

and the κ_r's are the cumulants of the parent distribution of the differences. In our present example, δ is found to be 0.913. Thus, t^2 is approximately distributed as F with 0.913 and 12.78 (instead of 1 and 14) degrees of freedom. In particular, the significance level associated with the hypothesis that $\theta = 0$ against the alternative $\theta > 0$ is now 2.388% as compared with the previous value of 2.485%. The test of the hypothesis that the true difference is zero, using the t criterion, is thus very little affected by this major departure from Normality. Furthermore, confidence intervals and hence the confidence distribution, based on the t statistic, would be almost unchanged.

However, if we really knew that the parent distribution was rectangular, then the largest observation $y_{(n)}$ and the smallest observation $y_{(1)}$ would be jointly sufficient for (θ, σ); and, as mentioned earlier, we would, on classical sampling theory, be led to consider not the t criterion but the function

$$W = \frac{m - \theta}{h/(n-1)}, \tag{3.2.2a}$$

where $m = \frac{1}{2}(y_{(n)} + y_{(1)})$ is the midpoint, and $h = \frac{1}{2}(y_{(n)} - y_{(1)})$ is the half range. On the assumption that the parent is rectangular, the variate W is distributed as

$$p(W) = \frac{1}{2}\left(1 + \frac{|W|}{n-1}\right)^{-n}, \qquad -\infty < W < \infty, \tag{3.2.2b}$$

[Neyman and Pearson (1928); Carlton (1946)]. Thus, the variate $|W|$ has the F distribution with 2 and $2(n-1)$ degrees of freedom. The distribution of W in (3.2.2b) may be called a "double F distribution" with $[2, 2(n-1)]$ degrees of freedom, since it consists of two such F distributions standing "back to back."

The curve on the left in Fig. 3.2.1 is an appropriately scaled double F distribution with (2, 28) degrees of freedom centered at the sample midpoint, $m = 4.0$ with scale factor $h/(n-1) = 5.07$, where the sample half range is $h = 71$. Just as the right-hand curve in the figure exemplifies the inferential situation with the Normal assumption, so the curve on the left correspondingly exemplifies the situation with the rectangular assumption. As before, it can be interpreted either as a fiducial distribution, as a confidence distribution, or, on parallel assumptions to those previously used, as a posterior distribution. It is seen to be very different from the one appropriate to the Normal assumption. In particular, the sampling theory significance level associated with the hypothesis that $\theta = 0$ against the alternative $\theta > 0$ is now not 2.485%, but 23.215%. Thus, whichever form of derivation we favor, we see that, if we assume a rectangular parent distribution, the inferences to be drawn are very different from those appropriate for a Normal parent. This is so even though the t criterion itself is very little affected by even this large departure from Normality.

One reason for the large difference of the distributions in Fig. 3.2.1 is that one curve is centered at the sample mean $\bar{y} = 20.9$, and the other at the sample midpoint $m = 4.0$. The mean and the midpoint for Darwin's data are very different mainly because of two rather large negative differences, and for this data we have an example in which the criterion is robust but the inference is not.

As we have explained, it is not seriously suggested that the rectangular distribution is a reasonable choice for the parent. We wish only to emphasize that uncertainty in our knowledge of the parent distribution transmits itself rather forcefully into an uncertainty about the inferences we can make concerning θ, and the difficulty which this presents in our interpretation of the data is not avoided by knowledge of robustness under non-Normality of the criterion. The difficulty can be resolved by making provision for an appropriate state of uncertainty about the parent distribution in the formulation. Possible knowledge about the parent distribution is of two kinds, that coming from the sample itself and *a priori* knowledge which may come from familiarity with the physical setup. Both of these can be taken account of in an appropriate Bayesian formulation.

3.2.1 A Wider Choice of the Parent Distribution

If, in the analysis of Darwin's data, we supposed that the parent distributions of self-and cross-fertilized plants were identical except for location, then the distribution of the differences would certainly be symmetric. We will assume therefore that the parent distribution of differences is a member of a class of symmetric distributions which includes the Normal, together with other distributions on the one hand more leptokurtic, and on the other hand more platykurtic.

Now the standardized Normal distribution may be written

$$p(x) = k \exp\left(-\tfrac{1}{2}|x|^q\right) \qquad \text{with} \qquad q = 2.$$

By allowing q to take values other than two we obtain what may be called the class of exponential power distributions. These distributions were considered by Diananda (1949), Box (1953b), and Turner (1960); with $q = 2/(1 + \beta)$ they can be written in the general form

$$p(y \mid \theta, \phi, \beta) = k\phi^{-1} \exp\left(-\frac{1}{2}\left|\frac{y-\theta}{\phi}\right|^{2/(1+\beta)}\right), \qquad -\infty < y < \infty, \tag{3.2.3}$$

where

$$k^{-1} = \Gamma\left(1 + \frac{1+\beta}{2}\right)2^{1+\frac{1}{2}(1+\beta)} \qquad \text{and} \qquad \phi > 0, \quad -\infty < \theta < \infty, \quad -1 < \beta \leqslant 1.$$

In (3.2.3), θ is a location parameter and ϕ is a scale parameter.† It can be readily shown that

$$E(y) = \theta,$$

$$\operatorname{Var}(y) = \sigma^2 = 2^{(1+\beta)}\left\{\frac{\Gamma[\frac{3}{2}(1+\beta)]}{\Gamma[\frac{1}{2}(1+\beta)]}\right\}\phi^2. \tag{3.2.4}$$

We may alternatively express (3.2.3) as

$$p(y \mid \theta, \sigma, \beta) = \omega(\beta)\sigma^{-1} \exp\left[-c(\beta)\left|\frac{y-\theta}{\sigma}\right|^{2/(1+\beta)}\right], \qquad -\infty < y < \infty, \tag{3.2.5}$$

where

$$c(\beta) = \left\{\frac{\Gamma[\frac{3}{2}(1+\beta)]}{\Gamma[\frac{1}{2}(1+\beta)]}\right\}^{1/(1+\beta)}$$

and

$$\omega(\beta) = \frac{\{\Gamma[\frac{3}{2}(1+\beta)]\}^{1/2}}{(1+\beta)\{\Gamma[\frac{1}{2}(1+\beta)]\}^{3/2}}, \qquad \sigma > 0, \quad -\infty < \theta < \infty, \quad -1 < \beta \leqslant 1.$$

The parameters θ and σ are then the mean and the standard deviation of the population, respectively. The parameter β can be regarded as a measure of kurtosis indicating the extent of the "non-Normality" of the parent population. In particular, when $\beta = 0$, the distribution is Normal. When $\beta = 1$, the distribution is the double exponential

$$p(y \mid \theta, \sigma, \beta = 1) = \frac{1}{\sqrt{2}\sigma} \exp\left(-\sqrt{2}\left|\frac{y-\theta}{\sigma}\right|\right), \qquad -\infty < y < \infty. \tag{3.2.6a}$$

† In our earlier work (1962) the form (3.2.3) was employed, but there the symbol σ was used for the general scale parameter now called ϕ.

Finally, when β tends to -1, it can be shown that the distribution tends to the rectangular distribution,

$$\lim_{\beta \to -1} p(y \mid \beta, \theta, \sigma) = \frac{1}{2\sqrt{3}\,\sigma}, \qquad \theta - \sqrt{3}\,\sigma < y < \theta + \sqrt{3}\,\sigma. \tag{3.2.6b}$$

Figure 3.2.2 shows the exponential power distribution for various values of β. The distributions shown have common mean and standard deviation. We see that for $\beta > 0$ the distributions are leptokurtic and for $\beta < 0$ they are platykurtic.

To further illustrate the effect of β on the shape of the distribution, Table 3.2.2 gives the upper 100α percent points in units of σ for various choices of β with θ assumed zero. Except for β equals 0 and 1, and the limiting case $\beta \to -1$, the percentage points in the table were calculated by numerical integration on a computer [Tiao and Lund (1970)].

In (3.2.5) we have employed the non-Normality measure β which makes the double exponential and the rectangular distribution equally discrepant from the Normal. However, we might have used, for example, the familiar kurtosis measure $\gamma_2 = \kappa_4/\kappa_2^2$ for the class of exponential power distributions. It is readily shown that

$$\gamma_2 = \frac{\Gamma[\frac{5}{2}(1+\beta)]\,\Gamma[\frac{1}{2}(1+\beta)]}{\{\Gamma[\frac{3}{2}(1+\beta)]\}^2} - 3. \tag{3.2.7}$$

Table 3.2.3 gives the value of γ_2 for various values of β. In terms of γ_2, the double exponential distribution would appear as 3 and the rectangular distribution as -1.2. However, whether β, γ_2, or any similar measure of non-Normality is adopted, the analysis which follows will be much the same.

Table 3.2.2

Upper 100α percent points of the exponential power distribution for various values of β in units of standard deviation σ

	($\theta = 0$)						
β	$\alpha = 0.25$	0.10	0.05	0.025	0.01	0.005	0.001
−1.00	0.87	1.39	1.56	1.65	1.70	1.71	1.73
−0.75	0.84	1.36	1.57	1.71	1.84	1.91	2.05
−0.50	0.80	1.35	1.61	1.81	2.03	2.16	2.41
−0.25	0.73	1.31	1.63	1.89	2.18	2.37	2.75
0.00	0.67	1.28	1.64	1.96	2.33	2.58	3.09
0.25	0.62	1.25	1.65	2.02	2.46	2.77	3.43
0.50	0.58	1.22	1.65	2.06	2.58	2.94	3.76
0.75	0.53	1.18	1.64	2.09	2.68	3.10	4.08
1.00	0.49	1.14	1.63	2.12	2.77	3.28	4.39

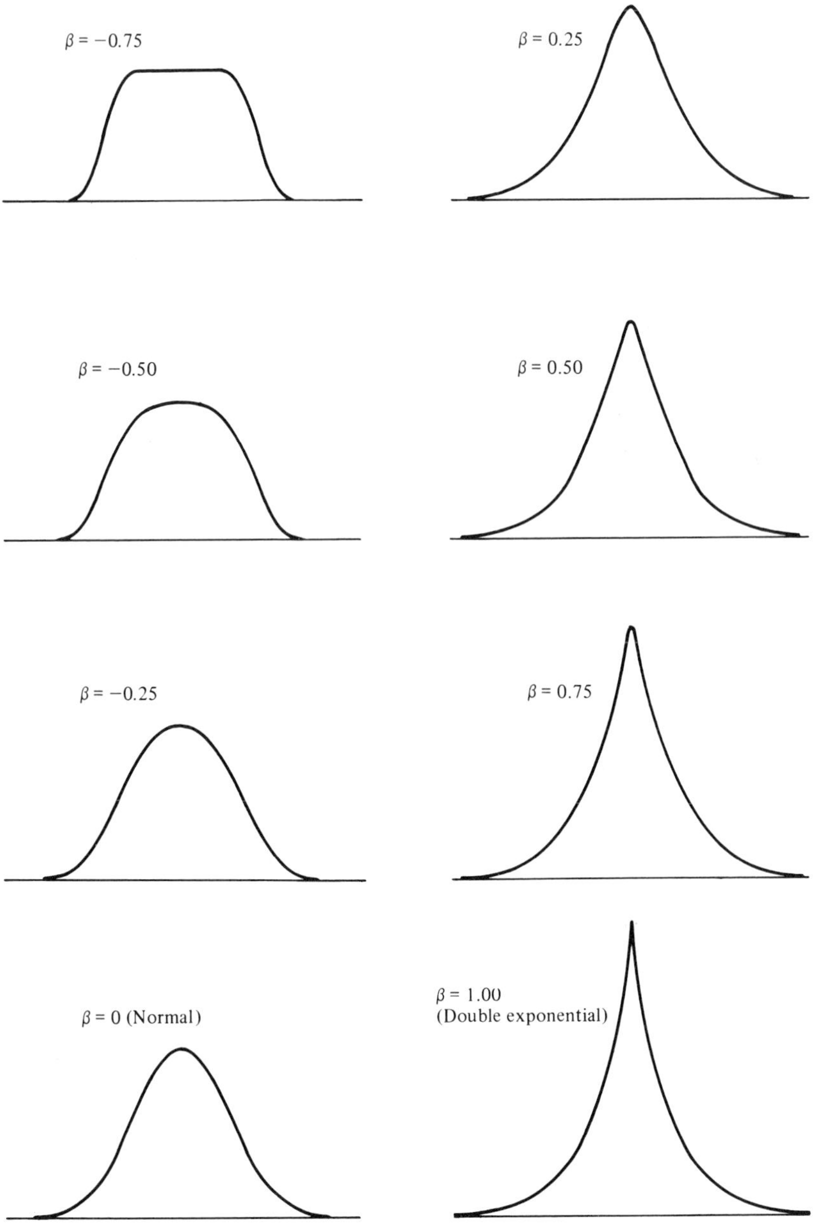

Fig. 3.2.2 Exponential power distributions with common standard deviation for various values of β.

Table 3.2.3

Relationship between (β, γ_2)

β	-1.0	-0.75	-0.50	-0.25	0.0	0.25	0.50	0.75	1.00
γ_2	-1.2	-1.07	-0.81	-0.45	0.0	0.55	1.22	2.03	3.00

3.2.2 Derivation of the Posterior Distribution of θ for a Specific Symmetric Parent

Given a sample of n independent observations from a member of the class of distributions in (3.2.5), the likelihood function is

$$l(\theta, \sigma, \beta \mid \mathbf{y}) \propto [\omega(\beta)]^n \sigma^{-n} \exp\left[-c(\beta) \sum_{i=1}^{n} \left|\frac{y_i - \theta}{\sigma}\right|^{2/(1+\beta)}\right]. \tag{3.2.8}$$

Consider now the posterior distribution of θ assuming that the parameter β is known. For a fixed β, the likelihood function is

$$l(\theta, \sigma \mid \beta, \mathbf{y}) \propto \sigma^{-n} \exp\left[-c(\beta) \sum_{i} \left|\frac{y_i - \theta}{\sigma}\right|^{2/(1+\beta)}\right]. \tag{3.2.9}$$

In addition, the noninformative reference prior for θ and σ is such that locally

$$p(\theta, \sigma \mid \beta) \propto \sigma^{-1}. \tag{3.2.10}$$

[See the discussion in Section 1.3 concerning the location-scale family (1.3.57), of which (3.2.5) is a special case, and the discussion leading to (1.3.105).]

The joint posterior distribution of (θ, σ) is then

$$p(\theta, \sigma \mid \beta, \mathbf{y}) \propto \sigma^{-(n+1)} \exp\left[-c(\beta) \sum_{i} \left|\frac{y_i - \theta}{\sigma}\right|^{2/(1+\beta)}\right], \qquad -\infty < \theta < \infty, \quad \sigma > 0. \tag{3.2.11}$$

Employing (A2.1.6) in Appendix A2.1 to integrate out σ, the posterior distribution of θ is then

$$p(\theta \mid \beta, \mathbf{y}) = [J(\beta)]^{-1}[M(\theta)]^{-\frac{1}{2}n(1+\beta)}, \qquad -\infty < \theta < \infty, \tag{3.2.12}$$

where

$$M(\theta) = \sum_{i} |y_i - \theta|^{2/(1+\beta)}$$

and $[J(\beta)]^{-1}$ is the appropriate normalizing constant. Thus, for any fixed β, $p(\theta \mid \beta, \mathbf{y})$ is simply proportional to a power of $M(\theta)/n$, the absolute moment of order $2/(1+\beta)$ of the observations about θ. The constant $[J(\beta)]^{-1}$ is such that

$$J(\beta) = \int_{-\infty}^{\infty} [M(\theta)]^{-\frac{1}{2}n(1+\beta)}\, d\theta, \tag{3.2.13}$$

and is merely a normalizing factor which ensures that the total area under the distribution is unity. This integral cannot usually be expressed as a simple function of the observations; it can, of course, always be computed numerically, and with the availability of electronic computers, this presents no particular difficulty. If all that is required is to draw the posterior distribution so that it can be presented to the investigator, it is seldom necessary to bother with the value of the normalizing constant at all. He appreciates that a probability distribution must have unit area and for most inferential purposes the ordinates need not even be marked. Using (3.2.12), posterior distributions computed from Darwin's data for various values of β are shown in Fig. 3.2.3.

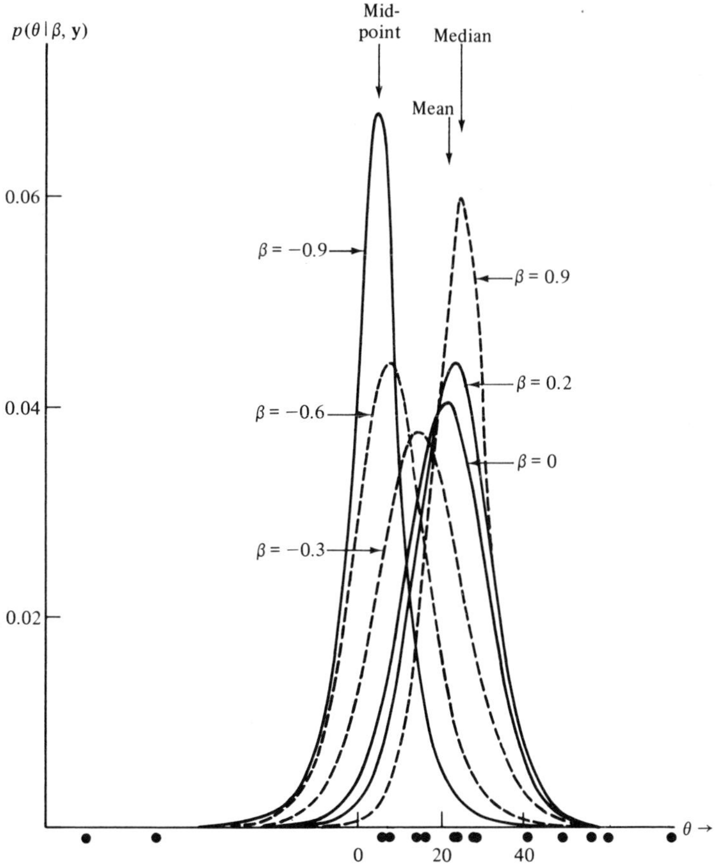

Fig. 3.2.3 Posterior distribution of θ for various choices of β: Darwin's data.

3.2.3 Properties of the Posterior Distribution of θ for a Fixed β

Since $p(\theta \mid \beta, \mathbf{y})$ is a monotonic function of $M(\theta)$, we find (see Appendix A3.1):

1. $p(\theta \mid \mathbf{y}, \beta)$ is continuous, and, for $-1 < \beta < 1$, is differentiable and unimodal, although not necessarily symmetric; the mode being attained in the interval $[y_{(1)}, y_{(n)}]$. The modal value is in fact the maximum likelihood estimate of θ. However, it should be noted that we are not concerned with the distribution of this maximum likelihood estimate; rather, we are considering the distribution of θ given the data.

2. When $\beta = 0$, $M(\theta) = \Sigma(y_i - \theta)^2 = (n-1)s^2 + n(\bar{y} - \theta)^2$, and making the necessary substitutions we obtain, for the posterior distribution of θ,

$$p\left(t = \frac{\theta - \bar{y}}{s/\sqrt{n}} \,\middle|\, \beta = 0, \mathbf{y}\right) = p(t_{n-1}), \tag{3.2.14}$$

where $p(t_{n-1})$ is the density of $t(0, 1, n-1)$ distribution.

3. When β approaches -1,

$$\lim_{\beta \to -1} [M(\theta)]^{(\beta+1)/2} = (h + |m - \theta|),$$

where m and h are as in (3.2.2a). Making the necessary substitutions, we find

$$\lim_{\beta \to -1} p(\theta \mid \beta, \mathbf{y}) = \frac{1}{2}\left(\frac{h}{n-1}\right)^{-1}\left[1 + \frac{|W|}{n-1}\right]^{-n}, \qquad -\infty < \theta < \infty, \tag{3.2.15}$$

where

$$W = \frac{\theta - m}{h/(n-1)}.$$

Thus,

$$\lim_{\beta \to -1} p\left(F = \frac{|\theta - m|}{h/(n-1)} \,\middle|\, \beta, \mathbf{y}\right) = p[F_{2,2(n-1)}], \qquad F > 0, \tag{3.2.16}$$

where $p[F_{2,2(n-1)}]$ is the density of an F variable with $[2, 2(n-1)]$ degrees of freedom. For Darwin's data, (3.2.15) is the reference distribution for θ, shown by the curve on the left in Fig. 3.2.1, but now derived as a limiting posterior distribution.

Thus, we see that, when the parent is Normal ($\beta = 0$), our expression (3.2.12) yields the t distribution as expected, and when the parent approaches the rectangular ($\beta \to -1$), again as expected, (3.2.12) tends to the double F distribution with 2 and $2(n-1)$ degrees of freedom. In each case, the posterior distribution can be expressed in terms of simple functions of the observations which are minimal sufficient statistics for θ and σ.

4. When β approaches 1, the distribution in (3.2.12) is not expressible in terms of simple functions of the observations. However, in the limit the mode of the posterior distribution is the median of the observations if n is odd, and is some unique value between the values of the middle two observations if n is even. When $\beta = 1$ and n is even, the density is, in fact, constant for values of θ between the middle two observations.

5. In certain other cases, it is possible to express the posterior distribution of θ in terms of a fixed number of functions of the observations. For instance, letting

$$\beta = (2 - q)/q, \qquad \text{then for} \qquad q = 2, 4, \ldots,$$

we have

$$p(\theta, \sigma \mid \beta, \mathbf{y}) \propto \sigma^{-(n+1)} \exp\left[-c(\beta)\sigma^{-q} \sum_{r=0}^{q} (-1)^r \binom{q}{r} \theta^r S_{q-r} \right],$$

$$-\infty < \theta < \infty, \qquad \sigma > 0, \qquad (3.2.17)$$

and

$$p(\theta \mid \beta, \mathbf{y}) \propto \left[\sum_{r=0}^{q} (-1)^r \binom{q}{r} \theta^r s_{q-r} \right]^{-n/q}, \qquad -\infty < \theta < \infty, \qquad (3.2.18)$$

where

$$S_r = \sum_i y_i^r.$$

It is readily seen that the set of q functions, $S_1, S_2, \ldots, S_q$, of the observations are jointly sufficient for θ and σ.

In general, however, the posterior distribution cannot be expressed in terms of a few simple functions of the observations. If we wish to think in terms of sufficiency and information as defined by Fisher (1922, 1925), our posterior distribution always, of course, employs a complete set of sufficient statistics, namely, the observations themselves. Consequently, no matter what is the value of β, no information is lost.

From the family of distributions for various values of β, shown in Fig. 3.2.3, we see that very different inferences would be drawn concerning θ, depending upon which value of β was assumed. The chief reason for this wide discrepancy is the fact that in Darwin's data, the center of the posterior distribution changes markedly as β is changed. In particular, for this sample, the median, mean, and the midpoint are 24.0, 20.9, 4.0 respectively; and these are the modes of the posterior distributions for the double exponential, Normal, and rectangular parent, respectively.

3.2.4 Posterior Distribution of θ and β when β is Regarded as a Random Variable

Because of the wide differences which occur in the posterior distribution of θ depending on which parent distribution we employ, it might be thought that there would be considerable uncertainty about what could be inferred from this set of data. It turns out that this is not the case when appropriate evidence concerning the value of β is put to use. There are two possible sources of information about the value of β, one from the data itself and the other from knowledge *a priori*. Both types of evidence can be injected into our analysis by allowing β itself to be a random variable associated with a prior distribution $p(\beta)$.

Joint Distribution of θ and β

Let us assume tentatively that, *a priori*, β is distributed independently of the mean θ and the standard deviation σ so that

$$p(\theta, \sigma, \beta) = p(\beta)p(\theta, \sigma). \tag{3.2.19}$$

As before, we adopt the noninformative reference prior for (θ, σ)

$$p(\theta, \sigma) \propto \sigma^{-1}. \tag{3.2.20}$$

Then, the joint posterior distribution of (θ, σ, β) is

$$p(\theta, \sigma, \beta \mid \mathbf{y}) \propto \sigma^{-1} p(\beta) l(\theta, \sigma, \beta \mid \mathbf{y}), \qquad -\infty < \theta < \infty, \quad \sigma > 0, \quad -1 < \beta \leqslant 1, \tag{3.2.21}$$

where $l(\theta, \sigma, \beta \mid \mathbf{y})$ is the likelihood function given in (3.2.8). After eliminating the standard deviation σ, we obtain the joint posterior distribution of (θ, β) as

$$p(\theta, \beta \mid \mathbf{y}) \propto p(\beta)[M(\theta)]^{-\frac{1}{2}n(1+\beta)}\, \Gamma[1 + \tfrac{1}{2}n(1+\beta)]\{\Gamma[1 + \tfrac{1}{2}(1+\beta)]\}^{-n},$$
$$-\infty < \theta < \infty, \quad -1 < \beta \leqslant 1. \tag{3.2.22}$$

Assumptions about locally uniform priors and particularly about independence of the component parameters have to be made with caution. Although it seems reasonable that θ should be assumed to be independent of σ and β *a priori*, we might have second thoughts about the independence of the scale parameter σ and the non-Normality parameter β. The definition of a scale parameter is always arbitrary to the extent of a multiplicative constant, and in particular, if we write the distribution in the form (3.2.3), using ϕ as the scale parameter, then

$$\phi = f(\beta)\sigma,$$

where $f(\beta)$ is some function of β. It might be supposed, therefore, that if β and σ were independent then β and ϕ could not be. It would then follow that the marginal posterior distribution $p(\theta, \beta \mid \mathbf{y})$ would be different if it was assumed [as in our earlier work (1962)] that $\log \phi$ was locally uniform and independent of β instead of assuming, as we have done here, that $\log \sigma$ was locally uniform and independent of β. In fact, the results obtained are approximately the same whatever function $f(\beta)$ is used. To see this, let us suppose that

$$p(\log \sigma, \beta) = p(\log \sigma)\, p(\beta), \tag{3.2.23}$$

with $p(\log \sigma) \propto c$. Since $\log \phi = \log f(\beta) + \log \sigma$, it follows that, for given β, locally,

$$p(\log \phi) \propto c. \tag{3.2.24}$$

Further, we can see from Appendix A3.2 that if $\log \sigma$ is locally uniform and independent of β, then $\log \phi$ and β will be approximately independent.

A Reference Distribution for β

It will turn out from our subsequent discussion that a useful reference prior distribution for β is a uniform distribution over the range $(-1, 1)$. If we denote the posterior distribution of θ and β based on this choice by $p_u(\theta, \beta \mid \mathbf{y})$, then

$$p_u(\theta, \beta \mid \mathbf{y}) \propto [M(\theta)]^{-\frac{1}{2}n(1+\beta)}\, \Gamma[1 + \tfrac{1}{2}n(1 + \beta)]\{\Gamma[1 + \tfrac{1}{2}(1 + \beta)]\}^{-n} \tag{3.2.25}$$

and

$$p(\theta, \beta \mid \mathbf{y}) \propto p_u(\theta, \beta \mid \mathbf{y}) p(\beta), \qquad -\infty < \theta < \infty, \quad -1 < \beta \leqslant 1, \tag{3.2.26}$$

where $p(\beta)$ is *any* appropriate prior distribution of β.

The joint distribution $p_u(\theta, \beta \mid \mathbf{y})$ actually obtained from the Darwin data is shown in Fig. 3.2.4. The sections of the joint distribution for various fixed values

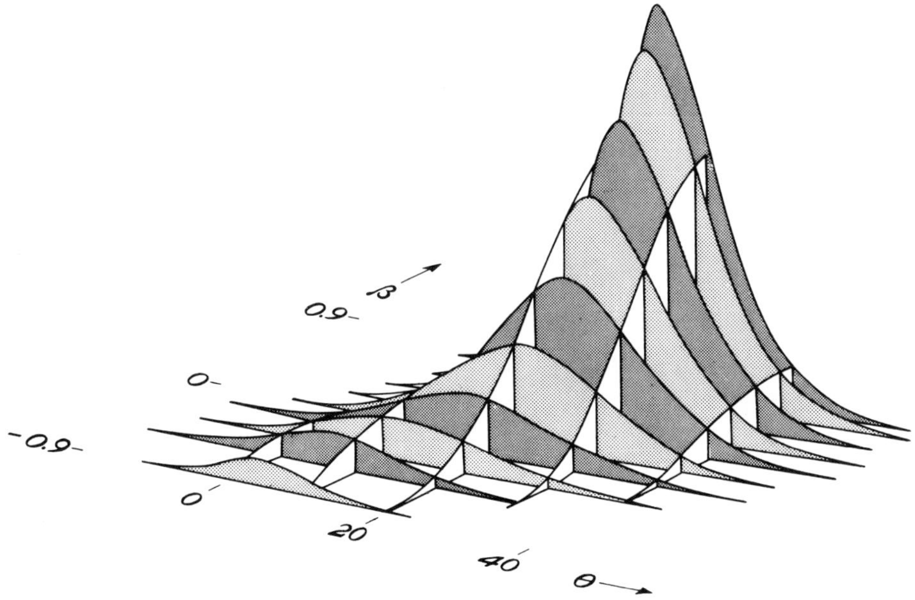

Fig. 3.2.4 The joint distribution $p_u(\theta, \beta \mid \mathbf{y})$: Darwin's data.

of β are, apart from a weighting factor, the conditional distributions $p(\theta \mid \beta, \mathbf{y})$ already sketched in Fig. 3.2.3. The weighting factor is of course $p_u(\beta \mid \mathbf{y})$ and it is evident that this factor attaches little credibility to platykurtic parenthood.

3.2.5 Marginal Distribution of β

If we integrate (3.2.25) over θ we obtain the weighting factor mentioned above which is the posterior marginal distribution of β for a uniform reference prior in β,

$$p_u(\beta \mid \mathbf{y}) \propto \Gamma[1 + \tfrac{1}{2}n(1+\beta)]\{\Gamma[1 + \tfrac{1}{2}(1+\beta)]\}^{-n} J(\beta), \quad -1 < \beta \leqslant 1, \qquad (3.2.27)$$

where $J(\beta)$ is given in (3.2.13). This distribution, shown by the solid curve in Fig. 3.2.5 for $a = 1$, allows appropriate inferences about β to be drawn in the light of the data for a uniform prior over the range $-1 < \beta \leqslant 1$. Now, in practice, there will be few instances where this prior would actually coincide with the attitude of the investigator, nevertheless, its use as a reference provides a valuable

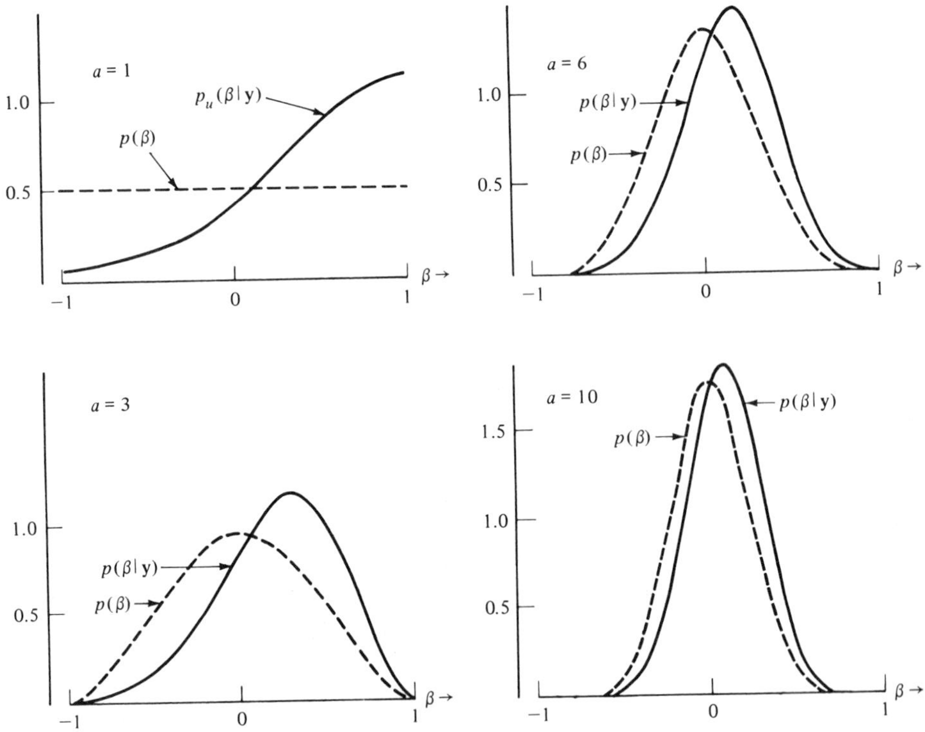

Fig. 3.2.5 Prior and posterior distributions of β for various choices of the parameter "a": Darwin's data.

intermediate step to more realistic choices. For, since

$$p(\beta \mid \mathbf{y}) \propto p_u(\beta \mid \mathbf{y})p(\beta), \tag{3.2.28}$$

we can produce the posterior distribution of β for *any* prior $p(\beta)$ from $p_u(\beta \mid \mathbf{y})$ merely by multiplication and renormalizing.

Choice of a Prior Distribution for β

In problems like the analysis of Darwin's data, usually some central limit effect would be expected. While this does not warrant an outright assumption of Normality which can be represented in the present context by making $p(\beta)$ a δ function at $\beta = 0$, it does mean that the investigator would want to associate high prior probability with distributions in the neighbourhood of the Normal. He could represent this attitude by choosing a prior distribution for β having a single mode at $\beta = 0$. A convenient distribution for this purpose is a symmetric beta distribution having mean zero and extending from -1 to $+1$ with one adjustable parameter which we call "a". Specifically, we assume that†

$$p(\beta) = w(1 - \beta^2)^{a-1}, \qquad -1 < \beta \leqslant 1, \tag{3.2.29}$$

where

$$w = \Gamma(2a)[\Gamma(a)]^{-2}\, 2^{-(2a-1)}, \qquad a \geqslant 1.$$

When $a = 1$, the distribution is uniform. With $a > 1$, it is a symmetric distribution having its mode at the normal theory value $\beta = 0$, and it becomes more and more concentrated about $\beta = 0$ as "a" is increased. When "a" tends to infinity, $p(\beta)$ approaches a delta function, representing an assumption of exact Normality.

The dotted curves in Fig. 3.2.5 show this distribution for $a = 1, 3, 6$, and 10. The corresponding posterior distributions $p(\beta \mid \mathbf{y})$ are shown by the solid curves in Fig. 3.2.5. The Figure shows how increasing prior certainty about Normality tends to override the information from the sample. When "a" tends to infinity, $p(\beta \mid \mathbf{y})$ will approach a δ function at $\beta = 0$ and Normality will be assumed no matter what the information from the sample.

3.2.6 Marginal Distribution of θ

The posterior distribution of θ is obtained by integrating out β from the joint distribution of θ and β, yielding

$$p(\theta \mid \mathbf{y}) = \int_{-1}^{1} p(\theta, \beta \mid \mathbf{y})\, d\beta, \qquad -\infty < \theta < \infty. \tag{3.2.30}$$

---

† Letting $x = (1 + \beta)/2$ and $p = q = a$, we have the usual beta distribution

$$p(x) = \frac{\Gamma(p + q)}{\Gamma(p)\Gamma(q)} x^{p-1}(1 - x)^{q-1}, \qquad 0 < x < 1.$$

Also, we may write

$$p(\theta, \beta \mid \mathbf{y}) = p(\theta \mid \beta, \mathbf{y})p(\beta \mid \mathbf{y}). \tag{3.2.31}$$

The posterior distribution of θ,

$$p(\theta \mid \mathbf{y}) = \int_{-1}^{1} p(\theta \mid \beta, \mathbf{y})p(\beta \mid \mathbf{y})\, d\beta, \qquad -\infty < \theta < \infty, \tag{3.2.32}$$

can thus be thought of as a weighted average of the "t-like" distributions $p(\theta \mid \beta, \mathbf{y})$ of Fig. 3.2.3, with a weight function $p(\beta \mid \mathbf{y})$ given by a solid curve of Fig. 3.2.5.

Since $p(\beta \mid \mathbf{y}) \propto p_u(\beta \mid \mathbf{y})\, p(\beta)$ we have also

$$p(\theta \mid \mathbf{y}) \propto \int_{-1}^{1} p(\theta \mid \beta, \mathbf{y})p_u(\beta \mid \mathbf{y})p(\beta)\, d\beta, \qquad -\infty < \theta < \infty. \tag{3.2.33}$$

Thus, we are averaging the t-like distributions with a weight function which depends partly on $p(\beta)$, representing prior information, and partly on $p_u(\beta \mid \mathbf{y})$,† representing information which is independent of the prior.

Finally we can write

$$p(\theta \mid \mathbf{y}) \propto \int_{-1}^{1} p_u(\theta, \beta \mid \mathbf{y})p(\beta)\, d\beta \tag{3.2.34}$$

so that $p(\theta \mid \mathbf{y})$ is the marginal distribution obtained when $p_u(\theta, \beta \mid \mathbf{y})$ shown in the three-dimensional diagram of Fig. 3.2.4 is averaged over the weight function $p(\beta)$.

The marginal distributions $p(\theta \mid \mathbf{y})$ for $a = 1$, 3, 6 and ∞ are drawn in Fig. 3.2.6. They show to what extent inferences about θ depend on the strength of central limit effect assumed. The curve for $a \to \infty$ is a t distribution corresponding to an outright assumption of parent Normality. In view of the very large differences exhibited by the conditional distributions $p(\theta \mid \beta, \mathbf{y})$, it is remarkable how little the marginal distribution $p(\theta \mid \mathbf{y})$ is affected by choices of "a" covering a range representing exact Normality ($a \to \infty$) to "no central limit effect" ($a = 1$). The reason for this is that widely discrepant conditional distributions generated by parents which approach the uniform ($\beta \to -1$) are almost ruled out by information coming from the sample itself.

The curve for $a = 10$ is not shown in Fig. 3.2.6 because it is almost identical to the t distribution obtained when $a \to \infty$. This is interesting because, as seen from Fig. 3.2.5, the central limit effect implied even by $a = 10$ is not an overwhelmingly strong one. For instance, with this distribution the probability *a priori* that $-0.33 < \beta < 0.33$ (that is, $q = 2/(1 + \beta)$ is between 1.5 and 3) is only 87 percent.

---

† We are employing noninformative prior distributions for θ and σ; a change in the prior distribution $p(\theta, \sigma)$ will of course change $p_u(\beta \mid \mathbf{y})$.

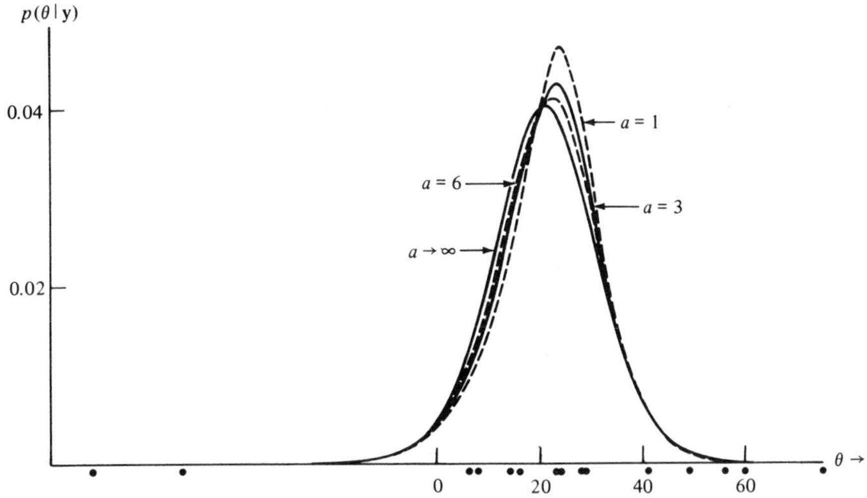

Fig. 3.2.6 Posterior distribution of θ for various choices of the parameter "a": Darwin's data.

3.2.7 Information Concerning the Nature of the Parent Distribution Coming from the Sample

In the past, the Normality (or otherwise) of a sample has often been assessed by inspecting the empirical distribution, by applying goodness-of-fit tests such as the χ^2 test and the Kolmogoroff–Smirnoff test (1941), and by checking measures of skewness and kurtosis. In cases such as the present one, where it is reasonable to assume symmetry, the calculation of $p_u(\beta \mid \mathbf{y})$ would provide another way of expressing sample information about the nature of the parent distribution. However, with this approach more can be done than merely "test" the assumption of Normality and then, in the absence of a "significant" result, assume it. The information about β coming from the sample is appropriately *used* in making inferences about θ. In particular, for Darwin's data it plays an important role in virtually eliminating the influence of unlikely parent distributions having extreme negative kurtosis.

3.2.8 Relationship to the General Bayesian Framework for Robustness Studies

It will now be seen how Darwin's example illustrates the general approach introduced in Section 1.6. After integrating out σ, we have the joint distribution of the mean θ and the non-Normality parameter β, which correspond to $\boldsymbol{\theta}_1$ and $\boldsymbol{\theta}_2$ respectively, in the general discussion. While it is true that we can obtain the marginal posterior distribution of θ immediately by integrating out β from $p(\theta, \beta \mid \mathbf{y})$, it is informative to write the integral in the form

$$p(\theta \mid \mathbf{y}) = \int_{-1}^{1} p(\theta \mid \beta, \mathbf{y}) p(\beta \mid \mathbf{y})\, d\beta, \tag{3.2.35}$$

and to say that $p(\beta \mid \mathbf{y})$ serves as a weight function acting on the various conditional posterior distributions $p(\theta \mid \beta, \mathbf{y})$. A study of the conditional distribution $p(\theta \mid \beta, \mathbf{y})$ for various β, together with the weight function $p(\beta \mid \mathbf{y})$ as was done in Sections 3.2.2 through 3.2.5, corresponds precisely to a study of the component distributions $p(\boldsymbol{\theta}_1 \mid \boldsymbol{\theta}_2, \mathbf{y})$ and $p(\boldsymbol{\theta}_2 \mid \mathbf{y})$ in (1.6.3) of the general approach.

Indeed, the attractiveness of this approach is further increased by the fact that if we let $p(\boldsymbol{\theta}_1)$ and $p(\boldsymbol{\theta}_2)$ be the prior distributions for $\boldsymbol{\theta}_1$ and $\boldsymbol{\theta}_2$ assumed independent, and let $l(\boldsymbol{\theta}_1, \boldsymbol{\theta}_2 \mid \mathbf{y})$ represent the joint likelihood, we may then write (1.6.3) in the form

$$p(\boldsymbol{\theta}_1 \mid \mathbf{y}) = c_1 \int_R p(\boldsymbol{\theta}_1 \mid \boldsymbol{\theta}_2, \mathbf{y}) p_u(\boldsymbol{\theta}_2 \mid \mathbf{y}) p(\boldsymbol{\theta}_2)\, d\boldsymbol{\theta}_2, \qquad (3.2.36)$$

where

$$p_u(\boldsymbol{\theta}_2 \mid \mathbf{y}) = c_2 \int p(\boldsymbol{\theta}_1) l(\boldsymbol{\theta}_1, \boldsymbol{\theta}_2 \mid \mathbf{y})\, d\boldsymbol{\theta}_1 \qquad (3.2.37)$$

is the likelihood integrated over a weight function $p(\boldsymbol{\theta}_1)$ and c_1 and c_2 are normalizing constants. The marginal distribution $p(\boldsymbol{\theta}_2 \mid \mathbf{y})$, which is proportional to the product,

$$p(\boldsymbol{\theta}_2 \mid \mathbf{y}) \propto p_u(\boldsymbol{\theta}_2 \mid \mathbf{y}) p(\boldsymbol{\theta}_2), \qquad (3.2.38)$$

is separated into the prior distribution of $\boldsymbol{\theta}_2$ and a part which is independent of this prior distribution. The function $p_u(\boldsymbol{\theta}_2 \mid \mathbf{y})$ can be regarded as a posterior distribution on the basis of a uniform reference prior for $\boldsymbol{\theta}_2$. It provides a basic reference posterior density which can be converted into a posterior distribution with an appropriate prior $p(\boldsymbol{\theta}_2)$ by multiplication. It is informative to consider the effect of varying $p(\boldsymbol{\theta}_2)$ to see how sensitive is the final result to changes in prior assumptions, and also to study $p_u(\boldsymbol{\theta}_2 \mid \mathbf{y})$ itself.

Thus, for the present example, the weight function $p(\beta \mid \mathbf{y})$ in (3.2.35) can be written

$$p(\beta \mid \mathbf{y}) \propto p(\beta) p_u(\beta \mid \mathbf{y}), \qquad (3.2.39)$$

with $p(\beta)$ the prior distribution of β, corresponding to $p(\boldsymbol{\theta}_2)$ in the general framework. Also

$$p_u(\beta \mid \mathbf{y}) \propto \int p_u(\beta, \theta \mid \mathbf{y})\, d\theta \qquad (3.2.40)$$

corresponds to $p_u(\boldsymbol{\theta}_2 \mid \mathbf{y})$ in the general formulation and can be thought of as the likelihood integrated over a weight function uniform in θ and $\log \sigma$.

3.3 APPROXIMATIONS TO THE POSTERIOR DISTRIBUTION $p(\theta \mid \beta, \mathbf{y})$†

In this section, we consider again the posterior distribution of θ for given β, $p(\theta \mid \beta, \mathbf{y})$ in (3.2.12), and discuss a method which can be used to approximate it.

3.3.1 Motivation for the Approximation

It turns out that over a wide range of values of β (say, from $\beta = -0.75$ to $\beta = 0.75$) the distribution $p(\theta \mid \beta, \mathbf{y})$ can be satisfactorily approximated by a t distribution.

† Much of the material in this section and Section 3.4 are taken from D. R. Lund's Ph.D. thesis (1967).

Specifically, we shall demonstrate that

$$p(\theta \mid \beta, \mathbf{y}) \mathrel{\dot{\propto}} [M(\hat{\theta}) + d(\theta - \hat{\theta})^2]^{-\frac{1}{2}n(1+\beta)}, \qquad -\infty < \theta < \infty, \tag{3.3.1}$$

where

$$M(\theta) = \sum_{u=1}^{n} |y_u - \theta|^{2/(1+\beta)},$$

d is an appropriately chosen constant, and $\hat{\theta}$ is the mode of $p(\theta \mid \beta, \mathbf{y})$. The symbol $\dot{\propto}$ means "approximately proportional to." To this degree of approximation, θ is distributed as the t distribution $t\{\hat{\theta}, M(\hat{\theta})/d[n(1 + \beta) - 1], n(1 + \beta) - 1\}$.

This type of approximation has to be justified in somewhat different ways, depending on whether β is negative or positive. In particular, for positive β the approximating process itself can be somewhat tedious. We shall, therefore, make clear why this kind of approximation is important.

Certainly, for many purposes of inference all that we really need is to be able to compute the posterior density function of θ and present its graph to the investigator. This computation is best made using the exact form

$$p(\theta \mid \beta, \mathbf{y}) \propto [M(\theta)]^{-\frac{1}{2}n(1+\beta)}, \qquad -\infty < \theta < \infty, \tag{3.3.2}$$

as given in (3.2.12).

The approximation (3.3.1) has two uses:

a) In some instances we may wish to calculate probability integrals. Although we can do this directly by integrating the density function (3.3.2) numerically, it is an advantage to be able to make use of the already tabled t integrals.

b) In problems discussed in Chapter 4, we shall often need to be able to integrate out θ from some posterior distribution involving other parameters (for example, the variance σ^2).

Specifically, we shall be involved with evaluating expressions of the form

$$\frac{\int_{-\infty}^{\infty} [M(\theta)]^{-c_1} d\theta}{\int_{-\infty}^{\infty} [M(\theta)]^{-c_2} d\theta}, \qquad c_1 > 0, \quad c_2 > 0. \tag{3.3.3}$$

Using (3.3.1), we have

$$\frac{\int_{-\infty}^{\infty} [M(\theta)]^{-c_1} d\theta}{\int_{-\infty}^{\infty} [M(\theta)]^{-c_2} d\theta} \doteq \frac{\int_{-\infty}^{\infty} [M(\hat{\theta}) + d(\theta - \hat{\theta})^2]^{-c_1} d\theta}{\int_{-\infty}^{\infty} [M(\hat{\theta}) + d(\theta - \hat{\theta})^2]^{-c_2} d\theta}$$

$$= [M(\hat{\theta})]^{c_2 - c_1} \frac{\Gamma(c_2)}{\Gamma(c_2 - \frac{1}{2})} \frac{\Gamma(c_1 - \frac{1}{2})}{\Gamma(c_1)}. \tag{3.3.4}$$

Note that, for this approximation, actual calculation of the value of d is not necessary since it cancels out on integration. So far as this more important application is concerned, what follows is mainly intended to show that an approximation of the form of (3.3.1) can be quite accurate and that, therefore, the approximate integration (3.3.4) can be used.

3.3.2 Quadratic Approximation to $M(\theta)$

Since $p(\theta \mid \beta, \mathbf{y})$ is a monotonic decreasing function of $M(\theta)$, we shall first conduct the discussion in terms of $M(\theta)$. Now, $M(\theta)$ is a convex function of θ and, for $-1 < \beta < 1$, possesses a unique minimum $\hat{\theta}$ [which is, of course, the mode of $p(\theta \mid \beta, \mathbf{y})$]. Further, for $\beta = 0$

$$M(\theta) = \Sigma\,(y_i - \bar{y})^2 + n(\theta - \bar{y})^2. \tag{3.3.5}$$

Now if for $\beta \neq 0$ we can use an approximation of the form

$$M(\theta) \doteq M(\hat{\theta}) + d(\theta - \hat{\theta})^2, \tag{3.3.6}$$

where d is some constant to be determined, then the corresponding approximate distribution of θ can be expressed in terms of the familiar Student's t form. In what follows, we first discuss the problem of determining the mode and then the question of finding the constant d.

Determination of the Mode $\hat{\theta}$

Two methods may be employed to obtain $\hat{\theta}$. The first of these makes use of the fact that $M'(\theta)$ is monotonically increasing in θ so that $M'(\theta) = 0$ has only one solution, $\hat{\theta}$. Thus, $M'(\theta) \gtrless 0$ implies that $\theta \gtrless \hat{\theta}$. By calculating $M'(\theta)$ at a point θ, we, therefore, know in which direction to proceed toward $\hat{\theta}$. We then increase or decrease the value of θ, whichever is required, in steps until $M'(\theta)$ changes sign. Repeating this process, but using steps which are halved after each sign change, we can obtain $\hat{\theta}$ to any desired degree of accuracy.

The second method employs Newton's iteration procedure for extracting roots of an equation. Starting with an initial guess value θ_0, we can write

$$M'(\theta) \doteq M'(\theta_0) + M''(\theta_0)(\theta - \theta_0) \tag{3.3.7}$$

provided the second derivative $M''(\theta)$ exists. Setting the right-hand side of (3.3.7) to zero, we obtain a new estimate, θ_1 of $\hat{\theta}$, such that

$$\theta_1 = \theta_0 - \frac{M'(\theta_0)}{M''(\theta_0)}. \tag{3.3.8}$$

Repeating the procedure with θ_1 in the role of θ_0, we obtain a second new estimate θ_2, and so on. We continue the process until $|\theta_{i+1} - \theta_i|/|\theta_i| < 0.001$, say, where θ_{i+1} is the value obtained in the $(i+1)$th iteration. Now

$$M'(\theta) = \frac{-2}{1+\beta} \sum_{u=1}^{n} |y_u - \theta|^{-2\beta/(1+\beta)} (y_u - \theta) \tag{3.3.9}$$

and

$$M''(\theta) = \frac{2(1-\beta)}{(1+\beta)^2} \sum_{u=1}^{n} |y_u - \theta|^{-2\beta/(1+\beta)}, \tag{3.3.10}$$

so that

$$\theta_{i+1} = \theta_i - \frac{M'(\theta_i)}{M''(\theta_i)}$$

$$= \left(\frac{-2\beta}{1-\beta}\right)\theta_i + \left(\frac{1+\beta}{1-\beta}\right)\frac{\sum_{u=1}^n |y_u - \theta_i|^{-2\beta/(1+\beta)}y_u}{\sum_{t=1}^n |y_t - \theta_i|^{-2\beta/(1+\beta)}}. \tag{3.3.11}$$

When the iteration process is terminated, we have, by setting $\hat{\theta} = \theta_{i+1} = \theta_i$,

$$\hat{\theta} = \sum_{u=1}^n a_u y_u, \tag{3.3.12}$$

where

$$a_u = \frac{|y_u - \hat{\theta}|^{-2\beta/(1+\beta)}}{\sum_{t=1}^n |y_t - \hat{\theta}|^{-2\beta/(1+\beta)}}.$$

Interpretation of Results

It is interesting to note that (3.3.12) says that modal value $\hat{\theta}$ is in the form of a weighted average of the observations. When $\beta \to 0$, all of the weights a_u tend to $1/n$ as they should, yielding $\hat{\theta} = \bar{y}$ in the limit. As β decreases below zero, observations corresponding to the residuals $y_u - \hat{\theta}$ with large absolute values tend to receive more weight. This is an intuitively pleasing result because for $\beta < 0$, the parent distributions have tails shorter than those of the Normal, and one would expect that extreme observations would thus provide more information about the location of the distribution. We have already seen that as $\beta \to -1$ in the limit, $\hat{\theta} = (y_{(1)} + y_{(n)})/2$ (the sample midpoint), so that all of the weight is divided equally between the largest and smallest observations, while all of the intermediate observations get zero weights. Indeed, the weights in (3.3.12) tend to this limit. The iteration procedure will, however, converge very slowly for β near -1 because the factor $(1 + \beta)/(1 - \beta)$ in (3.3.11) is small and θ_{i+1} is determined primarily by θ_i.

As β increases above zero, more weight is given to observations corresponding to the residuals $y_u - \hat{\theta}$ of smaller absolute value. This is again intuitively appealing because the parent then has longer tails than those of the Normal, and one would now expect that the "extreme" observations would provide relatively little information about the location of the parent distribution.

When $\beta < 0$, θ_{i+1} in (3.3.11) is a weighted average of two quantities θ_i and $\tilde{\theta}_i$, where

$$\tilde{\theta}_i = \frac{\sum_{u=1}^n |y_u - \theta_i|^{-2\beta/(1+\beta)}y_u}{\sum_{t=1}^n |y_t - \theta_i|^{-2\beta/(1+\beta)}}, \tag{3.3.13}$$

with weights equal to $-2\beta/(1 - \beta)$ and $(1 + \beta)/(1 - \beta)$, respectively. The quantity θ_i is the previous estimate of $\hat{\theta}$, and the quantity $\tilde{\theta}_i$ is itself a weighted average of the observations. Clearly θ_{i+1} lies between θ_i and $\tilde{\theta}_i$.

However, when $\beta > 0$, θ_{i+1}, instead of lying between θ_i and $\tilde{\theta}_i$, becomes a weighted difference of these two quantities and thus could be distinct from either one. Indeed, as

$\beta \to 1$, θ_{i+1} becomes a large multiple of $\theta_i - \tilde{\theta}_i$. The iteration procedure (3.3.11) may, therefore, be very erratic and unpredictable for $\beta > 0$, especially for β near 1. In practice, it may proceed toward the solution, but accelerate too fast and overshoot. It may then reverse direction and begin to diverge.

Thus, the second method should be used for $\beta < 0$ only and the first method used when $\beta > 0$. In practice, the second method often converges for $0 < \beta \leqslant 0.25$, and in our experience, the final estimate in such cases has always been the same from both methods. The first method, however, seems more reliable.

3.3.3 Approximation of $p(\theta \mid \mathbf{y})$

For $\beta \leqslant 0$, the second derivative $M''(\theta)$ exists for all θ and we can employ Taylor's theorem to write

$$M(\theta) \doteq M(\hat{\theta}) + \frac{M''(\hat{\theta})}{2}(\theta - \hat{\theta})^2. \qquad (3.3.14)$$

This expression is, of course, exact for $\beta = 0$. To this degree of approximation, the posterior distribution $p(\theta \mid \beta, \mathbf{y})$ is

$$p(\theta \mid \beta, \mathbf{y}) \propto [M(\hat{\theta}) + \tfrac{1}{2} M''(\hat{\theta})(\theta - \hat{\theta})^2]^{-\frac{1}{2}n(1+\beta)}. \qquad (3.3.15)$$

That is, for $\beta < 0$, θ is approximately distributed as

$$t\{\hat{\theta}, 2M(\hat{\theta})/M''(\hat{\theta})[n(1+\beta) - 1], n(1+\beta) - 1\}.$$

We have found in practice that (3.3.15) gives very close approximation in the range $-0.5 \leqslant \beta \leqslant 0$. To illustrate, Fig. 3.3.1 shows, for Darwin's data, the exact distribution (solid curve) and the corresponding approximation (broken curve) from (3.3.15) for $\beta = -0.25$, -0.50, and -0.75, respectively. In the first two instances, the agreement is very close. The approximation becomes somewhat less satisfactory for $\beta = -0.75$. This is to be expected, since the limiting distribution of θ as $\beta \to -1$ is not of the t form [see (3.2.15)].

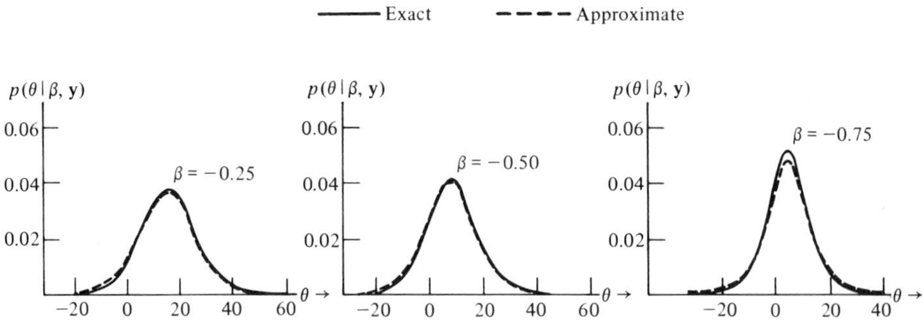

Fig. 3.3.1 Comparison of exact and approximate distributions of θ for several negative values of β: Darwin's data.

For $\beta > 0$, $M''(\hat{\theta})$ does not exist if $\hat{\theta} = y_u$, $(u = 1, \ldots, n)$, so that the approximation in (3.3.14) may break down. In practice, even in cases where $\hat{\theta}$ is not equal to any of the observations, good approximations are only rarely obtained through the use of this method. There is, however, reason to believe that the form (3.3.1) with d appropriately chosen will produce satisfactory results for $\beta > 0$. In our earlier work (1964a), this form was employed to obtain very close approximation to the moments of the variance ratio σ_2^2/σ_1^2 in the problem of comparing the variances of two exponential power distributions (this is discussed in Chapter 4). We now describe a method for determining d which does not depend on the existence of $M''(\hat{\theta})$.

This method employs the idea of least squares. For a set of suitably chosen values of θ, say, $\theta_1, \theta_2, \ldots, \theta_m$, we determine d so as to minimize the quantity

$$S = \sum_{j=1}^{m} \{M(\theta_j) - [M(\hat{\theta}) + d(\theta_j - \hat{\theta})^2]\}^2. \tag{3.3.16}$$

Solving $\partial S/\partial d = 0$, the solution is

$$d = \frac{\sum_{j=1}^{m} [M(\theta_j) - M(\hat{\theta})](\theta_j - \hat{\theta})^2}{\sum_{j=1}^{m} (\theta_j - \hat{\theta})^4}. \tag{3.3.17}$$

All that need be determined now is where to take the m points θ_j. If m is large, say forty or more, the exact location of the points should not be critical so long as they are spread out over the range for which the density of θ is appreciable. In particular, if m is *odd* and if the θ_j's are uniformly spaced at intervals of length c, with the center point at $\theta = \hat{\theta}$, then (3.3.17) can be simplified to

$$d = \frac{240\,\{[\sum_{j=1}^{m} M(\theta_j)(\theta_j - \hat{\theta})^2] - M(\hat{\theta})c^2(m-1)m(m+1)/12\}}{c^4(m-1)m(m+1)(3m^2-7)}. \tag{3.3.18}$$

To this degree of approximation, the posterior distribution of θ is

$$p(\theta \mid \beta, \mathbf{y}) \propto [M(\hat{\theta}) + d(\theta - \hat{\theta})^2]^{-\frac{1}{2}n(1+\beta)}, \tag{3.3.19}$$

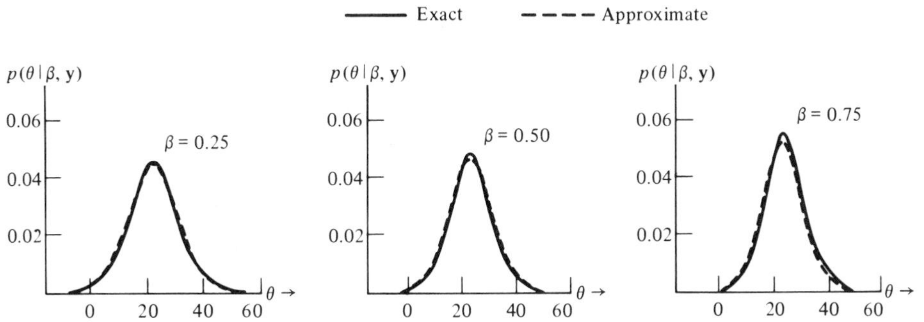

Fig. 3.3.2 Comparison of exact and approximate distributions of θ for several positive values of β: Darwin's data.

that is, a

$$t\{\hat{\theta}, M(\hat{\theta})/d[n(1+\beta)-1], n(1+\beta)-1\}$$

distribution.

For illustration, Fig. 3.3.2 shows, for $\beta = 0.25$, 0.50, and 0.75, a comparison of the actual distribution (solid curve) and the approximation (broken curve) using the present method for Darwin's data. In calculating the approximating form, the value of d was determined by using a set of 101 equally spaced points spread over four Normal theory standard deviations $[\Sigma(y_i - \bar{y})^2/n(n-1)]^{1/2}$ centered at $\theta = \hat{\theta}$. The agreement is remarkably close for the cases $\beta = 0.25$ and $\beta = 0.50$. The result for the case $\beta = 0.75$ is less satisfactory. In general, one would not expect the distribution of θ to be well approximated by that of a t form when β is near unity. Indeed, when $\beta = 1$ and n an even integer, the density will be constant between the middle two observations. Nevertheless, for Darwin's data, the approximation for $\beta = 0.75$ still seems to be close enough for practical purposes.

The process described above is admittedly somewhat tedious. However, as we have pointed out, the object of this section is not primarily to provide means of calculating d. Rather, it is to show that an approximation of the *form* $M(\theta) \doteq M(\hat{\theta}) + d(\theta - \hat{\theta})^2$ for *some* d can be used, and so to provide an easy means of integrating out θ from functions containing $M(\theta)$. In our subsequent applications d will cancel and need not be computed.

3.4 GENERALIZATION TO THE LINEAR MODEL

The analysis in Section 3.2 can be readily extended to the general linear model

$$\mathbf{y} = \mathbf{X}\boldsymbol{\theta} + \boldsymbol{\varepsilon}, \tag{3.4.1}$$

where $\mathbf{y}$ is a $n \times 1$ vector of observations, $\mathbf{X}$ a $n \times k$ full rank matrix of fixed elements, $\boldsymbol{\theta}$ a $k \times 1$ vector of unknown regression coefficients, and $\boldsymbol{\varepsilon}$ a $n \times 1$ vector of random errors. When this model was considered in Section 2.7, it was assumed that $\boldsymbol{\varepsilon}$ had the multivariate spherically Normal distribution $N_n(\mathbf{0}, \mathbf{I}\sigma^2)$. We now relax the assumption of Normality and suppose that the elements of $\boldsymbol{\varepsilon}$ are independent, each having the exponential power distribution

$$p(\varepsilon \mid \sigma, \beta) = \omega(\beta)\sigma^{-1} \exp\left[-c(\beta)\left|\frac{\varepsilon}{\sigma}\right|^{2/(1+\beta)}\right], \qquad -\infty < \varepsilon < \infty, \tag{3.4.2}$$

where $\omega(\beta)$ and $c(\beta)$ are given in (3.2.5). Our objective will be to study the effect of departures from Normality of this kind on inferences about the regression coefficients $\boldsymbol{\theta}$.

The likelihood function of $(\boldsymbol{\theta}, \sigma, \beta)$ is

$$l(\boldsymbol{\theta}, \sigma, \beta \mid \mathbf{y}) \propto [\omega(\beta)]^n \sigma^{-n} \exp\left[-c(\beta) \sum_{i=1}^{n} \left|\frac{y_i - \mathbf{x}'_{(i)}\boldsymbol{\theta}}{\sigma}\right|^{2/(1+\beta)}\right], \tag{3.4.3}$$

where $\mathbf{x}'_{(i)}$ is the ith row of $\mathbf{X}$. Following arguments similar to those leading to (3.2.19) and (3.2.20), suppose *a priori* that

$$p(\boldsymbol{\theta}, \sigma, \beta) = p(\beta)p(\boldsymbol{\theta}, \sigma) \tag{3.4.4}$$

with

$$p(\boldsymbol{\theta}, \sigma) \propto \sigma^{-1}$$

and $p(\beta)$ temporarily unspecified. Combining (3.4.3) and (3.4.4) and integrating out σ, we obtain the joint posterior distribution of $(\boldsymbol{\theta}, \beta)$, which can be written as the product

$$p(\boldsymbol{\theta}, \beta \mid \mathbf{y}) = p(\boldsymbol{\theta} \mid \beta, \mathbf{y})p(\beta \mid \mathbf{y}). \tag{3.4.5}$$

The conditional distribution $p(\boldsymbol{\theta} \mid \beta, \mathbf{y})$

In (3.4.5), the conditional posterior distribution $p(\boldsymbol{\theta} \mid \beta, \mathbf{y})$ is

$$p(\boldsymbol{\theta} \mid \beta, \mathbf{y}) = J(\beta)^{-1}[M(\boldsymbol{\theta})]^{-\frac{1}{2}n(1+\beta)}, \qquad -\infty < \theta_j < \infty, \quad j = 1, \ldots, k, \tag{3.4.6}$$

where

$$M(\boldsymbol{\theta}) = \sum_{i=1}^{n} |y_i - \mathbf{x}'_{(i)}\boldsymbol{\theta}|^{2/(1+\beta)},$$

$$J(\beta) = \int_R [M(\boldsymbol{\theta})]^{-\frac{1}{2}n(1+\beta)}\, d\boldsymbol{\theta}, \tag{3.4.7}$$

and $R: (-\infty < \theta_j < \infty, j = 1, \ldots, k)$. If $\boldsymbol{\theta}$ consists of a single element and $\mathbf{X}$ is a column of ones, the distribution in (3.4.6) reduces to that in (3.2.12) and much of the analysis which follows parallels that for the single mean.

In general, by studying the distribution $p(\boldsymbol{\theta} \mid \beta, \mathbf{y})$ as a function of β, we can determine how sensitive inferences about $\boldsymbol{\theta}$ are, to departures from Normality of the type postulated. In particular, when $\beta = 0$ (exact Normality),

$$M(\boldsymbol{\theta}) = (\mathbf{y} - \mathbf{X}\boldsymbol{\theta})'\,(\mathbf{y} - \mathbf{X}\boldsymbol{\theta}), \tag{3.4.8}$$

so that the distribution in (3.4.6) reduces to the $t_k[\hat{\boldsymbol{\theta}}, s^2\,(\mathbf{X}'\mathbf{X})^{-1}, \nu]$ distribution obtained in (2.7.20). For other values of β, it does not seem possible to express the distribution in terms of simple functions of the observations except when $\beta \to -1$ and $\mathbf{X}$ assumes special forms [see Lund (1967)], However, when the number of parameters in $\boldsymbol{\theta}$ is small, say $k = 2$ or 3, contours of $M(\boldsymbol{\theta})$ can be plotted. By investigating the changes in the location and shape of the contours for different values of β, one obtains a good appreciation of the effect of changes in β on inferences about $\boldsymbol{\theta}$. An illustrative example will be given later in this section.

For $-0.5 < \beta < 0.5$, Lund (1967) has demonstrated that the distribution (3.4.6) can be satisfactorily approximated by a multivariate t distribution.

The Marginal Distributions

The posterior distribution $p(\beta \mid \mathbf{y})$ in (3.4.5) can be written

$$p(\beta \mid \mathbf{y}) \propto p_u(\beta \mid \mathbf{y})p(\beta), \qquad -1 < \beta \leqslant 1, \tag{3.4.9}$$

where

$$p_u(\beta \mid \mathbf{y}) \propto \Gamma[1 + \tfrac{1}{2}n(1 + \beta)]\{\Gamma[1 + \tfrac{1}{2}(1 + \beta)]\}^{-n} J(\beta) \tag{3.4.10}$$

is the posterior distribution of β for a uniform reference prior in the range $-1 < \beta \leqslant 1$. From (3.4.5), the posterior distribution of $\boldsymbol{\theta}$ is

$$p(\boldsymbol{\theta} \mid \mathbf{y}) = \int_{-1}^{1} p(\boldsymbol{\theta} \mid \beta, \mathbf{y})p(\beta \mid \mathbf{y})\, d\beta$$

$$\propto \int_{-1}^{1} p(\boldsymbol{\theta} \mid \beta, \mathbf{y})p_u(\beta \mid \mathbf{y})p(\beta)\, d\beta, \qquad -\infty < \theta_j < \infty, \quad j = 1, \ldots, k. \tag{3.4.11}$$

As in the case of a single parameter θ, we are here averaging the conditional distributions $p(\boldsymbol{\theta} \mid \beta, \mathbf{y})$ for various β by a weight function $p(\beta \mid \mathbf{y})$ which is the product of $p_u(\beta \mid \mathbf{y})$ and $p(\beta)$.

3.4.1 An Illustrative Example

The data of Table 3.4.1 refers to an experiment relating the rate K of a chemical reaction to the absolute temperature T at which the experiment was conducted. The dependence of K on T is expected to be represented by the Arrhenius law

$$\log K = \log A - \frac{E}{R}\frac{1}{T}, \tag{3.4.12}$$

where R is the known gas constant; A and E are constants to be estimated. The twenty experimental runs were performed in random order, and the reasonable assumption was made that over the ranges of temperature employed $\log K$ had constant variance. We, therefore, consider the simple linear model

$$y_i = \theta_1 + \theta_2 x_i + \varepsilon_i, \qquad i = 1, 2, \ldots, 20, \tag{3.4.13}$$

where

$$y_i = \log K_i, \qquad \theta_1 = \log A - \frac{E}{R}\overline{T^{-1}}, \qquad \theta_2 = \frac{E}{50{,}000R},$$

$$x_i = (T_i^{-1} - \overline{T^{-1}}) \times 50{,}000, \qquad \text{and} \qquad \overline{T^{-1}} = \frac{1}{20}\sum_{i=1}^{20} T_i^{-1}.$$

The "Arrhenius plot" of y against x is shown in Fig. 3.4.1. In terms of the linear model in (3.4.1), the (20×2) matrix $\mathbf{X}$ has two columns, the first is a column of

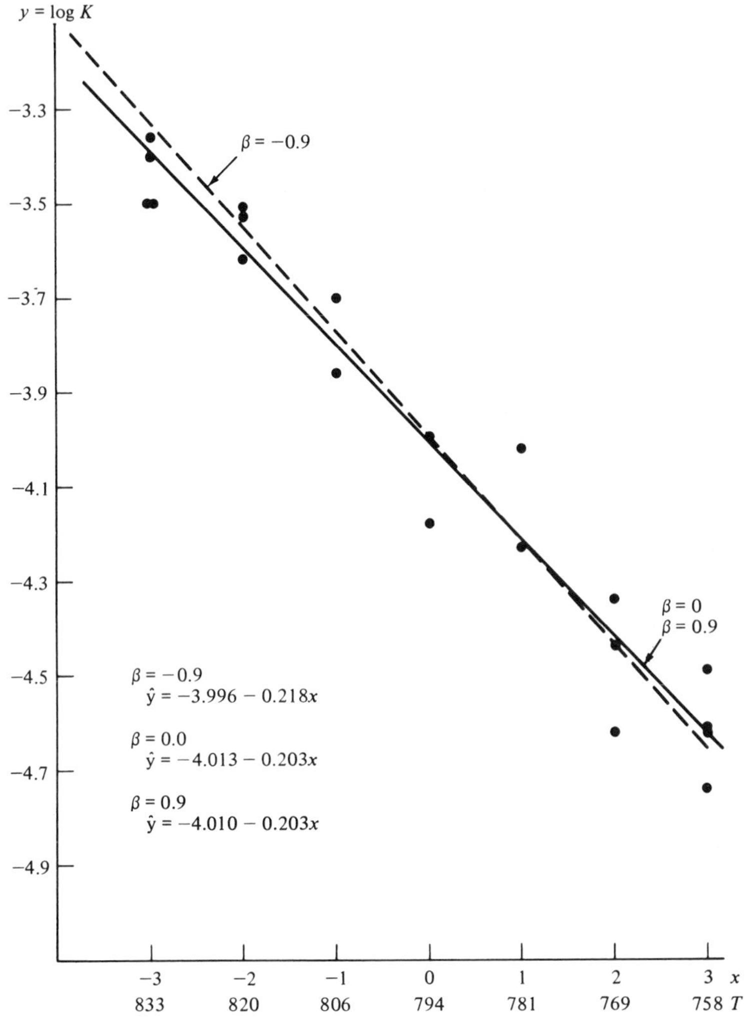

Fig. 3.4.1 Plot of reaction rate data with fitted straight lines for $\beta = -0.9, 0.0, 0.9$.

ones and the elements of the second are the values of x as defined above. The two columns are, of course, orthogonal. From (3.4.6), the posterior distribution of (θ_1, θ_2) for fixed β is

$$p(\theta_1, \theta_2 \mid \beta, \mathbf{y}) \propto [M(\boldsymbol{\theta})]^{-10(1+\beta)}, \qquad -\infty < \theta_1 < \infty, \quad -\infty < \theta_2 < \infty, \tag{3.4.14}$$

where

$$M(\boldsymbol{\theta}) = \sum_{i=1}^{20} |y_i - \theta_1 - x_i\theta_2|^{2/(1+\beta)}.$$

Table 3.4.1

Reaction rate data

T	x	$y = \log K$
833	-3	-3.50, -3.40, -3.36, -3.50
820	-2	-3.51, -3.62, -3.53
806	-1	-3.86, -3.70
794	0	-3.99, -4.18
781	1	-4.02, -4.23
769	2	-4.44, -4.62, -4.34
758	3	-4.74, -4.49, -4.61, -4.62

Table 3.4.2 gives the mode of the distribution, $\hat{\boldsymbol{\theta}} = (\hat{\theta}_1, \hat{\theta}_2)$, for $\beta = -0.9(0.3)0.9$. The fitted straight lines which result from using $\hat{\boldsymbol{\theta}}$ associated with $\beta = -0.9$, 0.0, and 0.9 are shown in Fig. 3.4.1.

Figure 3.4.2 shows, for each of the seven values of β considered, three contours A, B, and C of the corresponding posterior distribution of (θ_1, θ_2) together with the mode $(\hat{\theta}_1, \hat{\theta}_2)$. The density levels of the three contours are:

$$A:\quad p(\theta_1, \theta_2 \mid \beta, \mathbf{y}) = 0.5\, p(\hat{\theta}_1, \hat{\theta}_2 \mid \beta, \mathbf{y}),$$

$$B:\quad p(\theta_1, \theta_2 \mid \beta, \mathbf{y}) = 0.25\, p(\hat{\theta}_1, \hat{\theta}_2 \mid \beta, \mathbf{y}),$$

$$C:\quad p(\theta_1, \theta_2 \mid \beta, \mathbf{y}) = 0.05\, p(\hat{\theta}_1, \hat{\theta}_2 \mid \beta, \mathbf{y}).$$

Table 3.4.2

Values of the mode of $\boldsymbol{\theta}$ as a function of β: reaction rate data

β	-0.9	-0.6	-0.3	0.0	0.3	0.6	0.9
$\hat{\theta}_1$	-3.996	-4.017	-4.016	-4.013	-4.011	-4.010	-4.010
$\hat{\theta}_2$	-0.218	-0.205	-0.204	-0.203	-0.203	-0.203	-0.203

Using the bivariate Normal approximation, the contours would very roughly be the boundaries of the 50, 75, and 95 percent H.P.D. regions, respectively.

The contours, together with the modal values in Table 3.4.2, show how inferences about (θ_1, θ_2) are affected by changes in β in the range considered. For this example, both the location and the shape of the contours change appreciably as β decreases below -0.3. In contrast, the effect of β is relatively small in the range from -0.3 to 0.9. Since the two columns in $\mathbf{X}$ are orthogonal, the parameters θ_1 and θ_2 are *a posteriori* uncorrelated when $\beta = 0$. Figure 3.4.2 shows that this "orthogonality" property is gradually lost as β moves away from zero. In particular, the parameters θ_1 and θ_2 become increasingly negatively correlated and the dispersion tends to increase as β decreases from -0.3, namely, as the parent distribution becomes more and more platykurtic.

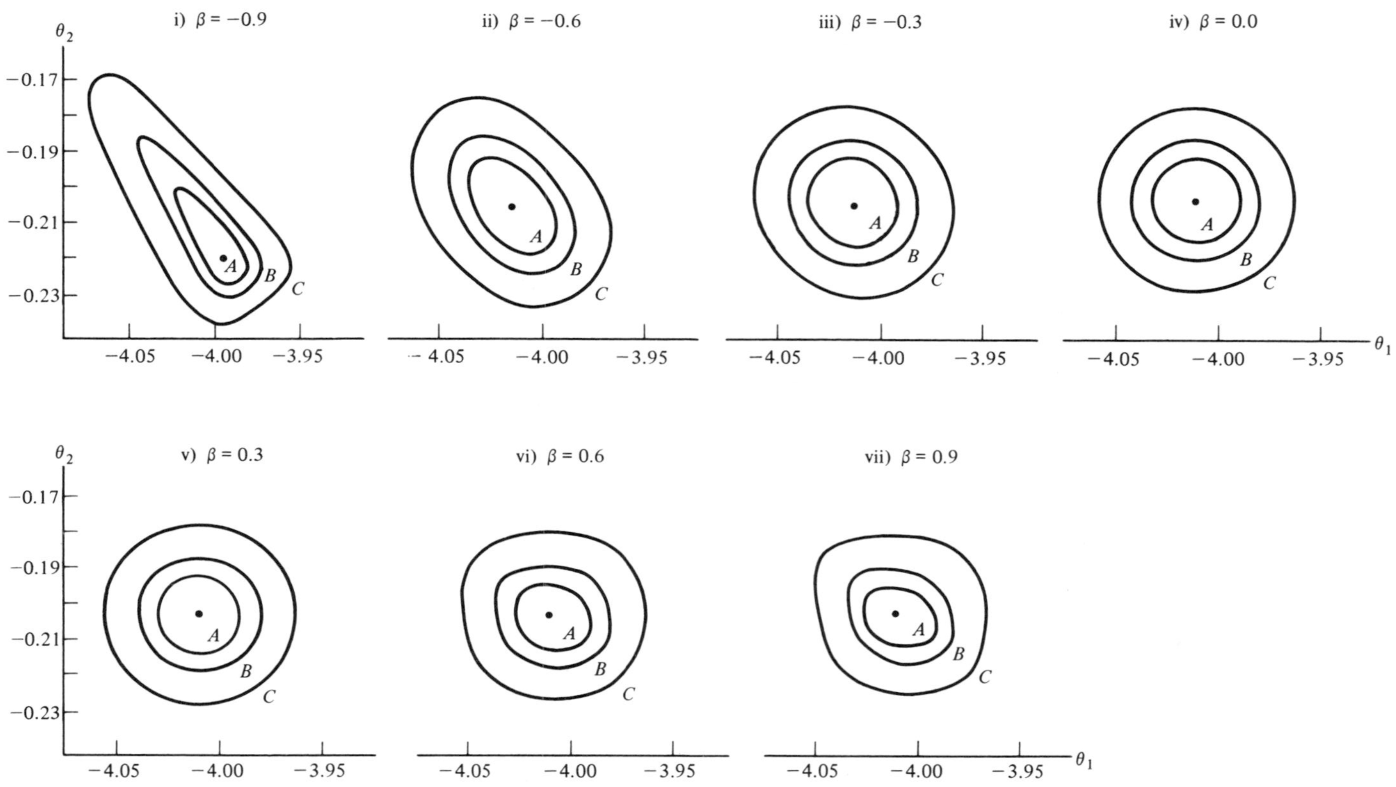

Fig. 3.4.2(i)–(vii) Contours of the posterior distribution of (θ_1, θ_2) for various values of β: reaction rate data (the density levels are such that $p(\theta_1, \theta_2 \mid \mathbf{y})/p(\hat{\theta}_1, \hat{\theta}_2 \mid \mathbf{y}) = (0.5, 0.25, 0.05)$ for (A, B, C), respectively).

The posterior distribution of either parameter, conditional on β, can be obtained from (3.4.14) by integrating out the other,

$$p(\theta_j \mid \beta, \mathbf{y}) = \int_{-\infty}^{\infty} p(\theta_1, \theta_2 \mid \beta, \mathbf{y})\, d\theta_l,$$

$$-\infty < \theta_j < \infty, \quad j = 1, 2, \quad l = 1, 2, \quad \text{and} \quad j \neq l. \qquad (3.4.15)$$

Figures 3.4.3 and 3.4.4 show respectively the marginal distributions of θ_1 and of θ_2 for the seven values of β considered. As expected, both the location and the spread of the marginal distributions are quite sensitive to changes in β in the range $-0.9 \leqslant \beta \leqslant -0.3$. Also, the sensitivity seems to be somewhat greater for θ_2 than for θ_1.

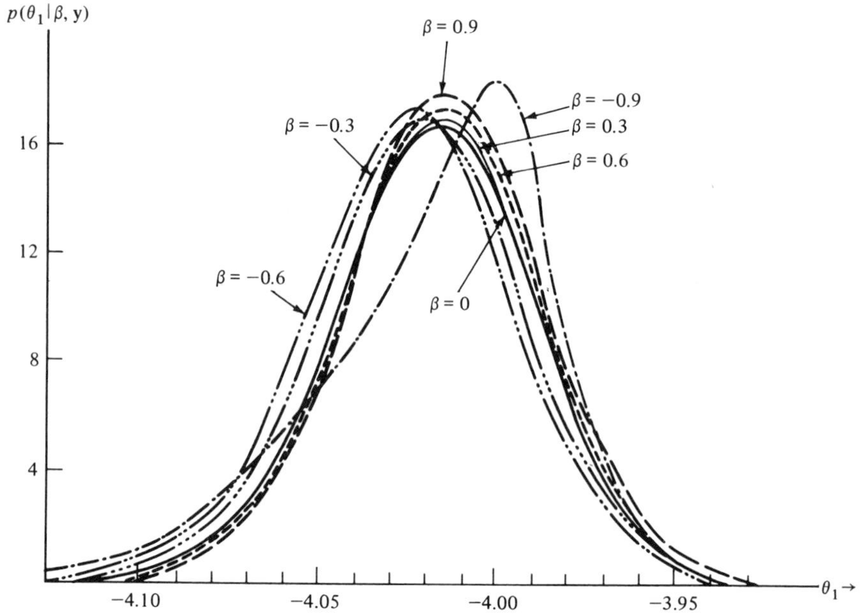

Fig. 3.4.3 Posterior distribution of θ_1 for various choices of β: reaction rate data.

Thus, so far as inferences about θ_1 and θ_2 are concerned, the assumption of Normality would not, for the present example, lead us much astray if the true parent were leptokurtic. On the other hand, such an assumption could lead to rather erroneous conclusions if the true parent distribution were close to the rectangular.

Proceeding as before, overall inferences about θ_1 and θ_2 can be made by averaging the posterior distributions in (3.4.14) and (3.4.15) over the posterior

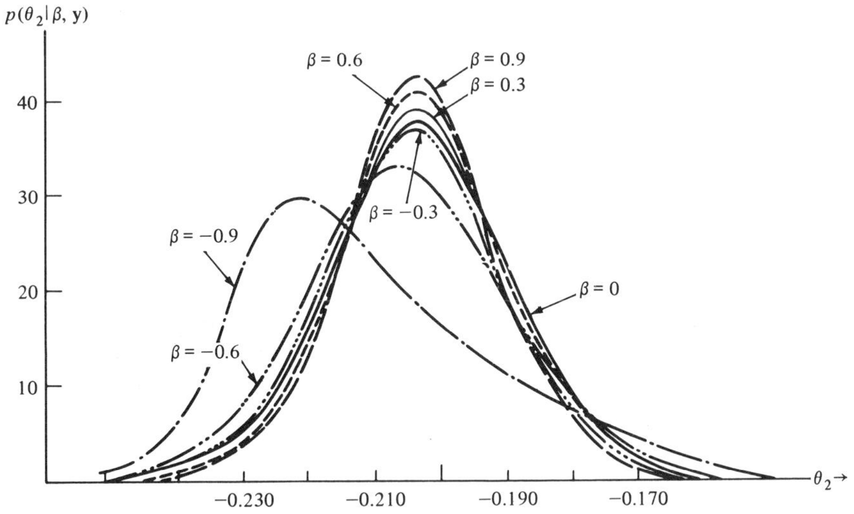

Fig. 3.4.4 Posterior distribution of θ_2 for various choices of β: reaction rate data.

distribution of β. From (3.4.11), the joint distribution of (θ_1, θ_2) is

$$p(\theta_1, \theta_2 \mid \mathbf{y}) \propto \int_{-1}^{1} p(\theta_1, \theta_2 \mid \beta, \mathbf{y}) p_u(\beta \mid \mathbf{y}) p(\beta)\, d\beta, \tag{3.4.16}$$

and from (3.4.15) the marginal distributions are

$$p(\theta_j \mid \mathbf{y}) \propto \int_{=1}^{1} p(\theta_j \mid \beta, \mathbf{y})\, p_u(\beta \mid \mathbf{y})\, p(\beta)\, d\beta,$$

$$-\infty < \theta_j < \infty, \quad j = 1, 2, \tag{3.4.17}$$

where $p_u(\beta \mid \mathbf{y})$ is the posterior distribution of β on the basis of a uniform *reference* prior for β, and $p(\beta)$ is the appropriate prior distribution.

The distribution

$$p_u(\beta \mid \mathbf{y}) \propto \Gamma[1 + 10(1 + \beta)]\{\Gamma[1 + \tfrac{1}{2}(1 + \beta)]\}^{-20} J(\beta), \qquad -1 < \beta \leqslant 1, \tag{3.4.18}$$

where

$$J(\beta) = \int_{-\infty}^{\infty}\int_{-\infty}^{\infty} [\Sigma\, |y_i - \theta_1 - \theta_2 x_i|^{2/(1+\beta)}]^{-10(1+\beta)}\, d\theta_1\, d\theta_2$$

is shown in Fig. 3.4.5 ($a = 1$). We see that the mode is near $\beta = 0$, indicating that the Normal distribution is a plausible parent. However, the evidence is not very strong. There is a probability of about 17% that $\beta \geqslant 0.6$ and a probability of about 7% that $\beta \leqslant -0.6$. Thus, parents which are quite different from the Normal cannot be ruled out from the evidence provided by $p_u(\beta \mid \mathbf{y})$ alone.

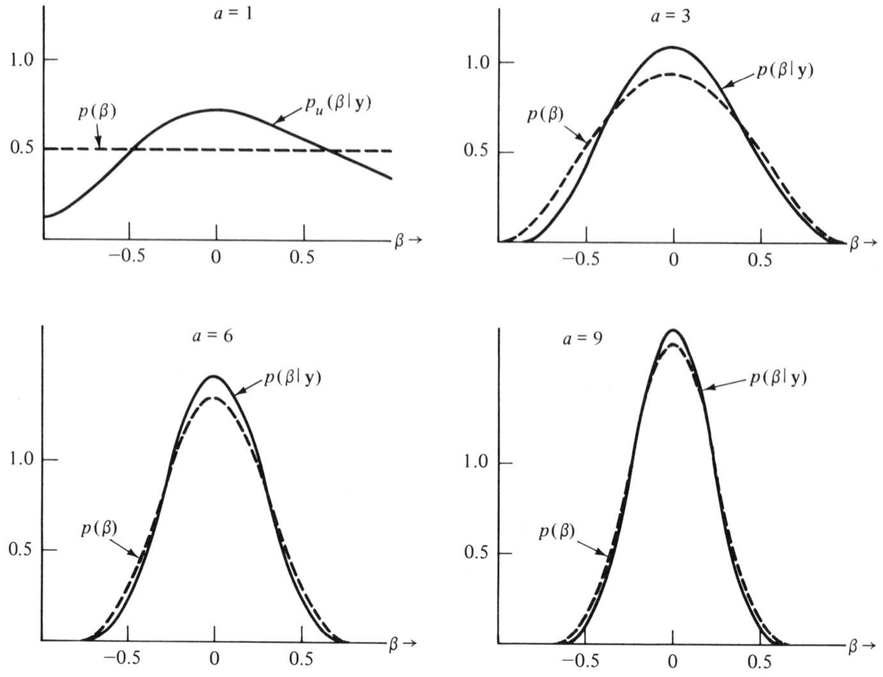

Fig. 3.4.5 Prior and posterior distributions of β for various choices of the parameter "a": reaction rate data.

In experiments of this kind a tendency toward Normality is to be expected. We represent this by supposing as before that

$$p(\beta) \propto (1 - \beta^2)^{a-1}, \qquad -1 < \beta \leqslant 1, \tag{3.4.19}$$

where the parameter "a" can be adjusted to represent a greater or lesser central limit effect. The result of varying "a" is shown in Fig. 3.4.5, where $p(\beta)$ and the corresponding posterior distribution of β

$$p(\beta \mid \mathbf{y}) \propto p_u(\beta \mid \mathbf{y})p(\beta), \qquad -1 < \beta \leqslant 1. \tag{3.4.20}$$

are plotted for $a = 1$, 3, 6, and 9, respectively. For $a \geqslant 6$, the distribution practically reproduces the prior.

For any specific value of "a", overall inference about (θ_1, θ_2) can now be made by averaging the conditional distributions $p(\theta_j \mid \beta, \mathbf{y})$ over the appropriate $p(\beta \mid \mathbf{y})$. We have computed the marginal distributions of θ_1 and θ_2 in (3.4.17) for $a = 1$, 3, 6, and 9, and specimens of the densities are shown in Tables 3.4.3, and 3.4.4, respectively. Also shown in the same tables are densities of the conditional distributions $p(\theta_j \mid \beta = 0, \mathbf{y})$ which correspond to $a \to \infty$, that is, to the assumption of exact Normality. These distributions are very little affected by the choice

Table 3.4.3

Posterior distribution of θ_1 for various choices of "a": reaction rate data

	$p(\theta_1 \mid \mathbf{y})$				
θ_1	$a = 1$	$a = 3$	$a = 6$	$a = 9$	$a \to \infty$
−4.090	0.19	0.19	0.19	0.19	0.19
−4.078	0.55	0.56	0.57	0.57	0.56
−4.066	1.51	1.55	1.56	1.56	1.54
−4.054	3.70	3.78	3.79	3.79	3.76
−4.042	7.66	7.81	7.82	7.81	7.73
−4.030	12.88	13.02	13.00	12.97	12.82
−4.018	16.72	16.71	16.65	16.61	16.49
−4.012	17.10	17.01	16.95	16.94	16.88
−4.006	16.30	16.13	16.10	16.10	16.12
−3.994	12.12	11.87	11.85	11.87	11.98
−3.982	6.80	6.74	6.78	6.81	6.95
−3.970	3.10	3.12	3.16	3.18	3.27
−3.958	1.23	1.25	1.27	1.28	1.31
−3.946	0.44	0.45	0.46	0.46	0.47
−3.934	0.15	0.15	0.15	0.15	0.16

Table 3.4.4

Posterior distribution of θ_2 for various choices of "a": reaction rate data

	$p(\theta_2 \mid \mathbf{y})$				
θ_2	$a = 1$	$a = 3$	$a = 6$	$a = 9$	$a \to \infty$
−0.2420	0.18	0.16	0.16	0.16	0.15
−0.2364	0.57	0.53	0.51	0.50	0.49
−0.2308	1.74	1.60	1.56	1.54	1.51
−0.2252	4.77	4.43	4.34	4.31	4.25
−0.2196	11.03	10.62	10.50	10.46	10.36
−0.2140	21.21	21.02	20.92	20.87	20.72
−0.2084	32.61	32.61	32.54	32.49	32.29
−0.2028	37.68	37.83	37.85	37.84	37.75
−0 1972	31.57	32.11	32.33	32.42	32.64
−0.1916	20.03	20.53	20.75	20.85	21.15
−0.1860	10.25	10.41	10.48	10.51	10.66
−0.1804	4.45	4.40	4.38	4.38	4.40
−0.1748	1.72	1.63	1.60	1.59	1.57
−0.1692	0.61	0.55	0.53	0.52	0.51
−0.1636	0.20	0.18	0.17	0.16	0.16

of "a". Even for the extreme case $a = 1$ where $p(\beta \mid \mathbf{y}) = p_u(\beta \mid \mathbf{y})$, the respective marginal distributions of θ_j are quite close to the limiting $p(\theta_j \mid \beta = 0, \mathbf{y})$ distributions.

For this data, then, where weak evidence from the sample would support a supposition of approximate Normality, analysis with wider assumptions confirms the Normal theory analysis.

3.5 FURTHER EXTENSION TO NONLINEAR MODELS

The object of much experimentation is to study the relationship between a response or output variable y subject to error and input variables $\xi_1, \xi_2, \ldots, \xi_p$ the levels of which are suitably chosen by the experimenter. Suppose the model can be written

$$y = \eta(\boldsymbol{\xi} \mid \boldsymbol{\theta}) + \varepsilon, \tag{3.5.1}$$

where $\boldsymbol{\xi}' = (\xi_1, \ldots, \xi_p)$, ε is the error term with zero expectation and $\boldsymbol{\theta}' = (\theta_1, \ldots, \theta_k)$ is the vector of k unknown parameters to be estimated. A model is linear in the parameters $\boldsymbol{\theta}$ if

$$\eta(\boldsymbol{\xi} \mid \boldsymbol{\theta}) = x_1\theta_1 + \cdots + x_k\theta_k,$$

where $x_1, \ldots, x_k$ are functions of $\xi_1, \ldots, \xi_p$, Otherwise, the model is called "nonlinear" (in the parameters $\boldsymbol{\theta}$). Linear models were considered in the last section. We now show how the approach is extended to include nonlinear models.

In n experiments, let y_i be the ith observation, $\boldsymbol{\xi}_i' = (\xi_{i1}, \ldots, \xi_{ip})$ the setting of the p input variables, and ε_i the corresponding error term. We then have

$$y_i = \eta(\boldsymbol{\xi}_i \mid \boldsymbol{\theta}) + \varepsilon_i, \qquad i = 1, \ldots, n, \tag{3.5.2}$$

where $E(\varepsilon_i) = 0$ or equivalently $E(y_i) = \eta(\boldsymbol{\xi}_i \mid \boldsymbol{\theta})$. We shall suppose that the errors $(\varepsilon_1, \ldots, \varepsilon_n)$ are independent, each having the exponential power distribution in (3.4.2). Thus, irrespective of whether the model (3.5.2) is linear or nonlinear in the parameters $\boldsymbol{\theta}$, the likelihood functions of $(\boldsymbol{\theta}, \sigma, \beta)$ is

$$l(\boldsymbol{\theta}, \sigma, \beta \mid \mathbf{y}) \propto [\omega(\beta)]^n \sigma^{-n} \exp\left[-c(\beta) \sum_{i=1}^{n} \left|\frac{\varepsilon_i}{\sigma}\right|^{2/(1+\beta)}\right], \tag{3.5.3}$$

where

$$\varepsilon_i = y_i - E(y_i) = y_i - \eta(\boldsymbol{\xi}_i \mid \boldsymbol{\theta}),$$

and $\omega(\beta)$ and $c(\beta)$ are given in (3.2.5). If we now suppose that $\boldsymbol{\theta}$ and $\log \sigma$ are approximately independent and locally uniform *a priori*, then for given β the joint posterior distribution of $\boldsymbol{\theta}$ and σ is

$$p(\boldsymbol{\theta}, \sigma \mid \beta, \mathbf{y}) \propto \sigma^{-(n+1)} \exp\left[-c(\beta) \sum_{i=1}^{n} \left|\frac{\varepsilon_i}{\sigma}\right|^{2/(1+\beta)}\right], \qquad \sigma > 0, \quad -\infty < \theta_j < \infty,$$

$$j = 1, \ldots, k. \tag{3.5.4}$$

On integrating out σ we obtain the joint distribution of $\boldsymbol{\theta}$ in the simple form

$$p(\boldsymbol{\theta} \mid \beta, \mathbf{y}) = J(\beta)^{-1}[M(\boldsymbol{\theta})]^{-\frac{1}{2}n(1+\beta)}, \qquad -\infty < \theta_j < \infty, \quad j = 1, \ldots, k, \qquad (3.5.5)$$

where

$$M(\boldsymbol{\theta}) = \sum_{i=1}^{n} |y_i - \eta(\boldsymbol{\xi}_i \mid \boldsymbol{\theta})|^{2/(1+\beta)}$$

and $J(\beta)$ is the normalizing integral

$$J(\beta) = \int_R [M(\boldsymbol{\theta})]^{-\frac{1}{2}n(1+\beta)}\, d\boldsymbol{\theta},$$

with

$$R: (-\infty < \theta_j < \infty, \ j = 1, \ldots, k).$$

Also if β is regarded as a random variable independent of $(\boldsymbol{\theta}, \sigma)$ *a priori*, the posterior distribution of β for a given prior $p(\beta)$ will be

$$p(\beta \mid \mathbf{y}) \propto p_u(\beta \mid \mathbf{y}) p(\beta), \qquad -1 < \beta \leqslant 1,$$

where

$$p_u(\beta \mid \mathbf{y}) \propto \frac{\Gamma[1 + \frac{1}{2}n(1+\beta)]}{\{\Gamma[1 + \frac{1}{2}(1+\beta)]\}^n} J(\beta) \qquad (3.5.6)$$

is the posterior distribution of β on the basis of a uniform reference prior over the range $(-1, 1)$. Finally, when β is not known the marginal posterior distribution of $\boldsymbol{\theta}$ is

$$p(\boldsymbol{\theta} \mid \mathbf{y}) = \int_{-1}^{1} p(\boldsymbol{\theta} \mid \beta, \mathbf{y}) p(\beta \mid \mathbf{y})\, d\beta$$

$$\propto \int_{-1}^{1} p(\boldsymbol{\theta} \mid \beta, \mathbf{y}) p_u(\beta \mid \mathbf{y}) p(\beta)\, d\beta, \qquad -\infty < \theta_j < \infty, \quad j = 1, \ldots, k. \qquad (3.5.7)$$

It will be noted that in the above we have not needed to assume either that

a) the model was linear in $\boldsymbol{\theta}$; or that

b) the error distribution was Normal.

3.5.1 An Illustrative Example

The following simple example, which serves to show the generality of the approach, concerns an experiment in a continuous chemical reactor. The object was to estimate the rate constant K of a particular chemical reaction. The quantity y is the observed concentration of a reactant A in the outlet from the reactor, expressed as a fraction of its concentration at the inlet. The inlet concentration was held fixed throughout the investigation but the flow rate was varied from one experimental run to another. If the reaction is first order with respect to A, it is possible to show that the outlet concentration is given by $\eta = 1/(1 + \theta\xi)$, where θ is a known multiple of the rate constant K and ξ is the ratio of the inlet concentration to the flow rate. Our problem then is to make inferences about θ, given data in the form of n pairs of values of y and ξ, and given the model

$$E(y_i) = \eta(\xi_i \mid \theta) = \frac{1}{1 + \theta\xi_i}, \qquad i = 1, \ldots, n. \tag{3.5.8}$$

In eight runs the following results were obtained:

Table 3.5.1

Chemical reactor data

ξ	50	75	100	125	150	200	250	300
y	0.78	0.69	0.54	0.57	0.54	0.37	0.40	0.32

We shall calculate, to a sufficient approximation, posterior densities of θ, $p(\theta|\beta, \mathbf{y})$, for $\beta \to -1$ and $\beta = -0.5, 0.0, 0.5, 1.0$. The corresponding five parent distributions are, then, the uniform, the fourth-power distribution, the Normal distribution, the 4/3 power distribution, and the double exponential distribution. For clarity we shall set out the calculations in some detail.

Table 3.5.2 shows values of the residuals $r_i = y_i - \eta(\xi_i \mid \theta)$ for a suitable range of values of θ at intervals of 0.025. From this table we calculate for each value of β the quantities:

$$r_L^{-8}, \qquad (\Sigma r_i^4)^{-2}, \qquad (\Sigma r_i^2)^{-4}, \qquad (\Sigma\, |r_i|^{4/3})^{-6}, \qquad (\Sigma \mid r_i \mid)^{-8}$$

(where r_L is the deviate largest in absolute magnitude).

These five quantities provide unstandardized ordinates of the conditional posterior distributions of θ, $p(\theta \mid \beta, \mathbf{y})$, for $\beta \to -1.0$ and $\beta = -0.5, 0.0, 0.5, 1.0$. For any given value of β, the quantity $\hat{J}(\beta) =$ (sum of these unstandardized ordinates $\times$ interval in θ) provides, for our present purpose,† an approximation sufficiently close to the normalizing integral $J(\beta)$. On dividing the unstandardized ordinates through by $\hat{J}(\beta)$ we obtain the approximate ordinates of the posterior distribution of θ for the five selected values of β. These calculations are shown in Table 3.5.3 (on page 192).

Using these values the posterior distributions of θ are plotted in Fig. 3.5.1. For this example, the distributions differ quite considerably from one another, both in their location and shape. Inferences about θ thus depend rather heavily upon the assumed form of the parent population.

It is of interest to consider the modal values of θ for various values of β, that is, the values $\hat{\theta}(\beta)$ giving maximum posterior density.

† A considerable improvement can be obtained, of course, by using a simple numerical integration procedure.

Table 3.5.2

Values of residuals $r_i = y_i - \eta(\xi_i \mid \theta)$ for various values of θ computed from the chemical reactor data†

ξ θ	0.500	0.525	0.550	0.575	0.600	0.625	0.650	0.675
50	−20	−13	−4	4	11	19	26	23
75	−37	−27	−17	−8	0	9	18	26
100	−127	−116	−105	−95	−85	−76	−66	−57
125	−45	−35	−23	−12	−1	8	18	28
150	−31	−20	−8	4	14	24	34	43
200	−130	−118	−106	−96	−85	−74	−65	−56
250	−44	−33	−21	−10	0	10	19	28
300	−80	−68	−57	−47	−37	−28	−19	−11

ξ θ	0.700	0.725	0.750	0.775	0.800	0.825	0.850
50	39	47	53	60	66	72	78
75	34	43	50	58	65	72	78
100	−48	−40	−31	−23	−16	−9	−2
125	37	45	54	62	70	78	84
150	52	61	69	78	85	92	100
200	−47	−38	−30	−22	−15	−8	0
250	36	44	52	60	67	74	80
300	−3	6	12	19	26	32	37

† The entries in the table are values of $r_i \times 100$, where $r_i = y_i - (1 + \theta\xi_i)^{-1}$.

Reading from the graphs we find approximately:

Parent distribution	Rectangular	Fourth power	Normal	$\frac{4}{3}$ power	Double exponential
Value of β	-1	$-\frac{1}{2}$	0	$\frac{1}{2}$	1
Value of $\hat{\theta}(\beta)$	0.70	0.69	0.665	0.63	0.595

Thus, the value

$\hat{\theta}(-1) = 0.70$ (rectangular parent) minimizes the maximum absolute deviations from the fitted curve,

$\hat{\theta}(0) = 0.665$ (Normal parent) is the least squares value and minimizes the sum of the squared deviations from the fitted curve, and

$\hat{\theta}(1) = 0.595$ (double exponential parent) minimizes the sum of the absolute deviations from the fitted curve.

The original data and fitted functions for these three values of θ are shown in Fig. 3.5.2.

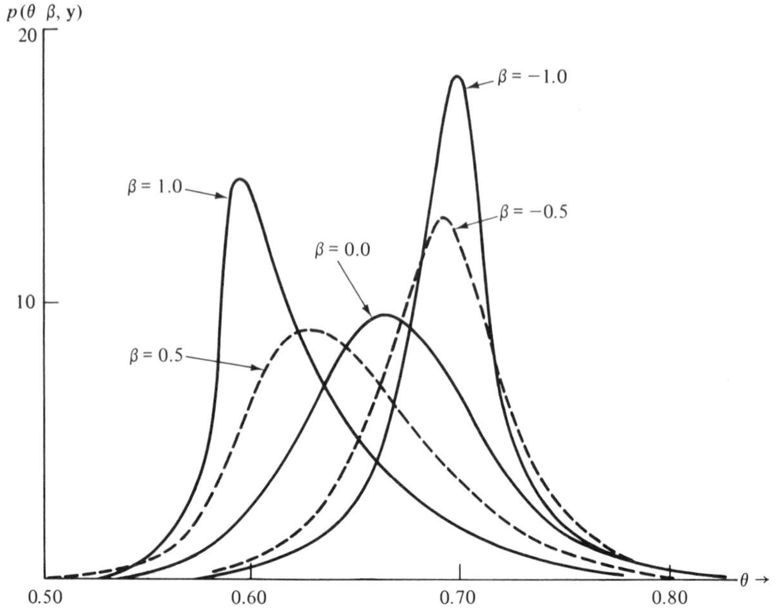

Fig. 3.5.1 Posterior distribution of θ for various values of β: chemical reactor data.

Information About β

Approximate unstandardized ordinates for $p_u(\beta \mid \mathbf{y})$ can be obtained by replacing $J(\beta)$ in (3.5.6) with $\hat{J}(\beta)$, so that

$$p_u(\beta \mid \mathbf{y}) \mathrel{\dot{\propto}} \frac{\Gamma[1 + \frac{1}{2}n(1 + \beta)]}{\{\Gamma[1 + \frac{1}{2}(1 + \beta)]\}^n} \hat{J}(\beta),$$

which are shown in the last row of Table 3.5.3 and plotted in Fig. 3.5.3. From this sample of only 8 observations, as is to be expected, there is very little evidence about the form of the parent distribution. In particular, the highest ordinate of $p_u(\beta \mid \mathbf{y})$ is only about three times the lowest. To see how slight is this apparent preference for negative values of β (for platykurtic parent distributions), we can temporarily consider $p_u(\beta \mid \mathbf{y})$ as a genuine posterior distribution (that is, consider it as if there *were* a uniform prior for β). Then the appropriate function $p_u(\beta \mid \mathbf{y})$, obtained in Fig. 3.5.3 by joining the ordinates with a dotted curve, would indicate a probability of about 1/3 that β was in fact *positive* (the parent was leptokurtic).

To obtain the marginal distribution of θ, one must integrate $p(\theta \mid \beta, \mathbf{y})$ with $p(\beta \mid \mathbf{y})$ as the weight function. Now the weight function is given by $p(\beta \mid \mathbf{y}) \propto p_u(\beta \mid \mathbf{y})p(\beta)$ and, for this example, it is the prior $p(\beta)$ which dominates rather than $p_u(\beta \mid \mathbf{y})$. Moreover, $p(\theta \mid \beta, \mathbf{y})$ changes drastically for different β. Here then

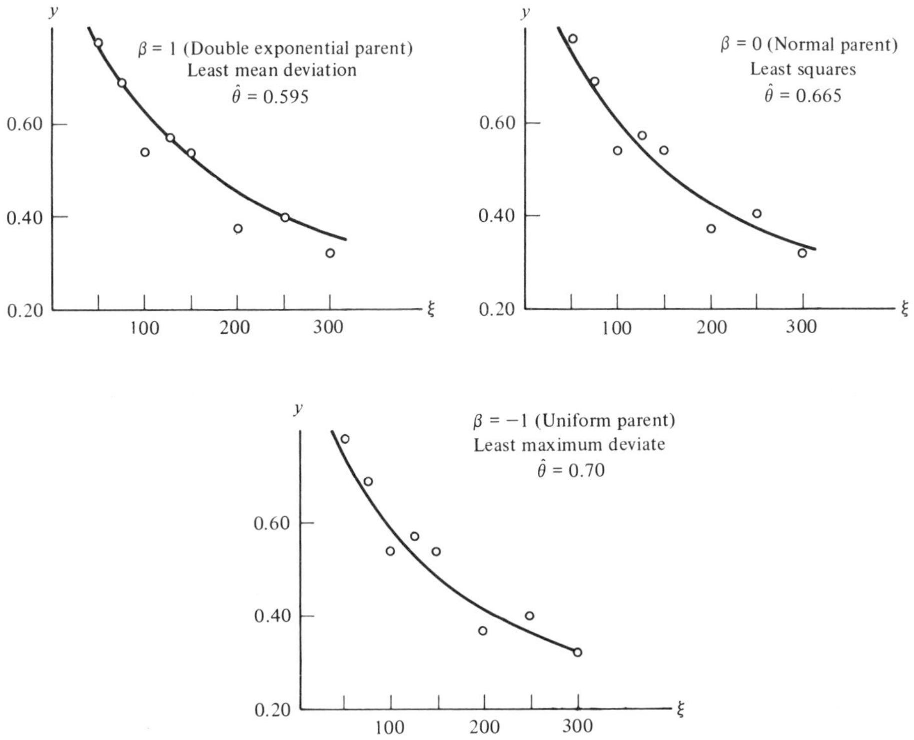

Fig. 3.5.2 Fitted functions using modal values of θ: chemical reactor data.

we are put on notice by the Bayes analysis that, for this example, the conclusions depend critically on what is assumed *a priori* about β and hence about the nature of the parent distribution. The data analyst is made aware that he is in a situation where prior assumptions about β (about Normality) must be given very careful attention, the data being too few to supply much information on their own account. If extensive records exist of similar experiments, he will be led to examine these to determine whether or not they suggest approximately Normal error structure. He will also carefully consider whether the hypothesis of many error sources, none of which dominate, is sensible for this experimental set-up.

If he decides that it is sensible to use for $p(\beta)$ a distribution centered at $\beta = 0$ as in Fig. 3.5.4 . Then for this data $p(\beta \mid \mathbf{y})$ will be much the same as $p(\beta)$. Averaging $p(\theta \mid \beta, \mathbf{y})$ with this weight function will be nearly the same as averaging with the prior $p(\beta)$. Thus the averaging process would lead approximately to the marginal distribution of θ, $p(\theta \mid \mathbf{y}) \doteq p(\theta \mid \beta = 0, \mathbf{y})$, and to Normal theory as an approximation.

The above kind of argument could also be used if his investigation led the data analyst to some central distribution other than the Normal. For example,

Table 3.5.3

Approximate ordinates of posterior distribution of θ for various values of β: chemical reactor data

	β				
	-1	$-\frac{1}{2}$	0	$\frac{1}{2}$	1
Value of θ	$r_L^{-8}/\hat{J}(-1)$	$(\Sigma r^4)^{-2}/\hat{J}(-\frac{1}{2})$	$(\Sigma r^2)^{-4}/\hat{J}(0)$	$(\Sigma\lvert r\rvert^{4/3})^{-6}/\hat{J}(\frac{1}{2})$	$(\Sigma\lvert r\rvert)^{-8}/\hat{J}(1)$
0.500	0.0	0.0	0.0	0.0	0.0
0.525	0.0	0.0	0.1	0.1	0.1
0.550	0.1	0.1	0.3	0.6	0.7
0.575	0.1	0.2	0.9	2.3	3.7
0.600	0.4	0.6	2.4	6.7	14.3
0.625	1.1	1.8	5.4	9.0	8.7
0.650	2.7	4.8	8.9	8.2	5.1
0.675	8.8	9.8	8.9	5.8	3.1
0.700	18.3	12.3	6.7	3.6	2.1
0.725	5.1	6.2	3.4	1.8	1.0
0.750	1.9	2.5	1.7	1.0	0.5
0.775	0.8	0.9	0.7	0.5	0.3
0.800	0.4	0.4	0.4	0.2	0.2
0.825	0.2	0.2	0.2	0.1	0.1
0.850	0.1	0.1	0.1	0.1	0.0
$\hat{J}(\beta)$	102.0	13.52	0.6045	0.02319	0.000804
$\dfrac{\Gamma[1+\frac{1}{2}n(1+\beta)]}{\{\Gamma[1+\frac{1}{2}(1+\beta)]\}^n}$	1	4.390	63.07	1,414.4	40,320
Approximate $p_u(\beta \mid \mathbf{y})$ (unstandardized)	102.0	59.4	38.1	32.8	32.4

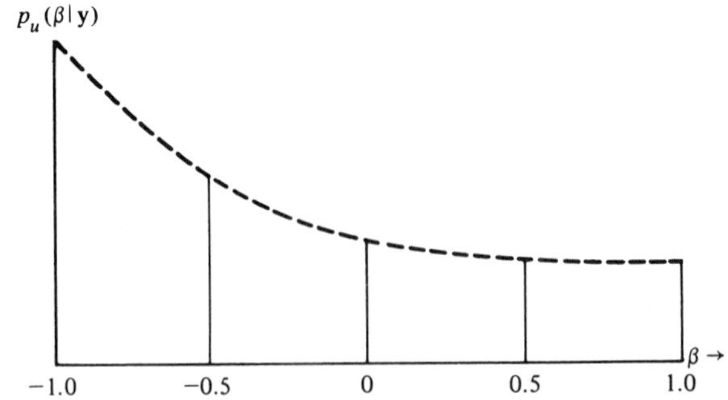

Fig. 3.5.3 Approximate ordinates of $p_u(\beta \mid \mathbf{y})$: chemical reactor data.

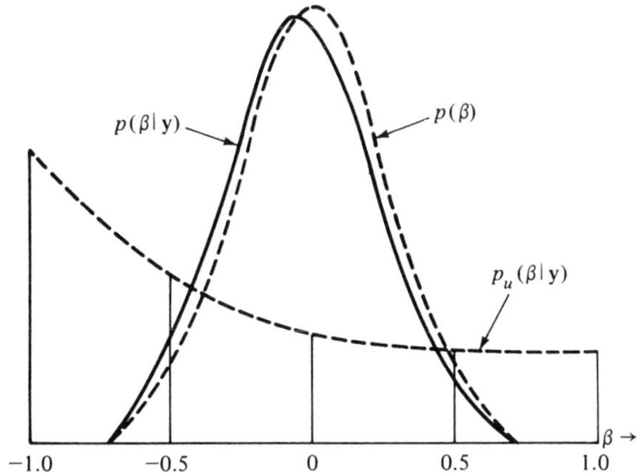

Fig. 3.5.4 Combination of the function $p_u(\beta \mid \mathbf{y})$ with a moderately informative prior $p(\beta)$: chemical reactor data.

we have said that breaking-strength test measurements often have a markedly leptokurtic distribution, so that in an experiment to determine the mean breaking strength from a sample of test pieces in which almost the only source of experimental error was the test itself, the analyst might employ a weight function $p(\beta)$ centered not on zero but on say $\beta = 0.8$. The resulting distribution $p(\theta \mid \mathbf{y})$ could then be considerably different from that obtained if the Normal were regarded as the most likely parent.

It is an advantage of the Bayesian analysis that when, as in this case, there is little information from the data itself concerning the value of some nuisance parameter such as β, then we *are put on notice* by evidence like that set out in Fig. 3.5.1 that assumptions about β ought not to be made lightly. If we really do have some external evidence about the parent distribution then we should use it; but if we do not, then the uncertainty about θ must be correspondingly increased. This small sample situation may be contrasted with that which applies when there is a larger number of observations. In this case the *sample* evidence about β is much stronger and there is correspondingly less need for strong prior evidence.

3.6 SUMMARY AND DISCUSSION

It seems pertinent at this point to summarize and discuss the results obtained in this chapter. The models employed in the analysis of the three sets of data in this chapter are all special cases of

$$y = \eta(\boldsymbol{\xi} \mid \boldsymbol{\theta}) + \varepsilon \tag{3.6.1}$$

given earlier in (3.5.1), with y the observed output, $\boldsymbol{\xi}$ a set of inputs and $\boldsymbol{\theta}$ a set of parameters. Specifically, the models and their special features are as follows:

Darwin's data (Section 3.2)	$y = \theta + \varepsilon$	Parameter θ is the population mean.
Reaction rate data (Section 3.4)	$y = \theta_1 + \theta_2 x + \varepsilon$	Parameters θ_1 and θ_2 appear *linearly* in regression equations.
Chemical reactor data (Section 3.5)	$y = 1/(1 + \theta\xi) + \varepsilon$	Parameter θ appears non-linearly.

The analysis of such sets of data would often be made on the assumption that the error ε was Normally distributed. Bayesian analysis makes it possible to study the effect of wider distributional assumptions, and to illustrate this, we have supposed that the parent error distribution was a member of the symmetric family

$$p(\varepsilon) = \omega(\beta)\sigma^{-1} \exp\left[-c(\beta)\left|\frac{\varepsilon}{\sigma}\right|^{2/(1+\beta)}\right] \tag{3.6.2}$$

in (3.4.2), which includes the Normal as a special case. By changing β we are able to study the effect of changing the kurtosis of the parent distribution in a particular way.

1. For the general class of models represented by (3.6.2), and on reasonable assumptions which include the idea that there is no appreciable prior knowledge concerning θ or σ, the conditional posterior distribution $p(\theta \mid \beta, \mathbf{y})$ has the remarkably simple general form

$$p(\boldsymbol{\theta} \mid \beta, \mathbf{y}) \propto \left[\sum_{i=1}^{n} |y_i - \eta(\boldsymbol{\xi}_i \mid \boldsymbol{\theta})|^{2(1+\beta)}\right]^{-\frac{1}{2}n(1+\beta)}. \tag{3.6.3}$$

2. In all the examples studied, $p(\boldsymbol{\theta} \mid \beta, \mathbf{y})$ changes quite markedly as β is changed. Consequently, the inferences which would be made about $\boldsymbol{\theta}$ could differ materially depending on the value of β. We have contrasted this lack of *inference* robustness with the *criterion* robustness enjoyed by the model under certain circumstances [see, for example, Box and Watson (1962)].

3. Inferences about the value of β itself can be made by studying the distribution

$$p(\beta \mid \mathbf{y}) \propto p_u(\beta \mid \mathbf{y})p(\beta) \tag{3.6.4}$$

where $p_u(\beta \mid \mathbf{y})$ is the distribution of β in relation to a uniform reference prior.

4. For small samples $p_u(\beta \mid \mathbf{y})$ will be rather flat. This indicates that, as might be expected, little information about the nature of a parent distribution is contained in a small sample of observations.

5. The marginal posterior distribution

$$p(\boldsymbol{\theta} \mid \mathbf{y}) \propto \int_{-1}^{1} p(\boldsymbol{\theta} \mid \beta, \mathbf{y}) p_u(\beta \mid \mathbf{y}) p(\beta) \, d\beta \tag{3.6.5}$$

summarizes what overall inferences might be made about $\boldsymbol{\theta}$. In this expression $p(\beta \mid \mathbf{y}) \propto p_u(\beta \mid \mathbf{y})\, p(\beta)$ acts as a weight function.

6. Some caution is needed in interpreting the integration in (3.6.5) if we need to rely heavily upon the form of $p(\beta)$. However, in many examples where a central limit tendency is expected, the precise form of $p(\beta)$ will not be of much importance. In the Darwin example, even though $p_u(\beta \mid \mathbf{y})$ is far from sharp, it is sufficient to ensure that extreme conditional distributions $p(\boldsymbol{\theta} \mid \beta, \mathbf{y})$ enter with little weight. Thus, even though the conditional distributions $p(\boldsymbol{\theta} \mid \beta, \mathbf{y})$ differ markedly over the entire range of β, the inferences to be drawn about $\boldsymbol{\theta}$ do not change very much for reasonable choices of $p(\beta)$. The same effect can be seen for the reaction rate data in Section 3.4.

7. An important advantage of the Bayesian approach is that it makes a deeper analysis of inference problems possible. It does this by showing

a) how sensitive the conditional distribution $p(\boldsymbol{\theta} \mid \beta, \mathbf{y})$ is to changes in the nuisance parameter β,

b) what the data can tell us about the nuisance parameter β.

Various situations could occur:

a) $p(\boldsymbol{\theta} \mid \beta, \mathbf{y})$ might be very *insensitive* to β, in which case it would be apparent that exact knowledge of the nuisance parameter β was not needed.

b) $p(\boldsymbol{\theta} \mid \beta, \mathbf{y})$ might be sensitive to changes in β, in which case the integration of this conditional distribution with weight function $p(\beta \mid \mathbf{y}) \propto p_u(\beta \mid \mathbf{y}) p(\beta)$ would need to be studied further.

c) If $p_u(\beta \mid \mathbf{y})$ was sharp (as it would be if the number of observations was large), then inferences would rest primarily on the information coming from the sample, as represented by $p_u(\beta \mid \mathbf{y})$, and very little on the choices of $p(\beta)$.

d) If, on the other hand, $p_u(\beta \mid \mathbf{y})$ was rather flat, as is particularly true in the third example of the chemical reactor data in Section 3.5, then any integration we make *would* in this case depend critically on the choice of $p(\beta)$, In this case then, where the inference *was* dependent to a large degree on prior assumptions, this fact would be clearly pointed out to us. Suppose for instance, that in the third example we made an outright assumption of Normality (corresponding to choosing $p(\beta)$ to be a δ function at $\beta = 0$), then the Bayesian analysis would make it perfectly clear that, *for this particular example*, the conclusions rested heavily on that choice, and that we ought to look to its justification.

In general, in order to draw conclusions on topics of importance to him, the experimenter needs information about primary parameters (for example, $\boldsymbol{\theta}$) in the context of a model structure about which he is uncertain. Such uncertainties can sometimes be represented as uncertainties about nuisance parameters. Whatever system of inference the investigator embraces, possible information about structure can either appear as prior assumptions or as information from the sample itself. In some instances reliance on prior assumptions could be greatly lessened by making proper use of appropriate information in the sample as can be done with Bayesian analysis. The investigator is often tempted to make whatever assumptions he thinks he needs to allow him to make inferences about primary parameters. Unless he has some means of keeping track of the consequences of such assumptions, he runs the risk of being misled.

Bayesian analysis helps to bring out into the open how much we need to assume and about what. Assumptions may be

a) necessary or unnecessary,

b) well founded or ill founded,

and inference procedures may be

c) sensitive or insensitive to assumptions.

Bayesian analysis supplies information in (a) and (c) and can draw our attention to the necessity for making up our minds about (b).

3.7 A SUMMARY OF FORMULAS FOR POSTERIOR DISTRIBUTIONS

Table 3.7.1 provides a short summary of the formulas for various prior and posterior distributions discussed in this chapter.

Table 3.7.1

A summary of prior and posterior distributions

---

1. Suppose the data $\mathbf{y}' = (y_1, \ldots, y_n)$ are drawn independently from the exponential power distribution in (3.2.5),

$$p(y \mid \theta, \sigma, \beta) = \omega(\beta)\,\sigma^{-1} \exp\left[-c(\beta)\left|\frac{y-\theta}{\sigma}\right|^{2/(1+\beta)}\right], \qquad -\infty < y < \infty,$$

where

$$c(\beta) = \left\{\frac{\Gamma[\frac{3}{2}(1+\beta)]}{\Gamma[\frac{1}{2}(1+\beta)]}\right\}^{1/(1+\beta)}$$

and

$$\omega(\beta) = \frac{\{\Gamma[\frac{3}{2}(1+\beta)]\}^{1/2}}{(1+\beta)\,\{\Gamma[\frac{1}{2}(1+\beta)]\}^{3/2}}.$$

Table 3.7.1 *Continued*

2. Suppose further that the prior distribution is, from (3.2.20) and (3.2.29),

$$p(\theta, \sigma, \beta) = p(\theta, \sigma)p(\beta),$$

with

$$p(\theta, \sigma) \propto \sigma^{-1} \qquad \text{and} \qquad p(\beta) \propto (1 - \beta^2)^{a-1}.$$

3. Given β, the conditional posterior distribution of θ is in (3.2.12),

$$p(\theta \mid \beta, \mathbf{y}) = [J(\beta)]^{-1}[M(\theta)]^{-\frac{1}{2}n(1+\beta)}, \qquad -\infty < \theta < \infty,$$

where

$$M(\theta) = \sum_{i=1}^{n} |y_i - \theta|^{2/(1+\beta)} \qquad \text{and} \qquad J(\beta) = \int_{-\infty}^{\infty} [M(\theta)]^{-\frac{1}{2}n(1+\beta)}\, d\theta.$$

In particular, for $\beta = 0$.

$$\frac{\theta - \bar{y}}{s/\sqrt{n}} \sim t_{n-1},$$

where

$$\bar{y} = \frac{1}{n}\Sigma\, y_i \qquad \text{and} \qquad s^2 = \frac{1}{n-1}\Sigma\,(y_i - \bar{y})^2.$$

Also, for $\beta \to -1$

$$\frac{|\theta - m|}{h/(n-1)} \sim F_{2,2(n-1)},$$

where $m = \frac{1}{2}(y_{(1)} + y_{(n)})$ and $h = \frac{1}{2}(y_{(n)} - y_{(1)})$.

4. The marginal distribution of β is in (3.2.28),

$$p(\beta \mid \mathbf{y}) \propto p(\beta)p_u(\beta \mid \mathbf{y}), \qquad -1 < \beta \leqslant 1,$$

where

$$p_u(\beta \mid \mathbf{y}) \propto J(\beta)\Gamma\left[1 + \frac{n}{2}(1 + \beta)\right]\{\Gamma[1 + \tfrac{1}{2}(1 + \beta)]\}^{-n}.$$

5. The marginal distribution of θ is from (3.2.32),

$$p(\theta \mid \mathbf{y}) = \int_{-1}^{1} p(\theta \mid \beta, \mathbf{y})\, p(\beta \mid \mathbf{y})\, d\beta$$

$$\propto \int_{-1}^{1} p(\theta \mid \beta, \mathbf{y})\, p_u(\beta \mid \mathbf{y})\, p(\beta)\, d\beta, \qquad -\infty < \theta < \infty.$$

Table 3.7.1 *Continued*

6. The conditional distribution $p(\theta \mid \beta, \mathbf{y})$, for β not close to -1, can be approximated by the t distribution

$$t\{\hat{\theta}, M(\hat{\theta})/d[n(1+\beta)-1], n(1+\beta)-1\},$$

where $\hat{\theta}$ is the mode and d is a positive constant. Procedurse for computing $\hat{\theta}$ and d are given in Section 3.3.

7. For the linear model $\mathbf{y} = \mathbf{X}\boldsymbol{\theta} + \boldsymbol{\varepsilon}$ in (3.4.1), where $\mathbf{X}$ is an $n \times k$ matrix, $\boldsymbol{\theta}' = (\theta_1, \ldots, \theta_k)$, and $\boldsymbol{\varepsilon}' = (\varepsilon_1, \ldots, \varepsilon_n)$, the formulae are of the same form as those given in (2) through (6) above. The only substitutions needed are
   a) replace θ by $\boldsymbol{\theta}$,
   b) replace $M(\theta)$ by

$$M(\boldsymbol{\theta}) = \sum_{i=1}^{n} |y_i - \mathbf{x}'_{(i)}\boldsymbol{\theta}|^{2/(1+\beta)},$$

   where $\mathbf{x}'_{(i)}$ is the ith row of $\mathbf{X}$, and
   c) redefine $J(\beta)$ as

$$J(\beta) = \int_R [M(\boldsymbol{\theta})]^{-\frac{1}{2}n(1+\beta)}\, d\boldsymbol{\theta},$$

   where $R: (-\infty < \theta_j < \infty,\ j = 1, \ldots, k)$.

8. For the nonlinear model in (3.5.2)

$$y_i = \eta(\boldsymbol{\xi}_i \mid \boldsymbol{\theta}) + \varepsilon_i,\ i = 1, \ldots, n,$$

where $\boldsymbol{\xi}'_i = (\xi_{i1}, \ldots, \xi_{ip})$. The substitutions required for the formulae in (2) through (6) are
   a) replace θ by $\boldsymbol{\theta}$,
   b) replace $M(\theta)$ by

$$M(\boldsymbol{\theta}) = \sum_{i=1}^{n} |y_i - \eta(\boldsymbol{\xi}_i \mid \boldsymbol{\theta})|^{2/(1+\beta)},$$

   c) redefine

$$J(\beta) = \int_R [M(\boldsymbol{\theta})]^{-\frac{1}{2}n(1+\beta)}\, d\boldsymbol{\theta},$$

   where $R: (-\infty < \theta_j < \infty,\ j = 1, \ldots, k)$.

APPENDIX A3.1

SOME PROPERTIES OF THE POSTERIOR DISTRIBUTION $p(\theta \mid \beta, \mathbf{y})$

In Section 3.2.3 we have asserted certain properties of the posterior distribution $p(\theta \mid \beta, \mathbf{y})$ in the permissible range of β. These properties follow from work on the sample median by Jackson (1921). For the class of parent distributions given in (3.2.3), the maximum likelihood estimate of θ for fixed β, which is also the mode of $p(\theta \mid \beta, \mathbf{y})$, has been considered for certain specific choices of β by Turner (1960). In our notation, consider the function

$$M(\theta) = \sum_{i=1}^{n} |y_i - \theta|^{2/(1+\beta)}, \qquad -\infty < \theta < \infty, \quad -1 < \beta \leqslant 1. \tag{A3.1.1}$$

For convenience, let us denote $q = 2/(1+\beta)$, so that

$$M(\theta) = \sum_{i=1}^{n} |y_i - \theta|^{q}, \qquad q \geqslant 1. \tag{A3.1.2}$$

1. We first show that

a) $M(\theta)$ is convex, continuous, and for $q > 1$, has continuous first derivative;

b) for $q > 1$, $M(\theta)$ has a unique minimum which is attained in the interval $[y_{(1)}, y_{(n)}]$.

To see (a), consider

$$g_i(\theta) = |\theta - y_i|^{q}, \qquad i = 1, 2, \ldots, n. \tag{A3.1.3}$$

Clearly $g_i(\theta)$ is convex and continuous everywhere. Now, suppose that $q > 1$. Then for $\theta < y_i$,

$$g_i'(\theta) = -q(y_i - \theta)^{q-1}, \tag{A3.1.4}$$

for $\theta > y_i$,

$$g_i'(\theta) = q(\theta - y_i)^{q-1},$$

and as θ approaches y_i from both directions

$$\lim_{\theta \uparrow y_i} g_i'(\theta) = \lim_{\theta \downarrow y_i} g_i'(\theta) = 0, \tag{A3.1.5}$$

which implies that $g_i'(y_i) = 0$. Since $q - 1 > 0$, $g_i'(\theta)$ exists and is continuous everywhere. Assertion (a) follows since $M(\theta)$ is the sum of all $g_i(\theta)$.

Let us now consider $M'(\theta)$. We see that for $q > 1$,

$$M'(\theta) = q \sum_{i=1}^{n} |y_{(i)} - \theta|^{q} \, \delta_i, \tag{A3.1.6a}$$

where

$$\delta_i = \begin{cases} -1 & \text{for} \quad \theta \leqslant y_{(i)} \\ 1 & \text{for} \quad \theta > y_{(i)} \end{cases}$$

and $y_{(i)}$, $(i = 1, \ldots, n)$ are the ordered observations. Thus, for $\theta < y_{(i)}$,

$$M'(\theta) = -q \sum_{i=1}^{n} (y_i - \theta)^{q-1} < 0 \tag{A3.1.6b}$$

and for $\theta > y_{(n)}$,

$$M'(\theta) = q \sum_{i=1}^{n} (\theta - y_i)^{q-1} > 0. \tag{A3.1.6c}$$

Thus, by properties of continuous function, there exists at least one θ_0, $y_{(1)} \leqslant \theta_0 \leqslant y_{(n)}$, such that $M'(\theta_0) = 0$.

Further, it is easy to see from (A3.1.6a) that $M'(\theta)$ is monotonically increasing in θ, so that $M'(\theta)$ can vanish once (and only once) and that the extreme value of $M(\theta)$ must be a minimum. This demonstrates assertion (b).

2. It has been shown by Jackson that when q approaches 1, in the limit the value of θ which minimizes $M(\theta)$ is the median of the y_i's, if n is odd; and, if n is even, is some unique value between the middle two of the y_i's.

3. When $q = 1$, it is readily seen that $M(\theta)$ is minimized at the median of the y's, for n odd, and is constant between the middle two observations for n even.

4. We now show that, when q is arbitrarily large

$$\lim_{q \to \infty} [M(\theta)]^{1/q} = (h + |m - \theta|), \tag{A3.1.7}$$

where

$$m = \tfrac{1}{2}(y_{(1)} + y_{(n)}), \qquad h = \tfrac{1}{2}(y_{(n)} - y_{(1)}).$$

Proof. Consider a finite sequence of monotone increasing positive numbers $\{a_n\}$ and a number S, such that

$$S = \left(\sum_{i=1}^{n} a^q \right)^{1/q}$$

$$= a_n \left[\sum_{i=1}^{n} \left(\frac{a_i}{a_n} \right)^q \right]^{1/q}, \tag{A3.1.8}$$

where

$$\frac{a_i}{a_n} \leqslant 1 \qquad \text{for all } i.$$

Hence

$$\left(\frac{S}{a_n}\right) = \left[\sum_{i=1}^{n}\left(\frac{a_i}{a_n}\right)^q\right]^{1/q}, \tag{A3.1.9}$$

so that

$$\log\left(\frac{S}{a_n}\right) = \frac{1}{q}\log\left[\sum_{i=1}^{n}\left(\frac{a_i}{a_n}\right)^q\right]. \tag{A3.1.10}$$

When $q \to \infty$,

$$\lim_{q\to\infty}\log\left[\sum_{i=1}^{n}\left(\frac{a_i}{a_n}\right)^q\right] = \log r, \tag{A3.1.11}$$

where $1 \leqslant r \leqslant n$. But this implies that

$$\lim_{q\to\infty}\log\left(\frac{S}{a_n}\right) = 0, \tag{A3.1.12}$$

from this

$$\lim_{q\to\infty} S = a_n.$$

Thus, for any given value of θ, when q is arbitrarily large,

$$\begin{aligned}\lim_{q\to\infty}[M(\theta)]^{1/q} &= \max_i |y_i - \theta| \\ &= \max[|\theta - y_{(1)}|, |\theta - y_{(n)}|] \\ &= \begin{cases} h + (m - \theta) & \text{for } \theta < m, \\ h & \text{for } \theta = m, \\ h + (\theta - m) & \text{for } \theta > m. \end{cases}\end{aligned} \tag{A3.1.13}$$

Hence, $\lim_{q\to\infty}[M(\theta)]^{1/q} = (h + |m - \theta|)$ and the assertion follows.

APPENDIX A3.2

A PROPERTY OF LOCALLY UNIFORM DISTRIBUTIONS

Let y_1 and y_2 be two independently distributed random variables. Suppose y_1 has a distribution such that the density is approximately uniform,

$$p(y_1) \doteq c,$$

over a range of y_1 which is wide relative to the range in which the density of y_2 is appreciable. Then the variable $X = y_1 + y_2$ has a locally uniform distribution and is distributed approximately independently of y_2.

Proof. The conditional density of X given y_2 is

$$\begin{aligned} f(X \mid y_2) &= f(y_1 = X - y_2 \mid y_2) \\ &= p(y_1 = X - y_2) \\ &\doteq c. \end{aligned}$$

It follows that X is approximately independent of y_2 and its marginal density is

$$f(X) \doteq c.$$

CHAPTER 4

BAYESIAN ASSESSMENT OF ASSUMPTIONS 2. COMPARISON OF VARIANCES

4.1 INTRODUCTION

In the preceding chapter, we analyzed the problem of making inferences about location parameters when the assumption of Normality was relaxed. We now extend the analysis to the problem of comparing variances.

As was shown by Geary (1947), Gayen (1950), Box (1953a), and others, the usual Normal sampling theory tests to compare variances are not criterion-robust under non-Normality. That is to say, the sampling distributions of Normal theory criteria such as the variance ratio and Bartlett's statistic to compare several variances can be seriously affected when the parent distribution is moderately non-Normal. The difficulty arises because of "confounding" of the effect of inequality of variances and the effect of kurtosis. In particular, critical regions for tests on variances and tests of kurtosis overlap substantially, so that it is usually difficult to know whether an observed discrepancy results from one or the other, or a mixture of both [see Box, (1953b)].

Within the Bayesian framework we are not limited by the necessity to consider "test criteria." However, as we have seen, the question of the sensitivity of the *inference* does arise. Therefore, following our earlier work (1964a), we now consider the comparison of variances in a wider Bayesian framework in which the parent distributions are permitted to exhibit kurtosis. We again employ as parent distributions the class of exponential power distributions in (3.2.5), which are characterized by three parameters (β, θ, σ).

The densities for samples from k such populations will thus depend in general upon $3k$ unknown parameters,

$$(\beta_1, \theta_1, \sigma_1), (\beta_2, \theta_2, \sigma_2), \ldots, (\beta_k, \theta_k, \sigma_k).$$

It will often be reasonable to suppose that the parameters $\beta_1, \ldots, \beta_k$ are essentially the same, and we shall make this assumption. In what follows, the case of two populations will be considered first, and the analysis is then extended to $k \geqslant 2$ populations.

4.2 COMPARISON OF TWO VARIANCES

For simplicity, the analysis is first carried through on the assumption that the location parameters θ_1 and θ_2 of the two populations to be compared, are known.

Later this assumption is relaxed. Proceeding as before, the inference robustness of the variance ratio σ_2^2/σ_1^2 may be studied for any particular sample $\mathbf{y}$ by computing the conditional posterior distribution $p(\sigma_2^2/\sigma_1^2 \mid \beta, \mathbf{y})$ for various choices of β. The marginal posterior distribution, which can be written as the product $p(\beta \mid \mathbf{y}) \propto p(\beta)p_u(\beta \mid \mathbf{y})$, indicates the plausiblity of the various choices. In particular, a prior distribution $p(\beta)$, concentrated about $\beta = 0$ to a greater or lesser degree, can represent different degrees of central limit tendency, and marginal posterior distributions $p(\sigma_2^2/\sigma_1^2 \mid \mathbf{y})$ appropriate to these different choices of $p(\beta)$ may be computed.

To relax the assumption that the location parameters θ_1 and θ_2 are known involves two further integrations, and proves to be laborious even on a fast electronic computer. However, we shall show that a close approximation to the integral is obtained by replacing the unknown means θ_1 and θ_2 by their modal values in the integrand, and changing the "degrees of freedom" by one unit. An example is worked out in detail.

4.2.1 Posterior Distribution of σ_2^2/σ_1^2 for Fixed Values of $(\theta_1, \theta_2, \beta)$

The likelihood function of $(\sigma_1, \sigma_2, \theta_1, \theta_2, \beta)$ given the two samples

$$\mathbf{y}_1 = (y_{11}, y_{12}, \ldots, y_{1n_1})' \qquad \text{and} \qquad \mathbf{y}_2 = (y_{21}, y_{22}, \ldots, y_{2n_2})'$$

is

$$l(\sigma_1, \sigma_2, \theta_1, \theta_2, \beta \mid \mathbf{y}) = [\omega(\beta)]^{(n_1+n_2)}\, \sigma_1^{-n_1} \sigma_2^{-n_2} \exp\left[-\tfrac{1}{2}c(\beta) \sum_{i=1}^{2} \frac{n_i s_i(\beta, \theta_i)}{\sigma_i^{2/(1+\beta)}} \right], \qquad (4.2.1)$$

where

$$\mathbf{y} = (\mathbf{y}_1, \mathbf{y}_2), \qquad s_i(\beta, \theta_i) = \frac{1}{n_i} \sum_{j=1}^{n_i} |y_{ij} - \theta_i|^{2/(1+\beta)}$$

and $c(\beta)$ and $\omega(\beta)$ are given in (3.2.5) as

$$c(\beta) = \left\{ \frac{\Gamma[\frac{3}{2}(1+\beta)]}{\Gamma[\frac{1}{2}(1+\beta)]} \right\}^{1/(1+\beta)},$$

$$\omega(\beta) = \frac{\{\Gamma[\frac{3}{2}(1+\beta)]\}^{1/2}}{(1+\beta)\{\Gamma[\frac{1}{2}(1+\beta)]\}^{3/2}}.$$

We assume, as before, that the means (θ_1, θ_2) and the logarithms of the standard deviations (σ_1, σ_2) are approximately independent and locally uniform *a priori*, so that

$$p(\theta_i) \propto c_1, \qquad (4.2.2)$$

$$p(\log \sigma_i) \propto c_2 \qquad \text{or} \qquad p(\sigma_i) \propto \frac{1}{\sigma_i}, \qquad i = 1, 2. \qquad (4.2.3)$$

Further, we suppose that these location and scale parameters are distributed independently of the non-Normality parameter β. It follows that for given values of $\boldsymbol{\theta} = (\theta_1, \theta_2)$ the joint posterior distribution of σ_1, σ_2, and β can then be written

$$p(\sigma_1, \sigma_2, \beta \mid \boldsymbol{\theta}, \mathbf{y}) \propto p(\sigma_1)p(\sigma_2)p(\beta)l(\sigma_1, \sigma_2, \beta \mid \boldsymbol{\theta}, \mathbf{y}) \tag{4.2.4a}$$

where $p(\beta)$ is the prior distribution of β and $l(\sigma_1, \sigma_2, \beta \mid \boldsymbol{\theta}, \mathbf{y})$ is the likelihood function in (4.2.1) when (θ_1, θ_2) are regarded as fixed. Now, we may write the joint distribution as the product

$$p(\sigma_1, \sigma_2, \beta \mid \boldsymbol{\theta}, \mathbf{y}) = p(\beta \mid \boldsymbol{\theta}, \mathbf{y})p(\sigma_1, \sigma_2 \mid \beta, \boldsymbol{\theta}, \mathbf{y}_i). \tag{4.2.4b}$$

For any given value of β, the conditional posterior distribution of σ_1 and σ_2 is then

$$p(\sigma_1, \sigma_2 \mid \beta, \boldsymbol{\theta}, \mathbf{y}) = \prod_{i=1}^{2} p(\sigma_i \mid \beta, \theta_i, \mathbf{y}_i), \tag{4.2.5}$$

where

$$p(\sigma_i \mid \beta, \theta_i, \mathbf{y}_i) \propto \sigma_i^{-(n_i+1)} \exp\left[-\frac{\frac{1}{2}c(\beta)n_i s_i(\beta, \theta_i)}{\sigma_i^{2/(1+\beta)}}\right], \qquad \sigma_i > 0.$$

This has the form of the product of two independent inverted gamma distributions. By making the transformation $V = \sigma_2^2/\sigma_1^2$ and $W = \sigma_1$, and integrating out W, we find that the posterior distribution of σ_2^2/σ_1^2 is

$$p(V \mid \beta, \boldsymbol{\theta}, \mathbf{y}) = \omega(\beta, \boldsymbol{\theta})V^{\frac{1}{2}n_i - 1}\left[1 + \frac{n_1 s_1(\beta, \theta_1)}{n_2 s_2(\beta, \theta_2)} V^{1/(1+\beta)}\right]^{-\frac{1}{2}(n_1+n_2)(1+\beta)}, \qquad V > 0, \tag{4.2.6}$$

where

$$\omega(\beta, \boldsymbol{\theta}) = \left(\frac{1}{1+\beta}\right)\frac{\Gamma[\frac{1}{2}(n_1 + n_2)(1+\beta)]}{\prod_{i=1}^{2}\Gamma[\frac{1}{2}n_i(1+\beta)]}\left[\frac{n_1 s_1(\beta, \theta_1)}{n_2 s_2(\beta, \theta_2)}\right]^{\frac{1}{2}n_1(1+\beta)}$$

It is convenient to consider the quantity $[s_1(\beta, \theta_1)/s_2(\beta, \theta_2)]V^{1/(1+\beta)}$, where it is to be remembered that $V = \sigma_2^2/\sigma_1^2$ is a random variable and $s_1(\beta, \theta_1)/s_2(\beta, \theta_2)$ is a constant calculated from the observations. We then have that

$$p\left[F = \frac{s_1(\beta, \theta_1)}{s_2(\beta, \theta_2)} V^{1/(1+\beta)} \,\middle|\, \beta, \boldsymbol{\theta}, \mathbf{y}\right] = p[F_{n_1(1+\beta), n_2(1+\beta)}], \tag{4.2.7}$$

where $p[F_{n_1(1+\beta), n_2(1+\beta)}]$ is the density of an F variable with $n_1(1+\beta)$ and $n_2(1+\beta)$ degrees of freedom. In particular, when $\beta = 0$, the quantity

$$V\frac{\sum_{j=1}^{n_1}(y_{1j} - \theta_1)^2/n_1}{\sum_{j=1}^{n_2}(y_{2j} - \theta_2)^2/n_2} \tag{4.2.8}$$

is distributed as F with n_1 and n_2 degrees of freedom. Further, when the value of

β tends to -1 so that the parent distributions tend to the rectangular form, the quantity

$$u = V\left(\frac{h_1 + |m_1 - \theta_1|}{h_2 + |m_2 - \theta_2|}\right)^2, \tag{4.2.9}$$

where (h_1, h_2) and (m_1, m_2) are the half-ranges and mid-points of the first and second samples, respectively, has the limiting distribution (see Appendix A4.1)

$$\lim_{\beta \to -1} p(u \mid \beta, \boldsymbol{\theta}, \mathbf{y}) = \begin{cases} \dfrac{n_1 n_2}{2(n_1 + n_2)} u^{\frac{1}{2}n_1 - 1} & \text{for} \quad 0 < u < 1, \\[2ex] \dfrac{n_1 n_2}{2(n_1 + n_2)} u^{-\frac{1}{2}n_2 - 1} & \text{for} \quad 1 < u < \infty. \end{cases} \tag{4.2.10}$$

Thus, for given β not close to -1, probability levels of V can be obtained from the F-table. In particular, the probability *a posteriori* that the variance ratio V is less than unity is simply

$$\Pr\{V < 1 \mid \beta, \boldsymbol{\theta}, \mathbf{y}) = \Pr\left\{F_{n_1(1+\beta), n_2(1+\beta)} < \frac{s_1(\beta, \theta_1)}{s_2(\beta, \theta_2)}\right\}. \tag{4.2.11}$$

4.2.2 Relationship Between the Posterior Distribution $p(V \mid \beta, \boldsymbol{\theta}, \mathbf{y})$ and Sampling Theory Procedures

In this simplified situation where θ_1, θ_2, and β are assumed known, a parallel result can be obtained from sampling theory. It is readily shown that the two power sums $n_1 s_1(\beta, \theta_1)$ and $n_2 s_2(\beta, \theta_2)$ in (4.2.1), when regarded as functions of the random variables $\mathbf{y}_1$ and $\mathbf{y}_2$, are sufficient statistics for (σ_1, σ_2) and have their joint moment generating function

$$M_{\mathbf{z}}(t_1, t_2) = \prod_{i=1}^{2}\left[1 - 2t_i \frac{\sigma_i^{2/(1+\beta)}}{c(\beta)}\right]^{-\frac{1}{2}n_i(1+\beta)}, \tag{4.2.12}$$

where

$$\mathbf{z} = [n_1 s_1(\beta, \theta_1), n_2 s_2(\beta, \theta_2)].$$

Thus, letting

$$\mathbf{z}^* = \left[\frac{c(\beta) n_1 s_1(\beta, \theta_1)}{\sigma_1^{2/(1+\beta)}}, \frac{c(\beta) n_2 s_2(\beta, \theta_2)}{\sigma_2^{2/(1+\beta)}}\right],$$

we obtain

$$M_{\mathbf{z}^*}(t_1, t_2) = \prod_{i=1}^{2} (1 - 2t_i)^{-\frac{1}{2}n_i(1+\beta)} \tag{4.2.13}$$

which is the product of the moment-generating functions of two independently distributed χ^2 distributions, with $n_1(1+\beta)$ and $n_2(1+\beta)$ degrees of freedom, respectively. Thus, on the hypothesis that $\sigma_2^2/\sigma_1^2 = 1$, the criterion $s_2(\beta, \theta_1)/s_1(\beta, \theta_2)$

is distributed as F with $n_2(1+\beta)$ and $n_1(1+\beta)$ degrees of freedom and, in fact, provides a uniformly most powerful similar test for this hypothesis against the alternative that $\sigma_2^2/\sigma_1^2 > 1$. The significance level associated with the observed $s_2(\beta, \theta_1)/s_1(\beta, \theta_2)$ is

$$\Pr\left\{F_{n_2(1+\beta), n_1(1+\beta)} > \frac{s_2(\beta, \theta_2)}{s_1(\beta, \theta_1)}\right\} = \Pr\left\{F_{n_1(1+\beta), n_2(1+\beta)} < \frac{s_1(\beta, \theta_1)}{s_2(\beta, \theta_2)}\right\} \qquad (4.2.14)$$

and is numerically equal to the probability for $V < 1$ given in (4.2.11). The level of significance associated with the observed ratio $s_2(\beta, \theta_1)/s_1(\beta, \theta_2)$ which can be derived from sampling theory is, therefore, precisely the probability *a posteriori* that the variance ratio σ_2^2/σ_1^2 is less than unity when $\log \sigma_1$ and $\log \sigma_2$ are supposed locally uniform *a priori*.

Table 4.2.1

Results from analyses of identical samples
[y = (percent of carbon -4.50) $\times$ 100]

Analyst A_1 Sample No.	y_1	Analyst A_2 Sample No.	y_2
1	−8	1	−10
2	−3	2	16
3	20	3	−8
4	22	4	9
5	3	5	5
6	5	6	−5
7	10	7	5
8	14	8	−11
9	−21	9	25
10	2	10	22
11	7	11	16
12	8	12	3
13	16	13	40
Mean	5.77	14	0
		15	−5
		16	16
		17	30
		18	−14
		19	25
		20	−28
		Mean	6.55

4.2.3 Inference Robustness on Bayesian Theory and Sampling Theory

The above problem of comparing two variances with the means (θ_1, θ_2) known, is an unusual one because we have the extraordinary situation where sufficient statistics $[n_1 s_1(\beta, \theta_1), n_2 s_2(\beta, \theta_2)]$ for σ_1^2 and σ_2^2 exist for each value of the nuisance parameter β. Inference robustness under changes of β can in this special case, therefore, be studied using *sampling theory* and a direct parallel can be drawn with the Bayesian analysis. We now illustrate this phenomenon with an example.

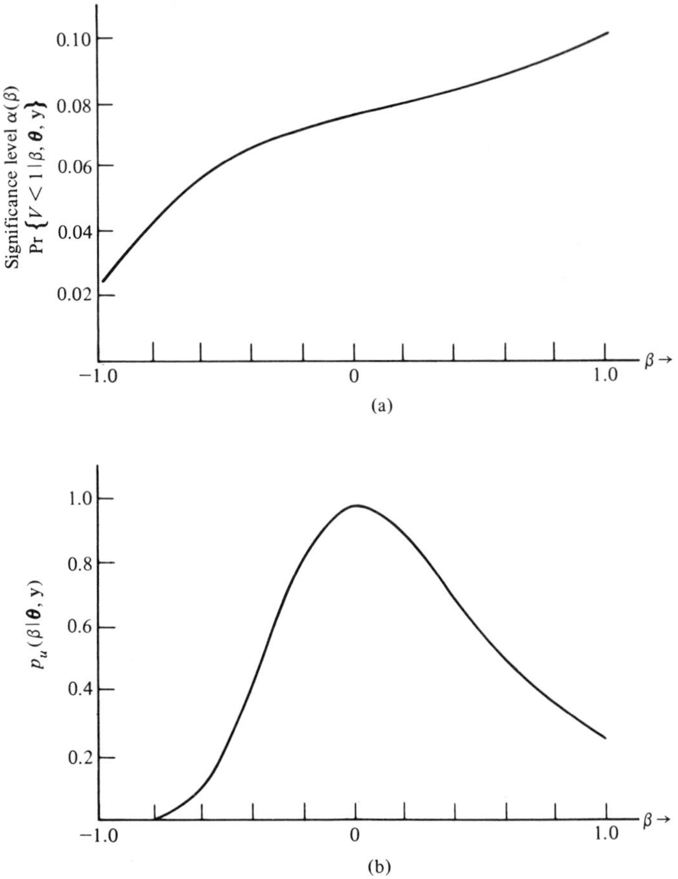

Fig. 4.2.1 Inference robustness: the analyst data.

a) The sampling theory significance level for various choices of β. This is numerically equal to the posterior probability that V is less than unity.

b) Posterior distribution of β for given θ_1 and θ_2 when the prior distribution of β is taken as uniform ($a = 1$).

The data in Table 4.2.1 were taken from Davies (1949). These data were collected to compare the accuracy of an experienced analyst A_1 and an inexperienced analyst A_2 in their assay of carbon in a mixed powder. We shall let σ_1^2 and σ_2^2 be the variances of A_1 and A_2, respectively. The population means θ_1 and θ_2 are, of course, unknown but for illustration we temporarily assume them known and equal to the sample means 5.77 and 6.55, respectively. In the original Normal sampling theory analysis, a test of the null hypothesis $\sigma_2^2/\sigma_1^2 = 1$ against the alternative $\sigma_2^2/\sigma_1^2 > 1$, yielded a significance level of 7.6%.

In Fig. 4.2.1(a), for each value of β, a significance level $\alpha(\beta)$ based on the uniformly most powerful similar criterion has been calculated. The figure shows how, for the wider class of parent distributions, the significance level changes from about 2.1% for β close to -1 (rectangular parent) to 9.9% when β is close to 1 (double exponential parent). In the sampling theory framework the *Normal theory* variance ratio lacks criterion robustness to kurtosis, but the inference robustness now considered is of a different character in which the criterion itself is changed appropriately as the parent distribution is changed.

As mentioned earlier, the significance level $\alpha(\beta)$ has a Bayesian interpretation. It is the posterior probability, given β, that the variance ratio V is less than unity.

4.2.4 Posterior Distribution of σ_2^2/σ_1^2 when β is Regarded as a Random Variable

Using the Bayesian approach, we can write the joint distribution of $V = \sigma_2^2/\sigma_1^2$ and β as the product

$$p(V, \beta \mid \boldsymbol{\theta}, \mathbf{y}) = p(V \mid \beta, \boldsymbol{\theta}, \mathbf{y})p(\beta \mid \boldsymbol{\theta}, \mathbf{y}). \tag{4.2.15}$$

The posterior distribution of V is then

$$p(V \mid \boldsymbol{\theta}, \mathbf{y}) = \int_{-1}^{+1} p(V \mid \beta, \boldsymbol{\theta}, \mathbf{y})p(\beta \mid \boldsymbol{\theta}, \mathbf{y})\, d\beta, \qquad V > 0. \tag{4.2.16}$$

As before, the marginal distribution of β which acts as the weight function can be written

$$p(\beta \mid \boldsymbol{\theta}, \mathbf{y}) \propto p(\beta)p_u(\beta \mid \boldsymbol{\theta}, \mathbf{y}), \qquad -1 < \beta \leqslant 1. \tag{4.2.17}$$

From (4.2.4a), we obtain

$$p_u(\beta \mid \boldsymbol{\theta}, \mathbf{y}) \propto \{\Gamma[1 + \tfrac{1}{2}(1 + \beta)]\}^{-(n_1+n_2)} \prod_{i=1}^{2} \Gamma[1 + \tfrac{1}{2}n_i(1 + \beta)]\, [n_i s_i(\beta, \theta_i)]^{-\frac{1}{2}n_i(1+\beta)} \tag{4.2.18}$$

which is the posterior distribution of β for a uniform reference prior.

It should be remembered that this reference prior is not necessarily one which represents one's genuine prior opinion; it does, however, provide a useful "way-stage" allowing us to compute $p_u(\beta \mid \boldsymbol{\theta}, \mathbf{y})$ which, when multiplied by any prior $p(\beta)$ of genuine interest, yields the appropriate posterior distribution. The function $p_u(\beta \mid \boldsymbol{\theta}, \mathbf{y})$ can also be thought of as the likelihood integrated over the weight function $(\sigma_1 \sigma_2)^{-1}$, which corresponds to the assumption of a uniform prior for $\log \sigma_1$ and $\log \sigma_2$. As was seen earlier, this choice is appropriate to the common situation where there is little prior information about these scale parameters relative to that which will be obtained from the data. It is important to notice, however, that $p_u(\beta \mid \boldsymbol{\theta}, \mathbf{y})$ *is* dependent on the choice of priors for σ_1 and σ_2.

For the analyst example, the function $p_u(\beta \mid \boldsymbol{\theta}, \mathbf{y})$ is shown in Fig. 4.2.1(b).

4.2.5 Inferences About V with β Eliminated

Now although sampling theory provides, in this exceptional instance, a basis for determining inference robustness in the sense that we can interpret Fig. 4.2.1(a) as showing the change in significance level that occurs for different values of β, it does not provide us with any satisfactory basis for using sample information about β. On the other hand, on Bayesian theory the overall probability

$$\Pr\{V < 1 \mid \boldsymbol{\theta}, \mathbf{y}\} = \int_{-1}^{+1} \Pr\{V < 1 \mid \beta, \boldsymbol{\theta}, \mathbf{y}\} p(\beta \mid \boldsymbol{\theta}, \mathbf{y})\, d\beta \qquad (4.2.19)$$

can be obtained by integrating the ordinates $\Pr\{V < 1 \mid \beta, \boldsymbol{\theta}, \mathbf{y}\}$ of Fig. 4.2.1(a) with the weight function $p(\beta \mid \boldsymbol{\theta}, \mathbf{y}) \propto p_u(\beta \mid \boldsymbol{\theta}, \mathbf{y}) p(\beta)$. Now $p_u(\beta \mid \boldsymbol{\theta}, \mathbf{y})$ is given for the example in Fig. 4.2.1(b), so that to obtain $\Pr\{V < 1 \mid \boldsymbol{\theta}, \mathbf{y}\}$ we have first to multiply the distribution in Fig. 4.2.1(b) by an appropriate prior $p(\beta)$, and then use the normalized product as a weight function in the integration of the ordinates of Fig. 4.2.1(a).

Certainly, the experimental situation here is one where we might expect a central limit effect, for it is likely that a large number of different components of error, associated with the various manipulations of the analysts, will make contributions of comparable importance to the overall error. Thus, as before in (3.2.29), we introduce the flexible prior distribution for β having a mode at $\beta = 0$,

$$p(\beta) = w(1 - \beta^2)^{a-1}, \qquad -1 < \beta \leqslant 1,$$

where (4.2.20)

$$w = \Gamma(2a)[\Gamma(a)]^{-2} 2^{-(2a-1)}, \qquad a \geqslant 1.$$

The dotted curves of Fig. 4.2.2 show $p(\beta)$ for various choices of "a." Multiplication of $p(\beta)$ by $p_u(\beta \mid \boldsymbol{\theta}, \mathbf{y})$ then produces $p(\beta \mid \boldsymbol{\theta}, \mathbf{y})$ shown by the solid curves in Fig. 4.2.2.

With $p(\beta)$ uniform ($a = 1$) we find $\Pr\{V < 1 \mid \boldsymbol{\theta}, \mathbf{y}\} = 7.91\%$, which happens to agree fairly closely with the value 7.59% obtained using Normal theory. This is not surprising because the distribution $p_u(\beta \mid \boldsymbol{\theta}, \mathbf{y})$ is, for this data, roughly centered about the value zero. The averaging of (4.2.19) can, therefore, be expected to yield a result close to that for Normal theory. Any injection of a sharper prior obtained by setting $a > 1$ and representing a prior expectation of central limit tendency could be expected to bring the probability even closer to that for Normal theory.

To facilitate comparison with sampling theory, we have so far calculated only the posterior probability that $V < 1$, which parallels directly the significance level. In most problems the whole posterior density function would be of interest. This fuller analysis will be illustrated below when the unrealistic assumption that θ_1 and θ_2 are known *a priori* is relaxed.

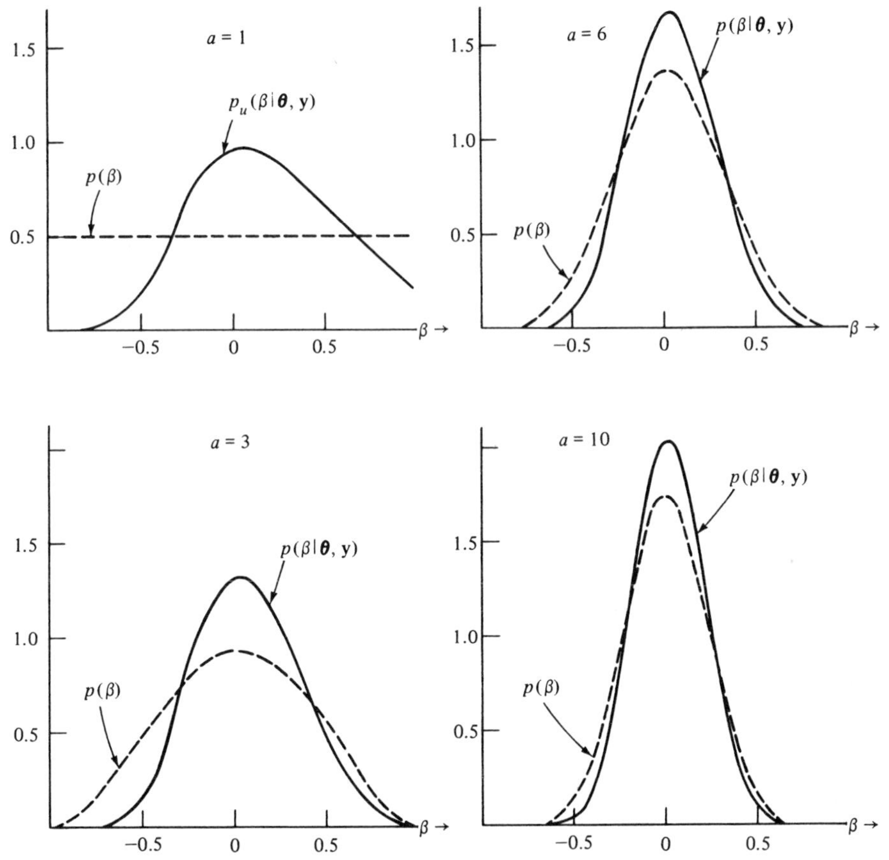

Fig. 4.2.2 Prior and posterior distributions of β for various choices of the parameter "a": the analyst data.

4.2.6 Posterior Distribution of V when θ_1 and θ_2 are Regarded as Random Variables

To concentrate attention on important issues, we supposed initially that the means $\boldsymbol{\theta} = (\theta_1, \theta_2)$ were known. Within the Bayesian formulation no difficulties of principle are associated with the elimination of such nuisance parameters.

With the prior assumptions of (4.2.2), (4.2.3) and (4.2.20), the joint posterior distribution of $(\boldsymbol{\theta}, \sigma_1, \sigma_2, \beta)$ is

$$p(\boldsymbol{\theta}, \sigma_1, \sigma_2, \beta \mid \mathbf{y}) \propto \sigma_1^{-1}\sigma_2^{-1}p(\beta)l(\sigma_1, \sigma_2, \theta_1, \theta_2, \beta \mid \mathbf{y}),$$

$$-\infty < \theta_i < \infty, \quad \sigma_i > 0, \quad i = 1, 2, \quad \text{and} \quad -1 < \beta \leqslant 1, \tag{4.2.21}$$

where $l(\sigma_1, \sigma_2, \theta_1, \theta_2, \beta \mid \mathbf{y})$ is given in (4.2.1).

Upon making the transformation from $(\boldsymbol{\theta}, \sigma_1, \sigma_2, \beta)$ to $(\boldsymbol{\theta}, V, \sigma_1, \beta)$ and integrating out $(\boldsymbol{\theta}, \sigma_1)$, the posterior distribution of β and V can be written as

$$p(V, \beta \mid \mathbf{y}) = p(V \mid \beta, \mathbf{y})p(\beta \mid \mathbf{y}). \tag{4.2.22}$$

The conditional posterior distribution of V for a fixed value of β is

$$p(V \mid \beta, \mathbf{y}) = \int_{-\infty}^{\infty}\int_{-\infty}^{\infty} p(V \mid \beta, \boldsymbol{\theta}, \mathbf{y})p(\boldsymbol{\theta} \mid \beta, \mathbf{y})\, d\theta_1\, d\theta_2, \qquad V > 0, \tag{4.2.23}$$

where $p(V \mid \beta, \boldsymbol{\theta}, \mathbf{y})$ is the distribution given in (4.2.6) and $p(\boldsymbol{\theta} \mid \beta, \mathbf{y})$, the joint distribution of θ_1 and θ_2, is

$$p(\boldsymbol{\theta} \mid \beta, \mathbf{y}) = \prod_{i=1}^{2} p(\theta_i \mid \beta, \mathbf{y}_i)$$

$$\prod_{i=1}^{2} \frac{[n_i s_i(\beta, \theta_i)]^{-\frac{1}{2}n_i(1+\beta)}}{\int_{-\infty}^{\infty} [n_i s_i(\beta, \theta_i)]^{-\frac{1}{2}n_i(1+\beta)}\, d\theta_i} \qquad -\infty < \theta_1 < \infty, \quad -\infty < \theta_2 < \infty, \tag{4.2.24}$$

with $s_i(\beta, \theta_i)$ $(i = 1, 2)$ given in (4.2.1). That is,

$$p(V \mid \beta, \mathbf{y}) = k(\beta)V^{\frac{1}{2}n_1 - 1}\int_{-\infty}^{\infty}\int_{-\infty}^{\infty} [n_2 s_2(\beta, \theta_2)$$

$$+ V^{1/(1+\beta)} n_1 s_1(\beta, \theta_1)]^{-\frac{1}{2}(n_1+n_2)(1+\beta)}\, d\theta_1\, d\theta_2, \qquad V > 0 \tag{4.2.25}$$

where

$$k(\beta)^{-1} = \frac{(1 + \beta)}{\Gamma[\frac{1}{2}(n_1 + n_2)(1 + \beta)]} \prod_{i=1}^{2} \Gamma[\tfrac{1}{2}n_i(1 + \beta)] \int_{-\infty}^{\infty} [n_i s_i(\beta, \theta_i)]^{-\frac{1}{2}n_i(1+\beta)}\, d\theta_i.$$

When the parents are Normal ($\beta = 0$), the quantity

$$V\frac{\Sigma\, (y_{1j} - \bar{y}_1)^2/(n_1 - 1)}{\Sigma\, (y_{2j} - \bar{y}_2)^2/(n_2 - 1)} \tag{4.2.26}$$

is distributed as F with $(n_1 - 1)$ and $(n_2 - 1)$ degrees of freedom. Also, when the parents tend to the rectangular form $(\beta \to -1)$, the quantity $u_* = V(h_1/h_2)^2$, where, as before, h_1 and h_2 are the half ranges of the two samples $\mathbf{y}_1$ and $\mathbf{y}_2$, has the limiting distribution (see Appendix A4.1)

$$\lim_{\beta\to -1} p(u_* \mid \beta, \mathbf{y})$$

$$= \begin{cases} k u_*^{\frac{1}{2}(n_1-1)-1}\left[(n_1+n_2)-(n_1+n_2-2)u_*^{1/2}\right] & \text{for} \quad 0 < u_* < 1, \\ k u_*^{-\frac{1}{2}(n_2-1)-1}\left[(n_1+n_2)-(n_1+n_2-2)u_*^{-1/2}\right] & \text{for} \quad 1 < u_* < \infty, \end{cases} \tag{4.2.27}$$

with

$$k = \frac{n_1 n_2}{2(n_1+n_2)} \frac{(n_1-1)(n_2-1)}{(n_1+n_2-1)(n_1+n_2-2)}.$$

For other choice of parents, it does not appear possible to express the posterior distribution of V in terms of simple functions. Methods are now derived which yield a close approximation to this distribution.

4.2.7 Computational Procedures for the Posterior Distribution $p(V \mid \beta, \mathbf{y})$

Numerical evaluation of (4.2.25) involves, among other things, computing a double integral for each value of V. This is laborious even on a fast electronic computer. However, a general expression for the rth moment of V for $(r < n_2/2)$ is readily obtained as

$$E(V^r \mid \beta, \mathbf{y}) = k(\beta) \frac{\int_{-\infty}^{\infty} [n_1 s_1(\beta, \theta_1)]^{-\frac{1}{2}(n_1+2r)(1+\beta)}\, d\theta_1}{\int_{-\infty}^{\infty} [n_1 s_1(\beta, \theta_1)]^{-\frac{1}{2}n_1(1+\beta)}\, d\theta_1}$$

$$\times \frac{\int_{-\infty}^{\infty} [n_2 s_2(\beta, \theta_2)]^{-\frac{1}{2}(n_2-2r)(1+\beta)}\, d\theta_2}{\int_{-\infty}^{\infty} [n_2 s_2(\beta, \theta_2)]^{-\frac{1}{2}n_2(1+\beta)}\, d\theta_2}, \tag{4.2.28}$$

with

$$k(\beta) = \Gamma[\tfrac{1}{2}(n_1+2r)(1+\beta)]\Gamma[\tfrac{1}{2}(n_2-2r)(1+\beta)] \Big/ \prod_{i=1}^{2} \Gamma[\tfrac{1}{2}n_i(1+\beta)].$$

Computation of the moments thus only involves the comparatively simple evaluation of one-dimensional integrals. It would now be possible to proceed by evaluating these moments and fitting appropriate forms of distributions suggested by the exact results obtainable when $\beta = 0$ and $\beta \to -1$. However, a simpler and more intuitively satisfying procedure is as follows.

Employing the approximating form (3.3.6) we may write

$$n_i s_i(\beta, \theta_i) \doteq n_i s_i(\beta, \hat{\theta}_i) + d_i(\theta_i - \hat{\theta}_i)^2, \qquad i = 1, 2 \tag{4.2.29}$$

where $\hat{\theta}_i$ is the value which minimizes $n_i s_i(\beta, \theta_i)$ and is, therefore, also the mode of the posterior distribution of θ_i in (4.2.24). Substituting (4.2.29) into (4.2.28), the integrals in the numerator and the denominator can then be explicitly evaluated.

Specifically, for the integrals in the numerator, we obtain

$$\int_{-\infty}^{\infty} [n_1 s_1(\beta, \theta_1)]^{-\frac{1}{2}(n_1+2r)(1+\beta)}\, d\theta_1 \int_{-\infty}^{\infty} [n_2 s_2(\beta, \theta_2)]^{-\frac{1}{2}(n_2-2r)(1+\beta)}\, d\theta_2$$

$$\doteq [n_1 s_1(\beta, \hat{\theta}_1)]^{-\frac{1}{2}[(n_1+2r)(1+\beta)-1]}\, d_1^{-1/2} B\{\tfrac{1}{2}, \tfrac{1}{2}[(n_1+2r)(1+\beta)-1]\}$$
$$\times [n_2 s_2(\beta, \hat{\theta}_2)]^{-\frac{1}{2}[(n_2-2r)(1+\beta)-1]}\, d_2^{-1/2} B\{\tfrac{1}{2}, \tfrac{1}{2}[(n_2-2r)(1+\beta)-1]\}, \qquad (4.2.30)$$

and for those in the denominator, we obtain

$$\prod_{i=1}^{2} \int_{-\infty}^{\infty} [n_i s_i(\beta, \theta_i)]^{-\frac{1}{2}n_i(1+\beta)}\, d\theta_i \doteq \prod_{i=1}^{2} [n_i s_i(\beta, \hat{\theta}_i)]^{-\frac{1}{2}[n_i(1+\beta)-1]}$$
$$\times\, d_i^{-1/2} B\{\tfrac{1}{2}, \tfrac{1}{2}[n_i(1+\beta)-1]\}, \qquad (4.2.31)$$

where $B(p, q) = \Gamma(p)\Gamma(q)/\Gamma(p+q)$.

By taking the ratio of (4.2.30) and (4.2.31), we see that the quantities d_1 and d_2 are *cancelled*. The rth moment, $(r < \frac{1}{2}n_2 - \frac{1}{2}(1+\beta)^{-1})$, of V is, approximately,

$$E(V^r \mid \beta, \mathbf{y}) \doteq \frac{\Gamma\{\frac{1}{2}[(n_1+2r)(1+\beta)-1]\}\Gamma\{\frac{1}{2}[(n_2-2r)(1+\beta)-1]\}}{\prod_{i=1}^{2} \Gamma\{\frac{1}{2}[n_i(1+\beta)-1]\}}$$
$$\times \left[\frac{n_2 s_2(\beta, \hat{\theta}_2)}{n_1 s_1(\beta, \hat{\theta}_1)}\right]^{r(1+\beta)}. \qquad (4.2.32)$$

That is, the moments of

$$V^{1/(1+\beta)} \frac{n_1 s_1(\beta, \hat{\theta}_1)/[n_1(1+\beta)-1]}{n_2 s_2(\beta, \hat{\theta}_2)/[n_2(1+\beta)-1]} \qquad (4.2.33)$$

are approximately the same as those of an F variable with $n_1(1+\beta)-1$ and $n_2(1+\beta)-1$ degrees of freedom.

Table 4.2.2

Comparison of exact and approximate moments of the variance ratio V for the analyst data with various choices of β

	β	−0.8	−0.6	−0.4	−0.2	0.2	0.4	0.6	0.8
μ_1'	exact	2.16	2.03	2.07	2.20	2.55	2.72	2.89	3.05
	approx.	2.11	2.02	2.06	2.20	2.55	2.72	2.89	3.04
μ_2'	exact	5.12	4.71	5.13	6.10	9.00	10.79	12.77	14.97
	approx.	4.99	4.69	5.11	6.08	9.01	10.79	12.74	14.84
μ_3'	exact	13.31	12.50	15.23	21.24	44.02	62.14	86.19	117.63
	approx.	13.14	12.49	15.18	21.17	44.07	62.14	85.69	115.96
μ_4'	exact	38.04	38.16	54.53	93.54	300.60	525.11	896.08	1497.80
	approx.	39.00	38.44	54.51	93.36	301.04	524.67	888.07	1465.83

Table 4.2.2 shows, for the analyst example, the exact and approximate first four moments of V for various values of β. The exact moments were obtained by direct evaluation of (4.2.28), using Simpson's rule, and the approximate moments were computed using (4.2.32).

The moments approximation discussed above suggests that the quantity in (4.2.33) is approximately distributed as an F variable with $n_1(1+\beta)-1$ and $n_2(1+\beta)-1$ degrees of freedom. This means that in (4.2.23) the process of averaging $p(V \mid \beta, \boldsymbol{\theta}, \mathbf{y})$ over the weight function $p(\boldsymbol{\theta} \mid \beta, \mathbf{y})$ may be approximated by simply replacing the unknown (θ_1, θ_2) in $p(V \mid \beta, \boldsymbol{\theta}, \mathbf{y})$ by their modal values and reducing the degrees of freedom by one unit. That is

$$p(V \mid \beta, \mathbf{y}) \doteq \omega(\beta, \hat{\boldsymbol{\theta}}) V^{\frac{1}{2}[n_1-(1+\beta)^{-1}]-1}\left[1+\frac{n_1 s_1(\beta, \hat{\theta}_1)}{n_2 s_2(\beta, \hat{\theta}_2)} V^{1/(1+\beta)}\right]^{-\frac{1}{2}[(n_1+n_2)(1+\beta)-2]} \qquad (4.2.34)$$

where

$$\omega(\beta, \hat{\boldsymbol{\theta}}) = \left(\frac{1}{1+\beta}\right) \frac{\Gamma[\frac{1}{2}(n_1+n_2)(1+\beta)-1]}{\prod_{i=1}^{2} \Gamma\{\frac{1}{2}[n_i(1+\beta)-1]\}} \left[\frac{n_1 s_1(\beta, \hat{\theta}_1)}{n_2 s_2(\beta, \hat{\theta}_2)}\right]^{\frac{1}{2}[n_1(1+\beta)-1]}.$$

This approximating distribution can be justified directly by using the expansion (4.2.29) to represent the quantities $n_1 s_1(\beta, \theta_1)$ and $n_2 s_2(\beta, \theta_2)$ in (4.2.25) and integrating directly. We have followed through the moment approach here because the exact moments can be expressed in rather simple form and, as we pointed out, could be used, if desired, to fit appropriate distributions.

It should be noted that in obtaining the approximating moments (4.2.32) and the distribution (4.2.34) through the use of the expansion (4.2.29), it is only necessary to determine the modes $\hat{\theta}_1$ and $\hat{\theta}_2$. The more difficult problem of finding suitable values for d_1 and d_2 (see Section 3.3) does not arise here because of the cancellation in the integration process. Although calculating the modal values $(\hat{\theta}_1, \hat{\theta}_2)$ still requires the use of numerical methods such as those discussed in Section 3.3, this is a procedure of great simplicity compared with the exact evaluation of multiple integrals for each V in (4.2.25).

Table 4.2.3 shows exact and approximate densities of V for several values of β using the analyst data. The exact densities are obtained by direct evaluation of (4.2.25) using Simpson's rule, and the approximate densities are calculated using (4.2.34). Some discrepancies occur when $\beta = -0.8$ (that is, as β approaches -1) but for the remainder of the β range the agreement is very close.

4.2.8 Posterior Distribution of V for Fixed β with θ_1 and θ_2 Eliminated

For the analyst data, the graphs of $p(V \mid \beta, \mathbf{y})$ are shown in Fig. 4.2.3 for various choices of β, where, this time, the simplifying assumption that the location parameters θ_1 and θ_2 are known and equal to the sample means was not made. In computing the posterior distributions of V for $\beta = -0.6$ and $\beta = 0.99$, θ_1 and

θ_2 were eliminated by actual integration and also by the approximate procedure mentioned above. In each of these two cases, the curves obtained by the two methods are practically indistinguishable. For the cases $\beta = 0$ and $\beta = -1$, the exact distributions were calculated directly using (4.2.26) and (4.2.27), respectively.

Table 4.2.3

Specimen probability densities of the variance ratio V for various values of β: the analyst data

V	$\beta = -0.8$		$\beta = -0.6$		$\beta = -0.2$	
	exact	approx.	exact	approx.	exact	approx.
0.5	0.005	0.020	0.017	0.021	0.048	0.049
1.0	0.099	0.159	0.229	0.245	0.299	0.301
1.2	0.209	0.264	0.378	0.388	0.389	0.390
1.4	0.353	0.388	0.507	0.510	0.444	0.444
1.6	0.503	0.510	0.583	0.580	0.461	0.461
1.8	0.619	0.597	0.594	0.588	0.449	0.448
2.2	0.628	0.579	0.474	0.468	0.372	0.371
2.6	0.412	0.380	0.301	0.297	0.276	0.275
3.0	0.206	0.194	0.168	0.167	0.192	0.192
4.0	0.026	0.028	0.032	0.032	0.070	0.069
V	$\beta = 0.2$		$\beta = 0.6$		$\beta = 0.8$	
	exact	approx.	exact	approx.	exact	approx.
0.5	0.070	0.069	0.090	0.089	0.100	0.100
1.0	0.274	0.274	0.258	0.257	0.253	0.253
1.2	0.329	0.329	0.294	0.293	0.282	0.283
1.4	0.359	0.359	0.311	0.311	0.294	0.295
1.6	0.367	0.368	0.313	0.313	0.295	0.296
1.8	0.359	0.360	0.305	0.306	0.287	0.287
2.2	0.314	0.315	0.272	0.273	0.256	0.257
2.6	0.255	0.256	0.231	0.231	0.219	0.219
3.0	0.200	0.200	0.190	0.190	0.183	0.183
4.0	0.100	0.100	0.110	0.110	0.111	0.111

We see from these graphs how the posterior distribution of the variance ratio V changes as the value of β is changed. When the parents are rectangular ($\beta \to -1$), the distribution is sharply concentrated around its modal value. As β is made larger so that the parent distribution passes through the Normal to become leptokurtic, the spread of the distribution of V increases, the modal value becomes smaller, and the distribution becomes more and more positively skewed. It is evident in this example that inferences concerning the variance ratio depend heavily upon the form of the parent distributions.

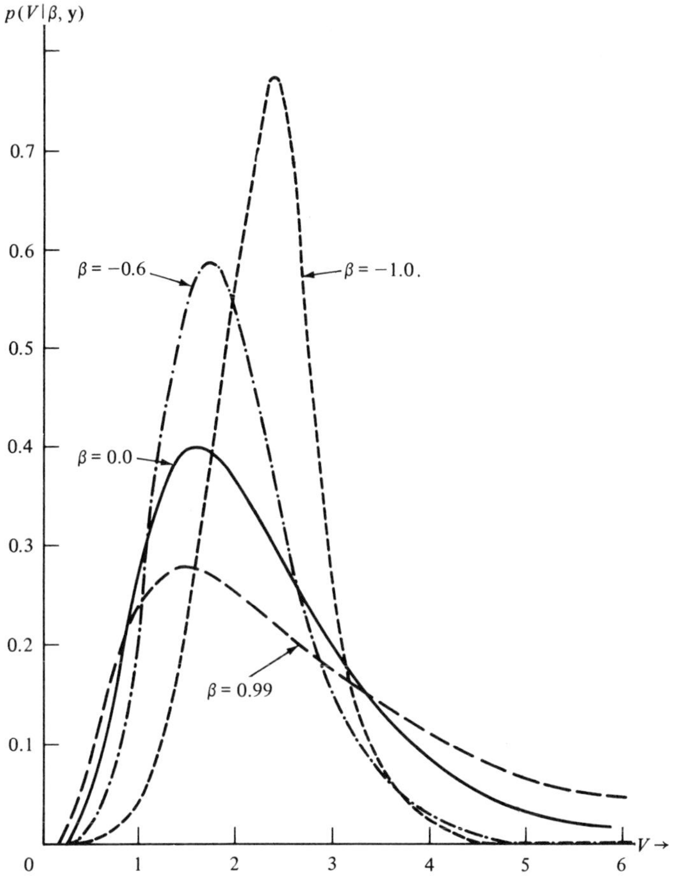

Fig. 4.2.3 Posterior distribution of V for various choices of β: the analyst data.

4.2.9 Marginal Distribution of β

The marginal posterior distribution of β, $p(\beta \mid \mathbf{y})$, with $(\boldsymbol{\theta}, \sigma_1, \sigma_2)$ eliminated may, as before, be written as the product

$$p(\beta \mid \mathbf{y}) \propto p_u(\beta \mid \mathbf{y})p(\beta), \qquad -1 < \beta \leqslant 1, \tag{4.2.35}$$

where

$$p_u(\beta \mid \mathbf{y}) \propto \frac{\prod_{i=1}^{2} \Gamma[1 + \frac{1}{2}n_i(1+\beta)]}{\{\Gamma[1 + \frac{1}{2}(1+\beta)]\}^{n_1+n_2}} \prod_{i=1}^{2} \int_{-\infty}^{\infty} [n_i s_i(\beta, \theta_i)]^{-\frac{1}{2}n_i(1+\beta)} \, d\theta_i$$

is the posterior distribution of β relative to a uniform prior for $-1 < \beta \leqslant 1$. For the analyst example, the distribution $p_u(\beta \mid \mathbf{y})$ shown in Fig. 4.2.4 is very little different from $p_u(\beta \mid \boldsymbol{\theta}, \mathbf{y})$ shown in Fig. 4.2.1(b). As before, the mode of the distribution is close to $\beta = 0$.

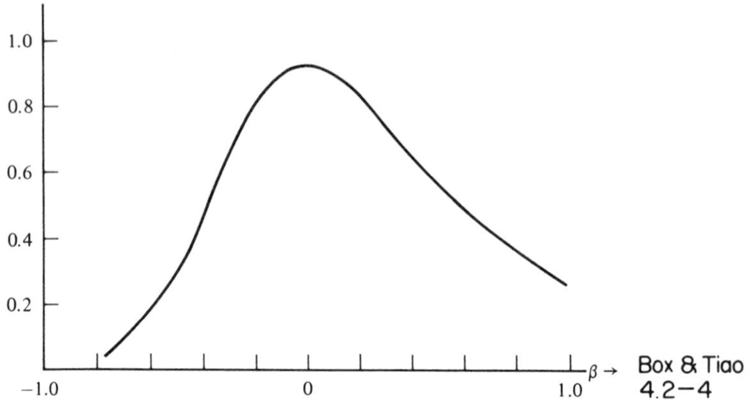

Fig. 4.2.4 Posterior distribution of β for a uniform reference prior: the analyst data.

4.2.10 Marginal Distribution of V

The main object of the analyst investigation was to make inferences about the variance ratio V. This can be done by eliminating β from the posterior distribution of V by integrating the conditional distribution $p(V \mid \beta, \mathbf{y})$ with the weight function $p(\beta \mid \mathbf{y})$,

$$p(V \mid \mathbf{y}) = \int_{-1}^{+1} p(V \mid \beta, \mathbf{y}) p(\beta \mid \mathbf{y})\, d\beta, \qquad V > 0. \tag{4.2.36}$$

Fig. 4.2.5 shows the distribution of V for a uniform reference prior in β, that is, with the weight function $p_u(\beta \mid \mathbf{y})$ in (4.2.35). Also shown by the dotted curve is the posterior distribution on the assumption of exact Normality. Any prior on β which is more peaked at $\beta = 0$ can be expected to produce even *greater* agreement with the Normal theory posterior. For this particular set of data, therefore, the Normal theory inference evidently approximates very closely that which would be appropriate for a wide choice of "a" in (4.2.20). This agreement is to be expected because, in this particular instance, the sample evidence expressed through $p_u(\beta \mid \mathbf{y})$ (see Fig. 4.2.4) indicates the Normal to be about the most probable parent distribution. However, such agreement might not be found if the sample evidence favored some other form of parent.

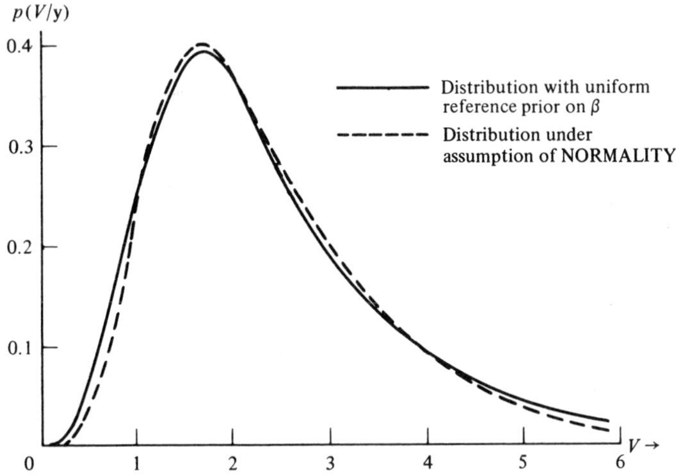

Fig. 4.2.5 Posterior distribution of V for a uniform reference prior in β: the analyst data.

4.3 COMPARISON OF THE VARIANCES OF k DISTRIBUTIONS

We now consider the problem of comparing the variances of k exponential power distributions when k can be greater than two. As mentioned earlier, we shall assume that the k populations have the same non-Normality parameter β, but possibly different values of θ and σ. When independent samples of size $n_1, \ldots, n_k$ are drawn from the k populations, the likelihood function is

$$l(\boldsymbol{\sigma}, \boldsymbol{\theta}, \beta \mid \mathbf{y}) = [\omega(\beta)]^n \prod_{i=1}^{k} \sigma_i^{-n_i} \exp\left[-\frac{\frac{1}{2}c(\beta)n_i s_i(\beta, \theta_i)}{\sigma_i^{2/(1+\beta)}}\right], \tag{4.3.1}$$

where

$$s_i(\beta, \theta_i) = \frac{1}{n_i} \sum_{j=1}^{n_i} |y_{ij} - \theta_i|^{2/(1+\beta)},$$

$$\boldsymbol{\theta} = (\theta_1, \ldots, \theta_k), \qquad \boldsymbol{\sigma} = (\sigma_1, \ldots, \sigma_k), \qquad \mathbf{y} = (\mathbf{y}_1, \ldots, \mathbf{y}_k),$$

$$\mathbf{y}_i = (y_{i1}, \ldots, y_{in_i}), \qquad n = \sum_{i=1}^{k} n_i,$$

and $c(\beta)$ and $\omega(\beta)$ are given in (4.2.1).

4.3.1 Comparison of k Variances for Fixed Values of $(\boldsymbol{\theta}, \beta)$

At first, we suppose that β and the location parameter θ in each population are known. With

$$p(\log \sigma_i) \propto c \qquad \text{or} \qquad p(\sigma_i) \propto \frac{1}{\sigma_i}, \qquad i = 1, \ldots, k \tag{4.3.2}$$

and for *given* $(\boldsymbol{\theta}, \beta)$, the posterior distribution of $\boldsymbol{\sigma}$ is

$$p(\boldsymbol{\sigma} \mid \boldsymbol{\theta}, \beta, \mathbf{y}) = \prod_{i=1}^{k} p(\sigma_i \mid \theta_i, \beta, \mathbf{y}_i), \qquad \sigma_i > 0, \quad i = 1, ..., k, \tag{4.3.3}$$

where

$$p(\sigma_i \mid \theta_i, \beta, \mathbf{y}_i) \propto \sigma_i^{-(n_i+1)} \exp\left[-\tfrac{1}{2}c(\beta)n_i s_i(\beta, \theta_i)/\sigma_i^{2/(1+\beta)}\right].$$

The joint distribution for $(\sigma_1, ..., \sigma_k)$ is thus a product of k independent inverted gamma distributions. In a certain academic sense, the obtaining of this joint distribution solves the inference problem. However, there is still the practical difficulty of *appreciating* what this k dimensional distribution has to tell us.

One question of interest is whether a particular parameter point is, or is not, included in an H.P.D. region of given probability mass. For instance, we may be interested in considering the plausibility of *equality* of variances. Following our earlier argument in Section 2.12, we consider the joint distribution of the $(k - 1)$ contrasts $\boldsymbol{\phi}' = (\phi_1, ..., \phi_{k-1})$ in the logarithms of the standard deviations $\boldsymbol{\sigma}$,

$$\phi_i = 2(\log \sigma_i - \log \sigma_k), \qquad i = 1, ..., k - 1. \tag{4.3.4}$$

It is straightforward to verify that the posterior distribution of $\boldsymbol{\phi}$ is

$$p(\boldsymbol{\phi} \mid \boldsymbol{\theta}, \beta, \mathbf{y}) = \frac{\Gamma[\frac{1}{2}n(1+\beta)]}{(1+\beta)^{(k-1)} \prod_{i=1}^{k} \Gamma[\frac{1}{2}n_i(1+\beta)]}$$
$$\times\, U_1^{\frac{1}{2}n_1(1+\beta)} \cdots U_{k-1}^{\frac{1}{2}(n_{k-1})(1+\beta)} (1 + U_1 + \cdots + U_{k-1})^{-\frac{1}{2}n(1+\beta)}$$
$$-\infty < \phi_1 < \infty, ..., -\infty < \phi_{k-1} < \infty, \tag{4.3.5}$$

where

$$U_i = \frac{n_i s_i(\beta, \theta_i)}{n_k s_k(\beta, \theta_k)} \exp\left[-\phi_i/(1+\beta)\right], \qquad i = 1, ..., (k-1).$$

Except for the factor $(1 + \beta)^{-1}$, this distribution is in precisely the same form as the distribution in (2.12.6), with U_i, $n_i(1 + \beta)$, and $s_i(\beta, \theta_i)$ replacing T_i, v_i, and s_i^2, respectively. From the developments in Section 2.12 it follows, in particular, that the distribution (4.3.5) has a unique mode at

$$\hat{\phi}_i(\beta) = (1+\beta)[\log s_i(\beta, \theta_i) - \log s_k(\beta, \theta_k)], \qquad i = 1, ..., k-1. \tag{4.3.6}$$

Further, in deciding if a particular point $\boldsymbol{\phi}_0$ is included in the H.P.D. region of content $(1 - \alpha)$, we may make use of the fact that $p(\boldsymbol{\phi} \mid \boldsymbol{\theta}, \beta, \mathbf{y})$ is a monotonic decreasing function of M, where

$$M = -2 \log W, \tag{4.3.7}$$

with

$$W = \frac{n^{\frac{1}{2}n(1+\beta)}}{\prod_{i=1}^{k} n_i^{\frac{1}{2}n_i(1+\beta)}} \left[\prod_{i=1}^{k-1} U_i^{\frac{1}{2}n_i(1+\beta)}\right] (1 + U_1 + \cdots + U_{k-1})^{-\frac{1}{2}n(1+\beta)} \tag{4.3.8}$$

The cumulant generating function of M, is, for $\beta \neq -1$,

$$\kappa_M(t) = a - \frac{k-1}{2}\log(1-2t) + \sum_{r=1}^{\infty} \alpha_r(1-2t)^{-(2r-1)}, \tag{4.3.9}$$

with

$$\alpha_r = \frac{B_{2r}}{2r(2r-1)}\left(\frac{2}{1+\beta}\right)^{2r-1}\left[\sum_{i=1}^{k} n_i^{-(2r-1)} - n^{-(2r-1)}\right]$$

and "a" is a constant. Bartlett's method of approximation discussed in Section 2.12 can be extended here so that

$$\Pr\{p(\boldsymbol{\phi} \mid \boldsymbol{\theta}, \beta, \mathbf{y}) > p(\boldsymbol{\phi}_0 \mid \boldsymbol{\theta}, \beta, \mathbf{y}) \mid \mathbf{y}\} = \Pr\{M < -2\log W_0\}$$

$$\doteq \Pr\left\{\chi^2_{k-1} < \frac{-2\log W_0}{1+A}\right\}, \tag{4.3.10}$$

where

$$A = \frac{1}{3(k-1)(1+\beta)}\left[\sum_{i=1}^{k} n_i^{-1} - n^{-1}\right]$$

and W_0 is obtained by inserting a particular parameter point of interest $\boldsymbol{\phi}_0$ in (4.3.8). In particular, if $\boldsymbol{\phi}_0 = \mathbf{0}$, so that $\sigma_1 = \sigma_2 = \cdots = \sigma_k$, then $-2\log W_0$ reduces to

$$-2\log W_0 = -\sum_{i=1}^{k} n_i(1+\beta)[\log s_i(\beta, \theta_i) - \log \bar{s}(\beta, \boldsymbol{\theta})], \tag{4.3.11}$$

with

$$\bar{s}(\beta, \boldsymbol{\theta}) = \frac{1}{n}\sum_j n_j s_j(\beta, \theta_j).$$

In this special case where the θ's are assumed known, exactly parallel results can be obtained within the sampling theory framework. In fact, the likelihood ratio criterion for testing the null hypothesis (H_0: $\sigma_1 = \sigma_2 = \cdots = \sigma_k$) against the alternative H_1 that they are not all equal, is

$$\lambda(\beta) = \prod_{i=1}^{k}\left[\frac{s_i(\beta, \theta_i)}{\bar{s}(\beta, \boldsymbol{\theta})}\right]^{-\frac{1}{2}n_i(1+\beta)}. \tag{4.3.12}$$

Thus $-2\log\lambda(\beta)$ is given by (4.3.11) and on the null hypothesis, the cumulant generating function of the sampling distribution of $-2\log\lambda(\beta)$ is precisely that given by the right-hand side of (4.3.9). It follows that the complement of the probability (4.3.10) is numerically equivalent to the significance level associated with the observed likelihood ratio statistic $\lambda(\beta)$.

An Example

For illustration, consider the three samples of 30 observations set out in Table 4.3.1. These observations were, in fact, generated from exponential power distributions with common mean $\theta = 0$ and $\beta = -0.5$. Their standard deviations σ_i were chosen to be 0.69, 0.59, 0.76, respectively. The analysis will be made assuming the means to be known.

Table 4.3.1

Data for three samples generated from the symmetric exponential power distribution with common $\beta = -0.5$ and $\theta = 0$

Observation number	Sample One ($\sigma = 0.69$)	Sample Two ($\sigma = 0.59$)	Sample Three ($\sigma = 0.76$)
1	0.23	−0.84	0.90
2	0.24	0.61	−0.26
3	−0.26	0.20	−0.03
4	−1.13	−0.83	0.04
5	−0.72	0.46	0.73
6	−0.52	−0.38	−0.06
7	0.65	0.13	−0.91
8	−1.16	0.33	−1.38
9	0.25	−0.20	1.24
10	−0.83	−0.45	−0.09
11	0.90	1.16	−0.40
12	−0.28	−0.55	−1.32
13	0.19	0.42	−0.57
14	−0.43	−0.73	−1.34
15	−1.12	0.37	0.97
16	−0.93	−0.37	0.68
17	−0.58	−0.50	0.70
18	0.62	−0.50	−0.52
19	0.61	−0.59	0.34
20	0.42	−0.77	−1.00
21	−1.34	−0.99	−1.28
22	−0.13	0.06	0.21
23	−0.92	0.46	1.42
24	−1.24	0.03	0.31
25	−1.02	−1.01	−0.31
26	−0.40	0.68	−0.14
27	−0.70	−0.28	0.99
28	0.10	−0.62	1.14
29	0.38	−0.70	0.21
30	−0.73	−0.32	−0.76

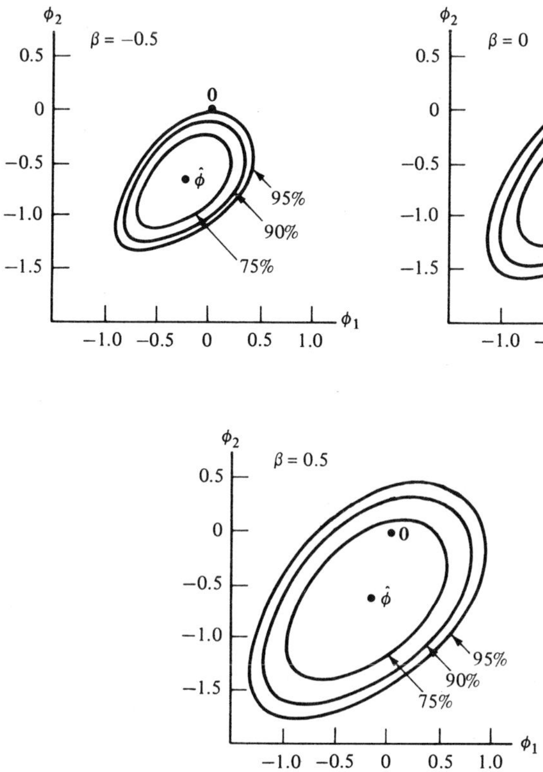

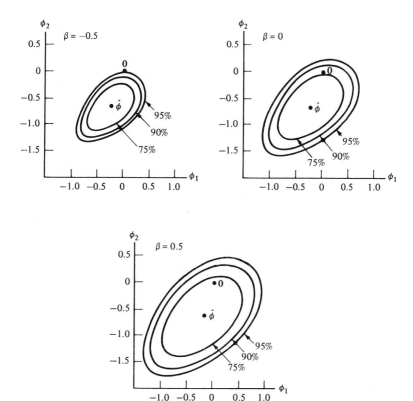

Fig. 4.3.1 Contours of the posterior distribution of (ϕ_1, ϕ_2) for various values of β: the generated example.

The first object is to study how inferences about the comparative values of σ_i would be affected by changes in the value of β. For fixed β, the posterior distribution of the two contrasts

$$\phi_1 = 2(\log \sigma_1 - \log \sigma_3) \quad \text{and} \quad \phi_2 = 2(\log \sigma_2 - \log \sigma_3) \tag{4.3.13}$$

is that given in (4.3.5) with $k = 3$. Figure 4.3.1 shows for $\beta = -0.5$, 0.0, and 0.5 the mode and three contours of the posterior distribution for ϕ_1 and ϕ_2. These contours, which correspond approximately to the 75, 90, and 95% H.P.D. regions, were drawn such that

$$M = -2 \log W = (1 + A)\chi^2(2, \alpha), \tag{4.3.14}$$

for $\alpha = 0.25, 0.10, 0.05$, respectively. For this example,

$$\begin{aligned} M = -90(1+\beta)\log 3 + 90(1+\beta)\log\big(1 + \exp\{-(1+\beta)^{-1}[\phi_1 - \hat{\phi}_1(\beta)]\} \\ + \exp\{-(1+\beta)^{-1}[\phi_2 - \hat{\phi}_2(\beta)]\}\big) \\ + 30\{[\phi_1 - \hat{\phi}_1(\beta)] + [\phi_2 - \hat{\phi}_2(\beta)]\} \end{aligned} \qquad (4.3.15)$$

and $1 + A = 1 + (1+\beta)^{-1}(0.0148)$.

Inspection of Fig. 4.3.1 shows that in all cases ϕ_1 and ϕ_2 are positively correlated. This might be expected, for although σ_1^2, σ_2^2, and σ_3^2 are independent *a posteriori*, $\phi_1 = \log\sigma_1^2 - \log\sigma_3^2$ and $\phi_2 = \log\sigma_2^2 - \log\sigma_3^2$ contain $\log\sigma_3^2$ in common. Also, for this particular set of data, the mode of the posterior distribution is not much affected by the choice of β, and the roughly elliptical shape of the contours is similar. The distributions differ markedly, however, in their spread; the dispersion becoming much larger as β is increased. Inferences about the relative values of the variances are for this reason very sensitive to the choice of β.

The possibility that the variances are all equal so that $\boldsymbol{\phi} = \mathbf{0}$, is often of special interest to the investigator. Figure 4.3.1 shows that the parameter point $\boldsymbol{\phi} = \mathbf{0}$ is just excluded by the approximate 95% H.P.D. region for $\beta = -0.5$, but lies well inside the 90% region for $\beta = 0$, and the 75% region for $\beta = 0.5$. To present a more complete picture, we may calculate, as a function of β, the probability associated with the region

$$p(\boldsymbol{\phi} \mid \boldsymbol{\theta}, \beta, \mathbf{y}) < p(\boldsymbol{\phi} = \mathbf{0} \mid \boldsymbol{\theta}, \beta, \mathbf{y}). \qquad (4.3.16)$$

For this example, we have from (4.3.10) and (4.3.11), with $k = 3$,

$$\Pr\{p(\boldsymbol{\phi} \mid \boldsymbol{\theta}, \beta, \mathbf{y}) < p(\boldsymbol{\phi} = \mathbf{0} \mid \boldsymbol{\theta}, \beta, \mathbf{y}) \mid \mathbf{y}\} \doteq \exp\left\{\frac{-\log W_0}{1 + (1+\beta)^{-1}(0.0148)}\right\}, \qquad (4.3.17)$$

with

$$\log W_0 = 15(1+\beta)\sum_{i=1}^{3}[\log s_i(\beta, \theta_i) - \log \bar{s}(\beta, \boldsymbol{\theta})].$$

Figure 4.3.2 shows this probability for various values of β ranging from -0.9 to 1.0. It is less than 1% for $\beta = -0.9$, monotonically increasing to 21% for $\beta = 0$, and to almost 58% for $\beta = 1$. As mentioned earlier, for a fixed value of β, (4.3.17) is numerically identical to the significance level associated with the likelihood ratio criterion for testing the null hypothesis that the variances are equal. Inferences about the equality of the variances are thus very sensitive to changes in β, irrespective of whether we adopt the sampling or the Bayesian theory.

4.3.2 Posterior Distribution of β and $\boldsymbol{\Phi}$

As before, the non-Normality parameter β can, in the Bayesian framework, be included in the formulation as a variable parameter. Assuming that β and the standard deviations $\boldsymbol{\sigma}$ are *a priori* independent, we may write

$$p(\beta, \boldsymbol{\sigma}) = p(\beta)p(\boldsymbol{\sigma}), \qquad (4.3.18)$$

where $p(\beta)$ is the prior of β and $p(\boldsymbol{\sigma})$ is the distribution in (4.3.2). Combining (4.3.18) with the likelihood function in (4.3.1) and integrating out $\boldsymbol{\sigma}$, for fixed $\boldsymbol{\theta}$ the posterior distribution of β is obtained as

$$p(\beta \mid \boldsymbol{\theta}, \mathbf{y}) \propto p(\beta) p_u(\beta \mid \boldsymbol{\theta}, \mathbf{y}), \qquad -1 < \beta \leqslant 1. \tag{4.3.19}$$

In this expression

$$p_u(\beta \mid \boldsymbol{\theta}, \mathbf{y}) \propto \left[\Gamma\left(1 + \frac{1+\beta}{2}\right)\right]^{-n} \prod_{i=1}^{k} \Gamma\left[1 + \frac{n_i}{2}(1+\beta)\right] [n_i s_i(\beta, \theta_i)]^{-\frac{1}{2}n_i(1+\beta)} \tag{4.3.20}$$

is the posterior distribution of β corresponding to a uniform reference prior.

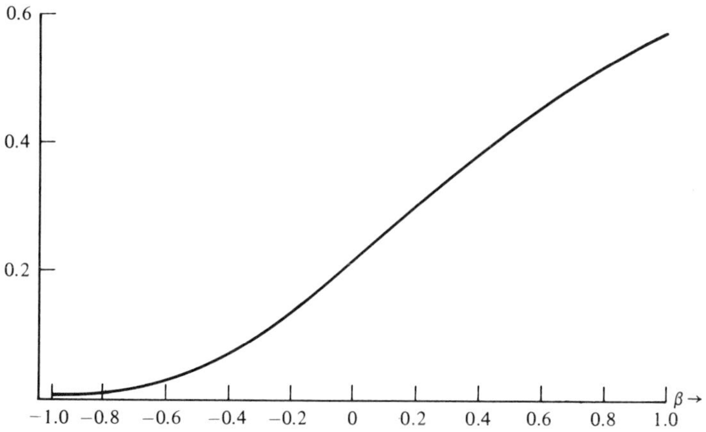

Fig. 4.3.2 The posterior probability $\Pr\{p(\boldsymbol{\phi} \mid \boldsymbol{\theta}, \beta, \mathbf{y}) < p(\boldsymbol{\phi} = \mathbf{0} \mid \boldsymbol{\theta}, \beta, \mathbf{y}) \mid \mathbf{y}\}$ as a function of β for the generated example; the probability is numerically equivalent to the sampling theory significance level.

Figure 4.3.3 for $a = 1$ shows the distribution $p_u(\beta \mid \boldsymbol{\theta}, \mathbf{y})$ for the data of Table 4.3.1. Its close concentration about the mode $\beta \doteq -0.7$, shows it to be quite informative about β and this is to be expected because of the large number of observations ($n = 90$). Even after multiplication by a prior distribution $p(\beta)$ which indicated a rather strong belief in a central limit tendency, inferences about β would still not be much different from that based upon $p_u(\beta \mid \boldsymbol{\theta}, \mathbf{y})$. In each of the four sets of graphs in Fig. 4.3.3, the dotted curve is the assumed prior and the solid curve the corresponding posterior distribution. The prior distributions were taken from (4.2.20) with $a = 1$, 3, 6, 10, respectively. Thus with "a" as high as 10, the information from $p_u(\beta \mid \boldsymbol{\theta}, \mathbf{y})$ still dominates the "prior" information. This figure may be contrasted with Fig. 3.2.5 for which only 15 observations (differences) were available and the information from $p_u(\beta \mid \mathbf{y})$ had much less weight.

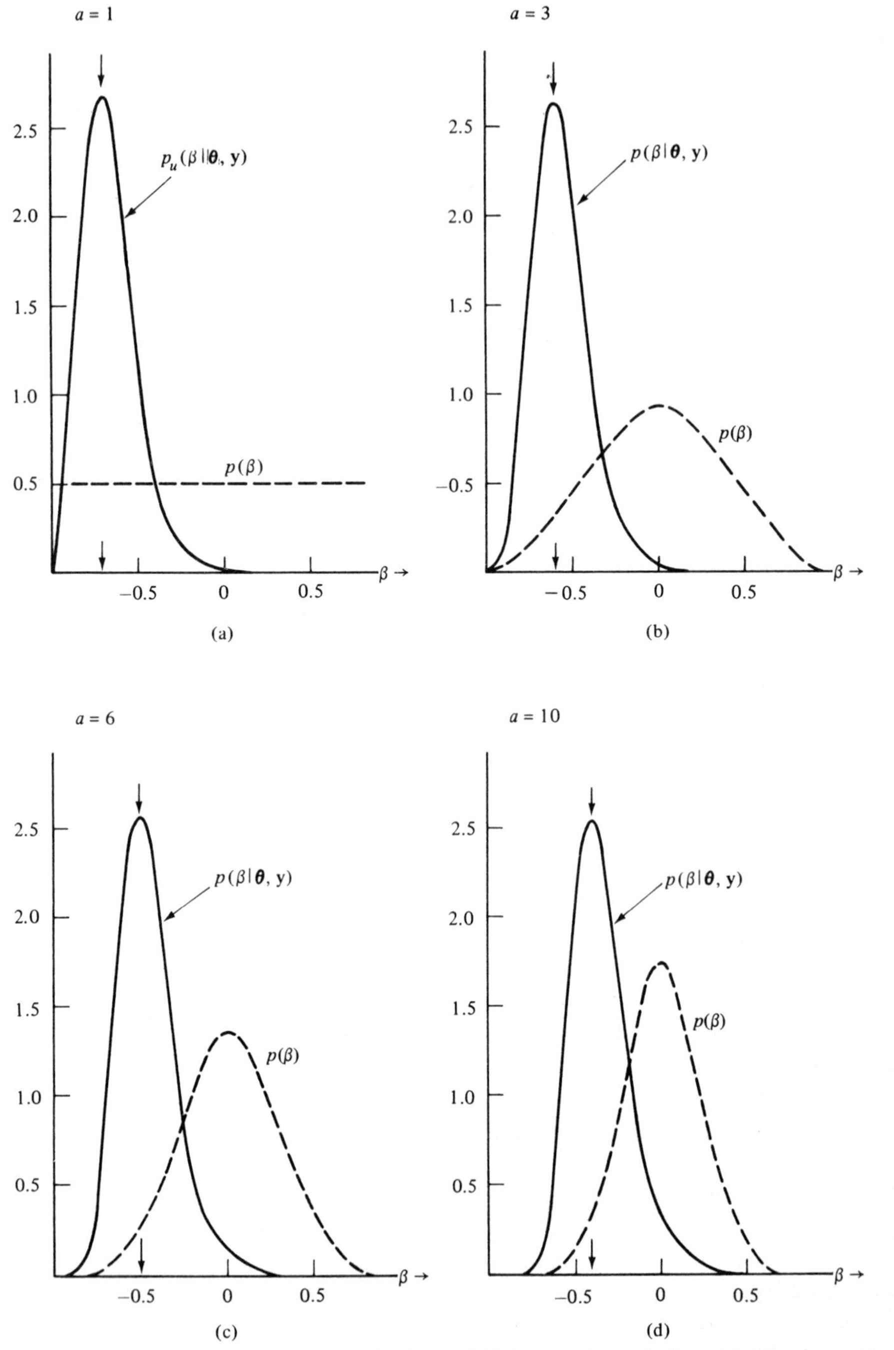

Fig. 4.3.3 Prior and posterior distribution of β for various choices of the parameter "a": the generated example.

Posterior Distribution of $\boldsymbol{\phi}$ *and its Approximation*

In the Bayesian framework, uncertainty in the inferences about $\boldsymbol{\phi}$ due to changes in β can be removed by considering the marginal posterior distribution of $\boldsymbol{\phi}$. From the distribution of $\boldsymbol{\phi}$, given β, in (4.3.5) and that of β in (4.3.19), we may write

$$p(\boldsymbol{\phi} \mid \boldsymbol{\theta}, \mathbf{y}) = \int_{-1}^{1} p(\boldsymbol{\phi} \mid \boldsymbol{\theta}, \beta, \mathbf{y}) p(\beta \mid \boldsymbol{\theta}, \mathbf{y})\, d\beta, \qquad -\infty < \phi_i < \infty, \quad i = 1, \ldots, k-1. \tag{4.3.21}$$

Although both the distribution $p(\beta \mid \boldsymbol{\theta}, \mathbf{y})$ and the conditional distribution $p(\boldsymbol{\phi} \mid \boldsymbol{\theta}, \beta, \mathbf{y})$ involve only simple functions of the observations, it does not seem possible to express the marginal distribution in a simple closed form. For $k \geqslant 3$, complete evaluation of the distribution would be very burdensome even on a fast computer. However, in situations in which the distribution of β is sharp and nearly symmetrical, we can write, approximately,

$$p(\boldsymbol{\phi} \mid \boldsymbol{\theta}, \mathbf{y}) \doteq p(\boldsymbol{\phi} \mid \boldsymbol{\theta}, \hat{\beta}, \mathbf{y}) \tag{4.3.22}$$

where $\hat{\beta}$ is the mode of $p(\beta \mid \boldsymbol{\theta}, \mathbf{y})$. For instance, the posterior distribution $p_u(\beta \mid \boldsymbol{\theta}, \mathbf{y})$ based upon a reference uniform prior in β has its mode close to $\beta = -0.7$ for the present example. Using the approximation in (4.3.22), the marginal posterior distribution is thus nearly $p(\boldsymbol{\phi} \mid \boldsymbol{\theta}, \beta = -0.7, \mathbf{y})$. Contours of this approximate distribution are shown in Fig. 4.3.4.

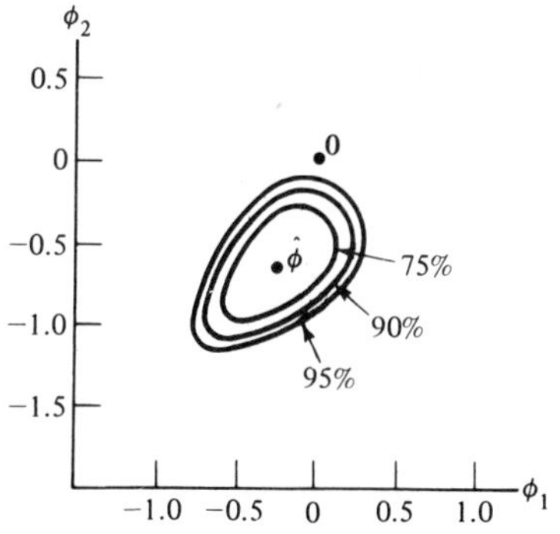

Fig. 4.3.4 Contours of the approximate posterior distribution of (ϕ_1, ϕ_2), $p(\phi_1, \phi_2 \mid \boldsymbol{\theta}, \mathbf{y}) \doteq p(\phi_1, \phi_2 \mid \boldsymbol{\theta}, \hat{\beta} = -0.7, \mathbf{y})$: the generated example.

Accuracy of the Approximation

To check the accuracy of the modal approximation, the exact marginal distributions may be compared with the approximate marginals implied by (4.3.22). From

(4.3.5), it is straightforward to verify that, given β, the distribution of ϕ_i is

$$p(\phi_i \mid \boldsymbol{\theta}, \beta, \mathbf{y}) = \frac{\Gamma[\frac{1}{2}(n_i + n_k)(1+\beta)]}{(1+\beta)\Gamma[\frac{1}{2}n_i(1+\beta)]\Gamma[\frac{1}{2}n_k(1+\beta)]} U_i^{\frac{1}{2}n_i(1+\beta)} (1 + U_i)^{-\frac{1}{2}(n_i+n_k)(1+\beta)}$$

$$-\infty < \phi_i < \infty, \qquad (4.3.23)$$

where

$$U_i = \frac{n_i s_i(\beta, \theta_i)}{n_k s_k(\beta, \theta_k)} \exp[-(1+\beta)^{-1}\phi_i], \qquad i = 1, \ldots, k-1.$$

Thus, the marginal distribution of ϕ_i is

$$p(\phi_i \mid \boldsymbol{\theta}, \mathbf{y}) = \int_{-1}^{1} p(\phi_i \mid \boldsymbol{\theta}, \beta, \mathbf{y}) p(\beta \mid \boldsymbol{\theta}, \mathbf{y})\, d\beta \qquad (4.3.24)$$

which, by adopting the same argument leading to (4.3.22), is approximately

$$p(\phi_i \mid \boldsymbol{\theta}, \mathbf{y}) \doteq p(\phi_i \mid \boldsymbol{\theta}, \hat{\beta}, \mathbf{y}). \qquad (4.3.25)$$

For example, the solid curve in Fig. 4.3.5 shows the exact distribution of ϕ_1 for the generated data based upon $p_u(\beta \mid \boldsymbol{\theta}, \mathbf{y})$, and the dotted curve in the same figure is the corresponding approximate distribution. The agreement between the two is fairly close.

4.3.3 The Situation When $\theta_1, \ldots, \theta_k$ Are Not Known

In the above, we have assumed that the location parameters $\theta_1, \ldots, \theta_k$ are known. The more common situation is when they are not known. With the assumption that $\boldsymbol{\theta}$, $\boldsymbol{\sigma}$, and β are *a priori* approximately independent, and that

$$p(\boldsymbol{\theta}) = \prod_{i=1}^{k} p(\theta_i) \qquad \text{where} \qquad p(\theta_i) \propto c, \qquad (4.3.26)$$

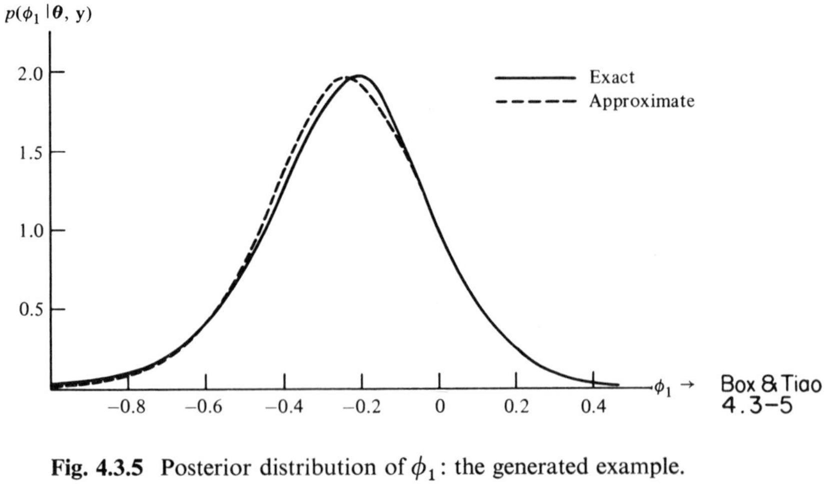

Fig. 4.3.5 Posterior distribution of ϕ_1: the generated example.

the posterior distribution of $\boldsymbol{\theta}$, for a given β, is

$$p(\boldsymbol{\theta} \mid \beta, \mathbf{y}) = \prod_{i=1}^{k} p(\theta_i \mid \beta, \mathbf{y}_i), \qquad -\infty < \theta_i < \infty, \quad i = 1, \ldots, k, \qquad (4.3.27)$$

where

$$p(\theta_i \mid \beta, \mathbf{y}_i) = \frac{[n_i s_i(\beta, \theta_i)]^{-\frac{1}{2}n_i(1+\beta)}}{\int_{-\infty}^{\infty} [n_i s_i(\beta, \theta_i)]^{-\frac{1}{2}n_i(1+\beta)}\, d\theta_i}.$$

Consequently, for a given β, the posterior distribution of the $(k-1)$ contrasts $\boldsymbol{\phi}$ defined in (4.3.4) becomes,

$$p(\boldsymbol{\phi} \mid \beta, \mathbf{y}) = \int_{-\infty}^{\infty} \cdots \int_{-\infty}^{\infty} p(\boldsymbol{\phi} \mid \boldsymbol{\theta}, \beta, \mathbf{y}) p(\boldsymbol{\theta} \mid \beta, \mathbf{y})\, d\theta_1 \cdots d\theta_k,$$

$$-\infty < \phi_i < \infty, \quad i = 1, \ldots, k-1 \qquad (4.3.28)$$

where the first factor in the integrand is given by (4.3.5). In the special case where the populations are Normal, it is readily verified that the integral can be evaluated exactly yielding,

$$p(\boldsymbol{\phi} \mid \beta = 0, \mathbf{y}) = \frac{\Gamma(\frac{1}{2}\nu)}{\prod_{i=1}^{k} \Gamma(\frac{1}{2}\nu_i)}\, T_1^{\frac{1}{2}\nu_1} \cdots T_{k-1}^{\frac{1}{2}\nu_{k-1}}(1 + T_1 + \cdots + T_{k-1})^{-\frac{1}{2}\nu},$$

$$-\infty < \phi_i < \infty, \ldots, -\infty < \phi_{k-1} < \infty, \qquad (4.3.29)$$

where

$$\nu_i = n_i - 1, \qquad \nu = n - k, \qquad \text{and} \qquad T_i = \frac{\sum^{n_i} (y_{ij} - \bar{y}_i)^2}{\sum^{n_k} (y_{kj} - \bar{y}_k)^2}\, e^{-\phi_i},$$

which is, of course, the same distribution obtained earlier in (2.12.6).

In the more general situation when the parent populations are not necessarily Normal, it does not seem possible to express the integral exactly in terms of simple functions. However, when dealing with the ratio of two variances, it was demonstrated that the effect of integrating over the posterior distribution of (θ_1, θ_2) was essentially to replace the θ's by their corresponding modal values and to reduce the "degrees of freedom," $n_i(1+\beta)$, $(i = 1, 2)$, by one unit. We can extend this approximation to the more general case where again it is exact if $\beta = 0$. As in (3.3.6) we write

$$n_i s_i(\beta, \theta_i) \doteq n_i s_i(\beta, \hat{\theta}_i) + d_i(\theta_i - \hat{\theta}_i)^2, \qquad i = 1, \ldots, k, \qquad (4.3.30)$$

where $\hat{\theta}_i$ is the value of θ_i minimizing $n_i s_i(\beta, \theta_i)$ and, therefore, is also the mode of the posterior distribution $p(\theta_i \mid \beta, \mathbf{y}_i)$ in (4.3.27). Substituting (4.3.30) into

(4.3.28) and integrating out $\boldsymbol{\theta}$, the posterior distribution $p(\boldsymbol{\phi} \mid \beta, \mathbf{y})$ is then approximately

$$p(\boldsymbol{\phi} \mid \beta, \mathbf{y}) \doteq (1+\beta)^{-(k-1)} \frac{\Gamma(\frac{1}{2}m)}{\prod_{i=1}^{k} \Gamma(\frac{1}{2}m_i)} \gamma_1^{\frac{1}{2}m_1} \cdots \gamma_{k-1}^{\frac{1}{2}m_{k-1}} (1+\gamma_1+\cdots+\gamma_{k-1})^{-\frac{1}{2}m},$$

$$-\infty < \phi_1 < \infty, \ldots, -\infty < \phi_{k-1} < \infty, \qquad (4.3.31)$$

where

$$m_i = n_i(1+\beta) - 1, \qquad m = n(1+\beta) - k,$$

and

$$\gamma_i = \frac{n_i s_i(\beta, \hat{\theta}_i)}{n_k s_k(\beta, \hat{\theta}_k)} \exp\left[-\frac{\phi_i}{(1+\beta)}\right], \qquad i = 1, \ldots, k-1.$$

As in (4.3.24) for the case of the comparison of two variances, in obtaining the approximate distribution (4.3.31) through the use of (4.3.30), it is only necessary to determine the modal values $\hat{\theta}_i$, the quantities d_i being cancelled in the integration process. The distribution (4.3.31) is of exactly the same form as that in (4.3.5). Consequently, to this degree of approximation, the cumulant generating function of

$$M^* = -2 \log W^* \qquad (4.3.32)$$

where

$$W^* = \frac{m^{\frac{1}{2}m}}{\prod_{i=1}^{k} m_i^{\frac{1}{2}m_i}} \gamma_1^{\frac{1}{2}m_1} \cdots \gamma_{k-1}^{\frac{1}{2}m_{k-1}} (1+\gamma_1+\cdots+\gamma_{k-1})^{-\frac{1}{2}m}$$

is given by

$$\kappa_{M^*}(t) = a - \frac{k-1}{2} \log(1-2t) + \sum_{r=1}^{\infty} \alpha_r (1-2t)^{-(2r-1)}, \qquad (4.3.33)$$

where

$$\alpha_r = \frac{B_{2r}}{2r(2r-1)} 2^{2r-1} \left[\sum_{i=1}^{k} m_i^{-(2r-1)} - m^{-(2r-1)}\right].$$

Hence the distribution of M^* can, as before, be approximated using Bartlett's method.

To decide whether a parameter point $\boldsymbol{\phi}_0$ is included in the H.P.D. region of content $(1-\alpha)$, we calculate

$$\Pr\{M^* > -2 \log W_0^*\} = \Pr\left\{\chi_{k-1}^2 < \frac{-2 \log W_0^*}{1+A^*}\right\}, \qquad (4.3.34)$$

where W_0^* is obtained by substituting $\boldsymbol{\phi}_0$ into (4.3.32) and

$$A^* = \frac{1}{3(k-1)} \left[\sum_{i=1}^{k} m_i^{-1} - m^{-1}\right].$$

In particular, for $\boldsymbol{\phi}_0 = \mathbf{0}$, $-2 \log W_0^*$ reduces to

$$-2 \log W_0^* = -\sum_{i=1}^{k} m_i \left[\log \frac{n_i s_i(\beta, \hat{\theta}_i)}{m_i} - \log \bar{s}(\beta, \hat{\boldsymbol{\theta}})\right], \qquad (4.3.35)$$

with

$$\bar{s}(\beta, \hat{\boldsymbol{\theta}}) = \frac{1}{m} \sum_{i=1}^{k} n_i s_i(\beta, \hat{\theta}_i).$$

In the case $\beta = 0$, we obtain precisely the Normal theory results already discussed.

Using Bartlett's method of approximation the somewhat remarkable result is obtained, therefore, that for *any* known value of β (not close to -1), the decision as to whether the point $\boldsymbol{\phi}_0$ lies inside or outside the $(1 - \alpha)$ H.P.D. region is made by referring $-2 \log W_0^*$ to a scaled χ^2 distribution. In the case $\boldsymbol{\phi}_0 = \mathbf{0}$ which corresponds to $\sigma_1 = \cdots = \sigma_k$, the quantity $-2 \log W_0^*$ is in exactly the same form as Bartlett's modified form of the likelihood ratio statistic for $\beta = 0$, except that n_i is replaced by $n_i(1 + \beta)$ and the sample variances s_i^2 by the quantities

$$\frac{n_i s_i(\beta, \hat{\theta}_i)}{n_i(1 + \beta) - 1} = \frac{\Sigma\, |y_{ij} - \hat{\theta}_i|^{2/(1+\beta)}}{n_i(1 + \beta) - 1}.$$

We are thus able to obtain, for each β, the approximate probability content of an H.P.D. region which would just exclude the point $\boldsymbol{\phi} = \mathbf{0}$. For a given set of data, we can, therefore, study how inferences about equality of the standard deviations $\boldsymbol{\sigma}$ may be affected by the departure from Normality in the parent populations. If β is assumed equal to zero and/or $\boldsymbol{\theta}$ are assumed known, sampling results can be obtained which exactly match the Bayesian results. But in the more general situation when $\beta \neq 0$ and the $\boldsymbol{\theta}$ unknown, no corresponding sampling result is available.

Marginal Posterior Distributions of β and $\boldsymbol{\phi}$

Combining the likelihood function (4.3.1) with the priors of $(\boldsymbol{\sigma}, \beta, \boldsymbol{\theta})$ in (4.3.2), (4.2.20), and (4.3.26), and integrating out $\boldsymbol{\theta}$ and $\boldsymbol{\sigma}$, we obtain the posterior distribution of β

$$p(\beta \mid \mathbf{y}) \propto p(\beta) p_u(\beta \mid \mathbf{y}), \qquad -1 < \beta \leqslant 1, \qquad (4.3.36)$$

where

$$p_u(\beta \mid \mathbf{y}) \propto \left[\Gamma\left(1 + \frac{1 + \beta}{2}\right)\right]^{-n} \prod_{i=1}^{k} \Gamma\left[1 + \frac{n_i}{2}(1 + \beta)\right] \int_{-\infty}^{\infty} [n_i s_i(\beta, \theta_i)]^{-\frac{1}{2}n_i(1+\beta)}\, d\theta_i$$

is the posterior distribution of β, corresponding to a uniform reference prior distribution.

Now, we can write the marginal distribution of $\boldsymbol{\phi}$ as

$$p(\boldsymbol{\phi} \mid \mathbf{y}) = \int_{-1}^{1} p(\boldsymbol{\phi} \mid \beta, \mathbf{y}) p(\beta \mid \mathbf{y})\, d\beta, \qquad (4.3.37)$$

where the first factor of the integrand is given by (4.3.28). In obtaining this distribution, the unknown location parameters $\boldsymbol{\theta}$ and the non-Normality parameter β are eliminated by integration. The distribution thus provides the final overall inferences about the linear contrasts of the logarithms of the variances of the k populations.

In practice, numerical integration of β and $\boldsymbol{\theta}$, and particularly $\boldsymbol{\theta}$, would be exceedingly burdensome. However, the factor $p(\boldsymbol{\phi} \mid \beta, \mathbf{y})$ can be approximated by (4.3.31) for values of β not close to -1, and when $p(\beta \mid \mathbf{y})$ is concentrated about some modal value $\hat{\beta}$ we can employ the further approximation

$$p(\boldsymbol{\phi} \mid \mathbf{y}) \doteq p(\boldsymbol{\phi} \mid \hat{\beta}, \mathbf{y}). \tag{4.3.38}$$

Thus in appropriate circumstances $p(\boldsymbol{\phi} \mid \mathbf{y})$ may be approximated by the right-hand side of (4.3.31) with $\hat{\beta}$ substituted for the unknown β. Although evaluation of $\hat{\beta}$ for (4.3.36) still requires numerical integration of one-dimensional integrals as well as numerical determination of a maximum, these are simple processes compared with evaluation of the exact distribution.

4.4 INFERENCE ROBUSTNESS AND CRITERION ROBUSTNESS

Using the notation of Section 1.6 suppose that $\boldsymbol{\theta}_1$ are a set of parameters of interest and $\boldsymbol{\theta}_2$ a set of nuisance parameters measuring discrepancies from "ideal" conditions $\boldsymbol{\theta}_{20}$. Then:

a) robustness of inferences about $\boldsymbol{\theta}_1$ to departures from the ideal conditions, may be studied by considering how the conditional posterior distribution $p(\boldsymbol{\theta}_1 \mid \boldsymbol{\theta}_2, \mathbf{y})$ changes, as the elements of $\boldsymbol{\theta}_2$ are moved away from the values $\boldsymbol{\theta}_{20}$,

b) at the same time, the marginal posterior distribution $p(\boldsymbol{\theta}_2 \mid \mathbf{y})$ measures the plausibility of various choices for $\boldsymbol{\theta}_2$, and integration of $p(\boldsymbol{\theta}_1 \mid \boldsymbol{\theta}_2, \mathbf{y})$ with $p(\boldsymbol{\theta}_2 \mid \mathbf{y})$ as weight function yields $p(\boldsymbol{\theta}_1 \mid \mathbf{y})$, which, in suitable circumstances, shows what overall inferences can be made about $\boldsymbol{\theta}_1$.

The ideas have been illustrated, in this chapter and the previous one, with $\boldsymbol{\theta}_2 = \beta$ measuring a departure from Normality and with the elements of $\boldsymbol{\theta}_1$ being in turn means, regression coefficients and variances. Since in any given instance we never know for certain how elaborate assumptions need to be, the general possibility of embedding a tentative parsimonious subset of assumptions in a more prodigal set and studying the sensitivity of the conditional inference can provide a highly informative technique of preliminary data analysis.

Such studies of inference robustness are, as has been said, of a different character from the customary criterion robustness studies of sampling theory. However, inference robustness *can* be interpreted in terms of sampling theory although usually the limitations of that theory make it difficult to apply.

In the sampling theory study of criterion robustness, an "optimal" criterion $C_{\boldsymbol{\theta}_{20}}(\mathbf{y})$ is selected which is appropriate for making inferences about the parameters

$\boldsymbol{\theta}_1$ for some fixed $\boldsymbol{\theta}_2 = \boldsymbol{\theta}_{20}$ corresponding to the "ideal" assumptions. The change in the sampling distribution of $C_{\boldsymbol{\theta}_{20}}(\mathbf{y})$ is then studied as the parameters $\boldsymbol{\theta}_2$ are changed from $\boldsymbol{\theta}_{20}$.

By contrast, to study inference robustness in the sampling framework, we need to study the sampling distribution not of $C_{\boldsymbol{\theta}_{20}}(\mathbf{y})$ but of $C_{\boldsymbol{\theta}_2}(\mathbf{y})$ as $\boldsymbol{\theta}_2$ are changed.

It is an extraordinary fact that for a sample drawn from an exponential power distribution of which the mean is assumed known, sufficient statistics exist for the variance over the *entire range* of the kurtosis parameter β. In this unusual circumstance therefore, it is possible with sampling theory to study not only criterion robustness but also inference robustness. We find in this exceptional case, where study of inference robustness is accessible to sampling theory, that it parallels exactly the Bayesian result.

4.4.1 The Analyst Example

As a specific illustration, we may again consider the variance comparison data of Table 4.2.1 consisting of 13 independent observations made by an Analyst A_1 and 20 made by an Analyst A_2. Although population means were unknown, for the purpose of this demonstration we shall assume them known and equal to the sample means. On the Normal theory test of equality of variances, the significance level, that is, the probability of exceeding the observed variance ratio by chance, is 7.6%.

Table 4.4.1

Changes in percentage significance level induced by departures of the population β from the β_0 defining the criterion $C_{\beta 0}(\mathbf{y})$: the analyst data

	β / β_0	Parent population −0.6	−0.4	−0.2	0.0	0.2	0.4	0.6
Criterion	−0.6	**6.0**	11.0	—	—	—	—	—
	−0.4	3.5	**6.5**	9.0	14.0	—	—	—
	−0.2	3.0	4.8	**7.0**	9.8	11.5	17.0	22.0
	0.0	2.8	4.5	6.0	**7.6**	9.5	12.0	14.5
	0.2	2.5	4.0	5.0	6.0	**8.0**	10.0	12.0
	0.4	2.4	3.5	4.8	6.0	8.0	**8.6**	9.5
	0.6	2.4	4.0	4.8	5.0	6.5	7.5	**9.2**

Table 4.4.1 shows the percentage significance levels calculated on standard sampling theory for this example under a number of different circumstances. The details of how these calculations were made are given later. In calculating the table, we have assumed the two parent distributions to be members of the class of exponential power distributions defined in (3.2.5) having common value of β and known means. Then, for any fixed value of β, say $\beta = \beta_0$, the uniformly

most powerful similar test of the hypothesis $H_0: \sigma_2^2/\sigma_1^2 = 1$ against the alternative $H_1: \sigma_2^2/\sigma_1^2 > 1$ is provided by the ratio

$$C_{\beta_0}(\mathbf{y}) = \frac{s_2(\beta_0, \theta_2)}{s_1(\beta_0, \theta_1)} = \frac{\sum^{n_2} |y_{2j} - \theta_2|^{2/(1+\beta_0)}/n_2}{\sum^{n_1} |y_{1j} - \theta_1|^{2/(1+\beta_0)}/n_1}, \tag{4.4.1}$$

in which the numerator and the denominator are sufficient statistics for the variances σ_2^2 and σ_1^2, respectively. In particular, when $\beta_0 = 0$ the criterion $C_0(\mathbf{y})$ is the usual F statistic. Consider for example the row in the table for $\beta_0 = 0$. The entries show how the percentage significance levels for the F criterion change as the β value of the parent population changes. For instance, 7.6% is the F criterion significance level for a Normal parent ($\beta = 0$) and the value 9.5% immediately to the right of this shows the significance level for a somewhat more leptokurtic parent distribution ($\beta = 0.2$) but with the same Normal theory F criterion $C_0(\mathbf{y})$. Similarly, if we take the values corresponding to the next row for $\beta_0 = 0.2$, we have the corresponding probabilities for the criterion

$$C_{0.2}(\mathbf{y}) = \frac{n_1}{n_2}\left[\frac{\sum^{n_2} |y_{2j} - \theta_2|^{2/1.2}}{\sum^{n_1} |y_{1j} - \theta_1|^{2/1.2}}\right], \tag{4.4.2}$$

which will provide a uniformly most powerful similar test for a parent with $\beta = 0.2$. In particular, the significance level for this parent population is 8.0%.

Now, however, consider the change from 7.6% to 8.0% in the diagonal. This gives a measure of inference robustness. Specifically, it shows how much the significance level changes when the parent distribution *and* the appropriate criterion are changed together.

Thus, while the familiar criterion robustness is measured by the changes occurring horizontally across the table, inference robustness is measured by changes occurring in the diagonal elements which are printed in bold type. It is noticeable that the changes which occur horizontally are, for this data, considerably greater than those which occur diagonally. In fact, whereas the probability of the error of the first kind for the Normal theory criterion is changed by a factor of 5 (from 2.8% to 14.5%), in changing from a platykurtic distribution with $\beta = -0.6$ to a leptokurtic distribution with $\beta = 0.6$, it is changed only by a factor of 1.5 (from 6.0% to 9.2%) when appropriate modification is made in the criterion. While one cannot on this evidence draw any general conclusions, it is true that for these particular data, the inferential probabilities about the ratio of variances are much less affected by changes in β than are the probabilities associated with a particular criterion.

The above discussion has been conducted so far entirely in terms of sampling theory. It will be recalled from Section 4.2.3 that the diagonal elements of the table are precisely the *a posteriori* probabilities that the variance ratio σ_2^2/σ_1^2 is less than unity for the corresponding values of β_0 in the parents. The *inference robustness* study under sampling theory and the Bayesian robustness study thus give precisely parallel results.

4.4.2 Derivation of the Criteria

For any specific value of β, say $\beta = \beta_0$, the uniformly most powerful similar criterion $C_{\beta_0}(\mathbf{y})$ in (4.4.1) follows an F distribution with $n_2(1+\beta_0)$ and $n_1(1+\beta_0)$ degrees of freedom when the hypothesis $H_0: \sigma_2^2/\sigma_1^2 = 1$ is true. Equivalently, the statistic

$$W(\beta_0) = \frac{\sum^{n_1} |y_{1j} - \theta_1|^{2/(1+\beta_0)}}{\sum^{n_1} |y_{1i} - \theta_1|^{2/(1+\beta_0)} + \sum^{n_2} |y_{2j} - \theta_2|^{2/(1+\beta_0)}} \tag{4.4.3}$$

has a beta distribution with parameters $\frac{1}{2}n_1(1+\beta_0)$ and $\frac{1}{2}n_2(1+\beta_0)$.

Using this result, the exact probabilities in the diagonal of Table 4.4.1, namely

$$\Pr\{F_{n_2(1+\beta_0), n_1(1+\beta_0)} > C_{\beta_0}(\mathbf{y})\},$$

can be readily calculated. For the off-diagonal elements, we need to find the distribution of the criterion $C_{\beta_0}(\mathbf{y})$ when the parent β takes some value other than β_0. This can be approximated using permutation theory.

For the exponential power distribution (3.2.5), it is readily shown by employing (A2.1.5) in Appendix A2.1 that the variate

$$X = \left| \frac{y - \theta}{\sigma} \right|^{2/(1+\beta_0)} \tag{4.4.4}$$

has as its rth moment (about the origin)

$$\mu_r = \frac{\Gamma\{(1+\beta)[\frac{1}{2} + r/(1+\beta_0)]\}}{\Gamma[\frac{1}{2}(1+\beta)]} [c(\beta)]^{-r(1+\beta)/(1+\beta_0)} \qquad r = 1, 2, 3, \ldots. \tag{4.4.5}$$

Now write

$$X_j = \begin{cases} \left| \dfrac{y_{1j} - \theta_1}{\sigma_1} \right|^{2/(1+\beta_0)} & j = 0, 1, \ldots, n_1, \\ \left| \dfrac{y_{2(j-n_1)} - \theta_2}{\sigma_2} \right|^{2/(1+\beta_0)} & j = n_1 + 1, n_1 + 2, \ldots, n_1 + n_2. \end{cases}$$

Then, on the hypothesis $H_0: \sigma_2^2/\sigma_1^2 = 1$ and following the method in Box and Andersen (1955), the permutation moments of

$$W(\beta_0) = \frac{\sum_{i=1}^{n_1} X_j}{\sum_{j=1}^{n} X_j}, \qquad n = n_1 + n_2 \tag{4.4.6}$$

can be written

$$\underset{p}{E}[W(\beta_0)] = \frac{n_1}{n}, \quad \underset{p}{V}[W(\beta_0)] = \frac{2n_1n_2}{n^2(n+2)}\left[1 + \frac{1}{2}\frac{n}{n-1}(b_2 - 3)\right], \tag{4.4.7}$$

where

$$b_2 = (n+2)\frac{\sum^n X_j^2}{(\sum^n X_j)^2}.$$

By taking the expectation over all samples of the permutation moments, we obtain the ordinary moments of $W(\beta_0)$ as

$$E[W(\beta_0)] = \frac{n_1}{n}, \quad V[W(\beta_0)] = \frac{2n_1 n_2}{n^2(n+2)}\left[1 + \frac{1}{2}\frac{n}{n-1}E(b_2 - 3)\right]. \tag{4.4.8}$$

Expanding the denominator of b_2 around the mean of X and taking expectations, we find that, to order n^{-2}

$$E(b_2 - 3) = \frac{\mu_2}{\mu_1^2} - 3 + n^{-1}\left[\frac{\mu_2}{\mu_1^2} - \frac{2\mu_3}{\mu_1^3} + \frac{3\mu_2^2}{\mu_1^4}\right]$$

$$- n^{-2}\left[\frac{\mu_2}{\mu_1^2} - \frac{2\mu_3}{\mu_1^3} - \frac{3}{\mu_1^4}(\mu_4 - \mu_2^2) + \frac{16\mu_3\mu_2}{\mu_1^5} - \frac{15\mu_2^3}{\mu_1^6}\right], \tag{4.4.9}$$

where μ_r are given in (4.4.5).

For a specific value of β, it can be shown that the statistic $W(\beta_0)$ is approximately distributed as a beta variable with parameters $[\frac{1}{2}n_1(1+\beta_0)\delta, \frac{1}{2}n_2(1+\beta_0)\delta]$, where

$$\delta = [1 + \tfrac{1}{2}E(b_2 - 3)]^{-1} \tag{4.4.10}$$

represents the modification due to departure of β from β_0. Equivalently, $C_{\beta_0}(\mathbf{y})$ is approximately distributed as an F variable with $n_2(1+\beta_0)\delta$ and $n_1(1+\beta_0)\delta$ degrees of freedom from which the off-diagonal probabilities in Table 4.4.1 can be approximately determined. The result is, of course, exact when $\beta = 0$.

4.5 A SUMMARY OF FORMULAE FOR VARIOUS PRIOR AND POSTERIOR DISTRIBUTIONS

For convenience of the reader, Table 4.5.1 gives a summary of the formulae for various prior and posterior distributions concerning inferences about comparison of variances.

Table 4.5.1

A summary of prior and posterior distributions

1. Suppose k samples $\mathbf{y}_1' = (y_{11}, \ldots, y_{1n_1}), \ldots, \mathbf{y}_k' = (y_{k1}, \ldots, y_{kn_k})$ are drawn from possibly different members of the exponential power distribution with common β

$$p(y \mid \theta, \sigma, \beta) = \omega(\beta)\sigma^{-1}\exp\left[-c(\beta)\left|\frac{y-\theta}{\sigma}\right|^{2/(1+\beta)}\right], \qquad -\infty < y < \infty,$$

Table 4.5.1 *Continued*

where

$$c(\beta) = \left\{\frac{\Gamma[\frac{3}{2}(1+\beta)]}{\Gamma[\frac{1}{2}(1+\beta)]}\right\}^{1/(1+\beta)},$$

$$\omega(\beta) = \frac{\{\Gamma[\frac{3}{2}(1+\beta)]\}^{1/2}}{(1+\beta)\{\Gamma[\frac{1}{2}(1+\beta)]\}^{3/2}}.$$

2. Let $\boldsymbol{\theta} = (\theta_1, \ldots, \theta_k)$ and $\boldsymbol{\sigma} = (\sigma_1, \ldots, \sigma_k)$
The prior from (4.3.2), (4.3.18), and (4.3.26) is

$$p(\boldsymbol{\theta}, \boldsymbol{\sigma}, \beta) = p(\boldsymbol{\theta})p(\boldsymbol{\sigma})p(\beta)$$

with

$$p(\boldsymbol{\theta}) \propto c, \qquad p(\boldsymbol{\sigma}) \propto \prod_{i=1}^{k} \sigma_i^{-1}, \qquad p(\beta) \propto (1-\beta^2)^{a-1}.$$

The Case $k = 2$

3. Conditional on $(\boldsymbol{\theta}, \beta)$, the posterior distribution of $V = \sigma_2^2/\sigma_1^2$, $p(V \mid \beta, \boldsymbol{\theta}, \mathbf{y})$, is in (4.2.7),

$$V^{1/(1+\beta)} \frac{s_1(\beta, \theta_1)}{s_2(\beta, \theta_2)} \sim F_{n_1(1+\beta),\, n_2(1+\beta)},$$

where

$$s_i(\beta, \theta_i) = \frac{1}{n_i} \Sigma \, |y_{ij} - \theta_i|^{2/(1+\beta)}.$$

Thus

$$\Pr\{V < 1 \mid \beta, \boldsymbol{\theta}, \mathbf{y}\} = \Pr\left\{F_{n_1(1+\beta), n_2(1+\beta)} < \frac{s_1(\beta, \theta_1)}{s_2(\beta, \theta_2)}\right\}.$$

In particular, for $\beta = 0$

$$V \frac{\Sigma(y_{1j} - \theta_1)^2/n_1}{\Sigma(y_{2j} - \theta_2)^2/n_2} \sim F_{n_1,\, n_2}.$$

For $\beta \to -1$, the distribution of

$$u = V\left(\frac{h_1 + |m_1 - \theta_1|}{h_2 + |m_2 - \theta_2|}\right)^2$$

is, from (4.2.10),

$$p(u \mid \beta \to -1, \boldsymbol{\theta}, \mathbf{y}) = \begin{cases} \dfrac{n_1 n_2}{2(n_1+n_2)} u^{\frac{1}{2}n_1 - 1} & 0 < u < 1 \\[2ex] \dfrac{n_1 n_2}{2(n_1+n_2)} u^{-\frac{1}{2}n_2 - 1} & 1 < u < \infty, \end{cases}$$

where (h_1, h_2) and (m_1, m_2) are the half-ranges and the mid-points, respectively.

Table 4.5.1 *Continued*

4. Conditional on $\boldsymbol{\theta}$, the posterior distribution of β is, in (4.2.17),

$$p(\beta \mid \boldsymbol{\theta}, \mathbf{y}) \propto p(\beta) p_u(\beta \mid \boldsymbol{\theta}, \mathbf{y}), \qquad -1 < \beta \leqslant 1,$$

with

$$p_u(\beta \mid \boldsymbol{\theta}, \mathbf{y}) \propto \{\Gamma[1 + \tfrac{1}{2}(1+\beta)]\}^{-(n_1+n_2)} \times \prod_{i=1}^{2} \Gamma[1 + \tfrac{1}{2}n_i(1+\beta)][n_i s_i(\beta, \theta_i)]^{-\frac{1}{2}n_i(1+\beta)}$$

5. The posterior distribution of V given $\boldsymbol{\theta}$ is, from (4.2.16),

$$p(V \mid \boldsymbol{\theta}, \mathbf{y}) = \int_{-1}^{1} p(V \mid \beta, \boldsymbol{\theta}, \mathbf{y}) p(\beta \mid \boldsymbol{\theta}, \mathbf{y})\, d\beta, \qquad V > 0.$$

6. The posterior distribution of $\boldsymbol{\theta}$, given β, is, in (4.2.24),

$$p(\boldsymbol{\theta} \mid \beta, \mathbf{y}) = \prod_{i=1}^{2} \frac{[n_i s_i(\beta, \theta_i)]^{-\frac{1}{2}n_i(1+\beta)}}{\int_{-\infty}^{\infty} [n_i s_i(\beta, \theta_i)]^{-\frac{1}{2}n_i(1+\beta)}\, d\theta_i}, \qquad -\infty < \theta_i < \infty.$$

7. The posterior distribution of V, given β, with $\boldsymbol{\theta}$ eliminated is, in (4.2.23) and (4.2.25),

$$p(V \mid \beta, \mathbf{y}) = \int_{-\infty}^{\infty} \int_{-\infty}^{\infty} p(V \mid \beta, \boldsymbol{\theta}, \mathbf{y}) p(\boldsymbol{\theta} \mid \beta, \mathbf{y})\, d\theta_1\, d\theta_2.$$

In particular, for $\beta = 0$

$$V \frac{\Sigma(y_{1j} - \bar{y}_1)^2/(n_1 - 1)}{\Sigma(y_{2j} - \bar{y}_2)^2/(n_2 - 1)} \sim F_{n_1-1, n_2-1},$$

and when $\beta \to -1$ the distribution of $u_* = V(h_1/h_2)^2$ is, from (4.2.27),

$$p(u_* \mid \beta \to -1, \mathbf{y}) = \begin{cases} k u_*^{\frac{1}{2}(n_1-1)-1}[(n_1+n_2) - (n_1+n_2-2)u_*^{\frac{1}{2}}], & 0 < u_* < 1 \\ k u_*^{-\frac{1}{2}(n_2-1)-1}[(n_1+n_2) - (n_1+n_2-2)u_*^{-\frac{1}{2}}], & 1 < u_* < \infty \end{cases}$$

with

$$k = \frac{n_1 n_2}{2(n_1+n_2)} \frac{(n_1-1)(n_2-1)}{(n_1+n_2-1)(n_1+n_2-2)}.$$

For β not close to -1, from (4.2.34), we have the approximation

$$V^{1/(1+\beta)} \frac{n_1 s_1(\beta, \hat{\theta}_1)/[n_1(1+\beta) - 1]}{n_2 s_2(\beta, \hat{\theta}_2)/[n_2(1+\beta) - 1]} \mathrel{\dot{\sim}} F_{n_1(1+\beta)-1, n_2(1+\beta)-1},$$

where $(\hat{\theta}_1, \hat{\theta}_2)$ is the mode of $p(\boldsymbol{\theta} \mid \beta, \mathbf{y})$.

Table 4.5.1 *Continued*

8. With $\boldsymbol{\theta}$ eliminated, the posterior distribution of β is, in (4.2.35),

$$p(\beta \mid \mathbf{y}) \propto p(\beta) p_u(\beta \mid \mathbf{y}), \qquad -1 < \beta \leqslant 1$$

with

$$p_u(\beta \mid \mathbf{y}) \propto \{\Gamma[1 + \tfrac{1}{2}(1 + \beta)]\}^{-(n_1+n_2)}$$

$$\times \prod_{i=1}^{2} \Gamma\left[1 + \frac{n_i}{2}(1 + \beta)\right] \int_{-\infty}^{\infty} [n_i s_i(\beta, \theta_i)]^{-\frac{1}{2}n_i(1+\beta)}\, d\theta_i$$

9. With $\boldsymbol{\theta}$ and β eliminated, the final posterior distribution of V is, in (4.2.36),

$$p(V \mid \mathbf{y}) = \int_{-1}^{1} p(V \mid \beta, \mathbf{y}) p(\beta \mid \mathbf{y})\, d\beta.$$

The General Case $k \geqslant 2$

10. Conditional on $(\boldsymbol{\theta}, \beta)$, then from (4.3.5) the posterior distribution of the $(k - 1)$ contrasts

$$\phi_i = 2(\log \sigma_i - \log \sigma_k), \qquad i = 1, \ldots, k - 1$$

is

$$p(\boldsymbol{\phi} \mid \boldsymbol{\theta}, \beta, \mathbf{y}) = c \left[\prod_{i=1}^{k-1} U_i^{\frac{1}{2}n_i(1+\beta)}\right] \left(1 + \sum_{i=1}^{k-1} U_i\right)^{-\frac{1}{2}n(1+\beta)},$$

$$-\infty < \phi_i < \infty, \quad i = 1, \ldots, k-1$$

where

$$U_i = \frac{n_i s_i(\beta, \theta_i)}{n_k s_k(\beta, \theta_k)}\, e^{-\phi_i/(1+\beta)}, \qquad n = \sum_{i=1}^{k} n_i,$$

and

$$c = \Gamma\left[\frac{n}{2}(1 + \beta)\right] (1 + \beta)^{-(k-1)} \left\{\prod_{i=1}^{k} \Gamma\left[\frac{n_i}{2}(1 + \beta)\right]\right\}^{-1}.$$

Thus, from (4.3.7) to (4.3.11), the approximate $(1 - \alpha)$ H.P.D. region of $\boldsymbol{\phi}$ is

$$\frac{-2 \log W}{1 + A} < \chi^2(k - 1, \alpha),$$

where

$$\frac{-2 \log W}{1 + \beta} = \sum_{i=1}^{k} n_i \log \frac{n_i}{n} - \sum_{i=1}^{k-1} U_i \log u_i + n \log\left(1 + \sum_{i=1}^{k-1} U_i\right),$$

$$A = \frac{1}{3(k - 1)(1 + \beta)} \left(\sum_{i=1}^{k} n_i^{-1} - n^{-1}\right),$$

Table 4.5.1 *Continued*

and $\chi^2(k-1, \alpha)$ is given in Table III (at the end of this book). In particular, for $\phi_i = 0$ $(i = 1, ..., k-1)$ (that is, $\sigma_1^2 = \cdots = \sigma_k^2$)

$$-2\log W = -\sum_{i=1}^{k} n_i(1+\beta)[\log s_i(\beta, \theta_i) - \log \bar{s}(\beta, \boldsymbol{\theta})],$$

with

$$\bar{s}(\beta, \boldsymbol{\theta}) = \frac{1}{n}\sum_{j=1}^{k} n_j s_j(\beta, \theta_j).$$

11. Conditional on $\boldsymbol{\theta}$, the posterior distribution of β is, in (4.3.19) and (4.3.20),

$$p(\beta \mid \boldsymbol{\theta}, \mathbf{y}) \propto p(\beta) p_u(\beta \mid \boldsymbol{\theta}, \mathbf{y}), \qquad -1 < \beta \leqslant 1,$$

with

$$p_u(\beta \mid \boldsymbol{\theta}, \mathbf{y}) \propto \{\Gamma[1 + \tfrac{1}{2}(1+\beta)]\}^{-n} \prod_{i=1}^{k} \Gamma\left[1 + \frac{n_i}{2}(1+\beta)\right] [n_i s_i(\beta, \theta_i)]^{-\frac{1}{2}n_i(1+\beta)}.$$

12. With β eliminated, the posterior distribution of $\boldsymbol{\phi}' = (\phi_1, ..., \phi_{k-1})$, given $\boldsymbol{\theta}$, is from (4.3.22), approximately

$$p(\boldsymbol{\phi} \mid \boldsymbol{\theta}, \mathbf{y}) \doteq p(\boldsymbol{\phi} \mid \boldsymbol{\theta}, \hat{\beta}, \mathbf{y}),$$

where $\hat{\beta}$ is the mode of $p(\beta \mid \boldsymbol{\theta}, \mathbf{y})$.

13. Conditional on β, the posterior distribution of $\boldsymbol{\theta}$ is, in (4.3.27),

$$p(\boldsymbol{\theta} \mid \beta, \mathbf{y}) \propto \prod_{i=1}^{k} \left\{ [n_i s_i(\beta, \theta_i)]^{-\frac{1}{2}n_i(1+\beta)} \bigg/ \int_{-\infty}^{\infty} [n_i s_i(\beta, \theta_i)]^{-\frac{1}{2}n_i(1+\beta)} \, d\theta_i \right\},$$

$$-\infty < \theta_i < \infty, \quad i = 1, ..., k.$$

14. Conditional on β, with $\boldsymbol{\theta}$ eliminated and for β not close to -1, the posterior distribution of $\boldsymbol{\phi}$ is approximately, from (4.3.31),

$$p(\boldsymbol{\phi} \mid \beta, \mathbf{y}) \doteq c' \left[\prod_{i=1}^{k-1} \gamma_i^{\frac{1}{2}m_i}\right] \left(1 + \sum_{i=1}^{k-1} \gamma_i\right)^{-\frac{1}{2}m}, \qquad -\infty < \phi_i < \infty, \quad i = 1, ..., k-1,$$

where

$$m = \sum_{i=1}^{k} m_i, \qquad m_i = n_i(1+\beta) - 1, \qquad \gamma_i = \frac{n_i s_i(\beta, \hat{\theta}_i)}{n_k s_k(\beta, \hat{\theta}_k)} \exp[-\phi_i/(1+\beta)],$$

$$c' = \Gamma\left(\frac{m}{2}\right)(1+\beta)^{-(k-1)} \left[\prod_{i=1}^{k} \Gamma\left(\frac{m_i}{2}\right)\right]^{-1},$$

Table 4.5.1 *Continued*

and $(\hat{\theta}_1, \ldots, \hat{\theta}_k)$ is the mode of $p(\boldsymbol{\theta} \mid \beta, \mathbf{y})$. Thus, from (4.3.32) to (4.3.35) the approximate $(1 - \alpha)$ H.P.D. region of $\boldsymbol{\phi}$ is given by

$$\frac{-2 \log W^*}{1 + A^*} < \chi^2(k - 1, \alpha),$$

where

$$-2 \log W^* = \sum_{i=1}^{k} m_i \log \frac{m_i}{m} - \sum_{i=1}^{k-1} m_i \log \gamma_i + m \log \left(1 + \sum_{i=1}^{k-1} \gamma_i\right)$$

and

$$A^* = \frac{1}{3(k-1)} \left(\sum_{i=1}^{k} m_i^{-1} - m^{-1}\right).$$

In particular, for the point $\boldsymbol{\phi} = \mathbf{0}$ $(\sigma_1^2 = \cdots = \sigma_k^2)$

$$-2 \log W^* = -\sum_{i=1}^{k} m_i \left[\log \frac{n_i s_i(\beta, \hat{\theta}_i)}{m_i} - \log \bar{s}(\beta, \hat{\boldsymbol{\theta}})\right]$$

with

$$\bar{s}(\beta, \hat{\boldsymbol{\theta}}) = \frac{1}{m} \sum_{i=1}^{k} n_i s_i(\beta, \hat{\theta}_i).$$

15. With $\boldsymbol{\theta}$ eliminated, the posterior distribution of β is, in (4.3.36),

$$p(\beta \mid \mathbf{y}) \propto p(\beta)\, p_u(\beta \mid \mathbf{y}), \qquad -1 < \beta \leqslant 1,$$

where

$$p_u(\beta \mid \mathbf{y}) \propto \{\Gamma[1 + \tfrac{1}{2}(1 + \beta)]\}^{-n} \prod_{i=1}^{k} \Gamma[1 + \tfrac{1}{2} n_i (1 + \beta)]$$

$$\times \int_{-\infty}^{\infty} [n_i s_i(\beta, \theta_i)]^{-\frac{1}{2} n_i (1+\beta)}\, d\theta_i.$$

16. Finally, with $\boldsymbol{\theta}$ and β eliminated, the posterior distribution of $\boldsymbol{\phi}$ is approximately

$$p(\boldsymbol{\phi} \mid \mathbf{y}) \doteq p(\boldsymbol{\phi} \mid \hat{\beta}, \mathbf{y}),$$

where $\hat{\beta}$ is the mode of $p(\beta \mid \mathbf{y})$.

APPENDIX A4.1

LIMITING DISTRIBUTIONS FOR THE VARIANCE RATIO V WHEN β APPROACHES -1

We here sketch the derivation of the limiting distributions given in (4.2.10) and (4.2.27) for the variance ratio V when β approaches -1. These results follow readily by making use of (A3.1.7) in Appendix A3.1.

Specifically, for the distribution of u in (4.2.10), we first make the transformation in (4.2.9),

$$u = V\left(\frac{h_1 + |m_1\,\theta_1| -}{h_2 + |m_2 - \theta_2|}\right)^2, \tag{A4.1.1}$$

so that

$$p(u \mid \beta, \boldsymbol{\theta}, \mathbf{y}) = q(\beta)[r(\beta, \theta_1, \theta_2)]^{n_1} u^{\frac{1}{2}n_1 - 1}[1 + u^{1/(1+\beta)} r(\beta, \theta_1, \theta_2)^{2/(1+\beta)}]^{-\frac{1}{2}(n_1+n_2)(1+\beta)},$$

$$0 < u < \infty, \tag{A4.1.2}$$

where

$$r(\beta, \theta_1, \theta_2) = \frac{[n_1 s_1(\beta, \theta_1)]^{\frac{1}{2}(1+\beta)}}{h_1 + |m_1 - \theta_1|} \cdot \frac{h_2 + |m_2 - \theta_2|}{[n_2 s_2(\beta, \theta_2)]^{\frac{1}{2}(1+\beta)}}$$

and

$$q(\beta) = \frac{1}{1+\beta} \frac{\Gamma[\frac{1}{2}(n_1 + n_2)(1 + \beta)]}{\prod_{i=1}^{2} \Gamma[\frac{1}{2}n_i(1 + \beta)]}.$$

As $\beta \to -1$,

$$\lim_{\beta \to -1} q(\beta) = \frac{n_1 n_2}{2(n_1 + n_2)} = q, \tag{A4.1.3}$$

say, and by using (A3.1.7),

$$\lim_{\beta \to -1} r(\beta, \theta_1, \theta_2) = 1. \tag{A4.1.4}$$

Further, following the argument in (A3.1.8) through (A3.1.12), we have

$$\lim_{\beta \to -1} [1 + u^{1/(1+\beta)} r(\beta, \theta_1, \theta_2)^{2/(1+\beta)}]^{\frac{1}{2}(1+\beta)} = \begin{cases} u & \text{for} \quad 1 < u < \infty, \\ 1 & \text{for} \quad 0 < u \leqslant 1. \end{cases} \tag{A4.1.5}$$

It follows that

$$\lim_{\beta \to -1} p(u \mid \beta, \boldsymbol{\theta}, \mathbf{y}) = \begin{cases} q u^{\frac{1}{2}n_1 - 1} & \text{for} \quad 0 < u \leqslant 1, \\ q u^{-\frac{1}{2}n_2 - 1} & \text{for} \quad 1 < u < \infty, \end{cases} \tag{A4.1.6}$$

as given in (4.2.10).

To derive the distribution of u_* in (4.2.27), we first note that from the result in (3.2.15) as $\beta \to -1$, the joint distribution of (θ_1, θ_2) in (4.2.24) tends to

$$\lim_{\beta \to -1} p(\boldsymbol{\theta} \mid \beta, \mathbf{y}) = \prod_{i=1}^{2} \tfrac{1}{2}(n_i - 1) h^{(n_i - 1)} (h_i + |m_i - \theta_i|)^{-n_i},$$

$$-\infty < \theta_i < \infty, \quad i = 1, 2. \tag{A4.1.7}$$

It is readily verified from (A4.1.7) that the quantity

$$z = \left(1 + \left|\frac{m_2 - \theta_2}{h_2}\right|\right)^2 \Bigg/ \left(1 + \left|\frac{m_1 - \theta_1}{h_1}\right|\right)^2 \tag{A4.1.8}$$

is distributed *a posteriori* as

$$\lim_{\beta\to-1} p(z \mid \beta, \mathbf{y}) = \begin{cases} cz^{\frac{1}{2}(n_1-1)-1} & \text{for} \quad 0 < z \leqslant 1, \\ cz^{-\frac{1}{2}(n_2-1)-1} & \text{for} \quad 1 < z < \infty, \end{cases} \tag{A4.1.9}$$

where

$$c = \frac{(n_1 - 1)(n_2 - 1)}{2(n_1 + n_2 - 2)}.$$

Noting that

$$u_* = V\left(\frac{h_1}{h_2}\right)^2 = uz. \tag{A4.1.10}$$

We obtain from (A4.1.6) the conditional posterior distribution of u_* given z as

$$\lim_{\beta\to-1} p(u_* \mid z, \beta, \mathbf{y}) = \begin{cases} qz^{-\frac{1}{2}n_1} u_*^{\frac{1}{2}n_1-1} & \text{for} \quad 0 < u_* < z, \\ qz^{\frac{1}{2}n_2} u_*^{-\frac{1}{2}n_2-1} & \text{for} \quad z < u_* < \infty. \end{cases} \tag{A4.1.11}$$

It follows from (A4.1.9) and (A4.1.11) that for $0 < u_* \leqslant 1$,

$$\begin{aligned} \lim_{\beta\to-1} p(u_* \mid \beta, \mathbf{y}) &= qc\left[u_*^{-\frac{1}{2}n_2-1}\int_0^{u_*} z^{\frac{1}{2}(n_1+n_2-1)-1}\,dz + u_*^{\frac{1}{2}n_1-1}\int_{u_*}^{1} z^{-\frac{1}{2}-1}\,dz \right. \\ &\qquad \left. + u_*^{\frac{1}{2}n_1-1}\int_1^{\infty} z^{-\frac{1}{2}(n_1+n_2+1)}\,dz\right] \\ &= ku_*^{\frac{1}{2}(n_1-1)-1}\left[(n_1+n_2) - (n_1+n_2-2)u_*^{1/2}\right], \end{aligned} \tag{A4.1.12}$$

and for $1 < u_* < \infty$,

$$\begin{aligned} \lim_{\beta\to-1} p(u_* \mid \beta, \mathbf{y}) &= qc\left[u_*^{-\frac{1}{2}n_2-1}\int_0^{1} z^{\frac{1}{2}(n_1+n_2-1)-1}\,dz + u_*^{-\frac{1}{2}n_2-1}\int_1^{u_*} z^{\frac{1}{2}-1}\,dz \right. \\ &\qquad \left. + u_*^{\frac{1}{2}n_1-1}\int_{u_*}^{\infty} z^{-\frac{1}{2}(n_1+n_2+1)}\,dz\right] \\ &= ku_*^{-[\frac{1}{2}(n_2-1)+1]}\left[(n_1+n_2) - (n_1+n_2-2)u_*^{-1/2}\right], \end{aligned}$$

with $k = \dfrac{2qc}{n_1 + n_2 - 1}$ as given in (4.2.27).

CHAPTER 5

RANDOM EFFECT MODELS

5.1 INTRODUCTION

In previous chapters problems about means, variances, and regression coefficients were discussed. We now consider another important class of practical problems concerned with *variance components*.

To illustrate how variance component problems arise, consider a chemical process producing batches of material which are sampled and then analysed for some characteristic such as product yield. When the total variance σ^2 of the observed yield is large, the efficient operation of the process is hampered, control becomes difficult, and process improvement studies are hindered. So that effort can be effectively directed to reducing σ^2, it is necessary to discover the relative importance of the various sources from which variation may spring. It would for example be fruitless to devote effort to improving the analytical method if in fact an inadequate technique for *sampling* the material were the main cause of variation. With this in mind a special type of experiment may be conducted using what is called a hierarchical design.

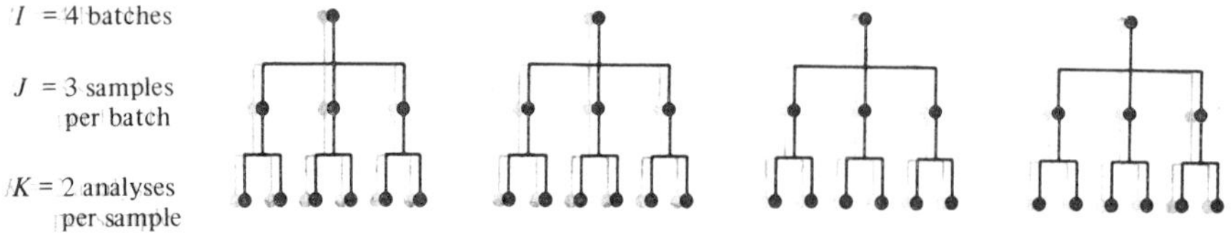

Fig. 5.1.1 Illustration of a hierarchical classification.

Figure 5.1.1 illustrates a hierarchical design for four batches, three samples per batch and two analyses per sample. More generally suppose that I batches of product are randomly taken, each batch is sampled J times and K analyses are performed on each sample. Then on the assumption that batches, samples, and analyses vary independently, and additively contribute errors e_i, e_{ij}, and e_{ijk}, we have the mathematical model

$$y_{ijk} = \theta + e_i + e_{ij} + e_{ijk}, \qquad i = 1, \ldots, I; \quad j = 1, \ldots, J; \quad k = 1, \ldots, K \qquad (5.1.1)$$

where y_{ijk} are the observations, θ a common location parameter, e_i, e_{ij} and e_{ijk}

are independently distributed random variables with zero means and variances $\text{Var}(e_{ijk}) = \sigma_1^2$, $\text{Var}(e_{ij}) = \sigma_2^2$ and $\text{Var}(e_i) = \sigma_3^2$. Thus the total variance of the (ijk)th observation y_{ijk} becomes $\sigma_1^2 + \sigma_2^2 + \sigma_3^2$ and the quantities $(\sigma_1^2, \sigma_2^2, \sigma_3^2)$ are called the variance components.

Obviously, more than three sources of variation may be involved and many different applications of this kind of study could be quoted.

The random variables e_j, e_{jk} are sometimes called random effects and the model in (5.1.1) is known as a *random effect* model. This is in contrast to the model in (2.7.7) for the comparison of Normal means considered in Section 2.11. In the sampling theory framework, these means are usually regarded as fixed constants and the model is therefore commonly called a *fixed effect* model. The relationship between these two types of models will be considered in more detail later in Chapters 6 and 7.

5.1.1 The Analysis of Variance Table

We have already seen that in the comparison of Normal means, certain calculations are conveniently set out in the form of an analysis of variance table. It so happens that such a table is also of value in analyzing variance components. An appropriate analysis of variance table for the present random effect model is shown in Table 5.1.1.

Table 5.1.1

Analysis of variance of hierarchical classification with three variance components

ource	Sum of squares (S.S.)	Degrees of freedom (d.f.)	Mean square (M.S.)	Sampling expectation of mean square (E.M.S.)
ue to tches	$S_3 = JK\Sigma(y_{i..} - y_{...})^2$	$\nu_3 = (I-1)$	$m_3 = S_3/\nu_3$	$\sigma_{123}^2 = \sigma_1^2 + K\sigma_2^2 + JK\sigma_3^2$
ue to mples	$S_2 = K\Sigma\Sigma(y_{ij.} - y_{i..})^2$	$\nu_2 = I(J-1)$	$m_2 = S_2/\nu_2$	$\sigma_{12}^2 = \sigma_1^2 + K\sigma_2^2$
ue to alyses	$S_1 = \Sigma\Sigma\Sigma(y_{ijk} - y_{ij.})^2$	$\nu_1 = IJ(K-1)$	$m_1 = S_1/\nu_1$	σ_1^2
tal	$\Sigma\Sigma\Sigma(y_{ijk} - y_{...})^2$	$IJK - 1$		

In Table 5.1.1 (and hereafter) we adopt the notation by which *a dot replacing a subscript indicates an average over that subscript.* Thus,

$$y_{ij.} = \frac{1}{K}\Sigma y_{ijk}, \qquad y_{i..} = \frac{1}{J}\Sigma y_{ij.}, \qquad \text{and} \qquad y_{...} = \frac{1}{I}\Sigma y_{i..}.$$

From the sampling theory point of view, by pooling the sample variances within the individual samples, we obtain an estimate of σ_1^2, the analytical variance. Then by pooling the sample variances of the sample means within each batch, we obtain an estimate of $\sigma_1^2 + K\sigma_2^2$ where σ_2^2 measures the variation due to sampling. Finally, from the variation of the batch means, we can obtain an estimate of $\sigma_1^2 + K\sigma_2^2 + JK\sigma_3^2$ where σ_3^2 is the batch variance. It is then customary to obtain estimates of the variance components $(\sigma_1^2, \sigma_2^2, \sigma_3^2)$ by solving the equations in the expectations. Thus, from Table 5.1.1 we have the relationships

$$\sigma_2^2 = (\sigma_{12}^2 - \sigma_1^2)/K, \qquad \sigma_3^2 = (\sigma_{123}^2 - \sigma_{12}^2)/JK, \tag{5.1.2a}$$

so that the usual estimates of the variance components are

$$\hat{\sigma}_1^2 = m_1, \qquad \hat{\sigma}_2^2 = (m_2 - m_1)/K, \qquad \text{and} \qquad \hat{\sigma}_3^2 = (m_3 - m_2)/JK. \tag{5.1.2b}$$

5.1.2 Two Examples

We will begin by illustrating the simple case where there are only two components of variance. The model is then

$$y_{jk} = \theta + e_j + e_{jk}, \qquad j = 1, \ldots, J; \quad k = 1, \ldots, K, \tag{5.1.3}$$

which is a special case of (5.1.1) when $I = 1$.

The first data set we consider is taken from Davies (1967, p. 105). The object of the experiment was to learn to what extent batch to batch variation in a certain raw material was responsible for variation in the final product yield. Five samples from each of six randomly chosen batches of raw material were taken and a single laboratory determination of product yield was made for each of the resulting 30 samples. The data are set out in Table 5.1.2 with an analysis of variance in Table 5.1.3. In this example $j = 1, \ldots, 6$ refers to the batches and for each j, $k = 1, \ldots, 5$ refers to the samples. The "within batches" component σ_1^2 will be referred to as the "analyses" component although it includes sampling errors as well as chemical

Table 5.1.2

Dyestuff data (Yield of dyestuff in grams of standard color)

Batch	1	2	3	4	5	6
Individual observations	1545	1540	1595	1445	1595	1520
	1440	1555	1550	1440	1630	1455
	1440	1490	1605	1595	1515	1450
	1520	1560	1510	1465	1635	1480
	1580	1495	1560	1545	1625	1445
$y_{j.}$	1505	1528	1564	1498	1600	1470
			$y_{..} = 1527.5$			

analysis errors. The variance component σ_2^2 associated with "batches" represents the variation associated with raw material quality changes when sampling and analytical errors are fully allowed for.

The data are characterized by the fact that the between-batches mean square is large compared with the within-batches mean square, strongly indicating that the component σ_2^2 associated with batches is nonzero.

Table 5.1.3

Analysis of variance for the dyestuff example

Source	S.S.	d.f.	M.S.	E.M.S.
Between batches	$S_2 =$ 56,357.5	$\nu_2 =$ 5	$m_2 =$ 11,271.50	$\sigma_1^2 + 5\sigma_2^2$
Within batches (analyses)	$S_1 =$ 58,830.0	$\nu_1 =$ 24	$m_1 =$ 2,451.25	σ_1^2
Total	115,187.5	29		

$\hat{\sigma}_1^2 = 2{,}451.25, \quad \hat{\sigma}_2^2 = (11{,}271.50 - 2{,}451.25)/5 = 1{,}764.05, \quad m_2/m_1 = 4.60.$

The second data set we consider illustrates the case where the between-batches mean square is less than the within-batches mean square. These data had to be constructed for although examples of this sort undoubtedly occur in practice they seem to be rarely published. The model in (5.1.3) was used to generate six groups of five observations each. The errors e_{jk} were drawn from a table of random Normal deviates with $\sigma_1 = 4$, the e_j were drawn from the same table but with $\sigma_2 = 2$, and the parameter θ was set to be equal to five. For convenience we shall refer to the components (σ_1^2, σ_2^2) in this second example associated with "analyses" and "batches" as we did in the first. The data are set out in Table 5.1.4 and the corresponding analysis of variance, in Table 5.1.5.

Table 5.1.4

Data generated from a table of random normal deviates for the two-component model (with $\theta = 5$, $\sigma_1 = 4$, $\sigma_2 = 2$)

Batch	1	2	3	4	5	6
Individual observations	7.298	5.220	0.110	2.212	0.282	1.722
	3.846	6.556	10.386	4.852	9.014	4.782
	2.434	0.608	13.434	7.092	4.458	8.106
	9.566	11.788	5.510	9.288	9.446	0.758
	7.990	−0.892	8.166	4.980	7.198	3.758
$y_{j.}$	6.2268	4.6560	7.5212	5.6848	6.0796	3.8252

$y_{..} = 5.6656$

Table 5.1.5

Analysis of variance for the generated example

Source	S.S.	d.f.	M.S.	E.M.S.
Between batches	$S_2 = 41.6816$	$\nu_2 = 5$	$m_2 = 8.3363$	$\sigma_1^2 + 5\sigma_2^2$
Within batches (analyses)	$S_1 = 358.7014$	$\nu_1 = 24$	$m_1 = 14.9459$	σ_1^2
Total	400.3830	29		

$\hat{\sigma}_1^2 = 14.9459, \quad \hat{\sigma}_2^2 = (8.3363 - 14.9459)/5 = -1.3219, \quad m_2/m_1 = 0.558$

5.1.3 Difficulties in the Sampling Theory Approach

The sampling theory approach to the problem of variance components encounters a number of snags. These have bothered statisticians for many years, as is evidenced by the great variety of attempts which have been made to resolve the problems.† The difficulties can conveniently be discussed in terms of the examples quoted above.

Negative Estimate of $\hat{\sigma}_2^2$

The most commonly used estimate of σ_2^2 is obtained from the difference of the two mean squares (m_1, m_2) and we have referred to this estimate as $\hat{\sigma}_2^2$. When the between batch mean square m_2 is smaller than the within batch mean square m_1 as it is in the second example above, $\hat{\sigma}_2^2$ will be negative. Since σ_2^2 must be nonnegative, this is generally regarded as objectionable.

Difficulties with Confidence Intervals

Even with the assumption that e_j and e_{jk} are Normally and independently distributed, the sampling distribution of $\hat{\sigma}_2^2$ is complicated and depends on the unknown variance ratio σ_2^2/σ_1^2. The problem of obtaining a confidence interval for σ_2^2 is complex, and no generally accepted procedure has been found. Also, in estimating the variance ratio σ_2^2/σ_1^2 the commonly used confidence intervals, which are based upon the sampling distribution of the mean square ratio m_2/m_1, may include negative values.

Intuitively, one would feel that the part of the interval associated with negative values of σ_2^2 ought to be discounted in some way, but attempts to do this within the sampling theory framework are difficult to justify.

† See for example, Daniels (1939), Crump (1946), Eisenhart (1947), Henderson (1953), Moriguti (1954), Tukey (1956), Graybill and Wortham (1956), Bulmer (1957), Searle (1958), Herbach (1959), Gates and Shine (1962), Thompson (1962, 1963), Gower (1962), Williams (1962), Bush and Anderson (1963), Wang (1967), Zacks (1967). For a review of recent work in this area, see Tan (1964) and Ali (1969).

Pooling of Estimates

When m_2 and m_1 are not very different, it has sometimes been argued that a pooled estimate $(\nu_1 m_1 + \nu_2 m_2)/(\nu_1 + \nu_2)$ of σ_1^2 should be used. But how does one decide when to pool and when not to pool? And how is the sampling distribution of the estimate of σ_1^2 affected by the fact that pooling will only be practiced when m_1 and m_2 are sufficiently close? Attempts have been made to get around these difficulties and to study the consequences of various proposed procedures, but the situation remains unsatisfactory.

Departures from Assumptions

Additional complications arise when we consider the effect of departures from the assumptions of Normality and independence. It is shown in Scheffé (1959) that non-Normality in e_j and lack of independence in e_{jk} will have serious effects on the distributions of the criteria which one uses to make inferences about (σ_1^2, σ_2^2). This further confuses an already chaotic situation.

In summary, then, traditional sampling theory methods have led to worrisome difficulties in the variance component problem to which no generally accepted set of solutions have been obtained. Our aim in this chapter is to reexamine these problems from a Bayesian standpoint, making the standard Normality and independence assumptions. Within the Bayesian framework, no difficulty in principle occurs in relaxing the Normality and independence assumptions. Indeed, Tiao and Tan (1966) and Hill (1967) have studied variance component problems while relaxing the independence assumption for the e_{jk}, and more recently Tiao and Ali (1971a) have considered the effect of departure from Normality of the e_j.

5.2 BAYESIAN ANALYSIS OF HIERARCHICAL CLASSIFICATIONS WITH TWO VARIANCE COMPONENTS

As a preliminary to the Bayesian analysis, a summary of notation and assumptions for the two-component random effect model is given below.

With J groups (batches) of K observations (analyses) we employ the model

$$y_{jk} = \theta + e_j + e_{jk}, \qquad j = 1, \ldots, J; \quad k = 1, \ldots, K, \tag{5.2.1}$$

where the e's are supposed Normally and independently distributed with

$$E(e_j) = E(e_{jk}) = 0, \qquad \text{Var}(e_j) = \sigma_2^2, \qquad \text{and} \qquad \text{Var}(e_{jk}) = \sigma_1^2. \tag{5.2.2}$$

Thus, $\text{Var}(y_{jk}) = \sigma_1^2 + \sigma_2^2$, and the traditional unbiased estimators for the variance components σ_1^2 and σ_2^2 are respectively

$$\hat{\sigma}_1^2 = m_1 \qquad \text{and} \qquad \hat{\sigma}_2^2 = \frac{(m_2 - m_1)}{K}, \tag{5.2.3}$$

with m_2 and m_1 the "between" and "within" mean squares defined in Table 5.2.1. It is to be noted that an important quantity occurring in the analysis is $\sigma_{12}^2 = \sigma_1^2 + K\sigma_2^2$.

Table 5.2.1

Analysis of variance of hierarchical classifications with two components

Source	S.S.	d.f.	M.S.	E.M.S.
Between batches	$S_2 = K\Sigma\,(y_{j.} - y_{..})^2$	$\nu_2 = (J-1)$	$m_2 = S_2/\nu_2$	$\sigma_{12}^2 = \sigma_1^2 + K\sigma_2^2$
Within batches	$S_1 = \Sigma\Sigma\,(y_{jk} - y_{j.})^2$	$\nu_1 = J(K-1)$	$m_1 = S_1/\nu_1$	σ_1^2
Total	$\Sigma\Sigma\,(y_{jk} - y_{..})^2$	$JK - 1$		

5.2.1 The Likelihood Function

To derive the likelihood function for random effect models, we first recall certain useful results summarized in the following theorem.

Theorem 5.2.1 Let $x_1, \ldots, x_n$ be n independent observations from a Normal distribution $N(0, \sigma^2)$. Let $\bar{x}$ be the sample mean and $(x_i - \bar{x})$ $(i = 1, \ldots, n)$, be the residuals. Then

1. $\bar{x}$ is distributed independently of the $x_i - \bar{x}$ as $N(0, \sigma^2/n)$,
2. $\Sigma\,(x_1 - \bar{x})^2 \sim \sigma^2\,\chi_{n-1}^2$, and
3. In so far as the $x_i - \bar{x}$ are concerned, $\Sigma\,(x_i - \bar{x})^2$ is a sufficient statistic for σ^2.

Turning to the model in (5.2.1), one convenient way to obtain the likelihood function is to work with the group means $y_{j.}$ and the residuals $y_{jk} - y_{j.}$. Clearly, in terms of (θ, e_j, e_{jk})

$$y_{j\cdot} = \theta + e_j + e_{j.} \qquad \text{and} \qquad y_{jk} - y_{j.} = e_{jk} - e_{j.}. \tag{5.2.4}$$

It follows from the Normality and independence assumptions of (e_j, e_{jk}) and the results in Theorem 5.2.1 that

a) the $y_{j.}$ are independent, each having a Normal distribution $N(\theta, \sigma_{12}^2/K)$,

b) $y_{j.}$ are distributed independently of $y_{jk} - y_{j.}$,

c) the quantity $\Sigma_k(y_{jk} - y_{j.})^2$ is distributed as $\sigma_1^2\chi_{(K-1)}^2$ so that the sum $\Sigma\Sigma\,(y_{jk} - y_{j.})^2 = S_1 = \nu_1 m_1$ is distributed as $\sigma_1^2\chi_{\nu_1}^2$, and finally

d) in so far as the $y_{jk} - y_{j.}$ are concerned, $\nu_1 m_1$ is a sufficient statistic for σ_1^2.

Thus, the likelihood function is

$$l(\theta, \sigma_1^2, \sigma_{12}^2 \mid \mathbf{y}) \propto (\sigma_1^2)^{-\nu_1/2} (\sigma_{12}^2)^{-(\nu_2+1)/2} \exp\left\{-\frac{1}{2}\left[\frac{K\Sigma(y_{j.}-\theta)^2}{\sigma_{12}^2} + \frac{\nu_1 m_1}{\sigma_1^2}\right]\right\} \tag{5.2.5}$$

where it is to be noted that $\sigma_{12}^2 > \sigma_1^2$. Alternatively, in terms of $(\theta, \sigma_1^2, \sigma_2^2)$ and noting that

$$K\Sigma(y_{j.}-\theta)^2 = \nu_2 m_2 + JK(y_{..}-\theta)^2, \tag{5.2.6}$$

we have

$$l(\theta, \sigma_1^2, \sigma_2^2 \mid \mathbf{y}) \propto (\sigma_1^2)^{-\nu_1/2} (\sigma_1^2 + K\sigma_2^2)^{-(\nu_2+1)/2} \exp\left\{-\frac{1}{2}\left[\frac{JK(y_{..}-\theta)^2}{\sigma_1^2 + K\sigma_2^2} + \frac{\nu_2 m_2}{\sigma_1^2 + K\sigma_2^2} + \frac{\nu_1 m_1}{\sigma_1^2}\right]\right\}. \tag{5.2.7}$$

5.2.2 Prior and Posterior Distribution of $(\theta, \sigma_1^2, \sigma_2^2)$

From (5.2.5) the likelihood function can be regarded as having arisen from J independent observations from a population $N(\theta, \sigma_{12}^2/K)$ and $J(K-1)$ further independent observations from a population $N(0, \sigma_1^2)$. Treating the location parameters θ separately from the variances $(\sigma_1^2, \sigma_{12}^2)$, we therefore take, as a noninformative reference prior, a distribution with $(\theta, \log\sigma_1^2, \log\sigma_{12}^2)$ locally uniform and locally independent. The fact that $\sigma_{12}^2 = \sigma_1^2 + K\sigma_2^2$ must be positive leads to the restriction $\sigma_{12}^2 > \sigma_1^2$ in the parametei space of $(\sigma_1^2, \sigma_{12}^2)$.

Thus the noninformative prior is defined by

$$p(\theta, \sigma_1^2, \sigma_{12}^2) = p(\theta)\, p(\sigma_1^2, \sigma_{12}^2) \tag{5.2.8}$$

with

$$p(\theta) \propto c$$

and

$$p(\sigma_1^2, \sigma_{12}^2) \propto \sigma_1^{-2}\sigma_{12}^{-2}, \qquad \sigma_{12}^2 > \sigma_1^2.$$

Alternatively, in terms of $(\theta, \sigma_1^2, \sigma_2^2)$, the prior distribution is

$$p(\theta, \sigma_1^2, \sigma_2^2) = p(\theta)\, p(\sigma_1^2, \sigma_2^2) \propto \sigma_1^{-2}(\sigma_1^2 + K\sigma_2^2)^{-1}. \tag{5.2.9}$$

This choice of the prior distribution has been criticized by Stone and Springer (1965). A discussion of their criticism is given in Appendix A5.5. An important issue raised by Klotz, Milton and Zacks (1969) and Portnoy (1971) in connection with sampling theory point estimators of (σ_1^2, σ_2^2) based on this prior will be considered in detail in Appendix A5.6.

The noninformative reference prior distribution for σ_1^2 and σ_2^2 may also be obtained directly by applying Jeffreys' rule discussed in Section 1.3. It can be readily shown that the information matrix for (σ_1^2, σ_2^2) is

$$\mathscr{I}(\sigma_1^2, \sigma_2^2) = E\left\{-\frac{\partial^2 \log l}{\partial\sigma_i^2\partial\sigma_j^2}\right\} = \frac{1}{2}\begin{bmatrix} \frac{\nu_2+1}{\sigma_{12}^4} + \frac{\nu_1}{2\sigma_1^4} & \frac{K(\nu_2+1)}{\sigma_{12}^4} \\ \frac{K(\nu_2+1)}{\sigma_{12}^4} & \frac{K^2(\nu_2+1)}{\sigma_{12}^4} \end{bmatrix}. \tag{5.2.10a}$$

Thus, the determinant is

$$|\mathscr{I}(\sigma_1^2, \sigma_2^2)| = \frac{K^2(\nu_2+1)\nu_1}{2\sigma_1^4\sigma_{12}^4} \tag{5.2.10b}$$

whence, as before, $p(\sigma_1^2, \sigma_2^2) \propto \sigma_1^{-2}\sigma_{12}^{-2}$.

By combining the prior in (5.2.9) with the likelihood in (5.2.7), the posterior distribution of $(\theta, \sigma_1^2, \sigma_2^2)$ is obtained as

$$p(\theta, \sigma_1^2, \sigma_2^2 \mid \mathbf{y}) \propto (\sigma_1^2)^{-(\frac{1}{2}\nu_1+1)}(\sigma_1^2 + K\sigma_2^2)^{-\frac{1}{2}(\nu_2+1)-1}$$
$$\exp\left\{-\frac{1}{2}\left[\frac{\nu_1 m_1}{\sigma_1^2} + \frac{\nu_2 m_2}{\sigma_1^2 + K\sigma_2^2} + \frac{JK(\bar{y}_{..} - \theta)^2}{\sigma_1^2 + K\sigma_2^2}\right]\right\},$$
$$-\infty < \theta < \infty, \quad \sigma_1^2 > 0, \quad \sigma_2^2 > 0. \tag{5.2.11}$$

To obtain the posterior distribution of the variance components (σ_1^2, σ_2^2), (5.2.11) is integrated over θ yielding

$$p(\sigma_1^2, \sigma_2^2 \mid \mathbf{y}) = \omega(\sigma_1^2)^{-(\frac{1}{2}\nu_1+1)}(\sigma_1^2 + K\sigma_2^2)^{-\frac{1}{2}(\nu_2+1)}\exp\left[-\frac{1}{2}\left(\frac{\nu_1 m_1}{\sigma_1^2} + \frac{\nu_2 m_2}{\sigma_1^2 + K\sigma_2^2}\right)\right]$$
$$\sigma_1^2 > 0, \quad \sigma_2^2 > 0, \tag{5.2.12}$$

where ω is the appropriate normalizing constant which, as it will transpire in the next section, is given by

$$\omega = \frac{K(\nu_1 m_1)^{\frac{1}{2}\nu_1}(\nu_2 m_2)^{\frac{1}{2}\nu_2}2^{-\frac{1}{2}(\nu_1+\nu_2)}}{\Gamma(\frac{1}{2}\nu_1)\Gamma(\frac{1}{2}\nu_2)\Pr\{F_{\nu_2,\nu_1} < m_2/m_1\}}. \tag{5.2.13}$$

From the definition of the χ^{-2} distribution, we may write the joint distribution of (σ_1^2, σ_2^2) as

$$p(\sigma_1^2, \sigma_2^2 \mid \mathbf{y}) = \frac{(\nu_1 m_1)^{-1} p\left(\chi_{\nu_1}^{-2} = \frac{\sigma_1^2}{\nu_1 m_1}\right) K(\nu_2 m_2)^{-1} p\left(\chi_{\nu_2}^{-2} = \frac{\sigma_1^2 + K\sigma_2^2}{\nu_2 m_2}\right)}{\Pr\left\{F_{\nu_2,\nu_1} < \frac{m_2}{m_1}\right\}},$$
$$\sigma_1^2 > 0, \quad \sigma_2^2 > 0, \tag{5.2.14}$$

where $p(\chi_\nu^{-2} = x)$ is the density of an χ^{-2} variable with ν degrees of freedom evaluated at x.

5.2.3 Posterior Distribution of σ_2^2/σ_1^2 and its Relationship to Sampling Theory Results

Before considering the distributions of σ_1^2 and σ_2^2 separately we discuss the distribution of their ratio. Problems can arise where this is the only quantity of interest. For example, in deciding how to allocate a *fixed* total effort to the sampling and analyzing of a chemical product, one is concerned only with σ_2^2/σ_1^2 and not with σ_1^2 and σ_2^2 individually.

For mathematical convenience, we work with $\sigma_{12}^2/\sigma_1^2 = 1 + K(\sigma_2^2/\sigma_1^2)$ rather than σ_2^2/σ_1^2 itself. In the joint distribution of (σ_1^2, σ_2^2) in (5.2.12), we make the transformation

$$W = \left(\frac{\sigma_{12}^2}{\sigma_1^2}\right) = 1 + K\left(\frac{\sigma_2^2}{\sigma_1^2}\right), \qquad V = \sigma_1^2,$$

and integrate over V to obtain

$$p(W \mid \mathbf{y}) = \omega K^{-1} W^{-\frac{1}{2}(\nu_2+2)} \int_0^\infty V^{-\frac{1}{2}(\nu_1+\nu_2+2)} \exp\left[-\frac{1}{2V}(\nu_1 m_1 + \nu_2 m_2/W)\right] dV. \tag{5.2.15}$$

For fixed W the integral in (5.2.15) is in the form of an inverted gamma function and can be evaluated explicitly using (A2.1.2) in Appendix A2.1. Upon normalizing, the expression for ω, already given in (5.2.13), is obtained. The distribution of W can then be written as

$$p(W \mid \mathbf{y}) = \left(\frac{m_1}{m_2}\right) p\left(F_{\nu_1,\nu_2} = \frac{m_1}{m_2} W\right) \Big/ \Pr\left\{F_{\nu_2,\nu_1} < \frac{m_2}{m_1}\right\}, \qquad W > 1 \tag{5.2.16}$$

where $p(F_{\nu_1,\nu_2} = x)$ is the density of an F variable with (ν_1, ν_2) degrees of freedom evaluated at x. Probability statements about the variance ratio σ_2^2/σ_1^2 can thus be made using an F table. In particular, for $0 < \eta_1 < \eta_2$, the probability that σ_2^2/σ_1^2 falls in the interval (η_1, η_2) is

$$\Pr\left\{\eta_1 < \frac{\sigma_2^2}{\sigma_1^2} < \eta_2 \mid \mathbf{y}\right\} = \frac{\Pr\left\{\dfrac{m_2}{m_1}(1 + K\eta_2)^{-1} < F_{\nu_2,\nu_1} < \dfrac{m_2}{m_1}(1 + K\eta_1)^{-1}\right\}}{\Pr\left\{F_{\nu_2,\nu_1} < \dfrac{m_2}{m_1}\right\}}. \tag{5.2.17}$$

Note that the distribution of W is defined over the range $W > 1$. Therefore, H.P.D. or other Bayesian posterior intervals will *not* extend over values of W less than one, that is, they will not cover negative values of σ_2^2/σ_1^2.

Now from (5.2.16) we can write for the distribution of $(m_2/m_1)W^{-1}$

$$p\left(\frac{m_2}{m_1}W^{-1}\,\middle|\,\mathbf{y}\right) = p\left(F_{\nu_2,\nu_1} = \frac{m_2}{m_1}W^{-1}\right)\Big/\Pr\left\{F_{\nu_2,\nu_1} < \frac{m_2}{m_1}\right\}, \qquad W > 1. \tag{5.2.18}$$

This result merits careful study. If σ_{12}^2 and σ_1^2 were unconstrained by the inequality $\sigma_{12}^2 > \sigma_1^2$, then using an asterisk to denote unconstrained probability densities, we should have (from the results in Section 2.6),

$$p^*\left(\frac{m_2}{m_1}W^{-1}\,\middle|\,\mathbf{y}\right) = p^*\left(F_{\nu_2,\nu_1} = \frac{m_2}{m_1}W^{-1}\right), \qquad W > 0, \tag{5.2.19}$$

where from the Bayesian viewpoint W is a random variable and m_2/m_1 is a ratio of fixed sample quantities. This probability distribution also has the interpretation that given the fixed ratio $W = \sigma_{12}^2/\sigma_1^2$, the sampling distribution of $(m_2/m_1)W^{-1}$ follows the F_{ν_2,ν_1} distribution. With this interpretation (5.2.19) gives the confidence distribution of W from sampling theory.

In contrast to the distributions in (5.2.16) and (5.2.18), the confidence distribution in (5.2.19) is defined over the entire range $W > 0$. Confidence intervals for W based on this distribution *could* thus cover values of W in the interval $0 < W < 1$, that is, $-K^{-1} < \sigma_2^2/\sigma_1^2 < 0$. Since σ_2^2/σ_1^2 must be nonnegative, this result is certainly objectionable.

In comparing (5.2.18) with (5.2.19), we see that the posterior distribution in (5.2.18) contains an additional factor

$$1\Big/\Pr\left\{F_{\nu_2,\nu_1} < \frac{m_2}{m_1}\right\}.$$

From (5.2.19),

$$\Pr\left\{F_{\nu_2,\nu_1} < \frac{m_2}{m_1}\right\} = \Pr^*\{W > 1\} = \Pr^*\{\sigma_{12}^2 > \sigma_1^2\}. \tag{5.2.20}$$

The Bayesian result can then be written

$$p\left(\frac{m_2}{m_1}W^{-1}\,\middle|\,\mathbf{y}\right) = \frac{p^*\left[F_{\nu_2,\nu_1} = (m_2/m_1)\,W^{-1}\right]}{\Pr^*\{W > 1\}}, \qquad W > 1. \tag{5.2.21}$$

In other words, the constrained distribution appropriate to the variance component problem is the unconstrained F distribution truncated to the left at $W = 1$ and normalized by dividing through by the total area of the admissible part. The truncation automatically rules out values of W less than one which are inadmissible *on a priori* grounds.

If we write the factor $\Pr\{F_{\nu_2,\nu_1} < m_2/m_1\}$ as $1 - \alpha$, then, on sampling theory, α may be interpreted as the significance level of a test of the null hypothesis that $\sigma_{12}^2/\sigma_1^2 = 1$ against the alternative that $\sigma_{12}^2/\sigma_1^2 > 1$. However, while the *interpre-*

tation of α as a significance level is clear enough, it seems difficult to produce any sampling theory justification for its use in the intuitively sensible manner required by the Bayesian analysis.

5.2.4 Joint Posterior Distribution of σ_1^2 and σ_2^2

Our *a posteriori* knowledge about the parameters σ_1^2 and σ_2^2 appropriate to the reference prior considered is, of course, contained in their joint distribution given by (5.2.14). From discussion in the preceding section, we can also regard this distribution as the *constrained* distribution of the variance components (σ_1^2, σ_2^2), the constraint being given by $\sigma_{12}^2 > \sigma_1^2$. If the constraint were not present, the posterior distribution of (σ_1^2, σ_2^2) would be

$$p^*(\sigma_1^2, \sigma_2^2 \mid \mathbf{y}) = (v_1 m_1)^{-1} p\left\{\chi_{v_1}^{-2} = \frac{\sigma_1^2}{v_1 m_1}\right\} K(v_2 m_2)^{-1} p\left\{\chi_{v_2}^{-2} = \frac{\sigma_1^2 + K\sigma_2^2}{v_2 m_2}\right\},$$

$$\sigma_1^2 > 0, \quad \sigma_2^2 > -\frac{1}{K}\sigma_1^2. \qquad (5.2.22)$$

Figures 5.2.1 and 5.2.2 show contours of the joint posterior distributions for the two examples introduced earlier in Section 5.1.2. Dotted lines indicate those parts of the distributions which have been truncated. The examples respectively illustrate situations in which $\hat{\sigma}_2^2$ has a positive and a negative value. Knowledge of the joint distribution of σ_1^2 and σ_2^2 clearly allows a much deeper appreciation of the situation than is possible by mere point estimates.

Modal Values of the Joint Posterior Distribution

On equating partial derivatives to zero, it is easily shown that the mode of the unconstrained distribution (5.2.22) is at

$$\sigma_{10}^2 = m_1\left(\frac{v_1}{v_1 + 2}\right), \qquad \sigma_{20}^2 = \left(m_2\frac{v_2}{v_2 + 2} - m_1\frac{v_1}{v_1 + 2}\right)\bigg/K. \qquad (5.2.23)$$

The mode is thus to the right of the line $\sigma_2^2 = 0$ provided that

$$\frac{m_2}{m_1} > \frac{v_1\,(v_2 + 2)}{v_2\,(v_1 + 2)}. \qquad (5.2.24)$$

In this case, the modes of the posterior distribution in (5.2.14) and of the unconstrained distribution are at the same point. For the dyestuff example, $m_2/m_1 = 4.6$ which is much greater that $v_1(v_2 + 2)/[v_2(v_1 + 2)] = 1.3$ so that from (5.2.23) the mode is at $(\sigma_{10}^2 = 2{,}262.7, \sigma_{20}^2 = 1{,}277.7)$, as shown by the point P in Fig. 5.2.1.

When the inequality (5.2.24) is reversed, it can be readily shown that the mode of the constrained distribution is on the line $\sigma_2^2 = 0$ and is at

$$\sigma_{10}^2 = \frac{v_1 m_1 + v_2 m_2}{v_1 + v_2 + 4}. \qquad (5.2.25)$$

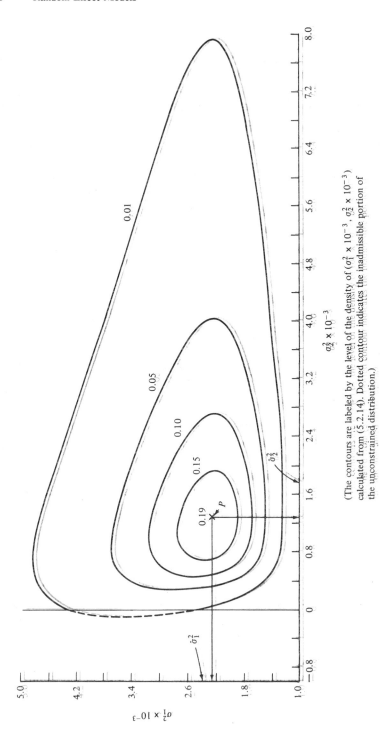

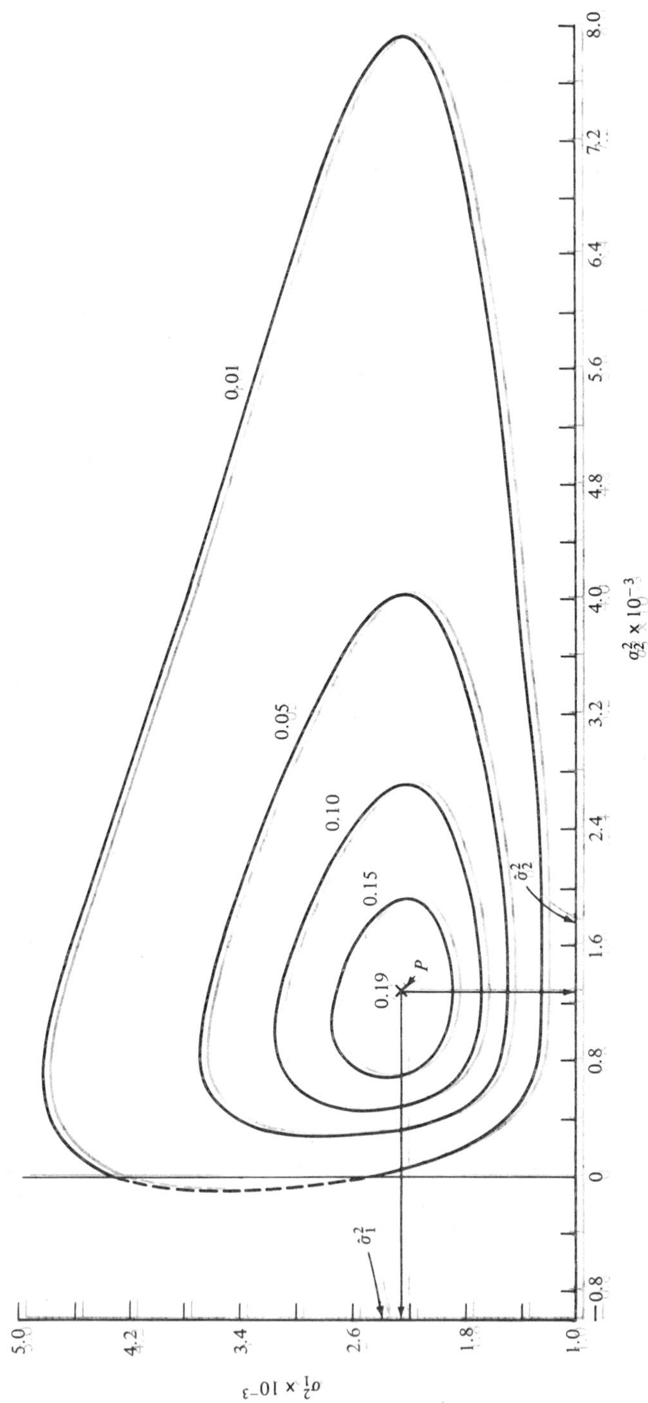

(The contours are labeled by the level of the density of $(\sigma_1^2 \times 10^{-3}, \sigma_2^2 \times 10^{-3})$ calculated from (5.2.14). Dotted contour indicates the inadmissible portion of the unconstrained distribution.)

Fig. 5.2.1 Contours of the joint distribution of the variance components (σ_1^2, σ_2^2): the dyestuff example.

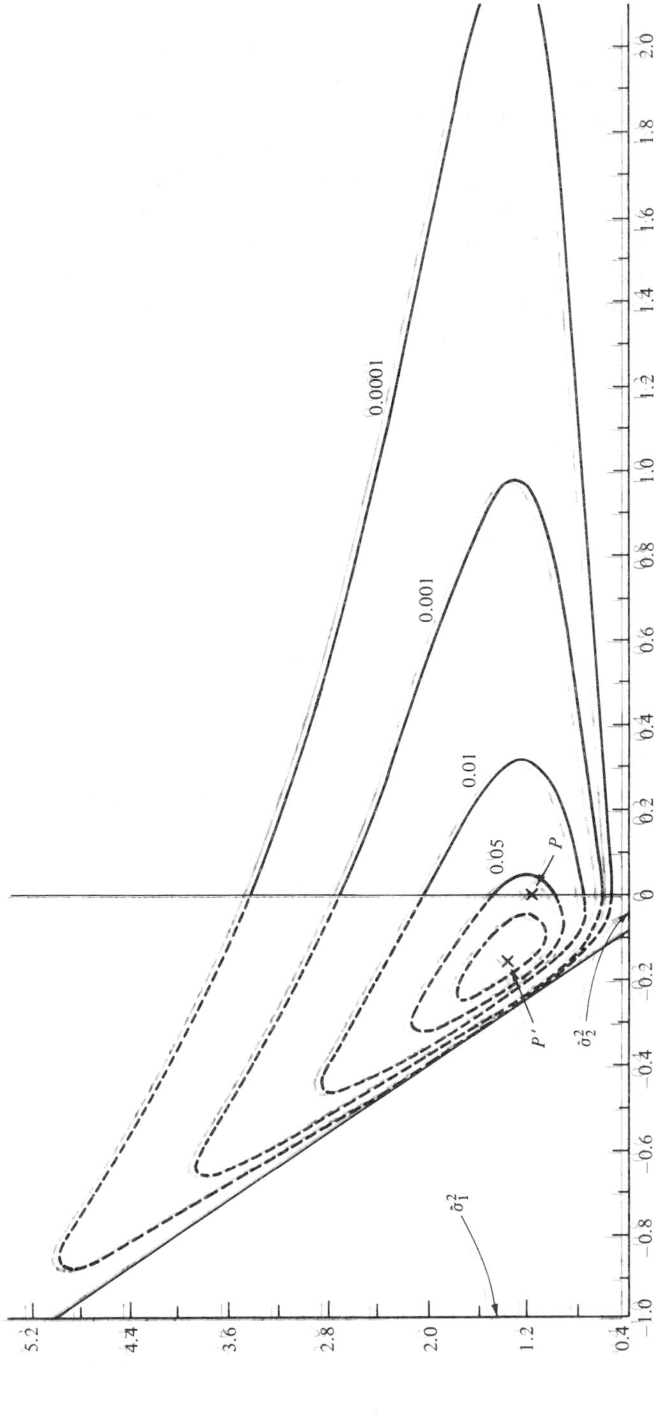

(The contours are labeled by the level of the density of (σ_1^2, σ_2^2) calculated from (5.2.14). Dotted contours indicate the inadmissible portion of the unconstrained distribution.)

Fig. 5.2.2 Contours of the joint distribution of the variance components (σ_1^2, σ_2^2): the generated example.

For the generated example $m_2/m_1 = 0.56$ which is less than $\nu_1(\nu_2+2)/[\nu_2(\nu_1+2)] = 1.3$ and consequently the mode is at $(\sigma_{10}^2 = 12.133, \sigma_{20}^2 = 0)$ as illustrated by the point P in Fig. 5.2.2. If we ignored the constraint $\sigma_2^2 > 0$, then from (5.2.23) the mode would be at $(\sigma_{10}^2 = 13.796, \sigma_{20}^2 = -1.568)$ as shown by the point P' in the same figure. A negative value of $\hat{\sigma}_2^2$ always leads to a distribution of (σ_1^2, σ_2^2) with its mode constrained on the line $\sigma_2^2 = 0$. This is because $\hat{\sigma}_2^2 = 0$ corresponds to $(m_2/m_1) < 1$ which, with $\nu_1 > \nu_2$, implies that $(m_2/m_1) < \nu_1(\nu_2+2)/[\nu_2(\nu_1+2)]$.

The mode reflects only a single aspect of the distribution. Inferences can best be made by plotting density contours, a task readily accomplished with electronic computation.

5.2.5 Distribution of σ_1^2

The posterior distribution of σ_1^2 is obtained by integrating out σ_2^2 from the joint distribution (5.2.14) to give

$$p(\sigma_1^2 \mid \mathbf{y}) = (\nu_1 m_1)^{-1} p(\chi_{\nu_1}^{-2} = \sigma_1^2/\nu_1 m_1) f(\sigma_1^2), \qquad \sigma_1^2 > 0, \qquad (5.2.26)$$

where

$$f(\sigma_1^2) = \frac{\Pr\{\chi_{\nu_2}^2 < \nu_2 m_2/\sigma_1^2\}}{\Pr\{F_{\nu_2,\nu_1} < m_2/m_1\}}.$$

We shall later in Section 5.2.11 discuss an alternative derivation by the method of constrained distributions discussed in Section 1.5. For the moment we notice that if, as before, we denote an unconstrained distribution by an asterisk, then from (5.2.22)

$$\Pr\left\{\chi_{\nu_2}^2 < \frac{\nu_2 m_2}{\sigma_1^2}\right\} = \Pr{}^*\{\sigma_2^2 > 0 \mid \sigma_1^2, \mathbf{y}\} = \Pr{}^*\{C \mid \sigma_1^2, \mathbf{y}\} \qquad (5.2.27a)$$

and

$$\Pr\left\{F_{\nu_2,\nu_1} < \frac{m_2}{m_1}\right\} = \Pr{}^*\{\sigma_2^2 > 0 \mid \mathbf{y}\} = \Pr{}^*\{C \mid \mathbf{y}\}, \qquad (5.2.27b)$$

where C is the constraint $\sigma_{12}^2 > \sigma_1^2$ (or equivalently $\sigma_2^2 > 0$). We can thus write (5.2.26) as

$$p(\sigma_1^2 \mid \mathbf{y}) = p^*(\sigma_1^2 \mid \mathbf{y}) f(\sigma_1^2), \qquad \sigma_1^2 > 0, \qquad (5.2.27c)$$

where

$$f(\sigma_1^2) = \frac{\Pr{}^*\{C \mid \sigma_1^2, \mathbf{y}\}}{\Pr{}^*\{C \mid \mathbf{y}\}}.$$

Consider the first factor $(\nu_1 m_1)^{-1} p(\chi_{\nu_1}^{-2} = \sigma_1^2/\nu_1 m_1)$ on the right of (5.2.26). In this expression, σ_1^2 is a random variable and m_1 a fixed sample quantity, but since the sampling distribution of $\nu_1 m_1$ for any fixed value of σ_1^2 is $\sigma_1^2 \chi_{\nu_1}^2$, the same expression would also supply the usual confidence distribution for σ_1^2 within the sampling theory framework. In this instance, therefore, the confidence distribution does *not* correspond to the posterior distribution because the latter includes an additional modifying factor $f(\sigma_1^2)$.

Because this second factor has no counterpart in sampling theory, its presence in the Bayesian result is of special interest. It expresses the additional information about σ_1^2 *which comes from* m_2. That some information of this kind exists is obvious on commonsense grounds. For instance, in the dyestuff example, $m_2 =$ 11,271.5 is an estimate of $\sigma_1^2 + 5\sigma_2^2$. This tells us *without any reference* to m_1 that values of σ_1^2 say four times as large as 11,271.5 are unlikely. This intuitive argument is given a precise expression in the Bayesian analysis through the modifying factor $f(\sigma_1^2)$. The numerator of $f(\sigma_1^2)$ given by (5.2.27a) is a function of σ_1^2 depending only on m_2. It is $\Pr^*\{C \mid \sigma_1^2, \mathbf{y}\}$, the probability of the constraint $\sigma_{12}^2 > \sigma_1^2$ being true for each *specific value* of σ_1^2. The denominator, given by (5.2.27b), is $\Pr^*\{C \mid \mathbf{y}\}$, the probability of the same constraint being true over *all* values of σ_1^2. It is independent of σ_1^2 and is merely a normalizing constant. Figures 5.2.3 and 5.2.4 show the effect of the modifying factor for the two sets of data we have previously discussed. Shown in each case are:

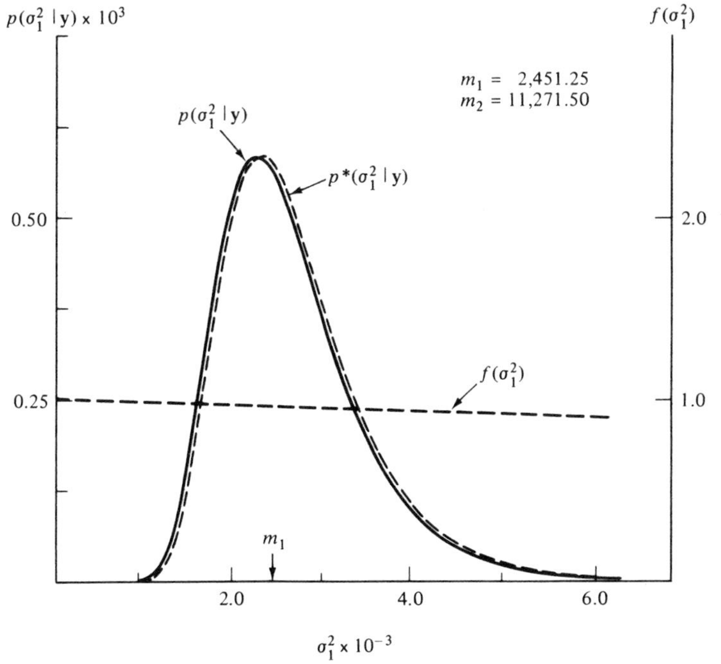

Fig. 5.2.3 Decomposition of the posterior distribution of σ_1^2: the dyestuff example.

a) the unconstrained distribution $p^*(\sigma_1^2 \mid \mathbf{y})$; this is also the confidence distribution of σ_1^2 which would be obtained from sampling theory,

b) the modifying factor $f(\sigma_1^2)$, and

c) the product of (a) and (b) which is the posterior distribution of σ_1^2.

The roles played by the two factors in determining the distribution of σ_1^2 depend critically on the relative size of m_1 and m_2. In the first example, m_2 is large compared with m_1, and we see that over the range in which $p^*(\sigma_1^2 \mid \mathbf{y})$ is appreciable, $f(\sigma_1^2)$ is relatively flat. Multiplication then produces a $p(\sigma_1^2 \mid \mathbf{y})$ which is close to $p^*(\sigma_1^2 \mid \mathbf{y})$ and the modifying factor has little effect on the distribution. For the second example, however, because m_2 is actually less than m_1, the factor $f(\sigma_1^2)$ falls off quite sharply over the range in which $p^*(\sigma_1^2 \mid \mathbf{y})$ is appreciable and multiplication produces a $p(\sigma_1^2 \mid \mathbf{y})$ which is considerably modified.

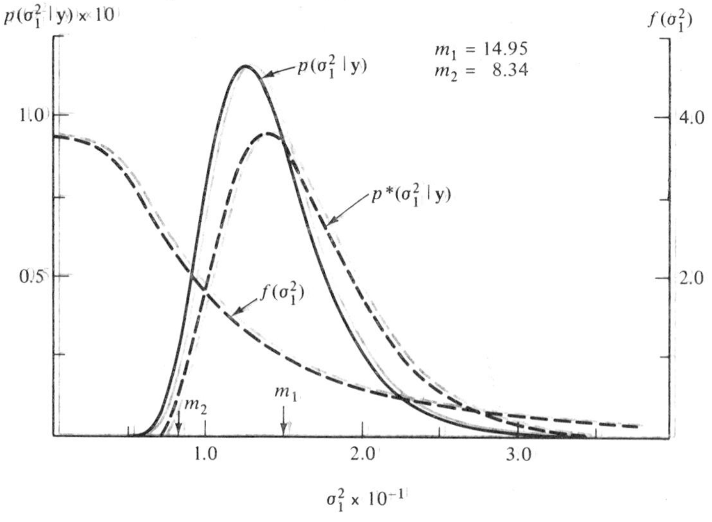

Fig. 5.2.4 Decomposition of the posterior distribution of σ_1^2: the generated example.

The decomposition illustrated in Figs. 5.2.3 and 5.2.4 of the posterior distribution of σ_1^2 into its two basic components is of interest for another reason. Examples are occasionally met where m_2 is significantly *smaller* than m_1. In such cases it has been suggested, for example by Anscombe (1948b) and Nelder (1954), in the sampling theory framework, that the model itself should be regarded as suspect. From the Bayesian point of view we are led to the same conclusion. For, when m_2 is very small compared with m_1, the factor $f(\sigma_1^2)$ will fall sharply and the information about σ_1^2 coming from the two components will be contradictory.

Effects of this kind can in particular be produced by serial correlation between the errors and, as has been pointed out earlier, Bayesian analysis which takes account of this correlation has been carried out.

Sampling Theory Interpretation of the Various Factors

Once more, while it does not seem possible to *justify* a similar analysis on sampling theory, the meaning which can be attached to the various factors in (5.2.26), from the sampling theory point of view, is interesting. From (5.2.26) we have

$$p\left(\frac{\nu_1 m_1}{\sigma_1^2}\,\middle|\,\mathbf{y}\right) = p\left(\chi_{\nu_1}^2 = \frac{\nu_1 m_1}{\sigma_1^2}\right)\frac{\Pr\{\chi_{\nu_2}^2 < \nu_2 m_2/\sigma_1^2\}}{\Pr\{F_{\nu_2,\nu_1} < m_2/m_1\}}. \tag{5.2.28}$$

Sampling theory inferences about σ_1^2 are usually made using the fact that

$$p\left(\frac{\nu_1 m_1}{\sigma_1^2}\right) = p\left(\chi_{\nu_1}^2 = \frac{\nu_1 m_1}{\sigma_1^2}\right), \tag{5.2.29}$$

where $\nu_1 m_1$ is the random variable and σ_1^2 is the fixed but unknown parameter. The second factor of (5.2.28) may be written

$$\frac{\Pr\{\chi_{\nu_2}^2 < \nu_2 m_2/\sigma_1^2\}}{\Pr\{F_{\nu_2,\nu_1} < m_2/m_1\}} = \frac{1-\alpha(\sigma_1^2)}{1-\alpha}. \tag{5.2.30}$$

In this expression, the quantity $\alpha(\sigma_1^2)$, which is *a function of* σ_1^2, is the significance level associated with the test in which the mean square m_2 is employed to test the hypothesis that $\sigma_{12}^2 = \sigma_1^2$ against the alternative $\sigma_{12}^2 > \sigma_1^2$ for each specified value of σ_1^2. As mentioned earlier in Section 5.2.3, the quantity α is the significance level associated with the *over-all test*, based on the mean square ratio m_2/m_1, of the hypothesis that $\sigma_{12}^2 = \sigma_1^2$ against the alternative that $\sigma_{12}^2 > \sigma_1^2$ where σ_1^2 is not specified.

Direct Calculation of $p(\sigma_1^2 \mid \mathbf{y})$

The distribution of σ_1^2 in (5.2.26) is equal to the product of two factors, the first being the density of an χ^{-2} variable with ν_1 degrees of freedom and the second, the ratio of a χ^2 probability integral and a F integral. The density function is readily calculated therefore using tables or charts of the χ^2 density and integral, and of the F integral. The distribution can also be expressed as a weighted series of χ^2 densities. Details are given in Appendix (A5.1).

5.2.6 A Scaled χ^{-2} Approximation to the Distribution of σ_1^2

Although the density function of σ_1^2 can be calculated directly from (5.2.26), for many practical purposes and in particular to study the problem of "pooling of variance estimates," it is useful to use an approximation involving only a single χ^2 variable.

Writing

$$z = \frac{\nu_1 m_1}{\sigma_1^2} \tag{5.2.31}$$

and applying the identity (A5.2.1) in Appendix A5.2, we find directly from (5.2.12) that the rth moment of z is

$$E(z^r \mid \mathbf{y}) = E\left[\left(\frac{\nu_1 m_1}{\sigma_1^2}\right)^r \Bigg| \mathbf{y}\right] = 2^r \frac{\Gamma\left(\dfrac{\nu_1}{2} + r\right) \Pr\left\{F_{\nu_2, \nu_1 + 2r} < \dfrac{(\nu_1 + 2r)}{\nu_1}\dfrac{m_2}{m_1}\right\}}{\Gamma\left(\dfrac{\nu_1}{2}\right) \Pr\left\{F_{\nu_2, \nu_1} < \dfrac{m_2}{m_1}\right\}}. \tag{5.2.32}$$

Further, the moment generating function is

$$M_z(t) = (1 - 2t)^{-\frac{1}{2}\nu_1} \frac{\Pr\left\{F_{\nu_2, \nu_1} < \dfrac{m_2/m_1}{1 - 2t}\right\}}{\Pr\left\{F_{\nu_2, \nu_1} < \dfrac{m_2}{m_1}\right\}}, \qquad |t| < \tfrac{1}{2}. \tag{5.2.33}$$

Consider the two extremes case when m_2/m_1 is very large and when it is close to zero. In the first case, $M_z(t)$ tends in the limit to $(1 - 2t)^{-\frac{1}{2}\nu_1}$ so that the distribution of z tends to the χ^2 distribution with ν_1 degrees of freedom. When m_2/m_1 tends to zero, it is easy to verify by applying L'Hospital's rule that

$$\lim_{m_2/m_1 \to 0} M_z(t) = (1 - 2t)^{-\frac{1}{2}(\nu_1 + \nu_2)} \tag{5.2.34}$$

so that the distribution of z again tends to the χ^2 distribution but with ν_2 additional degrees of freedom.

The preceding discussion suggests that the distribution of z could be well approximated by a scaled χ^2 variable. By equating the first two moments of z to those of $a\chi_b^2$, we find

$$a = \left(\frac{\nu_1}{2} + 1\right) \frac{I_x(\frac{1}{2}\nu_2, \frac{1}{2}\nu_1 + 2)}{I_x(\frac{1}{2}\nu_2, \frac{1}{2}\nu_1 + 1)} - \frac{\nu_1}{2} \frac{I_x(\frac{1}{2}\nu_2, \frac{1}{2}\nu_1 + 1)}{I_x(\frac{1}{2}\nu_2, \frac{1}{2}\nu_1)}$$

$$b = \frac{\nu_1}{a} \frac{I_x(\frac{1}{2}\nu_2, \frac{1}{2}\nu_1 + 1)}{I_x(\frac{1}{2}\nu_2, \frac{1}{2}\nu_1)} \tag{5.2.35}$$

where

$$x = \frac{\nu_2 m_2}{\nu_1 m_1 + \nu_2 m_2}$$

and $I_x(p, q)$ is the incomplete beta integral. To this degree of approximation, then

$$\frac{\nu_1 m_1}{\sigma_1^2} \mathrel{\dot\sim} a\chi_b^2, \tag{5.2.36}$$

where, as before, the symbol "$\overset{\cdot}{\sim}$" means "approximately distributed as". In terms of σ_1^2 we thus have

$$\sigma_1^2 \overset{\cdot}{\sim} \left(\frac{v_1 m_1}{a}\right) \chi_b^{-2}. \tag{5.2.37}$$

It can be verified that $0 < a < 1$ and $b > v_1$. To illustrate this approximation, we show below in Table 5.2.2 the exact and approximate densities for the two sets of data introduced earlier. Using Pearson's *Tables of the Incomplete Beta–Function* (1934) we find that for the dyestuff example that $a = 0.9912$ and $b = 24.26$ and for the generated example $a = 0.9426$ and $b = 28.67$. The agreement between the exact and the approximate values is exceedingly close for both examples.

Table 5.2.2

Comparison of the exact density $p(\sigma_1^2 \mid \mathbf{y})$ with approximate density obtained from $v_1 m_1/\sigma_1^2 \overset{\cdot}{\sim} a\chi_b^2$

Dyestuff example			Generated example		
$\sigma_1^2 \times 10^{-3}$	Exact density $\times 10^3$	Approximate density $\times 10^3$	$\sigma_1^2 \times 10^{-1}$	Exact density $\times 10$	Approximate density $\times 10$
1.00	0.0018	0.0017	0.60	0.0064	0.0062
1.25	0.0350	0.0345	0.80	0.2099	0.2097
1.50	0.1652	0.1647	1.00	0.7959	0.7975
1.75	0.3666	0.3674	1.20	1.1605	1.1610
2.00	0.5282	0.5300	1.40	1.0530	1.0518
2.25	0.5854	0.5870	1.60	0.7432	0.7418
2.50	0.5499	0.5504	1.80	0.4573	0.4566
2.75	0.4639	0.4633	2.00	0.2611	0.2611
3.00	0.3646	0.3633	2.20	0.1435	0.1437
3.25	0.2733	0.2718	2.40	0.0775	0.0778
3.50	0.1986	0.1973	2.60	0.0416	0.0419
3.75	0.1413	0.1403	2.80	0.0225	0.0227
4.00	0.0992	0.0986	3.00	0.0122	0.0124
4.25	0.0691	0.0688	3.20	0.0067	0.0068
4.50	0.0479	0.0479	3.40	0.0038	0.0038
4.75	0.0332	0.0333	3.60	0.0021	0.0022
5.00	0.0230	0.0232	3.80	0.0012	0.0013
5.25	0.0160	0.0162	4.00	0.0007	0.0007
5.50	0.0111	0.0114			

It will be recalled from (5.2.26) that if the additional information coming from m_2 were ignored, then for each set of data we would have

$$\frac{\nu_1 m_1}{\sigma_1^2} \sim \chi^2_{24}.$$

In fact we find

Dyestuff example	*Generated example*
$\dfrac{\nu_1 m_1}{\sigma_1^2} \mathrel{\dot\sim} 0.99\chi^2_{24.3}$	$\dfrac{\nu_1 m_1}{\sigma_1^2} \mathrel{\dot\sim} 0.94\chi^2_{28.7}$

As expected, the modification which occurs in the dyestuff example is very slight, whereas that for the generated example is considerably greater.

5.2.7 A Bayesian Solution to the Pooling Dilemma

On sampling theory, when m_1 and m_2 are of about the same size they are often pooled. The idea is that *if it can be assumed that* σ_2^2 *is zero*, the appropriate estimate of σ_1^2 is $(\nu_1 m_1 + \nu_2 m_2)/(\nu_1 + \nu_2)$ and not m_1.

Thus:

a) if m_2/m_1 is large, the estimate of σ_1^2 is taken to be m_1, distributed as $\sigma_1^2 \chi^2_{\nu_1}/\nu_1$,

b) if m_2/m_1 is close to one, the estimate of σ_1^2 is taken to be $(\nu_1 m_1 + \nu_2 m_2)/(\nu_1 + \nu_2)$, distributed as $\sigma_1^2 \chi^2_{\nu_1+\nu_2}/(\nu_1 + \nu_2)$.

Another way of saying this would be, that the estimate is taken to be

$$\frac{\nu_1 m_1 + \lambda \nu_2 m_2}{\nu_1 + \omega \nu_2} \quad \text{distributed as} \quad \sigma_1^2 \chi^2_{\nu_1+\omega\nu_2}/(\nu_1 + \omega\nu_2)$$

with the weights λ and ω such that

a) $\lambda = \omega = 0$ if m_2/m_1 is large

b) $\lambda = \omega = 1$ if m_2/m_1 is close to one

Now, from the Bayesian viewpoint the distribution of σ_1^2 may be approximated, as in (5.2.36), by $(\nu_1 m_1)/\sigma_1^2 \mathrel{\dot\sim} a\chi^2_b$. This approximation may equally well be written in terms of the pooled estimate as

$$\frac{1}{\sigma_1^2} \frac{\nu_1 m_1 + \lambda \nu_2 m_2}{\nu_1 + \omega \nu_2} \mathrel{\dot\sim} \frac{\chi^2_{\nu_1+\omega\nu_2}}{\nu_1 + \omega\nu_2}, \tag{5.2.38a}$$

or in terms of sums of squares as

$$\frac{1}{\sigma_1^2}(S_1 + \lambda S_2) \mathrel{\dot\sim} \chi^2_{\nu_1+\omega\nu_2}, \tag{5.2.38b}$$

where

$$\lambda = \frac{\nu_1[(1/a) - 1]}{\nu_2(m_2/m_1)}$$

$$\omega = \frac{(b - \nu_1)}{\nu_2}. \tag{5.2.38c}$$

Since $0 < a < 1$ and $b > \nu_1$, the weights λ and ω are both positive.

For illustration, consider first the dyestuff example. Here the mean square ratio $(m_2/m_1 = 4.6)$ is quite large and the corresponding weights $(\lambda = 0.009, \omega = 0.052)$ are therefore small. The contribution of m_2 for this example may thus be ignored for practical purposes, and the posterior distribution and the sampling theory confidence distribution essentially coincide. For the generated example, however, the mean square ratio $(m_2/m_1 = 0.56)$ is small and the corresponding weights $(\lambda = 0.524, \omega = 0.934)$ are no longer negligible. In this case then, the contribution of m_2 to the posterior distribution of σ_1^2 is considerable. These two posterior distributions have of course already been shown in Figs. 5.2.3 and 5.2.4 and their derivation discussed. The present excursion merely provides an alternative but interesting means of viewing them.

Using (5.2.35) with (5.2.38c), λ and ω may be plotted as functions of m_2/m_1 for any J and K. Figure 5.2.5 shows such graphs for the case $J = 6$ and $K = 5$ corresponding to the examples we have considered. As might be expected, for large values of m_2/m_1, λ and ω approach zero but, as m_2/m_1 becomes smaller, larger values of λ and ω calling for increasing degrees of pooling occur.

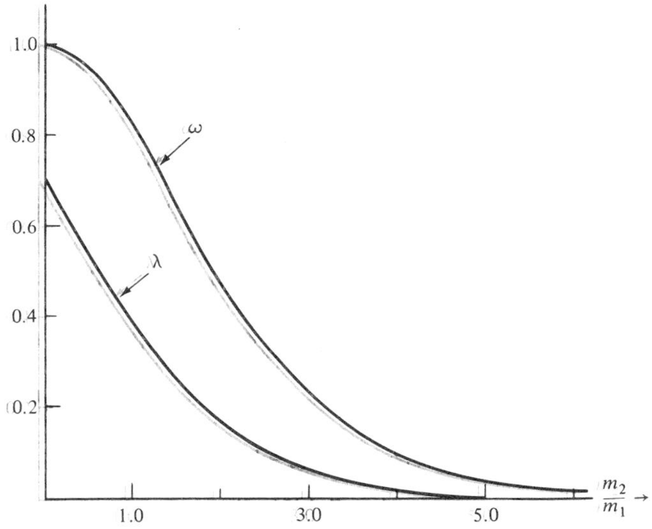

Fig. 5.2.5 Values of (λ, ω) as functions of m_2/m_1 $(J = 6, K = 5)$.

It will be noticed that ω is larger than λ for all values of m_2/m_1. To see the reason for this, consider the case where m_1 and m_2 are equal so that $m_2/m_1 = 1$. While m_1 provides an estimate of σ_1^2 alone, m_2 is an estimate of $\sigma_{12}^2 = \sigma_1^2 + K\sigma_2^2$ and σ_2^2 must be *positive* or zero. Thus with $m_2/m_1 = 1$ the evidence coming from m_2 and m_1 together implies a *smaller* value for σ_1^2 than would have been suggested by m_1 alone. This is consonant with the fact that, for $m_2/m_1 = 1$, ω is larger than λ, and corresponding compensation would be expected, and is found, for other values of m_2/m_1.

Thus, in the Bayesian context there is no dilemma "to pool or not to pool." The distribution of σ_1^2 *always* depends on m_2 as well as on m_1, as obviously it should. No "decision" needs to be made, instead there is a steady pooling transition which depends on the evidence supplied by the sample.

5.2.8 Posterior Distribution of σ_2^2

We now turn to the problem of making inferences about σ_2^2 which in the context of sampling theory is accompanied by many difficulties. The posterior distribution of σ_2^2 is obtained by integrating (5.2.14) over σ_1^2 to yield

$$p(\sigma_2^2 \mid \mathbf{y}) = \frac{K(\nu_1 m_1)^{-1}(\nu_2 m_2)^{-1}}{\Pr\{F_{\nu_2,\nu_1} < m_2/m_1\}} \int_0^\infty p\left(\chi_{\nu_2}^{-2} = \frac{\sigma_1^2 + K\sigma_2^2}{\nu_2 m_2}\right) p\left(\chi_{\nu_1}^{-2} = \frac{\sigma_1^2}{\nu_1 m_1}\right) d\sigma_1^2,$$

$$\sigma_2^2 > 0. \qquad (5.2.39)$$

For the noninformative reference prior distribution (5.2.9), the distribution $p(\sigma_2^2 \mid \mathbf{y})$ summarizes all our *a posteriori* knowledge about σ_2^2. It is defined over the range $(0, \infty)$ and thus no problem of "negative variance estimate" will ever arise. The exact posterior distributions of σ_2^2 for the dyestuff example and for the generated example are given by the solid curves in Figs. 5.2.6 and 5.2.7 respectively.

Some properties of the distribution of σ_2^2 are discussed in Appendix A5.3. In particular, it is shown that

a) if
$$\frac{m_2}{m_1} > \frac{\nu_2 + 2}{\nu_2} \frac{\nu_1}{\nu_1 + 2},$$

then the distribution is monotonically increasing in the interval $0 < \sigma_2^2 < c_1$, and monotonically decreasing in the interval $\sigma_2^2 > c_2$, where

$$c_1 = K^{-1}\left(\frac{\nu_1 + \nu_2 + 4}{\nu_1 + 2\nu_2 + 10}\right)\left(\frac{\nu_2}{\nu_2 + 2} m_2 - \frac{\nu_1 m_1 + \nu_2 m_2}{\nu_1 + \nu_2 + 4}\right),$$

$$c_2 = K^{-1}\left(\frac{\nu_2}{\nu_2 + 2} m_2 - \frac{\nu_1}{\nu_1 + \nu_2 + 4} m_1\right), \qquad 0 < c_1 < c_2. \qquad (5.2.40)$$

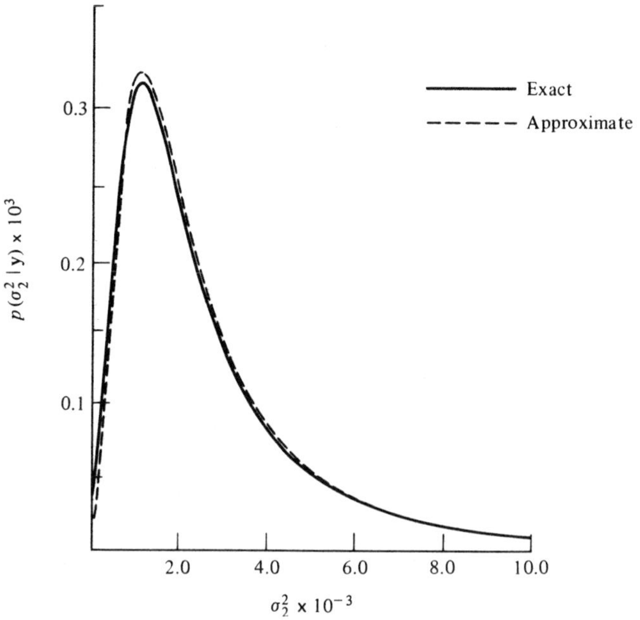

Fig. 5.2.6 Posterior distribution of σ_2^2: the dyestuff example.

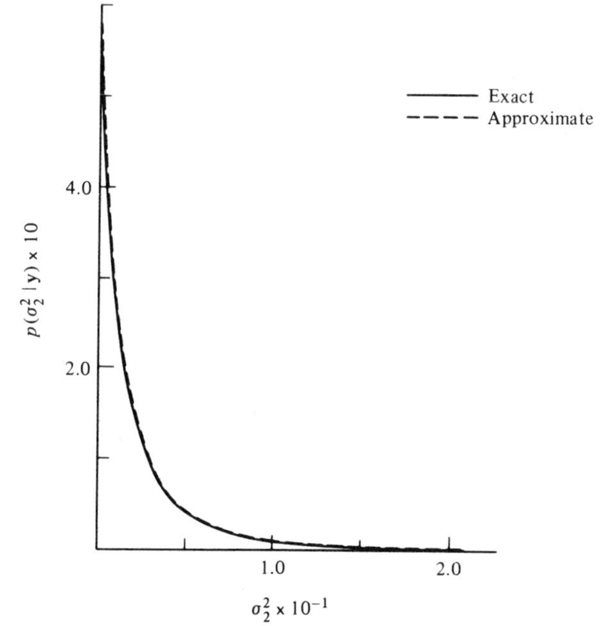

Fig. 5.2.7 Posterior distribution of σ_2^2: the generated example.

so that the mode must lie in the interval $c_1 < \sigma^2 < c_2$, and

b) if

$$\frac{m_2}{m_1} < \left(\frac{\nu_2 + 2}{\nu_2}\right)\left(\frac{\nu_1}{\nu_1 + \nu_2 + 4}\right),$$

then the distribution is monotonically decreasing in the entire range $\sigma_2^2 > 0$ and the mode is at the origin.

For the dyestuff example, $m_2/m_1 \doteq 4.60$, which is greater than

$$\left(\frac{\nu_1}{\nu_1 + 2}\right)\left(\frac{\nu_2 + 2}{\nu_2}\right) \doteq 1.29$$

so that the mode must lie in the interval $c_1 < \sigma^2 < c_2$ where $c_1 = 684.08$ and $c_2 = 1{,}253.67$. Inspection of Fig. 5.2.6 shows that the mode is approximately at $\sigma_2^2 = 1{,}100$. For the generated example, $m_2/m_1 = 0.558$ which is less than

$$\left(\frac{\nu_2 + 2}{\nu_2}\right)\left(\frac{\nu_1}{\nu_1 + \nu_2 + 4}\right) = 1.02.$$

Thus, as shown in Fig. 5.2.7, the distribution is J shaped, having its mode at the origin and monotonically decreasing in the entire range $\sigma_2^2 > 0$.

5.2.9 Computation of $p(\sigma_2^2 \mid \mathbf{y})$

It does not appear possible to express the distribution of σ_2^2 in (5.2.39) in terms of simple functions. The density for each value of σ_2^2 can, however, be obtained to any desired accuracy by numerical calculation of the one-dimensional integral. This was the method used to obtain the exact distributions of Figs. 5.2.6 and 5.2.7. Approximations which require less labour are now described.

A Simple χ^2 Approximation

It is seen from the posterior distribution of σ_1^2 and σ_2^2 in (5.2.14) that if σ_1^2 were known, say $\sigma_1^2 = \sigma_{10}^2$, then the conditional distribution of σ_2^2 would be in the form of a *truncated* "inverted" χ^2 distribution. That is,

$$p(\sigma_2^2 \mid \sigma_1^2 = \sigma_{10}^2, \mathbf{y}) = \frac{K(\nu_2 m_2)^{-1}\, p\left(\chi_{\nu_2}^{-2} = \dfrac{\sigma_{10}^2 + K\sigma_2^2}{\nu_2 m_2}\right)}{\Pr\left(\chi_{\nu_2}^{-2} > \dfrac{\sigma_{10}^2}{\nu_2 m_2}\right)}, \qquad \sigma_2^2 > 0, \qquad (5.2.41)$$

so that the random variable

$$x = \frac{\sigma_{10}^2 + K\sigma_2^2}{\nu_2 m_2} \qquad (5.2.42)$$

is distributed as χ^{-2} with ν_2 degrees of freedom truncated from $\sigma_{10}^2/(\nu_2 m_2)$. Although the marginal posterior distribution $p(\sigma_2^2 \mid \mathbf{y})$ cannot be expressed exactly in terms of such a simple form, nevertheless, the expression of the integral in (5.2.39) does suggest that, for large $\nu_1/2$, we can write

$$\int_0^\infty p\left(\chi_{\nu_2}^{-2} = \frac{\sigma_1^2 + K\sigma_2^2}{\nu_2 m_2}\right) p\left(\chi_{\nu_1}^{-2} = \frac{\sigma_1^2}{\nu_1 m_1}\right) d\left(\frac{\sigma_1^2}{\nu_1 m_1}\right)$$

$$= \frac{1}{\Gamma(t+1)} \int_{-\infty}^{\infty} p\left(\chi_{\nu_2}^{-2} = \frac{\nu_1 m_1 (2z)^{-1} + K\sigma_2^2}{\nu_2 m_2}\right) z^{(\nu_1/2)} e^{-z} \, d(\log z)$$

$$\doteq p\left(\chi_{\nu_2}^{-2} = \frac{\nu_1 m_1 (2t)^{-1} + K\sigma_2^2}{\nu_2 m_2}\right), \tag{5.2.43}$$

where the substitution $z = \nu_1 m_1/(2\sigma_1^2)$ is made and $t = \nu_1/2$ is the value of z which maximizes the factor $z^{(\nu_1/2)} e^{-z}$ appearing in the integrand. Making use of (5.2.43) and upon renormalizing, we thus arrive at a simple truncated inverted χ^2 approximating form for the posterior distribution of σ_2^2,

$$p(\sigma_2^2 \mid \mathbf{y}) \doteq \frac{K(\nu_2 m_2)^{-1} p\left(\chi_{\nu_2}^{-2} = \dfrac{m_1 + K\sigma_2^2}{\nu_2 m_2}\right)}{\Pr\left\{\chi_{\nu_2}^{-2} > \dfrac{m_1}{\nu_2 m_2}\right\}}, \qquad \sigma_2^2 > 0. \tag{5.2.44}$$

or equivalently,

$$p\left(\frac{m_1 + K\sigma_2^2}{\nu_2 m_2} \,\middle|\, \mathbf{y}\right) \doteq \frac{p\left(\chi_{\nu_2}^{-2} = \dfrac{m_1 + K\sigma_2^2}{\nu_2 m_2}\right)}{\Pr\left\{\chi_{\nu_2}^{-2} > \dfrac{m_1}{\nu_2 m_2}\right\}}, \qquad \frac{m_1 + K\sigma_2^2}{\nu_2 m_2} > \frac{m_1}{\nu_2 m_2}. \tag{5.2.45}$$

In words, we have that $(m_1 + K\sigma_2^2)/(\nu_2 m_2)$ is approximately distributed as χ^{-2} with ν_2 degrees of freedom truncated from below at $m_1/(\nu_2 m_2)$. Or equivalently, $\nu_2 m_2/(m_1 + K\sigma_2^2)$ is distributed approximately as χ^2 with ν_2 degrees of freedom truncated from above at $\nu_2 m_2/m_1$. Posterior probabilities for σ_2^2 can thus be approximately determined from an ordinary χ^2 table. In particular, for $\eta > 0$

$$\Pr\{0 < \sigma_2^2 < \eta \mid \mathbf{y}\} \doteq \frac{\Pr\left\{\dfrac{\nu_2 m_2}{m_1 + K\eta} < \chi_{\nu_2}^2 < \dfrac{\nu_2 m_2}{m_1}\right\}}{\Pr\left\{\chi_{\nu_2}^2 < \dfrac{\nu_2 m_2}{m_1}\right\}}. \tag{5.2.46}$$

The posterior densities for σ_2^2 for each of the two examples obtained from this approximation are shown by the dotted curves in Figs. 5.2.6 and 5.2.7. It is seen that this simple approximation gives satisfactory results even for the rather small sample sizes considered.

Table 5.2.3

Comparison of exact and approximate posterior density of σ_2^2†

a) *Dyestuff Example*

	$p(\sigma_2^2 \mid \mathbf{y}) \times 10^3$			
	(1)	(2)	(3)	(4)
$\sigma_2^2 \times 10^{-3}$	t^0	t^{-1}	t^{-2}	Exact
0.0	0.0070	0.0246	0.0332	0.0290
0.2	0.0591	0.0865	0.0851	0.0841
0.4	0.1518	0.1694	0.1665	0.1667
0.6	0.2386	0.2455	0.2433	0.2440
0.8	0.2948	0.2952	0.2938	0.2946
1.0	0.3200	0.3169	0.3161	0.3170
1.1	0.3233	0.3193	0.3187	0.3196
1.2	0.3221	0.3175	0.3170	0.3179
1.3	0.3173	0.3123	0.3120	0.3128
1.4	0.3099	0.3047	0.3045	0.3053
1.5	0.3006	0.2953	0.2952	0.2959
1.6	0.2899	0.2847	0.2846	0.2853
1.7	0.2784	0.2733	0.2732	0.2740
1.8	0.2664	0.2615	0.2615	0.2621
1.9	0.2543	0.2495	0.2495	0.2502
2.0	0.2422	0.2376	0.2376	0.2383
2.2	0.2187	0.2145	0.2146	0.2151
2.4	0.1967	0.1930	0.1930	0.1935
2.6	0.1766	0.1732	0.1733	0.1737
2.8	0.1584	0.1554	0.1555	0.1559
3.0	0.1421	0.1395	0.1395	0.1399
3.5	0.1089	0.1070	0.1070	0.1073
4.0	0.0843	0.0829	0.0829	0.0831
4.5	0.0661	0.0650	0.0650	0.0652
5.0	0.0524	0.0516	0.0516	0.0518
5.5	0.0421	0.0415	0.0415	0.0416
6.0	0.0342	0.0337	0.0337	0.0338
6.5	0.0281	0.0277	0.0277	0.0278
7.0	0.0232	0.0230	0.0230	0.0230
7.5	0.0194	0.0192	0.0192	0.0193
8.0	0.0164	0.0162	0.0162	0.0162

Table 5.2.3 (*continued*)

b) *Generated Example*

	$p(\sigma_2^2 \mid \mathbf{y}) \times 10$			
	(1)	(2)	(3)	(4)
$\sigma_2^2 \times 10^{-1}$	t^0	t^{-1}	t^{-2}	Exact
0.00	5.3566	5.5170	5.4635	5.4795
0.05	3.8069	3.8943	3.8829	3.8897
0.10	2.7671	2.7997	2.8003	2.8039
0.15	2.0563	2.0581	2.0623	2.0644
0.20	1.5595	1.5463	1.5514	1.5527
0.25	1.2046	1.1853	1.1902	1.1911
0.30	0.9460	0.9251	0.9294	0.9301
0.35	0.7538	0.7336	0.7373	0.7379
0.40	0.6087	0.5901	0.5932	0.5937
0.45	0.4974	0.4807	0.4833	0.4837
0.50	0.4107	0.3961	0.3982	0.3986
0.60	0.2881	0.2769	0.2784	0.2787
0.70	0.2086	0.2002	0.2012	0.2014
0.80	0.1552	0.1488	0.1495	0.1497
0.90	0.1181	0.1132	0.1137	0.1138
1.00	0.0916	0.0878	0.0882	0.0883
1.20	0.0579	0.0556	0.0558	0.0559
1.40	0.0386	0.0371	0.0372	0.0373
1.60	0.0268	0.0258	0.0259	0.0259
1.80	0.0193	0.0186	0.0187	0.0187
2.00	0.0143	0.0138	0.0138	0.0138

† Accuracy of various approximations [see expression (A5.4.7) in Appendix A5.4 for the formulas].
(1) Asymptotic expansion with the leading term.
(2) Asymptotic expansion with terms to order t^{-1}.
(3) Asymptotic expansion with terms to order t^{-2}.
(4) Exact evaluation by numerical integration.

An Asymptotic Expansion

The argument which led to (5.2.44) can be further exploited to yield an asymptotic expansion of $p(\sigma_2^2 \mid \mathbf{y})$ in powers of $t^{-1} = (\nu_1/2)^{-1}$ following the method discussed for example in Jeffreys and Swirlee (1956) and De Bruijn (1961). Details of the derivation of the asymptotic series are given in Appendix A5.4. The series is such that the simple approximation in (5.2.44) is the leading term. Table 5.2.3 shows, for the two examples, the degree of improvement that can be obtained by including terms up to order t^{-1} and up to order t^{-2}, respectively.

The Distribution of σ_2^2 When the Mean Square Ratio is Close to Zero

Finally, we remark here that in the extreme situation when m_2/m_1 approaches zero the posterior distribution of $K\sigma_2^2/m_2$ becomes diffuse, a result noted by Hill (1965). This behavior is reflected by the approximating form (5.2.44) from which it will be seen that as m_2/m_1 tends to zero, the distribution of $K\sigma_2^2/m_2$ approaches a horizontal line. In the limit, then, all values of $K\sigma_2^2/m_2$ become equally probable and no information about σ_2^2 is gained. It has been noted in Section 5.2.5 that if m_2/m_1 is close to zero, this will throw doubt on the adequacy of the model. In a Bayesian analysis the contradiction between model and data is evidenced by the conflict between $p^*(\sigma_1^2 \mid \mathbf{y})$ and $f(\sigma_1^2)$. The distribution of $K\sigma_2^2/m_2$ thus becomes diffuse precisely as overwhelming evidence becomes available that the basis on which this distribution is computed is unsound. If assumptions are persisted in, in spite of evidence that they are false, bizarre conclusions can be expected.

Table 5.2.4

Summarized calculations for approximating the posterior distributions of (σ_1^2, σ_2^2) (illustrated using dyestuff data)

---

1. Use standard analysis of variance (Table 5.2.1) to compute

$$m_1 = 2{,}451.25, \qquad m_2 = 11{,}271.50, \qquad \nu_1 = 24, \qquad \nu_2 = 5, \qquad K = 5.$$

2. Use (5.2.35) to compute

$$x = 0.4893, \qquad a = 0.9912, \qquad b = 24.26.$$

3. Then, using (5.2.37) the posterior distribution of the "within" component σ_1^2 is given by

$$\sigma_1^2 \mathrel{\dot\sim} \nu_1 m_1 a^{-1} \chi_b^{-2},$$

and for this example

$$\sigma_1^2 \mathrel{\dot\sim} 59{,}352.30 \chi_{24.26}^{-2}.$$

4. Also from (5.2.45) posterior distribution of the "between" component σ_2^2 is given by

$$\frac{m_1 + K\sigma_2^2}{\nu_2 m_2} \mathrel{\dot\sim} \chi_{\nu_2}^{-2}$$

truncated from below at

$$\frac{m_1}{\nu_2 m_2}.$$

For this example,

$$0.043 + (0.89 \times 10^{-4} \sigma_2^2) \mathrel{\dot\sim} \chi_5^{-2}$$

truncated from below at 0.043.

5.2.10 A Summary of Approximations to the Posterior Distribution of (σ_1^2, σ_2^2)

Using the dyestuff data for illustration Table 5.2.4. above provides a summary of the calculations needed to approximate the posterior distributions of σ_1^2 and σ_2^2.

5.2.11 Derivation of the Posterior Distribution of Variance Components Using Constraints Directly

It is of some interest to deduce directly the distributions of the variance components by employing the result (1.5.3) on constrained distributions. We have seen that the data can be thought of as consisting of J independent observations from a population $N(\theta, \sigma_{12}^2/K)$ and $J(K-1)$ further independent observations from a population $N(\theta, \sigma_1^2)$. Suppose that, as before, we assume that the joint prior distribution $p(\theta, \log\sigma_1^2, \log\sigma_{12}^2)$ is locally uniform but now we temporarily omit the constraint $C\colon \sigma_{12}^2 > \sigma_1^2$. Then σ_1^2 and σ_{12}^2 would be independent *a posteriori*. Also, if, as before, $p^*(x)$ denotes the unconstrained density of x, then

$$p^*(\sigma_1^2 \mid \mathbf{y}) = (\nu_1 m_1)^{-1}\, p\left(\chi_{\nu_1}^{-2} = \frac{\sigma_1^2}{\nu_1 m_1}\right), \qquad \sigma_1^2 > 0, \tag{5.2.47}$$

$$p^*(\sigma_{12}^2 \mid \mathbf{y}) = (\nu_2 m_2)^{-1}\, p\left(\chi_{\nu_2}^{-2} = \frac{\sigma_{12}^2}{\nu_2 m_2}\right), \qquad \sigma_{12}^2 > 0, \tag{5.2.48}$$

$$p^*(W \mid \mathbf{y}) = \left(\frac{m_1}{m_2}\right) p\left(F_{\nu_1,\nu_2} = \frac{m_1}{m_2} W\right), \qquad W > 0, \tag{5.2.49}$$

where $W = \sigma_{12}^2/\sigma_1^2$. It follows that

$$\Pr{}^*\,(\sigma_{12}^2 > \sigma_1^2 \mid \sigma_1^2, \mathbf{y}) = \Pr\left\{\chi_{\nu_2}^2 < \frac{\nu_2 m_2}{\sigma_1^2}\right\}, \tag{5.2.50}$$

$$\Pr{}^*\,(\sigma_{12}^2 > \sigma_1^2 \mid \sigma_{12}^2, \mathbf{y}) = \Pr\left\{\chi_{\nu_1}^2 > \frac{\nu_1 m_1}{\sigma_{12}^2}\right\}, \tag{5.2.51}$$

and

$$\Pr{}^*\,(\sigma_{12}^2 > \sigma_1^2 \mid \mathbf{y}) = \Pr\left\{F_{\nu_2,\nu_1} < \frac{m_2}{m_1}\right\}. \tag{5.2.52}$$

Using (1.5.3) the joint posterior distribution of $(\sigma_1^2, \sigma_{12}^2)$ given the constraint C is

$$p(\sigma_1^2, \sigma_{12}^2 \mid \mathbf{y}) = \frac{p^*(\sigma_1^2 \mid \mathbf{y})\, p^*(\sigma_{12}^2 \mid \mathbf{y})\, \Pr{}^*\,\{C \mid \sigma_1^2, \sigma_{12}^2, \mathbf{y}\}}{\Pr{}^*\,(C \mid \mathbf{y})}, \qquad \sigma_{12}^2 > \sigma_1^2 > 0. \tag{5.2.53}$$

Noting that given $(\sigma_1^2, \sigma_{12}^2)$ such that $\sigma_{12}^2 > \sigma_1^2$, $\Pr^*\{C \mid \sigma_1^2, \sigma_{12}^2, \mathbf{y}\} = 1$, we have

$$p(\sigma_1^2, \sigma_{12}^2 \mid \mathbf{y}) = \frac{(\nu_1 m_1)^{-1}\, p\,[\chi_{\nu_1}^{-2} = \sigma_1^2/(\nu_1 m_1)]\, (\nu_2 m_2)^{-1}\, p\,[\chi_{\nu_2}^{-2} = \sigma_{12}^2/(\nu_2 m_2)]}{\Pr\{F_{\nu_2,\nu_1} < m_2/m_1\}},$$

$$\sigma_{12}^2 > \sigma_1^2 > 0, \qquad (5.2.54)$$

which is, of course, equivalent to the posterior distribution of (σ_1^2, σ_2^2) in (5.2.14) if we remember that $\sigma_{12}^2 = \sigma_1^2 + K\sigma_2^2$. The distribution of the ratio W given C is, by a similar argument,

$$p(W \mid \mathbf{y}) = \frac{p^*(W \mid \mathbf{y}) \Pr^*\{C \mid W, \mathbf{y}\}}{\Pr^*\{C \mid \mathbf{y}\}}$$

$$= \frac{(m_1/m_2)\, p(F_{\nu_1,\nu_2} = (m_1/m_2)\, W)}{\Pr\{F_{\nu_2,\nu_1} < m_2/m_1\}}, \qquad W > 1, \qquad (5.2.55)$$

where $\Pr^*(C \mid W, \mathbf{y}) = 1$. This is the same distribution obtained earlier in (5.2.16). Further, the distribution of σ_1^2 subject to the constraint C is

$$p(\sigma_1^2 \mid \mathbf{y}) = \frac{p^*(\sigma_1^2 \mid \mathbf{y}) \Pr^*\{C \mid \sigma_1^2, \mathbf{y}\}}{\Pr^*\{C \mid \mathbf{y}\}}$$

$$= \frac{(\nu_1 m_1)^{-1}\, p\left(\chi_{\nu_1}^{-2} = \dfrac{\sigma_1^2}{\nu_1 m_1}\right) \Pr\left\{\chi_{\nu_2}^2 < \dfrac{\nu_2 m_2}{\sigma_1^2}\right\}}{\Pr\left\{F_{\nu_2,\nu_1} < \dfrac{m_2}{m_1}\right\}}, \qquad \sigma_1^2 > 0, \qquad (5.2.56)$$

which is identical to expression (5.2.26). Finally, by the same argument, the distribution of σ_{12}^2 given C is

$$p(\sigma_{12}^2 \mid \mathbf{y}) = \frac{p^*(\sigma_{12}^2 \mid \mathbf{y}) \Pr^*\{C \mid \sigma_{12}^2, \mathbf{y}\}}{\Pr^*\{C \mid \mathbf{y}\}}$$

$$= \frac{(\nu_2 m_2)^{-1}\, p\left(\chi_{\nu_2}^{-2} = \dfrac{\sigma_{12}^2}{\nu_2 m_2}\right) \Pr\left\{\chi_{\nu_1}^2 > \dfrac{\nu_1 m_1}{\sigma_{12}^2}\right\}}{\Pr\left\{F_{\nu_2,\nu_1} < \dfrac{m_2}{m_1}\right\}}, \qquad \sigma_{12}^2 > 0. \qquad (5.2.57)$$

Certain properties of this distribution are discussed in the next section.

5.2.12 Posterior Distribution of σ_{12}^2 and a Scaled χ^{-2} Approximation

Earlier we saw that not only m_1 but also m_2 contained information about the component σ_1^2. In a similar way we find that both m_1 and m_2 contain information about σ_{12}^2. While this parameter is not usually of direct interest for the two-component model, we need to study its distribution and methods of approximating it for more complicated random effect models. From (5.2.57) it is seen that the posterior distribution of σ_{12}^2 is proportional to the product of two factors. The first factor is a χ^{-2} density function while the second factor is the right tail area of a χ^2 distribution. This is in contrast to the posterior distribution of σ_1^2 in (5.2.56), where the probability integral is a left tail probability of a χ^2 variable. Using the identity (A5.2.1) in Appendix A5.2, the rth moment of $z_1 = v_2 m_2/\sigma_{12}^2$ is

$$E(z_1^r) = 2^r \frac{\Gamma\left(\frac{v_2}{2} + r\right) \Pr\left\{F_{v_2+2r, v_1} < \frac{v_2}{v_2 + 2r}\frac{m_2}{m_1}\right\}}{\Gamma\left(\frac{v_2}{2}\right) \Pr\left\{F_{v_2, v_1} < \frac{m_2}{m_1}\right\}}, \tag{5.2.58}$$

and the moment generating function is

$$M_{z_1}(t) = (1 - 2t)^{-\frac{1}{2}v_2} \frac{\Pr\{F_{v_2, v_1} < (m_2/m_1)(1 - 2t)\}}{\Pr\{F_{v_2, v_1} < m_2/m_1\}}, \qquad |t| < \tfrac{1}{2}. \tag{5.2.59}$$

When m_2/m_1 approaches infinity, $M_{z_1}(t)$ tends to $(1 - 2t)^{-\frac{1}{2}v_2}$ so that the distribution of z_1 tends to the χ^2 distribution with v_2 degrees of freedom. This suggests that, for moderately large values of m_2/m_1, the distribution of $z_1 = v_2 m_2/\sigma_{12}^2$ might well be approximated by that of a scaled χ^2 variable, say $c\chi_d^2$. By equating the first two moments, we find

$$c = \left(\frac{v_2}{2} + 1\right) \frac{I_x(\frac{1}{2}v_2 + 2, \frac{1}{2}v_1)}{I_x(\frac{1}{2}v_2 + 1, \frac{1}{2}v_1)} - \frac{v_2}{2} \frac{I_x(\frac{1}{2}v_2 + 1, \frac{1}{2}v_1)}{I_x(\frac{1}{2}v_2, \frac{1}{2}v_1)}$$

$$d = \frac{v_2}{c} \frac{I_x(\frac{1}{2}v_2 + 1, \frac{1}{2}v_1)}{I_x(\frac{1}{2}v_2, \frac{1}{2}v_1)}, \tag{5.2.60}$$

where we recall from (5.2.35) that $x = v_2 m_2/(v_1 m_1 + v_2 m_2)$. Thus, to this degree of approximation,

$$\frac{v_2 m_2}{\sigma_{12}^2} \mathrel{\dot\sim} c\chi_d^2 \qquad \text{or} \qquad \sigma_{12}^2 \mathrel{\dot\sim} \frac{v_2 m_2}{c} \chi_d^{-2} \tag{5.2.61}$$

and values of c and d can be readily determined from tables of the incomplete beta function.

In practice, agreement between the exact distribution and the approximation is found to be close provided m_2/m_1 is not too small. Table 5.2.5 gives specimens of the exact and the approximate densities for the dyestuff and the generated examples. The agreement for the dyestuff example is excellent. For the generated example, for which $m_2/m_1 = 0.56$, the approximation is rather poor. The reason for this can be seen by studying the limiting behavior of the moment generating function of z_1 in (5.2.59) when $m_2/m_1 \to 0$. Applying L'Hospital's rule, we find

$$\lim_{m_2/m_1 \to 0} M_{z_1}(t) = 1,$$

which is far from the moment generating function of a χ^2 distribution. Thus, we would not expect the approximation (5.2.61) to work for small values of m_2/m_1.

Table 5.2.5

Comparison of the exact density $p(\sigma_{12}^2 \mid \mathbf{y})$ with approximate density obtained from $\nu_2 m_2/\sigma_{12}^2 \sim c\chi_d^2$

Dyestuff example			Generated example		
$\sigma_{12}^2/1000$	Exact	Approximate	$\sigma_{12}^2/10$	Exact	Approximate
3.0	0.0041	0.0047	0.8	0.0172	0.0396
5.0	0.0399	0.0385	1.0	0.1233	0.1435
7.0	0.0626	0.0619	1.2	0.3074	0.2774
9.0	0.0640	0.0639	1.4	0.4543	0.3852
11.0	0.0557	0.0564	1.6	0.5101	0.4427
13.0	0.0461	0.0468	1.8	0.4936	0.4541
15.0	0.0372	0.0379	2.0	0.4409	0.4336
17.0	0.0300	0.0304	2.2	0.3776	0.3955
19.0	0.0242	0.0245	2.4	0.3170	0.3501
21.0	0.0196	0.0198	3.0	0.1836	0.2217
25.0	0.0132	0.0133	3.5	0.1198	0.1456
30.0	0.0084	0.0084	4.0	0.0812	0.0955
35.0	0.0054	0.0053	4.5	0.0571	0.0633

5.2.13 Use of Features of Posterior Distribution to Supply Sampling Theory Estimates

In recent years sampling theorists have used certain features of Bayesian posterior distribution to produce "point estimators" which are then judged on their sampling properties. We briefly discuss this approach in Appendix A5.6.

5.3 A THREE-COMPONENT HIERARCHICAL DESIGN MODEL

The results obtained in the preceding sections may be generalized in various ways. In the remainder of this chapter an extension from the two-component to the three-component hierarchical design model is studied. In Chapters 6 and 7, certain two-way cross classification random effect models and mixed models are considered.

The following development of the three-component model is broadly similar to that for two components. However, the reader's attention is particularly directed to the study in Section 5.3.4 of the *relative* contribution of variance components. Here new features arise and study of an example leads to somewhat surprising and disturbing conclusions which underline the danger of using point estimates.

For convenient reference we repeat here the specification of the three-component model already given at the beginning of this chapter.

It is supposed that

$$y_{ijk} = \theta + e_i + e_{ij} + e_{ijk}, \qquad i = 1, \ldots, I; \quad j = 1, \ldots, J; \quad k = 1, \ldots, K, \tag{5.3.1}$$

where y_{ijk} are the observations, θ is a common location parameter, e_i, e_{ij} and e_{ijk} are three different kinds of random effects. We further assume that the random effects (e_i, e_{ij}, e_{ijk}) are all independent and that

$$e_i \sim N(0, \sigma_3^2), \qquad e_{ij} \sim N(0, \sigma_2^2), \qquad \text{and} \qquad e_{ijk} \sim N(0, \sigma_1^2). \tag{5.3.2}$$

It follows in particular that $\text{Var}(y_{ijk}) = \sigma_1^2 + \sigma_2^2 + \sigma_3^2$ so that the parameters $(\sigma_1^2, \sigma_2^2, \sigma_3^2)$ are the variance components.

As previously mentioned, a hierarchical design of this type might be used in an industrial experiment to investigate the variation in the quality of a product from a batch chemical process. One might randomly select I batches of material, take J samples from each batch, and perform K repeated analyses on each sample. A particular result y_{ijk} would then be subject to three sources of variation, that due to batches e_i, that due to samples e_{ij} and that due to analyses e_{ijk}. The purpose of an investigation of the kind could be (a) to make inferences about the *relative* contribution of each source of variation to the total variance or (b) to make inferences about the variance components individually.

5.3.1 The Likelihood Function

To obtain the likelihood function, it is convenient to work with the transformed variables $y_{i..}$, $y_{ij.} - y_{i..}$, and $y_{ijk} - y_{ij.}$ rather than the observations y_{ijk} themselves. By repeated application of Theorem 5.2.1 (on page 250) the quantities

$$\begin{aligned} y_{i..} &= \theta + e_i + e_{i.} + e_{i..}, \\ y_{ij.} - y_{i..} &= (e_{ij} - e_{i\cdot}) + (e_{ij.} - e_{i..}), \\ y_{ijk} - y_{ij.} &= e_{ijk} - e_{ij.}\,, \end{aligned} \tag{5.3.3}$$

are distributed independently and

$$\begin{aligned} y_{i..} &\sim N(0, \sigma_{123}^2/JK), \\ K\Sigma\Sigma\,(y_{ij.} - y_{i..})^2 &\sim \sigma_{12}^2\,\chi^2_{I(J-1)}, \\ \Sigma\Sigma\Sigma\,(y_{ijk} - y_{ij.})^2 &\sim \sigma_1^2\,\chi^2_{IJ(K-1)}, \end{aligned} \tag{5.3.4}$$

where

$$\sigma_{12}^2 = \sigma_1^2 + K\sigma_2^2 \qquad \text{and} \qquad \sigma_{123}^2 = \sigma_1^2 + K\sigma_2^2 + JK\sigma_3^2. \tag{5.3.5}$$

as defined in Table 5.1.1.

Further, insofar as $y_{ijk} - y_{ij.}$ and $y_{ij.} - y_{i..}$ are concerned, $\Sigma\Sigma\Sigma(y_{ijk} - y_{ij.})^2$ and $K\Sigma\Sigma(y_{ij.} - y_{i..})^2$ are sufficient for σ_1^2 and σ_{12}^2, respectively. Thus, the likelihood function is

$$l(\theta, \sigma_1^2, \sigma_{12}^2, \sigma_{123}^2 \mid \mathbf{y}) \propto (\sigma_1^2)^{-\nu_1/2} (\sigma_{12}^2)^{-\nu_2/2} (\sigma_{123}^2)^{-(\nu_3+1)/2}$$
$$\times \exp\left\{-\frac{1}{2}\left[\frac{JK\Sigma(y_{i..} - \theta)^2}{\sigma_{123}^2} + \frac{\nu_2 m_2}{\sigma_{12}^2} + \frac{\nu_1 m_1}{\sigma_1^2}\right]\right\}$$
$$\propto (\sigma_1^2)^{-\nu_1/2} (\sigma_{12}^2)^{-\nu_2/2} (\sigma_{123}^2)^{-(\nu_3+1)/2}$$
$$\times \exp\left[-\frac{1}{2}\left(\frac{IJK(y_{...} - \theta)^2}{\sigma_{123}^2} + \frac{\nu_3 m_3}{\sigma_{123}^2} + \frac{\nu_2 m_2}{\sigma_{12}^2} + \frac{\nu_1 m_1}{\sigma_1^2}\right)\right], \tag{5.3.6a}$$

where (m_1, m_2, m_3) and (ν_1, ν_2, ν_3) are the mean squares and degrees of freedom given in Table 5.1.1, that is,

$$\nu_1 = IJ(K-1), \qquad \nu_2 = I(J-1), \qquad \nu_3 = (I-1),$$
$$m_1 = \frac{S_1}{\nu_1} = \frac{\Sigma\Sigma\Sigma(y_{ijk} - y_{ij.})^2}{\nu_1},$$
$$m_2 = \frac{S_2}{\nu_2} = \frac{K\Sigma\Sigma(y_{ij.} - y_{i..})^2}{\nu_2} \tag{5.3.6b}$$

and

$$m_3 = \frac{S_3}{\nu_3} = \frac{JK\Sigma(y_{i..} - y_{...})^2}{\nu_3}.$$

The likelihood function in (5.3.6a) can be regarded as having arisen from I independent observations from $N(\theta, \sigma_{123}^2/JK)$, together with $I(J-1)$ independent observations from $N(0, \sigma_{12}^2/K)$ and a further $IJ(K-1)$ independent observations from $N(0, \sigma_1^2)$. From (5.3.5) the parameters $(\sigma_1^2, \sigma_{12}^2, \sigma_{123}^2)$ are subject to the constraint

$$C: \sigma_{123}^2 > \sigma_{12}^2 > \sigma_1^2. \tag{5.3.7}$$

On standard sampling theory, the unbiased estimators for $(\sigma_1^2, \sigma_2^2, \sigma_3^2)$ are

$$\hat\sigma_1^2 = m_1, \qquad \hat\sigma_2^2 = \frac{(m_2 - m_1)}{K}, \qquad \hat\sigma_3^2 = \frac{(m_3 - m_2)}{JK}. \tag{5.3.8}$$

5.3.2 Joint Posterior Distribution of $(\sigma_1^2, \sigma_2^2, \sigma_3^2)$

As before, the posterior distribution of the parameters is obtained relative to the noninformative prior distribution $p(\theta, \log\sigma_1^2, \log\sigma_{12}^2, \log\sigma_{123}^2) \propto$ constant. If there were no constraint, then, *a posteriori* $(\sigma_1^2, \sigma_{12}^2, \sigma_{123}^2)$ would be independently distributed as $(\nu_1 m_1)\chi_{\nu_1}^{-2}$, $(\nu_2 m_2)\chi_{\nu_2}^{-2}$, and $(\nu_3 m_3)\chi_{\nu_3}^{-2}$, respectively. Using (1.5.3), with the constraint $C: \sigma_1^2 < \sigma_{12}^2 < \sigma_{123}^2$, the joint posterior distribution is

$$
\begin{aligned}
p(\sigma_1^2, \sigma_{12}^2, \sigma_{123}^2 \mid \mathbf{y}) &= p^*(\sigma_1^2, \sigma_{12}^2, \sigma_{123}^2 \mid C, \mathbf{y}) \\
&= p^*(\sigma_1^2, \sigma_{12}^2, \sigma_{123}^2 \mid \mathbf{y}) \frac{\Pr^* \{C \mid \sigma_1^2, \sigma_{12}^2, \sigma_{123}^2, \mathbf{y}\}}{\Pr^* \{C \mid \mathbf{y}\}} \\
&= p^*(\sigma_1^2 \mid \mathbf{y})\, p^*(\sigma_{12}^2 \mid \mathbf{y})\, p^*(\sigma_{123}^2 \mid \mathbf{y}) \frac{\Pr^* \{C \mid \sigma_1^2, \sigma_{12}^2, \sigma_{123}^2, \mathbf{y}\}}{\Pr^* \{C \mid \mathbf{y}\}},
\end{aligned}
$$

$$\sigma_{123}^2 > \sigma_{12}^2 > \sigma_1^2 > 0, \qquad (5.3.9)$$

where, as before, an asterisk indicates an unconstrained distribution. Since the factor $\Pr^* \{C \mid \sigma_1^2, \sigma_{12}^2, \sigma_{123}^2, \mathbf{y}\}$ takes the value unity when the parameters satisfy the constraint and zero otherwise, it serves to restrict the distribution to that region in the parameter space where the constraint is satisfied. Thus,

$$
\begin{aligned}
&p(\sigma_1^2, \sigma_{12}^2, \sigma_{123}^2 \mid \mathbf{y}) \\
&= \frac{p^*(\sigma_1^2 \mid \mathbf{y})\, p^*(\sigma_{12}^2 \mid \mathbf{y})\, p^*(\sigma_{123}^2 \mid \mathbf{y})}{\Pr^* \{C \mid \mathbf{y}\}} \\
&= \frac{(v_1 m_1)^{-1} p\left(\chi_{v_1}^{-2} = \dfrac{\sigma_1^2}{v_1 m_1}\right) (v_2 m_2)^{-1} p\left(\chi_{v_2}^{-2} = \dfrac{\sigma_{12}^2}{v_2 m_2}\right) (v_3 m_3)^{-1} p\left(\chi_{v_3}^{-2} = \dfrac{\sigma_{123}^2}{v_3 m_3}\right)}{\Pr\left\{\dfrac{\chi_{v_1}^2}{\chi_{v_2}^2} > \dfrac{v_1 m_1}{v_2 m_2}, \dfrac{\chi_{v_2}^2}{\chi_{v_3}^2} > \dfrac{v_2 m_2}{v_3 m_3}\right\}},
\end{aligned}
$$

$$\sigma_{123}^2 > \sigma_{12}^2 > \sigma_1^2 > 0, \qquad (5.3.10)$$

where in the denominator $(\chi_{v_1}^2, \chi_{v_2}^2, \chi_{v_3}^2)$ are independently distributed χ^2 variables with (v_1, v_2, v_3) degrees of freedom respectively.

From (5.3.10), the joint posterior distribution of the variance components $(\sigma_1^2, \sigma_2^2, \sigma_3^2)$ is

$$
\begin{aligned}
&p(\sigma_1^2, \sigma_2^2, \sigma_3^2 \mid \mathbf{y}) \\
&\quad = \gamma\, p\left(\chi_{v_1}^{-2} = \frac{\sigma_1^2}{v_1 m_1}\right) p\left(\chi_{v_2}^{-2} = \frac{\sigma_1^2 + K\sigma_2^2}{v_2 m_2}\right) p\left(\chi_{v_3}^{-2} = \frac{\sigma_1^2 + K\sigma_2^2 + JK\sigma_3^2}{v_3 m_3}\right),
\end{aligned}
$$

$$\sigma_1^2 > 0, \sigma_2^2 > 0, \sigma_3^2 > 0, \qquad (5.3.11)$$

where

$$\gamma^{-1} = (JK^2)^{-1} (v_1 m_1)(v_2 m_2)(v_3 m_3) \Pr\left\{\frac{\chi_{v_1}^2}{\chi_{v_2}^2} > \frac{v_1 m_1}{v_2 m_2}, \frac{\chi_{v_2}^2}{\chi_{v_3}^2} > \frac{v_2 m_2}{v_3 m_3}\right\}.$$

It can be verified that

$$\Pr\left\{\frac{\chi^2_{\nu_1}}{\chi^2_{\nu_2}} > \frac{\nu_1 m_1}{\nu_2 m_2}, \frac{\chi^2_{\nu_2}}{\chi^2_{\nu_3}} > \frac{\nu_2 m_2}{\nu_3 m_3}\right\}$$

$$= I_{\nu_3 m_3/(\nu_3 m_3 + \nu_2 m_2)}\left(\frac{\nu_3}{2}, \frac{\nu_2}{2}\right) - \left[B\left(\frac{\nu_1}{2}, \frac{\nu_2}{2}\right) B\left(\frac{\nu_1+\nu_2}{2}, \frac{\nu_3}{2}\right)\right]^{-1}$$

$$\times \int_0^{\nu_1 m_1/\nu_2 m_2} \int_0^{\nu_3 m_3/\nu_2 m_2} t_1^{(\nu_1/2)-1} t_2^{(\nu_3/2)-1} (1 + t_1 + t_2)^{-(\nu_1+\nu_2+\nu_3)/2}\, dt_2\, dt_1, \tag{5.3.12}$$

where $I_x(p,q)$ and $B(p,q)$ are the usual incomplete and complete beta functions, respectively.

We now discuss various aspects of the variance component distribution.

5.3.3 Posterior Distribution of the Ratio σ_{12}^2/σ_1^2

In deriving the distributions in (5.3.10) and (5.3.11), use has been made of the fact that the variances $(\sigma_1^2, \sigma_{12}^2, \sigma_{123}^2)$ are subject to the constraint $C: \sigma_{123}^2 > \sigma_{12}^2 > \sigma_1^2$. To illustrate the role played by this constraint, we now obtain the posterior distribution of the ratio σ_{12}^2/σ_1^2 and show its connection with sampling theory results. For the two-component model the posterior distribution of the ratio σ_{12}^2/σ_1^2, which we there called W, was discussed in Section 5.2.3. In the persent case the posterior distribution of σ_{12}^2/σ_1^2 can be obtained from (5.3.10) by making a transformation from $(\sigma_1^2, \sigma_{12}^2, \sigma_{123}^2)$ to $(\sigma_{12}^2/\sigma_1^2, \sigma_1^2, \sigma_{123}^2)$ and integrating out σ_1^2 and σ_{123}^2. Alternatively, since

$$p\left(\frac{\sigma_{12}^2}{\sigma_1^2}\,\middle|\, \mathbf{y}\right) = p^*\left(\frac{\sigma_{12}^2}{\sigma_1^2}\,\middle|\, \mathbf{y}\right) \frac{\Pr^*\{C \mid \sigma_{12}^2/\sigma_1^2, \mathbf{y}\}}{\Pr^*\{C \mid \mathbf{y}\}}, \tag{5.3.13}$$

it follows that

$$p\left(\frac{\sigma_{12}^2}{\sigma_1^2}\,\middle|\, \mathbf{y}\right) = \left(\frac{m_1}{m_2}\right) p\left(F_{\nu_1,\nu_2} = \frac{\sigma_{12}^2}{\sigma_1^2}\frac{m_1}{m_2}\right) H\left(\frac{\sigma_{12}^2}{\sigma_1^2}\right), \quad \frac{\sigma_{12}^2}{\sigma_1^2} > 0, \tag{5.3.14}$$

where

$$H\left(\frac{\sigma_{12}^2}{\sigma_1^2}\right) = \frac{\Pr\left\{F_{\nu_3,\nu_1+\nu_2} < \dfrac{m_3}{[\nu_2 m_2 + (\sigma_{12}^2/\sigma_1^2)\nu_1 m_1]/(\nu_1+\nu_2)}\right\}}{\Pr\left\{\dfrac{\chi^2_{\nu_1}}{\chi^2_{\nu_2}} > \dfrac{\nu_1 m_1}{\nu_2 m_2}, \dfrac{\chi^2_{\nu_2}}{\chi^2_{\nu_3}} > \dfrac{\nu_2 m_2}{\nu_3 m_3}\right\}}.$$

The first factor $(m_1/m_2)p[F_{\nu_1,\nu_2} = (\sigma_{12}^2/\sigma_1^2)(m_1/m_2)]$ is the unconstrained distribution of σ_{12}^2/σ_1^2 and the factor $H(\sigma_{12}^2/\sigma_1^2)$ represents the effect of the constraint

C. Now, we can regard the C as the intersection of the two constraints $C_1 : \sigma_{12}^2 > \sigma_1^2$ and $C_2 : \sigma_{132}^2 > \sigma_{12}^2$. Thus, we can write

$$\frac{\Pr^*\left\{C \,\middle|\, \dfrac{\sigma_{12}^2}{\sigma_1^2}, \mathbf{y}\right\}}{\Pr^*\{C \mid \mathbf{y}\}} = \frac{\Pr^*\left\{C_1 \,\middle|\, \dfrac{\sigma_{12}^2}{\sigma_1^2}, \mathbf{y}\right\}}{\Pr^*\{C_1 \mid \mathbf{y}\}} \; \frac{\Pr^*\left\{C_2 \,\middle|\, C_1, \dfrac{\sigma_{12}^2}{\sigma_1^2}, \mathbf{y}\right\}}{\Pr^*\{C_2 \mid C_1, \mathbf{y}\}} \tag{5.3.15}$$

from which it can be verified that

$$p\left(\frac{\sigma_{12}^2}{\sigma_1^2} \,\middle|\, \mathbf{y}\right) = \left(\frac{m_1}{m_2}\right) p\left(F_{\nu_1,\nu_2} = \frac{\sigma_{12}^2}{\sigma_1^2}\frac{m_1}{m_2}\right) H_1\left(\frac{\sigma_{12}^2}{\sigma_1^2}\right) H_{2\cdot 1}\left(\frac{\sigma_{12}^2}{\sigma_1^2}\right) \tag{5.3.16}$$

with

$$H_1\left(\frac{\sigma_{12}^2}{\sigma_1^2}\right) = \frac{1}{\Pr\{F_{\nu_2,\nu_1} < (m_2/m_1)\}}$$

and

$$H_{2\cdot 1}\left(\frac{\sigma_{12}^2}{\sigma_1^2}\right)$$

$$= \frac{\Pr\{F_{\nu_2,\nu_1} < (m_2/m_1)\} \Pr\left\{F_{\nu_3,\nu_1+\nu_2} < \dfrac{m_3}{[\nu_2 m_2 + (\sigma_{12}^2/\sigma_1^2)\nu_1 m_1]/(\nu_1+\nu_2)}\right\}}{\Pr\left\{\dfrac{\chi_{\nu_1}^2}{\chi_{\nu_2}^2} > \dfrac{\nu_1 m_1}{\nu_2 m_2}, \dfrac{\chi_{\nu_2}^2}{\chi_{\nu_3}^2} > \dfrac{\nu_2 m_2}{\nu_3 m_3}\right\}}.$$

As in the two-component model, the effect of the constraint $C_1 : \sigma_1^2 < \sigma_{12}^2$ is to truncate the unconstrained distribution from below at the point $\sigma_{12}^2/\sigma_1^2 = 1$ and $H_1(\sigma_{12}^2/\sigma_1^2)$ is precisely the normalizing constant induced by the truncation. The distribution in (5.3.16), however, contains a further factor $H_{2\cdot 1}(\sigma_{12}^2/\sigma_1^2)$ which is a monotonic decreasing function of the ratio σ_{12}^2/σ_1^2. Thus, the "additional" effect of the second constraint $C_2 : \sigma_{12}^2 < \sigma_{123}^2$ is to pull the posterior distribution of σ_{12}^2/σ_1^2 towards the left of the distribution for which C_2 is ignored and to reduce the spread. In other words, for $\eta > 1$

$$\Pr\left\{1 < \frac{\sigma_{12}^2}{\sigma_1^2} < \eta \,\middle|\, \mathbf{y}\right\} \geqslant \Pr\left\{\frac{m_1}{m_2} < F_{\nu_1,\nu_2} < \eta\frac{m_1}{m_2}\right\} \Big/ \Pr\left\{F_{\nu_2,\nu_1} < \frac{m_2}{m_1}\right\}. \tag{5.3.17}$$

The extent of the influence of H_1 and $H_{2\cdot 1}$ depends of course upon the two mean squares ratios m_2/m_1 and m_3/m_2. In general, when m_2/m_1 is large, the effect of H_1 will be small and for given m_2/m_1, a large m_3/m_2 will produce a small effect from H_{2D1}. On the other hand, when the values of m_2/m_1 and m_3/m_2 are moderate, the combined effect can be appreciable.

For illustration, consider the set of data given in Table 5.3.1. The observations were generated from a table of random Normal deviates using the model (5.3.1) with ($\theta = 0$, $\sigma_1^2 = 1.0$, $\sigma_2^2 = 4.0$, $\sigma_3^2 = 2.25$, $I = 10$, $J = K = 2$). The relevant sample quantities are summarized in Table 5.3.2.

Table 5.3.1

Data generated from a table of random normal deviates for the three-component model

$$\theta = 0, \quad \sigma_1^2 = 1.0, \quad \sigma_2^2 = 4.0, \quad \sigma_3^2 = 2.25, \quad I = 10, \quad J = K = 2$$

so that

$$\sigma_{12}^2 = 9.0, \quad \sigma_{123}^2 = 18, \quad \sigma_1^2/(\sigma_1^2 + \sigma_2^2 + \sigma_3^2) = 13.8\%,$$

$$\sigma_2^2/(\sigma_1^2 + \sigma_2^2 + \sigma_3^2) = 55.2\% \quad \sigma_3^2/(\sigma_1^2 + \sigma_2^2 + \sigma_3^2) = 31.0\%.$$

	$j = 1$		$j = 2$	
I	$k = 1$	$k = 2$	$k = 1$	$k = 2$
1	2.004	2.713	0.603	0.252
2	4.342	4.229	3.344	3.057
3	0.869	−2.621	−3.896	−3.696
4	3.531	4.185	1.722	0.380
5	2.579	4.271	−2.101	0.651
6	−1.404	−1.003	−0.775	−2.202
7	−1.676	−0.208	−9.139	−8.653
8	1.670	2.426	1.834	1.200
9	2.141	3.527	0.462	0.665
10	−1.229	−0.596	4.471	1.606

Table 5.3.2

Analysis of variance for the generated data

S.S.	d.f.	M.S.
$S_3 = 240.28$	$\nu_3 = 9$	$m_3 = 26.70$
$S_2 = 122.43$	$\nu_2 = 10$	$m_2 = 12.29$
$S_1 = 20.87$	$\nu_1 = 20$	$m_1 = 1.04$
$\hat{\sigma}_1^2 = 1.04$	$\hat{\sigma}_2^2 = 5.62$	$\hat{\sigma}_3^2 = 3.60$
$\hat{\sigma}_1^2 + \hat{\sigma}_2^2 + \hat{\sigma}_3^2 = 10.26$		
$\dfrac{1.04}{10.26} = 10.1$ percent	$\dfrac{5.62}{10.26} = 54.7$ percent	$\dfrac{3.60}{10.26} = 35.1$ percent

The solid curve in Fig. 5.3.1 shows for this example the posterior distribution of σ_{12}^2/σ_1^2 when C_1 and C_2 are both included. The broken curve in the same figure is the unconstrained posterior distribution. For this example, m_2/m_1 is large and

$$\Pr\{F_{\nu_2,\nu_1} < (m_2/m_1)\} = 0.9999963$$

so that the effect of C_1 is negligible and the broken curve is essentially also the posterior distribution of σ_{12}^2/σ_1^2 with the constraint C_1. The effect of C_2 is, however, more appreciable. The moderate size of m_3/m_2 implies that σ_{12}^2/σ_1^2 may be slightly smaller than would otherwise be expected.

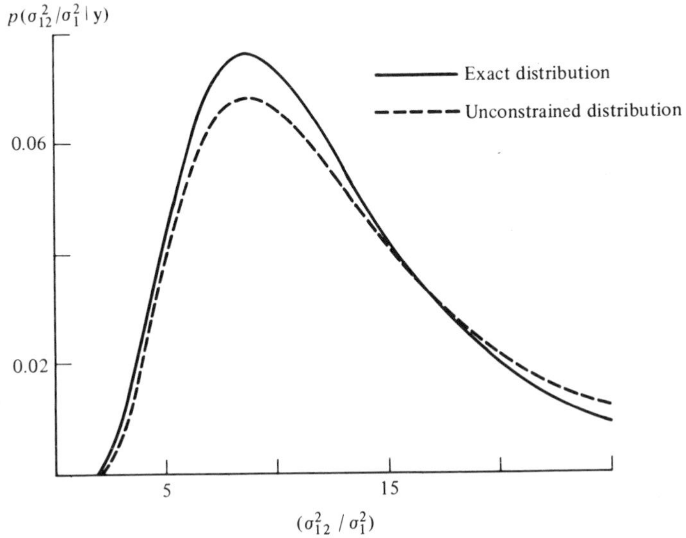

Fig. 5.3.1 Posterior distribution of σ_{12}^2/σ_2^1: the generated data.

We can now relate the above result to the sampling theory solution to this problem. It is well known that the sample quantity (m_1/m_2) is distributed as $(\sigma_1^2/\sigma_{12}^2)F_{\nu_1,\nu_2}$. Thus, the *unconstrained* posterior distribution

$$(m_1/m_2)\,p[F_{\nu_1,\nu_2} = (\sigma_{12}^2/\sigma_1^2)(m_1/m_2)]$$

is also the confidence distribution of σ_{12}^2/σ_1^2. Inference procedures based upon this confidence distribution are not satisfactory. In the first place the distribution extends from the origin to infinity so that the lower confidence limit for $(\sigma_{12}^2/\sigma_1^2) = 1 + K(\sigma_2^2/\sigma_1^2)$ can be smaller than unity and the corresponding limit for σ_2^2/σ_1^2 less than zero. In addition, the sampling procedure fails to take into account the information coming from the distribution of the mean square m_3, which, in the Bayesian approach, is included through the constraint $\sigma_{123}^2 > \sigma_{12}^2$.

5.3.4 Relative Contribution of Variance Components

One feature of the posterior distribution of $(\sigma_1^2, \sigma_2^2, \sigma_3^2)$, which is often of interest to the investigator, is the *relative contributions* of the variance components to the total variance of y_{ijk}. Inferences can be drawn by considering the joint distribution of any two of the three ratios

$$r_1 = \frac{\sigma_1^2}{\sigma_1^2 + \sigma_2^2 + \sigma_3^2}, \qquad r_2 = \frac{\sigma_2^2}{\sigma_1^2 + \sigma_2^2 + \sigma_3^2}, \qquad r_3 = \frac{\sigma_3^2}{\sigma_1^2 + \sigma_2^2 + \sigma_3^2}. \tag{5.3.18}$$

Since $r_1 = 1 - r_2 - r_3$, we make the transformation in (5.3.11) from $(\sigma_1^2, \sigma_2^2, \sigma_3^2)$ to (σ_1^2, r_2, r_3) and integrate out σ_1^2 to obtain

$$p(r_1, r_2, r_3 \mid \mathbf{y}) = N[x_1 x_2 (J-1) + x_1\phi_2 + x_2\phi_1 J(K-1)]^3\, x_1^{(\nu_1/2)-2}\, x_2^{(\nu_2/2)-2} \times (1 + x_1 + x_2)^{-(\nu_1+\nu_2+\nu_3)/2},$$

$$r_1 > 0, \quad r_2 > 0, \quad r_3 > 0, \quad r_1 + r_2 + r_3 = 1, \tag{5.3.19}$$

where

$$x_1 = \phi_1\left(\frac{r_1 + Kr_2}{r_1}\right), \qquad x_2 = \phi_2\left(\frac{r_1 + Kr_2}{r_1 + Kr_2 + JKr_3}\right),$$

$$\phi_1 = \frac{\nu_1 m_1}{\nu_2 m_2}, \qquad \phi_2 = \frac{\nu_3 m_3}{\nu_2 m_2},$$

and N is the normalizing constant

$$N^{-1} = \Pr\left\{\frac{\chi_{\nu_1}^2}{\chi_{\nu_2}^2} > \frac{\nu_1 m_1}{\nu_2 m_2}, \frac{\chi_{\nu_2}^2}{\chi_{\nu_3}^2} > \frac{\nu_2 m_2}{\nu_3 m_3}\right\} B(\tfrac{1}{2}\nu_1, \tfrac{1}{2}\nu_2)\, B(\tfrac{1}{2}\nu_1 + \tfrac{1}{2}\nu_2, \tfrac{1}{2}\nu_3) J^2 K \phi_1 \phi_2.$$

This distribution is analytically complicated. However, for a given set of data, contours of the density function can be plotted. Using the example in Table 5.3.1, three contours of the posterior distribution of (r_1, r_2, r_3) are shown in Fig. 5.3.2. To allow for the constraint $r_1 + r_2 + r_3 = 1$, the contours were drawn as a tri-coordinate diagram. The mode of the distribution is at approximately the point $P = (r_{10} = 0.075, r_{20} = 0.425, r_{30} = 0.500)$. The 50%, 70% and 90% in the figure were drawn such that

$$50\%\colon \log p(r_{10}, r_{20}, r_{30} \mid \mathbf{y}) - \log p(r_1, r_2, r_3 \mid \mathbf{y}) \doteq 0.69 = \tfrac{1}{2}\chi^2(2, 0.5),$$

$$70\%\colon \log p(r_{10}, r_{20}, r_{30} \mid \mathbf{y}) - \log p(r_1, r_2, r_3 \mid \mathbf{y}) \doteq 1.20 = \tfrac{1}{2}\chi^2(2, 0.3),$$

$$90\%\colon \log p(r_{10}, r_{20}, r_{30} \mid \mathbf{y}) - \log p(r_1, r_2, r_3 \mid \mathbf{y}) \doteq 2.30 = \tfrac{1}{2}\chi^2(2, 0.1),$$

and that the density of every point included exceeds that of every point excluded. That is, they are the boundaries of H.P.D. regions with approximate probability contents 50, 70, and 90 percent respectively.

Figure 5.3.2 provides us with a very illuminating picture of the inferential situation for this example. It is seen that on the one hand rather precise inferences are possible about r_1 (the percentage contribution of σ_1^2). In particular, nearly 90% of the probability mass is contained in the interval $0.05 < r_1 < 0.25$. On the other hand, however, the experiment has provided remarkably little information

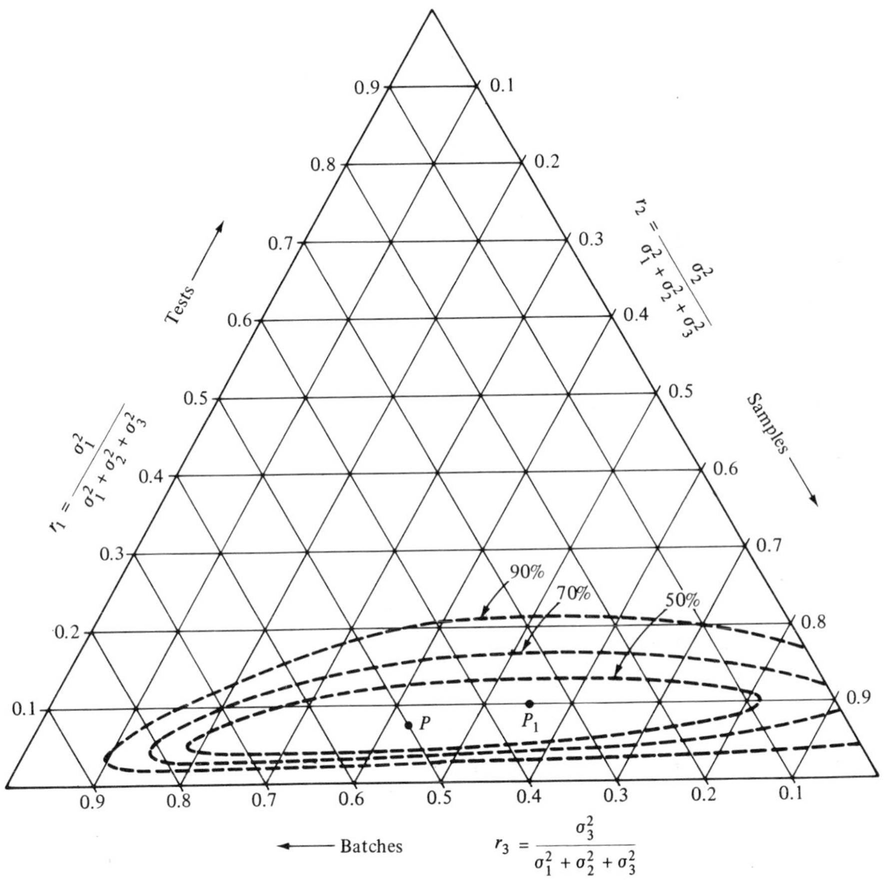

Fig. 5.3.2 Contours of the posterior distribution of (r_1, r_2, r_3): the generated data.

on the relative importance of σ_2^2 and σ_3^2 in accounting for the remaining variation Thus values of the ratio $r_2/r_3 = \sigma_2^2/\sigma_3^2$ ranging from $\frac{1}{9}$ to 9 are included in the 90 percent H.P.D. region. The diagram clearly shows that while we are on reasonably firm ground in making statements about the relative contributions of σ_1^2 on the one hand and $(\sigma_2^2 + \sigma_3^2)$ on the other to the total variance, we can say very little about the relative contributions of σ_2^2 and σ_3^2 to *their* total contribution $(\sigma_2^2 + \sigma_3^2)$. There would be need for additional data before any useful conclusions could be drawn on the relative contribution of these two components. The result is surprising and disturbing because the data comes from what we would have considered a reasonably well designed experiment. It underlines the danger we run into if we follow the common practice of considering only point estimates of variance components. Within the sampling theory framework such estimates of the relative contributions of $(\sigma_1^2, \sigma_2^2, \sigma_3^2)$ may be obtained by taking the ratios of the unbiased estimators $(\hat\sigma_1^2, \hat\sigma_2^2, \hat\sigma_3^2)$,

$$\hat r_1 = \frac{\hat\sigma_1^2}{\hat\sigma_1^2 + \hat\sigma_2^2 + \hat\sigma_3^2}, \qquad \hat r_2 = \frac{\hat\sigma_2^2}{\hat\sigma_1^2 + \hat\sigma_2^2 + \hat\sigma_3^2}, \qquad \hat r_3 = \frac{\hat\sigma_3^2}{\hat\sigma_1^2 + \hat\sigma_2^2 + \hat\sigma_3^2}. \tag{5.3.20}$$

The resulting ratios correspond to the point P_1 on the diagram. For large samples, the point P_1 would tend to the maximum likelihood estimate and the asymptotic confidence regions obtained from Normal theory would tend to the H.P.D. regions in the Bayesian framework. The small sample properties of the estimates $(\hat r_1, \hat r_2, \hat r_3)$ are, however, far from clear.

5.3.5 Posterior Distribution of σ_1^2

We now begin to study the marginal posterior distributions of the variance components $(\sigma_1^2, \sigma_2^2, \sigma_3^2)$ from which inferences about the individual components can be drawn. We shall develop an approximation method by which all the distributions involved here can be reduced to the χ^{-2} forms discussed earlier for the two-component model. This method not only gives an approximate solution to the present three-component hierarchical model, but also can be applied to the general q-component hierarchical models where q is any positive integer.

For the posterior distribution of σ_1^2, we have from (5.3.10) or (5.3.11)

$$p(\sigma_1^2 \mid \mathbf{y}) = \xi(\nu_1 m_1)^{-1}\, p\left(\chi_{\nu_1}^{-2} = \frac{\sigma_1^2}{\nu_1 m_1}\right) \int_{\sigma_1^2}^{\infty} (\nu_2 m_2)^{-1}\, p\left(\chi_{\nu_2}^{-2} = \frac{\sigma_{12}^2}{\nu_2 m_2}\right)$$

$$\times \Pr\left\{\chi_{\nu_3}^2 < \frac{\nu_3 m_3}{\sigma_{12}^2}\right\} d\sigma_{12}^2, \qquad \sigma_1^2 > 0, \tag{5.3.21}$$

where

$$\xi^{-1} = \Pr\left\{\frac{\chi_{\nu_1}^2}{\chi_{\nu_2}^2} > \frac{\nu_1 m_1}{\nu_2 m_2}, \frac{\chi_{\nu_2}^2}{\chi_{\nu_3}^2} > \frac{\nu_2 m_2}{\nu_3 m_3}\right\}.$$

The distribution of σ_1^2 is thus proportional to the product of two factors, the first being a χ^{-2} density with ν_1 degrees of freedom and the other a double integral of χ^2 variables. The first factor $(\nu_1 m_1)^{-1} p(\chi_{\nu_1}^{-2} = \sigma_1^2/\nu_1 m_1)$ is the unconstrained distribution of σ_1^2 and the integral on the right of (5.3.21) is induced by the constraint C. Because this integral is a monotonic decreasing function of σ_1^2, the effect of the constraint is, as is to be expected, to pull the distribution towards the left of the unconstrained χ^{-2} distribution. Exact evaluation of the distribution is tedious even on an electronic computer. But to approximate the distribution, one may employ the scaled χ^2 approach by equating the first two moments of

$\nu_1 m_1/\sigma_1^2$ to that of an $a\chi_b^2$ variable. Using the identity (A5.2.2) in Appendix A5.2 we find that the rth moment of $(\nu_1 m_1/\sigma_1^2)$ is

$$E\left[\left(\frac{\nu_1 m_1}{\sigma_1^2}\right)^r \bigg| \mathbf{y}\right] = 2^r \frac{\Gamma\left(\frac{\nu_1}{2}+r\right)}{\Gamma\left(\frac{\nu_1}{2}\right)} \frac{\Pr\left\{\frac{\chi^2_{\nu_1+2r}}{\chi^2_{\nu_2}} > \frac{\nu_1 m_1}{\nu_2 m_2}, \frac{\chi^2_{\nu_2}}{\chi^2_{\nu_3}} > \frac{\nu_2 m_2}{\nu_3 m_3}\right\}}{\Pr\left\{\frac{\chi^2_{\nu_1}}{\chi^2_{\nu_2}} > \frac{\nu_1 m_1}{\nu_2 m_2}, \frac{\chi^2_{\nu_2}}{\chi^2_{\nu_3}} > \frac{\nu_2 m_2}{\nu_3 m_3}\right\}} \tag{5.3.22}$$

where $(\chi^2_{\nu_1+2r}, \chi^2_{\nu_2}, \chi^2_{\nu_3})$ in the numerator are independent χ^2 variables with appropriate degrees of freedom. From (5.3.12), evaluation of (a, b) would thus involve calculating bivariate Dirichlet integrals. Although this approach simplifies the problem somewhat, existing methods for approximating such integrals, see for example, Tiao and Guttman (1965), still seem to be too complicated for routine practical use.

A Two-stage Scaled χ^{-2} Approximation

We now describe a two-stage scaled χ^{-2} approximation method. First, consider the joint distribution of $(\sigma_1^2, \sigma_{12}^2)$. Integrating out σ_{123}^2 in (5.3.10), we get

$$p(\sigma_1^2, \sigma_{12}^2 \mid \mathbf{y})$$

$$= \xi(\nu_1 m_1)^{-1} p\left(\chi_{\nu_1}^{-2} = \frac{\sigma_1^2}{\nu_1 m_1}\right)(\nu_2 m_2)^{-1} p\left(\chi_{\nu_2}^{-2} = \frac{\sigma_{12}^2}{\nu_2 m_2}\right) \Pr\left(\chi^2_{\nu_3} < \frac{\nu_3 m_3}{\sigma_{12}^2}\right)$$

$$= \xi'(\nu_1 m_1)^{-1} p\left(\chi_{\nu_1}^{-2} = \frac{\sigma_1^2}{\nu_1 m_1}\right) G(\sigma_{12}^2), \tag{5.3.23}$$

where

$$G(\sigma_{12}^2) = \frac{(\nu_2 m_2)^{-1} p\left(\chi_{\nu_2}^{-2} = \frac{\sigma_{12}^2}{\nu_2 m_2}\right) \Pr\left\{\chi^2_{\nu_3} < \frac{\nu_3 m_3}{\sigma_{12}^2}\right\}}{\Pr\left\{F_{\nu_3,\nu_2} < \frac{m_3}{m_2}\right\}}$$

$$\xi' = \xi \Pr\left\{F_{\nu_3,\nu_2} < \frac{m_3}{m_2}\right\},$$

and ξ is defined in (5.3.21). The quantity $G(\sigma_{12}^2)$ is in precisely the same form as the posterior distribution of σ_1^2 in (5.2.26) for the two-component model. Regarding the function $G(\sigma_{12}^2)$ as if it were the distribution of σ_{12}^2, we can then employ the scaled χ^{-2} method developed in Section 5.2.6 to approximate it by the distribution of $\nu_2' m_2' \chi_{\nu_2'}^{-2}$ where

$$\nu_2' = \frac{\nu_2}{a_1} \frac{I_{x_1}(\frac{1}{2}\nu_3, \frac{1}{2}\nu_2 + 1)}{I_{x_1}(\frac{1}{2}\nu_3, \frac{1}{2}\nu_2)}, \qquad m_2' = \frac{\nu_2 m_2}{a_1 \nu_2'},$$

$$a_1 = \left(\frac{\nu_2}{2} + 1\right) \frac{I_{x_1}(\frac{1}{2}\nu_3, \frac{1}{2}\nu_2 + 2)}{I_{x_1}(\frac{1}{2}\nu_3, \frac{1}{2}\nu_2 + 1)} - \frac{\nu_2}{2} \frac{I_{x_1}(\frac{1}{2}\nu_3, \frac{1}{2}\nu_2 + 1)}{I_{x_1}(\frac{1}{2}\nu_3, \frac{1}{2}\nu_2)}, \tag{5.3.24}$$

and

$$x_1 = \frac{\nu_3 m_3}{\nu_2 m_2 + \nu_3 m_3}.$$

To this degree of approximation, the distribution in (5.3.23) can be written

$$p(\sigma_1^2, \sigma_{12}^2 \mid \mathbf{y}) \doteq \xi'(\nu_1 m_1)^{-1} p\left(\chi_{\nu_1}^{-2} = \frac{\sigma_1^2}{\nu_1 m_1}\right)(\nu_2' m_2')^{-1} p\left(\chi_{\nu_2'}^{-2} = \frac{\sigma_{12}^2}{\nu_2' m_2'}\right),$$

$$\sigma_{12}^2 > \sigma_1^2 > 0. \qquad (5.3.25)$$

The effect of integration is thus essentially to change the mean square m_2 to m_2' and the degrees of freedom ν_2 to ν_2'. Noting that $\sigma_{12}^2 = \sigma_1^2 + K\sigma_2^2$, it follows that

$$p(\sigma_1^2, \sigma_2^2 \mid \mathbf{y}) \doteq \frac{\xi'' K(\nu_1 m_1)^{-1} p\left(\chi_{\nu_1}^{-2} = \dfrac{\sigma_1^2}{\nu_1 m_1}\right)(\nu_2' m_2')^{-1} p\left(\chi_{\nu_2'}^{-2} = \dfrac{\sigma_1^2 + K\sigma_2^2}{\nu_2' m_2'}\right)}{\Pr\left\{F_{\nu_2',\nu_1} < \dfrac{\nu_2' m_2'}{\nu_1 m_1}\right\}},$$

$$\sigma_1^2 > 0, \quad \sigma_2^2 > 0, \qquad (5.3.26)$$

where

$$\xi'' = \xi' \Pr\left\{F_{\nu_2',\nu_1} < \frac{\nu_2' m_2'}{\nu_1 m_1}\right\}.$$

If we ignore the constant ξ'', the distribution in (5.3.26) is of exactly the same form as the posterior distribution in (5.2.14). We can therefore apply the results for the two-component model to deduce the marginal posterior distributions of σ_1^2 and σ_2^2 for our three-component model. In particular, for the error variance σ_1^2, we may employ a second scaled χ^{-2} approximation so that σ_1^2 is approximately distributed as $\nu_1' m_1' \chi_{\nu_1'}^{-2}$ where

$$\nu_1' = \frac{\nu_1}{a_2} \frac{I_{x_2}(\frac{1}{2}\nu_2', \frac{1}{2}\nu_1 + 1)}{I_{x_2}(\frac{1}{2}\nu_2', \frac{1}{2}\nu_1)}, \qquad m_1' = \frac{\nu_1 m_1}{a_2 \nu_1'},$$

$$a_2 = \left(\frac{\nu_1}{2} + 1\right) \frac{I_{x_2}(\frac{1}{2}\nu_2', \frac{1}{2}\nu_1 + 2)}{I_{x_2}(\frac{1}{2}\nu_2', \frac{1}{2}\nu_1 + 1)} - \frac{\nu_1}{2} \frac{I_{x_2}(\frac{1}{2}\nu_2', \frac{1}{2}\nu_1 + 1)}{I_{x_2}(\frac{1}{2}\nu_2', \frac{1}{2}\nu_1)} \qquad (5.3.27)$$

and

$$x_2 = \frac{\nu_2' m_2'}{\nu_2' m_2' + \nu_1 m_1}.$$

Thus, making use of an incomplete beta function table, the quantities $(a_1, \nu_2', m_2';\ a_2, \nu_1', m_1')$ can be conveniently calculated from which the posterior distribution of σ_1^2 is approximately determined. For the generated example, we find $(a_1 = 0.865, \nu_2' = 12.35, m_2' = 11.51;\ a_2 = 1.0, \nu_1' = 20.0, m_1' = 1.04)$ so that *a posteriori* the variance σ_1^2 is approximately distributed as $(20.87)\chi^{-2}$ with 20 degrees of freedom. Since for this example $(\nu_1 = 20, m_1 = 1.04)$, the effect of the constraint $C: \sigma_{123}^2 > \sigma_{12}^2 > \sigma_1^2$ is thus negligible and inference about σ_1^2 can be based upon the unconstrained distribution $(\nu_1 m_1)^{-1} p(\chi_{\nu_1}^{-2} = \sigma_1^2/\nu_1 m_1)$. This is to be expected because for this example the mean square m_2 is much larger than the mean square m_1.

Pooling of Variance Estimates for the Three-component Model

The two-stage scaled χ^{-2} approximation leading to (5.3.27) can alternatively be put in the form

$$\frac{1}{\sigma_1^2} \frac{\nu_1 m_1 + \lambda_1 \nu_2 m_2 + \lambda_2 \nu_3 m_3}{\nu_1 + \omega_1 \nu_2 + \omega_2 \nu_3} \mathrel{\dot\sim} \frac{\chi^2_{\nu_1 + \omega_1 \nu_2 + \omega_2 \nu_3}}{\nu_1 + \omega_1 \nu_2 + \omega_2 \nu_3} \tag{5.3.28}$$

where

$$\lambda_1 = \left(\frac{1}{a_2} - 1\right) \frac{\nu_1 m_1}{\nu_2' m_2'}, \qquad \omega_1 = \frac{\nu_1' - \nu_1}{\nu_2'}$$

$$\lambda_2 = \left(\frac{1}{a_2} - 1\right)(1 - a_1) \frac{\nu_1 m_1}{\nu_3 m_3}, \qquad \omega_2 = \left(\frac{\nu_1' - \nu_1}{\nu_3}\right)\left(1 - \frac{\nu_2}{\nu_2'}\right).$$

This expression can be thought of as the Bayesian solution to the "pooling" of the three mean squares (m_1, m_2, m_3) in estimating the component σ_1^2. For illustration, suppose we have an example for which $I = 10$, $J = K = 2$ and the three squares (m_1, m_2, m_3) are equal. We find $\lambda_1 = 0.48$, $\lambda_2 = 0.157$, $\omega_1 = 0.754$ and $\omega_2 = 0.48$. Thus,

$$\frac{1}{\sigma_1^2} \frac{\nu_1 m_1 + \lambda_1 \nu_2 m_2 + \lambda_2 \nu_3 m_3}{\nu_1 + \omega_1 \nu_2 + \omega_2 \nu_3} = \frac{m_1}{\sigma_1^2} \frac{\nu_1 + 0.48\, \nu_2 + 0.157\, \nu_3}{\nu_1 + 0.754\, \nu_2 + 0.48\, \nu_3} = \frac{m_1}{\sigma_1^2} 0.82,$$

so that

$$\frac{m_1}{\sigma_1^2}(0.82) \mathrel{\dot\sim} \chi^2_{32.06}/32.06, \qquad \text{that is,} \qquad \frac{\sigma_1^2}{m_1} \mathrel{\dot\sim} (0.82)(32.06)\chi^{-2}_{32.06}.$$

In particular, the mean and variance of σ_1^2/m_1 are

$$E(\sigma_1^2/m_1) \doteq 0.875 \qquad \text{and} \qquad \operatorname{Var}(\sigma_1^2/m_1) \doteq 0.072.$$

By contrast, the evidence from m_1 *alone* would have given

$$(m_1/\sigma_1^2) \sim \chi^2_{20}/20, \qquad \text{that is,} \qquad (\sigma_1^2/m_1) \sim 20\chi_{20}^{-2},$$

so that

$$E(\sigma_1^2/m_1) = 1.111 \qquad \text{and} \qquad V(\sigma_1^2/m_1) = 0.154.$$

Thus, the additional evidence about σ_1^2 coming from m_2 and m_3 indicates that σ_1^2 is considerably smaller than would have been expected using m_1 alone. In addition, the variance of the distribution of σ_1^2/m_1 is seen to be only about half of what it would have been had only the evidence from m_1 been used.

5.3.6 Posterior Distribution of σ_2^2

Using the approximation leading to (5.2.44), it follows from the joint distribution of (σ_1^2, σ_2^2) in (5.3.26) that the posterior distribution of σ_2^2 is approximately

$$p(\sigma_2^2 \mid \mathbf{y}) \doteq \frac{K(\nu_2' m_2')^{-1} p[\chi_{\nu_2'}^{-2} = (m_1 + K\sigma_2^2)/(\nu_2' m_2')]}{\Pr\{\chi_{\nu_2'}^2 < \nu_2' m_2'/m_1\}}, \qquad \sigma_2^2 > 0. \tag{5.3.29}$$

To this degree of approximation, then, the quantity $(m_1 + K\sigma_2^2)/\nu_2' m_2'$ has the χ^{-2} distribution with ν_2' degrees of freedom truncated from below at $m_1/(\nu_2' m_2')$. For the example in Table 5.3.1, the quantity $0.007 + 0.014\,\sigma_2^2$ thus behaves like an χ^{-2} variable with 12.35 degrees of freedom truncated from below at 0.007. Posterior intervals about σ_2^2 can then be calculated from a table of χ^2 probabilities.

5.3.7 Posterior Distribution of σ_3^2

From the joint posterior distribution of $(\sigma_1^2, \sigma_2^2, \sigma_3^2)$ in (5.3.11), we may integrate out (σ_1^2, σ_2^2) to get the posterior distribution of σ_3^2,

$$p(\sigma_3^2 \mid \mathbf{y}) = \gamma \int_0^\infty \int_0^\infty p\left(\chi_{\nu_1}^{-2} = \frac{\sigma_1^2}{\nu_1 m_1}\right) p\left(\chi_{\nu_2}^{-2} = \frac{\sigma_1^2 + K\sigma_2^2}{\nu_2 m_2}\right)$$

$$\times\, p\left(\chi_{\nu_3}^{-2} = \frac{\sigma_1^2 + K\sigma_2^2 + JK\sigma_3^2}{\nu_3 m_3}\right) d\sigma_1^2\, d\sigma_2^2, \qquad \sigma_3^2 > 0. \tag{5.3.30}$$

The density function is a double integral which does not seem expressible in terms of simple forms. To obtain an approximation, we shall first consider the joint posterior distribution of $(\sigma_{12}^2, \sigma_{123}^2)$ and then deduce the result by making use of the fact that $\sigma_{123}^2 = \sigma_{12}^2 + JK\sigma_3^2$. From (5.3.10), we obtain

$$p(\sigma_{12}^2, \sigma_{123}^2 \mid \mathbf{y})$$

$$= \xi(\nu_3 m_3)^{-1} p\left(\chi_{\nu_3}^{-2} = \frac{\sigma_{123}^2}{\nu_3 m_3}\right)(\nu_2 m_2)^{-1} p\left(\chi_{\nu_2}^{-2} = \frac{\sigma_{12}^2}{\nu_2 m_2}\right) \Pr\left(\chi_{\nu_1}^2 > \frac{\nu_1 m_1}{\sigma_{12}^2}\right)$$

$$= \xi'''(\nu_3 m_3)^{-1} p\left(\chi_{\nu_3}^{-2} = \frac{\sigma_{123}^2}{\nu_3 m_3}\right) G_1(\sigma_{12}^2), \qquad \sigma_{123}^2 > \sigma_{12}^2 > 0, \tag{5.3.31}$$

where

$$G_1(\sigma_{12}^2) = (\nu_2 m_2)^{-1} p\left(\chi_{\nu_2}^{-2} = \frac{\sigma_{12}^2}{\nu_2 m_2}\right) \Pr\left\{\chi_{\nu_1}^2 > \frac{\nu_1 m_1}{\sigma_{12}^2}\right\} \Big/ \Pr\left\{F_{\nu_2,\nu_1} < \frac{m_2}{m_1}\right\},$$

$$\xi''' = \xi \Pr\{F_{\nu_2,\nu_1} < m_2/m_1\},$$

and ξ is given in (5.3.21). The function $G_1(\sigma_{12}^2)$ is in exactly the same form as the distribution of σ_{12}^2 in (5.2.57) for the two-component model. Provided m_2/m_1 is not too small, we can make use of the approximation method in (5.2.60) to get

$$p(\sigma_{12}^2, \sigma_{123}^2 \mid \mathbf{y}) \doteq \frac{(v_3 m_3)^{-1}\, p\left(\chi_{v_3}^{-2} = \dfrac{\sigma_{123}^2}{v_3 m_3}\right)(v_2'' m_2'')^{-1}\, p\left(\chi_{v_2''}^{-2} = \dfrac{\sigma_{12}^2}{v_2'' m_2''}\right)}{\Pr\left\{F_{v_3, v_2''} < \dfrac{m_3}{m_2}\right\}},$$

$$\sigma_{123}^2 > \sigma_{12}^2 > 0, \qquad (5.3.32)$$

where

$$v_2'' = \frac{v_2}{a_3}\frac{I_{x_3}(\frac{1}{2}v_2 + 1, \frac{1}{2}v_1)}{I_{x_3}(\frac{1}{2}v_2, \frac{1}{2}v_1)}, \qquad m_2'' = \frac{v_2 m_2}{a_3 v_2''},$$

$$a_3 = \left(\frac{v_2}{2} + 1\right)\frac{I_{x_3}(\frac{1}{2}v_2 + 2, \frac{1}{2}v_1)}{I_{x_3}(\frac{1}{2}v_2 + 1, \frac{1}{2}v_1)} - \frac{v_2}{2}\frac{I_{x_3}(\frac{1}{2}v_2 + 1, \frac{1}{2}v_1)}{I_{x_3}(\frac{1}{2}v_2, \frac{1}{2}v_1)}, \qquad (5.3.33)$$

and

$$x_3 = \frac{v_2 m_2}{v_1 m_1 + v_2 m_2}.$$

It follows that the posterior distribution of σ_3^2 is approximately

$$p(\sigma_3^2 \mid \mathbf{y}) \doteq \frac{JK(v_3 m_3)^{-1}\,(v_2'' m_2'')^{-1}}{\Pr\{F_{v_3, v_2''} < m_3/m_2''\}} \int_0^\infty p\left(\chi_{v_3}^{-2} = \frac{\sigma_{12}^2 + JK\sigma_3^2}{v_3 m_3}\right)$$

$$\times\, p\left(\chi_{v_2''}^{-2} = \frac{\sigma_{12}^2}{v_2'' m_2''}\right) d\sigma_{12}^2, \qquad \sigma_3^2 > 0. \qquad (5.3.34)$$

The reader will note that the distribution in (5.3.34) is in precisely the same form as the posterior distribution of σ_2^2 in (5.2.39) for the two-component model. Provided v_2'' is moderately large, we can thus employ (5.2.44) to express the distribution of σ_3^2 approximately in the form of a truncated χ^{-2} distribution

$$p(\sigma_3^2 \mid \mathbf{y}) \doteq \frac{JK(v_3 m_3)^{-1}\, p[\chi_{v_3}^{-2} = (m_2'' + JK\sigma_3^2)/(v_3 m_3)]}{\Pr\{\chi_{v_3}^{-2} > m_2''/(v_3 m_3)\}}, \qquad \sigma_3^2 > 0. \qquad (5.3.35)$$

To this degree of approximation, the quantity $(m_2'' + JK\sigma_3^2)/(v_3 m_3)$ is distributed as an χ^{-2} variable with v_3 degrees of freedom truncated from below at $m_2''/(v_3 m_3)$, from which Bayesian intervals for σ_3^2 can be determined. For the set of data in Table 5.3.1, we find ($a_3 = 1.0$, $v_2'' = 10.0$, $m_2'' = 12.29$) so that the quantity 0.051 + 0.017 σ_3^2 is approximately distributed as χ^{-2} with 9 degrees of freedom truncated from below at the point 0.051.

5.3.8 A Summary of the Approximations to the Posterior Distributions of $(\sigma_1^2, \sigma_2^2, \sigma_3^2)$

For convenience in calculation, Table 5.3.3 provides a short summary of the quantities needed for approximating the individual posterior distributions of the variance components $(\sigma_1^2, \sigma_2^2, \sigma_3^2)$. The numerical values shown are those for the set of data introduced in Table 5.3.1.

Table 5.3.3

Summarized Calculations for Approximating the Posterior Distributions of $(\sigma_1^2, \sigma_2^2, \sigma_3^2)$ (illustrated using the data in Table 5.3.1)

---

1. Use Table 5.1.1, or expression (5.3.6b), to compute

$$m_1 = 1.04, \quad m_2 = 12.29, \quad m_3 = 26.70,$$
$$\nu_1 = 20, \quad \nu_2 = 10, \quad \nu_3 = 9.$$
$$K = 2, \quad JK = 4,$$

2. Use (5.3.24) to determine

$$x_1 = 0.662, \quad a_1 = 0.865,$$
$$\nu_2' = 12.35, \quad m_2' = 11.51.$$

3. Use (5.3.27) to determine

$$x_2 = 0.872, \quad a_2 = 1.0,$$
$$\nu_1' = 20.0 \quad m_1' = 1.04.$$

4. Use (5.3.33) to determine

$$x_3 = 0.854, \quad a_3 = 1.0,$$
$$\nu_2'' = 10.0, \quad m_2'' = 12.29.$$

5. Then, for inference about σ_1^2

$$\sigma_1^2 \mathrel{\dot\sim} \nu_1' m_1' \chi_{\nu_1'}^{-2}$$
$$\sigma_1^2 \mathrel{\dot\sim} (20.8)\chi_{20}^{-2}.$$

6. For inference about σ_2^2

$$\frac{m_1 + K\sigma_2^2}{\nu_2' m_2'} \mathrel{\dot\sim} \chi_{\nu_2'}^{-2} \qquad \text{truncated from below at } m_1/(\nu_2' m_2'),$$

$$0.007 + 0.014\sigma_2^2 \mathrel{\dot\sim} \chi_{12.35}^{-2} \qquad \text{truncated from below at } 0.007.$$

7. For inference about σ_3^2

$$\frac{m_2'' + JK\sigma_3^2}{\nu_3 m_3} \mathrel{\dot\sim} \chi_{\nu_3}^{-2} \qquad \text{truncated from below at } m_2''/(\nu_3 m_3),$$

$$0.051 + 0.017\sigma_3^2 \mathrel{\dot\sim} \chi_9^{-2} \qquad \text{truncated from below at } 0.051.$$

5.4 GENERALIZATION TO q-COMPONENT HIERARCHICAL DESIGN MODEL

The preceding analysis of the three-component model and in particular the approximation methods for the posterior distributions of the individual variance components can be readily extended to the general q-component hierarchical model

$$y_{i_q \dots i_1} = \theta + e_{i_q} + e_{i_q i_{q-1}} + \cdots + e_{i_q \dots i_t} + \cdots + e_{i_q \dots i_1},$$
$$i_q = 1, \dots, I_q; \quad \dots; \quad i_1 = 1, \dots, I_1, \qquad (5.4.1)$$

where $y_{i_q \dots i_1}$ are the observations, θ is a common location parameter, and $e_{i_q}, \dots, e_{i_q \dots i_1}$ are q different kinds of random effects. Assuming that these effects are Normally and independently distributed with zero means and variances $(\sigma_q^2, \sigma_{q-1}^2, \dots, \sigma_t^2, \dots, \sigma_1^2)$ and following the argument in Sections 5.2 and 5.3, it is readily shown that the joint posterior distribution of the variance components is

$$p(\sigma_1^2, \dots, \sigma_q^2 \mid \mathbf{y}) = (I_1^{q-1} I_2^{q-2} \cdots I_{q-1}) p(\sigma_1^2, \sigma_{12}^2, \dots, \sigma_{1\dots t}^2, \dots, \sigma_{1\dots q}^2 \mid \mathbf{y}),$$
$$\sigma_1^2 > 0, \dots, \sigma_q^2 > 0, \qquad (5.4.2)$$

where

$$\sigma_{12}^2 = (\sigma_1^2 + I_1 \sigma_2^2), \dots, \sigma_{1\dots t}^2 = (\sigma_{1\dots(t-1)}^2 + I_1 \cdots I_{t-1} \sigma_t^2), \dots,$$
$$\sigma_{1\dots q}^2 = (\sigma_{1\dots(q-1)}^2 + I_1 \cdots I_{q-1} \sigma_q^2),$$

$$p(\sigma_1^2, \dots, \sigma_{1\dots t}^2, \dots, \sigma_{1\dots q}^2 \mid \mathbf{y})$$
$$= (\nu_1 m_1)^{-1} p\left(\chi_{\nu_1}^{-2} = \frac{\sigma_1^2}{\nu_1 m_1}\right) \cdots (\nu_t m_t)^{-1} p\left(\chi_{\nu_t}^{-2} = \frac{\sigma_{1\dots t}^2}{\nu_t m_t}\right) \cdots$$
$$\cdots (\nu_q m_q)^{-1} p\left(\chi_{\nu_q}^{-2} = \frac{\sigma_{1\dots q}^2}{\nu_q m_q}\right)$$
$$\times \left[\Pr\left\{\frac{\chi_{\nu_1}^2}{\chi_{\nu_2}^2} > \frac{\nu_1 m_1}{\nu_2 m_2}, \dots, \frac{\chi_{\nu_{t-1}}^2}{\chi_{\nu_t}^2} > \frac{\nu_{t-1} m_{t-1}}{\nu_t m_t}, \dots, \frac{\chi_{\nu_{q-1}}^2}{\chi_{\nu_q}^2} > \frac{\nu_{q-1} m_{q-1}}{\nu_q m_q}\right\}\right]^{-1},$$
$$0 < \sigma_1^2 < \sigma_{12}^2 < \cdots < \sigma_{1\dots t}^2 < \cdots < \sigma_{1\dots q}^2, \qquad (5.4.3)$$

and the m_t's and the ν_t's are the corresponding mean squares and the degrees of freedom. The distributions in (5.4.3) and (5.4.2) parallel exactly those in (5.3.10) and (5.3.11).

In principle, the marginal posterior distribution of a particular variance component, say, σ_t^2, can be obtained from (5.4.2) simply by integrating out $\sigma_1^2, \dots, \sigma_{t-1}^2, \sigma_{t+1}^2, \dots, \sigma_q^2$. In practice, however, this involves calculation of a $(q-1)$ dimensional integral for each value of σ_t^2 and would be difficult even on a fast computer. A simple approximation to the distribution of σ_t^2 can be obtained by first considering the joint distribution $p(\sigma_{1\dots(t-1)}^2, \sigma_{1\dots t}^2 \mid \mathbf{y})$. The latter distribution is obtained by integrating (5.4.3) over $(\sigma_{1\dots(t+1)}^2, \dots, \sigma_{1\dots q}^2)$ and $(\sigma_1^2, \dots, \sigma_{1\dots(t-2)}^2)$. It is clear that the set $(\sigma_{1\dots(t+1)}^2, \dots, \sigma_{1\dots q}^2)$ can be eliminated in reverse order by repeated applications of the $a\chi_b^2$ approximation in (5.2.35). The effect of each integration is merely to change the values of the mean square m and the degrees of freedom ν of the succeeding variable. Similarly, the set $(\sigma_1^2, \dots, \sigma_{1\dots(t-2)}^2)$ can be approximately integrated out upon repeated applications of the $c\chi_d^2$ method in (5.2.60). Thus, the distribution of $(\sigma_{1\ldots(t-1)}^2, \sigma_{1\ldots t}^2)$ takes the approximate form

$$p(\sigma_{1\ldots(t-1)}^2, \sigma_{1\ldots t}^2 \mid \mathbf{y}) \doteq \frac{(v'_{t-1}m'_{t-1})\,p\left(\chi_{v'_{t-1}}^{-2} = \dfrac{\sigma_{1\ldots(t-1)}^2}{v'_{t-1}m'_{t-1}}\right)(v'_t m'_t)^{-1}\,p\left(\chi_{v'_t}^{-2} = \dfrac{\sigma_{1\ldots t}^2}{v'_t m'_t}\right)}{\Pr\left\{F_{v'_t, v'_{t-1}} < \dfrac{m'_t}{m'_{t-1}}\right\}},$$

$$\sigma_{1\ldots t}^2 > \sigma_{1\ldots(t-1)}^2 > 0. \qquad (5.4.4)$$

Noting that $\sigma_{1\ldots t}^2 = \sigma_{1\ldots(t-1)}^2 + I_1 \cdots I_{t-1}\sigma_t^2$, the corresponding posterior distribution of $(\sigma_{1\ldots(t-1)}^2, \sigma_t^2)$ is then of exactly the same form as the posterior distribution of (σ_1^2, σ_2^2) in (5.2.14) for the two-component model. The marginal distribution of σ_t^2 can then be approximated using the truncated χ^{-2} form (5.2.44) or the associated asymptotic formulas. Finally, in the case $t = 2$, so that $v_1 = v'_1$, $m_1 = m'_1$ and the posterior distribution of the variance σ_1^2 can be approximated by that of an inverted χ^2 distribution as in the two-component model.

APPENDIX A5.1

The Posterior Density of σ_1^2 for the Two-Component Model Expressed as a χ^{-2} Series

We now express the distribution of σ_1^2 in (5.2.26) as a χ^{-2} series. Using the Mittag–Leffler formula [see Milne–Thomson (1960), p. 331], we can write

$$\Pr\{\chi_{v_2}^2 < v_2 m_2/\sigma_1^2\} = \exp\left[-\left(\frac{v_2 m_2}{2\sigma_1^2}\right)\right] \sum_{r=0}^{\infty} \frac{(v_2 m_2/2\sigma_1^2)^{(v_2/2)+r}}{\Gamma(\frac{1}{2}v_2 + r + 1)}. \qquad (A5.1.1)$$

Thus, the distribution of $\sigma_1^2/(v_1 m_1 + v_2 m_2)$ can be alternatively expressed as a weighted series of the densities of χ^{-2} variables,

$$p\left(\frac{\sigma_1^2}{v_1 m_1 + v_2 m_2}\,\middle|\,\mathbf{y}\right) = \sum_{r=0}^{\infty} \omega_r\, p\left(\chi_{v_1+v_2+2r}^{-2} = \frac{\sigma_1^2}{v_1 m_1 + v_2 m_2}\right), \qquad (A5.1.2)$$

with the weights ω_r given by

$$\omega_r = \frac{[v_2 m_2/(v_1 m_1)]^{(v_2/2)+r}\,[1 + v_2 m_2/(v_1 m_1)]^{\frac{1}{2}(v_1+v_2)+r}}{\Pr\{F_{v_2,v_1} < m_2/m_1\}}\,\frac{\Gamma[\frac{1}{2}(v_1 + v_2) + r]}{\Gamma(\frac{1}{2}v_2 + r + 1)\Gamma(\frac{1}{2}v_1)}.$$

We note that when v_2 is even, the right-hand side of (A5.1.1) can, of course, be expressed as a finite Poisson series so that

$$p(\sigma_1^2 \mid \mathbf{y}) = \frac{(v_1 m_1)^{-1}\,p[\chi_{v_1}^{-2} = \sigma_1^2/(v_1 m_1)]}{\Pr\{F_{v_2,v_1} < m_2/m_1\}}$$

$$\times \left\{1 - \exp[-v_2 m_2/(2\sigma_1^2)] \sum_{r=0}^{\frac{1}{2}v_2 - 1} \frac{[v_2 m_2/(2\sigma_1^2)]^r}{r!}\right\}. \qquad (A5.1.3)$$

In this case, posterior probabilities of σ_1^2 can be expressed as a finite series of χ^2 probabilities, that is,

$$\Pr\{d_1 < \sigma_1^2 < d_2 \mid \mathbf{y}\} = \frac{\Pr\{\nu_1 m_1/d_2 < \chi^2_{\nu_1} < \nu_1 m_1/d_1\}}{\Pr\{F_{\nu_2,\nu_1} < m_2/m_1\}}$$

$$- \sum_{r=0}^{\frac{1}{2}\nu_2 - 1} \phi_r \Pr\left\{\frac{\nu_1 m_1 + \nu_2 m_2}{d_2} < \chi^2_{\nu_1+2r} < \frac{\nu_1 m_1 + \nu_2 m_2}{d_1}\right\}, \tag{A5.1.4}$$

where

$$\phi_r = \frac{\Gamma(\frac{1}{2}\nu_1 + r)[\nu_2 m_2/(\nu_1 m_1)]^r[1 + \nu_2 m_2/(\nu_1 m_1)]^{-(\nu_1/2)+r}}{r!\,\Gamma(\nu_1/2)\Pr\{F_{\nu_2,\nu_1} < m_2/m_1\}},$$

so that probability integrals of σ_1^2 can be evaluated exactly from an ordinary χ^2 table.

APPENDIX A5.2

Some Useful Integral Identities

The derivation of the posterior distribution in (5.2.12) leads to the following useful identity. For $p_1 > 0$, $p_2 > 0$, $a_1 > 0$, $a_2 > 0$ and $c > 0$

$$\int_0^\infty \int_0^\infty x^{-(p_1+1)}(x + cy)^{-(p_2+1)} \exp\left[-\left(\frac{a_1}{x} + \frac{a_2}{x + cy}\right)\right] dx\,dy$$

$$= \frac{\Gamma(p_1)\,\Gamma(p_2)}{c\,a_1^{p_1}\,a_2^{p_2}} I_{a_2/(a_1+a_2)}(p_2, p_1) \tag{A5.2.1}$$

where $I_x(p, q)$ is the usual incomplete beta function.

Similarly, the results in (5.3.10) for the three-component model implies the identity: for $p_i > 0$, $a_i > 0$, $i = 1, 2, 3$ and $c_1 > 0$, $c_2 > 0$.

$$\int_0^\infty \int_0^\infty \int_0^\infty x_1^{-(p_1+1)}(x_1 + c_1 x_2)^{-(p_2+1)}(x_1 + c_1 x_2 + c_2 x_3)^{-(p_3+1)}$$

$$\times \exp\left[-\left(\frac{a_1}{x_1} + \frac{a_2}{x_1 + c_1 x_2} + \frac{a_3}{x_1 + c_1 x_2 + c_2 x_3}\right)\right] dx_1\,dx_2\,dx_3$$

$$= (c_1 c_2)^{-1}\left(\prod_{i=1}^{3} \Gamma(p_i) a_i^{-p_i}\right) \Pr\left\{\frac{\chi^2_{2p_1}}{\chi^2_{2p_2}} > \frac{a_1}{a_2}, \frac{\chi^2_{2p_2}}{\chi^2_{2p_3}} > \frac{a_2}{a_3}\right\}, \tag{A5.2.2}$$

where $\chi^2_{2p_1}$, $\chi^2_{2p_2}$ and $\chi^2_{2p_3}$ are independent χ^2 variables with $2p_1$, $2p_2$, and $2p_3$ degrees of freedom, respectively.

We record here another integral identity the proof of which can be found in Tiao and Guttman (1965). For $a > 0$, $p > 0$ and n a positive integer

$$\int_0^a (1+x)^{-(p+n)} x^{n-1}\, dx = B(n,p) - \sum_{j=0}^{n-1} \binom{n-1}{j} a^j (1+a)^{-(p+j)} B(p+j, n-j), \tag{A5.2.3}$$

where $B(p,q)$ is the usual complete beta function.

APPENDIX A5.3

Some Properties of the Posterior Distribution of σ_2^2 for the Two-component Model

We now discuss some properties of the posterior distribution of σ_2^2 in (5.2.39).

Moments

By repeated application of the identity (A5.2.1) in Appendix A5.2, it can be verified from (5.2.12) that for $\nu_1 > \nu_2 > 2r$, the rth moment of σ_2^2 is

$$\mu_r' = \left(\frac{\nu_2 m_2}{2K}\right)^r \sum_{i=0}^{r} \binom{r}{i} \left(-\frac{\nu_2 m_2}{\nu_1 m_1}\right)^i \frac{I_x[(\nu_2/2) - n + i, (\nu_1/2) - i]}{I_x(\nu_2/2, \nu_1/2)}$$

$$\times \frac{\Gamma[(\nu_2/2) - n + i]\, \Gamma[(\nu_1/2) - i]}{\Gamma(\nu_2/2)\, \Gamma(\nu_1/2)},$$

$$x = \frac{\nu_2 m_2}{\nu_1 m_1 + \nu_2 m_2}. \tag{A5.3.1}$$

Making use of an incomplete beta function table, the first few moments of σ_2^2, when they exist, can be conveniently calculated and will, of course, provide the investigator with valuable information about the shape of the distribution.

Mode

To obtain the mode of the distribution of σ_2^2, it is convenient to make the transformation

$$u = \frac{2K\sigma_2^2}{\nu_2 m_2}. \tag{A5.3.2}$$

so that from (5.2.39)

$$p(u \mid \mathbf{y}) = M \int_0^\infty [(\phi z)^{-1} + u]^{-(\frac{1}{2}\nu_2 + 1)}$$

$$\times \exp\left\{-[(\phi z)^{-1} + u]^{-1}\right\} z^{\frac{1}{2}\nu_1 - 1} \exp(-z)\, dz \tag{A5.3.3}$$

where M is a positive constant, $\phi = \nu_2 m_2/(\nu_1 m_1)$ and the substitution $z = \nu_1 m_1/(2\sigma_1^2)$ is made.

Upon differentiating (A5.3.3) with respect to u, we find

$$\frac{\partial p(u \mid \mathbf{y})}{\partial u} = M\left\{\left[1 - \left(\frac{v_2}{2}+1\right)u\right]\int_0^\infty g(u,z)\,dz - \left(\frac{v_2}{2}+1\right)\phi^{-1}\int_0^\infty g(u,z)z^{-1}\,dz\right\}, \tag{A5.3.4}$$

where

$$g(u,z) = [(\phi z)^{-1} + u]^{-(\frac{1}{2}v_2+3)}\, z^{\frac{1}{2}v_1-1}\exp\left(-\{z + [(\phi z)^{-1}+u]^{-1}\}\right).$$

Applying integration by parts

$$\int x\,dv = xv - \int v\,dx \tag{A5.3.5}$$

and setting

$$dv = z^{\frac{1}{2}(v_1+v_2)+1},\quad x = [\phi^{-1}+uz]^{-(\frac{1}{2}v_2+3)}\exp\left(-\{z+[(\phi z)^{-1}+u]^{-1}\}\right), \tag{A5.3.6}$$

the second integral on the right-hand side of (A5.3.4) can be written

$$\int_0^\infty g(u,z)z^{-1}\,dz = \left(\frac{2}{v_1+v_2+4}\right)\int_0^\infty g(u,z)[1+N(u,z)]\,dz, \tag{A5.3.7}$$

where

$$N(u,z) = \left(\frac{v_2}{2}+3\right)(\phi^{-1}+uz)^{-1}u + \phi^{-1}(\phi^{-1}+uz)^{-2}.$$

Substituting the right-hand side of (A5.3.7) into (A5.3.4), we have

$$M^{-1}\frac{\partial p(u \mid \mathbf{y})}{\partial u} = \left[1 - \left(\frac{v_2}{2}+1\right)\left(u - \frac{2\phi^{-1}}{v_2+v_1+4}\right)\right]\int_0^\infty g(u,z)\,dz$$

$$- \left(\frac{v_2}{2}+1\right)\left(\frac{2\phi^{-1}}{v_1+v_2+4}\right)\int_0^\infty g(u,z)N(u,z)\,dz. \tag{A5.3.8}$$

Since $N(u,z) \geqslant 0$, it follows that

$$\frac{\partial p(u \mid \mathbf{y})}{\partial u} < 0 \quad \text{if} \quad u > u^*, \tag{A5.3.9}$$

where

$$u^* = \left(\frac{v_2}{2}+1\right)^{-1} - \frac{2\phi^{-1}}{v_1+v_2+4}.$$

Also since,

$$N(u,z) \geqslant \left(\frac{v_2}{2}+3\right)\phi u + \phi, \tag{A5.3.10}$$

therefore,

$$\frac{\partial p(u \mid \mathbf{y})}{\partial u} > 0 \quad \text{if} \quad u < u_*, \tag{A5.3.11}$$

where

$$u_* = \left(1 + \frac{\nu_2 + 6}{\nu_1 + \nu_2 + 4}\right)^{-1} \left[\left(\frac{\nu_2}{2} + 1\right)^{-1} - \frac{2(\phi^{-1} + 1)}{\nu_1 + \nu_2 + 4}\right] \quad \text{and} \quad u_* < u^*.$$

Hence,

a) if $u_* > 0$, then $p(u \mid \mathbf{y})$ is monotonically increasing in the interval $0 < u < u^*$ and monotonically decreasing in the interval $u > u^*$ so that the mode must lie in the interval $u_* < u < u^*$; and

b) if $u^* < 0$, then $p(u \mid \mathbf{y})$ is monotonically decreasing in the entire range $u > 0$ and the mode is at the origin.

In terms of σ_2^2 and the mean squares m_1 and m_2, we may conclude that:

a) If

$$\frac{m_2}{m_1} > \left(\frac{\nu_1}{\nu_1 + 2}\right)\left(\frac{\nu_2 + 2}{\nu_2}\right),$$

then $p(\sigma_2^2 \mid \mathbf{y})$ is monotonically increasing in the interval $0 < \sigma_2^2 < c_1$ and monotonically decreasing in the interval $\sigma_2^2 > c_2$ so that the mode lies in the interval $c_1 < \sigma_2^2 < c_2$, where

$$c_1 = K^{-1}\left(\frac{\nu_1 + \nu_2 + 4}{\nu_1 + 2\nu_2 + 10}\right)\left[\frac{\nu_2}{\nu_2 + 2} m_2 - \frac{1}{\nu_1 + \nu_2 + 4}(\nu_1 m_1 + \nu_2 m_2)\right]$$

and

$$c_2 = K^{-1}\left(\frac{\nu_2}{\nu_2 + 2} m_2 - \frac{\nu_1}{\nu_1 + \nu_2 + 4} m_1\right). \tag{A5.3.12}$$

b) If

$$\frac{m_2}{m_1} < \left(\frac{\nu_2 + 2}{\nu_2}\right)\left(\frac{\nu_1}{\nu_1 + \nu_2 + 4}\right),$$

then $p(\sigma_2^2 \mid \mathbf{y})$ is monotonically decreasing in the entire range $\sigma_2^2 > 0$ and the mode is at the origin.

APPENDIX A5.4

An Asymptotic Expansion of the Distribution of σ_2^2 for the Two-component Model

We here derive an asymptotic expansion of the distribution of σ_2^2 in (5.2.39) in powers of $t^{-1} = (\nu_1/2)^{-1}$.† To simplify writing, we shall again work with the

† In Tiao and Tan (1965), the distribution of σ_2^2 was expanded in powers of $[(\nu_2/2) - 1]^{-1}$ with slightly different results.

distribution of $u=2K\sigma_2^2/(\nu_2 m_2)$ rather than σ_2^2 itself. Upon making the substitution $z = \nu_1 m_1/(2\sigma_1^2)$, the distribution of u is

$$p(u \mid \mathbf{y}) = \frac{1}{\Pr\{F_{\nu_2,\nu_1} < m_2/m_1\}\Gamma(t)} \int_0^\infty h(u, z) z^{t-1} e^{-z}\, dz, \qquad \text{(A5.4.1)}$$

where

$$h(u, z) = \frac{1}{\Gamma(\nu_2/2)} [(\phi z)^{-1} + u]^{\frac{1}{2}(\nu_2+2)} \exp\{-[(\phi z)^{-1} + u]^{-1}\}, \qquad \text{(A5.4.2)}$$

$$\phi = \frac{\nu_2 m_2}{\nu_1 m_1}, \qquad \text{and} \qquad t = \frac{\nu_1}{2}.$$

For fixed u, the function $h(u, z)$ in the integral (A5.4.1) is clearly analytic in $0 < z < \infty$. Using Taylor's theorem, we can expand $h(u, z)$ around t. By reversing the order of integration, we obtain

$$p(u \mid \mathbf{y}) = \frac{1}{\Pr\{F_{\nu_2,\nu_1} < m_2/m_1\}} \sum_{r=0}^{\infty} \frac{1}{r!} h^{(r)}(u, t) \frac{1}{\Gamma(t)} \int_0^\infty (z - t)^r z^{t-1} e^{-z}\, dz, \qquad \text{(A5.4.3)}$$

where

$$h^{(0)}(u, t) = [\Gamma(\nu_2/2)]^{-1} (\lambda^{-1} + u)^{-\frac{1}{2}(\nu_2+2)} \exp[-(\lambda^{-1} + u)^{-1}],$$

$$h^{(1)}(u, t) = -t^{-1} h^{(0)}(u, t) R_1(\lambda, u),$$

$$h^{(2)}(u, t) = t^{-2} h^{(0)}(u, t)[R_2(\lambda, u) + 2R_1(\lambda, u)],$$

$$h^{(3)}(u, t) = -t^{-3} h^{(0)}(u, t)[R_3(\lambda, u) + 6R_2(\lambda, u) + 6R_1(\lambda, u)],$$

$$h^{(4)}(u, t) = t^{-4} h^{(0)}(u, t)[R_4(\lambda, u) + 12R_3(\lambda, u) + 36R_2(\lambda, u) + 24R_1(\lambda, u)],$$

....

$$R_1(\lambda, u) = \lambda^{-1}\left[w^2 - \left(\frac{\nu_2}{2} + 1\right)w\right],$$

$$R_2(\lambda, u) = \lambda^{-2}\left[w^4 - 2\left(\frac{\nu_2}{2} + 2\right)w^3 + \left(\frac{\nu_2}{2} + 1\right)\left(\frac{\nu_2}{2} + 2\right)w^2\right],$$

$$R_3(\lambda, u) = \lambda^{-3}\left[w^6 - 3\left(\frac{\nu_2}{2} + 3\right)w^5 + 3\left(\frac{\nu_2}{2} + 2\right)\left(\frac{\nu_2}{2} + 3\right)w^4 - \left(\frac{\nu_2}{2} + 1\right)\left(\frac{\nu_2}{2} + 2\right)\left(\frac{\nu_2}{2} + 3\right)w^3\right],$$

$$R_4(\lambda, u) = \lambda^{-4}\left[w^8 - 4\left(\frac{\nu_2}{2} + 4\right)w^7 + 6\left(\frac{\nu_2}{2} + 3\right)\left(\frac{\nu_2}{2} + 4\right)w^6\right.$$

$$- 4\left(\frac{\nu_2}{2} + 2\right)\left(\frac{\nu_2}{2} + 3\right)\left(\frac{\nu_2}{2} + 4\right)w^5$$

$$\left. + \left(\frac{\nu_2}{2} + 1\right)\left(\frac{\nu_2}{2} + 2\right)\left(\frac{\nu_2}{2} + 3\right)\left(\frac{\nu_2}{2} + 4\right)w^4\right],$$

$$\lambda = \phi t \qquad \text{and} \qquad w = (\lambda^{-1} + u)^{-1}.$$

It is to be noted that for fixed $\lambda = \phi t$, $h^{(1)}(u, t)$ is of order t^{-1}, $h^{(2)}(u, t)$ is of order t^{-2} and in general $h^{(r)}(u, t)$ is of order t^{-r}. From the asymptotic relationship between the gamma distribution and the normal distribution one can easily verify that the integral

$$\frac{1}{\Gamma(t)}\int_0^\infty (z - t)^r z^{t-1} e^{-z}\, dz \tag{A5.4.4}$$

is a polynomial in t of degree $[\frac{1}{2}(r - 1)]$, where $[q]$ is the smallest nonnegative integer greater than or equal to q. Thus, the expression in (A5.4.3) can be written as a power series of t^{-1}. Also, to obtain the coefficient of t^{-1}, we need only to evaluate the first three terms of the series in (A5.4.3) and to obtain that of t^{-2}, the first five terms of the series and so on.

Now the quantity $\Pr\{F_{\nu_2,\nu_1} < m_2/m_1\}$ in (A5.4.1) can be written

$$\Pr\{F_{\nu_2\,\nu_1} < m_2/m_1\}$$

$$= \frac{1}{B(\frac{1}{2}\nu_2, t)}\int_0^\phi x^{\frac{1}{2}\nu_2 - 1}(1 + x)^{-\frac{1}{2}\nu_2} \exp\{-t \log(1 + x)\}\, dx$$

$$= c(t)\int_0^\lambda z^{\frac{1}{2}\nu_2 - 1} \exp(-z)\left[\left(1 + \frac{z}{t}\right)^{\frac{1}{2}\nu_2} \exp\left\{\frac{z^2}{2t} - \frac{z^3}{3t^2} + \cdots\right\}\right] dz, \tag{A5.4.5}$$

where

$$c(t) = \frac{\Gamma(\frac{1}{2}\nu_2 + t)t^{-\frac{1}{2}\nu_2}}{\Gamma(\frac{1}{2}\nu_2)\Gamma(t)}.$$

Applying Stirling's series (see Appendix A2.2) to $c(t)$ and expanding the factor in the square bracket of the integrand in powers of t^{-1}, we find that for fixed λ,

$$\Pr\{F_{\nu_2,\nu_1} < m_2/m_1\} \doteq G_\lambda(\tfrac{1}{2}\nu_2) + \frac{1}{t}A_1(\lambda) + \frac{1}{t^2}A_2(\lambda) + O(t^{-3}), \tag{A5.4.6}$$

where

$$A_1(\lambda) = g_\lambda\left(\frac{v_2}{2}\right)\frac{1}{2}\left[\lambda\left(\frac{v_2}{2} - 1\right) - \lambda^2\right],$$

$$A_2(\lambda) = g_\lambda\left(\frac{v_2}{2}\right)\frac{1}{8}\left[\lambda\left(\frac{v_2}{2} - 1\right)\left(\frac{v_2^2}{4} - \frac{7v_2}{6} + \frac{2}{3}\right) - \left(\frac{v_2}{2} - 1\right)\left(\frac{3v_2}{2} - \frac{2}{3}\right)\lambda^2 + \left(\frac{3v_2}{2} - \frac{1}{3}\right)\lambda^3 - \lambda^4\right],$$

$G_\lambda(\frac{1}{2}v_2)$ is the cumulative distribution of a gamma variable with parameter $v_2/2$ evaluated at λ, and $g_\lambda(\frac{1}{2}v_2)$ is the corresponding density. From (A5.4.3) and (A5.4.6), we find that for fixed λ the distribution of u can be written

$$p(u \mid \mathbf{y}) \doteq \frac{(\lambda^{-1} + u)^{-(\frac{1}{2}v_2+1)}}{G_\lambda(\frac{1}{2}v_2)\,\Gamma(\frac{1}{2}v_2)} \exp\left[-(\lambda^{-1} + u)^{-1}\right] \times \left[1 + \frac{1}{t}B_1(u,\lambda) + \frac{1}{t^2}B_2(u,\lambda) + O(t^{-3})\right] \qquad \text{(A5.4.7)}$$

with

$$B_1(u, \lambda) = \tfrac{1}{2}R_2(\lambda, u) + R_1(\lambda, u) - \frac{A_1(\lambda)}{G_\lambda(\frac{1}{2}v_2)},$$

$$B_2(u, \lambda) = \tfrac{1}{8}R_4(\lambda, u) + \tfrac{7}{6}R_3(\lambda, u) + \tfrac{5}{2}R_2(\lambda, u) + R_1(\lambda, u) - \frac{A_2(\lambda)}{G(\frac{1}{2}v_2)} - \frac{A_1(\lambda)}{G(\frac{1}{2}v_2)}B_1(u, \lambda),$$

where $R_1(\lambda, u), \ldots, R_4(\lambda, u)$ are given in (A5.4.3), and $A_1(\lambda)$ and $A_2(\lambda)$ are given in (A5.4.6). The distribution of u is thus expressed as a series in powers of t^{-1}. The leading term is in the form of a truncated "inverted" gamma distribution and is equivalent to the distribution in (5.2.44). The coefficients $B_1(u, \lambda)$, $B_2(u, \lambda)$, etc. are polynomials in $(\lambda^{-1} + u)^{-1}$. It is straightforward to verify that when one integrates the distribution (A5.4.7) over u all terms except the leading one vanish so that (A5.4.7) defines a proper density function. By making the transformation $\chi^2 = 2/(\lambda^{-1} + u)$ one can, of course, alternatively express the distribution in (A5.4.7) as a series of χ^2 densities.

If one is interested in individual posterior probabilities of u, a corresponding asymptotic expansion of the probability integrals in powers of t^{-1} can be readily obtained. From (A5.4.7), it can be verified that for fixed λ,

$$\Pr\{u > u_0 \mid \mathbf{y}\} = P_0(\gamma_0) + \frac{1}{t}P_1(\gamma_0) + \frac{1}{t^2}P_2(\gamma_0) + O(t^{-3}), \qquad \text{(A5.4.8)}$$

where

$$\gamma_0^{-1} = \lambda^{-1} + u_0,$$

$$P_0(\gamma_0) = G_{\gamma_0}(\tfrac{1}{2}\nu_2)/G_\lambda(\tfrac{1}{2}\nu_2),$$

$$P_1(\gamma_0) = \frac{1}{G_\lambda(\tfrac{1}{2}\nu_2)}\left[\left(\frac{\gamma_0}{\lambda}\right)^2 A_1(\gamma_0) - \frac{G_{\gamma_0}(\tfrac{1}{2}\nu_2)}{G_\lambda(\tfrac{1}{2}\nu_2)} A_1(\lambda) + g_{\gamma_0}(\tfrac{1}{2}\nu_2)\gamma_0\left(\frac{\gamma_0}{\lambda}\right)\left(\frac{\gamma_0}{\lambda} - 1\right)\right]$$

and

$$P_2(\gamma_0) = -\frac{A_1(\lambda)}{G_\lambda(\tfrac{1}{2}\nu_2)} P_1(\gamma_0) - \frac{A_2(\lambda)}{G_\lambda(\tfrac{1}{2}\nu_2)}\frac{G_{\gamma_0}(\tfrac{1}{2}\nu_2)}{G_\lambda(\tfrac{1}{2}\nu_2)} - \frac{g_{\gamma_0}(\tfrac{1}{2}\nu_2)}{G_\lambda(\tfrac{1}{2}\nu_2)}$$

$$\times\left\{\left(\frac{\gamma_0}{\lambda}\right)\gamma_0 + \frac{5}{2}\left(\frac{\gamma_0}{\lambda}\right)^2\left[\gamma_0^2 - \gamma_0\left(\frac{\nu_2}{2}+1\right)\right] + \frac{7}{6}\left(\frac{\gamma_0}{\lambda}\right)^3\left[\gamma_0^3 - 2\gamma_0^2\left(\frac{\nu_2}{2}+2\right)\right.\right.$$

$$\left. + \gamma_0\left(\frac{\nu_2}{2}+1\right)\left(\frac{\nu_2}{2}+2\right)\right] + \frac{1}{8}\left(\frac{\gamma_0}{\lambda}\right)^4\left[\gamma_0^4 - 3\gamma_0^3\left(\frac{\nu_2}{2}+3\right)\right.$$

$$\left.\left. + 3\gamma_0^2\left(\frac{\nu_2}{2}+2\right)\left(\frac{\nu_2}{2}+3\right) - \gamma_0\left(\frac{\nu_2}{2}+1\right)\left(\frac{\nu_2}{2}+2\right)\left(\frac{\nu_2}{2}+3\right)\right]\right\}.$$

Posterior probabilities of u and therefore σ_2^2 can thus be conveniently calculated using (A5.4.8) in conjunction with a standard incomplete gamma function table or a χ^2 table.

While the asymptotic expansions of $p(u \mid \mathbf{y})$ and the related probability integrals as obtained above are justified for large t, their usefulness depends, of course, upon how close they approximate the exact distribution and probabilities for moderate values of t. We have already seen in Figs. 5.2.6 and 5.2.7 that the posterior distribution of σ_2^2 obtained just by employing the leading term of (A5.4.7) is in close agreement with the exact distribution for both the dyestuff and the generated example. Tables 5.2.3(a) and 5.2.3(b) illustrate how further improvement can be made by employing additional terms in the expansion (A5.4.7). The first columns of the two tables give, respectively, a specimen of the densities of σ_2^2 for the two examples calculated from (A5.4.7) with just the leading term. In the second and third columns of the same tables are shown densities of σ_2^2 which now include terms to order t^{-1} and t^{-2}, respectively, and in the fourth column the corresponding exact densities obtained from (5.2.39) by numerical integration are listed. The values of t for these two examples are not at all large. Further, the dyestuff example has a moderate value of m_2/m_1, while the generated example has a rather small m_2/m_1. Thus, the results for the two examples demonstrate that even with moderate values of t, the expansion formula in (A5.4.7) is useful both for small and for moderate values of m_2/m_1.

The main practical advantage of the expressions (A5.4.7) and (A5.4.8) is, of course, their computational simplicity. The leading terms of order t^{-1} in (A5.4.7) and (A5.4.8) are simple functions of χ^2 densities and probabilities which can be easily evaluated on a desk calculator in conjunction with a standard χ^2 table or an incomplete gamma function table. If terms of order t^{-2} or higher are included, the use of an electronic computer is then desirable. Nevertheless, the expansion gives an efficient computational method compared with the direct numerical integration of the exact expressions in (5.2.39) and its related probability integrals.

APPENDIX A5.5

A Criticism of the Prior Distribution of (σ_1^2, σ_2^2)

We now discuss an interesting theoretical point raised by Stone and Springer (1965) concerning the prior distribution of $(\sigma_1^2, \sigma_{12}^2)$ in (5.2.8)

$$p(\sigma_1^2, \sigma_{12}^2) \propto \sigma_1^{-2}\sigma_{12}^{-2}, \qquad \sigma_{12}^2 > \sigma_1^2. \tag{A5.5.1}$$

They argue that this distribution can be regarded as the limiting distribution of the following sequence of distributions

$$p(\sigma_1^2, \sigma_{12}^2) \propto (\sigma_1^2)^{-(1-\varepsilon)}\,\sigma_{12}^{-2}, \qquad \sigma_{12}^2 > \sigma_1^2, \quad \varepsilon > 0 \tag{A5.5.2}$$

as $\varepsilon \to 0$. Making the transformation from $(\sigma_1^2, \sigma_{12}^2)$ to

$$\rho = \frac{\sigma_1^2}{\sigma_{12}^2} \qquad \text{and} \qquad \sigma_{12}^2, \tag{A5.5.3}$$

they claim that the marginal prior distribution of ρ is

$$p(\rho) = \varepsilon\rho^{-(1-\varepsilon)}, \qquad 0 < \rho < 1, \quad \varepsilon > 0. \tag{A5.5.4}$$

Letting $\delta > 0$, the prior probability that $0 < \rho < \delta$ is

$$\Pr(0 < \rho < \delta) = \delta^{\varepsilon} \tag{A5.5.5}$$

which approaches 1 as $\varepsilon \to 0$. This means that the probability limit

$$\rho \lim \rho = 0. \tag{A5.5.6}$$

Now, consider the distribution of the mean squares ratio m_1/m_2. Since conditional on ρ

$$\frac{m_1}{m_2} \sim \rho\, F_{\nu_1,\nu_2} \tag{A5.5.7}$$

it follows from (A5.5.6) that, *a priori*

$$\rho \lim \frac{m_1}{m_2} = 0. \tag{A5.5.8}$$

Thus, the prior distribution in (A5.5.1) implies that *a priori*, one would expect m_1/m_2 to be very close to zero, and would therefore be surprised if a somewhat large value of m_1/m_2 were observed.

Consider now the posterior distribution of σ_1^2 in (5.2.26). If m_1/m_2 were small, (that is, m_2/m_1 were very large), then

$$\lim_{m_1/m_2 \to 0} f(\sigma_1^2) = 1.$$

The implication is, then, since *a prior* one would expect m_1/m_2 to be small in any event, it should be therefore unnecessary to consider this factor at all in the posterior inference. And if m_1/m_2 turned out to be somewhat large, it would be in conflict with the implied prior belief.

While the above argument sounds interesting, given the way Stone and Springer approach the problem, it has hardly any practical significance so far as the posterior inference is concerned. Consider the two situations $\varepsilon \to 0$ and $\varepsilon = 0.5$, and let $\delta = 0.2$. Then, *a priori*,

$$\lim_{\varepsilon \to 0} \Pr(0 < \rho < 0.2) = 1 \qquad \text{and} \qquad \Pr(0 < \rho < 0.2 \mid \varepsilon = 0.5) = \sqrt{0.2} = 0.447.$$

Thus, a large difference exists between the prior for which $\varepsilon = 0.5$ and the prior in (A5.5.1), and in the case $\varepsilon = 0.5$, Stone and Springer's objection would not arise. However, the use of $\varepsilon = 0.5$ will only serve to change, in the posterior distribution of (σ_1^2, σ_2^2) in (5.2.14), v_1 to $v_1^* = v_1 - 0.5$ and m_1 to $m_1^* = m_1(v_1/v_1^*)$. The reader can verify that for either the dyestuff or the generated example or any other example in which v_1 is of moderate size, the posterior inferences about (σ_1^2, σ_2^2) would be scarcely affected.

APPENDIX A5.6

"Bayes" Estimators

In this book we treat sampling theory inference and Bayesian inference as quite distinct concepts. In sampling theory, inferences about parameters, supposed *fixed*, are made by considering the sampling properties of specifically selected functions of the observations called estimators. In Bayesian inference, the parameters are considered as random variables and inferences are made by considering their distributions conditional on the *fixed* data. In recent years, some workers have used a dual approach which, in a sense, goes part of the way along the Bayes route and then switches to sampling theory. Specifically, the Bayesian posterior distribution is obtained, but not used to make inferences directly. Instead, it is employed to suggest appropriate functions of the data for use as estimators. Inferences are then made by considering the sampling properties of these "Bayes" estimators. Because this dual approach is now quite widespread—see, for example, Mood and Graybill (1963)—, we feel that we should say something about it here.

The present authors believe that while estimators undoubtedly can serve some useful purposes (for example, in the calculation of residuals in the *criticism* phase of statistical analysis), we do not believe their sampling properties have much relevance to scientific inference.

Description of Distributions (Including Posterior Distributions)

We have argued in this book that the *whole* of the appropriate posterior distribution provides the means of making all relevant inferences about a parameter or a set of parameters. We have further argued that, by using a noninformative prior, a posterior distribution is obtained appropriate to the situation where the information supplied by the data is large compared with that available *a priori*. If for some reason it was *demanded* that we characterized a continuous posterior distribution without using its mathematical form and without using diagrams, we could try to do this in terms of a few descriptive measures. The principles we would then have to employ and the difficulties that would then confront us would be the same as those we would meet in so describing any other continuous distribution or indeed in representing any infinite set by a finite set.

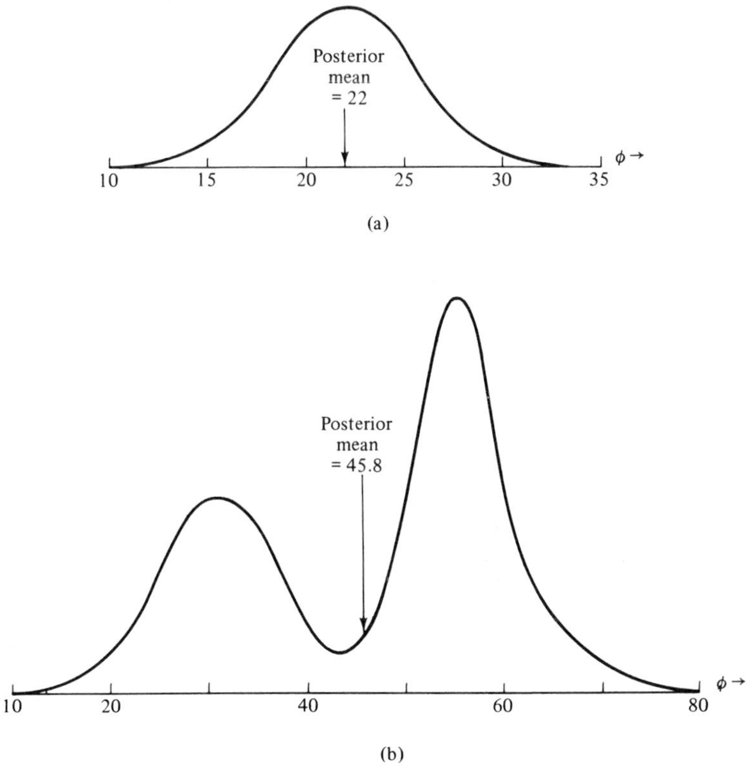

Fig. A5.6.1 Two hypothetical posterior distributions

For example, the description of the posterior distribution of Fig. A5.6.1(a) is simple. It is a Normal distribution with mean 22 and standard deviation 3.7 Thus, with knowledge of the Normal form, two quantities suffice to provide an exact description.

Description of the posterior distribution in Fig. A5.6.1(b) would be more difficult. We might describe it thus: "It is a bimodal distribution with most of its mass between 10 and 80. The left hand mode at 30 is about half the height of the right hand mode at 57. A local minimum occurs at 43 and is of about one tenth the height of the right hand mode". Alternatively, we could quote, say, 10 suitably spaced densities to give the main features of the distribution. However we proceeded we would need to quote five or more different numbers to provide even an approximate idea of what we were talking about. Now there is nothing unrealistic about this hypothetical posterior distribution. It would imply that values of the parameter around 57 and around 30 were plausible but that intermediate values around 43 were much less so. Practical examples of this sort are not uncommon and one such case will be discussed later in Section 8.2.

Posterior distributions met in practice are usually neither so simple as the Normal nor so complex as the bimodal form, just considered. But what *is* clear is that the number and kinds of measures needed to approximately *describe* the posterior distribution must depend on the type and complexity of that distribution. Questions about how to describe a posterior distribution cannot be (and do not need to be) decided until *after* that distribution is computed. Furthermore, they are decided simply on the basis of what would best convey the right picture to the investigator's mind.

Point Estimators

Over a very long period, attempts of one sort or another have been made to obtain by some *prior* recipe a *single number* calculated from the data which in some way best represents a parameter under study. Historically there has been considerable ambivalence about objectives. Classical writers would for example sometimes state the problem as that of finding the most "probable" value and sometimes that of finding a best "average" or "mean" value.

A Measure of Goodness of Estimators

Modern sampling theorists refer to the single representative quantity calculated from a sample $\mathbf{y}$ as a *point estimator* and they argue as follows. Suppose we are interested in a particular parameter ϕ. Then any function $\hat{\phi}(\mathbf{y})$ of the data which provides some idea of the value of ϕ may be called an *estimator*. Usually a very large number of possible estimators can be devised. For example, the variance of a distribution might be estimated by the sample variance, the sample range, the sample mean absolute deviation and so on. Therefore a criterion of goodness is needed. Using it the various estimators can be compared and the "best" selected. It is argued that goodness of an estimator ought to be measured by the average closeness of its values over all possible samples to the true value of ϕ. The criterion of average closeness most frequently used nowadays (but of course an arbitrary one) is the mean squared error (M.S.E.). Thus among a class of possible candidate estimators $\hat{\phi}_1(\mathbf{y}), \ldots, \hat{\phi}_j(\mathbf{y}), \ldots, \hat{\phi}_k(\mathbf{y})$, the goodness of a particular one, say, $\hat{\phi}_j(\mathbf{y})$ would be measured by the size of

$$\text{M.S.E.}(\hat{\phi}_j) = \underset{\mathbf{y}}{E}[\hat{\phi}_j(\mathbf{y}) - \phi]^2 = f_j(\phi),$$

where the expectation is taken over the sampling distribution of $\mathbf{y}$. The estimator $\hat{\phi}_i(\mathbf{y})$ would be "best" in the class considered if

$$f_i(\phi) \leqslant f_j(\phi), \qquad j = 1, \ldots, k; \quad j \neq i$$

for all values of ϕ, and would be called a minimum mean squared error (M.M.S.E.) estimator.

At first sight, the argument seems plausible, but the M.S.E. criterion is an arbitrary one and is easily shown to be unreliable. A simple example is the estimation of the reciprocal of the mean of the Normal distribution $N(\theta, 1)$. From a random sample $\mathbf{y}$, there is surely something to be said for the maximum likelihood estimator $\bar{y}^{-1}$ which is, after all, sufficient for θ^{-1}, but it so happens that the M.S.E. of $\bar{y}^{-1}$ is infinite.

The real weakness of the argument lies in that it requires a *universal* measure of goodness. Even granted that some measure of goodness may be desirable, it ought to depend on the probabilistic structure of the estimator as well as what the estimator is used for. But inherent in the theory is the supposition that, a criterion is to be arbitrarily chosen *a priori* and then applied universally to all estimators.

How to Obtain the Estimators

If we were to accept the above argument, we would have a means of ranking a set of estimators *once they were presented to us.* But how can we cope with the embarrassingly large variety of functions of the data which might have relevance as estimators? Is there some way in which we can simplify the problem so as to pick out, if not one, then at least a limited number of good estimators which may then be compared?

One early attempt to make the problem manageable was to arbitrarily limit the estimators to be (a) linear functions of $\mathbf{y}$ which were (b) unbiased, and then to find the one which had minimum variance. These requirements had the advantage that the "best" estimators could often be obtained from their definition by a simple minimization process. For example, in a random sample of size n from a population with finite mean ϕ and finite variance, the sample mean $\bar{y}$ is the "best" estimator of ϕ satisfying these requirements.

However, after it was realized that biased estimators often had smaller mean squared errors than unbiased ones, the requirement of unbiasedness was usually dropped and the criterion of minimum M.S.E. substituted for that of minimum variance. As an example, for the exponential distribution $f(y) = \phi^{-1} e^{-y/\phi}$, $(y > 0, \phi > 0)$. The linear function $n\bar{y}/(n+1)$, although biased, has smaller M.S.E. than that of the sample mean $\bar{y}$ in estimating ϕ. When it was further shown that non-linear estimators could be readily produced (from Bayesian sources) having smaller M.S.E. than the "best" linear ones for all values of ϕ, linear estimators could no longer be wholeheartedly supported.†

Possible ways of obtaining non-linear estimators *which still might minimize mean squared error* were therefore examined. Perhaps then the logical final step will be to abandon the arbitrary criterion of minimum M.S.E. With this step taken, it would seem best to give up the idea of point estimation altogether and use instead the posterior distribution for inferential purposes.

One method of finding estimators which are not necessarily linear in the observations is Fisher's method of maximum likelihood. Here the estimator is the mode of the likelihood, and, under certain conditions, asymptotically at least, the estimators have Normal distribution and smallest M.S.E. However, some investigators have felt that the likelihood method itself was too limited for their purposes. A few years ago therefore, the Bayesians were surprised by unexpected bed fellows anxious to examine their posterior distributions for interesting features. Whereas it was the *mode* of the likelihood function which had received almost exclusive attention, it now seemed to be the *mean* of the posterior distribution which became the main center of interest. Nevertheless the posterior mode and the median have also come in for some examination and each has been recommended or condemned on one context or another.

A Decision-theoretic Approach

Justification for this way of choosing estimators is sometimes attempted on formal decision-theoretic grounds. In this argument, let $\hat{\phi}$ be the decision-maker's "best" guess of the unknown ϕ relative to a loss function $L(\hat{\phi}, \phi)$. Then the so-called "Bayes estimator" is the guessed value which minimizes the posterior expected loss

$$\underset{\phi \mid \mathbf{y}}{E}\, L(\hat{\phi}, \phi) = \int_{-\infty}^{\infty} L(\hat{\phi}, \phi)\, p(\phi \mid \mathbf{y})\, d\phi.$$

In particular, if the loss function is the squared error

$$L_1(\hat{\phi}, \phi) = (\hat{\phi} - \phi)^2 \qquad \text{(A5.6.1a)}$$

then, provided it exists, the Bayes estimator is the posterior mean,‡ but by suitably

† An example of this kind will be discussed later in Section 7.2.

‡ When authors write of the "Bayes estimator" without further qualification they usually refer to the posterior mean. The naming of "Bayes" estimator seems particularly unfair to the Rev. Thomas Bayes who already has had much to answer for and should be cleared of all responsibility for fathering this particular foundling.

changing the loss function the posterior mode or median can also be produced. Thus, see Pratt, Raiffa and Schlaifer (1965), for

$$L_2(\hat{\phi}, \phi) = \begin{cases} 1 & \text{for} \quad |\hat{\phi} - \phi| > \varepsilon \\ 0 & \qquad\; |\hat{\phi} - \phi| < \varepsilon \end{cases} \tag{A5.6.1b}$$

where $\varepsilon > 0$ is an arbitrarily small constant, the Bayes estimator is a mode of the posterior distribution. Also, for

$$L_3(\hat{\phi}, \phi) = |\hat{\phi} - \phi|$$

the Bayes estimator is the posterior median. Indeed, if the loss function were

$$L_4 = \begin{cases} a_1|\hat{\phi} - \phi| & \hat{\phi} < \phi \\ a_2|\hat{\phi} - \phi| & \hat{\phi} > \phi \end{cases}$$

then by appropriately choosing a_1 and a_2, any fractile of the posterior distribution could be an estimator for ϕ.

In a proper Bayesian decision analysis, the loss function is supposed to represent a realistic economic penalty associated with the available actions. This justification, in our view, only serves to emphasize the weakness of the point estimation position. (a) In the context in which the estimator was to be employed, *if* the consequences associated with inaccurate estimation were really what the loss function said they were, then the resulting estimator *would* be best for *minimizing this loss* and this would be so *whether its* M.S.E. *in repeated sampling was large or small.* (b) It seems very dubious that a general loss function could be chosen in advance to meet all situations. In particular, it is easy to imagine practical situations where the squared error loss (A5.6.1a) would be inappropriate. In summary it seems to us that this Bayesian decision argument has little to do with point estimation for inferential purposes.

Are Point Estimates Useful for Inference?

The first question a Bayesian should ask is whether or not point estimates provide a useful tool for making inferences. Certainly, if he were presented with the Normal distribution of Fig. A5.6.1(a) and told he must pick a single number to represent ϕ he would have little hesitation in picking 22 which is both the mean and the mode of the distribution. Even here he would regret that he had not been allowed to say anything about the uncertainty associated with the parameter ϕ. But what single number should he pick to represent ϕ in Fig. A5.6.1(b)? If he chose the higher mode at 57, he would certainly feel unhappy about not mentioning the lower mode at 30. He would surely not choose the "Bayes" estimator—the posterior mean at 45.8 which is close to a local minimum density.

While it easy to demonstrate examples for which there can be no satisfactory point estimate, yet the idea is very strong among people in general and some statisticians in particular that there is a need for such a quantity. To the idea that people like to have a single number we answer that usually they shouldn't get it. Most people know they live in a statistical world and common parlance is full of words implying uncertainty. As in the case of weather forecasts, statements about uncertain quantities ought to be made in terms which reflect that uncertainty as nearly as possible.†

Should Inferences be Based on Sampling Distribution of Point Estimators?

Having computed the estimators from the posterior distribution, the sampling theorist often uses its *sampling distribution* in an attempt to make inferences about the unknown parameter. The bimodal example perhaps makes it easy to see why we are dubious about the logic of this practice. First we feel that having gone to the trouble of computing the posterior distribution, it may as well be put to use to make appropriate inferences about the parameters. Second we cannot see what relevance the *sampling distribution* of, say, the mean of the posterior distribution would have in making inferences about ϕ.

Why Mean Squared Error and Squared Error Loss?

We have mentioned that the M.S.E. criterion is arbitrary (as are alternative measures of average closeness). We have also said that in the decision theoretic framework the quadratic loss function leading to the posterior mean is arbitrary (as are alternative loss functions yielding different features of the posterior distribution). The question remains as to why many sampling theorists seem to cling rather tenaciously to the mean squared error criterion and the quadratic loss function. The reasons seem to be that (a) for their theory to work, they must cling to *something*, arbitrary though it be and (b) given this, it is best to cling to something that works well for the Normal distribution.

The latent belief in universal near-Normality of estimators is detectable in the general public's idea that the "most probable" and the "average" are almost synonymous, implying a belief, if not in Normality, at least that "mean equals mode". Many statisticians familiar with ideas of asymptotic Normality of maximum likelihood and other estimators have a similar tendency.

This has led to a curious reaction by some workers when the application of these arbitrary principles led to displeasing results. The Bayesian approach or the Bayesian prior have been blamed and not the arbitrary features, curious mechanism and dubious relevance of the point estimation method itself.

† The public is ill served when they are supplied with single numbers published in newspapers and elsewhere with no measure of their uncertainty. For example, most figures in a financial statement, balanced down to the last penny, are in fact merely estimates. When asked whether it would be a good idea to publish error limits for those figures, an accounting professor replied "This is commonly done internally in a company, but it would be too much of a shock for the general public!"

Point Estimators for Variance Components

For illustration we consider some work on the variance component problem tackled from the standpoint of point estimation. Two papers are discussed, one by Klotz, Milton and Zacks (1969) and the other by Portnoy (1971) concerning point estimators of the component σ_2^2 for the two-component model.

M.S.E. of Various Estimators of σ_2^2

These authors computed the M.S.E. of a variety of estimators for (σ_1^2, σ_2^2). Specifically, for the component σ_2^2, they have considered for various values of J and K the following estimators:

a) the usual unbiased estimator $\hat{\sigma}_2^2$

b) the maximum likelihood estimator

c) the mean of the posterior distribution (5.2.39)—see (A5.3.1) in Appendix A5.3

d) the component for σ_2^2 of the mode of the joint distribution (5.2.14)

e) the component for σ_2^2 of the mode of the corresponding joint distribution of $(\sigma_1^{-2}, \sigma_2^2)$,

among some others. Their calculation shows that over all values of the ratio $\sigma_2^2/(\sigma_1^2 + \sigma_2^2)$

i) the M.S.E. of the estimators (b), (d) and (e) are comparable and are much smaller than that of (a),

ii) by far the worst estimator is the posterior mean (c). For example, when $J = 5$ and $K = 2$, the M.S.E. of (c) is at least 8 times as large as those of (b), (d) and (e).

The fact that the posterior mean has very large M.S.E. seems especially disturbing to these authors, since such a choice is motivated by the squared error loss (A5.6.1a). Thus, they concluded that

"The numerical results on the M.S.E. of the formal Bayes estimators strongly indicate that the practice of using posterior means may yield very inefficient estimators. The inefficiency appears due to a large bias caused by posteriors with heavy tails resulting *from the quasi prior distributions*. Other characteristics of the posterior distributions such as the mode or median can give more efficient point estimators". (*J. Amer. Statist. Assoc.*, **64**, p. 1401).

The italics in the above are ours and they draw attention to a point we cannot understand. The authors' conclusions and associated implications may be summarized in part as follows.

Table A5.6.1
Comparison of estimators of σ_2^2

Prior	Implied loss function [see (A5.6.1a,b)]	Feature of posterior distribution	Estimators judged	Criterion
Noninformative (5.2.9)	L_1	mean	bad	M.S.E.
Noninformative (5.2.9)	L_2	mode of σ_2^2 w.r.t. (σ_1^2, σ_2^2)	good	M.S.E.
Noninformative (5.2.9)	L_2	mode of σ_2^2 w.r.t. $(\sigma_1^{-2}, \sigma_2^2)$	good	M.S.E.

But since the *same prior* distribution is assumed in every case yielding the same posterior, how is the prior to blame for the alleged inefficiency of one of the estimators? Or why on this reasoning should it not be praised for the efficiency of the other two?

We are neither encouraged by the fact that based on the reference prior in (5.2.9), some characteristics of the posterior distribution in (5.2.14) such as the mode has desirable sampling properties, nor are we dismayed by the larger M.S.E. of the corresponding posterior mean of σ_2^2. One simply should not confuse point estimation with inference.

Inferences about a parameter are provided not by some arbitrarily chosen characteristic of the posterior distribution, but by the *entire* distribution itself. One of the first lessons we learn in Statistics is that the mean is but one possible measure of the location of a distribution. While it can be an informative measure in certain cases such as the Normal, it can be of negligible value for others.

Other Examples of the Inefficiency of the Posterior Mean

For the problem of the Normal mean with known variance σ^2, the posterior distribution of θ based on the uniform reference prior is $N(\bar{y}, \sigma^2/n)$. As mentioned earlier, this distribution allowed us to make inferences not only about θ but also about any transformation of θ such as e^θ or θ^{-1}. While the posterior mean of θ, $\bar{y}$, is a useful and important characteristic of the Normal posterior, the posterior mean of e^θ, $e^{\bar{y}+\frac{1}{2}\sigma^2/n}$, tells us much less about the log Normal posterior, and the posterior mean of θ^{-1}, since it fails to exist, tell us *nothing* about the distribution of θ^{-1}! Thus, so far as inference is concerned, it is irrelevant whether some *arbitrarily chosen* characteristic of the posterior has or does not have some allegedly desirable sampling property. In particular, it is irrelevant that $\bar{y}$ is the M.M.S.E.

estimator for θ, or that for e^θ, as can be readily shown, the posterior mean $e^{\bar{y}+\frac{1}{2}\sigma^2/n}$ has larger M.S.E. than that of the posterior mode $e^{\bar{y}-\sigma^2/n}$ (which, for that matter, has larger M.S.E. than that of the estimator $e^{\bar{y}-1.5\sigma^2/n}$). Furthermore, the noninformative prior should not be to blame for causing the heavy right tail of the log Normal posterior for e^θ, nor should the same prior be praised for producing the symmetric Normal posterior for θ. It would certainly seem illogical to "judge" the appropriateness of the posterior and therefore of the prior on such grounds.

Adjusting Prior to Produce "Efficient" Posterior Mean

In view of all this, the analysis given in Portnoy's paper we find even more perplexing. He proposes the scale invariant loss function

$$L(\hat{\phi}, \sigma_2^2) = \frac{(\hat{\phi} - \sigma_2^2)^2}{4(\sigma_1^2 + K\sigma_2^2)^2} \tag{A5.6.2}$$

and proceeds to obtain Bayes estimators of σ_2^2 with respect to the class of priors

$$p(\sigma_1^2, \sigma_2^2) \propto (\sigma_1^2)^{a-(b+1)} (\sigma_1^2 + K\sigma_2^2)^b. \tag{A5.6.3}$$

For given (a, b), the estimator is the ratio of the posterior expectations

$$\hat{\phi}_{(a,b)} = \underset{\sigma_2^2|\mathbf{y}}{E} \left\{ \frac{\sigma_2^2}{(\sigma_1^2 + K\sigma_2^2)^2} \right\} \bigg/ \underset{\sigma_2^2|\mathbf{y}}{E} (\sigma_1^2 + K\sigma_2^2)^{-2}.$$

Using the "admissibility" proofs similar to James and Stein (1961), he concludes that the best estimator is the one corresponding to $(a = -1, b = -1)$ in (A5.6.3) which is then *precisely our noninformative prior.* His computation shows that the M.S.E. of $\hat{\phi}_{(-1,-1)}$ is much smaller than the corresponding posterior mean.

This result is certainly not surprising if we consider the similar but algebraically much simpler problem of the Normal variance σ^2 with the mean θ known. As we have discussed in Section 2.3, the posterior distribution of σ^2 based on the noninformative prior $p(\sigma^2) \propto \sigma^{-2}$ is $ns^2\chi_n^{-2}$ where $ns^2 = \Sigma(y_i - \theta)^2$. It is readily seen that the Bayes estimator with respect to the scale invariant loss function

$$L(\hat{\phi}, \sigma^2) = \frac{(\hat{\phi} - \sigma^2)^2}{\sigma^4}$$

is then

$$\hat{\phi} = \underset{\sigma^2|\mathbf{y}}{E} (\sigma^{-2}) \bigg/ \underset{\sigma^2|\mathbf{y}}{E} (\sigma^{-4}) = \frac{ns^2}{n+2}$$

which, as is well known, is the minimum M.S.E. estimator of σ^2 among all function of s^2 having the form cs^2. In particular, the M.S.E. of $\hat{\phi}$ is smaller than that of the posterior mean $ns^2/(n-2)$.

What is extraordinary is the conclusions Portnoy draws from his study. Using our notation, he says that

"For the case of the present estimator, $\hat{\phi}_{(-1,-1)}$, we actually find that the prior distribution (for $a = -1$, $b = -1$) corresponds exactly to the Jeffreys' prior,

$\sigma_1^{-2}(\sigma_1^2 + K\sigma_2^2)^{-1}$. However, $\hat{\phi}_{(-1,-1)}$ is not the posterior mean, but the posterior expected value of $\sigma_2^2/(\sigma_1^2 + K\sigma_2^2)^2$ over the posterior expected value of $1/(\sigma_1^2 + K\sigma_2^2)^2$, the denominator coming from the normalizing factor in the loss function. This is the same as taking the posterior mean for the prior $\sigma_1^{-2}(\sigma_1^2 + K\sigma_2^2)^{-3}$. The reasonable size of the mean squared error of $\hat{\phi}_{(-1,-1)}$ shows that this latter posterior distribution is centred at least as well and has substantially smaller variance (on the average) than the posterior for the Jeffreys' prior. Thus, if one wishes to make inferences based on a posterior distribution, one can seriously recommend using the prior $\sigma_1^{-2}(\sigma_1^2 + K\sigma_2^2)^{-3}$ instead of the Jeffreys' prior. This serves to justify, in my opinion, the use of squared error as loss: by using squared error and by taking an appropriate limit of what might be called Bayes invariant priors, one is assured of finding a posterior distribution which, on the average, is reasonably centered and has reasonably small variance. As this example shows, the Jeffreys' prior can lead to posterior distributions with mean and variance far too large." (*Ann. Math. Statist.* **42**, p.1394).

This seems to be suggesting that in order to obtain a good estimator with respect to the scale invariant loss function in (A5.6.2), Jeffreys' prior is needed, but the use of such a prior is, at the same time, undesirable in making inferences because it leads to a posterior mean having large mean squared error. To conclude that this study serves to justify "the use of squared error loss" is indeed difficult to follow.

To see where this sort of argument leads, consider again the problem of the Normal variance σ^2 with known mean. Let the prior distribution be of the form

$$p(\sigma^2) \propto (\sigma^2)^{-\alpha}.$$

Suppose we wish to estimate some power of σ^2, say σ^{2p}. Now for different values of α, a family of posterior means for σ^{2p} can be generated. If we were to insist on a squared error loss function and hence on the posterior mean as the "best" point estimator of σ^{2p}, then it is shown in Appendix 5.7 that the estimator which minimizes the M.S.E. is the one for which the prior would be such that

$$\alpha = 2p + 1,$$

that is,

$$p(\sigma^2) \propto (\sigma^2)^{-(2p+1)} \qquad \text{(A5.6.4)}$$

provided

$$n > -4p.$$

The implications would then be: (i) if we were interested in σ^2 ($p = 1$), the prior should be proportional to $(\sigma^2)^{-3}$ which is similar to the prior suggested by Portnoy for estimating σ_2^2; (ii) on the other hand, if $p = 0$ corresponding to $\log \sigma$, then we would use the noninformative prior σ^{-2}; but (iii) if we should be interested in σ^{-100} ($p = -50$), then we ought to employ a prior which is proportional to σ^{198}!

To put the matter in another way, the posterior distribution of σ^2 corresponding to (A5.6.4) would then be (for $n > -4p$)

$$p(\sigma^2 \mid \mathbf{y}) \propto (\sigma^2)^{-[\frac{1}{2}(n+4p)+1]} \exp\left\{-\frac{ns^2}{2\sigma^2}\right\}, \qquad \text{(A5.6.5)}$$

$$\propto (\sigma^2)^{-(\frac{1}{2}n'+1)} \exp\left\{-\frac{n's'^2}{2\sigma^2}\right\}, \qquad \sigma^2 > 0.$$

with $n' = n + 4p$ and $s'^2 = ns^2/n'$, which would be the same as that resulting from Jeffreys' noninformative prior but with $n' = n + 4p$ observations. Noting that $n's'^2/\sigma^2 \sim \chi^2_{n'}$, so that $(n's'^2)^p/\sigma^{2p} \sim \{\chi^2_{n'}\}^p$ this would mean for example that if we wish to estimate $(\sigma^2)^{10}$ from a sample of, say, $n = 8$ observations, we would have to base our inferences on a posterior distribution which would correspond to a *confidence distribution* appropriate for 48 observations. On the other hand, if we were interested in estimating $(\sigma^2)^{-10}$ from a sample of $n = 48$ observations, the posterior we would use would be numerically equivalent to a confidence distribution relative to only 8 observations.

Conclusions

The principle that the prior, once decided, should be consistent under transformation of the parameter seems to us sensible. By contrast, the suggestion that to allow for arbitrariness in the methods of obtaining point estimators, the principle be abandoned in favour of freely varying the prior wherever a different function is to be estimated seems less so.

APPENDIX A5.7

Mean Squared Error of the Posterior Means of σ^{2p}

In this appendix, we prove the result stated in (A5.6.4). That is, suppose we have a random sample of n observations from the Normal population $N(\theta, \sigma^2)$ where σ^2 is known, and supposed that the prior distribution of σ^2 takes the form

$$p(\sigma^2) \propto (\sigma^2)^{-\alpha}, \qquad \text{(A5.7.1)}$$

then among the class of estimators of σ^{2p} which are the posterior means of σ^{2p}, the estimator which minimizes the M.S.E. is obtained by setting the prior to be such that

$$\alpha = 2p + 1 \qquad \text{(A5.7.2)}$$

provided $n > -4p$.

Corresponding to the prior (A5.7.1), the posterior distribution of σ^2 is

$$p(\sigma^2 \mid \mathbf{y}) = \frac{(ns^2)^{\frac{1}{2}v}}{\Gamma(\frac{1}{2}v)2^{\frac{1}{2}v}} (\sigma^2)^{-(\frac{1}{2}v+1)} e^{-(ns^2/2\sigma^2)}, \qquad \sigma^2 > 0, \qquad \text{(A5.7.3)}$$

where
$$\nu = n + 2(\alpha - 1) > 0.$$

Thus, using (A2.1.2) in Appendix A2.1, the posterior mean of σ^{2p} is,
$$E(\sigma^{2p} \mid \mathbf{y}) = c(ns^2)^p,$$
where
$$c = \frac{\Gamma(\frac{1}{2}\nu - p)}{\Gamma(\frac{1}{2}\nu)} (\tfrac{1}{2})^p, \qquad \text{(A5.7.4)}$$
provided $\frac{1}{2}\nu - p > 0$. Now, the sampling distribution of ns^2 is $\sigma^2\chi_n^2$. It follows that the M.S.E. of $c(ns^2)^p$ is, for $(\frac{1}{2}n + 2p) > 0$
$$\text{M.S.E.} = \underset{\mathbf{y}}{E}\,[c(ns^2)^p - \sigma^{2p}]^2 = \sigma^{2p}\left[1 + c^2\,\frac{\Gamma(\frac{1}{2}n + 2p)2^{2p}}{\Gamma(\frac{1}{2}n)} - 2c\,\frac{\Gamma(\frac{1}{2}n + p)2^p}{\Gamma(\frac{1}{2}n)}\right]. \qquad \text{(A5.7.5)}$$

Letting
$$g(\nu) = \left(\frac{\text{M.S.E.}}{\sigma^{2p}} - 1\right)\Gamma(\tfrac{1}{2}n)$$
we have
$$g(\nu) = \left[\frac{\Gamma(\frac{1}{2}\nu - p)}{\Gamma(\frac{1}{2}\nu)}\right]^2 \Gamma(\tfrac{1}{2}n + 2p) - 2\frac{\Gamma(\frac{1}{2}\nu - p)}{\Gamma(\frac{1}{2}\nu)}\Gamma(\tfrac{1}{2}n + p). \qquad \text{(A5.7.6)}$$

Differentiating with respect to ν and setting the derivative to zero, we obtain
$$\Gamma(\tfrac{1}{2}n + 2p)\,\{\Gamma(\tfrac{1}{2}\nu - p)\,\Gamma'(\tfrac{1}{2}\nu - p) - [\Gamma(\tfrac{1}{2}\nu - p)]^2\,\Gamma'(\tfrac{1}{2}\nu)/\Gamma(\tfrac{1}{2}\nu)\}$$
$$- \Gamma(\tfrac{1}{2}n + p)[\Gamma(\tfrac{1}{2}\nu)\,\Gamma'(\tfrac{1}{2}\nu - p) - \Gamma(\tfrac{1}{2}\nu - p)\,\Gamma'(\tfrac{1}{2}\nu)] = 0, \qquad \text{(A5.7.7)}$$
where $\Gamma'(x)$ is the digamma function. The solution is
$$\nu = n + 4p \qquad \text{or} \qquad \alpha = 2p + 1.$$

By differentiating $g'(\nu)$ it can be verified that this solution minimizes the M.S.E.

CHAPTER 6

ANALYSIS OF CROSS CLASSIFICATION DESIGNS

6.1 INTRODUCTION

In the previous chapter we have discussed random effect models for hierarchical designs. In some situations it is appropriate to run *cross classification* designs and it may be appropriate to assume either

a) that all factors have random effects,
b) that all factors have fixed effects, or
c) that some factors have random effects and some have fixed effects.

We begin with a general discussion of the situation when there are just two factors and for illustration we suppose the factors to be car drivers and cars.

6.1.1 A Car–Driver Experiment: Three Different Classifications

Consider an experiment in which the response of interest was the performance, as measured by mileage per gallon of gasoline, achieved in driving a new Volkswagen sedan car. Suppose that

a) eight drivers and six cars were used in the investigation,
b) each driver drove each car twice, and
c) the resulting 96 trials were run in random order.

The above description of the investigation is, by itself, inadequate to determine an appropriate analysis for the results. To decide this a number of other things must be known. Most important are:

a) the *objectives* of the investigation,
b) the *design of the experiment* to achieve these objectives, in this example the method of selection of drivers and cars.

Objectives: The Particular or the General?

In one situation the objective may be to obtain information about the *mean* performance of each of the *particular* six cars and eight drivers included in the experiment. Comparisons between such means are sometimes called "fixed effects" and the analysis is conducted in terms of a *fixed effect* model.

In another situation the individual cars and/or drivers in the experiment may be of no interest in themselves. The objective may be to use them only to make inferences about the *populations* of cars and/or drivers from which they are presumed to be randomly drawn. In particular, the *variances* (and possibly certain covariances) of these populations will be of special interest. In this situation our analysis is conducted in terms of a *random effect* model.

In still another situation the objective may be to learn about the mean performance of each car as it performs in the hands of a *population* of drivers. In this case the analysis is conducted in terms of a *mixed* (random and fixed effect) model.

With the *two factor* cross classification then, three situations can be distinguished which are associated with specific models as follows:

a) both factors random—two-way random effect model,

b) one factor random, one factor fixed—two-way mixed model, and

c) both factors fixed—two-way fixed effect model.

Design Considerations

So far we have talked only of objectives but it is important that the experiment be conducted so that it is possible to achieve these objectives. In the random effect model, interest centers on the characteristics of *populations*, and in particular, on their variance components. To estimate such characteristics the particular cars or drivers included in the experiment must be *random selections* from the *relevant populations* of cars and drivers. For example, the two-way random effect setup would be appropriate if the six VW sedan cars and the eight drivers were randomly drawn from relevant and definable populations. The sampled population of cars might be all new VW sedans available for sale in the State of Wisconsin in a particular week. The sampled population of drivers might be all students at the University of Wisconsin-Madison prepared to take part in the experiment for a payment of $20.

It will often happen, that the population one would really like to make inferences about is different from the population that can be conveniently sampled. For instance, we may really be interested in the whole population of drivers in the United States, but may settle for the student drivers to produce a feasible experiment. In such a case it must be remembered that the *statistical* conclusions apply only to the limited population sampled. Any extrapolation to a wider population is based solely on an *opinion* that, for example, "student drivers are similar to other drivers."

Questions concerning the car–driver populations which this kind of experiment might answer are:

1. How large a variance in gas mileage might be experienced by different drivers operating new VW sedan cars.

2. What total proportion of this total variance was associated with: (a) variation in cars, (b) variation in drivers, (c) interaction between cars and drivers, and (d) unassignable variation.

By contrast, the fixed effect setup implies that our interest centers on the mean performance of each of the six particular cars or the eight particular drivers which may or may not be realistically considered as relevant random samples from specific populations.

The present discussion uses the familiar sampling-theory terms, "random" and "fixed" effects. As we shall see later, in the Bayesian framework, where of course all effects (parameters) are random variables, appropriate choice of model forms with suitable prior distributions achieves the different inferential objectives.

Classification of the Data in a Two-way Table

Suppose in general that the test conducted with I cars denoted by $(1, \ldots, i, \ldots, I)$ and J drivers denoted by $(1, \ldots, j, \ldots, J)$ and every driver operates every car K times $(K \geqslant 1)$. Let y_{ijk} be the kth $(k = 1, \ldots, K)$ observation for the jth driver on the ith car. The data can be set out in an $I \times J$ arrangement with K observations per cell as in Table 6.1.1. In the example considered above there are $I = 6$ cars, $J = 8$ drivers, and $K = 2$ replicate runs.

Table 6.1.1

Arrangement of observations in a two-way cross classification design

	Drivers 1	. . .	j	. . .	J
Cars 1					
. . .					
i			y_{ij1} ⋮ y_{ijk} ⋮ y_{ijK}		
. . .					
J					

Whichever setup is being considered we can always write

$$y_{ijk} = \theta + \delta_{ij} + e_{ijk}. \tag{6.1.1}$$

In this equation θ is the overall mean gas mileage, δ_{ij} is the mean increase or decrease in gas mileage achieved when the jth driver is operating the ith car, and e_{ijk} is the experimental error which is responsible for differences between replicate runs performed by the same driver in the same car. It will be supposed throughout that these experimental errors are distributed independently of each other, and of the δ_{ij}, with zero mean and variance σ_e^2.

6.1.2 Two-way Random Effect Model

We now consider the formulation of an appropriate model when both factors are random.

Classification of the Population in a Two-way Table

We can picture the whole population of possible combinations of cars and drivers as given by the cells in the $\mathscr{I} \times \mathscr{J}$ table of Table 6.1.2 having all possible columns

Table 6.1.2

Two-way population of increments δ_{ij}

Drivers

		1	.	.	j	.	.	.	$\mathscr{J}$	Row (car) means
Cars	1									r_1
	.									.
	.									.
	i				δ_{ij}					r_i
	.									.
	.									.
	$\mathscr{I}$									$r_{\mathscr{I}}$
Column (driver) means		c_1			c_j				$c_{\mathscr{J}}$	0

(drivers) indexed by $j = 1, 2, \ldots, \mathcal{J}$ and all possible rows (cars) indexed by $i = 1, 2, \ldots, \mathcal{I}$. Then δ_{ij} will be the mean increment associated with the ith car operated by the jth driver. That is, the amount by which the average gas mileage achieved by this car–driver combination exceeds or falls short of the overall population mean θ.

It is now convenient to define the row means and column means of the δ_{ij} in this *population* table as the (random) effects for cars and drivers respectively and the "nonadditive" increment as the (random) interaction effect.

Thus,

$$r_i = \delta_{i.} = \text{random effect for } i\text{th row (car)},$$

$$c_j = \delta_{.j} = \text{random effect for } j\text{th column (driver)},$$

$$t_{ij} = \delta_{ij} - \delta_{i.} - \delta_{.j} = \text{random interaction effect (car} \times \text{driver)},$$

$$\delta_{ij} = r_i + c_j + t_{ij},$$

where, as before, we use the notation that a dot replacing a subscript indicates the average over that subscript. The identity

$$\Sigma\Sigma\,\delta_{ij}^2 = \mathcal{J}\,\Sigma\,\delta_{i.}^2 + \mathcal{I}\,\Sigma\,\delta_{.j}^2 + \Sigma\Sigma\,(\delta_{ij} - \delta_{i.} - \delta_{.j})^2$$

can then be written

$$\Sigma\Sigma\,\delta_{ij}^2 = \mathcal{J}\,\Sigma\,r_i^2 + \mathcal{I}\,\Sigma\,c_j^2 + \Sigma\Sigma\,t_{ij}^2. \tag{6.1.2}$$

If we now define the variance components as

$$\sigma_r^2 = \frac{\Sigma\,r_i^2}{\mathcal{I}} \quad \text{row (car) component,}$$

$$\sigma_c^2 = \frac{\Sigma\,c_j^2}{\mathcal{J}} \quad \text{column (driver) component,} \tag{6.1.3}$$

$$\sigma_t^2 = \frac{\Sigma\,t_{ij}^2}{\mathcal{I}\mathcal{J}} \quad \text{interaction (car–driver) component.}$$

Then letting

$$\sigma^2 = \frac{\Sigma\Sigma\,\delta_{ij}^2}{\mathcal{I}\mathcal{J}}$$

we obtain

$$\sigma^2 = \sigma_r^2 + \sigma_c^2 + \sigma_t^2. \tag{6.1.4}$$

On this formulation, then, we can substitute $\delta_{ij} = r_i + c_j + t_{ij}$ in (6.1.1). The model for the actual test data in the 6×8 replicated array of Table 6.1.1 is thus

$$y_{ijk} = \theta + r_i + c_j + t_{ij} + e_{ijk}.$$

We now consider how the r_i, c_j, and t_{ij} in the equation are to be interpreted.

The random effect model assumes that the $I = 6$ rows (cars) were selected randomly from a population of $\mathscr{I}$ rows (cars) and the $J = 8$ columns (drivers) were similarly selected from a population of $\mathscr{J}$ columns (drivers). We show in the next section that, assuming the populations to be large, this random sampling ensures that the r_i, c_j, and t_{ij} may all be treated as *uncorrelated* random variables. Also the errors e_{ijk} are by assumption uncorrelated with the δ_{ij}.

Thus finally the two-way random effect model may be written

$$y_{ijk} = \theta + r_i + c_j + t_{ij} + e_{ijk}; \qquad i = 1, \ldots, I; \quad j = 1, \ldots, J; \quad k = 1, \ldots, K, \tag{6.1.5}$$

where r_i, c_j, t_{ij}, e_{ijk} are all uncorrelated random variables with zero means and

$$E(r_i^2) = \sigma_r^2, \qquad E(c_j^2) = \sigma_c^2, \qquad E(t_{ij}^2) = \sigma_t^2, \qquad E(e_{ijk}^2) = \sigma_e^2, \tag{6.1.6}$$

so that

$$\operatorname{Var}(y_{ijk}) = \sigma_r^2 + \sigma_c^2 + \sigma_t^2 + \sigma_e^2. \tag{6.1.7}$$

In the traditional sampling-theory analysis of the problem, this setup would be appropriate for making inferences about the behavior of an infinite population of drivers and cars, but not about the particular drivers and cars tested. The problem might arise for example in a general production study by a large manufacturer of motor cars. The conclusions drawn could indicate the degree of variability of gas mileage over the relevant populations of drivers and cars and the relative contribution of cars, drivers, and interaction to that overall variability.

Justification of the Two-way Random Effect Model

To show that the r_i, c_j, and t_{ij} in (6.1.5) are uncorrelated random variables with the properties just mentioned we regard the particular small sets of I rows (cars) and J columns (drivers), actually under study, as random samples (without replacement) from the $\mathscr{I}$ rows and $\mathscr{J}$ columns of Table 6.1.2. The quantities δ_{ij} in (6.1.1) are then random variables. It can be readily shown that

$$E(\delta_{ij}) = 0, \qquad E(\delta_{ij}^2) = \sigma^2,$$

$$E(\delta_{ij}\delta_{i\mathscr{j}}) = \sigma_r^2 + \frac{1}{(\mathscr{J} - 1)}(\sigma_r^2 - \sigma^2), \qquad j \neq \mathscr{j}, \tag{6.1.8}$$

$$E(\delta_{ij}\delta_{\mathscr{i}j}) = \sigma_c^2 + \frac{1}{(\mathscr{I} - 1)}(\sigma_c^2 - \sigma^2), \qquad i \neq \mathscr{i},$$

and

$$E(\delta_{ij}\delta_{\mathscr{i}\mathscr{j}}) = \frac{\sigma^2}{(\mathscr{I} - 1)(\mathscr{J} - 1)} - \frac{\mathscr{J}}{(\mathscr{I} - 1)(\mathscr{J} - 1)}\sigma_c^2 - \frac{\mathscr{I}}{(\mathscr{I} - 1)(\mathscr{J} - 1)}\sigma_r^2,$$

$$i \neq \mathscr{i},\ j \neq \mathscr{j}; \quad i, \mathscr{i} = 1, \ldots, I; \quad j, \mathscr{j} = 1, \ldots, J.$$

On the assumption that the populations are large, we have

$$E(\delta_{ij}\delta_{ij'}) = \sigma_r^2, \qquad E(\delta_{ij}\delta_{i'j}) = \sigma_c^2, \qquad \text{and} \qquad E(\delta_{ij}\delta_{i'j'}) = 0,. \tag{6.1.9}$$

If $\mathbf{I}_J$ is a $J \times J$ identity matrix and $\mathbf{1}_J$ is a $J \times 1$ vector of ones, then the $IJ \times IJ$ covariance matrix of the quantities δ_{ij} is given by

$$\operatorname{Cov}(\delta_{11}, \ldots, \delta_{1J}, \ldots, \delta_{I1}, \ldots, \delta_{IJ})$$

$$= \left[\begin{array}{c:c:c:c} \mathbf{I}_J(\sigma_c^2 + \sigma_t^2) + \mathbf{1}_J\mathbf{1}_J'\sigma_r^2 & \mathbf{I}_J\sigma_c^2 & \cdots & \mathbf{I}_J\sigma_c^2 \\ \hdashline \mathbf{I}_J\sigma_c^2 & \mathbf{I}_J(\sigma_c^2 + \sigma_t^2) + \mathbf{1}_J\mathbf{1}_J'\sigma_r^2 & \ddots & \vdots \\ \hdashline \vdots & \ddots & \ddots & \vdots \\ \hdashline \mathbf{I}_J\sigma_c^2 & \cdots & \ddots & \ddots \end{array}\right]. \tag{6.1.10}$$

The form of this matrix confirms the appropriateness of the formulation set out in (6.1.5) through (6.1.7).

6.1.3 Two-way Mixed Model

The same car–driver data could arise in quite different circumstances. Suppose a car rental agency has a particular set of $I = 6$ cars, not necessarily all the same model or make, and it is desired to estimate their individual mean performances when operated by a relevant population of drivers. To do this a test is conducted with $J = 8$ drivers randomly chosen from the population of interest. The underlying setup is represented by the elements in Table 6.1.3, which can be thought of as sub-population of Table 6.1.2 in which are selected the I *particular* rows (cars) of interest.

Table 6.1.3

A sub-population of I particular rows of the population of increments δ_{ij}

		DRIVERS $1 \; . \; . \; . \; \mathscr{j} \; . \; . \; . \; \mathscr{J}$	
	1	$\delta_{1\mathscr{j}}$	r_1
	⋮	⋮	⋮
CARS	i	$\delta_{i\mathscr{j}}$	r_i
	⋮	⋮	⋮
	I	$\delta_{I\mathscr{j}}$	r_I

The behavior of the increments of the I cars with, say, the $\tilde{j}$th driver, is thus represented by an I-variate observation $\boldsymbol{\delta}'_{\tilde{j}} = (\delta_{1\tilde{j}}, \ldots, \delta_{I\tilde{j}})$ from a large population of $\mathscr{J}$ drivers with

$$E(\boldsymbol{\delta}_{\tilde{j}}) = \mathbf{r} = (r_1, \ldots, r_I)',$$

$$\operatorname{Cov}(\boldsymbol{\delta}_{\tilde{j}}) = \mathbf{V}_I = \{\sigma_{i\tilde{i}}\}, \qquad i, \tilde{i} = 1, \ldots, I, \tag{6.1.11}$$

where

$$\sigma_{ii} = \frac{1}{\mathscr{J}} \Sigma(\delta_{i\tilde{j}} - r_i)^2,$$

$$\sigma_{i\tilde{i}} = \frac{1}{\mathscr{J}} \Sigma(\delta_{i\tilde{j}} - r_i)(\delta_{\tilde{i}\tilde{j}} - r_{\tilde{i}}).$$

Further, it can be verified that when $\mathscr{J} \to \infty$ the increments associated with two different drivers are uncorrelated, that is,

$$\operatorname{Cov}(\delta_{\tilde{i}j}, \delta_{i\tilde{j}}) = 0, \qquad j \neq \tilde{j}.$$

The model in (6.1.1) can then be written

$$\mathbf{y}_{jk} = \theta \mathbf{1}_I + \boldsymbol{\delta}_j + \mathbf{e}_{jk}, \qquad j = 1, \ldots, J; \quad k = 1, \ldots, K, \tag{6.1.12}$$

where $\mathbf{y}'_{jk} = (y_{1jk}, \ldots, y_{Ijk})$ and $\mathbf{e}'_{jk} = (e_{1jk}, \ldots, e_{Ijk})$. Letting

$$\boldsymbol{\theta}' = (\theta_1, \ldots, \theta_i, \ldots, \theta_I),$$

with

$$\theta_i = \theta + r_i, \tag{6.1.13}$$

we have finally

$$E(\mathbf{y}_{jk}) = \boldsymbol{\theta}, \qquad \operatorname{Cov}(\mathbf{y}_{jk}) = \mathbf{V}_I + \mathbf{I}\sigma_e^2. \tag{6.1.14}$$

The principal objective here is to make inferences about the mean vector $\boldsymbol{\theta}$ which determines the mean performances of the cars. However the meaning to be attached to the individual elements of the covariance matrix $\mathbf{V}_I$ of the δ's is also worth considering. The ith diagonal element σ_{ii} of this matrix represents the variation in performance when the ith car is operated by the whole population of drivers. The covariance $\sigma_{i\tilde{i}}$ may be written $\sigma_{i\tilde{i}} = \rho_{i\tilde{i}}(\sigma_{ii}\sigma_{\tilde{i}\tilde{i}})^{1/2}$ where $\rho_{i\tilde{i}}$ measures the correlation between performances achieved on cars i and $\tilde{i}$ by the population of drivers. If the two cars are similar, the driver who does well on one will usually do well on the other so that $\rho_{i\tilde{i}}$ would be positive. For two cars with widely different characteristics, $\rho_{i\tilde{i}}$ may be a quite small positive number or even negative. For example, if car i was easily manipulated by right-handed people but not by left-handed people while car $\tilde{i}$ was easily manipulated by left-handed people but not by right-handed people, then $\rho_{i\tilde{i}}$ could be negative. We shall see in the next section that by use of this correlation structure we can allow for "interactions" of a much more sophisticated kind than is possible with what has been the customary assumptions about "interaction variance." Indeed the customary assumptions correspond to a very restricted special case of the general model given here.

6.1.4 Special Cases of Mixed Models

The mixed model set out in (6.1.12) involves I unknown means θ_i, $\frac{1}{2}I(I-1)+I$ unknown elements in the covariance matrix $\mathbf{V}_I$ and an unknown σ_e^2, as parameters. Situations occur where simpler models are appropriate. Two special cases may be of interest.

1. *Additive Model*

What sometimes is called the *additive mixed model* is obtained if we assume that

$$\mathbf{V}_I = \sigma_c^2 \mathbf{1}_I \mathbf{1}_I'. \tag{6.1.15}$$

In terms of the entries in Table 6.1.3, this is saying that

$$\delta_{ij} - r_i = \delta_{i'j} - r_{i'} \quad \text{so that} \quad \sigma_{ii} = \sigma_{ii'} = \sigma_c^2, \quad \text{all} \quad (i, i').$$

In this case, the model (6.1.1) may be written in the form

$$y_{ijk} = \theta_i + c_j + e_{ijk}, \tag{6.1.16}$$

where c_j and e_{ijk} are uncorrelated random variables with zero means and $E(c_j^2) = \sigma_c^2$, $E(e_{ijk}^2) = \sigma_e^2$ so that $E(y_{ijk}) = \theta_i$ and $\text{Var}(y_{ijk}) = \sigma_c^2 + \sigma_e^2$. If for the moment we ignore the experimental error e_{ijk} the model can be represented as in Fig. 6.1.1 for $I = 2$, $J = 3$.

For the driver–car example, c_j is the incremental performance for the jth driver. It measures his ability to get a little more or little less gas mileage. The

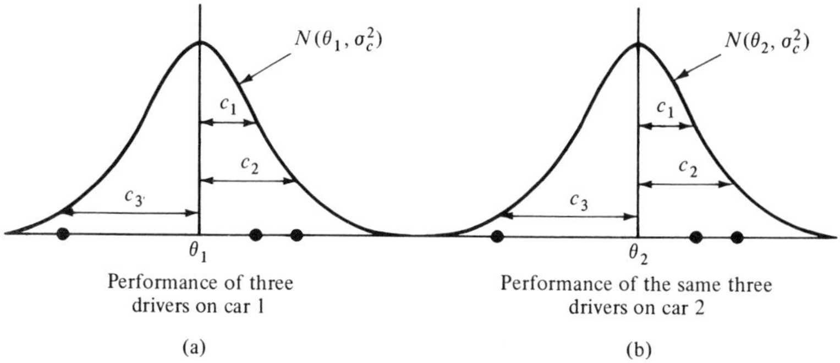

Fig. 6.1.1 Graphical illustration of an additive model.

additive model will be appropriate if, apart from experimental error, this incremental performance c_j associated with the jth driver remains the same for all I cars which he drives (that is, $c_j = \delta_{ij} - r_i = \delta_{i'j} - r_{i'}$). In choosing a random sample of drivers we will choose a random sample of incremental performance so that the $c_1, c_2, \ldots, c_J$ can be taken as a random sample from a distribution (for example, the Normal shown in Fig. 6.1.1) with zero mean and fixed variance σ_c^2.

This structure is sometimes adopted in modelling data generated by Fisher's randomized block designs. From this point of view we would here be testing I treatments (cars) in J randomly chosen blocks (drivers).

2. *Interaction Model*

Another special case arises when the covariance matrix $\mathbf{V}_I$ takes the form

$$\mathbf{V}_I = [(1 - \rho)\mathbf{I}_I + \rho \mathbf{1}_I \mathbf{1}_I']\phi^2 \tag{6.1.17}$$

where $|\rho| < 1$.

In terms of the entries in Table 6.1.3, this says that, the I elements $(\delta_{1j}, \ldots, \delta_{Ij})$ are equi-correlated and have common variance, that is,

$$\frac{1}{\mathscr{J}} \Sigma(\delta_{ij} - r_i)^2 = \phi^2,$$

$$\frac{1}{\mathscr{J}} \Sigma(\delta_{ij} - r_i)(\delta_{i'j} - r_{i'}) = \phi^2 \rho, \qquad i, i' = 1, \ldots, I; \quad i \neq i'. \tag{6.1.18}$$

If we further assume that $\rho > 0$, the covariance matrix in (6.1.17) can be written

$$\mathbf{V}_I = \sigma_c^2 \mathbf{1}_I \mathbf{1}_I' + \sigma_t^2 \mathbf{I}_I, \tag{6.1.19}$$

where $\sigma_c^2 = \phi^2 \rho$ and $\sigma_t^2 = \phi^2(1 - \rho)$. The model in (6.1.1) can be equivalently expressed as

$$y_{ijk} = \theta_i + c_j + t_{ij} + e_{ijk}, \tag{6.1.20}$$

where c_j, t_{ij}, and e_{ijk} are uncorrelated variables with zero means and

$$E(c_j^2) = \sigma_c^2, \qquad E(t_{ij}^2) = \sigma_t^2, \qquad \text{and} \qquad E(e_{ijk}^2) = \sigma_e^2$$

so that

$$E(y_{ijk}) = \theta_i \qquad \text{and} \qquad \operatorname{Var}(y_{ijk}) = \sigma_c^2 + \sigma_t^2 + \sigma_e^2.$$

This model is sometimes known as the *mixed model with interaction.* When $\sigma_t^2 = 0$, it reduces to the additive mixed model in (6.1.16).

The implications of this interaction model are illustrated in Fig. 6.1.2. Ignoring the measurement error e_{ijk} the two dots in Fig. 6.1.2(a) show the contribution $\theta_1 + c_1 + t_{11}$ and $\theta_1 + c_2 + t_{12}$ for drivers 1 and 2 using car 1. The quantities θ_1 and θ_2 are fixed constants, c_1 and c_2 are drawn from a Normal distribution

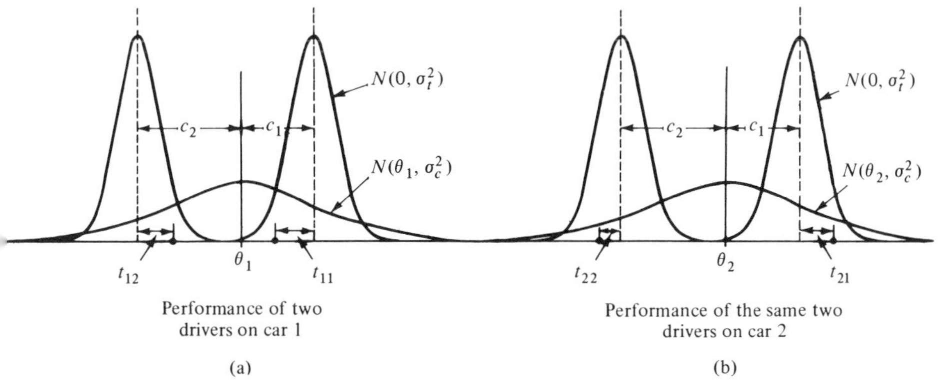

Fig. 6.1.2 Graphical illustration of an interaction model.

having variance σ_c^2 and t_{11} and t_{12} are random drawings from a Normal distribution having variance σ_t^2. Figure 6.1.2(b) shows the situation for the same two drivers operating a second car. Note that although a shift in mean for θ_1 to θ_2 has occured, c_1 and c_2 are assumed to remain the same as before but a different sample of t's is added.

This interaction model is less restrictive than the additive mixed model because it allows the possibility that an increment associated with a given driver will be different on different cars. However, because the additional increments t_{ij} are assumed to be independent and to have the same variance for each driver and each car, it is still too restrictive to represent many real situations.

For example, consider two particular cars, say car 1 and car 2, which were included in the experiment. According to the model, the increments associated with the J drivers would be as follows:

	Driver 1		Driver j		Driver J
Car 1	$c_1 + t_{11}$	...	$c_j + t_{1j}$	...	$c_J + t_{1J}$
Car 2	$c_1 + t_{21}$	...	$c_j + t_{2j}$	...	$c_J + t_{2J}$

Now suppose that the particular cars 1 and 2 were very much alike in some important respect but differed from the remaining cars. Thus, these particular cars might have less leg room for the driver than the remaining $I - 2$ cars which were tested. Then we would expect that the incremental performance c_j of a short legged driver j would be enhanced by a factor t_{1j} almost identical to t_{2j} while the incremental performance of a long legged driver j would be reduced almost equally and the negative contributions t_{1j} and t_{2j} would also be almost equal. Therefore, for these almost identical cars we would expect t_{1j} to be the same as t_{2j} for all j. In other words, we would expect t_1 and t_2 to be highly *correlated* within drivers and not uncorrelated as required by the model. The assumption of common variance for t_{ij} might also be unrealistic because, for example, the differences between drivers might be emphasized or de-emphasized by particular cars.

6.1.5 Two-way Fixed Effect Model

In some situations, no question would arise of attempting to deduce general properties of either the population of drivers or the population of cars. Interest might center on the performance of a particular set of I cars with a particular group of J drivers. In this case, setting

$$\alpha_i = \delta_{i.} - \delta_{..}, \qquad \beta_j = \delta_{.j} + \delta_{..}, \qquad \gamma_{ij} = \delta_{ij} - \delta_{i.} - \delta_{.j} + \delta_{..}, \qquad \bar{\theta} = \theta + \delta_{..}, \tag{6.1.21}$$

where

$$\delta_{i.} = \frac{1}{J}\,\Sigma\,\delta_{ij}, \qquad \delta_{.j} = \frac{1}{I}\,\Sigma\,\delta_{ij}, \qquad \delta_{..} = \frac{1}{IJ}\,\Sigma\,\Sigma\,\delta_{ij},$$

the model (6.1.1) becomes

$$y_{ijk} = \bar{\theta} + \alpha_i + \beta_j + \gamma_{ij} + e_{ijk}. \tag{6.1.22}$$

This is the two-way classification fixed effect model appropriate in this context for estimating the parameters α_i, β_j, γ_{ij}. As we have mentioned earlier, the term *fixed effect* is essentially a sampling theory concept because in this framework, the effects α_i, β_j and γ_{ij} are regarded as fixed but unknown constants. From the Bayesian viewpoint, all parameters are random variables and the appropriate inference procedure for these parameters depends upon the *nature of the prior distribution* used to represent the behavior of the factors corresponding to rows and columns. Problems could arise where a noninformative situation could be approximated by allowing these to take locally uniform and independent priors. We should then be dealing with a special case of the linear model discussed in Section 2.7. On the other hand and in particular for the driver–car example it might be more realistic to assume that the α's, β's, and γ's were themselves random samples from distributions with zero means and variances σ_r^2, σ_c^2, σ_t^2, respectively. The problem of making inferences about the parameters α_i, β_j, γ_{ij} would then parallel the approach given later in Section 7.2 for the estimation of means from a one-way classification random effect model.

In the above we have classified a number of different problems which can be pertinent in the analysis of cross classification designs. In what follows we shall not attempt a comprehensive treatment. Rather, a few familiar situations will be studied to see what new features come to light with the Bayesian approach.

We begin by considering the two-way random effect model in (6.1.5) and then devote the remainder of the chapter to the analysis of the additive mixed model in (6.1.16) and the mixed model with interaction in (6.1.20).

6.2 CROSS CLASSIFICATION RANDOM EFFECT MODEL

In this section the problem is considered of making inferences about the variance components in the two-way model

$$y_{ijk} = \theta + r_i + c_j + t_{ij} + e_{ijk}, \qquad i = 1, \ldots, I; \quad j = 1, \ldots, J; \quad k = 1, \ldots, K, \quad (6.2.1)$$

where y_{ijk} are the observations, θ is a location parameter, r_i, c_j, and t_{ij} are three different kinds of random effects and e_{ijk} are the residual errors. In the context of our previous discussion the random effects r_i, c_j, and t_{ij} could be associated respectively with cars, drivers, and car–driver interaction. The model would come naturally if the samples under study were randomly drawn from large populations of cars and drivers about which we wished to make inferences. It was shown that the r_i, c_j, t_{ij}, and e_{ijk} could then be taken to be uncorrelated with zero means and variances σ_r^2, σ_c^2, σ_t^2, and σ_e^2, respectively so that

$$E(y_{ijk}) = \theta, \qquad \operatorname{Var}(y_{ijk}) = \sigma_r^2 + \sigma_c^2 + \sigma_t^2 + \sigma_e^2. \quad (6.2.2)$$

The form of the model (6.2.1) is closely related to the three-component hierarchical design model discussed in Section 5.3. Indeed, expression (6.2.1) reduces to (5.3.1) with appropriate changes in notation if either σ_r^2 or σ_c^2 is known to be zero.

The relevant sample quantities are conveniently arranged in the analysis of variance shown in Table 6.2.1.

Using the table, we can write:

$$\hat{\sigma}_e^2 = m_e, \qquad \hat{\sigma}_t^2 = \frac{m_t - m_e}{K}, \qquad \hat{\sigma}_c^2 = \frac{m_c - m_t}{IK}, \qquad \hat{\sigma}_r^2 = \frac{m_r - m_t}{JK},$$

where these sampling theory point estimates of the components σ_r^2, σ_c^2, σ_t^2, and σ_e^2 are obtained by solving the equations in expectations of mean squares as in the case of hierarchical design models. The difficulties encountered are similar to those discussed earlier.

6.2.1 The Likelihood Function

To obtain the likelihood function, it will be useful to recall first a number of results related to the analysis of two-way classifications which are summarized in the following theorem.

Theorem 6.2.1 Let x_{ij} $(i = 1, \ldots, I, j = 1, \ldots, J)$ be IJ independent observations from a Normal distribution $N(0, \sigma^2)$ arranged in a two-way table of I rows and J columns. Let $x_{..}$, $x_{i.}$, $x_{.j}$, and $x_{ij} - x_{i.} - x_{.j} + x_{..}$ be respectively the grand mean, row means, column means, and residuals. Then

1. $\Sigma\Sigma x_{ij}^2 = IJx_{..}^2 + J\Sigma(x_{i.} - x_{..})^2 + I\Sigma(x_{.j} - x_{..})^2$
   $+ \Sigma\Sigma(x_{ij} - x_{i.} - x_{.j} + x_{..})^2,$

Table 6.2.1

Analysis of variance of two-way random effect models

Source	S.S.	d.f.	M.S.	E.M.S.
Grand mean	$IJK(y_{...}-\theta)^2$	1	$IJK(y_{...}-\theta)^2$	$\sigma_e^2+K\sigma_t^2+IK\sigma_c^2+JK\sigma_r^2$
Main effect "r"	$S_r=JK\Sigma(y_{i..}-y_{...})^2$	$v_r=(I-1)$	$m_r=S_r/v_r$	$\sigma_{etr}^2=\sigma_e^2+K\sigma_t^2+JK\sigma_r^2$
Main effect "c"	$S_c=IK(y_{.j.}-y_{...})^2$	$v_c=(J-1)$	$m_c=S_c/v_c$	$\sigma_{etc}^2=\sigma_e^2+K\sigma_t^2+IK\sigma_c^2$
Interaction $r\times c$	$S_t=K\Sigma\Sigma(y_{ij.}-y_{i..}-y_{.j.}+y_{...})^2$	$v_t=(I-1)(J-1)$	$m_t=S_t/v_t$	$\sigma_{et}^2=\sigma_e^2+K\sigma_t^2$
Residual	$S_e=\Sigma\Sigma\Sigma(y_{ijk}-y_{ij.})^2$	$v_e=IJ(K-1)$	$m_e=S_e/v_e$	σ_e^2
Total	$\Sigma\Sigma\Sigma(y_{ijk}-\theta)^2$	IJK		

2. The sets of quantities $x_{..}$, $\{x_{i.} - x_{..}\}$, $\{x_{.j} - x_{..}\}$, and $\{x_{ij} - x_{i.} - x_{.j} + x_{..}\}$ are distributed independently of one another,
3. $J\,\Sigma\,(x_{i.} - x_{..})^2 \sim \sigma^2\chi^2_{(I-1)}$, $I\,\Sigma\,(x_{.j} - x_{..})^2 \sim \sigma^2\chi^2_{(J-1)}$ and
$$\Sigma\,\Sigma\,(x_{ij} - x_{i.} - x_{.j} + x_{..})^2 \sim \sigma^2\chi^2_{(I-1)(J-1)},$$
4. So far as *each* of the sets $\{x_{i.} - x_{..}\}$, $\{x_{.j} - x_{..}\}$, and $\{x_{ij} - x_{i.} - x_{.j} + x_{..}\}$ is concerned, the corresponding sum of squares given in (3) is a sufficient statistic for σ^2.

In the theorem, (1) is an algebraic identity, (2) can be proved by showing that the four sets of variables are uncorrelated with one another, (3) follows from Cochran's theorem (1934) on distribution of quadratic forms, and (4) can be seen from inspection of the likelihood function.

For the model (6.2.1), we shall make the usual assumption that $(r_i, c_j, t_{ij}, e_{ijk})$ are Normally distributed. Since the effects are uncorrelated, the assumption of Normality implies that they are also independent. To obtain the likelihood function, it is convenient to work with the set of quantities $y_{...}$, $\{y_{i..} - y_{...}\}$, $\{y_{.j.} - y_{...}\}$, $\{y_{ij.} - y_{i..} - y_{.j.} + y_{...}\}$, and $\{y_{ijk} - y_{ij.}\}$. We have

$$\begin{aligned}
y_{...} &= \theta + r_. + c_. + t_{..} + e_{...},\\
y_{i..} - y_{...} &= (r_i - r_.) + (t_{i.} - t_{..}) + (e_{i..} - e_{...}),\\
y_{.j.} - y_{...} &= (c_j - c_.) + (t_{.j} - t_{..}) + (e_{.j.} - e_{...}),\\
y_{ij.} - y_{i..} - y_{.j.} + y_{...} &= (t_{ij} - t_{i.} - t_{.j} + t_{..}) + (e_{ij.} - e_{i..} - e_{.j.} + e_{...}),\\
y_{ijk} - y_{ij.} &= e_{ijk} - e_{ij.}.
\end{aligned} \tag{6.2.3a}$$

It follows from repeated use of Theorem 5.2.1 (on page 250) and Theorem 6.2.1 (on page 329) that the grand mean $y_{...}$ and the four mean squares (m_r, m_c, m_t, m_e) in Table 6.2.1 are jointly sufficient for $(\theta, \sigma_e^2, \sigma_{et}^2, \sigma_{etr}^2, \sigma_{etc}^2)$ and are independently distributed as

$$\begin{aligned}
&y_{...} \sim N[\theta, (\sigma_{etr}^2 + IK\sigma_c^2)/IJK], \quad v_r m_r \sim \sigma_{etr}^2\chi_{v_r}^2,\\
&v_c m_c \sim \sigma_{etc}^2\chi_{v_c}^2, \quad v_t m_t \sim \sigma_{et}^2\chi_{v_t}^2, \quad v_e m_e \sim \sigma_e^2\chi_{v_e}^2.
\end{aligned} \tag{6.2.3b}$$

The likelihood function is, therefore,

$$\begin{aligned}
&l(\theta, \sigma_e^2, \sigma_{et}^2, \sigma_{etr}^2, \sigma_{etc}^2 \mid \mathbf{y})\\
&\propto (\sigma_{etr}^2 + \sigma_{etc}^2 - \sigma_{et}^2)^{-1/2}(\sigma_e^2)^{-v_e/2}(\sigma_{et}^2)^{-v_t/2}(\sigma_{etr}^2)^{-v_r/2}(\sigma_{etc}^2)^{-v_c/2}\\
&\quad \times \exp -\left\{\frac{1}{2}\left[\frac{IJK(\theta - y_{...})^2}{\sigma_{etr}^2 + \sigma_{etc}^2 - \sigma_{et}^2} + \frac{v_e m_e}{\sigma_e^2} + \frac{v_t m_t}{\sigma_{et}^2} + \frac{v_r m_r}{\sigma_{etr}^2} + \frac{v_c m_c}{\sigma_{etc}^2}\right]\right\}.
\end{aligned} \tag{6.2.4}$$

The likelihood can alternatively be thought of as arising from one observation from a Normal distribution with mean θ and variance $(\sigma_{etr}^2 + \sigma_{etc}^2 - \sigma_{et}^2)/(IJK)$, together with $(\nu_e, \nu_t, \nu_r, \nu_c)$ further independent observations drawn from Normal distributions with zero means and variances $(\sigma_e^2, \sigma_t^2, \sigma_{etr}^2, \sigma_{etc}^2)$, respectively. The model supplies the additional information that the parameters $(\sigma_e^2, \sigma_{et}^2, \sigma_{etr}^2, \sigma_{etc}^2)$ are subject to the constraint

$$C: \begin{cases} \sigma_e^2 < \sigma_{et}^2 < \sigma_{etr}^2, \\ \sigma_e^2 < \sigma_{et}^2 < \sigma_{etc}^2. \end{cases} \tag{6.2.5}$$

6.2.2 Posterior Distribution of $(\sigma_e^2, \sigma_t^2, \sigma_r^2, \sigma_c^2)$

Adopting the argument in Section 1.3 and treating the location parameter θ separately from the variances $(\sigma_e^2, \sigma_{et}^2, \sigma_{etr}^2, \sigma_{etc}^2)$, we employ the noninformative reference prior distribution

$$p(\theta, \sigma_e^2, \sigma_{et}^2, \sigma_{etr}^2, \sigma_{etc}^2) \propto \sigma_e^{-2}\, \sigma_{et}^{-2}\, \sigma_{etr}^{-2}\, \sigma_{etc}^{-2} \tag{6.2.6}$$

subject to the constraint C. Upon integrating out θ, the joint posterior distribution of $(\sigma_e^2, \sigma_{et}^2, \sigma_{etr}^2, \sigma_{etc}^2)$ is

$$p(\sigma_e^2, \sigma_{et}^2, \sigma_{etr}^2, \sigma_{etc}^2 \mid \mathbf{y}) = w(\nu_e m_e)^{-1}\, p\left(\chi_{\nu_e}^{-2} = \frac{\sigma_e^2}{\nu_e m_e}\right) (\nu_t m_t)^{-1}\, p\left(\chi_{\nu_t}^{-2} = \frac{\sigma_{et}^2}{\nu_t m_t}\right)$$

$$\times\ (\nu_r m_r)^{-1}\, p\left(\chi_{\nu_r}^{-2} = \frac{\sigma_{etr}^2}{\nu_r m_r}\right) (\nu_c m_c)^{-1}\, p\left(\chi_{\nu_c}^{-2} = \frac{\sigma_{etc}^2}{\nu_c m_c}\right),$$

$$\sigma_e^2 > 0, \quad \sigma_{et}^2 > \sigma_e^2, \quad \sigma_{etr}^2 > \sigma_{et}^2, \quad \sigma_{etc}^2 > \sigma_{et}^2, \tag{6.2.7}$$

where

$$w^{-1} = \Pr{}^*(C \mid \mathbf{y}) = \Pr\left\{\frac{\chi_{\nu_e}^2}{\chi_{\nu_t}^2} > \frac{\nu_e m_e}{\nu_t m_t}, \frac{\chi_{\nu_r}^2}{\chi_{\nu_t}^2} < \frac{\nu_r m_r}{\nu_t m_t}, \frac{\chi_{\nu_c}^2}{\chi_{\nu_t}^2} < \frac{\nu_c m_c}{\nu_t m_t}\right\},$$

and $(\chi_{\nu_e}^2, \chi_{\nu_r}^2, \chi_{\nu_t}^2, \chi_{\nu_c}^2)$ are independently distributed χ^2 variables with $(\nu_e, \nu_r, \nu_t, \nu_c)$ degrees of freedom, respectively. It follows that the joint distribution of the variance components $(\sigma_e^2, \sigma_t^2, \sigma_r^2, \sigma_c^2)$ is

$$p(\sigma_e^2, \sigma_t^2, \sigma_r^2, \sigma_c^2 \mid \mathbf{y}) = w(\nu_e m_e)^{-1}\, p\left(\chi_{\nu_e}^{-2} = \frac{\sigma_e^2}{\nu_e m_e}\right) \left(\frac{\nu_t m_t}{K}\right)^{-1} p\left(\chi_{\nu_t}^{-2} = \frac{\sigma_e^2 + K\sigma_t^2}{\nu_t m_t}\right)$$

$$\times \left(\frac{\nu_r m_r}{JK}\right)^{-1} p\left(\chi_{\nu_r}^{-2} = \frac{\sigma_e^2 + K\sigma_t^2 + JK\sigma_r^2}{\nu_r m_r}\right) \left(\frac{\nu_c m_c}{IK}\right)^{-1}$$

$$\times\ p\left(\chi_{\nu_c}^{-2} = \frac{\sigma_e^2 + K\sigma_t^2 + IK\sigma_c^2}{\nu_c m_c}\right), \quad \sigma_e^2 > 0,\ \sigma_t^2 > 0,\ \sigma_r^2 > 0,\ \sigma_c^2 > 0. \tag{6.2.8}$$

As might be expected, this distribution is similar to the distribution in (5.3.11) for the three-component model. The scaled χ^{-2} approximation techniques developed in Sections 5.2.6 and 5.2.12 can now be employed to obtain marginal distribution of certain subsets of the components.

6.2.3 Distribution of $(\sigma_e^2, \sigma_t^2, \sigma_r^2)$

To illustrate, suppose we wish to make inferences jointly about $(\sigma_e^2, \sigma_t^2, \sigma_r^2)$. The corresponding posterior distribution can, of course, be obtained by integrating (6.2.8) over σ_c^2. Alternatively, we may use (6.2.7) and first obtain the distribution of $(\sigma_e^2, \sigma_{et}^2, \sigma_{etr}^2)$ by integrating out σ_{etc}^2 yielding

$$p(\sigma_e^2, \sigma_{et}^2, \sigma_{etr}^2 \mid \mathbf{y}) = w'(v_e m_e)^{-1} p\left(\chi_{v_e}^{-2} = \frac{\sigma_e^2}{v_e m_e}\right) (v_r m_r)^{-1} p\left(\chi_{v_r}^{-2} = \frac{\sigma_{etr}^2}{v_r m_r}\right) G(\sigma_{et}^2),$$

$$\sigma_{etr}^2 > \sigma_{et}^2 > \sigma_e^2 > 0, \qquad (6.2.9)$$

where

$$G(\sigma_{et}^2) = (v_t m_t)^{-1} p\left(\chi_{v_t}^{-2} = \frac{\sigma_{et}^2}{v_t m_t}\right) \frac{\Pr\{\chi_{v_c}^2 < v_c m_c/\sigma_{et}^2\}}{\Pr\{F_{v_c, v_t} < m_c/m_t\}}$$

and

$$w' = w \Pr\left\{F_{v_c, v_t} < \frac{m_c}{m_t}\right\}.$$

We see that the function $G(\sigma_{et}^2)$ is in precisely the same form as the posterior distribution of the variance σ_1^2 for the one-way model in (5.2.26). It follows by using the technique developed in Section 5.2.6 that $G(\sigma_{et}^2)$ is approximately

$$G(\sigma_{et}^2) \doteq (v_t' m_t')^{-1} p\left(\chi_{v_t'}^{-2} = \frac{\sigma_{et}^2}{v_t' m_t'}\right), \qquad (6.2.10)$$

where

$$v_t' = \frac{v_t}{a_1} \frac{I_{x_1}(\frac{1}{2}v_c, \frac{1}{2}v_t + 1)}{I_{x_1}(\frac{1}{2}v_c, \frac{1}{2}v_t)}, \qquad m_t' = \frac{v_t m_t}{a_1 v_t'},$$

$$a_1 = \left(\frac{v_t}{2} + 1\right) \frac{I_{x_1}(\frac{1}{2}v_c, \frac{1}{2}v_t + 2)}{I_{x_1}(\frac{1}{2}v_c, \frac{1}{2}v_t + 1)} - \frac{v_t}{2} \frac{I_{x_1}(\frac{1}{2}v_c, \frac{1}{2}v_t + 1)}{I_{x_1}(\frac{1}{2}v_c, \frac{1}{2}v_t)}, \qquad x_1 = \frac{v_c m_c}{v_c m_c + v_t m_t},$$

and, as before, $I_x(p, q)$ is the incomplete beta function. To this degree of approximation, the joint distribution of $(\sigma_e^2, \sigma_t^2, \sigma_r^2)$ is

$$p(\sigma_e^2, \sigma_t^2, \sigma_r^2 \mid \mathbf{y}) \doteq w_1 (v_e m_e)^{-1} p\left(\chi_{v_e}^{-2} = \frac{\sigma_e^2}{v_e m_e}\right) \left(\frac{v_t' m_t'}{K}\right)^{-1} p\left(\chi_{v_t'}^{-2} = \frac{\sigma_e^2 + K\sigma_t^2}{v_t' m_t'}\right)$$

$$\times \left(\frac{v_r m_r}{JK}\right)^{-1} p\left(\chi_{v_r}^{-2} = \frac{\sigma_e^2 + K\sigma_t^2 + JK\sigma_r^2}{v_r m_r}\right),$$

$$\sigma_e^2 > 0, \sigma_t^2 > 0, \quad \sigma_r^2 > 0, \qquad (6.2.11)$$

where

$$w_1^{-1} = \Pr\left\{\frac{\chi^2_{\nu_e}}{\chi^2_{\nu_t'}} > \frac{\nu_e m_e}{\nu_t' m_t'},\quad \frac{\chi^2_{\nu_t'}}{\chi_{\nu_r}} > \frac{\nu_t' m_t'}{\nu_r m_r}\right\},$$

which is exactly the same form as the distribution in (5.3.11) and can thus be analyzed as before.

It is clear that an analogous argument can be used to derive the joint distribution of $(\sigma_e^2, \sigma_t^2, \sigma_c^2)$ if desired.

6.2.4 Distribution of (σ_r^2, σ_c^2)

The derivation given above allows us to make inferences about the individual components $(\sigma_e^2, \sigma_t^2, \sigma_r^2, \sigma_c^2)$ and joint inferences about the sets $(\sigma_e^2, \sigma_t^2, \sigma_r^2)$ and $(\sigma_e^2, \sigma_t^2, \sigma_c^2)$. In cross classification designs, we are often interested mainly in the "main effect" variances (σ_r^2, σ_c^2). The corresponding joint posterior distribution is, of course, given by integrating (6.2.8) over σ_e^2 and σ_t^2,

$$p(\sigma_r^2, \sigma_c^2 \mid \mathbf{y}) = \int_0^\infty \int_0^\infty p(\sigma_e^2, \sigma_t^2, \sigma_r^2, \sigma_c^2 \mid \mathbf{y})\, d\sigma_e^2\, d\sigma_t^2, \qquad \sigma_r^2 > 0, \quad \sigma_c^2 > 0. \tag{6.2.12}$$

It does not seem possible to express the exact distribution in terms of simple functions and direct calculation of the distribution would require numerical evaluation of a double integral for every pair of values (σ_r^2, σ_c^2). We therefore introduce an approximation method which reduces (6.2.12) to a one-dimensional integral. First, we obtain the distribution of $(\sigma_{et}^2, \sigma_{etr}^2, \sigma_{etc}^2)$ from (6.2.7),

$$p(\sigma_{et}^2, \sigma_{etr}^2, \sigma_{etc}^2 \mid \mathbf{y}) = w''(\nu_r m_r)^{-1}\, p\left(\chi_{\nu_r}^{-2} = \frac{\sigma_{etr}^2}{\nu_r m_r}\right)(\nu_c m_c)^{-1}\, p\left(\chi_{\nu_c}^{-2} = \frac{\sigma_{etc}^2}{\nu_c m_c}\right) H(\sigma_{et}^2),$$

$$\sigma_{et}^2 > 0, \quad \sigma_{etr}^2 > \sigma_{et}^2, \quad \sigma_{etc}^2 > \sigma_{et}^2, \tag{6.2.13}$$

where

$$H(\sigma_{et}^2) = (\nu_t m_t)^{-1}\, p\left\{\chi_{\nu_t}^{-2} = \frac{\sigma_{et}^2}{\nu_t m_t}\right\} \frac{\Pr\{\chi^2_{\nu_e} > \nu_e m_e/\sigma_{et}^2\}}{\Pr\{F_{\nu_t,\nu_e} < m_t/m_e\}}$$

and

$$w'' = w \Pr\left\{F_{\nu_t,\nu_e} < \frac{m_t}{m_e}\right\}.$$

The function $H(\sigma_{et}^2)$ is in the same form as the distribution of σ_{12}^2 in (5.2.57) for the two-component model. It follows from the method in (5.2.60) that

$$H(\sigma_{et}^2) \doteq (\nu_t'' m_t'')^{-1}\, p\left\{\chi_{\nu_t''}^{-2} = \frac{\sigma_{et}^2}{\nu_t'' m_t''}\right\} \tag{6.2.14}$$

where

$$v_t'' = \frac{v_t}{a_2} \frac{I_{x_2}(\frac{1}{2}v_t + 1, \frac{1}{2}v_e)}{I_{x_2}(\frac{1}{2}v_t, \frac{1}{2}v_e)}, \qquad m_t'' = \frac{v_t m_t}{a_2 v_t''},$$

$$a_2 = \left(\frac{v_t}{2} + 1\right) \frac{I_{x_2}(\frac{1}{2}v_t + 2, \frac{1}{2}v_e)}{I_{x_2}(\frac{1}{2}v_t + 1, \frac{1}{2}v_e)} - \frac{v_t}{2} \frac{I_{x_2}(\frac{1}{2}v_t + 1, \frac{1}{2}v_e)}{I_{x_2}(\frac{1}{2}v_t, \frac{1}{2}v_e)},$$

and

$$x_2 = \frac{v_t m_t}{v_t m_t + v_e m_e}.$$

Hence,

$$p(\sigma_{et}^2, \sigma_{etr}^2, \sigma_{etc}^2 \mid \mathbf{y}) \doteq w_2 (v_r m_r)^{-1} p\left(\chi_{v_r}^{-2} = \frac{\sigma_{etr}^2}{v_r m_r}\right) (v_c m_c)^{-1} p\left(\chi_{v_c}^{-2} = \frac{\sigma_{etc}^2}{v_c m_c}\right)$$

$$\times (v_t'' m_t'')^{-1} p\left(\chi_{v_t''}^{-2} = \frac{\sigma_{et}^2}{v_t'' m_t''}\right), \qquad \sigma_{et}^2 > 0, \ \ \sigma_{etr}^2 > \sigma_{et}^2, \ \ \sigma_{etc}^2 > \sigma_{et}^2,$$

where

$$w_2 = \Pr\left\{\frac{\chi_{v_r}^2}{\chi_{v_t''}^2} < \frac{v_r m_r}{v_t'' m_t''}, \ \frac{\chi_{v_c}^2}{\chi_{v_t''}^2} < \frac{v_c m_c}{v_t'' m_t''}\right\}$$

so that

$$p(\sigma_r^2, \sigma_c^2 \mid \mathbf{y}) \doteq w_2 (v_t'' m_t'')^{-1} \left(\frac{v_r m_r}{JK}\right)^{-1} \left(\frac{v_c m_c}{IK}\right)^{-1}$$

$$\times \int_0^\infty p\left(\chi_{v_r}^{-2} = \frac{\sigma_{et}^2 + JK\sigma_r^2}{v_r m_r}\right) p\left(\chi_{v_c}^{-2} = \frac{\sigma_{et}^2 + IK\sigma_c^2}{v_c m_c}\right)$$

$$\times p\left(\chi_{v_t''}^{-2} = \frac{\sigma_{et}^2}{v_t'' m_t''}\right) d\sigma_{et}^2, \qquad \sigma_r^2 > 0, \ \ \sigma_c^2 > 0. \tag{6.2.15}$$

Calculation of the approximating distribution in (6.2.15) thus involves computing only one dimensional integrals, a task which is considerably simpler than exact evaluation of (6.2.12).

6.2.5 An Illustrative Example

To illustrate the analysis presented in this section, Table 6.2.2 gives the results for a randomized experiment. The data consists of 162 observations representing mileages per gallon of gasoline for 9 drivers on 9 cars with each driver making two runs with each car ($I = J = 9$, $K = 2$). We shall analyze the example by supposing that the underlying Normality and independence assumptions for the model (6.2.1) are appropriate and shall adopt the noninformative reference prior in (6.2.6) for making inferences.

Table 6.2.2 Gas mileage for 9 drivers driving 9 cars (duplicate runs)

Cars	Drivers 1	2	3	4	5	6	7	8	9
1	32.431	26.111	29.719	31.915	34.582	28.712	31.518	26.513	29.573
	31.709	26.941	29.218	32.183	36.490	27.091	30.448	28.440	29.464
2	26.356	22.652	25.966	27.856	31.090	25.956	23.375	25.329	28.648
	26.225	22.139	26.835	26.241	31.320	23.653	25.298	24.098	27.136
3	30.243	25.218	27.682	28.912	33.821	29.394	28.713	26.005	31.174
	31.785	27.189	27.521	28.059	34.462	27.859	29.302	27.020	28.003
4	29.830	25.192	25.962	28.717	34.619	27.663	27.511	26.145	26.834
	29.859	25.081	28.715	29.783	34.653	25.516	30.906	23.299	29.549
5	33.464	24.631	29.567	27.140	34.553	27.746	31.371	27.290	33.239
	30.307	25.930	28.368	30.818	35.337	26.210	21.495	24.689	32.319
6	28.313	21.809	28.030	28.447	31.432	26.551	28.073	24.575	28.026
	27.998	23.144	28.234	27.670	32.203	24.538	28.148	23.999	28.820
7	28.294	22.236	26.467	25.716	31.916	25.028	25.238	21.607	27.687
	27.363	22.245	27.115	25.059	31.541	26.296	25.083	21.900	29.357
8	29.864	24.542	27.103	25.051	32.282	25.096	27.655	25.038	26.414
	27.363	23.816	24.817	25.293	31.295	25.909	26.207	22.951	26.256
9	27.438	21.472	25.108	28.419	31.241	27.020	26.676	22.795	27.638
	27.486	23.130	27.589	25.941	32.459	24.445	25.738	23.299	27.385

The averages for each driver on each car as well as the driver means, car means, and grand mean are given in Table 6.2.3. The breakdown of the sum of squares and other relevant quantities for the analysis are summarized in Table 6.2.4 in the form of an analysis of variance table.

Table 6.2.3 Average mileage for 9 drivers on 9 cars

Cars	Drivers 1	2	3	4	5	6	7	8	9	Row means
1	32.0700	26.5260	29.4685	32.0490	35.5360	27.9015	30.9830	27.4765	29.5185	30.1699
2	26.2905	22.3955	26.4005	27.0485	31.2050	24.8045	24.3365	24.7135	27.8920	26.1207
3	31.0140	26.2035	27.6015	28.4855	34.1415	28.6265	29.0075	26.5125	29.5885	29.0201
4	29.8445	25.1365	27.3385	29.2500	34.6360	26.5895	29.2085	24.7220	28.1915	28.3241
5	31.8855	25.2805	28.9675	28.9790	34.9450	26.9780	30.4330	25.9895	32.7790	29.5819
6	28.1555	22.4765	28.1320	28.0585	31.8175	25.5445	28.1105	24.2870	28.4230	27.2228
7	27.8285	22.2405	26.7910	25.3875	31.7285	25.6620	25.1605	21.7535	28.5220	26.1193
8	28.6135	24.1790	25.9600	25.1720	31.7885	25.5025	26.9310	23.9945	26.3350	26.4973
9	27.4620	22.3010	26.3485	27.1800	31.8500	25.7325	26.2070	23.0470	27.5115	26.4044
Column	29.2404	24.0821	27.4453	27.9567	33.0720	26.3713	27.8197	24.7218	28.7512	27.7178

Table 6.2.4

Analysis of variance of 9 drivers on 9 cars

Source	S.S.	d.f.	M.S.	E.M.S.
Grand mean	$162(\theta-27.7178)^2$	1		
Main effect cars	362.0985	$v_r = 8$	$m_r = 45.2623$	$\sigma^2_{etr} = \sigma^2_e + 2\sigma^2_t + 18\sigma^2_r$
Main effect drivers	1,011.6550	$v_c = 8$	$m_c = 126.4569$	$\sigma^2_{etc} = \sigma^2_e + 2\sigma^2_t + 18\sigma^2_c$
Interaction	109.6123	$v_t = 64$	$m_t = 1.7127$	$\sigma^2_{et} = \sigma^2_e + 2\sigma^2_t$
Residuals	95.2460	$v_e = 81$	$m_e = 1.1759$	σ^2_e

$\hat{\sigma}^2_e = 1.1759,\quad \hat{\sigma}^2_t = 0.2684,\quad \hat{\sigma}^2_c = 6.9302,\quad \hat{\sigma}^2_r = 2.4194$

In examples of this kind, our main interests usually center on the "main effect" variances (σ_r^2, σ_c^2) and the interaction variance σ_t^2. However it seems clear from Table 6.2.4 [and can be readily confirmed by a fuller Bayesian analysis using, for example, (6.2.11)] that, for this example, σ_t^2 is small compared with σ_r^2 and σ_c^2.

If we wish to make inferences about the "main effect" variances (σ_r^2, σ_c^2) we may use the approximation (6.2.14) to eliminate σ_e^2 yielding

$$x_2 = \frac{109.6123}{204.8583} = 0.5351,$$

$$a_2 = 33 \times \frac{I_{x_2}(34, 40.5)}{I_{x_2}(33, 40.5)} - 32 \times \frac{I_{x_2}(33, 40.5)}{I_{x_2}(32, 40.5)},$$

$$= 33(0.98161) - 32(0.98461) = 0.88561,$$

$$v_t'' = \frac{64}{0.88561}(0.98461) = 71.1544,$$

$$m_t'' = \frac{109.6123}{(0.88561)(71.1544)} = 1.7395,$$

and

$$v_t'' m_t'' = 123.7704.$$

From (6.2.15), the approximate posterior distribution of (σ_r^2, σ_c^2) is then

$$p(\sigma_r^2, \sigma_c^2 \mid \mathbf{y}) \propto \int_0^\infty p\left(\chi_8^{-2} = \frac{\sigma_{et}^2 + 18\sigma_r^2}{362.0985}\right) p\left(\chi_8^{-2} = \frac{\sigma_{et}^2 + 18\sigma_c^2}{1{,}011.6550}\right)$$

$$\times\, p\left(\chi_{71.1544}^{-2} = \frac{\sigma_{et}^2}{123.7704}\right) d\sigma_{et}^2.$$

Figure 6.2.1 shows four contours of the distribution calculated by numerical integration together with the mode $(\sigma_{ro}^2, \sigma_{co}^2)$ which is at approximately (1.92, 5.52).

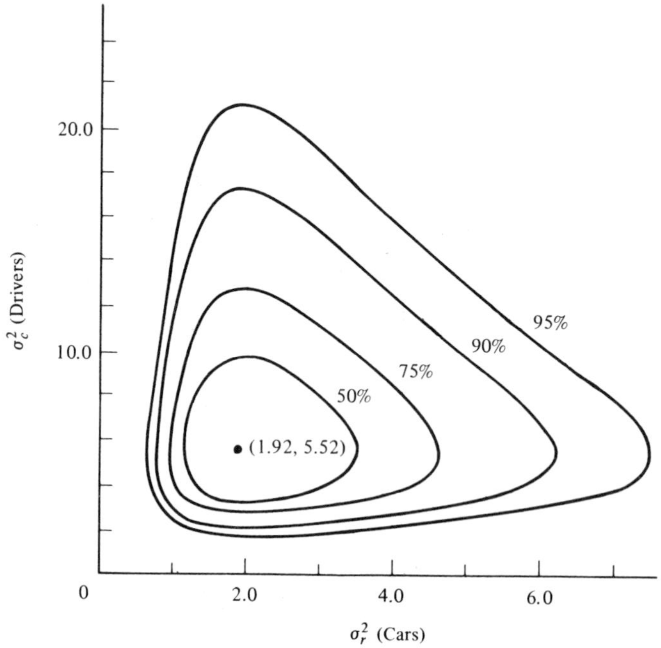

Fig. 6.2.1 Contours of posterior distribution of (σ_r^2, σ_c^2): the car–driver data.

The levels of the contours are respectively 50, 25, 10 and 5 per cent of the density at the mode. Using the asymptotic Normal approximation and since the χ^2 distribution with two degrees of freedom is an exponential, these contours are very roughly the boundaries of the 50, 75, 90 and 95 per cent H.P.D. regions. It will be seen that for this example both the main effect components are substantial and that they appear to be approximately independently distributed.

6.2.6 A Simple Approximation to the Distribution of (σ_r^2, σ_c^2)

When, as in this example, the modified degrees of freedom v_t'' is large, the last factor in the integrand of (6.2.15) will be sharply concentrated about m_t''. We may adopt an argument similar to the one leading to the modal approximation in (5.2.43) to write

$$p(\sigma_r^2, \sigma_c^2 \mid \mathbf{y}) \propto p\left(\chi_{v_r}^{-2} = \frac{m_t'' + JK\sigma_r^2}{v_r m_r}\right) p\left(\chi_{v_c}^{-2} = \frac{m_t'' + IK\sigma_c^2}{v_c m_c}\right). \quad (6.2.16)$$

To this degree of approximation, σ_r^2 and σ_c^2 are independent, each distributed like a truncated inverted χ^2 variable. More specifically,

$$\frac{m_t'' + JK\sigma_r^2}{v_r m_r} \mathrel{\dot\sim} \chi_{v_r}^{-2} \quad \text{truncated from below at} \quad \frac{m_t''}{v_r m_r}$$

and (6.2.17)

$$\frac{m_t'' + IK\sigma_c^2}{v_c m_c} \mathrel{\dot\sim} \chi_{v_c}^{-2} \quad \text{truncated from below at} \quad \frac{m_t''}{v_c m_c}.$$

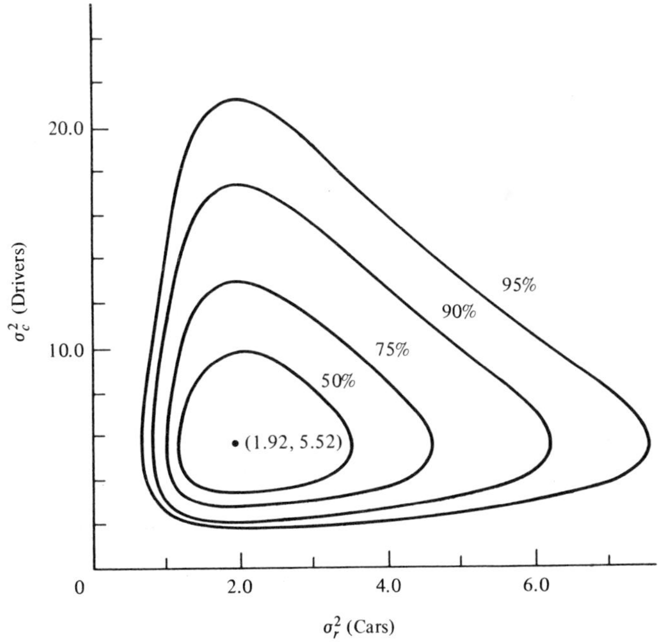

Fig. 6.2.2 Contours of the approximate posterior distribution of (σ_r^2, σ_c^2): the car–driver data.

Returning to our example, the posterior distribution of the variances (σ_r^2, σ_c^2) using the present approximation is

$$p(\sigma_r^2, \sigma_c^2 \mid \mathbf{y}) \propto p\left(\chi_8^{-2} = \frac{1.7395 + 18\sigma_r^2}{362.0985}\right) p\left(\chi_8^{-2} = \frac{1.7395 + 18\sigma_c^2}{1{,}011.6550}\right).$$

Figure 6.2.2 gives four contours corresponding to the 50, 75, 90 and 95 per cent H.P.D. regions, together with the mode $(\hat{\sigma}_r^2, \hat{\sigma}_c^2)$. They are very close to the contours in Fig. 6.2.1. In fact, by overlaying the two figures one finds that they almost completely coincide, showing that for this example (6.2.16) is an excellent approximation to (6.2.15).

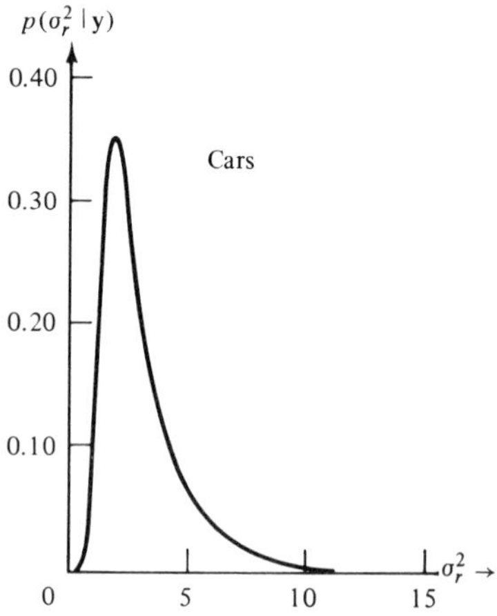

Fig. 6.2.3 Approximate posterior distribution of σ_r^2: the car–driver data.

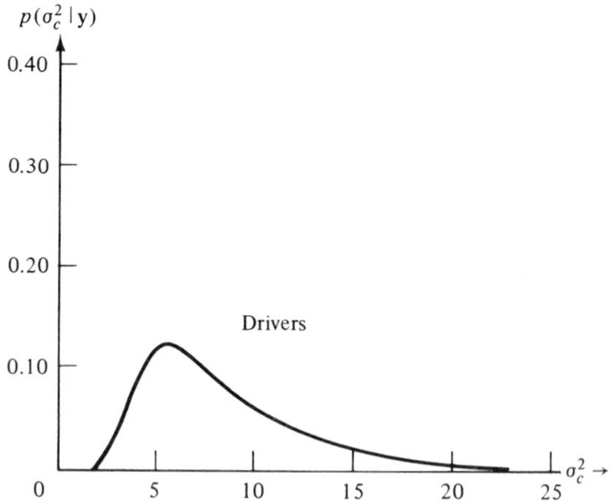

Fig. 6.2.4 Approximate posterior distribution of σ_c^2: the car–driver data.

Using (6.2.17), individual inferences about σ_r^2 can be made by referring the quantity $(1.7395 + 18\sigma_r^2)/362.0985$ to an inverted χ^2 with 8 degrees of freedom truncated from below at 0.0048 or equivalently $362.0985/(1.7395 + 18\sigma_r^2)$ to a χ^2 with 8 degrees of freedom truncated from above at 208.33. Similarly, inferences about σ_c^2 can be made by referring $(1.7395 + 18\sigma_c^2)/1{,}101.655$ to χ_8^{-2} truncated from below at 0.0016 or $1{,}101.655/(1.7395 + 18\sigma_c^2)$ to χ_8^2 truncated from above at 633.3. In both cases, the effect of truncation is negligible. The approximate posterior distributions of σ_r^2 and σ_c^2 are shown respectively in Figs 6.2.3 and 6.2.4.

6.3 THE ADDITIVE MIXED MODEL

We consider in this section the model

$$y_{ij} = \theta_i + c_j + e_{ij}, \qquad i = 1, \ldots, I; \quad j = 1, \ldots, J \tag{6.3.1}$$

which is the additive mixed model in (6.1.16) with $k = 1$. The random effects c_j and the errors e_{ij} are assumed to independently follow Normal distributions with zero means and variances σ_c^2 and σ_e^2, respectively. In terms of the previous example we would have I particular cars of interest tested on a random sample of J drivers. The assumption of additivity implies that the expected increment of performance c_j associated with a particular driver j is the same whichever of the I cars he drives. Though restrictive this assumption is adequate in some contexts. Suppose it was appropriate in the case of the car rental agency interested in the mean performances θ_i of $I = 6$ cars tested by a random sample of drivers. Then the variance component σ_c^2 would measure the variation ascribable to the differences in drivers, and σ_e^2 would measure the experimental error variation.

6.3.1 The Likelihood Function

For the likelihood function, it is convenient to transform the IJ observations into $\{y_{i.}\}$, $\{y_{.j} - y_{..}\}$ and $\{y_{ij} - y_{i.} - y_{.j} + y_{..}\}$. We have

$$\begin{aligned} y_{i.} &= \theta_i + c_{.} + e_{i.}, \\ y_{.j} - y_{..} &= c_j - c_{.} + e_{.j} - e_{..}, \\ y_{ij} - y_{i.} - y_{.j} + y_{..} &= e_{ij} - e_{i.} - e_{.j} + e_{..}. \end{aligned} \tag{6.3.2}$$

Using Theorem 5.2.1 (on page 250) and Theorem 6.2.1 (on page 329), we may conclude that:

1. the three sets of variables $\{y_{i.}\}$, $\{y_{.j} - y_{..}\}$, and $\{y_{ij} - y_{i.} - y_{.j} + y_{..}\}$ are independent one of another;
2. so far as the set $\{y_{.j} - y_{..}\}$ is concerned, the sum of squares $I\,\Sigma\,(y_{.j} - y_{..})^2$ is a sufficient statistic for $(\sigma_e^2 + I\sigma_c^2)$ and is distributed as $(\sigma_e^2 + I\sigma_c^2)\chi^2_{(J-1)}$;
3. so far as the set $\{y_{ij} - y_{i.} - y_{.j} + y_{..}\}$ is concerned, the sum of squares $\Sigma\,\Sigma\,(y_{ij} - y_{i.} - y_{.j} + y_{..})^2$ is a sufficient statistic for σ_e^2 and is distributed as $\sigma_e^2\chi^2_{(I-1)(J-1)}$; and
4. the vector of variables $\mathbf{y}'_{.} = (y_{1.}, \ldots, y_{I.})$ is distributed as the I-variate Normal $N_I(\boldsymbol{\theta}, \mathbf{V})$ where

$$\boldsymbol{\theta}' = (\theta_1, \ldots, \theta_I) \qquad \text{and} \qquad \mathbf{V} = J^{-1}[\sigma_e^2\mathbf{I} + \sigma_c^2\mathbf{1}_I\mathbf{1}_I']. \tag{6.3.3}$$

Noting that

$$|\mathbf{V}| = J^{-I}(\sigma_e^2)^{(I-1)}(\sigma_e^2 + I\sigma_c^2),$$

the likelihood function is

$$l(\boldsymbol{\theta}, \sigma_e^2, \sigma_c^2 \mid \mathbf{y}) \propto (\sigma_e^2)^{-[J(I-1)/2]} (\sigma_e^2 + I\sigma_c^2)^{-J/2}$$

$$\times \exp\left\{-\frac{1}{2}\left[\frac{I\,\Sigma(y_{.j} - y_{..})^2}{\sigma_e^2 + I\sigma_c^2} + \frac{\Sigma\,\Sigma\,(y_{ij} - y_{i.} - y_{.j} + y_{..})^2}{\sigma_e^2} + Q(\boldsymbol{\theta})\right]\right\}, \quad (6.3.4)$$

where

$$Q(\boldsymbol{\theta}) = (\mathbf{y}_{.} - \boldsymbol{\theta})'\mathbf{V}^{-1}(\mathbf{y}_{.} - \boldsymbol{\theta}).$$

In the analysis of additive mixed models, interest is usually centered on the comparative values of the location parameters θ_i rather than θ_i themselves. We may, therefore, work with

$$\phi_i = \theta_i - \bar{\theta}, \qquad i = 1, \ldots, I; \qquad \bar{\theta} = I^{-1}\,\Sigma\,\theta_i. \quad (6.3.5)$$

Since

$$\mathbf{V}^{-1} = J\sigma_e^{-2}\left[\mathbf{I} - \frac{\sigma_c^2}{\sigma_e^2 + I\sigma_c^2}\mathbf{1}_I\mathbf{1}_I'\right]$$

$$= J\sigma_e^{-2}\left(\mathbf{I} - \frac{1}{I}\mathbf{1}_I\mathbf{1}_I'\right) + JI^{-1}(\sigma_e^2 + I\sigma_c^2)^{-1}\mathbf{1}_I\mathbf{1}_I', \quad (6.3.6)$$

we may express the quadratic form $Q(\boldsymbol{\theta})$ as

$$Q(\boldsymbol{\theta}) = \sigma_e^{-2}J\,\Sigma\,(y_{i.} - y_{..} - \phi_i)^2 + (\sigma_e^2 + I\sigma_c^2)^{-1}IJ(y_{..} - \bar{\theta})^2. \quad (6.3.7)$$

The various sums of squares appearing in the likelihood function can be conveniently arranged in Table 6.3.1 in the form of an analysis of variance.

In Table 6.3.1, $\boldsymbol{\phi}' = (\phi_1, \ldots, \phi_{I-1})$ and note that $\phi_I = -(\phi_1 + \cdots + \phi_{I-1})$. We can express the likelihood function in (6.3.4) as

$$l(\bar{\theta}, \boldsymbol{\phi}, \sigma_{ce}^2, \sigma_e^2 \mid \mathbf{y}) = l_1(\bar{\theta}, \sigma_{ce}^2 \mid \mathbf{y})\,l_2(\boldsymbol{\phi}, \sigma_e^2 \mid \mathbf{y}), \quad (6.3.8)$$

where

$$l_1(\bar{\theta}, \sigma_{ce}^2 \mid \mathbf{y}) \propto (\sigma_{ce}^2)^{-(\nu_c+1)/2} \exp\left\{-\frac{1}{2\sigma_{ce}^2}[\nu_c m_c + IJ(\bar{\theta} - y_{..})^2]\right\}$$

and

$$l_2(\boldsymbol{\phi}, \sigma_e^2 \mid \mathbf{y}) \propto (\sigma_e^2)^{-(\nu_e+\nu_\phi)/2} \exp\left\{-\frac{1}{2\sigma_e^2}[S(\boldsymbol{\phi}) + \nu_e m_e]\right\}.$$

It is clear from the above expression that $l_1(\bar{\theta}, \sigma_{ce}^2 \mid \mathbf{y})$ can be regarded as the likelihood function of a sample consisting of one observation $y_{..}$ drawn from a Normal distribution $N(\bar{\theta}, \sigma_{ce}^2/IJ)$ and ν_c independent observations from a Normal population $N(0, \sigma_{ce}^2)$. Similarly $l_2(\boldsymbol{\phi}, \sigma_e^2 \mid \mathbf{y})$ can be taken as that of a sample of

Table 6.3.1

Analysis of variance for the additive mixed model

Source	S.S.	d.f.	M.S.	E.M.S.
Grand mean	$IJ(\bar{\theta}-y_{..})^2$	1	$IJ(\bar{\theta}-y_{..})^2$	$\sigma_{ce}^2=\sigma_e^2+I\sigma_c^2$
Fixed effect (cars)	$S(\boldsymbol{\phi})=J\,\Sigma[\phi_i-(y_{i.}-y_{..})]^2$	$\nu_\phi=(I-1)$	$m(\boldsymbol{\phi})=S(\boldsymbol{\phi})/\nu_\phi$	σ_e^2
Random effect (drivers)	$S_c=I\,\Sigma(y_{.j}-y_{..})^2$	$\nu_c=(J-1)$	$m_c=S_c/\nu_c$	σ_{ce}^2
Error	$S_e=\Sigma\Sigma(y_{ij}-y_{i.}-y_{.j}+y_{..})^2$	$\nu_e=(I-1)(J-1)$	$m_e=S_e/\nu_e$	σ_e^2

one observation from a $(I-1)$ dimensional Normal distribution with means $(\phi_1, \ldots, \phi_{I-1})$ and covariance matrix $(\sigma_e^2/J)[\mathbf{I}_{(I-1)} - (1/I)\mathbf{1}_{(I-1)}\mathbf{1}'_{(I-1)}]$ and of ν_e further observations from a Normal population $N(0, \sigma_e^2)$. Since $\sigma_{ce}^2 = \sigma_e^2 + I\sigma_c^2$, the parameters $(\sigma_{ce}^2, \sigma_e^2)$ are thus subject to the constraint

$$C: \sigma_e^2 < \sigma_{ce}^2. \tag{6.3.9}$$

As before, it is convenient to proceed first ignoring the constraint C. The posterior distribution of the various parameters appropriate to the model (6.3.1) are then conveniently derived from the unconstrained distributions using the result (1.5.3).

6.3.2 Posterior Distributions of $(\bar{\theta}, \boldsymbol{\phi}, \sigma_e^2, \sigma_{ce}^2)$ Ignoring the Constraint C

With the noninformative reference prior distribution†

$$p(\bar{\theta}, \boldsymbol{\phi}, \sigma_e^2, \sigma_{ce}^2) \propto \sigma_e^{-2}\sigma_{ce}^{-2} \tag{6.3.10}$$

the unconstrained posterior distribution of $(\bar{\theta}, \boldsymbol{\phi}, \sigma_e^2, \sigma_{ce}^2)$ is

$$p^*(\bar{\theta}, \boldsymbol{\phi}, \sigma_e^2, \sigma_{ce}^2 \mid \mathbf{y}) \propto (\sigma_e^2)^{-\frac{1}{2}(\nu_e + \nu_\phi) - 1}(\sigma_{ce}^2)^{-\frac{1}{2}(\nu_c - 1) - 1}$$

$$\times \exp\left\{-\frac{1}{2}\left[\frac{\nu_e m_e + S(\boldsymbol{\phi})}{\sigma_e^2} + \frac{\nu_c m_c + IJ(\bar{\theta} - y_{..})^2}{\sigma_{ce}^2}\right]\right\},$$

$$-\infty < \bar{\theta} < \infty, \quad -\infty < \phi_i < \infty, \quad i = 1, \ldots, (I-1), \quad \sigma_e^2 > 0, \sigma_{ce}^2 > 0, \tag{6.3.11}$$

from which for the *unconstrained* situation we deduce the following solutions which parallel known sampling results.

Given σ_e^2, the conditional posterior distribution of $\boldsymbol{\phi}$ is the $(I-1)$ dimensional multivariate Normal

$$N_{(I-1)}(\hat{\boldsymbol{\phi}}, \sigma_e^2 \boldsymbol{\Sigma}) \quad \text{with} \quad \hat{\boldsymbol{\phi}}' = (y_{1.} - y_{..}, \ldots, y_{(I-1).} - y_{..}) \tag{6.3.12a}$$

and

$$\boldsymbol{\Sigma} = \frac{1}{J}\left[\mathbf{I}_{(I-1)} - \frac{1}{I}\mathbf{1}_{(I-1)}\mathbf{1}'_{(I-1)}\right].$$

In particular, given σ_e^2, the conditional distribution of the difference $\delta = \theta_i - \theta_{i'} = \phi_i - \phi_{i'}$ between two particular means of interest is Normal $N(y_{i.} - y_{i'.}, 2\sigma_e^2/J)$. (6.3.12b)

The marginal posterior distribution of $\boldsymbol{\phi}$ is the $(I-1)$ dimensional multivariate t distribution $t_{(I-1)}(\hat{\boldsymbol{\phi}}, m_e\boldsymbol{\Sigma}, \nu_e)$. (6.3.13a)

† When the number of means I is large, the assumption that the θ's are locally uniform can be an inadequate approximation, and a more appropriate analysis may follow the lines given later in Section 7.2.

Hence the difference δ is distributed as $t(y_{i.} - y_{\acute{\iota}.}, 2m_e/J, \nu_e$ so that the quantity $\tau = [\delta - (y_{i.} - y_{\acute{\iota}.})]/(2m_e/J)^{1/2}$ is distributed as $t(0, 1, \nu_e)$. (6.3.13b)

Also, the quantity $V = m(\boldsymbol{\phi})/m_e$ is distributed as F with ν_ϕ and ν_e degrees of freedom. (6.3.14)

The quantity $\sigma_e^2/\nu_e m_e$ has the $\chi_{\nu_e}^{-2}$ distribution. (6.3.15)

Given $\boldsymbol{\phi}$, the conditional posterior distribution of σ_e^2 is such that $\sigma_e^2/[\nu_e m_e + \nu_\phi m(\boldsymbol{\phi})]$ has the χ^{-2} distribution with $\nu_e + \nu_\phi$ degrees of freedom. (6.3.16a)

It follows in particular that given δ, $\sigma_e^2/[m_e(\nu_e + \tau^2)]$ is distributed as $\chi_{\nu_e+1}^{-2}$. (6.3.16b)

The quantity $\sigma_{ce}^2/(\nu_c m_c)$ has the χ^{-2} distribution with ν_c degrees of freedom. (6.3.17)

The ratio $m_c\sigma_e^2/(m_e\sigma_{ce}^2)$ has the F distribution with ν_c and ν_e degrees of freedom. (6.3.18)

Given $\boldsymbol{\phi}$, the conditional distribution of the ratio σ_e^2/σ_{ce}^2 is such that the quantity

$$\frac{\sigma_e^2}{\sigma_{ce}^2} \frac{m_c}{[\nu_e m_e + \nu_\phi m(\boldsymbol{\phi})]/(\nu_e + \nu_\phi)}$$

is distributed as F with ν_c and $\nu_e + \nu_\phi$ degrees of freedom. (6.3.19a)

In particular, if just δ is given, the quantity

$$\frac{\sigma_e^2}{\sigma_{ce}^2} \frac{m_c}{m_e(\nu_e + \tau^2)/(\nu_e + 1)}$$

follows the F distribution with ν_c and $\nu_e + 1$ degrees of freedom. (6.3.19b)

6.3.3 Posterior Distribution of σ_e^2 and σ_c^2

The unconstrained posterior distributions given above are, of course, not themselves the appropriate distributions for the model (6.3.1), but provide a useful stepping stone from which they may be reached. When interest centers on the variance components (σ_e^2, σ_c^2), we may first employ the result (1.5.3) to obtain from (6.3.15) and (6.3.17) the distribution of $(\sigma_e^2, \sigma_{ce}^2)$,

$$p(\sigma_e^2, \sigma_{ce}^2 \mid \mathbf{y}) = (\nu_e m_e)^{-1} p\left(\chi_{\nu_e}^{-2} = \frac{\sigma_e^2}{\nu_e m_e}\right) (\nu_c m_c)^{-1} p\left(\chi_{\nu_c}^{-2} = \frac{\sigma_{ce}^2}{\nu_c m_c}\right)$$

$$\times \frac{\Pr^*\{C \mid \sigma_e^2, \sigma_{ce}^2, \mathbf{y}\}}{\Pr^*\{C \mid \mathbf{y}\}}, \qquad \sigma_{ce}^2 > \sigma_e^2 > 0. \qquad (6.3.20a)$$

Using (6.3.18), the overall probability of the constraint C: $\sigma_e^2 < \sigma_{ce}^2$ is

$$\Pr^*\{C \mid \mathbf{y}\} = \Pr\left\{F_{\nu_c,\nu_e} < \frac{m_c}{m_e}\right\}. \tag{6.3.20b}$$

In addition, given σ_e^2 and σ_{ce}^2, $\Pr^*\{C \mid \sigma_e^2, \sigma_{ce}^2, \mathbf{y}\}$ is clearly unity for $\sigma_e^2 < \sigma_{ce}^2$ and zero otherwise. Thus, the posterior distribution of (σ_e^2, σ_c^2) is

$$p(\sigma_e^2, \sigma_c^2 \mid \mathbf{y}) = (\nu_e m_e)^{-1} p\left(\chi_{\nu_e}^{-2} = \frac{\sigma_e^2}{\nu_e m_e}\right)\left(\frac{\nu_c m_c}{I}\right)^{-1}$$

$$\times\, p\left(\chi_{\nu_c}^{-2} = \frac{\sigma_e^2 + I\sigma_c^2}{\nu_c m_c}\right)\Big/\Pr\left\{F_{\nu_c,\nu_e} < \frac{m_c}{m_e}\right\}, \quad \sigma_c^2 > 0, \quad \sigma_e^2 > 0. \tag{6.3.21}$$

The reader will note that if we set $I = K$, $\nu_1 = \nu_e$, $\nu_2 = \nu_c$, $m_1 = m_e$, and $m_2 = m_c$, the distribution in (6.3.21) is precisely the same as that obtained in (5.2.14). Inferences about (σ_e^2, σ_c^2) can thus be made using the results obtained earlier.

6.3.4 Inferences About Fixed Effects for the Mixed Model

In many problems, the items of principal interest are the comparisons $\phi_i = \theta_i - \bar{\theta}$. For instance, for the rental car agency the aim of an analysis would most likely be to determine the relative performance of the cars. We would then have an example of the analysis of a randomized block design using a mixed model with σ_c^2 a variance component associated with blocks (drivers), and σ_e^2 a component due to error, and with the ϕ_i representing comparisons of the treatments (cars).

6.3.5 Comparison of Two Means

Consider the problem of comparing two particular means, say those for the ith and the $\acute{\imath}$th treatment. To make inferences about $\delta = \theta_i - \theta_{\acute{\imath}}$ it is convenient to work with the quantity

$$\tau = [\delta - (y_{i.} - y_{\acute{\imath}.})]\Big/\left(\frac{2m_e}{J}\right)^{1/2}. \tag{6.3.22}$$

From the unconstrained distribution of τ in (6.3.13b), we again employ the result (1.5.3) to obtain the posterior distribution of τ as

$$p(\tau \mid \mathbf{y}) = p(t_{\nu_e} = \tau)\,\frac{\Pr^*\{C \mid \tau, \mathbf{y}\}}{\Pr^*\{C \mid \mathbf{y}\}}, \qquad -\infty < \tau < \infty, \tag{6.3.23}$$

where $\Pr^*\{C \mid \mathbf{y}\}$ is given in (6.3.20b). From (6.3.19b), the conditional posterior probability of C, given τ, is

$$\Pr^*\{C \mid \tau, \mathbf{y}\} = \Pr\left\{F_{\nu_c,\nu_e+1} < \frac{m_c(\nu_e + 1)}{m_e(\nu_e + \tau^2)}\right\}. \tag{6.3.24}$$

The posterior distribution of τ is, therefore,

$$p(\tau \mid \mathbf{y}) = p(t_{\nu_e} = \tau)\, g\left(\tau^2 \,\middle|\, \frac{m_c}{m_e}\right), \qquad -\infty < \tau < \infty, \tag{6.3.25}$$

a Student's t distribution with ν_e degrees of freedom modified by a factor $g(\tau^2 \mid m_c/m_e)$ which is the ratio of two F probability integrals

$$g\left(\tau^2 \,\middle|\, \frac{m_c}{m_e}\right) = \frac{\Pr\left\{F_{\nu_c, \nu_e+1} < \dfrac{m_c}{m_e}\left(\dfrac{\nu_e+1}{\nu_e}\right)\left(1+\dfrac{\tau^2}{\nu_e}\right)^{-1}\right\}}{\Pr\left\{F_{\nu_c, \nu_e} < \dfrac{m_c}{m_e}\right\}}. \tag{6.3.26}$$

We now discuss some properties of this modified t distribution. Since both $p(t_{\nu_e} = \tau)$ and $g(\tau^2 \mid m_c/m_e)$ are functions of τ^2, the distribution is symmetric about the origin. It follows that all odd moments when they exist are zero. Using the identity (A5.2.1) in Appendix A5.2, the $2r$th ($r < \nu_e/2$) moment of τ is

$$E(\tau^{2r} \mid \mathbf{y}) = \nu_e^r \frac{B[(\frac{1}{2}\nu_e) - r, r]}{B(\frac{1}{2}, r)} \frac{\Pr\{F_{\nu_c, \nu_e - 2r} < (m_c/m_e)(\nu_e - 2r)/\nu_e\}}{\Pr\{F_{\nu_c, \nu_e} < m_c/m_e\}} \tag{6.3.27}$$

where, as before, $B(p, q)$ is the complete beta function. The function $g(\tau^2 \mid m_c/m_e)$ is monotonically decreasing in τ^2. This implies that the distribution $p(\tau \mid \mathbf{y})$ is uniformly more concentrated about the origin than $p(t_{\nu_e} = \tau)$. That is, for an arbitrary constant $d > 0$,

$$\Pr\{|\tau| < d \mid \mathbf{y}\} \geqslant \Pr\{|t_{\nu_e}| < d\}. \tag{6.3.28}$$

The above result is indeed a sensible one. For, the distribution $p(\tau \mid \mathbf{y})$ is obtained from the distribution $p(t_{\nu_e} = \tau)$ by imposing the constraint C: $\sigma_e^2 < \sigma_{ce}^2$. We see in (6.3.12b) that σ_e^2, if known, is proportional to the conditional variance of δ. Thus, it is not at all surprising that the posterior distribution of τ with the constraint C is more concentrated about the origin than the distribution of τ without the constraint. We have here in fact a further example of Bayesian pooling which is discussed in more detail in Section 6.3.7.

An Example

In the particular case $I = 2$, the randomized block arrangement results in pairs of observations. On standard sampling theory, the comparison of the two means is usually carried out by considering the differences $z_j = y_{2j} - y_{1j}$ $(j = 1, \ldots, J)$ between the pairs and making use of the fact that the sampling distribution of

$$\tau = \frac{[\delta - (y_{2.} - y_{1.})]}{(2m_e/J)^{1/2}} = \frac{J^{1/2}(\delta - \bar{z})}{[\Sigma(z_j - \bar{z})^2/(J-1)]^{1/2}}, \tag{6.3.29}$$

Table 6.3.2

Measurement of percentage of ammonia by two analysts

	Days (blocks)	1	2	3	4	5	6	7	8
(% Ammonia–12) × 100	*Analyst* 1	37	35	43	34	36	48	33	33
	Analyst 2	37	38	36	47	48	57	28	42

where $\bar{z} = J^{-1} \Sigma z_j$, is the $t(0, 1, \nu_e)$ distribution. In this analysis, however, any "inter-block information" provided by m_c about σ_e^2 is not taken account of.

We consider this "paired t" problem in some detail using for illustration an example quoted in Davies (1967, p. 57). The data shown in Table 6.3.2 are determinations made by two analysts of the percentage of ammonia in a plant gas made on eight different days. The primary purpose of the experiment was to make inferences about a possible systematic mean difference δ, that is, a bias between results from the two analysts. The example illustrates the use of the additive mixed model (6.3.1) with $I = 2$, $J = 8$. It should be borne in mind that we assume that (i) day to day variations of the true ammonia percentage follow a Normal distribution having variance σ_c^2, (ii) the eight particular daily samples considered may be regarded as a random sample from this distribution, (iii) there is no interaction between analysts and samples, and (iv) the Normally distributed analytical error has the same variance σ_e^2 for both analysts. The sample quantities needed in the analysis are given in Table 6.3.3 in the form of an analysis of variance.

Table 6.3.3

Analysis of variance for the analyst data

Source	S.S.	d.f.	M.S.
Fixed effect (bias between analysts)	$S(\delta)=\frac{1}{2}J(\delta-\bar{z})^2 = 4(\delta-4.25)^2$	1	$4(\delta-4.25)^2$
Random block effect (days)	$S_c=553.0$	$\nu_c=7$	$m_c=79.0$
Error	$S_e=206.75$	$\nu_e=7$	$m_e=29.54$
	Average for analyst 1	$y_{1.}=37.38$	
	Average for analyst 2	$y_{2.}=41.63$	
	$m_c/m_e=2.67$		

The solid curve in Fig. 6.3.1 is the posterior distribution of the bias $\delta = \theta_2 - \theta_1$ calculated from expression (6.3.25), corresponding to a noninformative reference prior. This distribution is seen to be more concentrated about its mean, $\bar{z} = y_{2.} - y_{1.} = 4.25$, than the unconstrained posterior distribution of δ shown by the dotted curve in the same figure. This latter curve (which is also the confidence distribution of δ) is a scaled t distribution centered at the same mean with 7 degrees of freedom. In particular, an H.P.D. interval for δ is shown with content 91.2%. This would have the somewhat smaller content 90.0% if the unconstrained distribution were used. The broken curve shown in the figure is the posterior distribution of δ obtained by taking $\sigma_c^2 = 0$ in (6.3.1), the implication of which will be discussed later in this section.

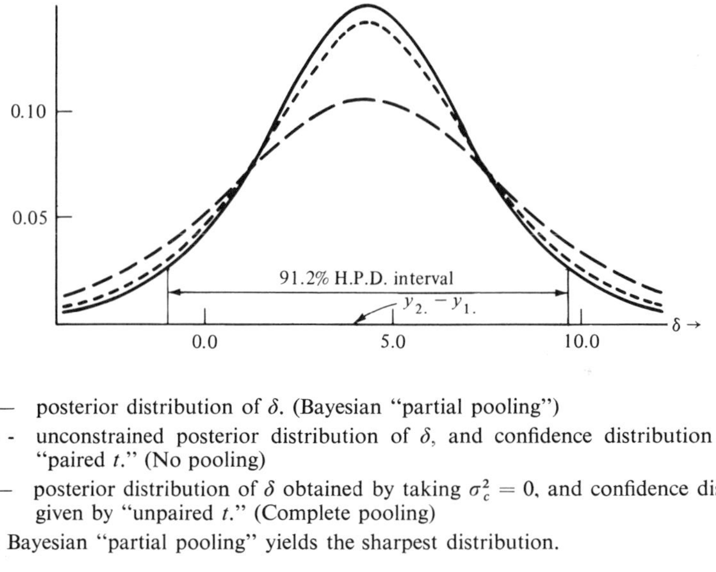

———— posterior distribution of δ. (Bayesian "partial pooling")

- - - - - - - - unconstrained posterior distribution of δ, and confidence distribution given by "paired t." (No pooling)

— — — — posterior distribution of δ obtained by taking $\sigma_c^2 = 0$, and confidence distribution given by "unpaired t." (Complete pooling)

Note that Bayesian "partial pooling" yields the sharpest distribution.

Fig. 6.3.1 Posterior distribution of the analytical bias δ.

Effect of m_c/m_e on the Distribution of δ

In the example considered, the posterior distribution of the bias δ coming from the Bayesian analysis is not very different from the unconstrained distribution which parallels the sampling results. This is because the ratio $m_c/m_e = 2.67$ is fairly large so that the influence of the factor $g(\tau^2 \mid m_c/m_e)$ is relatively mild. However, when m_c/m_e is close to or less than unity, the effect of $g(\tau^2 \mid m_c/m_e)$ will be much greater.

For illustration, suppose that in the analyst example the pair (48, 57) is excluded. One then obtains the analysis of variance presented in Table 6.3.4.

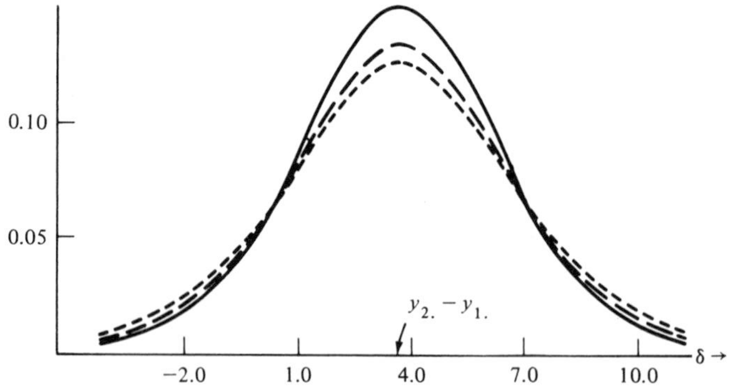

———— posterior distribution of δ. (Bayesian "partial pooling")

- - - - - unconstrained posterior distribution of δ, and confidence distribution given by "paired t." (No pooling)

— — — posterior distribution of δ obtained by taking $\sigma_c^2 = 0$, and confidence distribution given by "unpaired t." (Complete pooling)

Note that Bayesian "partial pooling" yields the sharpest distribution.

Fig. 6.3.2 Posterior distribution of the analytical bias δ for data excluding the pair (48, 57).

Here the ratio $m_c/m_e = 0.86$ is less than unity and, as shown in Fig. 6.3.2, the posterior distribution of δ is markedly different from the unconstrained distribution. It will be seen later in Section 6.3.7 that $g(\tau^2 \mid m_c/m_e)$, in fact, reflects the effect of pooling the variance estimates m_c and m_e in the estimation of σ_e^2.

Table 6.3.4

Analysis of variance for the analysts data excluding the pair (48, 57)

Source	S.S.	d.f.	M.S.
Fixed effect (bias between analysts)	$S(\delta) = \frac{1}{2}J(\delta - \bar{z})^2 = 3.5(\delta - 3.57)^2$	1	$3.5(\delta - 3.57)^2$
Random block effect (days)	$S_c = 166.71$	$\nu_c = 6$	$m_c = 27.79$
Error	$S_e = 193.86$	$\nu_e = 6$	$m_e = 32.31$
	$y_{1.} = 35.86,$	$y_{2.} = 39.43$	
	$m_c/m_e = 0.86$		

For given τ, the modifying factor $g(\tau^2 \mid m_c/m_e)$ is a function of m_c/m_e. It is instructive to consider the behavior of $p(\tau \mid \mathbf{y})$ in the extreme cases $m_c/m_e \to \infty$ and $m_c/m_e \to 0$. When $m_c/m_e \to \infty$, $g(\tau^2 \mid m_c/m_e)$ approaches unity for any fixed τ, so that the distribution $p(\tau \mid \mathbf{y})$ tends in the limit to the unconstrained distribution $p(t_\nu = \tau)$. On the other hand, when $m_c/m_e \to 0$, applying L'Hospital's rule, one finds that

$$\lim_{(m_c/m_e)\to 0} g\left(\tau^2 \,\middle|\, \frac{m_c}{m_e}\right) = \frac{\Gamma[\frac{1}{2}(\nu_e + \nu_c + 1)]\,\Gamma(\frac{1}{2}\nu_e)}{\Gamma[\frac{1}{2}(\nu_e + \nu_c)]\,\Gamma[\frac{1}{2}(\nu_e + 1)]}\left(1 + \frac{\tau^2}{\nu_e}\right)^{-\nu_c/2}. \qquad (6.3.30)$$

Hence in thelim it the distribution of the random variable $[I/(I-1)]^{1/2}\,\tau$ approaches the $t(0, 1, \nu_e + \nu_c)$ distribution. Further, it can be verified that the mixed derivative

$$\frac{\partial^2 \log g(\tau^2 \mid m_c/m_e)}{\partial\tau^2\,\partial(m_c/m_e)}$$

is

$$\frac{\partial^2 \log g\,(\tau^2 \mid m_c/m_e)}{\partial\tau^2\,\partial\,(m_c/m_e)} = N\left(\frac{m_c}{m_e}\right) H(a) \qquad (6.3.31)$$

where

$$H(a) = \frac{1}{2}\left(\nu_e + 1 - \frac{\nu_e + \nu_c + 1}{1 + a}\right)\int_0^a x^{\frac{1}{2}\nu_c - 1}\,(1 + x)^{-\frac{1}{2}(\nu_c + \nu_e + 1)}\,dx$$
$$+\, a^{\frac{1}{2}\nu_c}\,(1 + a)^{-\frac{1}{2}(\nu_c + \nu_e + 1)},$$

$$a = \frac{\nu_c m_c}{\nu_e m_e}\left(1 + \frac{\tau^2}{\nu_e}\right)^{-1},$$

and $N(m_c/m_e)$ is a non-negative function af m_c/m_e. Clearly $H(0) = 0$ and upon differentiating we find that $H'(a) > 0$. Thus

$$\frac{\partial^2 \log g(\tau^2 \mid m_c/m_e)}{\partial\tau^2\,\partial(m_c/m_e)} \geqslant 0 \qquad (6.3.32)$$

for all values of m_c/m_e and τ^2. This implies that the posterior distribution of τ^2 has the "monotone likelihood ratio" property—see e.g. Lehman (1959). It follows that for $d > 0$, the probability $\Pr\{|\tau| < d \mid \mathbf{y}\}$ is a monotonically decreasing function of m_c/m_e. This, together with (6.3.28) and (6.3.30), shows that, in obvious notation

$$\Pr\{|t_{\nu_e}| < d\} \leqslant \Pr\{|\tau| < d \mid \mathbf{y}\} \leqslant \Pr\{|t_{\nu_e + \nu_c}| < [I/(I-1)]^{1/2} d\} \qquad (6.3.33)$$

which provides an upper and a lower bound for the probability $\Pr\{|\tau| < d \mid \mathbf{y}\}$.

6.3.6 Approximations to the Distribution $p(\tau \mid \mathbf{y})$

In this section, we discuss certain methods which can be used to evaluate the posterior distribution of τ.

When $\nu_c = (J-1)$ is a positive even integer, we can use the identity (A5.2.3) in Appendix A5.2 to expand the numerator of $g(\tau^2 \mid m_c/m_e)$ into

$$\Pr\left\{F_{\nu_c,\,\nu_e+1} < \frac{m_c}{m_e}\left(\frac{\nu_e+1}{\nu_e}\right)\left(1+\frac{\tau^2}{\nu_e}\right)^{-1}\right\} = 1 - \left[(1-x)\left(1+\frac{\tau^2}{\nu_e}\right)\right]^{\frac{1}{2}(\nu_e+1)}$$
$$\times \sum_{j=1}^{\frac{1}{2}\nu_c-1}\binom{\frac{1}{2}\nu_c-1}{j}\frac{B[\frac{1}{2}\nu_c-j,\frac{1}{2}(\nu_e+1)+j]}{B[\frac{1}{2}\nu_c,\frac{1}{2}(\nu_e+1)]}x^j\left[1+(1-x)\frac{\tau^2}{\nu_e}\right]^{-\frac{1}{2}(\nu_e+1)-j} \tag{6.3.34}$$

where

$$x = \frac{\nu_c m_c}{\nu_e m_e + \nu_c m_c},$$

Substituting (6.3.34) into (6.3.25), we obtain

$$p(\tau \mid \mathbf{y}) = \frac{p(t_{\nu_e} = \tau)}{\Pr\{F_{\nu_c,\nu_e} < m_c/m_e\}} - \sum_{j=0}^{\frac{1}{2}\nu_c-1}\omega_j\gamma_j\,p(t_{\nu_e+2j} = \gamma_j\tau), \qquad -\infty < \tau < \infty, \tag{6.3.35}$$

where

$$\gamma_j = (1-x)^{1/2}(1+2j/\nu_e)^{1/2} \quad \text{and} \quad \omega_j = \frac{x^j(1-x)^{\frac{1}{2}\nu_e}}{j\,B(\frac{1}{2}\nu_e, j)\Pr\{F_{\nu_c,\nu_e} < m_c/m_e\}}.$$

It follows that for $d > 0$,

$$\Pr\{|\tau| < d \mid \mathbf{y}\} = \frac{\Pr\{|t_{\nu_e}| < d\}}{\Pr\{F_{\nu_c,\nu_e} < m_c/m_e\}} - \sum_{j=0}^{\frac{1}{2}\nu_c-1}\omega_j\Pr\{|t_{\nu_e+2j}| < \gamma_j d\} \tag{6.3.36}$$

which can be used to calculate probabilities to any desired degree of accuracy using a standard t table in conjunction with an incomplete beta function table. When ν_c is large, evaluation of the probability of τ from (6.3.36) would be rather laborious and for odd values of ν_c, this formula is not applicable. When appropriate it can, however, be used to check the usefulness of simpler approximations.

A Scaled t Approximation

We now show that a scaled t distribution can provide a simple and overall satisfactory approximation to the posterior distribution of τ. This result is to be expected because we have seen that in the two extreme cases when $m_c/m_e \to \infty$ and $m_c/m_e \to 0$, τ follows a scaled t distribution exactly. We can write (6.3.25) as

$$p(\tau \mid \mathbf{y}) = \underset{\sigma_e^2 \mid \mathbf{y}}{E}\; p(\tau \mid \sigma_e^2, \mathbf{y})$$
$$= \int_0^\infty p(\tau \mid \sigma_e^2, \mathbf{y})\,p(\sigma_e^2 \mid \mathbf{y})\,d\sigma_e^2, \qquad -\infty < \tau < \infty. \tag{6.3.37}$$

From (6.3.12b), the unconstrained conditional distribution of τ, given σ_e^2, is Normal $N(0, \sigma_e^2/m_e)$. Once σ_e^2 is given, the constraint $C: \sigma_e^2 < \sigma_{ce}^2$ has no effect on the distribution of τ. It follows that the first factor in the integrand, $p(\tau \mid \sigma_e^2, \mathbf{y})$, is the Normal distribution $N(0, \sigma_e^2/m_e)$. Now from (6.3.21) the posterior distribution $p(\sigma_e^2 \mid \mathbf{y})$ in the integrand is

$$p(\sigma_e^2 \mid \mathbf{y}) = (v_e m_e)^{-1}\, p\left(\chi_{v_e}^{-2} = \frac{\sigma_e^2}{v_e m_e}\right) \frac{\Pr\{\chi_{v_c}^2 < v_c m_c/\sigma_e^2\}}{\Pr\{F_{v_c, v_e} < m_c/m_e\}}, \qquad \sigma_e^2 > 0. \tag{6.3.38}$$

This distribution is of exactly the same form as the distribution of σ_1^2 in (5.2.26) for the two-component random model effect. Thus, employing the method developed in Section 5.2.6, the distribution of σ_e^2 can be closely approximated by that of a scaled χ^{-2} variable. Using the resulting approximation, we obtain

$$(m_e/m_e')^{1/2}\, \tau \mathrel{\dot\sim} t_{v_e'} \tag{6.3.39}$$

where

$$v_e' = \frac{v_e}{a} \frac{I_x(\frac{1}{2}v_c, \frac{1}{2}v_e + 1)}{I_x(\frac{1}{2}v_c, \frac{1}{2}v_e)}, \qquad m_e' = \frac{v_e m_e}{a v_e'},$$

$$a = \left(\frac{v_e}{2} + 1\right) \frac{I_x(\frac{1}{2}v_c, \frac{1}{2}v_e + 2)}{I_x(\frac{1}{2}v_c, \frac{1}{2}v_e + 1)} - \frac{v_e}{2} \frac{I_x(\frac{1}{2}v_c, \frac{1}{2}v_e + 1)}{I_x(\frac{1}{2}v_c, \frac{1}{2}v_e)} \quad \text{and} \quad x = \frac{v_c m_c}{v_e m_e + v_c m_c}.$$

That is, the quantity

$$\frac{\delta - (y_{i.} - y_{\acute{\imath}.})}{(2m_e'/J)^{1/2}} \tag{6.3.40}$$

is approximately distributed as $t(0, 1, v_e')$, or equivalently, δ is approximately distributed as $t(y_{i.} - y_{\acute{\imath}.}, 2m_e'/J, v_e')$. For the complete analysts data, we have $(v_e' = 8.57, m_e' = 27.55)$. On the other hand, excluding the pair (48, 57) one would get $(v_e' = 10.48, m_e' = 23.80)$. Tables 6.3.5a and 6.3.5b give, respectively, for these two cases specimens of the posterior densities of δ obtained from exact evaluation of (6.3.25) and from the above approximation. The agreement is very close.

6.3.7 Relationship to Some Sampling Theory Results and the Problem of Pooling

The standard sampling theory "paired t" analysis depends on the fact that the sampling distribution of $\tau = [\delta - (y_{i.} - y_{\acute{\imath}.})]/(2m_e/J)^{1/2}$ is the $t(0, 1, v_e)$ distribution. The resulting confidence distribution of δ is numerically equivalent to the unconstrained posterior distribution of δ. Now $E(m_e) = \sigma_e^2$ and $E(m_c) = 2\sigma_c^2 + \sigma_e^2$. In obtaining this confidence distribution, therefore, one may feel intuitively that some information about the variance component σ_e^2 is lost by not taking m_c into account because its sampling distribution also involves σ_e^2. What has sometimes been done is to use the ratio m_c/m_e to test the hypothesis $\sigma_c^2 = 0$

Table 6.3.5 Comparison of exact and approximate density of δ

(a) Data of Table 6.3.2

δ	*Exact*	*Approximate*
4.25	0.147830	0.147659
4.45	0.147353	0.147181
4.65	0.145932	0.145759
4.85	0.143600	0.143425
5.05	0.140411	0.140232
5.25	0.136437	0.136253
5.45	0.134182	0.133995
5.65	0.126489	0.126292
5.85	0.120718	0.120512
6.05	0.114559	0.114343
6.25	0.108120	0.107894
6.65	0.094818	0.094571
7.05	0.081560	0.081299
7.45	0.068940	0.068675
7.85	0.057370	0.057115
8.25	0.047087	0.046858
8.65	0.038185	0.037995
9.05	0.030645	0.030505
9.45	0.024376	0.024292
9.85	0.019244	0.019216
10.25	0.015097	0.015122

(b) Data of Table 6.3.2 excluding the pair (48, 57)

δ	*Exact*	*Approximate*
3.57	0.149453	0.149380
3.77	0.148973	0.148900
3.97	0.147543	0.147470
4.17	0.145197	0.145122
4.37	0.141985	0.141908
4.57	0.137977	0.137898
4.77	0.133258	0.133177
4.97	0.127925	0.127839
5.17	0.122079	0.121990
5.37	0.115830	0.115736
5.57	0.109286	0.109187
5.97	0.095729	0.095621
6.37	0.082174	0.082060
6.77	0.069237	0.069124
7.17	0.057359	0.057253
7.57	0.046800	0.046711
7.97	0.037672	0.037604
8.37	0.029966	0.029924
8.77	0.023591	0.023576
9.17	0.018407	0.018419
9.57	0.014256	0.014290
9.97	0.010971	0.011024

and, in the absence of a significant result, to run an "unpaired t" analysis. That is, to employ the quantity $(v_e m_e + v_c m_c)/(v_c + v_e)$ as a pooled estimator of σ_e^2 with $(v_c + v_e)$ degrees of freedom. In this case, inferences about δ are made by referring the quantity

$$\tau_1 = \frac{\delta - (y_{i.} - y_{i.})}{\left(\dfrac{2}{J}\,\dfrac{v_e m_e + v_c m_c}{v_e + v_c}\right)^{1/2}} \tag{6.3.41}$$

to the $t(0, 1, v_c + v_e)$ distribution. From the Bayesian point of view, this parallels the case when σ_c^2 is known to be equal to zero. To see this, suppose we now take the prior distribution of $(\boldsymbol{\theta}, \sigma_e^2)$ to be, locally,

$$p(\boldsymbol{\theta}, \sigma_e^2) \propto \sigma_e^2. \tag{6.3.42}$$

Then, on combining this with the likelihood in (6.3.4) conditional on $\sigma_c^2 = 0$, it can be verified that *a posteriori* the quantity τ_1 follows the same $t(0, 1, v_e + v_c)$ distribution.

Note that

$$\tau = \frac{\delta - (y_{2.} - y_{1.})}{\{\Sigma(z_j - \bar{z})^2/[J(J-1)]\}^{1/2}} \quad \text{and} \quad \tau_1 = \frac{\delta - (y_{2.} - y_{1.})}{\{\Sigma\,\Sigma\,(y_{ij} - y_{i.})^2/[J(J-1)]\}^{1/2}} \tag{6.3.43}$$

where, as before in (6.3.29), $z_j = y_{2j} - y_{1j}$ and $\bar{z} = (1/J)\Sigma z_j$. The problem is thus the familiar one of deciding whether an unpaired t test or a paired t test should be adopted in analyzing paired data when it is felt that there might or might not be variation from pair to pair.

For illustration, consider again the complete set of analyst data. On sampling theory, testing the hypothesis that $\sigma_c^2 = 0$ against the alternative $\sigma_c^2 > 0$ (or equivalently testing $\sigma_{ce}^2/\sigma_e^2 = 1$ against $\sigma_{ce}^2/\sigma_e^2 > 1$) involves calculating

$$\Pr\left\{F_{v_c, v_e} > \frac{m_c}{m_e}\right\} = \Pr\{F_{7,7} > 2.67\} = 0.1095.$$

The result is thus not quite significant at the 10% level. The confidence distribution of δ shown by the dotted curve in Fig. 6.3.1 which is obtained from τ (paired t) is, however, appreciably different from that given by the broken curve in the same figure when τ_1 (unpaired t) is used. As mentioned earlier, from the Bayesian point of view, these two curves correspond, respectively, to the posterior distribution of δ when the constraint C: $\sigma_e^2 < \sigma_{ce}^2$ is ignored and that of δ when σ_c^2 is assumed zero. Both of these curves are different from the solid curve which is the appropriate posterior distribution of δ. In obtaining the latter distribution we do not make the outright assumption that $\sigma_c^2 = 0$, nor do we ignore the constraint $\sigma_e^2 < \sigma_{ce}^2$ given by the model.

The difficulty in the sampling theory approach in deciding whether to use τ or τ_1 is another example of the "pooling" dilemma discussed in Section 5.2.7. The posterior distribution $p(\tau \mid \mathbf{y})$ can be analyzed in this light. The discrepancy between $p(\tau \mid \mathbf{y})$ with the constraint C and the unconstrained distribution $p(t_{\nu_e} = \tau)$ can in fact be regarded as a direct consequence of "Bayesian pooling" of m_c and m_e in obtaining the former distribution. As seen in (6.3.37), the posterior distribution of τ can be written

$$p(\tau \mid \mathbf{y}) = \int_0^\infty p(\tau \mid \sigma_e^2, \mathbf{y})\, p(\sigma_e^2 \mid \mathbf{y})\, d\sigma_e^2, \qquad -\infty < \tau < \infty,$$

where $p(\tau \mid \sigma_e^2, \mathbf{y})$ is Normal $N(0, \sigma_e^2/m_e)$ and

$$p(\sigma_e^2 \mid \mathbf{y}) = (\nu_e m_e)^{-1}\, p\left(\chi_{\nu_e}^{-2} = \frac{\sigma_e^2}{\nu_e m_e}\right) \frac{\Pr\{\chi_{\nu_c}^2 < \nu_c m_c/\sigma_e^2\}}{\Pr\{F_{\nu_c,\nu_e} < m_c/m_e\}}, \qquad \sigma_e^2 > 0.$$

Thus, the departure of the posterior distribution of τ from the $t(0, 1, \nu_e)$ distribution depends entirely upon the departure of the distribution of σ_e^2 from that of $(\nu_e m_e)\chi_{\nu_e}^{-2}$.

For the complete analyst data, ($\nu_e = 7$, $m_e = 29.54$) and ($\nu_e' = 8.57$, $m_e' = 27.55$). Thus, the posterior distribution of τ (or equivalently of δ) was not much different from the unconstrained distribution of τ. In terms of the pooling discussion in Section 5.2.7, we can write

$$\frac{1}{\sigma_e^2} \frac{\nu_e m_e + \lambda \nu_c m_c}{\nu_e + \omega \nu_c} \mathrel{\dot\sim} \chi_{\nu_e + \omega\nu_c}^2/(\nu_e + \omega \nu_c) \tag{6.3.44}$$

where from (5.2.38),

$$\lambda = \nu_e\left(\frac{1}{a} - 1\right)\bigg/\nu_c\left(\frac{m_c}{m_e}\right) \qquad \text{and} \qquad \omega = (\nu_e' - \nu_e)/\nu_c.$$

For this example, $\lambda = 0.05$ and $\omega = 0.22$ so that

$$\nu_e m_e + \lambda \nu_c m_c = 206.75 + 0.05 \times 553.0 = 206.75 + 29.8 = 236.55$$

$$\nu_e + \omega \nu_c = 7 + 0.22 \times 7 = 7 + 1.6 = 8.6.$$

Thus, approximately

$$\frac{206.8 + 29.8}{\sigma_e^2} \mathrel{\dot\sim} \chi_{(7+1.6)}^2.$$

The combined effect of the addition of 29.8 to the sum of squares and of 1.6 to the degrees of freedom is small and the distribution of σ_e^2 is nearly the same as for the unconstrained situation which would give

$$\frac{206.8}{\sigma_1^2} \sim \chi_7^2.$$

The much larger effect when the pair (48, 57) is excluded—Fig. 6.3.2—arises from the much greater degree of pooling which occurs here. We have $\nu_e = 6$, $m_e = 32.21$, $\nu_e' = 10.48$, $m_e' = 23.80$ so that

$$\lambda = 0.33, \qquad \omega = 0.751.$$

Thus,

$$\frac{193.9 + 0.33 \times 166.7}{\sigma_e^2} \,\dot{\sim}\, \chi^2_{6+0.75\times 6}$$

or

$$\frac{193.9 + 55.7}{\sigma_e^2} \,\dot{\sim}\, \chi^2_{6+4.5}$$

as compared with the unconstrained distribution

$$\frac{193.9}{\sigma_e^2} \sim \chi_6^2.$$

In this example, therefore, m_c contributes significantly to the posterior distribution of σ_e^2. On sampling theory, because of the small size of the mean squares ratio $m_c/m_e = 0.86$, one might be tempted to pool m_c and m_e. Then

$$\frac{\nu_e m_e + \nu_c m_c}{\sigma_e^2} \sim \chi^2_{\nu_c+\nu_e} \qquad \text{so that} \qquad \frac{193.9 + 166.7}{\sigma_e^2} \sim \chi^2_{6+6}.$$

defines the confidence distribution of σ_e^2 for this set of data. The corresponding confidence distribution of δ which results from this complete pooling and which, from the Bayesian viewpoint, is the posterior distribution of δ on the assumption that $\sigma_c^2 = 0$, has been shown in Fig. 6.3.2. As in the analysis of the complete data (Fig. 6.3.1), the posterior distribution of δ corresponding to a Bayesian 'partial pooling' is *sharper than with "no pooling" or "complete pooling"*.

6.3.8 Comparison of I Means

When I ($I \geqslant 2$) means are to be compared, it is convenient to consider the $I - 1$ linearly independent contrasts $\phi_i = \theta_i - \bar{\theta}$, $i = 1, \ldots, (I-1)$. The sample quantities needed in the analysis were conveniently summarized in Table 6.3.1, the analysis of variance table for the present model. As we have already indicated, the model can be appropriate in the analysis of randomized block data in those cases where it is sensible to represent block contributions as random effects.

The posterior distribution of $\boldsymbol{\phi}' = (\phi_1, \ldots, \phi_{I-1})$ may be obtained from (6.3.13a), (6.3.18), and (6.3.19a) by applying the result (1.5.3) concerning constrained distributions. We thus obtain

$$p(\boldsymbol{\phi} \mid \mathbf{y}) = p(\mathbf{t}_{(I-1)} = \boldsymbol{\phi} - \hat{\boldsymbol{\phi}} \mid m_e \boldsymbol{\Sigma}, \nu_e)\, g(\boldsymbol{\phi}), \qquad -\infty < \phi_i < \infty, \quad i = 1, \ldots, I-1, \tag{6.3.45}$$

where the factor $p(\mathbf{t}_{(I-1)} = \boldsymbol{\phi} - \hat{\boldsymbol{\phi}} \mid m_e \boldsymbol{\Sigma}, \nu_e)$ is the density of a $t_{(I-1)}(\hat{\boldsymbol{\phi}}, m_e \boldsymbol{\Sigma}, \nu_e)$ distribution which is the unconstrained distribution of $\boldsymbol{\phi}$, and the modifying factor $g(\boldsymbol{\phi})$ is

$$g(\boldsymbol{\phi}) = \frac{\Pr^*\{C \mid \boldsymbol{\phi}, \mathbf{y}\}}{\Pr^*\{C \mid \mathbf{y}\}} = \frac{\Pr\{F_{\nu_c, \nu_e + \nu_\phi} < m_c/m_e(\boldsymbol{\phi})\}}{\Pr\{F_{\nu_c, \nu_e} < m_c/m_e\}} \tag{6.3.46}$$

with

$$m_e(\boldsymbol{\phi}) = [\nu_e m_e + S(\boldsymbol{\phi})]/(\nu_e + \nu_\phi).$$

Two "F" ratios occur in the modifying factor $g(\boldsymbol{\phi})$, $m_c/m_e = (S_c/\nu_c)/(S_e/\nu_e)$ is the usual ratio of random effect and error mean squares whereas $m_c/m_e(\boldsymbol{\phi})$ is the ratio of the random effect mean square to a modified "error" mean square $m_e(\boldsymbol{\phi}) = [S_e + S(\boldsymbol{\phi})]/(\nu_e + \nu_\phi)$ in which the treatment and error mean squares are pooled. Both the unconstrained distribution and the factor $g(\boldsymbol{\phi})$ are functions of $S(\boldsymbol{\phi})$ only, so that the posterior distribution $p(\boldsymbol{\phi} \mid \mathbf{y})$ has the same ellipsoidal contours as the unconstrained multivariate distribution. However, since $g(\boldsymbol{\phi})$ is monotonically decreasing in $S(\boldsymbol{\phi})$, the probability contained within any given ellipsoidal contour defined by $S(\boldsymbol{\phi}) = d$ is greater for $p(\boldsymbol{\phi} \mid \mathbf{y})$ than for the unconstrained distribution.

To study this effect more formally we notice that since $p(\boldsymbol{\phi} \mid \mathbf{y})$ is a function of $S(\boldsymbol{\phi})$ only, the probability content of any given contour may be obtained by considering the distribution of $S(\boldsymbol{\phi})$ or more conveniently of

$$V = \frac{m(\boldsymbol{\phi})}{m_e} = \frac{S(\boldsymbol{\phi})/\nu_\phi}{S_e/\nu_e}. \tag{6.3.47}$$

From (6.3.14), (6.3.18), and (6.3.19a) we have

$$p(V \mid \mathbf{y}) = p\,(F_{\nu_\phi, \nu_e} = V)\, g\left(V \,\middle|\, \frac{m_c}{m_e}\right), \qquad V > 0. \tag{6.3.48}$$

The first factor is the ordinary F density with (ν_ϕ, ν_e) degrees of freedom which would define the posterior distribution of V if there were no constraint. The modifying factor $g(V \mid m_c/m_e)$ is the same as $g(\boldsymbol{\phi})$ in (6.3.46) but now considered as a function of V. It is, therefore, monotonically decreasing in V.

It is readily shown that V is stochastically smaller than an F_{ν_ϕ, ν_e} variable. That is, for $V_0 > 0$

$$\Pr\{V < V_0 \mid \mathbf{y}\} \geqslant \Pr\{F_{\nu_\phi, \nu_e} < V_0\}. \tag{6.3.49}$$

To see this, since $p(V \mid \mathbf{y})$ is a probability density, we have

$$1 = \int_0^\infty p(V \mid \mathbf{y})\, dV = \int_0^\infty p(F_{\nu_\phi,\nu_e} = V)\, g\left(V \,\middle|\, \frac{m_c}{m_e}\right) dV = E g\left(V \,\middle|\, \frac{m_c}{m_e}\right) \tag{6.3.50}$$

where the expectation E on the extreme right is taken over the unconstrained F_{ν_ϕ,ν_e} distribution. Now $g(V \mid m_c/m_e)$ is monotonically decreasing in V so that there exists a value V' such that

$$g\left(V \,\middle|\, \frac{m_c}{m_e}\right) > 1 \quad \text{for} \quad V < V' \quad \text{and} \quad g\left(V \,\middle|\, \frac{m_c}{m_e}\right) < 1 \quad \text{for} \quad V > V'. \tag{6.3.51}$$

Consider the difference

$$f(V_0) = \Pr\{V < V_0 \mid \mathbf{y}\} - \Pr\{F_{\nu_\phi,\nu_e} < V_0\}. \tag{6.3.52}$$

Clearly $f(0) = f(\infty) = 0$. Upon differentiating, we have

$$f'(V_0) = p(F_{\nu_\phi,\nu_e} = V_0)\left[g\left(V_0 \,\middle|\, \frac{m_c}{m_e}\right) - 1\right]. \tag{6.3.53}$$

Expressions (6.3.51) and (6.3.53) together imply that $f(V_0)$ can never be negative, and (6.3.49) follows at once.

As in the case of comparing two means, when m_c/m_e tends to infinity, $g(V \mid m_c/m_e)$ approaches unity for all V and the posterior distribution $p(V \mid \mathbf{y})$ tends to the unconstrained $F_{(\nu_\phi,\nu_e)}$ distribution. It can be verified that the mixed derivative

$$\frac{\partial^2 \log g(V \mid m_c/m_e)}{\partial V\, \partial\,(m_c/m_e)} \geqslant 0 \tag{6.3.54}$$

for all V and m_c/m_e, so that for $V_0 > 0$, the probability $\Pr\{V < V_0 \mid \mathbf{y}\}$ is monotonically decreasing in m_c/m_e. Further as $m_c/m_e \to 0$, we obtain

$$\lim_{(m_c/m_e)\to 0} \Pr\{V < V_0 \mid \mathbf{y}\} = \Pr\left\{F_{\nu_\phi,\nu_e+\nu_c} < \left(\frac{I}{I-1}\right)V_0\right\}. \tag{6.3.55}$$

Thus, corresponding to (6.3.33),

$$\Pr\{F_{\nu_\phi,\nu_e} < V_0\} \leqslant \Pr\{V < V_0 \mid \mathbf{y}\} \leqslant \Pr\left\{F_{\nu_\phi,\nu_e+\nu_c} < \left(\frac{I}{I-1}\right)V_0\right\} \tag{6.3.56}$$

which provides an upper and a lower bound for $\Pr\{V < V_0 \mid \mathbf{y}\}$.

The relationship between these results and those of sampling theory is similar to that discussed earlier for the comparison of two means. In particular, the ellipsoid defined by $m(\boldsymbol{\phi})/m_e = F(\nu_\phi, \nu_e, \alpha)$ which encloses the smallest $(1-\alpha)$ confidence region will also be the $(1-\alpha)$ H.P.D. region for the unconstrained distribution of $\boldsymbol{\phi}$. The same ellipsoid will define an H.P.D. region for the posterior distribution of $\boldsymbol{\phi}$, but it is clear from (6.3.49) that the probability content will be greater than $(1 - \alpha)$.

6.3.9 Approximating the Posterior Distribution of $V = m(\boldsymbol{\phi})/m_e$

Using expression (6.3.48), it is computationally simple to calculate the exact probability density of V, the probability integrals involved being obtained from an F table or a table of incomplete beta functions.

Alternatively, as in the case of comparing two means, the distribution of V can be expressed as a finite series when v_c is even. Applying the identity (A5.2.3) in Appendix A5.2 to expand the numerator of $g(V \mid m_c/m_e) = g(\boldsymbol{\phi})$ in (6.3.46) and after some simplification, we get

$$p(V \mid \mathbf{y}) = \frac{p(F_{v_\phi, v_e} = V)}{\Pr\{F_{v_\phi, v_e} < m_c/m_e\}} - \sum_{j=0}^{\frac{1}{2}v_c - 1} \omega_j \gamma_j^2\, p(F_{v_\phi, v_e + 2j} = \gamma_j^2 V), \qquad V > 0, \qquad (6.3.57)$$

where ω_j and γ_j are given in (6.3.35). It follows that

$$\Pr\{V < V_0 \mid \mathbf{y}\} = \frac{\Pr\{F_{v_\phi, v_e} < V_0\}}{\Pr\{F_{v_c, v_e} < m_c/m_e\}} - \sum_{j=0}^{\frac{1}{2}v_c - 1} \omega_j \Pr\{F_{v_\phi, v_e + 2j} < \gamma_j^2 V_0\} \qquad (6.3.58)$$

which can be calculated using an F table or an incomplete beta function table. This expression is, however, not applicable for odd values of v_c and becomes rather complicated, computationally, for large v_c.

A Scaled F Approximation

Adopting an argument similar to that leading to the scaled t approximation for the individual comparison, we write

$$p(V \mid \mathbf{y}) = \int_0^\infty p(V \mid \sigma_e^2, \mathbf{y})\, p(\sigma_e^2 \mid \mathbf{y})\, d\sigma_e^2, \qquad V > 0. \qquad (6.3.59)$$

From the unconstrained joint distribution of $\boldsymbol{\phi}$ in (6.3.12a), it is clear that given σ_e^2, the quantity $V = m(\boldsymbol{\phi})/m_e$ has the $(\sigma_e^2/m_e)\chi^2_{v_\phi}/v_\phi$ distribution. Further, once σ_e^2 is given the constraint C does not affect the distribution of V. Thus the first factor in the integrand of (6.3.59) is, in fact, a $(\sigma_e^2/m_e)\chi^2_{v_\phi}/v_\phi$ distribution. The second factor is, of course, the same posterior distribution of σ_e^2 given in (6.3.38). Employing the scaled χ^{-2} approximation implied by (6.3.39) to the distribution of σ_e^2, we obtain

$$(m_e/m_e')V \sim F_{v_\phi, v_e'}. \qquad (6.3.60)$$

The probability integral of V can thus be approximately determined by using an F table or an incomplete beta function table.

6.3.10 Summarized Calculations

For the complete set of analyst data introduced in Table 6.3.2, Table 6.3.6 provides a summary of the approximating posterior distributions of the various parameters of interest for the additive mixed model.

Table 6.3.6

Summarized calculations for the various approximate posterior distributions for the additive mixed model (applied to the analyst data of Table 6.3.2)

1. Use Tables 6.3.1 and 6.3.3 to obtain

$$y_{1.}=37.38, \qquad y_{2.}=41.63, \qquad y_{..}=39.51$$

$$m_c=79.0, \qquad m_e=29.54, \qquad m(\boldsymbol{\phi})=8[(\phi_1+2.13)^2+(\phi_2-2.13)^2]$$

$$v_c=7, \qquad v_e=7, \qquad v_\phi=1$$

$$I=2, \qquad J=8$$

2. Use (6.3.39) to calculate

$$x=0.728, \qquad a=0.876$$

$$v_e'=8.57, \qquad m_e'=27.55$$

3. Then, for making inferences about σ_e^2

$$\sigma_e^2 \,\dot{\sim}\, v_e' m_e' \chi_{v_e'}^{-2} \qquad \sigma_e^2 \,\dot{\sim}\, (236.6)\chi_{8.6}^{-2}$$

4. For making inferences about σ_c^2

$$\frac{m_e+I\sigma_c^2}{v_c m_c} \,\dot{\sim}\, \chi_{v_c}^{-2} \quad \text{truncated from below at } \frac{m_e}{v_c m_c},$$

$$0.053+0.0036\,\sigma_c^2 \,\dot{\sim}\, \chi_7^{-2} \quad \text{truncated from below at } 0.053.$$

5. For comparison of two means

$$\delta=\theta_i-\theta_{i'}$$

$$\frac{\delta-(y_{i.}-y_{i'.})}{(2m_e'/J)^{1/2}} \,\dot{\sim}\, t(0,1,v_e')$$

$$\frac{\delta-4.25}{2.62} \,\dot{\sim}\, t(0,1,8.6)$$

6. For general comparison of I means (in this case $I=2$)

$$\frac{m(\boldsymbol{\phi})}{m_e'} \,\dot{\sim}\, F_{v_\phi,v_e'} \qquad \frac{m(\boldsymbol{\phi})}{27.55} \,\dot{\sim}\, F_{1,8.6}$$

6.4 THE INTERACTION MODEL

The method of analysis in the previous section can be easily extended to the so-called mixed model with interaction in (6.1.20),

$$y_{ijk} = \theta_i + c_j + t_{ij} + e_{ijk}, \qquad i = 1, \ldots, I; \quad j = 1, \ldots, J; \quad k = 1, \ldots, K \tag{6.4.1}$$

where y_{ijk} are the observations, θ_i are location parameters, c_j are random effects, t_{ij} are random "interaction" effects, and e_{ijk} are random errors. The IJK observations can be arranged in a two way table with I rows, J columns, and K observations per cell as in Table 6.1.1. Following our previous discussion, we assume that c_j, t_{ij}, and e_{ijk} are all independent and that

$$c_j \sim N(0, \sigma_c^2), \qquad t_{ij} \sim N(0, \sigma_t^2), \qquad \text{and} \qquad e_{ijk} \sim N(0, \sigma_e^2). \tag{6.4.2}$$

6.4.1 The Likelihood Function

To derive the likelihood function, it is convenient to define

$$\varepsilon_{ij} = t_{ij} + e_{ij.},$$

and write

$$y_{ijk} - y_{ij.} = e_{ijk} - e_{ij.}, \tag{6.4.3}$$

$$y_{ij.} = \theta_i + c_j + \varepsilon_{ij.} \tag{6.4.4}$$

From Theorem 5.2.1 (on page 250), the deviations $y_{ijk} - y_{ij.}$ are distributed independently to the cell means $y_{ij.}$ and also $S_e = \Sigma\Sigma\Sigma\,(y_{ijk} - y_{ij.})^2$ has the $\sigma_e^2\chi^2$ distribution with $IJ(K-1)$ degrees of freedom. Further, the model in (6.4.4) for the cell means $y_{ij.}$ is of exactly the same form as the additive model discussed in the preceding section with ε_{ij} having Normal distribution $N(0, \sigma_t^2 + K^{-1}\sigma_e^2)$. It follows that the likelihood function is

$$l(\bar{\theta}, \boldsymbol{\phi}, \sigma_e^2, \sigma_t^2, \sigma_c^2 \mid \mathbf{y}) \propto (\sigma_e^2)^{-\frac{1}{2}\nu_e} (\sigma_{et}^2)^{-\frac{1}{2}(\nu_t + \nu_\phi)} (\sigma_{etc}^2)^{-\frac{1}{2}(\nu_c + 1)}$$

$$\times \exp\left\{-\frac{1}{2}\left[\frac{\nu_e m_e}{\sigma_e^2} + \frac{\nu_t m_t + S(\boldsymbol{\phi})}{\sigma_{et}^2} + \frac{\nu_c m_c + IJK(\bar{\theta} - y_{...})^2}{\sigma_{etc}^2}\right]\right\}. \tag{6.4.5}$$

The quantities appearing in the likelihood function can be conveniently arranged in analysis of variance form as in Table 6.4.1.

In Table 6.4.1, $\bar{\theta} = (1/I)\,\Sigma\,\theta_i$, $\boldsymbol{\phi}' = (\phi_1, \ldots, \phi_{I-1})$, $\phi_i = \theta_i - \bar{\theta}$, and

$$\phi_I = -\sum_{i=1}^{I-1} \phi_i.$$

From the definitions of $(\sigma_e^2, \sigma_{et}^2, \sigma_{etc}^2)$, we see that these parameters are subject to the constraint

$$C: \sigma_e^2 < \sigma_{et}^2 < \sigma_{etc}^2. \tag{6.4.6}$$

Table 6.4.1

Analysis of variance for the interaction model

Source	S.S.	d.f.	M.S.	E.M.S.
Grand mean	$IJK(\bar{\theta}-y_{...})^2$	1	$IJK(\bar{\theta}-y_{...})^2$	$\sigma_{etc}^2=\sigma_e^2+K\sigma_t^2+IK\sigma_c^2$
Fixed effects	$S(\boldsymbol{\phi})=JK\Sigma[\phi_i-(y_{i..}-y_{...})]^2$	$\nu_\phi=I-1$	$m(\boldsymbol{\phi})=S(\boldsymbol{\phi})/\nu_\phi$	$\sigma_{et}^2=\sigma_e^2+K\sigma_t^2$
Random effects	$S_c=IK\Sigma(y_{.j.}-y_{...})^2$	$\nu_c=J-1$	$m_c=S_c/\nu_c$	σ_{etc}^2
Interaction	$S_t=K\Sigma\Sigma(y_{ij.}-y_{i..}-y_{.j.}+y_{...})^2$	$\nu_t=(I-1)(J-1)$	$m_t=S_t/\nu_t$	σ_{et}^2
Error	$S_e=\Sigma\Sigma\Sigma(y_{ijk}-y_{ij.})^2$	$\nu_e=IJ(K-1)$	$m_e=S_e/\nu_e$	σ_e^2

6.4.2 Posterior Distribution of $(\bar{\theta}, \boldsymbol{\phi}, \sigma_e^2, \sigma_{et}^2, \sigma_{etc}^2)$

Adopting an argument similar to that given in the previous section, we employ the noninformative reference prior distribution

$$p(\bar{\theta}, \boldsymbol{\phi}, \sigma_e^2, \sigma_{et}^2, \sigma_{etc}^2) \propto \sigma_e^{-2}\,\sigma_{et}^{-2}\,\sigma_{etc}^{-2} \tag{6.4.7}$$

with the constraint C, from which it follows that the posterior distribution can be written

$$p(\bar{\theta}, \boldsymbol{\phi}, \sigma_e^2, \sigma_{et}^2, \sigma_{etc}^2 \mid \mathbf{y}) = p(\bar{\theta} \mid \sigma_{etc}^2, \mathbf{y})\, p(\boldsymbol{\phi} \mid \sigma_{et}^2, \mathbf{y})\, p(\sigma_e^2, \sigma_{et}^2, \sigma_{etc}^2 \mid \mathbf{y}),$$

$$-\infty < \bar{\theta} < \infty; \quad -\infty < \phi_i < \infty, i = 1, \ldots, I-1; \quad \sigma_{etc}^2 > \sigma_{et}^2 > \sigma_e^2 > 0. \tag{6.4.8}$$

In (6.4.8), (i) the conditional distribution of $\bar{\theta}$ given σ_{etc}^2, $p(\bar{\theta} \mid \sigma_{etc}^2, \mathbf{y})$, is Normal $N(y_{\ldots}, \sigma_{etc}^2/IJK)$; (ii) the conditional distribution of $\boldsymbol{\phi}$ given σ_{et}^2, $p(\boldsymbol{\phi} \mid \sigma_{et}^2, \mathbf{y})$, is the $(I-1)$ dimensional multivariate Normal

$$N_{(I-1)}(\hat{\boldsymbol{\phi}}, \sigma_{et}^2\, \boldsymbol{\Sigma}) \tag{6.4.9}$$

with

$$\hat{\boldsymbol{\phi}}' = (y_{1..} - y_{\ldots}, \ldots, y_{(I-1)..} - y_{\ldots})$$

$$\boldsymbol{\Sigma} = \frac{1}{JK}\left[\mathbf{I}_{(I-1)} - \frac{1}{I}\mathbf{1}_{(I-1)}\mathbf{1}'_{(I-1)}\right];$$

and (iii) the marginal distribution of $(\sigma_e^2, \sigma_{et}^2, \sigma_{etc}^2)$ is

$$p(\sigma_e^2, \sigma_{et}^2, \sigma_{etc}^2 \mid \mathbf{y})$$

$$= \frac{(v_e m_e)^{-1}\, p\left(\chi_{v_e}^{-2} = \dfrac{\sigma_e^2}{v_e m_e}\right)(v_t m_t)^{-1}\, p\left(\chi_{v_t}^{-2} = \dfrac{\sigma_{et}^2}{v_t m_t}\right)(v_c m_c)^{-1}\, p\left(\chi_{v_c}^{-2} = \dfrac{\sigma_{etc}^2}{v_c m_c}\right)}{\Pr\left\{\dfrac{\chi_{v_e}^2}{\chi_{v_t}^2} > \dfrac{v_e m_e}{v_t m_t}, \dfrac{\chi_{v_t}^2}{\chi_{v_c}^2} > \dfrac{v_t m_t}{v_c m_c}\right\}},$$

$$\sigma_{etc}^2 > \sigma_{et}^2 > \sigma_e^2 > 0, \tag{6.4.10}$$

where $(\chi_{v_e}^2, \chi_{v_t}^2, \chi_{v_c}^2)$ are independent χ^2 variables with (v_e, v_t, v_c) degrees of freedom respectively. Expression (6.4.10) is of precisely the same form as the distribution in (5.3.10) for the three component hierarchical design model. Inferences about the variances $(\sigma_e^2, \sigma_t^2, \sigma_c^2)$ can thus be made using the corresponding results obtained earlier.

6.4.3 Comparison of I Means

Integrating out $(\bar{\theta}, \sigma_e^2, \sigma_{et}^2, \sigma_{etc}^2)$ from (6.4.8), the posterior distribution of the $(I-1)$ contrasts $\boldsymbol{\phi}$ is

$$p(\boldsymbol{\phi} \mid \mathbf{y}) = p(\mathbf{t}_{(I-1)} = \boldsymbol{\phi} - \hat{\boldsymbol{\phi}} \mid m_t \boldsymbol{\Sigma}, v_t)\, g(\boldsymbol{\phi}), \qquad -\infty < \phi_i < \infty, \quad i = 1, \ldots, I-1, \tag{6.4.11}$$

whose first factor is the density of the multivariate t-distribution $t_{(I-1)}(\hat{\boldsymbol{\phi}}, m_t\boldsymbol{\Sigma}, v_t)$ which would be the distribution of $\boldsymbol{\phi}$ if the constraint C in (6.4.6) were ignored. The modifying factor $g(\boldsymbol{\phi})$ in (6.4.11), which represents the effect of the constraint C in (6.4.6), is

$$g(\boldsymbol{\phi}) = \frac{\Pr\left\{\dfrac{\chi_{v_e}^2}{\chi_{v_t+v_\phi}^2} > \dfrac{v_e m_e}{v_t m_t + S(\boldsymbol{\phi})}, \dfrac{\chi_{v_t+v_\phi}^2}{\chi_{v_c}^2} > \dfrac{v_t m_t + S(\boldsymbol{\phi})}{v_c m_c}\right\}}{\Pr\left\{\dfrac{\chi_{v_e}^2}{\chi_{v_t}^2} > \dfrac{v_e m_e}{v_t m_t}, \dfrac{\chi_{v_t}^2}{\chi_{v_c}^2} > \dfrac{v_t m_t}{v_c m_c}\right\}} \tag{6.4.12}$$

where $\chi_{v_t+v_\phi}^2$ is an χ^2 variable with $v_t + v_\phi$ degrees of freedom independent of $\chi_{v_e}^2$ and $\chi_{v_c}^2$. Since $g(\boldsymbol{\phi})$ is a function of $S(\boldsymbol{\phi})$ only, it follows that the center of the distribution in (6.4.11) remains at $\hat{\boldsymbol{\phi}}$, and the density contours of $\boldsymbol{\phi}$ are ellipsoidal. It is easy to see that $\hat{\boldsymbol{\phi}}$ is the vector of the posterior means of $\boldsymbol{\phi}$ and is also the unique mode of the distribution. These properties are very similar to the distribution of $\boldsymbol{\phi}$ in (6.3.45) for the additive mixed model. However, unlike the latter distribution, $g(\boldsymbol{\phi})$ is no longer monotonically decreasing in $S(\boldsymbol{\phi})$ so that g does not necessarily make the distribution of $\boldsymbol{\phi}$ more concentrated about $\hat{\boldsymbol{\phi}}$. The reason is that the quantity σ_{et}^2 appearing in the covariance matrix of $\boldsymbol{\phi}$ for the conditional distribution $p(\boldsymbol{\phi} \mid \sigma_{et}^2, \mathbf{y})$ in (6.4.9) can be regarded as subject to two constraints, $\sigma_{et}^2 < \sigma_{etc}^2$ and $\sigma_{et}^2 > \sigma_e^2$. While the former constraint tends to reduce the spread of the distribution of $\boldsymbol{\phi}$, the latter tends to increase it. The combined effect of these two constraints depends upon the two mean square ratios m_e/m_t and m_c/m_t. In general the smaller the ratios m_e/m_t and m_c/m_t, the larger will be the reduction in the spread of the distribution of $\boldsymbol{\phi}$.

6.4.4 Approximating the Distribution $p(\boldsymbol{\phi} \mid \mathbf{y})$

For a given set of data, the specific effect of the constraint C can be approximately determined by the procedure we now develop. The posterior distribution of $\boldsymbol{\phi}$ can be written as the integral

$$p(\boldsymbol{\phi} \mid \mathbf{y}) = \int_0^\infty p(\boldsymbol{\phi} \mid \sigma_{et}^2, \mathbf{y})\, p(\sigma_{et}^2 \mid \mathbf{y})\, d\sigma_{et}^2, \qquad -\infty < \phi_i < \infty, \qquad i = 1, \ldots, I-1. \tag{6.4.13}$$

In the integrand, the weight function $p(\sigma_{et}^2 \mid \mathbf{y})$ is

$$p(\sigma_{et}^2 \mid \mathbf{y}) = \int_0^{\sigma_{et}^2} \int_{\sigma_{et}^2}^{\infty} p(\sigma_e^2, \sigma_{et}^2, \sigma_{etc}^2 \mid \mathbf{y})\, d\sigma_{etc}^2\, d\sigma_e^2, \qquad \sigma_{et}^2 > 0. \tag{6.4.14}$$

Adopting the scaled χ^2 approximation techniques developed in Sections 5.2.6 and 5.2.12, to first integrate out σ_{etc}^2 and then eliminate σ_e^2, we find

$$\sigma_{et}^2 \;\dot{\sim}\; (v_t' m_t')\chi_{v_t'}^{-2} \tag{6.4.15}$$

where

$$v_t' = \frac{v_t^*}{a_1} \frac{I_{x_1}(\frac{1}{2}v_t^* + 1, \frac{1}{2}v_e)}{I_{x_1}(\frac{1}{2}v_t^*, \frac{1}{2}v_e)}, \qquad m_t' = \frac{v_t m_t}{a_1 a_2 v_1'}$$

$$a_1 = \left(\frac{v_t^*}{2} + 1\right) \frac{I_{x_1}(\frac{1}{2}v_t^* + 2, \frac{1}{2}v_e)}{I_{x_1}(\frac{1}{2}v_t^* + 1, \frac{1}{2}v_e)} - \frac{v_t^*}{2} \frac{I_{x_1}(\frac{1}{2}v_t^* + 1, \frac{1}{2}v_e)}{I_{x_1}(\frac{1}{2}v_t^*, \frac{1}{2}v_e)}, \quad x_1 = \frac{v_t m_t}{a_2 v_e m_e + v_t m_t},$$

$$v_t^* = \frac{v_t}{a_2} \frac{I_{x_2}(\frac{1}{2}v_c, \frac{1}{2}v_t + 1)}{I_{x_2}(\frac{1}{2}v_c, \frac{1}{2}v_t)},$$

$$a_2 = \left(\frac{v_t}{2} + 1\right) \frac{I_{x_2}(\frac{1}{2}v_c, \frac{1}{2}v_t + 2)}{I_{x_2}(\frac{1}{2}v_c, \frac{1}{2}v_t + 1)} - \frac{v_t}{2} \frac{I_{x_2}(\frac{1}{2}v_c, \frac{1}{2}v_t + 1)}{I_{x_2}(\frac{1}{2}v_c, \frac{1}{2}v_t)},$$

and

$$x_2 = \frac{v_c m_c}{v_c m_c + v_t m_t}.$$

To this degree of approximation,

$$p(\boldsymbol{\phi} \mid \mathbf{y}) \doteq p(\mathbf{t}_{(I-1)} = \boldsymbol{\phi} - \hat{\boldsymbol{\phi}} \mid m_t' \boldsymbol{\Sigma}, v'), \qquad -\infty < \phi_i < \infty, \quad i = 1, \ldots, I-1, \tag{6.4.16}$$

i.e., a $t_{(I-1)}(\hat{\boldsymbol{\phi}}, m_t' \boldsymbol{\Sigma}, v_t')$ distribution. Thus, the approximate effect of the modifying factor $g(\boldsymbol{\phi})$ is the change in the scale factor from $m_t^{1/2}$ to $(m_t')^{1/2}$ and the "degrees of freedom" from v_t to v_t'.

In particular, if one is interested in comparing two means, say θ_i and $\theta_{i'}$, then the approximate posterior distribution of the difference in means $\delta = \theta_{i'} - \theta_i$ is $t(y_{i'..} - y_{i..}, (2/JK)m_t', v_t')$ and H.P.D. intervals for δ can be conveniently calculated. For an overall comparison of all I means $\boldsymbol{\theta}' = (\theta_1, \ldots, \theta_I)$, H.P.D. regions for the $(I-1)$ linearly independent contrasts $\boldsymbol{\phi}$ can be approximately determined by making use of the fact that the quantity

$$V = \frac{m(\boldsymbol{\phi})}{m_t'} = \frac{S(\boldsymbol{\phi})/v_\phi}{m_t'} \tag{6.4.17}$$

is approximately distributed as F with (v_ϕ, v_t') degrees of freedom.

6.4.5 An Example

Consider again the car–driver data introduced in Table 6.2.2, but now suppose that we are interested in comparing the performance of the 9 particular cars. The relevant sample quantities are:

$$m_c = 126.4569 \qquad m_t = 1.7127 \qquad m_e = 1.1759$$

$$v_c = 8 \qquad v_t = 64 \qquad v_e = 81$$

$$v_\phi = 8 \qquad J = 9 \qquad K = 2$$

Car	$\hat{\phi}_i = y_{i..} - y_{...}$	Car	$\hat{\phi}_i = y_{i..} - y_{...}$
1	2.4521	6	−0.4950
2	−1.5971	7	−1.5985
3	1.3023	8	−1.2205
4	0.6063	9	−1.3134
5	1.8641		

On the basis of the non-informative reference prior distribution in (6.4.7) the posterior distribution of the $I - 1 = 8$ independent contrasts $\boldsymbol{\phi} = (\phi_1, \ldots, \phi_8)'$ is that given in (6.4.11). To approximate this distribution, we find by using (6.4.15)

$$x_2 = \frac{1{,}011.655}{1{,}121.267} = 0.902,$$

$$I_{.902}(4, 34) = I_{.902}(4, 33) = I_{.902}(4, 32) \doteq 1.0$$

so that

$$v_t^* \doteq v_t = 64, \qquad a_2 \doteq 1.0.$$

Thus,

$$x_1 = \frac{109.6123}{204.858} = 0.5351$$

$$a_1 = 33 \times \frac{I_{x_1}(34, 40.5)}{I_{x_1}(33, 40.5)} - 32 \times \frac{I_{x_1}(33, 40.5)}{I_{x_1}(32, 40.5)}$$

$$= 33(0.9816) - 32(0.9846) = 0.8856$$

$$v_t' = \frac{64(0.9846)}{0.8856} = 71.15$$

and

$$m_t' = \frac{109.6123}{0.8856 \times 71.15} = 1.74.$$

It follows that $\boldsymbol{\phi}$ is approximately distributed as $t_8(\hat{\boldsymbol{\phi}}, 1.74\,\boldsymbol{\Sigma}, 71.15)$ where from (6.4.9),

$$\boldsymbol{\Sigma} = \tfrac{1}{18}[\mathbf{I}_8 - \tfrac{1}{9}\mathbf{1}_8\mathbf{1}_8'].$$

In particular, suppose we wish to compare the means of cars 1 and 2. Then

$$\delta = \theta_1 - \theta_2 = \phi_1 - \phi_2$$

is approximately distributed as $t(4.05, 0.1933, 71.15)$. Using the Normal approximation, limits of the 95% H.P.D. interval are (3.19, 4.91).

For overall comparison of the means of all nine cars, the $(1-\alpha)$ H.P.D. region of $\boldsymbol{\phi}$ is given by

$$\frac{m(\boldsymbol{\phi})}{1.74} \leqslant F(8, 71.15, \alpha)$$

where

$$m(\boldsymbol{\phi}) = \frac{18}{8}\sum_{i=1}^{9}(\phi_i - \hat{\phi}_i)^2.$$

In particular, one may wish to decide whether the parameter point $\boldsymbol{\phi} = \mathbf{0}$ (i.e., $\theta_1 = \cdots = \theta_9$) lies inside the 95% H.P.D. region. We find

$$\frac{m(\boldsymbol{\phi} = \mathbf{0})}{1.74} = \frac{45.26}{1.74} = 26.01 \quad \text{and} \quad F(8, 71.15, 0.05) \doteq 2.14$$

so that the point $\boldsymbol{\phi} = \mathbf{0}$ is excluded.

Finally, we note that if the constraint (6.4.6) were ignored, then the posterior distribution of $\boldsymbol{\phi}$ for the present example would be $t_8(\hat{\boldsymbol{\phi}}, 1.71\,\boldsymbol{\Sigma}, 64)$, compared with the approximating $t_8(\hat{\boldsymbol{\phi}}, 1.74\,\boldsymbol{\Sigma}, 71.15)$. The effect of the constraint is seen to be very slight.

In the above we have assumed that the nine means θ_i are locally uniform *a priori*. In some situations, especially when the nine cars were of the same make and year, they might be more realistically regarded as a random sample from a large population of cars. In these circumstances, a more appropriate analysis would follow the procedure developed in the following chapter.

CHAPTER 7

INFERENCE ABOUT MEANS WITH INFORMATION FROM MORE THAN ONE SOURCE: ONE-WAY CLASSIFICATION AND BLOCK DESIGNS

7.1 INTRODUCTION

Much scientific work has as its object the detection and measurement of possible changes in some observable response associated with qualitative change in the experimental conditions. To make it possible to assess such influences, experiments are frequently arranged so that the data can be *classified* in some convenient manner. Thus, in our earlier discussion, we have considered data coming from one-way classification arrangements, cross classifications, and hierarchical designs.

With sampling theory, when the object was to compare means, the analysis was ordinarily conducted using a "fixed effect" model, but when variance components were of interest the analysis employed a "random effect" model. Mixed models were employed when some factors were thought of as contributing fixed and others random effects. As we have seen in Chapters 5 and 6, a number of difficulties were associated with analysis of these models in the sampling theory approach.

By considering two other important problems in this general area we hope to show further, in the present chapter, how the Bayes approach illuminates and resolves difficulties. The two selected problems concern the comparison of means

i) for the one-way classification design,

ii) for balanced incomplete block designs.

For problem (i) the sampling theory difficulties are evidenced, for example, by the results of James and Stein (1961) which show the usual averages to be inadmissable estimators of the group means. For problem (ii) in sampling theory terms, one difficulty is that of appropriately "recovering inter-block information."

An inadequacy of the framework outlined above is that it does not provide for the common situation in which we wish to *compare means* which are nevertheless "*random effects.*" Similarly block effects are often best treated as random. These inadequacies are easily remedied in the Bayesian approach, and when this is used, it becomes clear that the problems encountered with sampling theory concern once more the difficulty, on that theory, of *appropriate pooling of information* from more than one source.

7.2 INFERENCES ABOUT MEANS FOR THE ONE-WAY RANDOM EFFECT MODEL

Suppose we have data y_{jk} arranged in a one-way classification with J groups and K observations per group. For example, consider again the dyestuff data [taken from Davies (1967, p. 105)] for $K = 5$ laboratory determinations made on samples from each of $J = 6$ batches. These data were previously considered in Sections 5.1 and 5.2. The observations and the batch averages are shown in Table 7.2.1.

Table 7.2.1

Yield of dyestuff in grams of standard color

	Batch					
	1	2	3	4	5	6
	145	140	195	45	195	120
Individual	40	155	150	49	230	55
observations	40	90	205	195	115	50
(yield–1400)	120	160	110	65	235	80
	180	95	160	145	225	45
Averages	105	128	164	98	200	70

7.2.1 Various Models Used in the Analysis of One-way Classification

In Section 2.11 we have considered the analysis of data arranged in this way in relation to a "fixed effect" model

$$y_{jk} = \theta_j + e_{jk}, \qquad j = 1, 2, \ldots, J: \quad k = 1, 2, \ldots, K, \tag{7.2.1}$$

with the errors distributed independently and Normally such that $e_{jk} \sim N(0, \sigma_1^2)$. In that formulation, the θ_j were supposed to have locally uniform reference priors. This would seem appropriate if the means θ_j were expected to bear no strong relationship one to another. Certainly, problems occasionally do occur where this is so. Thus, in comparing the laboratory yields for several different methods of making a particular chemical product, we could have a situation approximated by the supposition that any one of the methods could give yields anywhere within a wide range, *independently* of the others. Given this supposition, we have seen that *a posteriori* the θ_j would have a multivariate t distribution, and consideration as to whether a particular parameter point $\boldsymbol{\theta}$ was or was not included in a particular H.P.D. region could be decided by the use of the F distribution. These results all have exact parallels in the standard sampling theory analysis.

It is apparent, however, that locally uniform priors for the θ_j would be totally inappropriate for the dyestuff data. A model more likely to fit these circumstances would be one where the batch means θ_j were regarded as independent drawings from a distribution. These data have already been studied in connection with such a model in (5.2.1). Specifically, writing $\theta_j = \theta + e_j$, it was assumed that

$$e_j \sim N(0, \sigma_2^2), \qquad \text{that is,} \qquad \theta_j \sim N(\theta, \sigma_2^2), \tag{7.2.2}$$

but was supposed that we wished to learn about σ_2^2, the *variance* of the distribution of the θ_j. This "random effect" model could, however, equally well be appropriate when interest centered on the individual batch means $\theta_1, ..., \theta_6$, rather than on the variance σ_2^2.

As a further example, if the performance of six Volkswagen cars bought in the same year was tested on five different days, the main objective could be to compare the mean performance of the *particular* six cars being tested, even though the cars could reasonably be regarded as random drawings from a population of Volkswagens. As Lindley has pointed out in his discussion of the work of Stein (1962), it is common for a random effect model to be appropriate, and yet to wish to make inferences about means.

In what follows then, the random effect model of Section 5.2 is used but the object now is to make inferences about the individual batch means θ_j. We later consider in some detail the contrast between this random effect analysis of the means and the corresponding fixed effect analysis, and also its relation to certain results in sampling theory.

7.2.2 Use of the Random Effect Model in the Study of Means

Assuming then the random effect model defined in (7.2.1) and (7.2.2), the usual associated analysis of variance is given below in Table 7.2.2.

Table 7.2.2

Analysis of variance for one-way classification random effect model

Source	S.S.	d.f.	M.S.	Sampling expectation of M.S.
Between groups (batches)	$S_2 = K\Sigma(y_{j.} - y_{..})^2$	$\nu_2 = J-1$	$m_2 = S_2/\nu_2$	$\sigma_{12}^2 = \sigma_1^2 + K\sigma_2^2$
Within groups	$S_1 = \Sigma\Sigma(y_{jk} - y_{j.})^2$	$\nu_1 = J(K-1)$	$m_1 = S_1/\nu_1$	σ_1^2

As before, we have adopted the notation that a subscript replaced by a dot means the average over that subscript.

Prior and Posterior Distributions of $(\boldsymbol{\theta}, \theta, \sigma_1^2, \sigma_2^2)$

Let $\boldsymbol{\theta}$ be a vector whose elements are the group means $\theta_1, \theta_2, ..., \theta_J$. This vector must be distinguished from the scalar $\theta = E(\theta_j), j = 1, 2, ..., J$. The joint distribution of the JK observations $\mathbf{y}$ and the unknown parameters

$(\boldsymbol{\theta}, \sigma_1^2, \sigma_2^2, \theta)$ may be written

$$p(\mathbf{y}, \boldsymbol{\theta}, \sigma_1^2, \sigma_2^2, \theta) = p(\theta, \sigma_1^2, \sigma_2^2)\, p(\boldsymbol{\theta} \mid \sigma_2^2, \theta)\, p(\mathbf{y} \mid \boldsymbol{\theta}, \sigma_1^2) \tag{7.2.3}$$

where from (7.2.1)

$$p(\mathbf{y} \mid \boldsymbol{\theta}, \sigma_1^2) \propto (\sigma_1^2)^{-JK/2} \exp\left\{-\frac{1}{2\sigma_1^2}[\nu_1 m_1 + K\Sigma(\theta_j - y_{j.})^2]\right\},$$

$$-\infty < y_{jk} < \infty, \quad j = 1, \ldots, J, \quad k = 1, \ldots, K, \tag{7.2.4}$$

from (7.2.2)

$$p(\boldsymbol{\theta} \mid \sigma_2^2, \theta) \propto (\sigma_2^2)^{-J/2} \exp\left[-\frac{1}{2\sigma_2^2}\Sigma(\theta_j - \theta)^2\right], \qquad -\infty < \theta_j < \infty, \quad j = 1, \ldots, J \tag{7.2.5}$$

and $p(\theta, \sigma_1^2, \sigma_2^2)$ is the prior distribution of $(\theta, \sigma_1^2, \sigma_2^2)$.

For a noninformative reference prior of $(\theta, \sigma_1^2, \sigma_2^2)$, note that by combining (7.2.4) and (7.2.5) and integrating out $\boldsymbol{\theta}$, the joint distribution of $\mathbf{y}$ given $(\theta, \sigma_1^2, \sigma_2^2)$ is precisely proportional to the likelihood function in (5.2.7). Thus, given $\mathbf{y}$, the likelihood function of $(\theta, \sigma_1^2, \sigma_2^2)$ is

$$l(\theta, \sigma_1^2, \sigma_2^2 \mid \mathbf{y}) \propto (\sigma_1^2)^{-\nu_1/2} (\sigma_{12}^2)^{-(\nu_2+1)/2}$$
$$\times \exp\left\{-\frac{1}{2}\left[\frac{JK(\theta - y_{..})^2 + \nu_2 m_2}{\sigma_{12}^2} + \frac{\nu_1 m_1}{\sigma_1^2}\right]\right\}. \tag{7.2.6}$$

Thus, following the previous approach to variance component models, inferences are considered against the background of the noninformative reference prior

$$p(\sigma_1^2, \sigma_2^2, \theta) \propto \sigma_1^{-2} \sigma_{12}^{-2} \tag{7.2.7}$$

in which $\log \sigma_1^2$, $\log \sigma_{12}^2$, and θ are supposed locally uniform, and the fact that $\sigma_{12}^2 = \sigma_1^2 + K\sigma_2^2$ implies the constraint $\sigma_{12}^2 > \sigma_1^2$.

Given the sample $\mathbf{y}$, the joint posterior distribution of $(\boldsymbol{\theta}, \sigma_1^2, \sigma_2^2, \theta)$ is then

$$p(\boldsymbol{\theta}, \sigma_1^2, \sigma_2^2, \theta \mid \mathbf{y}) \propto \sigma_1^{-2}\sigma_{12}^{-2}\, p(\boldsymbol{\theta} \mid \sigma_2^2, \theta)\, p(\mathbf{y} \mid \boldsymbol{\theta}, \sigma_1^2) \propto \sigma_{12}^{-2} (\sigma_1^2)^{-(\frac{1}{2}JK+1)} (\sigma_2^2)^{-\frac{1}{2}J}$$
$$\times \exp\left\{-\frac{1}{2}\left[\frac{\nu_1 m_1 + K\Sigma(\theta_j - y_{j.})^2}{\sigma_1^2} + \frac{\Sigma(\theta_j - \theta)^2}{\sigma_2^2}\right]\right\},$$
$$\sigma_1^2 > 0, \quad \sigma_2^2 > 0, \quad -\infty < \theta_j < \infty, \quad j = 1, \ldots, J. \tag{7.2.8}$$

7.2.3 Posterior Distribution of $\boldsymbol{\theta}$ via Intermediate Distributions

Before obtaining the posterior distribution of $\boldsymbol{\theta}$, various intermediate distributions of $(\boldsymbol{\theta}, \sigma_1^2, \sigma_2^2, \theta)$ are first derived. This approach facilitates subsequent comparison of random and fixed effect models.

Conditional Posterior Distribution of $\boldsymbol{\theta}$ *Given* $(\theta, \sigma_1^2, \sigma_2^2)$

The joint distribution in (7.2.8) can be written as the product

$$p(\boldsymbol{\theta}, \sigma_1^2, \sigma_2^2, \theta \mid \mathbf{y}) = p(\boldsymbol{\theta} \mid \sigma_1^2, \sigma_2^2, \theta, \mathbf{y})\, p(\sigma_1^2, \sigma_2^2, \theta \mid \mathbf{y}). \tag{7.2.9}$$

Using (A.1.1.5) in Appendix A1.1, we can write

$$\frac{K}{\sigma_1^2}(\theta_j - y_{j.})^2 + \frac{1}{\sigma_2^2}(\theta_j - \theta)^2 = \frac{K}{\sigma_1^2(1-z)}(\theta_j - \tilde{\theta}_j)^2 + \frac{K}{\sigma_{12}^2}(\theta - y_{j.})^2, \tag{7.2.10}$$

where

$$\tilde{\theta}_j = (1 - z)y_{j.} + z\theta \tag{7.2.11}$$

and

$$z = \sigma_1^2/\sigma_{12}^2.$$

Note that z is the reciprocal of $W = \sigma_{12}^2/\sigma_1^2$, the quantity considered in (5.2.15).

It follows that conditional on $(\theta, \sigma_1^2, \sigma_2^2)$ the θ_j are independently distributed *a posteriori* as

$$\theta_j \sim N\left[\tilde{\theta}_j, \frac{\sigma_1^2}{K}(1-z)\right], \quad j = 1, \ldots, J. \tag{7.2.12}$$

Posterior Distribution of $(\theta, \sigma_1^2, \sigma_2^2)$

Also, the posterior distribution of $(\theta, \sigma_1^2, \sigma_2^2)$ is

$$p(\theta, \sigma_1^2, \sigma_2^2 \mid \mathbf{y}) \propto (\sigma_1^2)^{-(\frac{1}{2}\nu_1+1)}(\sigma_{12}^2)^{-(\frac{1}{2}J+1)} \exp\left\{\frac{1}{2}\left[\frac{\nu_1 m_1}{\sigma_1^2} + \frac{K\,\Sigma(\theta - y_{j.})^2}{\sigma_{12}^2}\right]\right\},$$

$$\sigma_1^2 > 0,\ \sigma_2^2 > 0, \quad -\infty < \theta < \infty, \tag{7.2.13}$$

which is equivalent to the distribution in (5.2.11). This implies that in particular

$$p(\sigma_2^2 \mid \theta, \sigma_1^2, \mathbf{y}) \propto (\sigma_{12}^2)^{-(\frac{1}{2}J+1)} \exp\left\{-\frac{K\,\Sigma\,(\theta - y_{j.})^2}{2\sigma_{12}^2}\right\}, \quad \sigma_2^2 > 0, \tag{7.2.14}$$

which is a truncated χ^{-2} distribution.

Posterior Distribution of $(\boldsymbol{\theta}, z)$

Using (7.2.10)

$$\frac{1}{2}\left[\frac{\nu_1 m_1 + K\,\Sigma\,(\theta_j - y_{j.})^2}{\sigma_1^2} + \frac{\Sigma\,(\theta_j - \theta)^2}{\sigma_2^2}\right] = \frac{1}{2\sigma_1^2}\left[\nu_1 m_1 + \frac{\sigma_1^2}{\sigma_2^2}J(\theta - \bar{\theta})^2 + KS(\boldsymbol{\theta}, z)\right], \tag{7.2.15}$$

where

$$S(\boldsymbol{\theta}, z) = \Sigma\,(\theta_j - y_{j.})^2 + \frac{z}{1-z}\,\Sigma\,(\theta_j - \bar{\theta})^2$$

and

$$\bar{\theta} = J^{-1}\,\Sigma\,\theta_j\ .$$

Making the transformation from $(\boldsymbol{\theta}, \sigma_1^2, \sigma_2^2, \theta)$ to $(\boldsymbol{\theta}, \sigma_1^2, z, \theta)$ and integrating out θ and σ_1^2, in (7.2.8), we obtain

$$p(\boldsymbol{\theta}, z \mid \mathbf{y}) \propto z^{\frac{1}{2}\nu_2 - 1}\,(1-z)^{-\frac{1}{2}\nu_2}\,[\nu_1 m_1 + KS(\boldsymbol{\theta}, z)]^{-\frac{1}{2}(\nu_1+\nu_2+J)},$$
$$0 < z < 1, \quad -\infty < \theta_j < \infty, \quad j = 1, \ldots, J. \qquad (7.2.16)$$

Now the quantity $S(\boldsymbol{\theta}, z)$ may be written

$$S(\boldsymbol{\theta}, z) = (\boldsymbol{\theta} - \mathbf{y}_{.})'(\boldsymbol{\theta} - \mathbf{y}_{.}) + \frac{z}{1-z}\,\boldsymbol{\theta}'(\mathbf{I} - J^{-1}\mathbf{1}\mathbf{1}')\boldsymbol{\theta} \qquad (7.2.17)$$

where $\mathbf{y}_{.}' = (y_{1.}, \ldots, y_{J.})$ and $\mathbf{1}$ is a $J \times 1$ vector of ones. By making use of the identity (A7.1.1) in Appendix A7.1, we find

$$KS(\boldsymbol{\theta}, z) = z\nu_2 m_2 + (1-z)^{-1}\,Q(\boldsymbol{\theta}, z), \qquad (7.2.18)$$

where

$$Q(\boldsymbol{\theta}, z) = K[\boldsymbol{\theta} - \hat{\boldsymbol{\theta}}(z)]'(\mathbf{I} - zJ^{-1}\mathbf{1}\mathbf{1}')[\boldsymbol{\theta} - \hat{\boldsymbol{\theta}}(z)],$$
$$\hat{\boldsymbol{\theta}}'(z) = [\hat{\theta}_1(z), \ldots, \hat{\theta}_J(z)],\ \hat{\theta}_j(z) = y_{j.} - z(y_{j.} - y_{..}), \quad j = 1, \ldots, J.$$

Thus,

$$p(\boldsymbol{\theta}, z \mid \mathbf{y}) \propto z^{\frac{1}{2}\nu_2 - 1}\,(1-z)^{-\frac{1}{2}\nu_2}\,[\nu_1 m_1 + z\nu_2 m_2 + (1-z)^{-1}\,Q(\boldsymbol{\theta}, z)]^{-\frac{1}{2}(\nu_1+\nu_2+J)},$$
$$0 < z < 1, \quad -\infty < \theta_j < \infty, \quad j = 1, \ldots, J. \qquad (7.2.19)$$

Posterior Distribution of $\boldsymbol{\theta}$ *Conditional on* z

From (7.2.19) it follows at once that, *conditional on* z,

a) the distribution of $\boldsymbol{\theta}$ is the J-dimensional t distribution

$$t_J[\hat{\boldsymbol{\theta}}(z), s^2(z)K^{-1}\,(\mathbf{I} - zJ^{-1}\mathbf{1}\mathbf{1}')^{-1}, \nu_1 + \nu_2], \qquad (7.2.20)$$

where

$$\hat{\boldsymbol{\theta}}'(z) = [\hat{\theta}_1(z), \ldots, \hat{\theta}_J(z)],$$
$$\hat{\theta}_j(z) = y_{j.} - z(y_{j.} - y_{..}) = (1-z)y_{j.} + zy_{..},$$
$$s^2(z) = \frac{\nu_1 m_1 + z\nu_2 m_2}{\nu_1 + \nu_2}\,(1-z)$$

and

$$(\mathbf{I} - zJ^{-1}\mathbf{1}\mathbf{1}')^{-1} = \mathbf{I} + J^{-1}\left(\frac{z}{1-z}\right)\mathbf{1}\mathbf{1}',$$

i.e.,

$$p(\boldsymbol{\theta} \mid z, \mathbf{y}) \propto \left[1 + \frac{Q(\boldsymbol{\theta}, z)}{(\nu_1 + \nu_2)s^2(z)}\right]^{-\frac{1}{2}(\nu_1+\nu_2+J)},$$

$$-\infty < \theta_j < \infty, \quad j = 1, \ldots, J,$$

b) the θ_j have means $\hat{\theta}_j(z)$, $j = 1, \ldots, J$, and they have common variance and covariance

$$\operatorname{Var}(\theta_j \mid z) = \frac{(\nu_1 + \nu_2)}{K(\nu_1 + \nu_2 - 2)}\left[1 + \frac{z}{J(1-z)}\right]s^2(z),$$

$$\operatorname{Cov}(\theta_i, \theta_j \mid z) = \frac{(\nu_1 + \nu_2)}{K(\nu_1 + \nu_2 - 2)}\left[\frac{z}{J(1-z)}\right]s^2(z), \tag{7.2.21}$$

c) the marginal distribution of θ_j is a t distribution having $\nu_1 + \nu_2$ degrees of freedom; specifically, the quantity

$$\frac{\theta_j - \theta_j(z)}{\left[\left(\dfrac{\nu_1 + \nu_2 - 2}{\nu_1 + \nu_2}\right)\operatorname{Var}(\theta_j \mid z)\right]^{1/2}}$$

is distributed as $t(0, 1, \nu_1 + \nu_2)$, (7.2.22)

d) more generally, the marginal distribution of a linear function of the means $\eta = \boldsymbol{l}'\boldsymbol{\theta}$, where $\boldsymbol{l}' = (l_1 \ldots, l_J)$, is such that the quantity

$$\frac{\eta - \hat{\eta}(z)}{\sigma(\boldsymbol{l})},$$

with

$$\hat{\eta}(z) = \boldsymbol{l}'\,\hat{\boldsymbol{\theta}}(z) \qquad \text{and} \qquad \sigma^2(\boldsymbol{l}) = s^2(z)K^{-1}\left[\boldsymbol{l}'\boldsymbol{l} + J^{-1}\left(\frac{z}{1-z}\right)(\boldsymbol{l}'\mathbf{1})^2\right],$$

is distributed as

$$t(0, 1, \nu_1 + \nu_2), \tag{7.2.23}$$

e) in particular, the marginal distribution of a particular difference $\theta_i - \theta_j$ is a t distribution having $\nu_1 + \nu_2$ degrees of freedom, namely, the quantity

$$\frac{(\theta_i - \theta_j) - (1 - z)(y_{i.} - y_{j.})}{[2K^{-1}s^2(z)]^{1/2}}$$

has the $t(0, 1, \nu_1 + \nu_2)$ distribution, (7.2.24)

f) the joint distribution of the $(J - 1)$ linearly independent contrasts $\boldsymbol{\phi} = (\phi_1, \ldots, \phi_{J-1})'$ where $\phi_j = \theta_j - \bar{\theta}$, $j = 1, \ldots, J - 1$, is given by the $J - 1$

dimensional t distribution having $\nu_1 + \nu_2$ degrees of freedom defined by

$$p(\boldsymbol{\phi} \mid z\, \mathbf{y}) \propto \left\{1 + \frac{K \sum_{j=1}^{J} [\phi_j - (1-z)(y_{j.} - y_{..})]^2}{(\nu_1 + \nu_2) s^2(z)}\right\}^{-\frac{1}{2}(\nu_1 + \nu_2 + J - 1)},$$

$$-\infty < \phi_j < \infty, \quad j = 1, \ldots, J-1, \tag{7.2.25}$$

where $\phi_J = -(\phi_1 + \cdots + \phi_{J-1})$.

Posterior Distribution of z
Now from (5.2.16) and remembering that $z = W^{-1}$, the marginal distribution of $z = \sigma_1^2/\sigma_{12}^2$ is given by

$$p(z \mid \mathbf{y}) = \frac{(m_2/m_1)\, p[F_{\nu_2,\nu_1} = (m_2/m_1)z]}{\Pr\{F_{\nu_2,\nu_1} < m_2/m_1\}}, \qquad 0 < z < 1, \tag{7.2.26}$$

where, as before, $p(F_{\nu_2,\nu_1} = c)$ is the density of an F variable with (ν_2, ν_1) degrees of freedom evaluated at c. The rth moment $(r < \frac{1}{2}\nu_1)$ of z is, therefore,

$$\mu'_r = E(z)^r = \frac{B(\frac{1}{2}\nu_2 + r, \frac{1}{2}\nu_1 - r)}{B(\frac{1}{2}\nu_2, \frac{1}{2}\nu_1)} \frac{I_x(\frac{1}{2}\nu_2 + r, \frac{1}{2}\nu_1 - r)}{I_x(\frac{1}{2}\nu_2, \frac{1}{2}\nu_1)} \left(\frac{\nu_1 m_1}{\nu_2 m_2}\right)^r, \tag{7.2.27}$$

where $x = \dfrac{\nu_2 m_2}{\nu_1 m_1 + \nu_2 m_2}$, and $B(p,q)$ and $I_x(p,q)$ are, respectively, the complete and the incomplete beta functions.

Posterior Distribution of $\boldsymbol{\theta}$
The unconditional distribution of the means $\boldsymbol{\theta}$ may now be obtained by integrating the conditional distribution of $\boldsymbol{\theta}$ for given z in (7.2.20) with $p(z \mid \mathbf{y})$ as weight function. Thus

$$p(\boldsymbol{\theta} \mid \mathbf{y}) = \int_0^1 p(\boldsymbol{\theta} \mid z, \mathbf{y})\, p(z \mid \mathbf{y}) dz, \qquad -\infty < \theta_j\, \infty, \quad j = 1, \ldots, J. \tag{7.2.28}$$

In particular, the unconditional distribution of any linear function $\eta = \boldsymbol{l}'\boldsymbol{\theta}$ of the elements of $\boldsymbol{\theta}$ can be similarly obtained by integrating the conditional distribution of η given z in (7.2.23) over the distribution $p\,(z \mid \mathbf{y})$. Also, to obtain the moments of θ_j, we can take the expected values over z of the conditional moments. In particular, we find

$$E(\theta_j \mid \mathbf{y}) = \hat{\theta}_j = \underset{z}{E}[\hat{\theta}_j(z)] = y_{j.} - \mu_1'(y_{j.} - y_{..})$$

$$= (1 - \mu_1')y_{j.} + \mu_1' y_{..}, \tag{7.2.29}$$

$$\operatorname{Var}(\theta_j \mid \mathbf{y}) = \underset{z}{E} \operatorname{Var}(\theta_j \mid z) + \underset{z}{E}[\hat{\theta}_j(z) - \hat{\theta}_j]^2$$

$$= \frac{J(\nu_1 m_1 + \mu_1' \nu_2 m_2) - (J-1)(\mu_1' \nu_1 m_1 + \mu_2' \nu_2 m_2)}{JK(\nu_1 + \nu_2 - 2)}$$

$$+ (y_{j.} - y_{..})^2 \mu_2, \tag{7.2.30}$$

and

$$\operatorname{Cov}(\theta_i, \theta_j \mid \mathbf{y}) = \underset{z}{E} \operatorname{Cov}(\theta_i, \theta_j \mid z) + \underset{z}{E}[\hat{\theta}_i(z) - \hat{\theta}_i][\hat{\theta}_j(z) - \hat{\theta}_j]$$

$$= \frac{\mu_1' v_1 m_1 + \mu_2' v_2 m_2}{JK(v_1 + v_2 - 2)} + (y_{i.} - y_{..})(y_{j.} - y_{..})\mu_2, \qquad (7.2.31)$$

where

$$\mu_2 = \mu_2' - (\mu_1')^2.$$

7.2.4 Random Effect and Fixed Effect Models for Inferences about Means

Inferences about the means θ_j resulting from the random effect formulation are different from those for the fixed effect formulation discussed earlier in Section 2.11. Specifically, in contrast to the results in (7.2.28) through (7.2.31), we recall from (2.11.1) that for the *fixed effect model, a posteriori*

$$\boldsymbol{\theta} \sim t_J(\mathbf{y}_{.}, m_1 K^{-1}\mathbf{I}, v_1) \qquad (7.2.32)$$

where $\mathbf{y}_{.}' = (y_{1.}, \ldots, y_{J.})$.

In particular, the marginal distribution of a linear function $\eta = \boldsymbol{l}'\boldsymbol{\theta}$ is

$$\eta \sim t(\boldsymbol{l}'\mathbf{y}_{.}, m_1 K^{-1}\boldsymbol{l}'\boldsymbol{l}, v_1) \qquad (7.2.33)$$

Also, the means, variances and covariances of the θ_j are

$$E(\theta_j \mid \mathbf{y}) = y_{j.}, \qquad \operatorname{Var}(\theta_j \mid \mathbf{y}) = \frac{m_1}{K} \frac{v_1}{v_1 - 2}, \qquad \text{and} \qquad \operatorname{Cov}(\theta_i, \theta_j \mid \mathbf{y}) = 0. \qquad (7.2.34)$$

An Illustrative Example

For illustration, consider the dyestuff data quoted in Table 7.2.1. In this example θ_j is the mean yield for the jth batch of dyestuff and

$$J = 6, K = 5, \quad v_1 = 24, \quad v_2 = 5, \quad m_1 = 2{,}451.25, \quad m_2 = 11{,}271.50$$

$$y_{..} = 127.5, \quad \mu_1' = 0.233, \quad \mu_2 = 0.026.$$

Figure 7.2.1 contrasts the posterior distributions of the θ_j from random and fixed effect models. The greater clustering of the distributions about $y_{..}$ which occurs with the random effect model is clearly seen. Table 7.2.3 contrasts the means and standard deviations of the distributions obtained from these two models. To facilitate comparison, we have arranged and numbered the groups in order of magnitude of the sample means $y_{j.}$. Inspection of the table shows the clustering of the posterior means and also the slight reduction in standard deviation of the distributions with the random effect models.

Figure 7.2.2 shows the distribution of $\theta_6 - \theta_1$ for the random model effect together with the corresponding distribution appropriate to the fixed effect model. In this extreme case, a very large difference is seen between the two distributions.

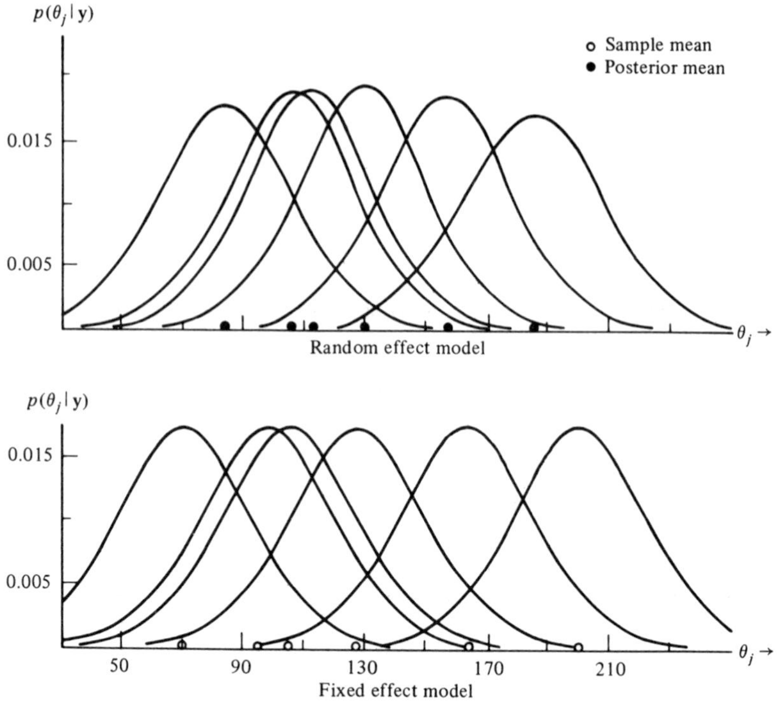

Fig. 7.2.1 Posterior distributions of θ_j for the dyestuff data.

Table 7.2.3

Means and standard deviations of θ_j for random effect and fixed effect models: the dyestuff data

Group (ordered by magnitude of mean)	Posterior means $E(\theta_j \mid \mathbf{y})$ Random effect $y_{j.} - 0.233(y_{j.} - y_{..})$	Fixed effect $y_{j.}$	Standard deviations $[\mathrm{Var}\,(\theta_j \mid \mathbf{y})]^{1/2}$ Random effect $[420.4 + 0.026(y_{j.} - y_{..})^2]^{1/2}$	Fixed effect $(534.8)^{1/2}$
1	83	70	22.5	23.1
2	105	98	21.1	23.1
3	110	105	20.8	23.1
4	128	128	20.5	23.1
5	156	164	21.3	23.1
6	183	200	23.6	23.1

Although for the fixed effect model the θ_j are uncorrelated, this is not so for the random effect model. The rather slight correlations that occur in the present example are shown in Table 7.2.4.

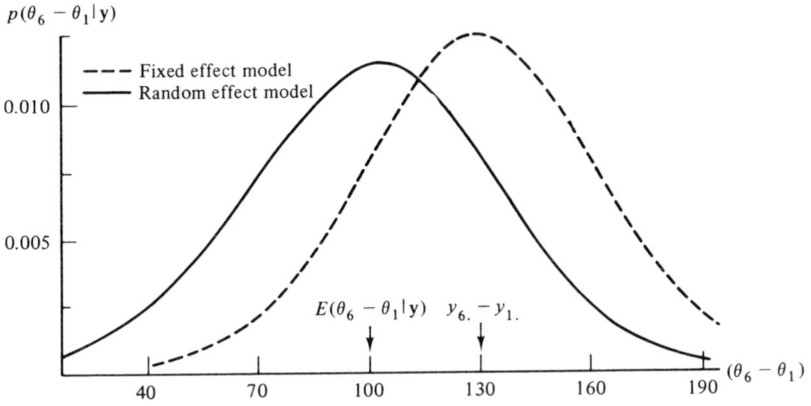

Fig. 7.2.2 Posterior distributions of $\theta_6 - \theta_1$ for the dyestuff data.

Table 7.2.4

Correlation matrix of θ_j for the random effect model: the dyestuff data
$\rho_{ij} = \text{Cov}(\theta_i, \theta_j | \mathbf{y}) / [\text{Var}(\theta_i | \mathbf{y}) \text{Var}(\theta_j | \mathbf{y})]^{1/2}$

Group	1	2	3	4	5	6
1	1	0.14	0.12	0.05	−0.07	−0.16
2		1	0.09	0.05	−0.01	−0.07
3			1	0.05	+0.00	−0.04
4				1	0.05	0.05
5					1	0.18
6						1

7.2.5 Random Effect Prior versus Fixed Effect Prior

In the Bayes framework there can, properly speaking, be no "fixed" effects since the parameters θ_j are regarded as having probability distributions in any case. The terminology "fixed effect" and "random effect" are sampling theory concepts which we have retained to make certain analogies clear. The distinctions arise from the different *prior* distributions appropriate in different circumstances. Specifically, corresponding to the "fixed effect" model, the appropriate non-informative prior is one where $p(\boldsymbol{\theta})$ is taken to be locally uniform so that

$$p(\boldsymbol{\theta}, \sigma_1^2) = p(\boldsymbol{\theta})\, p(\sigma_1^2) \tag{7.2.35}$$

with

$$p(\boldsymbol{\theta}) \propto c \qquad \text{and} \qquad p(\sigma_1^2) \propto \sigma_1^{-2}.$$

This will be called the "fixed effect" prior.

On the other hand, corresponding to the "random effect" model, the appropriate prior is

$$p(\boldsymbol{\theta}, \sigma_1^2, \sigma_2^2, \theta) = p(\theta, \sigma_1^2, \sigma_2^2)\, p(\boldsymbol{\theta} \mid \sigma_2^2, \theta), \tag{7.2.36}$$

where

$$p(\theta, \sigma_1^2, \sigma_2^2) \propto \sigma_1^{-2}\, \sigma_{12}^{-2}$$

is our noninformative prior for $(\theta, \sigma_1^2, \sigma_2^2)$ and $p(\boldsymbol{\theta} \mid \sigma_2^2, \theta)$ is supposed Normal as in (7.2.5). This we call the "random effect" prior.

In either case, the posterior distribution of $\boldsymbol{\theta}$ is obtained by combining the appropriate prior with the same likelihood

$$l(\boldsymbol{\theta}, \sigma_1^2 \mid \mathbf{y}) \propto p(\mathbf{y} \mid \boldsymbol{\theta}, \sigma_1^2), \tag{7.2.37}$$

where $p(\mathbf{y} \mid \boldsymbol{\theta}, \sigma_1^2)$ is given in (7.2.4). Thus, the posterior distributions of $\boldsymbol{\theta}$ are *different because the priors are.*

For purposes of comparison, both can be thought of in terms of a general model in which the prior for $\boldsymbol{\theta}$, σ_1^2, σ_2^2 and θ is written in the form

$$p(\boldsymbol{\theta}, \sigma_2^2, \sigma_1^2, \theta) = p(\boldsymbol{\theta}, \sigma_2^2 \mid \theta, \sigma_1^2)\, p(\theta)\, p(\sigma_1^2), \tag{7.2.38}$$

where

$$p(\sigma_1^2) \propto \sigma_1^{-2}, \qquad p(\theta) \propto c \tag{7.2.39}$$

and

$$p(\boldsymbol{\theta}, \sigma_2^2 \mid \theta, \sigma_1^2) = p(\sigma_2^2 \mid \sigma_1^2)\, p(\boldsymbol{\theta} \mid \sigma_2^2, \theta)$$

$$\propto p(\sigma_2^2 \mid \sigma_1^2)\, \sigma_2^{-J} \exp\left[-\frac{1}{2\sigma_2^2} \Sigma\, (\theta_j - \theta)^2\right]. \tag{7.2.40}$$

In both cases, the θ_j may be supposed to be a random sample from a Normal distribution $N(\Theta, \sigma_2^2)$. The crucial question concerns the choice of $p(\sigma_2^2 \mid \sigma_1^2)$.

The assumption of a locally uniform prior for the θ_j in the fixed effect formulation amounts to postulating that they are distributed about some unknown θ with a *large* variance σ_{20}^2. We can accomodate this assumption within the general framework by making $p(\sigma_2^2 \mid \sigma_1^2)$ in (7.2.40) a delta function at $\sigma_2^2 = \sigma_{20}^2$. On the other hand, by employing a noninformative prior $p(\sigma_2^2 \mid \sigma_1^2) \propto \sigma_{12}^{-2}$, we are led to the random effect analysis where, by allowing σ_2^2 to become a random variable, we let the data $\mathbf{y}$ determine what can be said about the spread of the θ_j.

Uniform Prior and Multiparameter Problem of High Dimension

We have seen that in the fixed effect formulation, the noninformative prior for $\boldsymbol{\theta}$ is supposedly locally uniform. It is natural to ask "Specifically what prior distribution for the θ_j is implied by the random effect prior of (7.2.36)." For simplicity, suppose σ_1^2 is known.

We have

$$p(\boldsymbol{\theta} \mid \theta) = \int_0^\infty p(\boldsymbol{\theta} \mid \sigma_2^2, \theta)\, p(\sigma_2^2)\, d\sigma_2^2, \tag{7.2.41}$$

where from (7.2.36) and (7.2.40)

$$p(\boldsymbol{\theta} \mid \sigma_2^2, \theta) \propto (\sigma_2^2)^{-J/2} \exp\left[-\frac{1}{2\sigma_2^2} \Sigma (\theta_j - \theta)^2\right] \tag{7.2.42}$$

and

$$p(\sigma_2^2) \propto \sigma_{12}^{-2} = (\sigma_1^2 + K\sigma_2^2)^{-1}. \tag{7.2.43}$$

Thus, for large J,

$$p(\boldsymbol{\theta} \mid \theta) \mathrel{\dot{\propto}} \left[1 + \frac{K\Sigma (\theta_j - \theta)^2}{J\sigma_1^2}\right]^{-1} \left[\Sigma (\theta_j - \theta)^2\right]^{-\frac{1}{2}(J-2)} \tag{7.2.44}$$

and, since the second factor dominates the first,

$$p(\boldsymbol{\theta} \mid \theta) \mathrel{\dot{\propto}} [\Sigma (\theta_j - \theta)^2]^{-\frac{1}{2}(J-2)}. \tag{7.2.45}$$

This distribution is very close to the prior for $\boldsymbol{\theta}$ suggested by Stein (1962) and by Anscombe (1963). Specifically, they proposed that an appropriate prior for *high* dimensional $\boldsymbol{\theta}$ is the multivariate t distribution

$$p(\boldsymbol{\theta} \mid \theta) \propto [\nu + \Sigma (\theta_j - \theta)^2]^{-\frac{1}{2}(J+\nu)}, \tag{7.2.46}$$

where ν is an arbitrarily small positive constant. The prior of $\boldsymbol{\theta}$ in (7.2.45) has its probability mass spread over a very wide region in the space of $\boldsymbol{\theta}$. In particular, it can be shown that the variance of any linear function of $\boldsymbol{\theta}$ is infinite.

To see the distinction between (7.2.45) and the locally uniform prior $p(\boldsymbol{\theta}) \propto c$, consider the quantity

$$\tilde{\sigma}_2^2 = J^{-1} \Sigma (\theta_j - \theta)^2 \tag{7.2.47}$$

which is, in a sense, an "estimate" of σ_2^2. In particular, consider what prior distribution for $\tilde{\sigma}_2^2$ is implied by the random effect prior for $\boldsymbol{\theta}$ in (7.2.45). First, conditional on σ_2^2, we obtain from (7.2.42) that

$$p(\tilde{\sigma}_2^2 \mid \sigma_2^2, \theta) \propto \sigma_2^{-J} \tilde{\sigma}_2^{J-2} \exp\left(-\frac{J\tilde{\sigma}_2^2}{2\sigma_2^2}\right). \tag{7.2.48}$$

By integrating over the distribution of σ_2^2 in (7.2.43), we then find that, for large J,

$$p(\tilde{\sigma}_2^2 \mid \theta) \mathrel{\dot{\propto}} (\sigma_1^2 + K\tilde{\sigma}_2^2)^{-1} \tag{7.2.49}$$

which is a decreasing function of $\tilde{\sigma}_2^2$ and, as might be expected, is of the same form as the noninformative prior for σ_2^2.

By contrast, consider what the prior of $\tilde{\sigma}_2^2$ would be if the θ_j were strictly uniformly distributed over some region R in the parameter space of $\boldsymbol{\theta}$. Now $\tilde{\sigma}_2^2 = \lambda$ defines a hyperspherical surface with radius proportional to $\lambda^{\frac{1}{2}}$ centered at $\theta\mathbf{1}$ where $\mathbf{1}$ is a $J \times 1$ vector of ones. Suppose this surface lies within R. Then the probability that $\tilde{\sigma}_2$ lies within $\lambda^{\frac{1}{2}}$ and $\lambda^{\frac{1}{2}} + \delta$, where δ is a small positive constant, is proportional to $\delta(\lambda^{\frac{1}{2}})^{J-1}$. Thus, a strictly uniform prior for $\boldsymbol{\theta}$ implies that

$$p(\tilde{\sigma}_2^2 \mid \theta) \propto (\tilde{\sigma}_2^2)^{\frac{1}{2}J-1} \tag{7.2.50}$$

which, for $J > 2$, is an increasing function of $\tilde{\sigma}_2^2$ and asserts that $\tilde{\sigma}_2^2$ is large with high probability. This is not surprising if we remember that a flat prior for $\boldsymbol{\theta}$ can be produced by allowing σ_2^2 to become large in the prior distribution (7.2.42). Thus, we obtain from (7.2.48)

$$p(\tilde{\sigma}_2^2 \mid \sigma_2^2 \to \infty, \theta) \propto (\tilde{\sigma}_2^2)^{\frac{1}{2}J-1}$$

as in (7.2.50) so that, for the "fixed effects" prior, $\tilde{\sigma}_2^2$ is large because σ_2^2 is tacitly assumed to be large.

One is thus led to contrast the two results

random effect (approximate result)	strictly uniform prior (or prior with $\sigma_2^2 \to \infty$)
$p(\tilde{\sigma}_2^2 \mid \theta) \propto (\sigma_1^2 + K\tilde{\sigma}_2^2)^{-1}$	$p(\tilde{\sigma}_2^2 \mid \theta) \propto (\tilde{\sigma}_2^2)^{\frac{1}{2}J-1}$

and we see that these two expressions become more and more discordant as the dimension J increases. As was pointed out by Anscombe, these results should lead one to approach with some caution the choice of noninformative prior distributions when the number of parameters is large.

One might wonder how can we justify two different priors for $\boldsymbol{\theta}$ both of which are supposed to be noninformative? The starting point for deriving a noninformative prior in Section 1.3 was that we wished to express the state of knowing little about the parameters *in a given model* relative to what the data would have to tell us. This implies, as we have seen already, that *different* priors will express this state when we contemplate different models. For this example the "fixed effect" prior expresses the idea directly that we know very little about any one of the parameters θ_j. By contrast, the "random effect" model says that the θ_j are random drawings from a Normal distribution $N(\theta, \sigma_2^2)$ and the prior expresses the fact that little is known about the mean and *variance* of *that distribution*. The implication here is that the means do cluster together and correctly implies a different prior for the θ_j.

Alternative Assumptions About the Prior

To discuss the choice of prior further, it is useful to reiterate the general role of assumptions in statistical methods. In practice, we know that all assumptions are false. For example, there never was an exactly straight line nor an exact Normal distribution. Logically, then, in selecting assumptions we should not ask (a) "Is this assumption true ?" nor only (b) "Does this assumption approximate the truth concerning the experimental setup ?" Rather we have to ask (c) "Does the use of this assumption *lead to* an approximately correct result?" This is so because the answer to (a) must always be "No" and so the question is irrelevant while the answer to (b) can be "Yes" when the answer to (c) is "No" and vice versa. The nature of the experimental setup should provide some guide, though not necessarily a decisive one, in what ought to be assumed. Therefore, in considering the choice of a prior distribution for a group of J means we shall consider both questions (b) and (c) above.

Priors for Different Experimental Set-ups

For the one-way classification model being considered, in rare instances it *might* be realistic to analyse the data against a noninformative reference prior which supposed that the treatment means θ_j were, approximately, independently and uniformly distributed over a wide range. Strictly speaking, this prior is only truly representative of a situation where the experimenter

i) knows little about any of the treatments,

ii) has no reason to believe any of the treatment will produce similar results.

In practice, however, (ii) will rarely represent his state of mind. Usually some of the treatments would be modifications of others and could be expected *a priori* to behave similarly. In particular, (ii) is clearly not appropriate in the example we have considered where the "treatments" are "batches." There, the alternative supposition is much closer to reality, whereby the θ_j are independent drawings from a Normal distribution $N(\theta, \sigma_2^2)$ with θ and σ_2^2 only vaguely known and hence approximately represented by noninformative priors.

Obviously the two alternative possibilities considered in this chapter are by no means exhaustive. When a large number of treatments are under consideration, the experimenter may for example be able to divide them into subgroups within which he expects similarities. Alternatively even if the "treatments" were different batches, he might not believe that the θ_j were independent but rather that they formed a time series represented perhaps by an autoregressive process.† In this case the prior would be defined in terms of the parameters of this process.

The Bayes approach has the advantage that any possibility of interest can be modelled and the appropriately chosen prior will invite the data to comment on relevant uncertain aspects (for instance on the variance of the population of the θ_j in the random effect model, or on the value of the autoregressive parameter if the θ_j are represented as a time series).

7.2.6 Effect of Different Prior Assumptions on the Posterior Distribution of $\boldsymbol{\theta}$

We have said that with data occurring in the form of a one-way classification, there are many prior assumptions that in different circumstances could make sense. The two priors for $\boldsymbol{\theta}$ examined here are the "fixed effect" uniform prior and the "random effect" prior.

As illustrated by the dyestuff example, the random effect prior results in a greater clustering of the posterior means about the grand average $y_{..}$ as well as an overall increase in the precision of the posterior distributions of the θ_j.

A Simpler Situation

A better intuitive understanding of how this happens is obtained by considering the simpler situation in which σ_1^2, σ_2^2 and θ are *all supposed known*.

---

† See, for example, Tiao and Ali (1971b).

Then, for the random effect model we would have two sources of information about a particular batch mean θ_j. First, we would know from (7.2.5) that the θ_j was drawn from a Normal distribution with mean θ and variance σ_2^2. Second, we would known from (7.2.4) that the sample mean $y_{j.}$ was distributed Normally about θ_j with variance σ_1^2/K. Combining these facts we see from (7.2.12) that *a posteriori* the θ_j would be Normally distributed with mean

$$\tilde{\theta}_j = E(\theta_j \mid \sigma_1^2, \sigma_{12}^2, \theta, y_{j.}) = (1 - z)\, y_{j.} + z\theta, \qquad (7.2.51)$$

where

$$z = \sigma_1^2/\sigma_{12}^2.$$

Thus, the posterior mean would be a linear interpolation between the sample mean $y_{j.}$ and the prior mean Θ. Also, the variance would be

$$\operatorname{Var}(\theta_j \mid \sigma_1^2, \sigma_{12}^2, \theta, y_{j.}) = \frac{\sigma_1^2}{K}(1 - z). \qquad (7.2.52)$$

Further, the θ_j would be distributed independently of one another. Note that $z^{-1} = \sigma_{12}^2/\sigma_1^2 = 1 + K(\sigma_2^2/\sigma_1^2)$ measures the relative variation of the group means θ_j compared with the within group variation of the data.

On the other hand, for the fixed effect model with σ_1^2 assumed known, we would only have the second source of information from (7.2.4) so that *a posteriori* the means θ_j would be Normally independently distributed with

$$E(\theta_j \mid \sigma_1^2, y_{j.}) = y_{j.} \quad \text{and} \quad \operatorname{Var}(\theta_j \mid \sigma_1^2, y_{j.}) = \sigma_1^2/K, \qquad j = 1, \ldots, J. \qquad (7.2.53)$$

Thus, the effect of the information from the random effect prior is to "pull" the posterior mean of θ_j towards the prior mean θ and to decrease the variance by a factor $(1 - z) = 1 - \sigma_1^2/\sigma_{12}^2$.

When σ_1^2, σ_2^2 and θ are not known, the posterior mean of θ_j for the random effect model becomes that in (7.2.29), that is,

$$E(\theta_j \mid \mathbf{y}) = \hat{\theta}_j = (1 - \mu_1')\, y_{j.} + \mu_1' y_{..},$$

where

$$\mu_1' = E(z \mid \mathbf{y}),$$

which is an interpolation between the sample mean $y_{j.}$ and the grand average $y_{..}$. Compared with (7.2.51), we are thus replacing θ by $y_{..}$ and z by its posterior expectation.

Situation When J is Large

Now, if the number of groups J is large so that ν_1 and ν_2 are both large, then in the integrand of (7.2.28) the conditional multivariate t distribution $p(\boldsymbol{\theta} \mid z, \mathbf{y})$ given in (7.2.20) approaches a J-dimensional multivariate Normal distribution. Also, the distribution $p(z \mid \mathbf{y})$ in (7.2.26) will become sharply concentrated about the reciprocal of the mean square ratio

$$\hat{z} = \frac{m_1}{m_2} = \frac{1}{F} \tag{7.2.54}$$

provided $F > 1$.

In this case, using the approximations $\mu_1' \doteq 1/F$, $\mu_2' \doteq (1/F)^2$ and $\mu_2 \doteq 0$ in (7.2.29) through (7.2.31) and noting that the integration process in (7.2.28) is essentially equivalent to replacing the unknown z in $p(\boldsymbol{\theta} \mid z, \mathbf{y})$ by $\hat{z}$, the θ_j are distributed approximately Normally and independently with

$$E(\theta_j \mid \mathbf{y}) \doteq \left(1 - \frac{1}{F}\right) y_{j.} + \frac{1}{F} y_{..} \qquad \text{and} \qquad \operatorname{Var}(\theta_j \mid \mathbf{y}) \doteq \frac{m_1}{K}\left(1 - \frac{1}{F}\right). \tag{7.2.55}$$

By contrast, for the fixed prior, it is clear from (7.3.32) that, for large J, the θ_j are approximately Normally and independently distributed such that

$$E(\theta_j \mid \mathbf{y}) = y_{j.} \qquad \text{and} \qquad \operatorname{Var}(\theta_j \mid \mathbf{y}) \doteq \frac{m_1}{K}. \tag{7.2.56}$$

Thus, by using the random effect prior (7.2.36)

a) the posterior means cluster more closely about $y_{..}$,
b) the variance of the distribution of θ_j tends to be reduced by a factor which, for large J, is approximately $1 - 1/F$.

It follows that when F is large so that $1/F$ is small, the clustering and increase in precision is negligible and the random effect prior yields what is essentially the fixed effect solution. On the other hand, when F is not large but takes a value such as 2 or 3, considerable modifications in the posterior means and variances occur leading to modifications in the inferences about $\boldsymbol{\theta}$.

The dyestuff data of Section 7.2.4 represent an intermediate situation. As we have seen, for this data, inferences about individual elements of $\boldsymbol{\theta}' = (\theta_1, \ldots, \theta_6)$ are not very different for the random and the fixed effect formulations. More generally, if $F = m_2/m_1$ is fairly large (say $F > 10$), the random effect analysis will not differ very markedly from the fixed effect analysis.

Now $F = m_2/m_1$ is a sample measure of

$$\frac{\sigma_{12}^2}{\sigma_1^2} = 1 + K\frac{\sigma_2^2}{\sigma_1^2}$$

indicating how large σ_2^2 is compared with σ_1^2/K. In the random effect analysis, then, we are as it were using the data (as reflected by the spread of the sample means $y_{j.}$ in relation to the within group variances m_1/K) to comment on the spread of the means and hence tell us to what extent the locally uniform prior assumption on θ_j is justified.

Size of H.D.P. regions

From (7.2.55) and (7.2.56), we see that for any individual mean θ_j, when J is large,

$$\frac{\text{length of } (1-\alpha) \text{ H.D.P. interval for random effect model}}{\text{length of } (1-\alpha) \text{ H.P.D. interval for fixed effect model}} = \left(1 - \frac{1}{F}\right)^{1/2}. \qquad (7.2.57)$$

On the other hand, for the θ_j considered jointly

$$\frac{\text{volume of } (1-\alpha) \text{ H.P.D. region for random effect model}}{\text{volume of } (1-\alpha) \text{ H.P.D. region for fixed effect model}} = \left(1 - \frac{1}{F}\right)^{J/2}. \qquad (7.2.58)$$

For large J, $(1 - 1/F)^{J/2}$ could be small even when F was large and $(1 - 1/F)^{1/2}$ was close to unity, so that it might seem at first that, for large J, a great increase in precision is to be expected from the use of internal evidence about σ_2^2. To set this result in proper perspective, however, it should be remembered that the situation is exactly as if the standard deviation σ_1 of the errors e_{jk} in (7.2.1) had been decreased by a factor $(1 - 1/F)^{1/2}$. This would also result in the *volume* of the H.P.D. region being reduced by the factor $(1 - 1/F)^{J/2}$.

Overall Analysis of Contrasts

In the sampling theory analysis of the one-way classification, a useful preliminary to more detailed study is the overall test of the hypothesis of equality of the group means $\boldsymbol{\theta}' = (\theta_1, \ldots, \theta_J)$. This is accomplished by referring the ratio of mean squares m_2/m_1 to a table of the F distribution. In Section 2.11 we have discussed the Bayesian analogue of this procedure. Given $\boldsymbol{\phi}' = (\phi_1, \ldots, \phi_{J-1})$, a set of $J - 1$ linearly independent contrasts among the J parameters $\boldsymbol{\theta}$, it was shown that the point $\boldsymbol{\phi} = \mathbf{0}$ was not included in the $(1 - \alpha)$ H.P.D. region for $\boldsymbol{\phi}$ if

$$\frac{m_2}{m_1} > F(J-1, \nu_1, \alpha) \qquad (7.2.59)$$

or, for ν_1 large, if

$$\frac{m_2}{m_1} > \frac{\chi^2(J-1, \alpha)}{J-1}$$

where, as before, $F(J - 1, \nu_1, \alpha)$ is the upper α percentage point of a F distribution with $[(J - 1), \nu_1]$ degrees of freedom and $\chi^2(J - 1, \alpha)$ is that of a χ^2 distribution with $(J - 1)$ degrees of freedom. This analysis was based on the assumption of a fixed effect prior.

Consider now the corresponding result for the random effect prior of (7.2.36) and for the moment let us take $\phi_1 = \theta_1 - \bar{\theta}, \ldots, \phi_{J-1} = \theta_{J-1} - \bar{\theta}$.

Using the result (7.2.25), conditional on z, we see that the quantity

$$\frac{K \sum_{j=1}^{J} (\phi_j - \hat{\phi}_j)^2}{(J-1)s^2(z)}, \qquad (7.2.60)$$

where

$$\hat{\phi}_j = (1 - z)(y_{j.} - y_{..}) \qquad \text{and} \qquad s^2(z) = \frac{v_1 m_1 + z v_2 m_2}{v_1 + v_2}(1 - z),$$

is distributed as F with $[(J - 1), (v_1 + v_2)]$ degrees of freedom. Thus, when J is large, z converges in probability to $1/F = m_1/m_2$ provided $F > 1$ so that

$$s^2(z) \doteq m_1\left(1 - \frac{1}{F}\right),$$
$$\hat{\phi}_j \doteq \left(1 - \frac{1}{F}\right)(y_{j.} - y_{..}), \tag{7.2.61}$$

and approximately

$$\frac{K \sum_{j=1}^{J} [\phi_j - (1 - 1/F)(y_{j.} - y_{..})]^2}{(J - 1)m_1(1 - 1/F)} \tag{7.2.62}$$

is distributed as F with $[(J - 1), (v_1 + v_2)]$ degrees of freedom. Thus, for the point $\boldsymbol{\phi} = \mathbf{0}$ not to be included in the $1 - \alpha$ H.P.D. region, we require approximately that

$$\frac{(1 - 1/F)\, K \sum_{j=1}^{J} (y_{j.} - y_{..})^2}{(J - 1)m_1} > F(J - 1, v_1 + v_2, \alpha), \tag{7.2.63}$$

that is,

$$\frac{m_2}{m_1} - 1 > F(J - 1, v_1 + v_2, \alpha)$$

or to a further approximation,

$$\frac{m_2}{m_1} - 1 > \frac{\chi^2(J - 1, \alpha)}{J - 1}.$$

Thus, for large J, the effect on inferences about the value $\boldsymbol{\phi} = \mathbf{0}$ of employing the random effect prior is approximately represented by subtracting one from the mean ratio before referring to the appropriate tables.

The situation is illustrated in Fig. 7.2.3. An H.P.D. region for two orthogonal contrasts† $\boldsymbol{\phi}^* = (\phi_1^*, \phi_2^*)'$ is shown for the fixed effect model centered about $\hat{\boldsymbol{\phi}}^*$. The region has radius r.

For the random effect model, the corresponding H.P.D. region is centered about $(1 - 1/F)\hat{\boldsymbol{\phi}}^*$ while the radius of the region is reduced by a factor $(1 - 1/F)^{1/2}$. In the situation illustrated, the point $\boldsymbol{\phi} = \mathbf{0}$ is just outside the fixed effect H.P.D. region but is inside the corresponding random effect region.

† The set of linearly independent contrasts $\phi_1 = \theta_1 - \bar{\theta}, \ldots, \phi_{J-1} = \theta_{J-1} - \bar{\theta}$ can always be transformed into a set of $(J - 1)$ orthogonal linear contrasts having spherical H.P.D. regions.

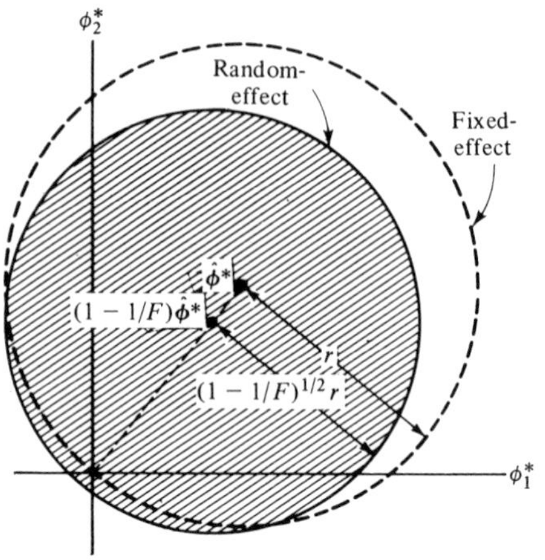

Fig. 7.2.3 H.P.D. regions for orthogonal contrasts ϕ_1^* and ϕ_2^* with "fixed-effect" and "random-effect" priors.

7.2.7 Relationship to Sampling Theory Results

We have seen earlier in Section 2.11 that, for the fixed-effect model, the standard sampling theory results parallel those obtained from a Bayesian analysis with a noninformative prior which is locally uniform in $\boldsymbol{\theta}$. By contrast, in this chapter a random effect Bayesian analysis of the means θ_j has been explored and has led to intuitively sensible results. It is interesting, therefore, that in this case also, parallel sampling theory results exist.

The "parallel" sampling theory comes out of the work of James and Stein (1961) who considered the problem of estimating a set of means $\theta_1,\ldots,\theta_J$ of the one=way model (7.2.1) so as to minimize mean squared error. In general, let $\tilde{\boldsymbol{\theta}}' = (\tilde{\theta}_1, \ldots, \tilde{\theta}_J)$ be a set of point estimators for $\boldsymbol{\theta}' = (\theta_1, \ldots, \theta_J)$ and let $J^{-1} \Sigma (\tilde{\theta}_j - \theta_j)^2$ be the appropriate squared error loss function. Then, as discussed in Appendix A5.6, the mean squared error of $\tilde{\boldsymbol{\theta}}$ is the expectation

$$\text{M.S.E.}(\tilde{\boldsymbol{\theta}}) = J^{-1} E\, \Sigma (\tilde{\theta}_j - \theta_j)^2, \tag{7.2.64}$$

where E is taken with respect to the distribution of the data $\mathbf{y}$.

For simplicity let us suppose that σ_1^2 is known. Then the analysis proceeds as follows. If considerations are limited to $\tilde{\boldsymbol{\theta}}$ which are *linear* functions of the data, then it is well known that the minimum mean squared error point estimators are the sample means $\mathbf{y}'_{.} = (y_{1.}, \ldots, y_{J.})$. Specifically,

$$\text{M.S.E.}(\mathbf{y}_{.}) = J^{-1} E\, \Sigma (y_{j.} - \theta_j)^2 = \frac{\sigma_1^2}{K}. \tag{7.2.65}$$

However, these authors relaxed the unnecessarily restrictive assumption of linearity and considered instead for $J \geqslant 2$ estimators $\hat{\theta}_j$ which are non-linear functions of the data such that

$$\hat{\theta}_j - \theta = \left[1 - \frac{(J-2)\sigma_1^2}{K\,\Sigma\,(y_{j.} - \theta)^2}\right](y_{j.} - \theta), \qquad j = 1, \ldots, J.$$

that is,

$$\hat{\theta}_j = \left(1 - \frac{1}{F'}\right)y_{j.} + \frac{1}{F'}\theta, \tag{7.2.66}$$

where

$$F' = \frac{K\,\Sigma\,(y_{j.} - \theta)^2}{(J-2)\sigma_1^2}$$

and θ is any arbitrary constant. They then showed that

$$\text{M.S.E.}(\hat{\boldsymbol{\theta}}) = J^{-1}\,E\,\Sigma\,(\hat{\theta}_j - \theta_j)^2 = \frac{\sigma_1^2}{K}\left(1 - \frac{1}{F''}\right), \tag{7.2.67}$$

where

$$F'' = f\left(J, \frac{K}{\sigma_1^2}, \tilde{\sigma}_2^2\right), \qquad \tilde{\sigma}_2^2 = J^{-1}\,\Sigma\,(\theta_j - \theta)^2$$

and $1/F''$ is such that

$$0 \leqslant \frac{1}{F''} < 1, \qquad \lim_{\tilde{\sigma}_2^2 \to \infty} \frac{1}{F''} = 0.$$

It was thus demonstrated that the non-linear estimators (7.2.66) have the remarkable sampling property that, whatever the true value of $\tilde{\sigma}_2^2$, M.S.E.$(\hat{\boldsymbol{\theta}})$ can never exceed M.S.E. $(\mathbf{y}_{.})$. Note that $\tilde{\sigma}_2^2$ is a measure of the spread of the means θ_j. Thus, the M.S.E. of $\hat{\boldsymbol{\theta}}$ will approach the M.S.E. of $\mathbf{y}_{.}$ when the means are spread out.

For comparison, we see that with the simplifying assumption that σ_1^2 and θ are known, the posterior distribution of $\boldsymbol{\theta}$ is obtained by integrating the conditional Normal distribution for $\boldsymbol{\theta}$ in (7.2.12) over the distribution of σ_2^2 in (7.2.14). Now for large J, the distribution of σ_2^2 is sharply concentrated about its mode $\hat{\sigma}_2^2$, where

$$\sigma_1^2 + K\hat{\sigma}_2^2 = \frac{K\,\Sigma\,(\theta - y_{j.})^2}{J+2}, \tag{7.2.68}$$

provided $\hat{\sigma}_2^2 > 0$. It follows that, for large J, the θ_j are approximately Normally distributed independent of one another with *posterior means* $\hat{\theta}_j$ such that

$$\hat{\theta}_j - \theta = \left[1 - \frac{(J+2)\sigma_1^2}{K\,\Sigma\,(y_{j.} - \theta)^2}\right](y_{j.} - \theta)$$

that is,

$$\hat{\theta}_j = \left(1 - \frac{1}{F^*}\right)y_{j.} + \frac{1}{F^*}\theta \qquad \text{where} \qquad F^* = \frac{K\,\Sigma\,(y_{j.} - \theta)^2}{(J+2)\sigma_1^2} \tag{7.2.69}$$

and with common variance

$$\frac{\sigma_1^2}{K}\left(1 - \frac{1}{F^*}\right). \tag{7.2.70}$$

Thus, the estimators in (7.2.66) which have remarkable sampling properties are of precisely the same form as the posterior means in (7.2.69) resulting from the Bayesian random effect analysis, the only difference is that $J - 2$ replaces $J + 2$ (that is, $F' = \dfrac{(J+2)}{(J-2)}F^*$). Indeed, Stein (1962) proposed that the appropriate non-linear functions to be used as estimators should be obtained from the posterior distribution with "suitably chosen" priors. For reasons explained in Appendix A5.6, we would not employ this approach but would feel that, having come so far towards the Bayesian position, one more step might as well be taken and the posterior distribution itself, rather than some arbitrary feature of it, be made the basis for inference.

7.2.8 Conclusions

1) Given an appropriate metric, a state of relative ignorance about a parameter can frequently be represented to an adequate approximation by supposing that, over the region where the likelihood is appreciable, the prior is constant, However, careful thought must be given to selecting the appropriate form of the model, and to deciding *which* of its characteristics it is desired to express ignorance about in this way. For example, suppose there are a number of parameters θ_j which (i) are expected to be random drawings from a distribution or (ii) are expected to be successive values if a time series. Then it would be the parameters of (i) the *distribution* or (ii) the *time series*, about which we would wish to express ignorance. It would not be appropriate in these cases to so express ignorance about the θ_j directly, for this would not take account of available information (i) that the θ_j clustered in a distribution or (ii) that the θ_j were members of a time series.

2) In this chapter we have discussed in some detail a situation of the first kind where the assumption of a uniform prior distribution applied directly to the θ_j might not be appropriate. Instead it was supposed *a priori* that the θ_j were randomly drawn from a Normal distribution with unknown mean θ and unknown variance σ_2^2. By employing noninformative priors for these parameters, the data themselves are induced to provide evidence about the location and *dispersion* of the θ_j. These ideas are of more general application and are akin to those employed by Mosteller and Wallace (1964) and provide a link with the empirical Bayes procedures of Robbins (1955, 1964).

3) Analysis in terms of a random effect prior is, of course, not necessarily appropriate (although, since clustering of means is so often to be expected, it is an important point of departure). Applications have been mentioned in which the means arranged in appropriate sequence might be expected to form a time series. In another situation two or more distinct clusters might be expected. By suitably

choosing the structure of the model and introducing suitable parameters associated with noninformative priors, the data are invited to comment on relevant aspects of the problem.

4) An important practical question is "How much does a correct choice of prior matter?" or equivalently "How sensitive (or robust) is the posterior distribution of the θ_j to change in the prior distribution?" The Bayes analysis makes it clear that the problem is one of combining, in a relevant way, information coming from the data in terms of comparisons between groups with information from within groups. Now suppose there are J groups of K observations. The different priors reflect different relations *between* the J groups. Thus if J is small compared with K one can expect that inferences will be robust with respect to choice of prior. However, when J is large compared with K, between group information will become of relative importance, and the precise choice of prior will be critical.

5) In general, Bayes' theorem says that

$$p(\boldsymbol{\theta} \mid \mathbf{y}) \propto p(\boldsymbol{\theta})\, l(\boldsymbol{\theta} \mid \mathbf{y})$$

where $\mathbf{y}$ represents a set of observations and $\boldsymbol{\theta}$ a set of unknown parameters. Thus, given the data $\mathbf{y}$, inferences about $\boldsymbol{\theta}$ depend upon what we assume about the prior distribution $p(\boldsymbol{\theta})$ as well as what is assumed about the data generating model leading to the likelihood function $l(\boldsymbol{\theta} \mid \mathbf{y})$. In Chapters 3 and 4, we studied the robustness of inferences about $\boldsymbol{\theta}$ by varying $l(\boldsymbol{\theta} \mid \mathbf{y})$ to determine to what extent such commonly made assumption as Normality can be justified. We can equally well, as has been done here, study the effect on the posterior distribution of varying the prior distribution $p(\boldsymbol{\theta})$. Examples of the latter type of inference robustness studies have in fact been given earlier in Section 1.2.3 and in Section 2.4.5.

7.3 INFERENCES ABOUT MEANS FOR MODELS WITH TWO RANDOM COMPONENTS

In Section 6.3 we considered the analysis of cross classification designs and in particular the analysis of a two-way layout using a model in which the row means were regarded as fixed effects and the column means as random effects. It is often appropriate to use such a two-way model in the analysis of randomized block designs where the blocks are associated with the random column effects and the treatments with the fixed row effects. Such a model contains two random components: one associated with the "within block error" and one with the "between block error."

In Section 6.3 the error variance component and the between columns variance were denoted by σ_e^2 and σ_c^2 respectively. In what follows where we associate the random column component with blocks we shall use σ_e^2 and σ_b^2 respectively for the error variance component and the between blocks variance component.

In practice, the blocks correspond to portions of experimental material which are expected to be more homogeneous than the whole aggregate of material.

Thus in an animal experiment the blocks might be groups of animals from the same litter. The randomized block design ensures that comparisons between the treatments can be made *within blocks* (for example, between litter mates within a litter) and so are not subject to the larger block to block errors.

When the available block size is smaller than the number of treatments, incomplete block designs may be employed—see, for example, Cochran and Cox (1950), Kempthorne (1952). Thus six pairs of identical twins might be used to compare four different methods of reading instruction in the balanced arrangement shown in Table 7.3.1.

Table 7.3.1

A balanced incomplete block design

		BLOCKS (Twin Pair) 1	2	3	4	5	6
TREATMENTS	A	×	×	×			
(method of	B	×			×	×	
reading	C		×		×		×
instruction)	D			×		×	×

The design is such that treatments A and B are randomly assigned to the first pair of twins, A and C to the second, and so on. In general, in a balanced incomplete block design (BIBD), *I treatments are examined in J blocks of equal size K. Each treatment appears r times and occurs λ times in a block with every other treatment.* In the reading instruction example $I = 4$, $J = 6$, $K = 2$, $r = 3$, and $\lambda = 1$.

Designs of this kind supply two different kinds of information about the treatment contrasts. The first uses within (intra) block comparisons and the second uses between (inter) block comparisons. Thus one within-block comparison is the difference in the scores of the first pair of twins, which yields information about the difference between treatments A and B. The difference in scores of the second pair of twins similarly yields information about the difference between A and C, and so on. All comparisons of this sort are affected only by the within block variance σ_e^2.

Analysis of incomplete block designs has sometimes been conducted as if these within block comparisons were the *only* source of information about the treatment contrasts. However, as was pointed out by Yates (1940), a second source of information about the treatment contrasts is supplied by comparisons *between* the block averages. Thus the average score for the first pair of twins, supplies an estimate of the average of the effects of treatments A and B; the average score for the second block supplies an estimate of the average of the effects of treatments A and C, and so on. Comparison of these block average scores which have an error variance $\sigma_b^2 + \frac{1}{2}\sigma_e^2$ thus supplies further information about the treatment contrasts.

On sampling theory, estimation of the treatment contrasts using either (i) within block information only or, (ii) between block information only, is readily achieved by a standard application of least squares. It is, however, far from clear how these estimates ought to be *combined* and resulting inferences drawn.

7.3.1 A General Linear Model with Two Random Components

There is no particular reason to limit the initial discussion to a particular kind of design and we now consider a general linear model appropriate to any design having J blocks with fixed block size K. Specifically, we consider the linear model

$$\mathbf{y}_j = \mathbf{A}_j\boldsymbol{\theta} + b_j\mathbf{1} + \mathbf{e}_j, \qquad j = 1, \ldots, J. \tag{7.3.1}$$

In this model $\mathbf{y}_j = (y_{1j}, \ldots, y_{Kj})$ is a $K \times 1$ vector (or block) of observations, and $\mathbf{A}_j$ is a $K \times I$ matrix of fixed elements. The quantity $\boldsymbol{\theta} = (\theta_1, \ldots, \theta_I)'$ is a $I \times 1$ vector of regression coefficients, b_j is the jth block (random) effect having mean zero and variance σ_b^2, $\mathbf{1}$ is a $K \times 1$ vector of ones, and $\mathbf{e}_j = (e_{1j}, \ldots, e_{Kj})'$ is a $K \times 1$ vector of random errors. In this model the vector $\mathbf{A}_j\boldsymbol{\theta}$ is a "fixed" component of the vector of observations $\mathbf{y}_j$ and the vectors $b_j\mathbf{1}$ and $\mathbf{e}_j$ are two distinct random components.

For an incomplete block design, $\boldsymbol{\theta}$ is the $I \times 1$ vector of treatment means and $\mathbf{A}_j$ consists of 1's and 0's, each row of which contains a single 1 indicating which treatments are included in the jth block. Thus for the reading instruction design

$$A_1 = \begin{bmatrix} 1 & 0 & 0 & 0 \\ 0 & 1 & 0 & 0 \end{bmatrix}, \quad A_2 = \begin{bmatrix} 1 & 0 & 0 & 0 \\ 0 & 0 & 1 & 0 \end{bmatrix}, \quad \ldots, \quad A_6 = \begin{bmatrix} 0 & 0 & 1 & 0 \\ 0 & 0 & 0 & 1 \end{bmatrix}.$$

7.3.2 The Likelihood Function

We assume that b_j and the elements of $\mathbf{e}_j, j = 1, \ldots, J$, are all independent and that

$$b_j \sim N(0, \sigma_b^2), \quad \mathbf{e}_j \sim N_K(\mathbf{0}, \sigma_e^2\mathbf{I}) \tag{7.3.2}$$

where $\mathbf{0}$ is a null vector and $\mathbf{I}$ is an identity matrix, both of size K. To derive the likelihood function of the parameters $(\boldsymbol{\theta}, \sigma_e^2, \sigma_b^2)$, it is convenient to work with the block averages and the within block residuals

$$y_{.j} = K^{-1}\mathbf{1}'\mathbf{y}_j = K^{-1}\sum_{k=1}^{K} y_{kj} \quad \text{and} \quad (\mathbf{I} - \mathbf{R})\mathbf{y}_j = \mathbf{y}_j - y_{.j}\mathbf{1}, \tag{7.3.3}$$

where $\mathbf{R} = K^{-1}\mathbf{1}\mathbf{1}'$.

The model in (7.3.1) can thus be alternatively expressed as

$$\begin{aligned} y_{.j} &= K^{-1}\mathbf{1}'\mathbf{A}_j\boldsymbol{\theta} + b_j + e_{.j} \\ (\mathbf{I} - \mathbf{R})\mathbf{y}_j &= (\mathbf{I} - \mathbf{R})(\mathbf{A}_j\boldsymbol{\theta} + \mathbf{e}_j), \end{aligned} \tag{7.3.4}$$

where

$$e_{.j} = K^{-1}\mathbf{1}'\mathbf{e}_j = K^{-1}\sum_{k=1}^{K} e_{kj}\ .$$

It follows from the Normality and independence assumptions of b_j and $\mathbf{e}_j$ that

a) the two sets of random variables $\{y_{.j}\}$, $\{(\mathbf{I}-\mathbf{R})\mathbf{y}_j\}$ are independent of each other,

b) $y_{.j}$ are independently distributed as Normal $N(K^{-1}\mathbf{1}'\mathbf{A}_j\boldsymbol{\theta}, \sigma_b^2 + K^{-1}\sigma_e^2)$, and

c) $(\mathbf{I}-\mathbf{R})\mathbf{y}_j$ are independently Normally distributed with mean vector $(\mathbf{I}-\mathbf{R})\,\mathbf{A}_j\boldsymbol{\theta}$ and a *singular* covariance matrix $\sigma_e^2(\mathbf{I}-\mathbf{R})$.

Thus, the likelihood function can be expressed as the product

$$l(\boldsymbol{\theta}, \sigma_e^2, \sigma_b^2 \mid \mathbf{y}) = l_b(\boldsymbol{\theta}, \sigma_{be}^2 \mid y_{.j}, j = 1, \ldots, J)\,l_e(\boldsymbol{\theta}, \sigma_e^2 \mid (\mathbf{I}-\mathbf{R})\mathbf{y}_j, j = 1, \ldots, J), \quad (7.3.5)$$

where the first factor is

$$l_b(\boldsymbol{\theta}, \sigma_{be}^2 \mid y_{.j}, j = 1, \ldots, J) \propto p(y_{.j}, j = 1, \ldots, J \mid \boldsymbol{\theta}, \sigma_{be}^2) \quad (7.3.6)$$

$$\propto (\sigma_{be}^2)^{-J/2} \exp\left[-\frac{1}{2}\frac{S_b(\boldsymbol{\theta})}{\sigma_{be}^2}\right]$$

with

$$\sigma_{be}^2 = \sigma_e^2 + K\sigma_b^2 \quad (7.3.7)$$

and

$$S_b(\boldsymbol{\theta}) = K\sum_{j=1}^{J}(y_{.j} - K^{-1}\mathbf{1}'\mathbf{A}_j\boldsymbol{\theta})^2. \quad (7.3.8)$$

For the second factor in (7.3.5), we first observe that $(\mathbf{I}-\mathbf{R})\mathbf{1} = \mathbf{0}$ and the matrix $(\mathbf{I}-\mathbf{R})$ is idempotent of rank $K-1$. Thus there exists a $K \times K$ orthogonal matrix $\mathbf{P}$, where

$$\mathbf{P} = [\overset{K-1}{\mathbf{P}^*} \mid \overset{1}{K^{-\frac{1}{2}}\mathbf{1}}]$$

such that

$$\mathbf{P}'(\mathbf{I}-\mathbf{R})\mathbf{P} = \begin{bmatrix} \mathbf{I}_{K-1} & \mathbf{0} \\ \mathbf{0}' & 0 \end{bmatrix}. \quad (7.3.9)$$

Let

$$\mathbf{x}_j = \mathbf{P}_*'(\mathbf{I}-\mathbf{R})(\mathbf{y}_j - \mathbf{A}_j\boldsymbol{\theta}). \quad (7.3.10)$$

Then the $(K-1)\times 1$ vector $\mathbf{x}_j$ is Normally distributed as $N_{K-1}(\mathbf{0}, \mathbf{I}_{k-1}\sigma_e^2)$. Noting that

$$\mathbf{x}_j'\mathbf{x}_j = (\mathbf{y}_j - \mathbf{A}_j\boldsymbol{\theta})'(\mathbf{I}-\mathbf{R})(\mathbf{y}_j - \mathbf{A}_j\boldsymbol{\theta}), \quad (7.3.11)$$

the second factor of the likelihood (7.3.5) is found to be

$$l_e(\boldsymbol{\theta}, \sigma_e^2 \mid (\mathbf{I}-\mathbf{R})\mathbf{y}_j, j = 1, \ldots, J) \propto p(\mathbf{x}_j, j = 1, \ldots, J \mid \sigma_e^2) \quad (7.3.12)$$

$$\propto (\sigma_e^2)^{-\frac{1}{2}J(K-1)} \exp\left[-\frac{1}{2}\frac{S_e(\boldsymbol{\theta})}{\sigma_e^2}\right],$$

where

$$S_e(\boldsymbol{\theta}) = \sum_{j=1}^{J} (\mathbf{y}_j - \mathbf{A}_j\boldsymbol{\theta})'(\mathbf{I} - \mathbf{R})(\mathbf{y}_j - \mathbf{A}_j\boldsymbol{\theta}). \tag{7.3.13}$$

Now, we can write

$$S_b(\boldsymbol{\theta}) = S_b + Q_b(\boldsymbol{\theta}), \tag{7.3.14}$$

where

$$S_b = S_b(\tilde{\boldsymbol{\theta}}), \qquad Q_b(\boldsymbol{\theta}) = (\boldsymbol{\theta} - \tilde{\boldsymbol{\theta}})'\left(\sum_{j=1}^{J} \mathbf{A}_j'\mathbf{R}\mathbf{A}_j\right)(\boldsymbol{\theta} - \tilde{\boldsymbol{\theta}})$$

and $\tilde{\boldsymbol{\theta}}$ satisfies the normal equations

$$\left(\sum_{j=1}^{J} \mathbf{A}_j\mathbf{R}\mathbf{A}_j\right)\tilde{\boldsymbol{\theta}} = \sum_{j=1}^{J} \mathbf{A}_j'\mathbf{1}y_{.j} \quad . \tag{7.3.15}$$

Similarly, we can write

$$S_e(\boldsymbol{\theta}) = S_e + Q_e(\boldsymbol{\theta}), \tag{7.3.16}$$

where

$$S_e = S_e(\hat{\boldsymbol{\theta}}), \qquad Q_e(\boldsymbol{\theta}) = (\boldsymbol{\theta} - \hat{\boldsymbol{\theta}})'\left[\sum_{j=1}^{J} \mathbf{A}_j'(\mathbf{I} - \mathbf{R})\mathbf{A}_j\right](\boldsymbol{\theta} - \hat{\boldsymbol{\theta}})$$

and $\hat{\boldsymbol{\theta}}$ satisfies the normal equations

$$\left[\sum_{j=1}^{J} \mathbf{A}_j'(\mathbf{I} - \mathbf{R})\mathbf{A}_j\right]\hat{\boldsymbol{\theta}} = \sum_{j=1}^{J} \mathbf{A}_j'(\mathbf{I} - \mathbf{R})\mathbf{y}_j \quad . \tag{7.3.17}$$

It is convenient to arrange the quantities $S_b, Q_b(\boldsymbol{\theta}), S_e,$ and $Q_e(\boldsymbol{\theta})$ and to indicate sampling distributional properties in the familiar form of an analysis of variance as in Table 7.3.2.

Table 7.3.2

Analysis of variance of the linear model with two random components

Sources	S.S.	d.f.	Sampling distribution of S.S. (all independent)
Between (inter) blocks	$Q_b(\boldsymbol{\theta})$	q_b	$\sigma_{be}^2\chi_{q_b}^2$
	S_b	$J-q_b$	$\sigma_{be}^2\chi_{J-q_b}^2$
Within (intra) blocks	$Q_e(\boldsymbol{\theta})$	q_e	$\sigma_e^2\chi_{q_e}^2$
	S_e	$J(K-1)-q_e$	$\sigma_e^2\chi_{J(K-1)-q_e}^2$

In Table 7.3.2, q_b and q_e are, respectively, the rank of the matrices

$$\sum_{j=1}^{J} \mathbf{A}_j'\mathbf{R}\mathbf{A}_j \qquad \text{and} \qquad \sum_{j=1}^{J} \mathbf{A}_j'(\mathbf{I} - \mathbf{R})\mathbf{A}_j.$$

7.3.3 Sampling Theory Estimation Problems Associated with the Model

In the sampling theory framework, the usual difficulties are associated with making inferences about the variance components σ_b^2 and σ_e^2. The unbiased estimator of σ_b^2 which has often been used

$$\hat{\sigma}_b^2 = \frac{S_b}{K(J - q_b)} - \frac{S_e}{K[J(K-1) - q_e]}. \tag{7.3.18}$$

is intuitively unsatisfactory because it can take negative values. In addition, the sampling distribution of $\hat{\sigma}_b^2$ involves the nuisance parameter σ_b^2/σ_e^2, and consequently leads to difficulties in constructing confidence intervals for σ_b^2. Finally, while inferences about σ_e^2 are often based upon S_e alone, yet if σ_b^2/σ_e^2 is not large, S_b can also contribute appreciable information about σ_e^2. Attempts to deal with this problem by "pooling variances" are accompanied by familiar problems previously discussed in Chapters 5 and 6.

Usually the parameters of principal interest are the elements of the vector $\boldsymbol{\theta}$ or linear functions of them. From Table 7.3.2 we see that if the inter-block information was ignored, confidence regions for $\boldsymbol{\theta}$ could be constructed by referring the quantity

$$\frac{Q_e(\boldsymbol{\theta})/q_e}{S_e/[J(K-1) - q_e]} \tag{7.3.19}$$

to an F distribution with $[q_e, J(K-1) - q_e]$ degrees of freedom. Similarly, if the intra-block information was ignored, a different set of confidence regions for $\boldsymbol{\theta}$ could be obtained by referring the quantity

$$\frac{Q_b(\boldsymbol{\theta})/q_b}{S_b/(J - q_b)} \tag{7.3.20}$$

to an F distribution with $(q_b, J - q_b)$ degrees of freedom.

In attempts to combine intra- and inter-block information the major problem for sampling theory is the difficulty of eliminating the nuisance parameter σ_b^2/σ_e^2, see, for example Yates (1940), Scheffé (1959, pp. 170–178), Graybill and Weeks (1959), and Seshadri (1966). We now consider the problems from a Bayesian viewpoint and discuss in detail the analysis of *balanced* incomplete block designs.

7.3.4 Prior and Posterior Distributions of $(\sigma_e^2, \sigma_{be}^2, \boldsymbol{\theta})$

From (7.3.5), (7.3.6) and (7.3.12) we see that information contained in the likelihood function can be regarded as coming from J independent observations $y_{j.}$ from Normal populations with variance σ_{be}^2/K, and $J(K-1)$ independent observations drawn from Normal populations with variance σ_e^2. The means of these observations are linear functions of the elements of the vector $\boldsymbol{\theta}$.

As in Chapters 5 and 6, we carry through the analysis using the noninformative reference prior distribution

$$p(\sigma_e^2, \sigma_{be}^2, \boldsymbol{\theta}) \propto \sigma_e^{-2}\sigma_{be}^{-2}, \tag{7.3.21}$$

where, since $\sigma_{be}^2 = \sigma_e^2 + K\sigma_b^2$, the constraint $\sigma_{be}^2 > \sigma_e^2$ is implied. Combining (7.3.5) with (7.3.21) we obtain the posterior distribution as

$$p(\sigma_e^2,\sigma_{be}^2,\boldsymbol{\theta} \mid \mathbf{y}) \propto (\sigma_{be}^2)^{-(\frac{1}{2}J+1)}\,(\sigma_e^2)^{-[\frac{1}{2}J(K-1)+1]}$$

$$\times \exp\left\{-\frac{1}{2}\left[\frac{S_b + Q_b(\boldsymbol{\theta})}{\sigma_{be}^2} + \frac{S_e + Q_e(\boldsymbol{\theta})}{\sigma_e^2}\right]\right\},$$

$$\sigma_{be}^2 > \sigma_e^2 > 0, \quad -\infty < \theta_i < \infty, \quad i = 1, \ldots, I. \tag{7.3.22}$$

We shall not proceed further with the general model but consider the especially important case of balanced incomplete block designs.

7.4 ANALYSIS OF BALANCED INCOMPLETE BLOCK DESIGNS

As we have noted earlier, in a balanced incomplete block design (BIBD), I treatments are examined in J blocks of equal size K. Each treatment appears r times and occurs λ times in a block with every other treatment. For any such design, we must have

$$J \geqslant I, \qquad JK = Ir, \qquad \lambda(I-1) = r(K-1). \tag{7.4.1}$$

In terms of the general model (7.3.1), $\mathbf{A}_j$ is a $K \times I$ matrix of 1's and 0's, each row of which contains a single 1 in the column corresponding to the treatment applied. Clearly $\mathbf{A}_j$ consists of orthogonal rows. The $I \times J$ matrix

$$[\mathbf{A}_1'\mathbf{1} \;\vdots\; \cdots \;\vdots\; \mathbf{A}_J'\mathbf{1}] \tag{7.4.2}$$

is known as the *incidence* matrix or the *design* matrix whose elements consist of 1's and 0's indicating the presence and absence of the I treatments in the J blocks. For the instruction design in Table 7.3.1, the incidence matrix is

$$\begin{array}{ll} & \quad\text{Block} \\ & \;\;1 \;\; 2 \;\; 3 \;\; 4 \;\; 5 \;\; J{=}6 \\ \text{Treatment} \begin{array}{r} 1 \\ 2 \\ 3 \\ I{=}4 \end{array} & \left[\begin{array}{cccccc} 1 & 1 & 1 & 0 & 0 & 0 \\ 1 & 0 & 0 & 1 & 1 & 0 \\ 0 & 1 & 0 & 1 & 0 & 1 \\ 0 & 0 & 1 & 0 & 1 & 1 \end{array}\right]. \end{array}$$

7.4.1 Properties of the BIBD Model

The BIBD model has the following properties:

$$\sum_{j=1}^{J} \mathbf{A}_j'\mathbf{A}_j = r\mathbf{I}_I \tag{7.4.3}$$

$$\sum_{j=1}^{J} \mathbf{A}_j'\mathbf{R}\mathbf{A}_j = \left(\frac{r-\lambda}{K}\right)\mathbf{I}_I + \frac{\lambda}{K}\mathbf{1}_I\mathbf{1}_I' \tag{7.4.4}$$

$$\left(\sum_{j=1}^{J} \mathbf{A}_j'\mathbf{R}\mathbf{A}_j\right)^{-1} = \left(\frac{K}{r-\lambda}\right)\left(\mathbf{I}_I - \frac{\lambda}{rK}\mathbf{1}_I\mathbf{1}_I'\right), \quad r > \lambda \tag{7.4.5}$$

$$\sum_{j=1}^{J} \mathbf{A}_j'(\mathbf{I}-\mathbf{R})\mathbf{A}_j = \frac{\lambda I}{K}(\mathbf{I}_I - I^{-1}\mathbf{1}_I\mathbf{1}_I') \tag{7.4.6}$$

$$\sum_{j=1}^{J} \mathbf{A}_j'\mathbf{1}_K y_{.j} = r\mathbf{B}, \tag{7.4.7}$$

$$\sum_{j=1}^{J} \mathbf{A}_j'(\mathbf{I}-\mathbf{R})\mathbf{y}_j = r(\mathbf{T}-\mathbf{B}), \tag{7.4.8}$$

$$\mathbf{1}_I'\mathbf{T} = \mathbf{1}_I'\mathbf{B} = I y_{..} \tag{7.4.9}$$

where $\mathbf{I}_I$ is an $I \times I$ identity matrix, $\mathbf{1}_I$ is an $I \times 1$ vector of ones, $\mathbf{B}' = (B_1, \ldots, B_I)$, B_i is the average of $y_{.j}$ for blocks j in which the ith treatment appears, $\mathbf{T}' = (T_1, \ldots, T_I)$ is the vector of treatment averages, and $y_{..}$ is the grand average.

Breakdown of $S_b(\boldsymbol{\theta})$

Using (7.4.4), (7.4.5), (7.4.7), and (7.4.9), the quantity $Q_b(\boldsymbol{\theta})$ in (7.3.14) can be written

$$Q_b(\boldsymbol{\theta}) = \left(\frac{r-\lambda}{K}\right)\sum_{i=1}^{I}(\theta_i - \tilde{\theta}_i)^2 + \frac{\lambda}{K}\left[\sum_{i=1}^{I}(\theta_i - \tilde{\theta}_i)\right]^2, \tag{7.4.10}$$

where

$$\tilde{\theta}_i = \frac{1}{(r-\lambda)}(rKB_i - \lambda I y_{..}), \qquad r > \lambda. \tag{7.4.11}$$

In a BIBD, it is the comparative values of the θ_i which are of primary interest. In this connection it is convenient to work with the contrasts

$$\phi_i = \theta_i - \bar{\theta}, \qquad i = 1, \ldots, I, \tag{7.4.12}$$

where $\quad \bar{\theta} = I^{-1}\sum_{i=1}^{I}\theta_i \quad$ so that $\quad \sum_{i=1}^{I}\phi_i = 0, \quad$ and to write

$$Q_b(\boldsymbol{\theta}) = \left(\frac{r-\lambda}{K}\right)\sum_{i=1}^{I}(\phi_i - \tilde{\phi}_i)^2 + rI(\bar{\theta} - y_{..})^2 \tag{7.4.13}$$

with

$$\tilde{\phi}_i = \frac{rK}{r-\lambda}(B_i - y_{..}), \qquad r > \lambda. \tag{7.4.14}$$

Since $S_b(\boldsymbol{\theta}) = S_b + Q_b(\boldsymbol{\theta})$ is true for all $\boldsymbol{\theta}$, by setting $\boldsymbol{\theta} = \mathbf{0}$ (which implies that $\phi_i = 0$, $i = 1, \ldots, I$), we obtain from (7.3.8) and (7.4.13) that

$$S_b = K \sum_{j=1}^{J} y_{.j}^2 - rIy_{..}^2 - \left(\frac{r-\lambda}{K}\right) \sum_{i=1}^{I} \tilde{\phi}_i^2$$

$$= K \sum_{j=1}^{J} (y_{.j} - y_{..})^2 - \left(\frac{r-\lambda}{K}\right) \sum_{i=1}^{I} \tilde{\phi}_i^2. \tag{7.4.15}$$

Breakdown of $S_e(\boldsymbol{\theta})$

From (7.4.6) the quantity $Q_e(\boldsymbol{\theta})$ in (7.3.16) can be written

$$Q_e(\boldsymbol{\theta}) = \frac{\lambda I}{K} \left\{ \sum_{i=1}^{I} (\theta_i - \hat{\theta}_i)^2 - I^{-1} \left[\sum_{i=1}^{I} (\theta_i - \hat{\theta}_i) \right]^2 \right\} \tag{7.4.16}$$

$$= \frac{\lambda I}{K} \sum_{i=1}^{I} (\phi_i - \hat{\phi}_i)^2,$$

where ϕ_i is defined in (7.4.12),

$$\hat{\phi}_i = \hat{\theta}_i - I^{-1} \sum_{l=1}^{I} \hat{\theta}_l \tag{7.4.17}$$

and $\hat{\boldsymbol{\theta}} = (\hat{\theta}_1, \ldots, \hat{\theta}_I)'$ satisfies the normal equations (7.3.17). It follows from (7.4.6), (7.4.8), and (7.4.9) that

$$\hat{\boldsymbol{\theta}} = \frac{rK}{\lambda I} (\mathbf{T} - \mathbf{B}) \tag{7.4.18}$$

is a particular solution of (7.3.17) for the BIBD and that

$$\hat{\phi}_i = \hat{\theta}_i = \frac{rK}{\lambda I} (T_i - B_i), \qquad i = 1, \ldots, I. \tag{7.4.19}$$

Consequently, we may write $S_e(\boldsymbol{\theta})$ in (7.3.16) as

$$S_e(\boldsymbol{\theta}) = S_e + \frac{\lambda I}{K} \sum_{i=1}^{I} (\phi_i - \hat{\phi}_i)^2 \tag{7.4.20}$$

with

$$S_e = \sum_{j=1}^{J} \sum_{k=1}^{K} y_{kj}^2 - K \sum_{j=1}^{J} y_{.j}^2 - \frac{\lambda I}{K} \sum_{i=1}^{I} \hat{\phi}_i^2$$

$$= \sum_{j=1}^{J} \sum_{k=1}^{K} (y_{kj} - y_{.j})^2 - \frac{\lambda I}{K} \sum_{i=1}^{I} \hat{\phi}_i^2. \tag{7.4.21}$$

The expressions for $\tilde{\phi}_i$, $\hat{\phi}_i$, S_b, and S_e given above are, of course, well known. We have sketched the development in order to relate the present results for the BIBD to those for the general linear model set out in Section 7.3.

The Analysis of Variance

The breakdowns of $S_b(\boldsymbol{\theta})$ and $S_e(\boldsymbol{\theta})$ are summarized in Table 7.4.1 in the usual analysis of variance form.

Table 7.4.1

Analysis of variance of a BIBD model

Sources	S.S.	d.f.	Sampling distribution of S.S. (all independent)
Between (inter) blocks	$Q_b(\boldsymbol{\theta})\begin{cases} rI(\bar{\theta}-y_{..})^2 \\ \dfrac{r-\lambda}{K}\sum_{i=1}^{I}(\phi_i-\tilde{\phi}_i)^2 \end{cases}$	1 $\delta(I-1)$	$\sigma_{be}^2\chi_1^2$ $\sigma_{be}^2\chi_{\delta(I-1)}^2$
	S_b	$(J-1)-\delta(I-1)$	$\sigma_{be}^2\chi_{(J-1)-\delta(I-1)}^2$
Within (intra) blocks	$Q_e(\boldsymbol{\theta})=\dfrac{\lambda I}{K}\sum_{i=1}^{I}(\phi_i-\hat{\phi}_i)^2$	$I-1$	$\sigma_e^2\chi_{(I-1)}^2$
	S_e	$J(K-1)-(I-1)$	$\sigma_e^2\chi_{J(K-1)-(I-1)}^2$

In Table 7.4.1, $\delta=1$ if $r>\lambda$ and $\delta=0$ if $r=\lambda$.

7.4.2 A Comparison of the One-way Classification, RCBD and BIBD Models

It is instructive to relate these results to those for the additive mixed model (Section 6.3) and those for the one-way classification (Sections 2.11 and 7.2). If, as we shall suppose, blocks may be represented as a "random effect," then the additive mixed model is appropriate for the analysis of the randomized block design. We shall call the latter the *randomized complete block design* (RCBD) to distinguish it from the BIBD.

For a BIBD, the model in (7.3.1) may be written

$$y_{kj}=E(y_{kj})+b_j+e_{kj} \qquad j=1,\ldots,J; \quad k=1,\ldots,K, \tag{7.4.22}$$

where y_{kj} is the kth observation in the jth block, b_j the block effect and e_{kj} the within block residual error. The expectation $E(y_{kj})$ is

$$E(y_{kj})=\theta_i \tag{7.4.23}$$

if the ith treatment, $i=1,\ldots,I$, is applied to y_{kj}. Alternatively, we may write the model as

$$y_{(im)}=\theta_i+\varepsilon_{(im)}, \qquad i=1,\ldots,I; \quad m=1,\ldots,r \tag{7.4.24}$$

where $y_{(im)}$ is the mth observation receiving the ith treatment and $\varepsilon_{(im)}$ is the error term such that

$$\varepsilon_{(im)}=b_j+e_{kj} \tag{7.4.25}$$

if $y_{(im)}$ happens to be in the kth position of the jth block. In other words, (7.4.22) and (7.4.24) show that we can classify the $JK = rI$ observations in two different ways, one according to the blocks and the other according to the treatments.

If there is no block effect, $b_j \equiv 0$, then the model in (7.4.24) is the one-way classification model discussed earlier with the appropriate change in notation. On the other hand, if the size of a block equals to the number of treatments, $K = I$, and every treatment appears in every block, so that $J = r = \lambda$, then the model in (7.4.22) is

$$y_{ij} = \theta_i + b_j + e_{ij}, \qquad i = 1, \dots, I; \quad j = 1, \dots, J \tag{7.4.26}$$

which is the RCBD model given in (6.3.1) upon substituting b_j for c_j.

Now, the sample averages needed to compute $\tilde{\phi}_i$, $\hat{\phi}_i$ and the sum of squares in Table (7.4.1) are

a) $y_{..} = y_{(..)}$ grand average

b) $T_i = y_{(i.)}$ treatment averages

c) $y_{.j}$ block averages

d) B_i average of block averages for blocks in which the ith treatment appears.

Note that when $b_j \equiv 0$ or $K = I, B_i \equiv y_{..}$, otherwise $I^{-1} \sum_i B_i = y_{..}$.

Table 7.4.2a compares the point estimates of the contrasts $\phi_i = \theta_i - \bar{\theta}, i = 1, \dots, I$, for the one-way classification, RCBD and BIBD models.

While the estimate $T_i - y_{..}$ is the same for both the one-way classification and the RCBD, it is regarded as an "intra-block" estimate for the latter because

$$T_i - y_{..} = \frac{1}{J} \sum_j \left(y_{ij} - y_{.j}\right) \tag{7.4.27}$$

that is, because it is the average of J within block contrasts among the observations. For the BIBD, there are two sources of information about ϕ_i, the "intra-block" estimate $\hat{\phi}_i$ and the "inter-block" estimate $\tilde{\phi}_i$. Let

$$y_{.m}^{(i)} \qquad m = 1, \dots, r \tag{7.4.28}$$

denote the r block averages in which the ith treatment appears. Then, the "intra-block" estimate $\hat{\phi}_i$ of ϕ_i is proportional to

$$T_i - B_i = \frac{1}{r} \sum_m (y_{(im)} - y_{.m}^{(i)}) \tag{7.4.29}$$

which is made up of r within block contrasts among the observations. On the other hand, the inter-block estimate $\tilde{\phi}_i$ is proportional to

$$B_i - y_{..}$$

representing a contrast among the block averages. The constants of proportionality f^{-1} and $(1-f)^{-1}$ are such that

$$f = \frac{\lambda I}{rK}, \qquad 1 - f = \left(\frac{rK - \lambda I}{rK}\right) = \left(\frac{r - \lambda}{rK}\right), \qquad 0 < f \leqslant 1. \tag{7.4.30}$$

The quantity f is sometimes called the "efficiency factor" of a BIBD. When $K = I$ and $r = \lambda, f = 1$ and the BIBD reduces to the RCBD. In this case

$$\hat{\phi}_i = T_i - B_i = T_i - y_{..}, \qquad \tilde{\phi} \equiv 0 \tag{7.4.31}$$

so that, as far as sampling theory is concerned, there is only one source of information about ϕ_i.

Table 7.4.2

Comparison of one-way classification, RCBD and BIBD models

a) Point estimates of the contrast $\phi_i = \theta_i - \bar{\theta}$:

	One-way classification	RCBD	BIBD
Inter-block	$T_i - y_{..}$	—	$\tilde{\phi}_i = \dfrac{1}{1-f}(B_i - y_{..})$
Intra-block		$T_i - y_{..}$	$\hat{\phi}_i = \dfrac{1}{f}(T_i - B_i)$

b) Decomposition of the sum of squares $\sum_i \sum_m (y_{(im)} - \theta_i)^2$:

Source		One-way classification	RCBD	BIBD
Grand mean		$rI(\bar{\theta} - y_{..})^2$	$rI(\bar{\theta} - y_{..})^2$	$rI(\bar{\theta} - y_{..})^2$
Treatment	inter-block	$r\Sigma[\phi_i - (T_i - y_{..})]^2$	—	$(1-f)r\Sigma(\phi_i - \tilde{\phi}_i)^2$
	intra-block		$r\Sigma[\phi_i - (T_i - y_{..})]^2$	$fr\Sigma(\phi_i - \hat{\phi}_i)^2$
Residuals	inter-block	$\Sigma\,\Sigma\,(y_{(im)} - T_i)^2$	$K\Sigma(y_{.j} - y_{..})^2$	$K\Sigma(y_{.j} - y_{..})^2 - (1-f)r\Sigma\tilde{\phi}^2$
	intra-block		$\Sigma\,\Sigma\,(y_{kj} - y_{.j})^2 - r\Sigma(T_i - y_{..})^2$	$\Sigma\,\Sigma\,(y_{kj} - y_{.j})^2 - fr\,\Sigma\,\hat{\phi}_i^2$

Table 7.4.2b gives a corresponding comparison of the decomposition of the sum of squares for the three models. The reader will have no difficulty in associating the entries in the table with the corresponding ones in Table 2.11.1 for the one-way classification, in Table 6.3.1 for the RCBD and in Table 7.4.1 for the BIBD.

7.4.3 Posterior Distribution of $\boldsymbol{\phi}=(\phi_1,\ldots,\phi_{I-1})^1$ for a BIBD

The quantities of principal interest are contrasts between the treatment parameters. If in (7.3.22) we make the transformation from $\theta_1, \ldots, \theta_I$ to $\bar{\theta}$ and the $I - 1$ linearly independent contrasts $\phi_i = \theta_i - \bar{\theta}, i = 1, \ldots, I - 1$, then upon substituting for $Q_b(\boldsymbol{\theta})$ and $Q_e(\boldsymbol{\theta})$ into (7.3.22) from Table 7.4.1 and integrating out $\bar{\theta}$, we obtain

$$p(\sigma_e^2, \sigma_{be}^2, \boldsymbol{\phi} \mid \mathbf{y}) \propto (\sigma_{be}^2)^{-[\frac{1}{2}(J-1)+1]} (\sigma_e^2)^{-[\frac{1}{2}J(K-1)+1]}$$
$$\times \exp\left\{-\frac{1}{2}\left[\frac{S_b + (1-f)r\sum_{i=1}^I (\phi_i - \tilde{\phi}_i)^2}{\sigma_{be}^2} + \frac{S_b + fr\sum_{i=1}^I (\phi_i - \hat{\phi}_i)^2}{\sigma_e^2}\right]\right\},$$
$$-\infty < \phi_i < \infty, \qquad i = 1, \ldots, I-1; \qquad 0 < \sigma_e^2 < \sigma_{be}^2 < \infty, \qquad (7.4.32)$$

where and in what follows it is to be remembered that $\phi_I = -(\phi_1 + \cdots + \phi_{I-1})$. In (7.4.32), use is made of the definition of the efficiency factor $f = \lambda I(rK)$ in (7.4.30). Applying the identity (A5.2.1) in Appendix A5.2 to integrate out σ_e^2 and σ_{be}^2, we have

$$p(\boldsymbol{\phi} \mid \mathbf{y}) \propto \left[S_e + fr\sum_{i=1}^{I} (\phi_i - \hat{\phi}_i)^2\right]^{-\frac{1}{2}J(K-1)}$$
$$\times \left[S_b + (1-f)r\sum_{i=1}^{I} (\phi_i - \tilde{\phi}_i)^2\right]^{-\frac{1}{2}(J-1)} I_{u(\boldsymbol{\phi})}\left[\frac{J-1}{2}, \frac{J(K-1)}{2}\right],$$
$$-\infty < \phi_i < \infty, \qquad i = 1, \ldots, I-1, \qquad (7.4.33)$$

where

$$u(\boldsymbol{\phi}) = \frac{S_b + (1-f)r\sum_{i=1}^I (\phi_i - \tilde{\phi}_i)^2}{S_e + S_b + fr\sum_{i=1}^I (\phi_i - \hat{\phi}_i)^2 + (1-f)r\sum_{i=1}^I (\phi_i - \tilde{\phi}_i)^2}$$

and, as before, $I_x(p,q)$ is the incomplete beta function. The posterior distribution is the product of three factors. The first of these is in the form of a $(I - 1)$ dimensional multivariate t distribution centered at $\hat{\boldsymbol{\phi}} = (\hat{\phi}_1, \ldots, \hat{\phi}_{I-1})'$, with covariance matrix proportional to $(\mathbf{I}_{I-1} - I^{-1}\mathbf{1}_{I-1}\mathbf{1}_{I-1}')$. For $r > \lambda$, the second factor again takes the form of an $(I - 1)$-dimensional multivariate t distribution centered at $\tilde{\boldsymbol{\phi}} = (\tilde{\phi}_1, \ldots, \tilde{\phi}_{I-1})'$, with covariance matrix also proportional to $(\mathbf{I}_{I-1} - I^{-1}\mathbf{1}_{I-1}\mathbf{1}_{I-1}')$. The third factor is an incomplete beta integral whose upper limit depends on $\boldsymbol{\phi}$. We note that the distribution $p(\boldsymbol{\phi} \mid \mathbf{y})$ is a multivariate generalization of a form of distribution obtained in Tiao and Tan (1966, Eqn. 7.2).

Relationship to Previous Results

The first factor in (7.4.33) is equivalent to the confidence distribution of $\boldsymbol{\phi}$ in the sampling theory framework if only the intra-block information is employed. It would also be the appropriate posterior distribution of $\boldsymbol{\phi}$ if only the portion of the likelihood corresponding to "within-block" was used in conjunction with a noninformative uniform prior in $\log \sigma_e$, $(\sigma_e > 0)$. For $r > \lambda$, the second factor is equivalent to the confidence distribution of $\boldsymbol{\phi}$ if only the inter-block information is employed. It would also be the appropriate posterior distribution of $\boldsymbol{\phi}$ if only the portion of the likelihood corresponding to "between-blocks" was used in conjunction with a noninformative uniform prior in $\log \sigma_{be}$, $(\sigma_{be} > 0)$.

The product of the first two factors provides a blending of intra- and inter-block information to form a posterior distribution for $\boldsymbol{\phi}$ which would be appropriate if the constraint $\sigma_{be}^2 > \sigma_e^2$ could be ignored. It is clear that the distribution of $\boldsymbol{\phi}$ based upon the first and second factors would be centered between the centers of the two separate multivariate t distributions.†

The third factor arises as a direct consequence of the constraint $\sigma_{be}^2 > \sigma_e^2$, the effect of which is to pull the distribution of $\boldsymbol{\phi}$ towards the multivariate t distribution given by the first factor.

To see this, consider the density constrained on *any* straight line in the space of the variables $\boldsymbol{\phi}$. Any straight line can be written in parametric form

$$\boldsymbol{\phi} = (1 - \mu)\boldsymbol{\phi}_1 + \mu\boldsymbol{\phi}_2,$$

where $\boldsymbol{\phi}_1$ and $\boldsymbol{\phi}_2$ are any two points. Substitution of $\boldsymbol{\phi}$ into (7.4.33) yields the form

$$p(\mu \mid \mathbf{y}) \propto [a_e + d_e(\mu - c_e)^2]^{-\frac{1}{2}J(K-1)}[a_b + d_b(\mu - c_b)^2]^{-\frac{1}{2}(J-1)}$$
$$\times\ I_{u(\mu)}\left[\frac{J-1}{2}, \frac{J(K-1)}{2}\right], \qquad -\infty < \mu < \infty, \qquad (7.4.34)$$

where

$$u(\mu) = \frac{a_b + d_b(\mu - c_b)^2}{a_e + d_e(\mu - c_e)^2 + a_b + d_b(\mu - c_b)^2}, \qquad a_e > 0,\ \ a_b > 0,\ \ d_e > 0,\ \ d_b > 0.$$

Now $\partial u(\mu)/\partial\mu \lessgtr 0$ in the interval between c_e and c_b as $c_e \lessgtr c_b$. Remembering that $I_{u(\mu)}[\frac{1}{2}(J-1), \frac{1}{2}J(K-1)]$ is monotonically increasing in $u(\mu)$, the effect of the this function is to give more weight to the first factor in determining the center of the complete distribution $p(\mu \mid \mathbf{y})$. Since the line we have chosen is arbitrary, it follows that the function $I_{u(\mu)}[\frac{1}{2}(J-1), \frac{1}{2}J(K-1)]$ gives more weight to the first factor of $p(\boldsymbol{\phi} \mid \mathbf{y})$ than to the second.

Finally, in the special case $r = \lambda$, namely, a randomized complete block design model, the distribution in (7.4.33) degenerates to that given in (6.3.45)

† Other results of the same kind will be considered further in Chapters 8 and 9 when combining the information about common regression coefficients from several sources, in cases where no variance restrictions exist.

with appropriate changes in notation. In this case, the second factor is a constant and the third factor, which is now monotonically decreasing in $\sum_{i=1}^{I}(\phi_i - \hat{\phi}_i)^2$, tends to make the distribution more concentrated about $\hat{\boldsymbol{\phi}}$ than it would be if only the first factor alone were considered.

7.4.4 An Illustrative Example

To illustrate the theory we consider an example involving $I = 3$ treatments, $J = 15$ blocks of size $K = 2$, each treatment replicated $r = 10$ times; thus $\lambda = 5$. Table 7.4.3 shows the data with some peripheral calculations. For convenience in presentation, we have associated the blocks (random effects) with rows and the treatment (fixed effects) with columns. The data was generated from a table of random Normal deviates using a BIBD model with $\theta_1 = 5$, $\theta_2 = \theta_3 = 2$, $\sigma_e^2 = 2.25$, and $\sigma_b^2 = 1$. For this example the distribution in (7.4.33) is, of

Table 7.4.3

Data from a BIBD ($I=3, J=15, K=2, r=10, \lambda=5, f=0.75$)

		Treatment 1	Treatment 2	Treatment 3	$y_{.j}$
Block	1	10.05	4.92		7.485
	2	10.10	5.49		7.795
	3	5.52	2.97		4.245
	4	9.90	3.72		6.810
	5	9.99	5.68		7.835
	6	5.93		0.24	3.085
	7	6.76		3.80	5.280
	8	7.94		2.90	5.420
	9	11.44		2.10	6.770
	10	9.74		3.94	6.840
	11		0.79	4.83	2.810
	12		6.09	4.41	5.250
	13		5.78	6.47	6.125
	14		3.22	5.30	4.260
	15		5.78	4.18	4.980
	T_i	8.737	4.444	3.817	$y_{..} = 5.666$
	B_i	6.1565	5.7595	5.082	$S_e = 29.1841$ (13 d.f.)
	$\hat{\phi}_i$	3.4406	-1.754	-1.6866	$S_b = 49.04448$ (12 d.f.)
	$\tilde{\phi}_i$	1.962	0.374	-2.336	

course, a bivariate one. Note that

$$\sum_{i=1}^{3} (\phi_i - \hat{\phi}_i)^2 = 2[(\phi_1 - 3.4406)^2 + (\phi_2 + 1.754)^2 + (\phi_1 - 3.4406)(\phi_2 + 1.754)]$$

$$\sum_{i=1}^{3} (\phi_i - \tilde{\phi}_i)^2 = 2[(\phi_1 - 1.962)^2 + (\phi_2 - 0.374)^2 + (\phi_1 - 1.962)(\phi_2 - 0.374)]$$

Figure 7.4.1 shows a number of contours related to the distribution (7.4.33).

1. The contour (1) centered at $P_1 = (3.44, -1.75)$ is the 95 per cent H.P.D. region derived from the first factor alone. It was calculated using the fact that, if the first factor alone were the posterior distribution, then the quadratic form

$$\frac{fr\sum_{i=1}^{I} (\phi_i - \hat{\phi}_i)^2/(I - 1)}{S_e/[J(K - 1) - (I - 1)]}$$

would be distributed as an F with $(I - 1)$ and $J(K - 1) - (I - 1)$ degrees of freedom.

2. The contour (2) centered at $P_2 = (1.96, 0.37)$ is the 95 per cent H.P.D. region derived from the second factor alone. It was calculated by referring the quantity

$$\frac{(1 - f)r\sum_{i=1}^{I} (\phi_i - \tilde{\phi}_i)^2/(I - 1)}{S_b/(J - I)}$$

to an F distribution with $(I - 1)$ and $(J - I)$ degrees of freedom.

3. The broken lines (3) are contours of the third factor, with the contour values shown.

4. The contour (4) centered at $P = (3.26, -1.50)$ defines the approximate 95 per cent H.P.D. region derived from the complete distribution (7.4.33). It was calculated from the formula

$$\log p\,(\phi_1^*, \phi_2^* \mid \mathbf{y}) - \log p\,(\phi_1, \phi_2 \mid \mathbf{y}) = 5.99 = \chi^2(2, 0.05)$$

where (ϕ_1^*, ϕ_2^*) are the coordinates of P, the center and maximum point. This is equivalent to

$$p(\phi_1, \phi_2 \mid \mathbf{y}) = 0.05p(\phi_1^*, \phi_2^* \mid \mathbf{y}).$$

We can see from Fig. 7.4.1 that, as expected, the overall distribution contour curve (4), is located between the contours (1) and (2) from the first and second factors respectively. In fact the center P is practically collinear with the other centers P_1 and P_2. Contours (1) and (2) are elliptical and have parallel axes, since the covariance matrices of the two corresponding bivariate t distributions are proportional. Contour (2) is much larger than contour (1) essentially because, for this example,

$$S_b/[(J - I)(1 - f)] \gg S_e/\{f[J(K - 1) - (I - 1)]\}$$

implying that the first factor will have a dominating role in determining the overall distribution. The lines (3) show that the third factor is increasing in a South-east direction confirming that this factor tends to pull the final distribution of $\boldsymbol{\phi}$ towards the bivariate t distribution given by the first factor. However, for this example, the effect is quite small, as is seen by the slight changes in the value of the function over the region in which the dominating first factor is appreciable, The shape of contour (4) is very nearly elliptical which suggests (as turns out to be the case) that the distribution might be approximated by a bivariate t distribution.

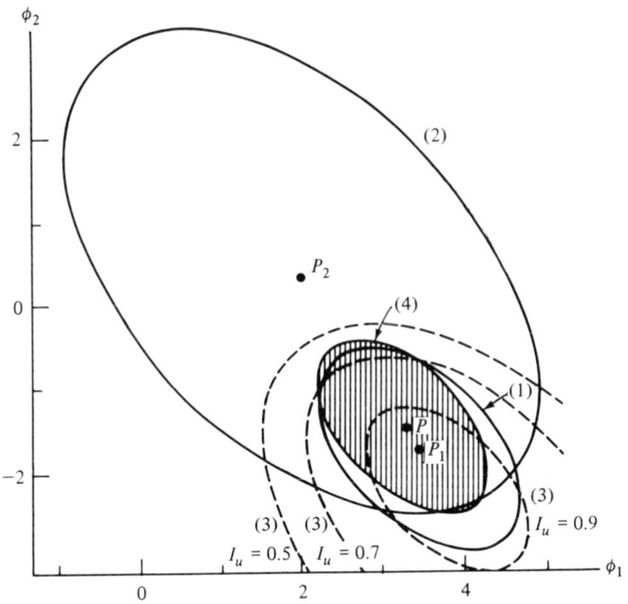

In the figure
1. 95 per cent contour of $\boldsymbol{\phi}$ based on first factor of (7.4.33)
2. 95 per cent contour of $\boldsymbol{\phi}$ based on second factor of (7.4.33)
3. Contours of $\boldsymbol{\phi}$ for various values of $I_{u(\boldsymbol{\phi})}\,[\frac{1}{2}(J-1), \frac{1}{2}J(K-1)]$ as marked
4. Approximate 95 per cent contour of the entire distribution of $\boldsymbol{\phi}$ in (7.4.33)

Fig. 7.4.1 Contours for the components of the posterior distribution of (ϕ_1, ϕ_2) for the BIBD data.

7.4.5 Posterior Distribution of σ_{be}^2/σ_e^2

In this section, we digress from our discussion of the problem of making inferences about the contrasts $\boldsymbol{\phi}$ to obtain the distribution of the variance ratio

$$w = \sigma_{be}^2/\sigma_e^2 \tag{7.4.35}$$

which plays an important role in approximating the distribution of $\boldsymbol{\phi}$.

In the joint distribution $p(\sigma_e^2, \sigma_{be}^2, \boldsymbol{\phi} \mid \mathbf{y})$ in (7.4.32) we may use the identity (A7.1.1) in Appendix A7.1 to write

$$\sigma_{be}^{-2}(1-f)r\sum_{i=1}^{I}(\phi_i-\tilde{\phi}_i)^2+\sigma_e^{-2}fr\sum_{i=1}^{I}(\phi_i-\hat{\phi}_i)^2$$

$$=\sigma_e^{-2}\left\{\left(f+\frac{1-f}{w}\right)r\sum_{i=1}^{I}[\phi_i-\bar{\phi}_i(w)]^2+\left(w+\frac{1-f}{f}\right)^{-1}S_t\right\}, \quad (7.4.36)$$

where

$$S_t=(1-f)r\sum_{i=1}^{I}(\tilde{\phi}_i-\hat{\phi}_i)^2 \quad (7.4.37)$$

and

$$\bar{\phi}_i(w)=\left(f+\frac{1-f}{w}\right)^{-1}\left[f\hat{\phi}_i+\left(\frac{1-f}{w}\right)\tilde{\phi}_i\right]. \quad (7.4.38)$$

Making the transformation from $(\sigma_{be}^2, \sigma_e^2)$ to (w, σ_e^2) and integrating out σ_e^2, the posterior distribution of $(w, \boldsymbol{\phi})$ is

$$p(w, \boldsymbol{\phi} \mid \mathbf{y}) \propto w^{-[\frac{1}{2}(J-1)+1]}$$

$$\times\left\{S_e+w^{-1}S_b+\left(w+\frac{1-f}{f}\right)^{-1}S_t+\left(f+\frac{1-f}{w}\right)r\sum_{i=1}^{I}[\phi_i-\bar{\phi}_i(w)]^2\right\}^{-[\frac{1}{2}(JK-1)]},$$

$$-\infty<\phi_i<\infty, \quad i=1,\ldots,I-1, \quad w>1. \quad (7.4.39)$$

Upon eliminating $\boldsymbol{\phi}$ by integration and recalling from Table 7.4.1 that $\delta = 1$ if $r > \lambda$ and $\delta = 0$ if $r = \lambda$, the posterior distribution of w is

$$p(w \mid \mathbf{y}) \propto H_1(w)H_2(w), \qquad w > 1, \quad (7.4.40)$$

with

$$H_1(w) \propto w^{\frac{1}{2}[J(K-1)-(I-1)]-1}\left(1+\frac{S_e}{S_b}w\right)^{-\frac{1}{2}[I(r-1)-\delta(I-1)]}$$

and

$$H_2(w) \propto \left(\frac{w}{w+(1-f)/f}\right)^{\frac{1}{2}(I-1)}\left(1+\frac{S_e}{S_b}w\right)^{-\frac{1}{2}\delta(I-1)}$$

$$\times\left[1+\left(\frac{w}{w+(1-f)/f}\right)\left(\frac{S_t}{S_b}\right)\left(1+\frac{S_e}{S_b}w\right)^{-1}\right]^{-\frac{1}{2}I(r-1)}.$$

This distribution is constrained in the interval $(1, \infty)$ and is proportional to the product of two factors $H_1(w)$ and $H_2(w)$. The constraint $w > 1$ arises from the inequality $\sigma_{be}^2 > \sigma_e^2$. If there were no constraint, the factor $H_1(w)$ would be proportional to an F distribution with $J(K-1)-(I-1)$ and $(J-1)-\delta(I-1)$ degrees of freedom and would represent a quantity proportional to the confidence distribution of w based on S_e and S_b. It would also be proportional to the posterior distribution of w based upon the portion of the likelihood relating to

S_b and S_e alone in conjunction with a uniform reference prior in $\log \sigma_{be}$ and $\log \sigma_e$.

For $r > \lambda$, the second factor $H_2(w)$ occurs because the quantities $\hat{\boldsymbol{\phi}}$ and $\tilde{\boldsymbol{\phi}}$ *are estimating the same parameters* $\boldsymbol{\phi}$. Thus, both the constraint $w > 1$ and the factor $H_2(w)$ can be thought of as resulting from the additional information supplied by the BIBD model.

To illustrate, consider again the data of Table 7.4.3. Curve (1) in Fig. 7.4.2 gives the posterior distribution of w in (7.4.40). Curve (2) in the same figure represents the confidence distribution of w based upon $H_1(w)$. Finally, the curve (3) would be the appropriate distribution of w if the constraint $w > 1$ were ignored. For this example, then, the effect of both the constraint and the second factor appears to be appreciable. In particular, if our inferences about w were based upon $H_1(w)$ alone, then the usual 95 per cent interval, Bayesian or otherwise, would have its lower limit less than unity which is *a priori* unacceptable. Finally, comparing Fig. 7.4.1 and Fig. 7.4.2 we see that while the effect of the constraint $w > 1$ is important insofar as inferences about w are concerned, it has much less effect on inferences about the contrasts $\boldsymbol{\phi}$.

7.4.6 Further Properties of, and Approximations to, $p(\boldsymbol{\phi} \mid \mathbf{y})$

We now return to our discussion of the posterior distribution $p(\boldsymbol{\phi} \mid \mathbf{y})$ in (7.4.33).

In the analysis of data from a balanced incomplete block design, we are often

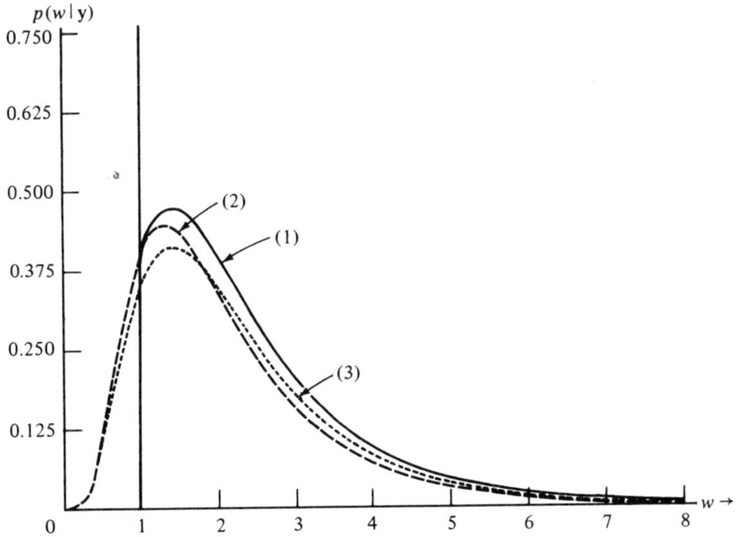

In the figure
1. Posterior distribution of w in (7.4.40)
2. Posterior distribution of w based on $H_1(w)$ alone, also a confidence distribution of w
3. Posterior distribution of w ignoring the constraint $w > 1$.

Fig. 7.4.2 Posterior distributions of w for the BIBD data.

concerned with the problem of making inferences about one or more linear contrasts of the treatment means θ_i. Since the ϕ_i are themselves linear contrasts, a contrast in the θ_t can be expressed as a contrast of $\phi_1, \ldots, \phi_I$. We are thus led to the problem of finding the marginal distributions of linear contrasts of $\phi_1, \ldots, \phi_I$.

From the joint distribution (7.4.33) it is clear that the conditional distribution of a subset of $\phi_1, \ldots, \phi_{I-1}$, given the remainder, is of exactly the same form as the original distribution. However, the marginal distribution of a subset is not To obtain the marginal distributions, it is helpful to write the joint distribution of $\boldsymbol{\phi}$ in the alternative form.

$$p(\boldsymbol{\phi} \mid \mathbf{y}) = \int_1^\infty p(\boldsymbol{\phi} \mid w, \mathbf{y})\, p(w \mid \mathbf{y})\, dw, \qquad -\infty < \phi_i < \infty, \qquad i = 1, \ldots, I-1, \tag{7.4.41}$$

where $p(w|\mathbf{y})$ is given in (7.4.40). From the joint distribution $p(\mathrm{w}, \boldsymbol{\phi}|\mathbf{y})$ in (7.4.39) it is readily seen that, given w, the conditional distribution $p(\boldsymbol{\phi} \mid w, \mathbf{y})$ is the $(I-1)$-dimensional multivariate t distribution

$$t_{(I-1)}\,[\bar{\boldsymbol{\phi}}(w), s^2(w)(\mathbf{I}_{I-1} - I^{-1}\mathbf{1}_{I-1}\mathbf{1}'_{I-1}), I(r-1)] \tag{7.4.42}$$

where

$$\bar{\boldsymbol{\phi}}(w) = [\bar{\phi}_1(w), \ldots, \bar{\phi}_{I-1}(w)]'$$

and

$$I(r-1)rf\left(w + \frac{1-f}{f}\right) s^2(w) = S_b + wS_e + w\left(w + \frac{1-f}{f}\right)^{-1} S_t. \tag{7.4.43}$$

Note that from (7.4.38), the conditional posterior mean of ϕ_i, given w,

$$\bar{\phi}_i(w) = \left(f + \frac{1-f}{w}\right)^{-1} \left[f\hat{\phi}_i + \left(\frac{1-f}{w}\right)\tilde{\phi}_i\right]$$

is a weighted average of the intra- and inter-block estimates $(\hat{\phi}_i, \tilde{\phi}_i)$ with weights proportional to f and $(1-f)/w$ respectively.

From (7.4.41) and properties of the multivariate t distribution, the marginal distribution of any set of p linearly independent contrasts $\boldsymbol{\eta} = \mathbf{L}\boldsymbol{\theta} = \mathbf{L}(\boldsymbol{\phi}' \,\vdots\, \phi_I)'$, where $\mathbf{L}$ is a $p \times I$ matrix of rank $p \leqslant (I-1)$ such that $\mathbf{L}\mathbf{1}_I = \mathbf{0}$, is

$$p(\boldsymbol{\eta} \mid \mathbf{y}) = \int_1^\infty p(\boldsymbol{\eta} \mid w, \mathbf{y})\, p(w \mid \mathbf{y})\, \mathrm{d}w, \qquad -\infty < \eta_i < \infty, \qquad i = 1, \ldots, p, \tag{7.4.44}$$

where $p(\boldsymbol{\eta} \mid w, \mathbf{y})$ is the $t_p[\hat{\boldsymbol{\eta}}(w), s^2(w)\mathbf{L}\mathbf{L}', I(r-1)]$ distribution with

$$\hat{\boldsymbol{\eta}}(w) = \mathbf{L}[\bar{\boldsymbol{\phi}}'(w) \,\vdots\, \bar{\phi}_I(w)]'.$$

In particular, the posterior distribution of a single contrast $\eta = \boldsymbol{\theta}'\boldsymbol{l} = (\boldsymbol{\phi}' \,\vdots\, \phi_I)\boldsymbol{l}$ is

$$p(\eta \mid \mathbf{y}) = \int_1^\infty p(\eta \mid w, \mathbf{y})\, p(w \mid \mathbf{y})\, dw, \qquad -\infty < \eta < \infty, \tag{7.4.45}$$

where, for given w, the quantity

$$\frac{\eta - \hat{\eta}(w)}{s(w)(\mathbf{l}'\mathbf{l})^{\frac{1}{2}}} \tag{7.4.46}$$

follows the $t[0, 1, I(r-1)]$ distribution.

Although the conditional distribution of $\boldsymbol{\eta}$ given w is of the multivariate t form, it does not seem possible to express the unconditional distribution in terms of simple functions when $p < (I-1)$. However, for given data, the unconditional distribution can be obtained by a one dimensional numerical integration.

The solid curves (1) in Figs. 7.4.3 and 7.4.4 show, respectively, the marginal distributions $p(\phi_1 \mid \mathbf{y})$ and $p(\phi_2 \mid \mathbf{y})$ for the data in Table 7.4.3. The distribution of ϕ_1 is derived from (7.4.45) by setting $\mathbf{l}' = (\frac{2}{3}, -\frac{1}{3}, -\frac{1}{3})$ so that $\eta = \phi_1$. The distribution of ϕ_2 is obtained similarly. These distributions are nearly symmetrical, centered about $\phi_1 = 3.25$ and $\phi_2 = -1.50$, respectively. Also shown in the same figures by the broken curves (2) centered at $\hat{\phi}_1 = 3.44$ and $\hat{\phi}_2 = -1.75$, are respectively the distributions of ϕ_1 and ϕ_2 based upon the first factor (intra-block) of (7.4.33) alone. These curves (2) are numerically equivalent to the corresponding confidence distributions of ϕ_1 and ϕ_2 in the sampling theory framework if inter-block information is ignored. By comparing the curves labeled

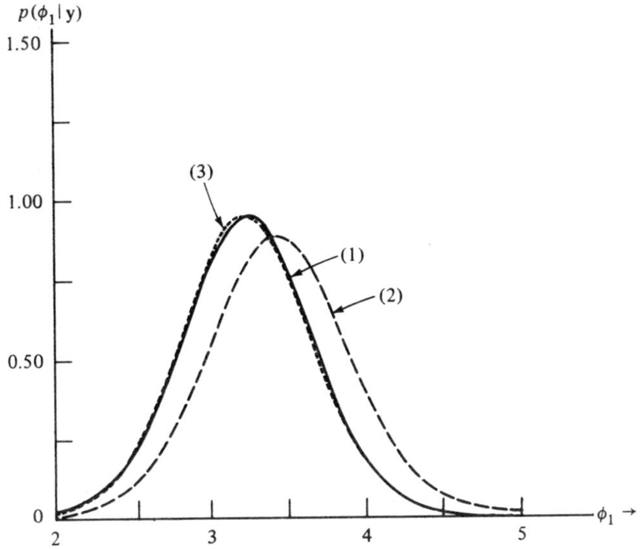

In the figure
1. Posterior distribution of ϕ_1 obtained from (7.4.45)
2. Posterior distribution of ϕ_1 based on the first factor of (7.4.33), also a confidence distribution of ϕ_1
3. Approximate distribution of ϕ_1, from (7.4.52)

Fig. 7.4.3 Posterior distributions of ϕ_1 for the BIBD data.

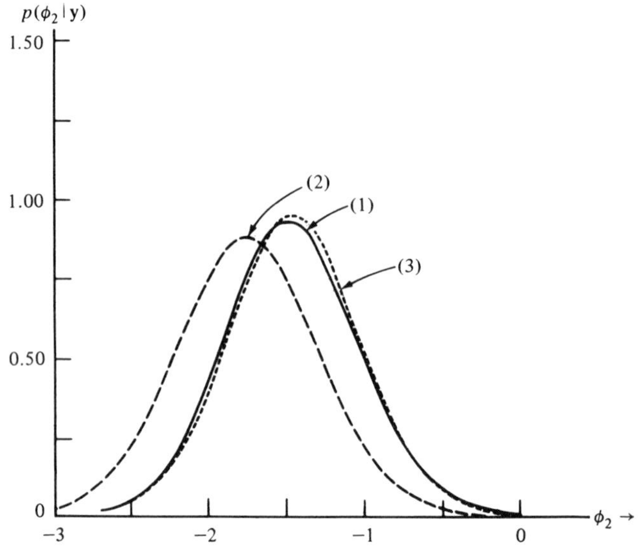

In the figure
1. Posterior distribution of ϕ_2 obtained from (7.4.45)
2. Posterior distribution of ϕ_2 based on the first factor of (7.4.33), also a confidence distribution of ϕ_2
3. Approximate distribution of ϕ_2, from (7.4.52)

Fig. 7.4.4 Posterior distributions of ϕ_2 for the BIBD data.

(1) and (2) in each figure, one sees how the inter-block information modifies the marginal inferences of the contrasts, and for the present example the modification is appreciable. The dotted curves (3) are approximations discussed below.

Approximations

In obtaining the marginal distributions of ϕ_1 and ϕ_2 shown in Figs. 7.4.3–4, numerical integration on a computer was necessary in both cases. However, both distributions are almost symmetrical and resemble Student's t distributions. In fact, close approximation to $p(\boldsymbol{\phi} \mid \mathbf{y})$ can be obtained by use of the multivariate t distribution. In general, we can write

$$p(\boldsymbol{\phi} \mid \mathbf{y}) = \underset{g(w)}{E}\, p(\boldsymbol{\phi} \mid w, \mathbf{y}), \qquad -\infty < \phi_i < \infty, \qquad i = 1, \ldots, I-1, \qquad (7.4.47)$$

where $p(\boldsymbol{\phi} \mid w, \mathbf{y})$ is given in (7.4.42) and the expectation is taken over the posterior distribution of $g(w)$ which is some monotonic function of w. If the conditional distribution $p(\boldsymbol{\phi} \mid w, \mathbf{y})$ is changing gently in the region of $g(w)$ where its density is appreciable, then

$$p(\boldsymbol{\phi} \mid \mathbf{y}) \doteq p(\boldsymbol{\phi} \mid \bar{w}, \mathbf{y}), \qquad -\infty < \phi_i < \infty, \qquad i = 1, \ldots, I-1 \qquad (7.4.48)$$

where $g(\bar{w}) = Eg(w)$. From the distribution of w in (7.4.40), we see that evaluation of the expectation $Eg(w)$ would necessitate numerical integration irrespective of the choice of g. Alternatively, we can write

$$p(\boldsymbol{\phi} \mid \mathbf{y}) \doteq p(\boldsymbol{\phi} \mid \hat{w}, \mathbf{y}) + h\, E[g(w) - g(\hat{w})] \tag{7.4.49}$$

where $g(\hat{w})$ is the mode of $g(w)$ and h is the first derivative of $p(\boldsymbol{\phi} \mid w, \mathbf{y})$ with respect to $g(w)$, evaluated at $\hat{w}$. If $g(w)$ is chosen so that its distribution is symmetrical, then

$$p(\boldsymbol{\phi} \mid \mathbf{y}) \doteq p(\boldsymbol{\phi} \mid \hat{w}, \mathbf{y}), \qquad -\infty < \phi_i < \infty, \qquad i = 1, ..., I-1. \tag{7.4.50}$$

To this degree of approximation, the posterior distribution of $\boldsymbol{\phi}$ is the multivariate t distribution

$$t_{(I-1)}\left[\bar{\boldsymbol{\phi}}(\hat{w}), s^2(\hat{w})\left(\mathbf{I}_{I-1} - \frac{1}{I}\mathbf{1}_{I-1}\mathbf{1}'_{I-1}\right), I(r-1)\right]$$

where $\bar{\boldsymbol{\phi}}(\hat{w})$ and $s^2(\hat{w})$ are obtained simply by substituting $\hat{w}$ into (7.4.42) and (7.4.43). To decide whether a particular parameter point $\boldsymbol{\phi} = \boldsymbol{\phi}_0$ lies inside or outside the $(1-\alpha)$ H.P.D. region, we may thus compare the quantity

$$\frac{\sum_{i=1}^{I} [\phi_i - \bar{\phi}_i(\hat{w})]^2}{(I-1)s^2(\hat{w})} \tag{7.4.51}$$

with the 100α upper percentage point of an F variable with $(I-1)$ and $I(r-1)$ degrees of freedom. Making use of the properties of the multivariate t distribution, any set of $p \leqslant (I-1)$ linearly independent contrasts will have an approximate p-dimensional multivariate t distribution. In particular, the quantity

$$\frac{\eta - \hat{\eta}(\hat{w})}{s(\hat{w})(\boldsymbol{l}'\boldsymbol{l})^{1/2}} \tag{7.4.52}$$

is approximately distributed as $t[0, 1, I(r-1)]$.

Now for the data of Table 7.4.3 the distribution of w, as whown in Fig. 7.4.2, is clearly not symmetrical. An F-like distribution with a long tail to the right (such as the one shown) can however be transformed into a more symmetrical distribution by a log transformation $g(w) = \log w$. Noting that $p(\log w \mid \mathbf{y}) = wp(w \mid \mathbf{y})$, it can be readily verified from (7.4.40) that for $r > \lambda$ the mode of $\log w$ is a root of the cubic equation,

$$w^3 + d_2 w^2 + d_1 w + d_0 = 0 \tag{7.4.53}$$

where

$$d_0 = -\frac{J(K-1)}{(J-1)}\left(\frac{1-f}{f}\right)^2 \frac{S_b}{S_e},$$

$$d_1 = \left(\frac{J-I}{J-1}\right)\left(\frac{1-f}{f}\right)^2 - \frac{1}{I-1}\left(\frac{1-f}{f}\right)[2J(K-1)-(I-1)]\frac{S_b}{S_e}$$

$$+ \left(\frac{J-I}{J-1}\right)\left(\frac{1-f}{f}\right)\frac{S_t}{S_e}$$

and

$$d_2 = \left(\frac{2J-I-1}{J-1}\right)\left(\frac{1-f}{f}\right) - \left[\frac{I(r-1)}{J-1} - 1\right]\left(\frac{S_b+S_t}{S_e}\right).$$

For the example, (7.4.53) becomes

$$w^3 - 1.509101\, w^2 - 0.850443\, w - 0.200062 = 0$$

and the appropriate root is $\hat{w} = 1.99$ to two decimal places. Using this value in (7.4.52) we find that, to this degree of approximation,

$$\frac{\phi_1 - 3.23}{0.42} \quad \text{and} \quad \frac{\phi_2 + 1.45}{0.42}$$

are distributed marginally as $t(0, 1, 27)$. The curves marked (3) in Figs. 7.4.3 and 7.4.4 represent the approximating distributions. In both cases, the agreement between the exact distribution obtained by numerical integration [curve (1)] and the approximating distributions seems to be sufficiently close for any practical inferential purpose.

To give further illustration of this approximation, Table 7.4.4 shows specimens of the exact and the approximate densities of $\phi_1 = \theta_1 - \bar{\theta}$ using the data in Federer (1955, pp. 419–422). In this example, $J = 15, I = 10, K = 4, r = 6, \lambda = 2$

$$S_e = 17.878935, \qquad S_b = 1.379463, \qquad S_t = 26.996222$$

$$\hat{\phi}_1 = 3.3450, \qquad \tilde{\phi}_1 = 1.90476$$

and the cubic equation in (7.4.53) becomes

$$w^3 - 3.809690\, w^2 + 0.032859\, w - 0.009920 = 0$$

of which the appropriate root is $w = 3.80$. The agreement between the exact and the approximate distributions is again close.

While we have illustrated the use of the log modal approximation only in the one dimensional case, i.e., a single contrast, the close agreement in the examples shown as well as the near elliptical shape of the contour of $p(\boldsymbol{\phi} \mid \mathbf{y})$ shown in Fig. 7.4.1 suggest that such approximations will also be useful in higher dimensions.

Table 7.4.4

Specimens of the exact and approximate densities of ϕ_1: Federer's example

	Exact	Approximate
2.37	0.02027	0.01580
2.47	0.04505	0.03770
2.57	0.09397	0.08332
2.67	0.18237	0.16919
2.77	0.32640	0.31316
2.87	0.53407	0.52445
2.97	0.79236	0.78959
3.07	1.05818	1.06303
3.17	1.26442	1.27463
3.27	1.34595†	1.35779†
3.37	1.27319	1.28383
3.47	1.06983	1.07832
3.57	0.79989	0.80649
3.67	0.53412	0.53922
3.77	0.32028	0.32400
3.87	0.17367	0.17608
3.97	0.08586	0.08718
4.07	0.03904	0.03964
4.17	0.01648	0.01668

† Density at the common mode.

7.4.7 Summarized Calculations for Approximating the Posterior Distributions of $\boldsymbol{\phi}$

Table 7.4.5 provides a summary of the calculations needed for approximating the posterior distributions of the contrasts $\phi_1, \ldots, \phi_I$. The numerical values shown are those associated with the data in Table 7.4.3.

Table 7.4.5

Summarized calculations for approximating the posterior distribution of $\boldsymbol{\phi}$

Data of Table 7.4.3.

1. $J=15, \quad K=2, \quad I=3, \quad r=10, \quad \lambda=5$

2. $y_{..}$ Grand average
   T_i Average of observations corresponding to treatment i, $i=1, \ldots, I$
   B_i Average of block averages $y_{.j}$ for the r blocks in which the ith treatment appears, $i=1, \ldots, I$

   $y_{..}=5.67$
   $T_1=8.74, \quad T_2=4.44, \quad T_3=3.82$
   $B_1=6.16, \quad B_2=5.67, \quad B_3=5.08$

Table 7.4.5 *Continued*

3. $\hat{\phi}_1 = 3.44,\quad \hat{\phi}_2 = -1.75,\quad \hat{\phi}_3 = -1.69$ Use (7.4.17–18)

$\tilde{\phi}_1 = 1.96,\quad \tilde{\phi}_2 = \ 0.37\quad \tilde{\phi}_3 = -2.34$ (7.4.14)

4. $S_e = 29.18$ Use (7.4.21)

$S_b = 49.04$ (7.4.15)

$S_t = 17.77$ (7.4.37)

5. $\hat{w} = 1.99$ Use (7.4.53)

6. $\bar{\phi}_1(\hat{w}) = 3.23\quad \bar{\phi}_2(\hat{w}) = -1.45\quad \bar{\phi}_3(\hat{w}) = -1.78$ Use (7.4.38)

7. $s^2(\hat{w}) = 0.26$ Use (7.4.43)

8. Marginal posterior distribution of ϕ_i

$$\frac{\phi_i - \bar{\phi}_i(\hat{w})}{[s^2(\hat{w})(I-1)/I]^{1/2}} \sim t[0, 1\ I(r-1)], \qquad i = 1, \ldots, I.$$

$$\frac{\phi_i - \bar{\phi}_i(\hat{w})}{0.42} \sim t(0, 1, 27), \qquad i = 1, 2, 3.$$

9. Comparison of two treatment means.

$$\eta = \theta_i - \theta_{i'},\ = \phi_i - \phi_{i'}$$

$$\frac{\eta - [\bar{\phi}_i(\hat{w}) - \bar{\phi}_{i'}(\hat{w})]}{[2s^2(\hat{w})]^{1/2}} \sim t[0, 1, I(r-1)]$$

e.g. $i = 1,\quad i' = 2$

$$\frac{\delta - 4.68}{0.71} \sim t(0, 1, 27)$$

10. Overall comparison of I treatment means.

$$\frac{\sum_{i=1}^{I} [\phi_i - \bar{\phi}_i(\hat{w})]^2}{(I-1)s^2(\hat{w})} \sim F_{(I-1), I(r-1)}$$

$$\frac{\sum_{i=1}^{3} [\phi_i - \bar{\phi}_i(\hat{w})]^2}{0.52} \sim F_{2,27}$$

7.4.8 Recovery of Interblock Information in Sampling Theory

We now consider to what extent the results in the preceding sections may be related to those of sampling theory. If the variance ratio $w = 1 + (K\sigma_b^2/\sigma_e^2)$ were known, then, on sampling theory, the quantity $\bar{\phi}_i(w)$ in (7.4.38) would be the natural estimator for ϕ_i and is, in fact, the minimum variance unbiased estimator—see Graybill and Weeks (1959).

In the case of a single contrast $\eta = \mathbf{l}'\boldsymbol{\theta}$, the *sampling* distribution of the quantity in (7.4.46) would be $t[0, 1, I(r - 1)]$. Thus, for given w, the conditional posterior distribution $p(\eta \mid w, \mathbf{y})$ in (7.4.45) is numerically equivalent to the confidence distribution of η.

In practice w is of course *not* known. Then in the sampling theory approach two problems arise; one concerns the choice of estimator for w and the other, the distribution of the resulting estimator for ϕ_i given that choice. Concerning the first problem, from Table 7.4.1 and using (7.4.36), it can be verified that for $r > \lambda$, the quantities (S_e, S_b, S_t) are independent and that

$$S_e \sim \sigma_e^2 \chi^2_{J(K-1)-(I-1)} \qquad S_b \sim \sigma_e^2 w \chi^2_{J-I}, \qquad S_t \sim \sigma_e^2 \left(w + \frac{1-f}{f}\right)\chi^2_{I-1}. \tag{7.4.54}$$

Clearly, w can be estimated in many different ways. In particular, the estimators of w proposed by Yates and by Graybill and Weeks are based upon separate unbiased estimators of σ_e^2 and σ_b^2, which are themselves linear combinations of (S_e, S_b, S_t). The appropriateness of these estimators seems questionable since it is the ratio w and not the individual components, on which inferences about ϕ_i depend. Further, granted that *separate* unbiased estimators of σ_e^2 and σ_b^2 are of interest, presumably one should use those with minimum variance. However, it is readily verified that among linear functions of (S_e, S_b, S_t), the UMV estimators of (σ_e^2, σ_b^2) depend upon the unknown w. Another approach would be to estimate w by the method of maximum likelihood. By substituting for $S_b(\boldsymbol{\theta})$ in (7.3.6) and $S_e(\boldsymbol{\theta})$ in (7.3.12) from (7.4.13), (7.4.15) and (7.4.20), (7.4.21), respectively, and by making use of (7.4.36), it can be verified that, for $w > 1$,

$$\begin{aligned} l(w \mid \mathbf{y}) &= \max_{\sigma_e^2, \bar{\theta}, \boldsymbol{\phi} \mid w} l(\sigma_e^2, \bar{\theta}, \boldsymbol{\phi}, w \mid \mathbf{y}) \\ &\propto w^{-J/2}\left[S_e + \frac{S_b}{w} + \left(w + \frac{1-f}{f}\right)^{-1} S_t\right]^{-\frac{1}{2}JK} \end{aligned} \tag{7.4.55}$$

It follows that the maximum likelihood estimator of w is the root of a cubic equation unless the maximum of $l(w \mid \mathbf{y})$ occurs at $w = 1$. Whether an estimator of w is obtained by using separate unbiased estimators of σ_e^2 and σ_b^2 or by maximum likelihood, the resulting distributional properties of the corresponding estimator of ϕ_i are unknown.

In the Bayesian approach, as we have seen, an estimator $\hat{w}$ of w *may* be used to obtain approximations to the distributions of the ϕ_i. However, no point estimate of w is actually necessary since, given the data, the actual distribution of the ϕ_i can always be obtained through a single integration as in (7.4.41). The integration in (7.4.41) averages the conditional distribution $p(\boldsymbol{\phi} \mid w, \mathbf{y})$ over a weight function $p(w \mid \mathbf{y})$ which reflects the information about w in the light of the data $\mathbf{y}$ and the prior assumption.

We have discussed in detail the problem of combining inter- and intra-block information about linear contrasts of treatment means θ_i only for the BIBD model. The methods can, however, be readily extended to the general linear model of (7.3.1).† In the above the prior distribution of θ_i was supposed locally uniform. It is often more appropriate to employ the random-effect prior in Section 7.2 to the present problem. For details of this analysis, see Afonja (1970).

APPENDIX A7.1

Some Useful Results in Combining Quadratic Forms

We here give two useful lemmas for combining quadratic forms.

Lemma 1. Let $\mathbf{x}$, $\mathbf{a}$ and $\mathbf{b}$ be $k \times 1$ vectors, and $\mathbf{A}$ and $\mathbf{B}$ be $k \times k$ symmetric matrices such that the inverse $(\mathbf{A}+\mathbf{B})^{-1}$ exists. Then,

$$(\mathbf{x}-\mathbf{a})'\mathbf{A}(\mathbf{x}-\mathbf{a}) + (\mathbf{x}-\mathbf{b})'\mathbf{B}(\mathbf{x}-\mathbf{b}) = (\mathbf{x}-\mathbf{c})'(\mathbf{A}+\mathbf{B})(\mathbf{x}-\mathbf{c}) + (\mathbf{a}-\mathbf{b})'\,\mathbf{A}(\mathbf{A}+\mathbf{B})^{-1}\mathbf{B}(\mathbf{a}-\mathbf{b}) \qquad \text{(A7.1.1)}$$

where

$$\mathbf{c} = (\mathbf{A}+\mathbf{B})^{-1}(\mathbf{A}\mathbf{a}+\mathbf{B}\mathbf{b}).$$

Proof:

$$\begin{aligned}(\mathbf{x}-\mathbf{a})'\mathbf{A}(\mathbf{x}-\mathbf{a}) &+ (\mathbf{x}-\mathbf{b})'\mathbf{B}(\mathbf{x}-\mathbf{b}) \\ &= \mathbf{x}'(\mathbf{A}+\mathbf{B})\mathbf{x} - 2\mathbf{x}'(\mathbf{A}\mathbf{a}+\mathbf{B}\mathbf{b}) + \mathbf{a}'\mathbf{A}\mathbf{a} + \mathbf{b}'\mathbf{B}\mathbf{b} \\ &= \mathbf{x}'(\mathbf{A}+\mathbf{B})\mathbf{x} - 2\mathbf{x}'(\mathbf{A}+\mathbf{B})\mathbf{c} + \mathbf{c}'(\mathbf{A}+\mathbf{B})\mathbf{c} + d \\ &= (\mathbf{x}-\mathbf{c})'(\mathbf{A}+\mathbf{B})(\mathbf{x}-\mathbf{c}) + d \end{aligned} \qquad \text{(A7.1.2)}$$

where

$$d = \mathbf{a}'\mathbf{A}\mathbf{a} + \mathbf{b}'\mathbf{B}\mathbf{b} - \mathbf{c}'(\mathbf{A}+\mathbf{B})\mathbf{c} \qquad \text{(A7.1.3)}$$

Now,

$$\begin{aligned}\mathbf{c}'(\mathbf{A}+\mathbf{B})\mathbf{c} &= (\mathbf{A}\mathbf{a}+\mathbf{B}\mathbf{b})'(\mathbf{A}+\mathbf{B})^{-1}(\mathbf{A}\mathbf{a}+\mathbf{B}\mathbf{b}) \\ &= [\mathbf{A}(\mathbf{a}-\mathbf{b}) + (\mathbf{A}+\mathbf{B})\mathbf{b}]'(\mathbf{A}+\mathbf{B})^{-1}[(\mathbf{A}+\mathbf{B})\mathbf{a} - \mathbf{B}(\mathbf{a}-\mathbf{b})] \\ &= -(\mathbf{a}-\mathbf{b})'\mathbf{A}(\mathbf{A}+\mathbf{B})^{-1}\mathbf{B}(\mathbf{a}-\mathbf{b}) + \mathbf{a}'\mathbf{A}\mathbf{a} + \mathbf{b}'\mathbf{B}\mathbf{b}\end{aligned} \qquad \text{(A7.1.4)}$$

† See Tiao and Draper (1968).

Substituting (A7.1.4) into (A7.1.3), the lemma follows at once. Note that if both **A** and **B** have inverses, then

$$\mathbf{A}(\mathbf{A} + \mathbf{B})^{-1}\mathbf{B} = (\mathbf{A}^{-1} + \mathbf{B}^{-1})^{-1}. \tag{A7.1.5}$$

It sometimes happens that we need to combine two quadratic forms for which the matrix $(\mathbf{A} + \mathbf{B})$ has no inverse. In this case, Lemma 1 may be modified as follows:

Lemma 2. Let **x**, **a** and **b** be $k \times 1$ vectors and **A** and **B** be two $k \times k$ positive semidefinite symmetric matrices. Suppose the rank of the matrix $\mathbf{A} + \mathbf{B}$ is $q(< k)$. Then, subject to the constraints $\mathbf{Gx} = \mathbf{0}$,

$$(\mathbf{x} - \mathbf{a})'\mathbf{A}(\mathbf{x} - \mathbf{a}) + (\mathbf{x} - \mathbf{b})'\mathbf{B}(\mathbf{x} - \mathbf{b}) = (\mathbf{x} - \mathbf{c}^*)'(\mathbf{A} + \mathbf{B} + \mathbf{M})(\mathbf{x} - \mathbf{c}^*)$$
$$+ (\mathbf{a} - \mathbf{b})'\mathbf{A}(\mathbf{A} + \mathbf{B} + \mathbf{M})^{-1}\mathbf{B}(\mathbf{a} - \mathbf{b}) \tag{A7.1.6}$$

where **G** is any $(k - q) \times k$ matrix of rank $k - q$ such that the rows of **G** are linearly independent of the rows of $\mathbf{A} + \mathbf{B}$, $\mathbf{M} = \mathbf{G}'\mathbf{G}$ and

$$\mathbf{c}^* = (\mathbf{A} + \mathbf{B} + \mathbf{M})^{-1}(\mathbf{Aa} + \mathbf{Bb}).$$

Proof. We shall first prove that

$$\mathbf{M}(\mathbf{A} + \mathbf{B} + \mathbf{M})^{-1}\mathbf{A} = \mathbf{M}(\mathbf{A} + \mathbf{B} + \mathbf{M})^{-1}\mathbf{B} = \mathbf{0}. \tag{A7.1.7}$$

Since $(\mathbf{A} + \mathbf{B})$ is of rank q, **G** is of rank $k - q$ and the rows of **G** are linearly independent of the rows of $(\mathbf{A} + \mathbf{B})$, there exists a $(k - q) \times k$ matrix **U** of rank $(k - q)$ such that $\mathbf{UG}'$ is non-singular and

$$\mathbf{U}(\mathbf{A} + \mathbf{B}) = \mathbf{0}. \tag{A7.1.8}$$

Now,

$$\mathbf{U}(\mathbf{A} + \mathbf{B} + \mathbf{M})(\mathbf{A} + \mathbf{B} + \mathbf{M})^{-1} = \mathbf{U}.$$

From (A7.1.8)

$$\mathbf{UM}(\mathbf{A} + \mathbf{B} + \mathbf{M})^{-1} = \mathbf{U}.$$

Postmultiplying both sides by $(\mathbf{A} + \mathbf{B})$, we get

$$\mathbf{UM}(\mathbf{A} + \mathbf{B} + \mathbf{M})^{-1}(\mathbf{A} + \mathbf{B}) = \mathbf{0}. \tag{A7.1.9}$$

Since $\mathbf{M} = \mathbf{G}'\mathbf{G}$ and $\mathbf{UG}'$ is non-singular, (A7.1.9) implies that

$$\mathbf{G}(\mathbf{A} + \mathbf{B} + \mathbf{M})^{-1}(\mathbf{A} + \mathbf{B}) = \mathbf{0}$$

so that

$$\mathbf{M}(\mathbf{A} + \mathbf{B} + \mathbf{M})^{-1}(\mathbf{A} + \mathbf{B}) = \mathbf{0}. \tag{A7.1.10a}$$

Postmultiplying both sides of (A7.1.10a) by $(\mathbf{A} + \mathbf{B} + \mathbf{M})^{-1}\mathbf{M}$, we obtain

$$\mathbf{M}(\mathbf{A} + \mathbf{B} + \mathbf{M})^{-1}\mathbf{A}(\mathbf{A} + \mathbf{B} + \mathbf{M})^{-1}\mathbf{M}$$
$$+ \mathbf{M}(\mathbf{A} + \mathbf{B} + \mathbf{M})^{-1}\mathbf{B}(\mathbf{A} + \mathbf{B} + \mathbf{M})^{-1}\mathbf{M} = \mathbf{0} \tag{A7.1.10b}$$

Since **A** and **B** are assumed positive semidefinite, it follows that both terms on the left of (A7.1.10b) are positive semidefinite matrices. Thus the equality implies that both must be null matrices. Writing $\mathbf{A} = \mathbf{C}'\mathbf{C}$ and $\mathbf{B} = \mathbf{D}'\mathbf{D}$ where **C** and **D** are $k \times k$ matrices we must then have

$$\mathbf{M}(\mathbf{A} + \mathbf{B} + \mathbf{M})^{-1}\mathbf{C}' = \mathbf{M}(\mathbf{A} + \mathbf{B} + \mathbf{M})^{-1}\mathbf{D}' = \mathbf{0}$$

and the assertion (A7.1.7) follows.

We now prove (A7.1.6). Using the fact that $\mathbf{Mx} = \mathbf{0}$, we have

$$\begin{aligned}(\mathbf{x} - \mathbf{a})'\mathbf{A}(\mathbf{x} - \mathbf{a}) + (\mathbf{x} - \mathbf{b})'\mathbf{B}(\mathbf{x} - \mathbf{b}) &= \mathbf{x}'(\mathbf{A} + \mathbf{B} + \mathbf{M})\mathbf{x} - 2\mathbf{x}'(\mathbf{A} + \mathbf{B} + \mathbf{M})\mathbf{c}^* \\ &\quad + \mathbf{c}^{*\prime}(\mathbf{A} + \mathbf{B} + \mathbf{M})\mathbf{c}^* + d_1 \\ &= (\mathbf{x} - \mathbf{c}^*)'(\mathbf{A} + \mathbf{B} + \mathbf{M})(\mathbf{x} - \mathbf{c}^*) + d_1\end{aligned} \tag{A7.1.11}$$

where

$$d_1 = \mathbf{a}'\mathbf{Aa} + \mathbf{b}'\mathbf{Bb} - \mathbf{c}^{*\prime}(\mathbf{A} + \mathbf{B} + \mathbf{M})\mathbf{c}^*. \tag{A7.1.12}$$

Now,

$$\begin{aligned}\mathbf{c}^{*\prime}(\mathbf{A} + \mathbf{B} + \mathbf{M})\mathbf{c}^* &= (\mathbf{Aa} + \mathbf{Bb})'(\mathbf{A} + \mathbf{B} + \mathbf{M})^{-1}(\mathbf{Aa} + \mathbf{Bb}) \\ &= [\mathbf{A}(\mathbf{a} - \mathbf{b}) + (\mathbf{A} + \mathbf{B})\mathbf{b}]'(\mathbf{A} + \mathbf{B} + \mathbf{M})^{-1}[(\mathbf{A} + \mathbf{B})\mathbf{a} - \mathbf{B}(\mathbf{a} - \mathbf{b})] \\ &= -(\mathbf{a} - \mathbf{b})'\mathbf{A}(\mathbf{A} + \mathbf{B} + \mathbf{M})^{-1}\mathbf{B}(\mathbf{a} - \mathbf{b}) \\ &\quad + \mathbf{b}'(\mathbf{A} + \mathbf{B})(\mathbf{A} + \mathbf{B} + \mathbf{M})^{-1}(\mathbf{A} + \mathbf{B})\mathbf{a} \\ &\quad + (\mathbf{a} - \mathbf{b})'\mathbf{A}(\mathbf{A} + \mathbf{B} + \mathbf{M})^{-1}(\mathbf{A} + \mathbf{B})\mathbf{a} \\ &\quad - \mathbf{b}'(\mathbf{A} + \mathbf{B})(\mathbf{A} + \mathbf{B} + \mathbf{M})^{-1}\mathbf{B}(\mathbf{a} - \mathbf{b})\end{aligned} \tag{A7.1.13}$$

Using (A7.1.10a), the second term on the extreme right of (A7.1.13) becomes

$$\mathbf{b}'(\mathbf{A} + \mathbf{B})(\mathbf{A} + \mathbf{B} + \mathbf{M})^{-1}(\mathbf{A} + \mathbf{B})\mathbf{a} = \mathbf{b}'(\mathbf{A} + \mathbf{B})\mathbf{a}. \tag{A7.1.14}$$

Applying (A7.1.7), the third and fourth terms are, respectively,

$$(\mathbf{a} - \mathbf{b})'\mathbf{A}(\mathbf{A} + \mathbf{B} + \mathbf{M})^{-1}(\mathbf{A} + \mathbf{B})\mathbf{a} = (\mathbf{a} - \mathbf{b})'\mathbf{Aa}, \tag{A7.1.15}$$

$$-\mathbf{b}'(\mathbf{A} + \mathbf{B})(\mathbf{A} + \mathbf{B} + \mathbf{M})^{-1}\mathbf{B}(\mathbf{a} - \mathbf{b}) = -(\mathbf{a} - \mathbf{b})'\mathbf{Bb}. \tag{A7.1.16}$$

Substituting (A7.1.14–16) into (A7.1.13), we get

$$\mathbf{c}^{*\prime}(\mathbf{A} + \mathbf{B} + \mathbf{M})\mathbf{c}^* = -(\mathbf{a} - \mathbf{b})'\mathbf{A}(\mathbf{A} + \mathbf{B} + \mathbf{M})^{-1}\mathbf{B}(\mathbf{a} - \mathbf{b}) + \mathbf{a}'\mathbf{Aa} + \mathbf{a}'\mathbf{Bb}$$

so that

$$d_1 = (\mathbf{a} - \mathbf{b})'\mathbf{A}(\mathbf{A} + \mathbf{B} + \mathbf{M})^{-1}\mathbf{B}(\mathbf{a} - \mathbf{b})$$

and the lemma is proved.

CHAPTER 8

SOME ASPECTS OF MULTIVARIATE ANALYSIS

8.1 INTRODUCTION

In all the problems we have so far considered, observations are made of a single unidimensional response or output y. The inference problems that result are called univariate problems. In this and the next chapter, we shall consider problems which arise when the output is multidimensional. Thus, in the study of a chemical process, at each experimental setting one might observe yield y_1, density y_2, and color y_3 of the product. Similarly, in a study of consumer behavior, for each household one might record spending on food y_1, spending on durables y_2, and spending on travel and entertainment y_3. We would then say that a three-dimensional output or response is observed. Inference problems which arise in the analysis of such data are called *multivariate*.

In this chapter, we shall begin by reviewing some *univariate* problems in a general setting which can be easily extended to the multivariate case.

8.1.1 A General Univariate Model

It is often desired to make inferences about *parameters* $\theta_1, \ldots, \theta_k$ contained in the relationship between a single observed *output variable* or *response* y subject to error and p *input invariables* $\xi_1, \ldots, \xi_p$ whose values are assumed exactly known. It should be understood that the inputs could include qualitative as well as quantitative variables. For example, ξ_i might take values of 0 or 1 depending on whether some particular quality was absent or present in which case ξ_i is called an *indicator variable* or less appropriately a *dummy* variable.

The Design Matrix

Suppose, in an investigation, n experimental "runs" are made, and the uth run consists of making an observation y_u at some fixed set of input conditions $\boldsymbol{\xi}_u' = (\xi_{u1}, \xi_{u2}, \ldots, \xi_{up})$. The $n \times p$ *design matrix*

$$\boldsymbol{\xi} = \begin{bmatrix} \boldsymbol{\xi}_1' \\ \vdots \\ \boldsymbol{\xi}_u' \\ \vdots \\ \boldsymbol{\xi}_n' \end{bmatrix} = \begin{bmatrix} \xi_{11} & \xi_{12} & \cdots & \xi_{1p} \\ \vdots & & & \\ \xi_{u1} & \xi_{u2} & \cdots & \xi_{up} \\ \vdots & & & \\ \xi_{n1} & \xi_{n2} & \cdots & \xi_{np} \end{bmatrix} \qquad (8.1.1)$$

lists the p input conditions to be used in each of the n projected runs and the uth row of $\boldsymbol{\xi}$ is the vector $\boldsymbol{\xi}'_u$. The phraseology "experimental run", "experimental design" is most natural in a situation in which a scientific experiment is being conducted and in which the levels of the inputs are at our choice. In some applications, however, and particularly in economic studies, it is often impossible to choose the experimental conditions. We have only historical data generated for us in circumstances beyond our control and often in a manner we would not choose. It is convenient here to extend the terminologies "experimental run" and "experimental design" to include experiments designed by nature, but we must, of course, bear in mind the limitations of such historical data.

To obtain a mathematical model for our set-up we need to link the n observations $\mathbf{y}' = (y_1, ..., y_n)$ with the inputs $\boldsymbol{\xi}$. This we do by defining two functions called respectively the *expectation function* and the *error function.*

The Expectation Function

The expected value $E(y_u)$ of the output from the uth run is assumed to be a known function η_u of the p fixed inputs $\boldsymbol{\xi}_u$ employed during that run, involving k unknown parameters $\boldsymbol{\theta}' = (\theta_1, ..., \theta_k)$,

$$E(y_u) = \eta_u = \eta(\boldsymbol{\xi}_u, \boldsymbol{\theta}). \tag{8.1.2}$$

The vector valued function $\boldsymbol{\eta} = \boldsymbol{\eta}(\boldsymbol{\xi}, \boldsymbol{\theta}), \boldsymbol{\eta}' = (\eta_1, ..., \eta_u, ..., \eta_n)$, is called the *expectation function.*

The Error Function

The expectation function links $E(y_u)$ to $\boldsymbol{\xi}_u$ and $\boldsymbol{\theta}$. We now have to link y_u to $E(y_u) = \eta_u$. This is done by means of an *error distribution function* in $\boldsymbol{\varepsilon}' = (\varepsilon_1, ..., \varepsilon_n)$. The n experimental errors $\boldsymbol{\varepsilon} = \mathbf{y} - \boldsymbol{\eta}$ which occur in making the runs are assumed to be random variables having zero means but in general not necessarily independently or Normally distributed. We denote the density function of the n errors by $p(\boldsymbol{\varepsilon} \mid \boldsymbol{\pi})$ where $\boldsymbol{\pi}$ is a set of error distribution parameters whose values are in general unknown.

Finally, then, the output in the form of the n observations $\mathbf{y}$ and the input in the form of the n sets of conditions $\boldsymbol{\xi}$ are linked together by a *mathematical model* containing the error function and the expectation function as follows

$$\boldsymbol{\xi} \xrightarrow{\boldsymbol{\eta} = \boldsymbol{\eta}(\boldsymbol{\xi}, \boldsymbol{\theta})} \boldsymbol{\eta} \xrightarrow{p(\mathbf{y} - \boldsymbol{\eta} \mid \boldsymbol{\pi})} \mathbf{y}. \tag{8.1.3}$$

This model involves a function

$$f(\mathbf{y}, \boldsymbol{\theta}, \boldsymbol{\pi}, \boldsymbol{\xi}) \tag{8.1.4}$$

of the observations $\mathbf{y}$, the parameters $\boldsymbol{\theta}$ of the expectation function, the parameters $\boldsymbol{\pi}$ of the error distribution and the design $\boldsymbol{\xi}$.

Data Generation Model

If we knew $\boldsymbol{\theta}$ and $\boldsymbol{\pi}$ and the design $\boldsymbol{\xi}$, we could use the function (8.1.4) to calculate the probability density associated with any particular set of data $\mathbf{y}$. This data generation model (which might, for example, be directly useful for simulation and Monte-Carlo studies) is the function $f(\mathbf{y}, \boldsymbol{\theta}, \boldsymbol{\pi}, \boldsymbol{\xi})$ with $\boldsymbol{\theta}, \boldsymbol{\pi}$ and $\boldsymbol{\xi}$ held fixed and we denote it by

$$p(\mathbf{y} \mid \boldsymbol{\theta}, \boldsymbol{\pi}, \boldsymbol{\xi}) = f(\mathbf{y}, \boldsymbol{\theta}, \boldsymbol{\pi}, \boldsymbol{\xi}), \tag{8.1.5}$$

which emphasizes that the density is a function of $\mathbf{y}$ alone for *fixed* $\boldsymbol{\theta}$, $\boldsymbol{\pi}$, and $\boldsymbol{\xi}$.

The Likelihood Function and the Posterior Distribution

In ordinary statistical practice, we are not directly interested in probabilities associated with *various* sets of data, given *fixed* values of the parameters $\boldsymbol{\theta}$ and $\boldsymbol{\pi}$. On the contrary, we are concerned with the probabilities associated with *various* sets of parameter values, given a *fixed* set of data which is known to have occurred.

After an experiment has been performed, $\mathbf{y}$ is known and fixed (as is $\boldsymbol{\xi}$) but $\boldsymbol{\theta}$ and $\boldsymbol{\pi}$ are unknown. The likelihood function has the same form as (8.1.4), but in it $\mathbf{y}$ and $\boldsymbol{\xi}$ are fixed and $\boldsymbol{\theta}$ and $\boldsymbol{\pi}$ are not to be regarded as variables. Thus, the likelihood may be written

$$l(\boldsymbol{\theta}, \boldsymbol{\pi} \mid \mathbf{y}, \boldsymbol{\xi}) = f(\mathbf{y}, \boldsymbol{\theta}, \boldsymbol{\pi}, \boldsymbol{\xi}). \tag{8.1.6}$$

In what follows we usually omit specific note of dependence on $\boldsymbol{\xi}$ and write $l(\boldsymbol{\theta}, \boldsymbol{\pi} \mid \mathbf{y}, \boldsymbol{\xi})$ as $l(\boldsymbol{\theta}, \boldsymbol{\pi} \mid \mathbf{y})$.

In the Bayesian framework, inferences about $\boldsymbol{\theta}$ and $\boldsymbol{\pi}$ can be made by suitable study of the posterior distribution $p(\boldsymbol{\theta}, \boldsymbol{\pi} \mid \mathbf{y})$ of $\boldsymbol{\theta}$ and $\boldsymbol{\pi}$ obtained by combining the likelihood with the appropriate prior distribution $p(\boldsymbol{\theta}, \boldsymbol{\pi})$,

$$p(\boldsymbol{\theta}, \boldsymbol{\pi} \mid \mathbf{y}) \propto l(\boldsymbol{\theta}, \boldsymbol{\pi} \mid \mathbf{y})\, p(\boldsymbol{\theta}, \boldsymbol{\pi}). \tag{8.1.7}$$

An example in which the expectation function is nonlinear and the error distribution is non-Normal was given in Section 3.5. In this chapter, we shall from now on assume Normality but will extend our general model to cover multivariate problems.

8.2 A GENERAL MULTIVARIATE NORMAL MODEL

Suppose now that a number of output responses are measured in each experimental run. Thus, in a chemical experiment, at each setting of the process conditions ξ_1 = temperature and ξ_2 = concentration, observations might be made on the output responses y_1 = yield of product A, y_2 = yield of product B, and y_3 = yield of product C. In general, then, from each experimental run the

m-variate observation

$$\mathbf{y}'_{(u)} = (y_{u1}, \ldots, y_{ui}, \ldots, y_{um})$$

would be available. There would now be m expectation functions

$$E(\mathbf{y}_{(u)}) = \boldsymbol{\eta}_{(u)} = (\eta_{u1}, \ldots, \eta_{um})'$$

where

$$\begin{aligned} E(y_{u1}) &= \eta_{u1} = \eta_1(\boldsymbol{\xi}_{u1}, \boldsymbol{\theta}_1) \\ &\vdots \\ E(y_{ui}) &= \eta_{ui} = \eta_i(\boldsymbol{\xi}_{ui}, \boldsymbol{\theta}_i) \\ &\vdots \\ E(y_{uj}) &= \eta_{uj} = \eta_j(\boldsymbol{\xi}_{uj}, \boldsymbol{\theta}_j) \\ &\vdots \\ E(y_{um}) &= \eta_{um} = \eta_m(\boldsymbol{\xi}_{um}, \boldsymbol{\theta}_m) \end{aligned} \tag{8.2.1}$$

where $\boldsymbol{\xi}_{ui}$ would contain p_i elements $(\xi_{u1i}, \ldots, \xi_{usi}, \ldots, \xi_{up_ii})$ and $\boldsymbol{\theta}_i$ would contain k_i elements $(\theta_{1i}, \ldots, \theta_{gi}, \ldots, \theta_{k_ii})$. The expectation functions η_{ui} might be linear or nonlinear both in the parameters $\boldsymbol{\theta}_i$ and the inputs $\boldsymbol{\xi}_{ui}$. Also, depending on the problem, some or all of the p_i elements of $\boldsymbol{\xi}_{ui}$ might be the same as those of $\boldsymbol{\xi}_{uj}$ and some or all of the elements of $\boldsymbol{\theta}_i$ might be the same as those of $\boldsymbol{\theta}_j$. That is to say, a given output would involve certain inputs and certain parameters which might or might not be shared by other outputs.

8.2.1 The Likelihood Function

Let us now consider the problem of making inferences about the $\boldsymbol{\theta}_i$ for a set of n m-variate observations. We assume that the error vector

$$\boldsymbol{\varepsilon}_{(u)} = \mathbf{y}_{(u)} - \boldsymbol{\eta}_{(u)} = (\varepsilon_{u1}, \ldots, \varepsilon_{um})', \qquad u = 1, \ldots, n, \tag{8.2.2}$$

is, for given $\boldsymbol{\theta}$ and Σ, distributed as the m-variate Normal $N_m(\mathbf{0}, \Sigma)$, and that the runs are made in such a way that it can be assumed that from run to run the observations are independent. Thus, in terms of the general framework of (8.1.4), $\boldsymbol{\Sigma} = \boldsymbol{\pi}$ are the parameters of an error distribution which is multivariate Normal. We first derive some very general results which apply to any model of this type, and then consider in detail the various important special cases that emerge if the expectation functions are supposed linear in the parameters $\boldsymbol{\theta}_i$.

The joint distribution of the n vectors of errors $\boldsymbol{\varepsilon} = (\boldsymbol{\varepsilon}_{(1)}, \ldots, \boldsymbol{\varepsilon}_{(u)}, \ldots, \boldsymbol{\varepsilon}_{(n)})'$ is

$$\begin{aligned} p(\boldsymbol{\varepsilon} \mid \boldsymbol{\Sigma}, \boldsymbol{\theta}) &= \prod_{u=1}^{n} p(\boldsymbol{\varepsilon}_{(u)} \mid \boldsymbol{\Sigma}, \boldsymbol{\theta}) \\ &= (2\pi)^{-mn/2} |\boldsymbol{\Sigma}|^{-n/2} \exp\left(-\frac{1}{2} \sum_{u=1}^{n} \boldsymbol{\varepsilon}'_{(u)} \boldsymbol{\Sigma}^{-1} \boldsymbol{\varepsilon}_{(u)}\right) \end{aligned}$$

$$-\infty < \varepsilon_{ui} < \infty, \quad i = 1, \ldots, m, \quad u = 1, \ldots, n, \tag{8.2.3}$$

where $\Sigma = \{\sigma_{ij}\}$ is the $m \times m$ covariance matrix, $\Sigma_{-1} = \{\sigma^{ij}\}$ its inverse and $\boldsymbol{\theta}$ refers to the complete set of all the $(k_1 + \cdots + k_m)$ parameters $\boldsymbol{\theta}_1, \ldots, \boldsymbol{\theta}_m$.

Denoting $\mathbf{S}(\boldsymbol{\theta})$ to be the $m \times m$ symmetric matrix

$$\mathbf{S}(\boldsymbol{\theta}) = \{S_{ij}(\boldsymbol{\theta}_i, \boldsymbol{\theta}_j)\}$$

with

$$S_{ij}(\boldsymbol{\theta}_i, \boldsymbol{\theta}_j) = \sum_{u=1}^{n} \varepsilon_{ui}\,\varepsilon_{uj} = \sum_{u=1}^{n} [y_{ui} - \eta_i(\boldsymbol{\xi}_{ui}, \boldsymbol{\theta}_i)]\,[y_{uj} - \eta_j(\boldsymbol{\xi}_{uj}, \boldsymbol{\theta}_j)], \qquad i, j = 1, \ldots, m, \tag{8.2.4}$$

then the exponent in (8.2.3) can be expressed as

$$\sum_{u=1}^{n} \boldsymbol{\varepsilon}'_{(u)} \boldsymbol{\Sigma}^{-1} \boldsymbol{\varepsilon}_{(u)} = \operatorname{tr} \mathbf{S}(\boldsymbol{\theta})\boldsymbol{\Sigma}^{-1} = \sum_{i=1}^{m} \sum_{j=1}^{m} \sigma^{ij} S_{ij}(\boldsymbol{\theta}_i, \boldsymbol{\theta}_j) \tag{8.2.5}$$

where $\operatorname{tr} \mathbf{A}$ means the trace of the matrix $\mathbf{A}$. Given the observations, the likelihood function can thus be written

$$\begin{aligned} l(\boldsymbol{\theta}, \boldsymbol{\Sigma} \mid \mathbf{y}) &\propto p(\boldsymbol{\varepsilon} \mid \boldsymbol{\Sigma}, \boldsymbol{\theta}) \\ &\propto |\boldsymbol{\Sigma}|^{-n/2} \exp\left[-\tfrac{1}{2} \operatorname{tr} \boldsymbol{\Sigma}^{-1} \mathbf{S}(\boldsymbol{\theta}) \right]. \end{aligned} \tag{8.2.6}$$

To clarify the notation, we emphasize that $\mathbf{y}$ refers to the $n \times m$ matrix of observations

$$\mathbf{y} = \begin{bmatrix} y_{11} & \cdots & y_{1i} & \cdots & y_{1m} \\ \vdots & & & & \\ y_{u1} & \cdots & y_{ui} & \cdots & y_{um} \\ \vdots & & & & \\ y_{n1} & \cdots & y_{ni} & \cdots & y_{nm} \end{bmatrix} = [\mathbf{y}_1, \ldots, \mathbf{y}_i, \ldots, \mathbf{y}_m] = \begin{bmatrix} \mathbf{y}'_{(1)} \\ \vdots \\ \mathbf{y}'_{(u)} \\ \vdots \\ \mathbf{y}'_{(n)} \end{bmatrix},$$

where $\mathbf{y}_i = (y_{1i}, \ldots, y_{ni})'$ is the vector of n observations corresponding to the ith response and $\mathbf{y}_{(u)} = (y_{u1}, \ldots, y_{um})'$ is the vector of m observations of the uth experimental run. Similarly, $\boldsymbol{\varepsilon}$ refers to the $n \times m$ matrix of errors

$$\boldsymbol{\varepsilon} = \begin{bmatrix} \varepsilon_{11} & \cdots & \varepsilon_{1i} & \cdots & \varepsilon_{1m} \\ \vdots & & & & \\ \varepsilon_{u1} & \cdots & \varepsilon_{ui} & \cdots & \varepsilon_{um} \\ \vdots & & & & \\ \varepsilon_{n1} & \cdots & \varepsilon_{ni} & \cdots & \varepsilon_{nm} \end{bmatrix} = [\boldsymbol{\varepsilon}_1, \ldots, \boldsymbol{\varepsilon}_i, \ldots, \boldsymbol{\varepsilon}_m] = \begin{bmatrix} \boldsymbol{\varepsilon}'_{(1)} \\ \vdots \\ \boldsymbol{\varepsilon}'_{(u)} \\ \vdots \\ \boldsymbol{\varepsilon}'_{(n)} \end{bmatrix}.$$

8.2.2 Prior Distribution of $(\boldsymbol{\theta}, \boldsymbol{\Sigma})$

For the prior distribution of the parameters $(\boldsymbol{\theta}, \boldsymbol{\Sigma})$, we shall first of all assume that $\boldsymbol{\theta}$ and $\boldsymbol{\Sigma}$ are approximately independent so that

$$p(\boldsymbol{\theta}, \boldsymbol{\Sigma}) \doteq p(\boldsymbol{\theta})\, p(\boldsymbol{\Sigma}). \tag{8.2.7}$$

We shall further suppose that the parameterization in terms of $\boldsymbol{\theta}$ is so chosen such that it is appropriate to take $\boldsymbol{\theta}$ as locally uniform,†

$$p(\boldsymbol{\theta}) \propto \text{constant}. \tag{8.2.8}$$

For the prior distribution of the $\frac{1}{2}m(m+1)$ distinct elements of $\boldsymbol{\Sigma}$, application of the argument in Section 1.3 for the multiparameter situation leads to the non informative reference prior

$$p(\boldsymbol{\Sigma}) \propto |\mathcal{I}(\boldsymbol{\Sigma})|^{1/2}. \tag{8.2.9}$$

Now,

$$|\mathcal{I}(\boldsymbol{\Sigma})| = |\mathcal{I}(\boldsymbol{\Sigma}^{-1})| \left|\frac{\partial \boldsymbol{\Sigma}}{\partial \boldsymbol{\Sigma}^{-1}}\right|^{-2}, \tag{8.2.10}$$

where

$$\left|\frac{\partial \boldsymbol{\Sigma}}{\partial \boldsymbol{\Sigma}^{-1}}\right| = \left|\frac{\partial(\sigma_{11}, \sigma_{12}, \ldots, \sigma_{mm})}{\partial(\sigma^{11}, \sigma^{12}, \ldots, \sigma^{mm})}\right| \tag{8.2.11}$$

is the Jacobian of the transformation from the elements σ_{ij} of $\boldsymbol{\Sigma}$ to the elements σ^{ij} of $\boldsymbol{\Sigma}^{-1}$. It is shown in Appendix A8.2 that

$$|\mathcal{I}(\boldsymbol{\Sigma}^{-1})| \propto \left|\frac{\partial \boldsymbol{\Sigma}}{\partial \boldsymbol{\Sigma}^{-1}}\right| \tag{8.2.12}$$

and that

$$\left|\frac{\partial \boldsymbol{\Sigma}}{\partial \boldsymbol{\Sigma}^{-1}}\right| = |\boldsymbol{\Sigma}|^{m+1}. \tag{8.2.13}$$

Thus,

$$p(\boldsymbol{\Sigma}) \propto |\boldsymbol{\Sigma}|^{-\frac{1}{2}(m+1)}. \tag{8.2.14}$$

In this special case $m = 1$, (8.2.14) reduces to

$$p(\sigma_{11}) \propto \frac{1}{\sigma_{11}} \tag{8.2.15}$$

which coincides with the usual assumption concerning a noninformative prior distribution for a single variance. Another special case of interest is when the errors $(\varepsilon_{u1}, \ldots, \varepsilon_{um})$ are uncorrelated, that is, $\sigma_{ij} = 0$ if $i \neq j$. In this case, the same argument leads to

$$p(\boldsymbol{\Sigma} \mid \sigma_{ij} = 0, i \neq j) = p(\sigma_{11}, \ldots, \sigma_{mm}) \propto \prod_{i=1}^{m} \sigma_{ii}^{-1}. \tag{8.2.16}$$

---

† As we have mentioned earlier, when the parameter space is of high dimension, the use of the locally uniform prior may be inappropriate and more careful considerations should be given to the structure of the model in selecting a noninformative prior.

8.2.3 Posterior Distribution of $(\boldsymbol{\theta}, \boldsymbol{\Sigma})$

Using (8.2.6), (8.2.8), and (8.2.14), the joint posterior distribution of $(\boldsymbol{\theta}, \boldsymbol{\Sigma})$ is

$$p(\boldsymbol{\theta}, \boldsymbol{\Sigma} \mid \mathbf{y}) \propto |\boldsymbol{\Sigma}|^{-\frac{1}{2}(n+m+1)} \exp\left[-\tfrac{1}{2}\operatorname{tr}\boldsymbol{\Sigma}^{-1}\mathbf{S}(\boldsymbol{\theta})\right], \qquad -\infty < \boldsymbol{\theta} < \infty, \quad \boldsymbol{\Sigma} > \mathbf{0}, \tag{8.2.17}$$

where the notation $-\infty < \boldsymbol{\theta} < \infty$ means that each element of the set of parameters $\boldsymbol{\theta}$ can vary from $-\infty$ to ∞, and the notation $\boldsymbol{\Sigma} > \mathbf{0}$ means that the $\frac{1}{2}m(m+1)$ elements σ_{ij} are such that the random matrix $\boldsymbol{\Sigma}$ is positive definite.

It is sometimes convenient to work with the elements of $\boldsymbol{\Sigma}^{-1} = \{\sigma^{ij}\}$ rather than the elements of $\boldsymbol{\Sigma}$. Since

$$p(\boldsymbol{\theta}, \boldsymbol{\Sigma}^{-1} \mid \mathbf{y}) = p(\boldsymbol{\theta}, \boldsymbol{\Sigma} \mid \mathbf{y}) \left|\frac{\partial \boldsymbol{\Sigma}}{\partial \boldsymbol{\Sigma}^{-1}}\right|, \tag{8.2.18}$$

it follows from (8.2.13) that the posterior distribution of $(\boldsymbol{\theta}, \boldsymbol{\Sigma}^{-1})$ is

$$p(\boldsymbol{\theta}, \boldsymbol{\Sigma}^{-1} \mid \mathbf{y}) \propto |\boldsymbol{\Sigma}^{-1}|^{\frac{1}{2}(n-m-1)} \exp\left[-\tfrac{1}{2}\operatorname{tr}\boldsymbol{\Sigma}^{-1}\mathbf{S}(\boldsymbol{\theta})\right],$$
$$-\infty < \boldsymbol{\theta} < \infty, \quad \boldsymbol{\Sigma}^{-1} > \mathbf{0}. \tag{8.2.19}$$

8.2.4 The Wishart Distribution

We now introduce a distribution which is basic in Normal theory multivariate problems. Let $\mathbf{Z}$ be a $m \times m$ positive definite symmetric random matrix which consists of $\frac{1}{2}m(m+1)$ distinct random variables z_{ij} $(i, j = 1, \ldots, m; i \geqslant j)$. Let $q > 0$, and $\mathbf{B}$ be a $m \times m$ positive definite symmetric matrix of fixed constants. The distribution of z_{ij},

$$p(\mathbf{Z}) \propto |\mathbf{Z}|^{\frac{1}{2}q-1} \exp\left(-\tfrac{1}{2}\operatorname{tr}\mathbf{ZB}\right), \quad \mathbf{Z} > \mathbf{0} \tag{8.2.20}$$

obtained by Wishart (1928), is a multivariate generalization of the χ^2 distribution. It can be shown that

$$\int_{\mathbf{Z}>0} |\mathbf{Z}|^{\frac{1}{2}q-1} \exp\left(-\tfrac{1}{2}\operatorname{tr}\mathbf{ZB}\right) d\mathbf{Z} = |\mathbf{B}|^{-\frac{1}{2}(q+m-1)}\, 2^{\frac{1}{2}m(q+m-1)}\, \Gamma_m\left(\frac{q+m-1}{2}\right) \tag{8.2.21}$$

where $\Gamma_p(b)$ is the generalized gamma function, Siegel (1935)

$$\Gamma_p(b) = [\Gamma(\tfrac{1}{2})]^{\frac{1}{2}p(p-1)} \prod_{\alpha=1}^{p} \Gamma\left(b + \frac{\alpha - p}{2}\right), \qquad b > \frac{p-1}{2}. \tag{8.2.22}$$

We shall denote the distribution (8.2.20) by $W_m(\mathbf{B}^{-1}, q)$ and say that $\mathbf{Z}$ is distributed as Wishart with q degrees of freedom and parameter matrix $\mathbf{B}^{-1}$. For a discussion of the properties of the Wishart distribution, see for example Anderson (1958). Note carefully that the parameterization used in (8.2.20) is different from the one used in Anderson in one respect. In his notation, the distribution in (8.2.20) is denoted as $W(\mathbf{B}^{-1}, \nu)$ where $\nu = q + m - 1$ is said to be the degrees of freedom.

As an application of the Wishart distribution, we see in (8.2.19) that, given $\boldsymbol{\theta}, \boldsymbol{\Sigma}^{-1}$ is distributed as $W_m[\mathbf{S}^{-1}(\boldsymbol{\theta}), n - m + 1]$ provided $n \geqslant m$.

8.2.5 Posterior Distribution of $\boldsymbol{\theta}$

Using the identity (8.2.21), we immediately obtain from (8.2.19) the marginal posterior distribution of $\boldsymbol{\theta}$ as

$$p(\boldsymbol{\theta} \mid \mathbf{y}) \propto |\mathbf{S}(\boldsymbol{\theta})|^{-n/2}, \qquad -\infty < \boldsymbol{\theta} < \infty, \tag{8.2.23}$$

provided $n \geqslant m$.

This extremely simple result is remarkable because of its generality. It will be noted that to reach it we have *not* had to assume either:

a) that any of the input variables ξ_{ui} were or were not common to more than one output, or
b) that the parameters $\boldsymbol{\theta}_i$ were or were not common to more than one output, or
c) that the expectation functions were linear or were nonlinear in the parameters.

This generality may be contrasted with the specification needed to obtain "nice" sampling theory results. For example, a common formulation assumes that the ξ_{ui} are common, that the $\boldsymbol{\theta}_i$ are *not*, and that the expectation functions are all linear in the parameters.

In the special case in which there is only one output response y, (8.2.23) reduces to

$$p(\boldsymbol{\theta} \mid \mathbf{y}) \propto [S(\boldsymbol{\theta})]^{-n/2}, \qquad -\infty < \boldsymbol{\theta} < \infty, \tag{8.2.24}$$

with $S(\boldsymbol{\theta}) = \sum_{u=1}^{n} [y_u - \eta(\xi_u, \boldsymbol{\theta})]^2$. As we have seen, this result can be regarded as supplying a Bayesian justification of least squares, since the modal values of $\boldsymbol{\theta}$ (those associated with maximum posterior density) are those which minimize S. The general result (8.2.23) supplies then, among other things, an appropriate Bayesian multivariate generalization of least squares. The "most probable" values of $\boldsymbol{\theta}$ being simply those which minimize the determinant $|\mathbf{S}(\boldsymbol{\theta})|$.

Finally, in the special case $\sigma_{ij} = 0, i \neq j$, combining (8.2.16) with (8.2.6) and integrating out $\sigma_{11}, \ldots, \sigma_{mm}$ yields

$$p(\boldsymbol{\theta} \mid \mathbf{y}) \propto \prod_{i=1}^{m} [S_{ii}(\boldsymbol{\theta}_i)]^{-n/2}, \qquad -\infty < \boldsymbol{\theta} < \infty. \tag{8.2.25}$$

8.2.6 Estimation of Common Parameters in a Nonlinear Multivariate Model

We now illustrate the general applicability of the result (8.2.23) by considering an example in which:

a) certain of the θ's are common to more than one output, and

b) the expectation functions are nonlinear in the parameters.

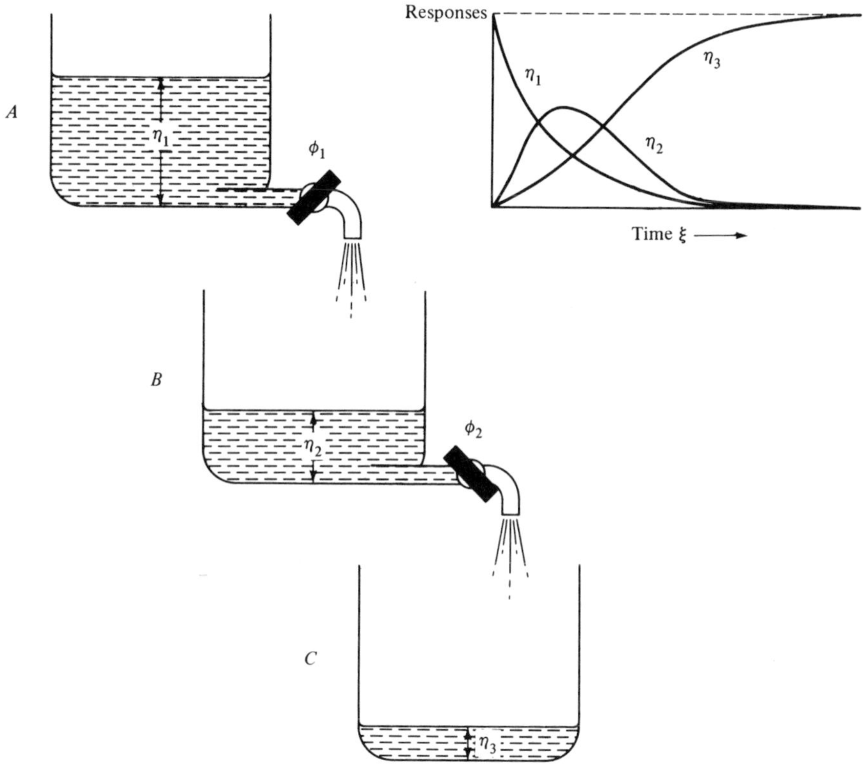

Fig. 8.2.1 Diagrammatic representation of a system $A \to B \to C$.

Suppose we have the consecutive system indicated in Fig. 8.2.1, which shows water running from a tank A via a tap opened an amount ϕ_1 into a tank B which then runs into a tank C via a tap opened an amount ϕ_2.

If η_1, η_2 and η_3 are the proportions of A, B, and C present at time ξ, with initial conditions $(\eta_1 = 1, \eta_2 = 0, \eta_3 = 0)$, the system can be described by the differential equations

$$\frac{d\eta_1}{d\xi} = -\phi_1\eta_1,$$

$$\frac{d\eta_2}{d\xi} = \phi_1\eta_1 - \phi_2\eta_2, \qquad (8.2.26)$$

$$\frac{d\eta_3}{d\xi} = \phi_2\eta_2.$$

Systems of this kind have many applications in engineering and in the physical and biological sciences. In particular, the equation (8.2.26) could represent a consecutive first-order chemical reaction in which a substance A decomposed to form B, which in turn decomposed to form C. The responses η_1, η_2, η_3 would then be the mole fractions of A, B, and C present at time ξ and the quantities ϕ_1 and ϕ_2 would then be rate constants associated with the first and second decompositions and would normally have to be estimated from data.

If we denote by y_1, y_2, and y_3 the *observed* values of η_1, η_2, and η_3, then, on integration of (8.2.26), we have the expectation functions

$$E(y_1) = \eta_1 = e^{-\phi_1 \xi}, \tag{8.2.27a}$$

$$E(y_2) = \eta_2 = (e^{-\phi_1 \xi} - e^{-\phi_2 \xi})\, \phi_1/(\phi_2 - \phi_1), \tag{8.2.27b}$$

$$E(y_3) = \eta_3 = 1 + (-\,\phi_2\, e^{-\phi_1 \xi} + \phi_1\, e^{-\phi_2 \xi})/(\phi_2 - \phi_1), \tag{8.2.27c}$$

and it is to be noted that for all ξ,

$$\eta_1 + \eta_2 + \eta_3 \equiv 1. \tag{8.2.27d}$$

Observations on y_1 could yield information only on ϕ_1, but observations on y_2 and y_3 could each provide information on both ϕ_1 and ϕ_2. If measurements of more than one of the quantities (y_1, y_2, y_3) were available, we should certainly expect to be able to estimate the parameters more precisely. The Bayesian approach allows us to pool the information from (y_1, y_2, y_3) and makes it easy

Table 8.2.1

Observations on the yield of three substances in a chemical reaction

Time $= \xi_u$	Yield of A y_{1u}	Yield of B y_{2u}	Yield of C y_{3u}
$\frac{1}{2}$	0.959	0.025	0.028
$\frac{1}{2}$	0.914	0.061	0.000
1	0.855	0.152	0.068
1	0.785	0.197	0.096
2	0.628	0.130	0.090
2	0.617	0.249	0.118
4	0.480	0.184	0.374
4	0.423	0.298	0.358
8†	0.166	0.147	0.651
8†	0.205	0.050	0.684
16†	0.034	0.000	0.899
16†	0.054	0.047	0.991

† These four runs are omitted in the second analysis.

to appreciate the contribution from each of the three responses. In this example ξ is the only input variable and is the elapsed time since the start of the reaction. We denote by $\mathbf{y}'_{(u)} = (y_{u1}, y_{u2}, y_{u3})$ a set of $m = 3$ observations made on $\eta_{1u}, \eta_{2u}, \eta_{3u}$ at time ξ_u. A typical set of such observations is shown in Table 8.2.1.

In some cases observations may not be available on all three of the outputs. Thus only the concentration y_2 of the product B might be observable, or y_2 and y_3 might be known, but there might be no independently measured observation y_1 of the concentration of A.†

We suppose that the observations of Table 8.2.1 may be treated as having arisen from 12 *independent* experimental runs, as might be appropriate if the runs were carried out in random order in sealed tubes, each reaction being terminated at the appropriate time by sudden cooling. Furthermore, we suppose that (y_1, y_2, y_3) are functionally independent so that the 3×3 matrix $\boldsymbol{\Sigma}$ may be assumed to be positive definite and contains three variances and three covariances, all unknown. It is perhaps most natural for the experimenter to think in terms of the logarithms $\theta_1 = \log \phi_1$ and $\theta_2 = \log \phi_2$ of the rate constants and to regard these as locally uniformally distributed *a priori*.‡ We shall, therefore, choose as our reference priors for (θ_1, θ_2) and $\boldsymbol{\Sigma}$ the distributions in (8.2.8) and (8.2.14), respectively.

---

† When the chemist has difficulty in determining one of the products he sometimes makes use of relations like (8.2.27d) to "obtain it by calculation." Thus he might "obtain" y_1 from the relation $y_1 = 1 - y_2 - y_3$. For the resulting data set, the 3×3 covariance matrix $\boldsymbol{\Sigma}$ will of course not be positive definite, and the analysis in terms of three-dimensional responses will be inappropriate. In particular, the determinant of the sums of squares and products which appears in (8.2.23) will be zero *whatever* the values of the parameters. The difficulty is of course overcome very simply. The quantity y_1 is not an observation and the *data* has two dimensions, not three. The analysis should be carried through with y_2 and y_3 which *have* actually been measured. For a fuller treatment of problems of this kind arising because of data dependence or near dependence, see Box, Erjavec, Hunter and MacGregor (1972).

‡ Suppose that (a) the expectation functions were *linear* in $\theta_1(\boldsymbol{\phi})$ and $\theta_2(\boldsymbol{\phi})$ where $\boldsymbol{\phi} = (\phi_1, \phi_2)$, (b) little was known *a priori* about either parameter compared with the information supplied by the data, and (c) any prior information about one parameter would supply essentially none about the other.

Then, arguing as in Section 1.3, a noninformative reference prior to $\boldsymbol{\theta}$ should be locally uniform.

Conditions (b) and (c) are likely to be applicable to this problem at least as approximations, but condition (a) is not, because the expectation functions are non-linear in ϕ_1 and ϕ_2 and no general linearizing transformation exists. However, [see for example Beale (1960), and Guttman and Meeter (1965)] the expectation functions are more "nearly linear" in $\theta_1 = \log \phi_1$ and $\theta_2 = \log \phi_2$. Thus, the assumption that θ_1 and θ_2 are locally uniform provides a better approximation to a noninformative prior for the rate constants. For reasons we have discussed earlier, the assumption is not critical and, if for example we assume ϕ_1 and ϕ_2 themselves to be locally uniform, the posterior distribution is not altered appreciably.

Expression (8.2.23) makes it possible to compute the posterior density for the parameters assuming observations are available on some or all of the products A, B, and C. Thus, we may consider the posterior distribution of $\boldsymbol{\theta} = (\theta_1, \theta_2)'$

a) if only yields $\mathbf{y}_2$ of product B are available

$$p(\boldsymbol{\theta} \mid \mathbf{y}_2) \propto [S_{22}(\boldsymbol{\theta})]^{-n/2}, \qquad -\infty < \boldsymbol{\theta} < \infty, \tag{8.2.28a}$$

b) if only yields $\mathbf{y}_3$ of product C are available,

$$p(\boldsymbol{\theta} \mid \mathbf{y}_3) \propto [S_{33}(\boldsymbol{\theta})]^{-n/2}, \qquad -\infty < \boldsymbol{\theta} < \infty, \tag{8.2.28b}$$

c) if only yields $\mathbf{y}_2$ and $\mathbf{y}_3$ of B and C are available

$$p(\boldsymbol{\theta} \mid \mathbf{y}_2, \mathbf{y}_3) \propto \begin{vmatrix} S_{22}(\boldsymbol{\theta}) & S_{23}(\boldsymbol{\theta}) \\ S_{23}(\boldsymbol{\theta}) & S_{33}(\boldsymbol{\theta}) \end{vmatrix}^{-n/2}, \qquad -\infty < \boldsymbol{\theta} < \infty, \tag{8.2.28c}$$

and

d) if yields $\mathbf{y}_1$, $\mathbf{y}_2$ and $\mathbf{y}_3$ of the products A, B and C are all available

$$p(\boldsymbol{\theta} \mid \mathbf{y}) \propto |\mathbf{S}(\boldsymbol{\theta})|^{-n/2}, \qquad -\infty < \boldsymbol{\theta} < \infty, \tag{8.2.28d}$$

where $\mathbf{S}(\boldsymbol{\theta}) = \{S_{ij}(\boldsymbol{\theta})\}$, $i, j = 1, 2, 3$.

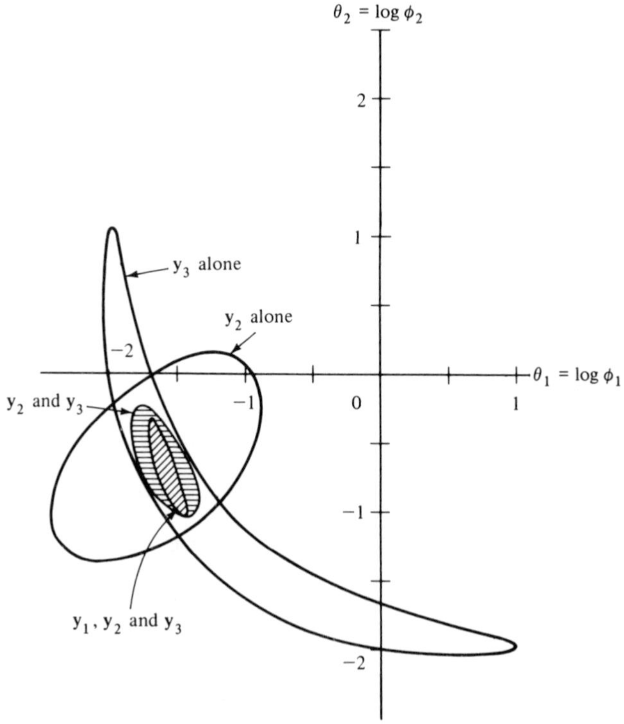

Fig. 8.2.2 99.75% H.P.D. regions for θ_1 and θ_2 for the chemical reaction data.

Since there are only two parameters θ_1 and θ_2, the posterior distributions can be represented by contour diagrams which may be superimposed to show the contributions made by the various output responses. Single contours are shown in Fig. 8.2.2 of the posterior distributions of θ_1 and θ_2 for (a) $\mathbf{y}_2$ alone, (b) $\mathbf{y}_3$ alone, (c) $\mathbf{y}_2$ and $\mathbf{y}_3$ jointly, and (d) $\mathbf{y}_1$, $\mathbf{y}_2$, and $\mathbf{y}_3$ jointly. The contours actually shown are those which should correspond to an H.P.D. region containing approximately 99.75% of the probability mass calculated from

$$\log p(\hat{\boldsymbol{\theta}} \mid \cdot) - \log p(\boldsymbol{\theta} \mid \cdot) = \tfrac{1}{2}\chi^2(2, \alpha), \qquad \alpha = 0.0025$$

where $p(\boldsymbol{\theta} \mid \cdot)$ refers to the appropriate distributions in (8.2.28a–d) and $\hat{\boldsymbol{\theta}}$, the corresponding modal values of $\boldsymbol{\theta}$. In this example, it is apparent, particularly for $\mathbf{y}_3$, that the posterior distributions are non-Normal. Nevertheless, the above very crude approximation will suffice for the purpose of the present discussion.

In studying Figure 8.2.2, we first consider the moon-shaped contour obtained from observations $\mathbf{y}_3$ on the end product C alone. In any sequential reaction $A \to B \to C \to \ldots$ etc., we should expect that observation of *only* the end product (C in this case) could provide little or no information about the *individual* parameters but only about some aggregate of these rate constants. A diagonally attenuated ridge-like surface is therefore to be expected. However, it should be further noted that since in this specific instance η_3 is *symmetric* in θ_1 and θ_2 [see expression (8.2.27c)], the posterior surface is *completely symmetric* about the line $\theta_1 = \theta_2$. In particular, if $(\hat{\theta}_1, \hat{\theta}_2)$ is a point of maximum density the point $(\hat{\theta}_2, \hat{\theta}_1)$ will also give the same maximum density. In general the surface will be bimodal and have two peaks of equal height symmetrically situated about the equi-angular line. Marginal distributions will thus display precisely the kind of behaviour shown in Fig. A5.6.1.

Figure 8.2.2 shows, how, for this data, the inevitable ambiguity arising when only observations $\mathbf{y}_3$ on product C are utilized, is resolved as soon as the additional information supplied by values $\mathbf{y}_2$ on the intermediate product B is considered. As can be expected, the nature of the evidence that the intermediate product $\mathbf{y}_2$ contributes, is preferentially concerned with the difference of the parameters. This is evidenced by the tendency of the region to be obliquely oriented approximately at right angles to that for $\mathbf{y}_3$. By combining information from the two sources we obtain a much smaller region contained within the intersection of the individual regions. Finally, information from $\mathbf{y}_1$ which casts further light on the value of θ_1, causes the region to be further reduced.

Data of this kind sometimes occur in which available observations trace only part of the reaction. To demonstrate the effect of this kind of inadequacy in the experimental design, the analysis is repeated omitting the last four observations in Table 8.2.1. As shown in Fig. 8.2.3, over the ranges studied, the contours for $\mathbf{y}_2$ alone and $\mathbf{y}_3$ alone do not now close. Nevertheless, quite precise estimation is possible using $\mathbf{y}_2$ and $\mathbf{y}_3$ together and the addition of $\mathbf{y}_1$ improves the estimation further.

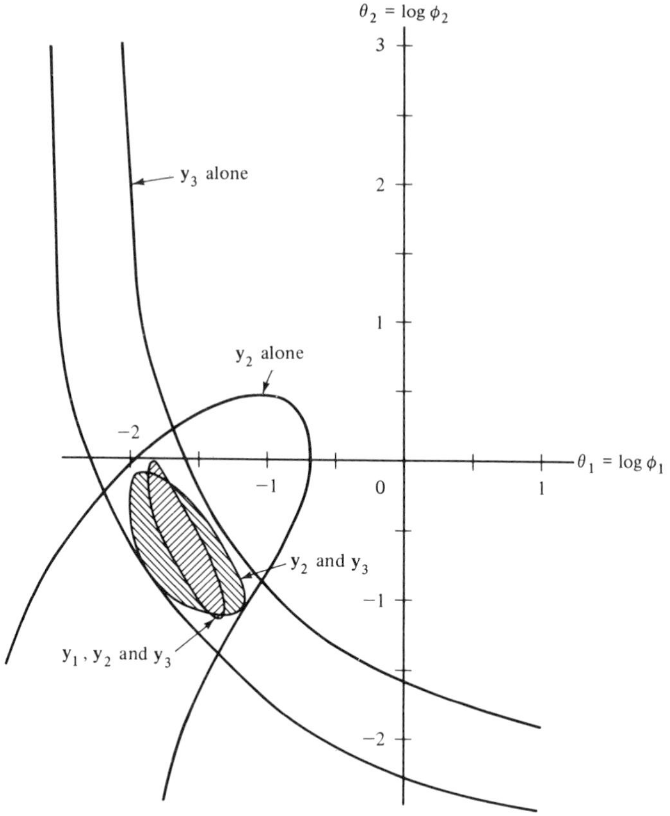

Fig. 8.2.3 99.75% H.P.D. regions for θ_1 and θ_2, excluding the last four observations.

Precautions in the Estimation of Common Parameters

Even in cases where only a single response is being considered, caution is needed in the fitting of functions. As explained in Section 1.1.4, fitting should be regarded as merely one element in the iterative model building process. The appropriate attitude is that when the model is initially fitted it is tentatively entertained rather than assumed. Careful checks on residuals are applied in a process of model criticism to see whether there is reason to doubt its applicability to the situation under consideration.

The importance of such precaution is even greater when several responses are considered. In multivariate problems, not only should each response model be checked individually but they must also be checked for overall consistency. The investigator should in practice *not* revert immediately to a joint analysis of responses. He should:

1) check the individual fit of each response,
2) compare posterior distributions to appraise the consistency of the information from the various responses (an aspect discussed in more detail in Chapter 9).

Only in those cases where he is satisfied with the individual fit and with the consistency shall he revert to the joint analysis.

8.3 LINEAR MULTIVARIATE MODELS

In discussing the general m-variate Normal model above, we have not needed to assume anything specific about the form of the m expectation functions $\boldsymbol{\eta}$. In particular, they need not be linear in the parameters† nor does it matter whether or not some parameters appear in more than one of the expectation functions. Many interesting and informative special cases arise if we suppose the expectation functions to be linear in the θ's. Moreover, as will be seen, the linear results can sometimes supply adequate local approximations for models non-linear in the parameters. From now on then we assume that

$$E(y_{ui}) = \eta_i(\boldsymbol{\xi}_{ui}, \boldsymbol{\theta}_i) = \mathbf{x}'_{(ui)}\boldsymbol{\theta}_i, \qquad i = 1, \dots, m, \qquad u = 1, \dots, n, \tag{8.3.1}$$

where

$$\boldsymbol{\theta}'_i = (\theta_{1i}, \dots, \theta_{gi}, \dots, \theta_{k_i i})$$

and

$$\mathbf{x}'_{(ui)} = (x_{u1i}, \dots, x_{ugi}, \dots, x_{uk_i i})$$

with

$$x_{ugi} = \frac{\partial \eta_i(\boldsymbol{\xi}_{ui}, \boldsymbol{\theta}_i)}{\partial \theta_{gi}}$$

independent of all the θ's.

The $n \times k_i$ matrix $\mathbf{X}_i$ whose uth row is $\mathbf{x}'_{(ui)}$ will be called the *derivative* matrix for the ith response.

Our linear m-variate model may now be written as

$$\begin{aligned} \mathbf{y}_1 &= \mathbf{X}_1\boldsymbol{\theta}_1 + \boldsymbol{\varepsilon}_1 \\ &\vdots \\ \mathbf{y}_i &= \mathbf{X}_i\boldsymbol{\theta}_i + \boldsymbol{\varepsilon}_i \\ &\vdots \\ \mathbf{y}_m &= \mathbf{X}_m\boldsymbol{\theta}_m + \boldsymbol{\varepsilon}_m. \end{aligned} \tag{8.3.2}$$

Certain characteristics of this mode of writing the model should be noted. In particular, it is clear that while the elements of $\mathbf{x}_{(ui)}$ will be functions of the

† Although, so that a uniform density can represent an approximately noninformative prior and also to assist local linear approximation, parameter transformations in terms of which the expectation function is more nearly linear, will often be employed.

elements of the vector input variables $\boldsymbol{\xi}_{ui}$, they will in general not be proportional to the elements of $\boldsymbol{\xi}_{ui}$ themselves. Thus if

$$E(y_{ui}) = \frac{\theta_1 \log \xi_{u1i} + \theta_2 \, \xi_{u1i} \, \xi_{u3i}}{\xi_{u2i}},$$

then

$$x_{u1i} = \frac{\log \xi_{u1i}}{\xi_{u2i}} \quad \text{and} \quad x_{u2i} = \frac{\xi_{u1i} \, \xi_{u3i}}{\xi_{u2i}}.$$

8.3.1 The Use of Linear Theory Approximations when the Expectation is Nonlinear in the Parameters

The specific form of posterior distributions which we shall obtain for the linear case will often provide reasonably close approximations even when the expectation functions $\boldsymbol{\eta}_{(u)}$ is *nonlinear* in $\boldsymbol{\theta}$. This is because we need only that the expectation functions are approximately linear *in the region of the parameter space covered by most of the posterior distribution*, say within the 95% H.P.D. region.† For moderate n, this can happen with functions that are highly nonlinear in the parameters when considered over their whole range. Then, in the region where the posterior probability mass is concentrated (say the 95% H.P.D. region), we may expand the expectation function around the mode $\hat{\boldsymbol{\theta}}_i$

$$E(y_{ui}) = \eta_{ui} \doteq \eta_i(\boldsymbol{\xi}_{ui}, \hat{\boldsymbol{\theta}}_i) + \sum_{g=1}^{k_i} x_{ugi} \, (\theta_{gi} - \hat{\theta}_{gi}), \tag{8.3.3}$$

where

$$x_{ugi} = \left. \frac{\partial \eta_i(\boldsymbol{\xi}_{ui}, \boldsymbol{\theta}_i)}{\partial \theta_{gi}} \right|_{\boldsymbol{\theta}_i = \hat{\boldsymbol{\theta}}_i},$$

which is, approximately, in the form of a linear model. Thus, the posterior distributions found from linear theory can, in many cases, provide close approximations to the true distributions. For example, in the univariate case ($m = 1$) with a single parameter θ, the posteriror distribution in (8.2.24) would be approximately

$$p(\theta \mid \mathbf{y}) \,\dot{\propto}\, [\nu s^2 + (\Sigma \, x_u^2) \, (\theta - \hat{\theta})^2]^{-n/2} \tag{8.3.4}$$

where

$$\nu = n - 1, \qquad s^2 = \frac{1}{\nu} \Sigma[y_u - \eta(\xi_u, \hat{\theta})]^2 \quad \text{and} \quad x_u = \left. \frac{\partial \eta(\xi_u, \theta)}{\partial \theta} \right|_{\theta = \hat{\theta}},$$

so that the quantity

$$\frac{\sqrt{\Sigma x_u^2}(\theta - \hat{\theta})}{s} \tag{8.3.5}$$

would be approximately distributed as $t(0, 1, \nu)$.

---

† A possibility that can be checked a posteriori for any specific case.

When, as in the case in which multivariate or univariate least squares is appropriate, a convenient method of calculating the $\hat{\theta}$'s is available for the linear case but not for the corresponding nonlinear situation, the linearization may be used iteratively to find the $\hat{\theta}$'s for the nonlinear situation.

For example, in the univariate model containing a single parameter θ, with a first guess θ_0, we can write approximately

$$E(z_{u0}) \doteq (\theta - \theta_0)x_u, \tag{8.3.6}$$

where

$$z_{u0} = y_u - \eta\,(\xi_u, \theta_0) \qquad \text{and} \qquad x_u = \left.\frac{\partial \eta(\xi_u, \theta)}{\partial \theta}\right|_{\theta=\theta_0}$$

Applying ordinary least squares to the model, we obtain an estimate of the correction $\theta - \theta_0$ and hence hopefully an improved "guess" θ_1 from

$$\theta_1 = \theta_0 + \frac{\Sigma z_{u0}\, x_u}{\Sigma x_u^2}. \tag{8.3.7}$$

This is the well-known Newton–Gauss method of iteration for nonlinear least squares, Box (1957, 1960), Hartley (1961), Marquardt (1963), and under favorable conditions the successive iterants will converge to $\hat{\theta}$.

8.3.2 Special Cases of the General Linear Multivariate Model

In general, the joint distribution of $\boldsymbol{\theta}$ and $\boldsymbol{\Sigma}$ is given by (8.2.17) and the marginal distribution of $\boldsymbol{\theta}$ is that in (8.2.23) quite independently of whether $\eta_i(\boldsymbol{\xi}_{ui}, \boldsymbol{\theta}_i)$ is linear in $\boldsymbol{\theta}_i$ or not. For practical purposes, however, it is of interest to consider a number of special cases.

For orientation we reconsider for a moment the linear *univariate* situation discussed earlier in Section 2.7,

$$\mathbf{y} = \mathbf{X}\boldsymbol{\theta} + \boldsymbol{\varepsilon}, \tag{8.3.8}$$

where $\mathbf{y}$ is a $n \times 1$ vector of observations, $\mathbf{X}$ a $n \times k$ matrix of fixed elements, $\boldsymbol{\theta}$ a $k \times 1$ vector of parameters and $\boldsymbol{\varepsilon}$ a $n \times 1$ vector of errors. In this case,

$$p(\boldsymbol{\theta}, \sigma^2 \mid \mathbf{y}) \propto (\sigma^2)^{-(\frac{1}{2}n+1)} \exp\left[-\frac{S(\boldsymbol{\theta})}{2\sigma^2}\right], \qquad \sigma^2 > 0, \quad -\infty < \boldsymbol{\theta} < \infty, \tag{8.3.9}$$

and

$$p(\boldsymbol{\theta} \mid \mathbf{y}) \propto [S(\boldsymbol{\theta})]^{-n/2}, \qquad -\infty < \boldsymbol{\theta} < \infty, \tag{8.3.10}$$

The determinant $|\mathbf{S}(\boldsymbol{\theta})|$ in (8.2.23) becomes the single sum of squares

$$S(\boldsymbol{\theta}) = (\mathbf{y} - \mathbf{X}\boldsymbol{\theta})'\,(\mathbf{y} - \mathbf{X}\boldsymbol{\theta}). \tag{8.3.11}$$

In this linear case, we may write

$$S(\boldsymbol{\theta}) = (n-k)\,s^2 + (\boldsymbol{\theta} - \hat{\boldsymbol{\theta}})'\,\mathbf{X}'\mathbf{X}(\boldsymbol{\theta} - \hat{\boldsymbol{\theta}}), \tag{8.3.12}$$

where

$$(n-k)s^2 = (\mathbf{y}-\hat{\mathbf{y}})'(\mathbf{y}-\hat{\mathbf{y}}) = (\mathbf{y}-\mathbf{X}\hat{\boldsymbol{\theta}})'(\mathbf{y}-\mathbf{X}\hat{\boldsymbol{\theta}})$$

and, assuming $\mathbf{X}$ is of rank k,

$$\hat{\boldsymbol{\theta}} = (\mathbf{X}'\mathbf{X})^{-1}\mathbf{X}'\mathbf{y}$$

so that, writing $\nu = n - k$,

$$p(\boldsymbol{\theta} \mid \mathbf{y}) \propto \left[1 + \frac{(\boldsymbol{\theta}-\hat{\boldsymbol{\theta}})'\mathbf{X}'\mathbf{X}(\boldsymbol{\theta}-\hat{\boldsymbol{\theta}})}{\nu s^2}\right]^{-\frac{1}{2}(\nu+k)}, \qquad -\infty < \boldsymbol{\theta} < \infty. \tag{8.3.13}$$

The posterior distribution of $\boldsymbol{\theta}$ is thus the k dimensional $t_k[\hat{\boldsymbol{\theta}}, s^2(\mathbf{X}'\mathbf{X})^{-1}, \nu]$ distribution. Further, integrating out $\boldsymbol{\theta}$ from (8.3.9) yields the distribution of σ^2,

$$p(\sigma^2 \mid \mathbf{y}) \propto (\sigma^2)^{-(\frac{1}{2}\nu+1)} \exp\left(-\frac{\nu s^2}{2\sigma^2}\right), \qquad \sigma^2 > 0, \tag{8.3.14}$$

so that $\sigma^2/(\nu s^2)$ has the χ_ν^{-2} distribution. All the above results have, of course, been already obtained earlier in Section 2.7.

It is clear that the general linear model (8.3.2) which can be regarded as the multivariate generalization of (8.3.8) need not be particularized in any way. The matrices $\mathbf{X}_1, \ldots, \mathbf{X}_m$ *may or may not* have elements in common; furthermore, the vectors of parameters $\boldsymbol{\theta}_1, \ldots, \boldsymbol{\theta}_m$ *may or may not* have elements in common. Using sampling theory, Zellner (1962, 1963) attempted to study the situation in which the $\mathbf{X}_i$ were not assumed to be identical. The main difficulty with his approach was that the minimum variance estimator for $\boldsymbol{\theta}$ involves the unknown $\boldsymbol{\Sigma}$, and the estimators proposed are "optimal" only in the asymptotic sense.

Cases of special interest which are associated with practical problems of importance and which relate to known results include:

a) when the derivative matrices $\mathbf{X}_1 = \ldots = \mathbf{X}_m = \mathbf{X}$ are common but the parameters $\boldsymbol{\theta}_1, \ldots, \boldsymbol{\theta}_m$ are not,

b) when $\boldsymbol{\theta}_1 = \ldots = \boldsymbol{\theta}_m$ but the matrices $\mathbf{X}_1, \ldots, \mathbf{X}_m$ are not, and

c) when $\boldsymbol{\theta}_1 = \ldots = \boldsymbol{\theta}_m$ *and* $\mathbf{X}_1 = \ldots = \mathbf{X}_m$.

In the remaining part of this chapter, we shall discuss case (a). The problem of estimating common parameters which includes (b) and (c) will be treated in the next chapter.

8.4 INFERENCES ABOUT $\boldsymbol{\theta}$ FOR THE CASE OF A COMMON DERIVATIVE MATRIX X

The model for which $\mathbf{X}_1 = \ldots = \mathbf{X}_m = \mathbf{X}$ (so that $k_1 = \ldots = k_m = k$) and $\boldsymbol{\theta}_1 \neq \ldots \neq \boldsymbol{\theta}_m$ has received most attention in the sampling theory framework —see, for example, Anderson (1958). From the Bayesian point of view, the problem has been studied by Savage (1961a), Geisser and Cornfield (1963), Geisser (1965a), Ando and Kaufman (1965), and others. In general, the multivariate model in (8.3.2) can now be written

$$\mathbf{y} = \mathbf{X}\boldsymbol{\theta} + \boldsymbol{\varepsilon} \tag{8.4.1}$$

$$\underset{n\times m}{[\,\mathbf{y}\,]} = \underset{n\times k}{[\,\mathbf{X}\,]}\ \underset{k\times m}{[\,\boldsymbol{\theta}\,]} + \underset{n\times m}{[\,\boldsymbol{\varepsilon}\,]}$$

where the notation beneath the matrices indicates that $\mathbf{y}$ is an $n\times m$ matrix of m-variate observations, $\boldsymbol{\theta}$ is a $k\times m$ matrix of parameters and $\boldsymbol{\varepsilon}$ an $n\times m$ matrix of errors.

The model would be appropriate for example if say a 2^p factorial experiment had been conducted on a chemical process and the output y_1 = product yield, y_2 = product purity, y_3 = product density had been measured. The elements of each column of the common matrix $\mathbf{X}$ would then be an appropriate sequence of $+1$'s and -1's corresponding to the experimental conditions and the "effect" parameters $\boldsymbol{\theta}_i$ would be different for each output. In econometrics, the model (8.4.1) is frequently encountered in the analysis of the reduced form of simultaneous equation systems.

We note that the $k\times m$ matrix of parameters

$$\boldsymbol{\theta} = \begin{bmatrix} \theta_{11} & \cdots & \theta_{1i} & \cdots & \theta_{1m} \\ \vdots & & \vdots & & \vdots \\ \theta_{g1} & \cdots & \theta_{gi} & \cdots & \theta_{gm} \\ \vdots & & \vdots & & \vdots \\ \theta_{k1} & \cdots & \theta_{ki} & \cdots & \theta_{km} \end{bmatrix} \tag{8.4.2}$$

can be written in the two alternative forms

$$\boldsymbol{\theta} = [\boldsymbol{\theta}_1, \ldots, \boldsymbol{\theta}_i, \ldots, \boldsymbol{\theta}_m] = \begin{bmatrix} \boldsymbol{\theta}'_{(1)} \\ \vdots \\ \boldsymbol{\theta}'_{(g)} \\ \vdots \\ \boldsymbol{\theta}'_{(k)} \end{bmatrix} \tag{8.4.3}$$

where $\boldsymbol{\theta}_i$ is the ith column vector and $\boldsymbol{\theta}'_{(g)}$ is the gth row vector of $\boldsymbol{\theta}$. For simplicity, we shall assume throughout the chapter that the rank of $\mathbf{X}$ is k.

8.4.1 Distribution of θ

Consider the elements of the $m\times m$ matrix $\mathbf{S}(\boldsymbol{\theta}) = \{S_{ij}(\boldsymbol{\theta}_i, \boldsymbol{\theta}_j)\}$ of (8.2.4). When $\mathbf{X}_1 = \ldots = \mathbf{X}_m = \mathbf{X}$, we can write

$$\begin{aligned} S_{ij}(\boldsymbol{\theta}_i,\boldsymbol{\theta}_j) &= (\mathbf{y}_i - \mathbf{X}\boldsymbol{\theta}_i)'\,(\mathbf{y}_j - \mathbf{X}\boldsymbol{\theta}_j) \\ &= (\mathbf{y}_i - \mathbf{X}\hat{\boldsymbol{\theta}}_i)'\,(\mathbf{y}_j - \mathbf{X}\hat{\boldsymbol{\theta}}_j) + (\boldsymbol{\theta}_i - \hat{\boldsymbol{\theta}}_i)'\,\mathbf{X}'\mathbf{X}(\boldsymbol{\theta}_j - \hat{\boldsymbol{\theta}}_j) \end{aligned} \tag{8.4.4}$$

where $\hat{\boldsymbol{\theta}}_i = (\mathbf{X}'\mathbf{X})^{-1}\mathbf{X}'\mathbf{y}_i$ is the least squares estimates of $\boldsymbol{\theta}_i, i = 1, \ldots, m$.

Consequently,

$$\mathbf{S}(\boldsymbol{\theta}) = \mathbf{A} + (\boldsymbol{\theta} - \hat{\boldsymbol{\theta}})' \mathbf{X}'\mathbf{X}(\boldsymbol{\theta} - \hat{\boldsymbol{\theta}}), \tag{8.4.5}$$

where $\hat{\boldsymbol{\theta}}$ is the $k \times m$ matrix of least squares estimates

$$\hat{\boldsymbol{\theta}} = [\hat{\boldsymbol{\theta}}_1, \ldots, \hat{\boldsymbol{\theta}}_i, \ldots, \hat{\boldsymbol{\theta}}_m] = \begin{bmatrix} \hat{\boldsymbol{\theta}}'_{(1)} \\ \vdots \\ \hat{\boldsymbol{\theta}}'_{(g)} \\ \vdots \\ \hat{\boldsymbol{\theta}}'_{(k)} \end{bmatrix}, \tag{8.4.6}$$

and $\mathbf{A}$ is the $m \times m$ matrix

$$\mathbf{A} = \{a_{ij}\}$$

with

$$a_{ij} = (\mathbf{y}_i - \mathbf{X}\hat{\boldsymbol{\theta}}_i)' (\mathbf{y}_j - \mathbf{X}\hat{\boldsymbol{\theta}}_j), \qquad i,j = 1, \ldots, m, \tag{8.4.7}$$

that is, $\mathbf{A}$ is proportional to the sample covariance matrix. For simplicity, we shall assume that $\mathbf{A}$ is positive definite. From the general result in (8.2.23), the posterior distribution of $\boldsymbol{\theta}$ is then

$$p(\boldsymbol{\theta} \mid \mathbf{y}) \propto |\mathbf{A} + (\boldsymbol{\theta} - \hat{\boldsymbol{\theta}})' \mathbf{X}'\mathbf{X}(\boldsymbol{\theta} - \hat{\boldsymbol{\theta}})|^{-n/2}, \qquad -\infty < \boldsymbol{\theta} < \infty. \tag{8.4.8}$$

As mentioned earlier, when there is a single output ($m = 1$), (8.4.8) is in the form of a k-dimensional multivariate t distribution. The distribution in (8.4.8) is a matric-variate generalization of the t distribution. It was first obtained by Kahirsagar (1960). A comprehensive discussion of its properties has been given by Dickey (1967b).

8.4.2 Posterior Distribution of the Means from a m-dimensional Normal Distribution

In the case $k = 1$ where each $\boldsymbol{\theta}_i$ consists of a single element and $\mathbf{X}$ is a $n \times 1$ vector of ones, expression (8.4.8) is the joint posterior distribution of the m means when sampling from an m-dimensional multivariate Normal distribution $N_m(\boldsymbol{\theta}, \boldsymbol{\Sigma})$. In this case

$$\boldsymbol{\theta} = (\theta_1, \ldots, \theta_i, \ldots, \theta_m), \quad \hat{\boldsymbol{\theta}} = (\bar{y}_1, \ldots, \bar{y}_i, \ldots, \bar{y}_m),$$

$$\mathbf{X}'\mathbf{X} = n \quad \text{and} \quad a_{ij} = \sum_{u=1}^{n} (y_{ui} - \bar{y}_i)(y_{uj} - \bar{y}_j), \tag{8.4.9}$$

where

$$\bar{y}_i = \frac{1}{n} \sum_{u=1}^{n} y_{ui}.$$

The posterior distribution of $\boldsymbol{\theta}$ can be written

$$p(\boldsymbol{\theta} \mid \mathbf{y}) \propto |\mathbf{A} + n(\boldsymbol{\theta} - \hat{\boldsymbol{\theta}})' (\boldsymbol{\theta} - \hat{\boldsymbol{\theta}})|^{-n/2}$$

$$\propto |\mathbf{I} + n\,\mathbf{A}^{-1} (\boldsymbol{\theta} - \hat{\boldsymbol{\theta}})' (\boldsymbol{\theta} - \hat{\boldsymbol{\theta}})|^{-n/2}, \qquad -\infty < \boldsymbol{\theta} < \infty. \tag{8.4.10}$$

We now make use of the fundamental identity

$$|\mathbf{I}_k - \mathbf{P}\,\mathbf{Q}| = |\mathbf{I}_l - \mathbf{Q}\,\mathbf{P}|, \tag{8.4.11}$$

where $\mathbf{I}_k$ and $\mathbf{I}_l$ are, respectively, $k \times k$ and a $l \times l$ identity matrices, $\mathbf{P}$ is a $k \times l$ matrix and $\mathbf{Q}$ is a $l \times k$ matrix. Noting that $(\boldsymbol{\theta} - \hat{\boldsymbol{\theta}})$ is a $1 \times m$ vector, we immediately obtain

$$p(\boldsymbol{\theta} \mid \mathbf{y}) \propto [1 + n(\boldsymbol{\theta} - \hat{\boldsymbol{\theta}})\,\mathbf{A}^{-1} (\boldsymbol{\theta} - \hat{\boldsymbol{\theta}})']^{-n/2}, \qquad -\infty < \theta_i < \infty, \quad i = 1, ..., m, \tag{8.4.12}$$

which is a m-dimensional $t_m[\hat{\boldsymbol{\theta}}', n^{-1}(n-m)^{-1}\mathbf{A}, n-m]$ distribution, a result first published by Geisser and Cornfield (1963). Thus, by comparing (8.4.12) with (8.3.13), we see that both when $m = 1$ and when $k = 1$, the distribution in (8.4.8) can be put in the multivariate t form.

8.4.3 Some Properties of the Posterior Matric-variate t Distribution of $\boldsymbol{\theta}$

When neither m nor k is equal to one, it is not possible to express the distribution of $\boldsymbol{\theta}$ as a multivariate t distribution. As we have mentioned, the distribution in (8.4.8) can be thought of as a matric-variate extension of the t distribution. We now discuss some properties of this distribution.

Two equivalent Representations of the Distribution of $\boldsymbol{\theta}$

It is shown in Appendix A8.3 that, for $\nu > 0$,

$$\int_{-\infty<\boldsymbol{\theta}<\infty} |\mathbf{I}_m + \mathbf{A}^{-1}(\boldsymbol{\theta} - \hat{\boldsymbol{\theta}})'\,\mathbf{X}'\mathbf{X}(\boldsymbol{\theta} - \hat{\boldsymbol{\theta}})|^{-\frac{1}{2}(\nu+k+m-1)}\,d\boldsymbol{\theta}$$

$$= c(m, k, \nu)\,|\mathbf{X}'\mathbf{X}|^{-m/2}\,|\mathbf{A}^{-1}|^{-k/2}, \tag{8.4.13}$$

where

$$\nu = n - (k + m) + 1,$$

$$c(m, k, \nu) = [\Gamma(\tfrac{1}{2})]^{mk}\,\frac{\Gamma_m[\frac{1}{2}(\nu + m - 1)]}{\Gamma_m[\frac{1}{2}(\nu + k + m - 1)]} \tag{8.4.14}$$

and $\Gamma_p(b)$ is the generalized Gamma function defined in (8.2.22). Thus,

$$p(\boldsymbol{\theta} \mid \mathbf{y}) = [c(m,k,\nu)]^{-1}\,|\mathbf{X}'\mathbf{X}|^{m/2}\,|\mathbf{A}^{-1}|^{k/2}\,|\mathbf{I}_m + \mathbf{A}^{-1}(\boldsymbol{\theta} - \hat{\boldsymbol{\theta}})'\,\mathbf{X}'\mathbf{X}(\boldsymbol{\theta} - \hat{\boldsymbol{\theta}})|^{-\frac{1}{2}(\nu+k+m-1)},$$

$$-\infty < \boldsymbol{\theta} < \infty. \tag{8.4.15}$$

and we shall say that the $k \times m$ matrix of parameters $\boldsymbol{\theta}$ is distributed as $t_{km}[\hat{\boldsymbol{\theta}}, (\mathbf{X}'\mathbf{X})^{-1}, \mathbf{A}, \nu]$. Note that by applying the identity (8.4.11), we can write

$$|\mathbf{I}_m + \mathbf{A}^{-1}(\boldsymbol{\theta} - \hat{\boldsymbol{\theta}})'\,\mathbf{X}'\mathbf{X}(\boldsymbol{\theta} - \hat{\boldsymbol{\theta}})| = |\mathbf{I}_k + (\mathbf{X}'\mathbf{X})\,(\boldsymbol{\theta} - \hat{\boldsymbol{\theta}})\mathbf{A}^{-1}\,(\boldsymbol{\theta} - \hat{\boldsymbol{\theta}})'| \quad (8.4.16)$$

so that in terms of the $m \times k$ matrix $\boldsymbol{\theta}'$ the roles of m and k on the one hand and of the matrices $(\mathbf{X}'\mathbf{X})^{-1}$ and $\mathbf{A}$ on the other are simultaneously interchanged. Thus, we may conclude that if

$$\boldsymbol{\theta} \sim t_{km}[\hat{\boldsymbol{\theta}}, (\mathbf{X}'\mathbf{X})^{-1}, \mathbf{A}, \nu],$$

then

$$\boldsymbol{\theta}' \sim t_{mk}[\hat{\boldsymbol{\theta}}', \mathbf{A}, (\mathbf{X}'\mathbf{X})^{-1}, \nu]. \quad (8.4.17)$$

It follows from these two equivalent representations that

$$c(k, m, \nu) \equiv c(m, k, \nu),$$

that is,

$$\frac{\Gamma_k\,[\frac{1}{2}(\nu + k - 1)]}{\Gamma_k\,[\frac{1}{2}(\nu + k + m - 1)]} \equiv \frac{\Gamma_m\,[\frac{1}{2}(\nu + m - 1)]}{\Gamma_m\,[\frac{1}{2}(\nu + k + m - 1)]}. \quad (8.4.18)$$

Marginal and Conditional Distributions of Subsets of Columns of $\boldsymbol{\theta}$

We now show that the marginal and conditional distributions of subsets of the m columns of $\boldsymbol{\theta}$ are also matric-variate t distributions. Let $m = m_1 + m_2$ and partition the matrices $\boldsymbol{\theta}$, $\hat{\boldsymbol{\theta}}$, and $\mathbf{A}$ into

$$\boldsymbol{\theta} = \overset{\quad m_1 \qquad m_2}{[\boldsymbol{\theta}_{1*} \;\vdots\; \boldsymbol{\theta}_{2*}\;]}\,{}_k, \qquad \hat{\boldsymbol{\theta}} = \overset{\quad m_1 \qquad m_2}{[\hat{\boldsymbol{\theta}}_{1*} \;\vdots\; \hat{\boldsymbol{\theta}}_{2*}\;]}\,{}_k,$$

$$\mathbf{A} = \overset{\quad m_1 \qquad m_2}{\left[\begin{array}{c:c} \mathbf{A}_{11} & \mathbf{A}_{12} \\ \hdashline \mathbf{A}_{21} & \mathbf{A}_{22} \end{array}\right]}\begin{matrix} m_1 \\ m_2 \end{matrix}. \quad (8.4.19)$$

Then:

a) conditional on $\boldsymbol{\theta}_{1*}$, the subset $\boldsymbol{\theta}_{2*}$ is distributed as

$$\boldsymbol{\theta}_{2*} \sim t_{km_2}(\tilde{\boldsymbol{\theta}}_{2*}, \mathbf{H}^{-1}, \mathbf{A}_{22\cdot 1}, \nu + m_1), \quad (8.4.20)$$

where

$$\mathbf{H}^{-1} = (\mathbf{X}'\mathbf{X})^{-1} + (\boldsymbol{\theta}_{1*} - \hat{\boldsymbol{\theta}}_{1*})\mathbf{A}_{11}^{-1}\,(\boldsymbol{\theta}_{1*} - \hat{\boldsymbol{\theta}}_{1*})',$$

$$\tilde{\boldsymbol{\theta}}_{2*} = \hat{\boldsymbol{\theta}}_{2*} + (\boldsymbol{\theta}_{1*} - \hat{\boldsymbol{\theta}}_{1*})\mathbf{A}_{11}^{-1}\,\mathbf{A}_{12},$$

$$\mathbf{A}_{22\cdot 1} = \mathbf{A}_{22} - \mathbf{A}_{21}\,\mathbf{A}_{11}^{-1}\,\mathbf{A}_{12}.$$

b) $\boldsymbol{\theta}_{1*}$ is distributed as

$$\boldsymbol{\theta}_{1*} \sim t_{km_1}\,[\hat{\boldsymbol{\theta}}_{1*}, (\mathbf{X}'\mathbf{X})^{-1}, \mathbf{A}_{11}, \nu]. \quad (8.4.21)$$

To prove these results, we can write

$$(\boldsymbol{\theta} - \hat{\boldsymbol{\theta}})\mathbf{A}^{-1}(\boldsymbol{\theta} - \hat{\boldsymbol{\theta}})' = (\boldsymbol{\theta}_{1*} - \hat{\boldsymbol{\theta}}_{1*})\mathbf{A}_{11}^{-1}(\boldsymbol{\theta}_{1*} - \hat{\boldsymbol{\theta}}_{1*})'$$
$$+ (\boldsymbol{\theta}_{2*} - \tilde{\boldsymbol{\theta}}_{2*})\mathbf{A}_{22\cdot 1}^{-1}(\boldsymbol{\theta}_{2*} - \tilde{\boldsymbol{\theta}}_{2*})'. \quad (8.4.22)$$

The determinant on the right-hand side of (8.4.16) can now be written

$$|\mathbf{I}_k + (\mathbf{X}'\mathbf{X})(\boldsymbol{\theta} - \hat{\boldsymbol{\theta}})\mathbf{A}^{-1}(\boldsymbol{\theta} - \hat{\boldsymbol{\theta}})'| = |\mathbf{I}_k + (\mathbf{X}'\mathbf{X})(\boldsymbol{\theta}_{1*} - \hat{\boldsymbol{\theta}}_{1*})\mathbf{A}_{11}^{-1}(\boldsymbol{\theta}_{1*} - \hat{\boldsymbol{\theta}}_{1*})'|$$
$$\times |\mathbf{I}_k + \mathbf{H}(\boldsymbol{\theta}_{2*} - \tilde{\boldsymbol{\theta}}_{2*})\mathbf{A}_{22\cdot 1}^{-1}(\boldsymbol{\theta}_{2*} - \tilde{\boldsymbol{\theta}}_{2*})'|. \qquad (8.4.23)$$

Substituting (8.4.23) into (8.4.15), we see that, given $\boldsymbol{\theta}_{1*}$, the conditional distribution of $\boldsymbol{\theta}_{2*}$ is

$$p(\boldsymbol{\theta}_{2*} | \boldsymbol{\theta}_{1*}, \mathbf{y}) \propto |\mathbf{I}_k + \mathbf{H}(\boldsymbol{\theta}_{2*} - \tilde{\boldsymbol{\theta}}_{2*})\mathbf{A}_{22\cdot 1}^{-1}(\boldsymbol{\theta}_{2*} - \tilde{\boldsymbol{\theta}}_{2*})'|^{-\frac{1}{2}(v+k+m-1)}$$
$$\propto |\mathbf{I}_{m_2} + \mathbf{A}_{22\cdot 1}^{-1}(\boldsymbol{\theta}_{2*} - \tilde{\boldsymbol{\theta}}_{2*})'\mathbf{H}(\boldsymbol{\theta}_{2*} - \tilde{\boldsymbol{\theta}}_{2*})|^{-\frac{1}{2}[(v+m_1)+k+m_2-1]}$$
$$-\infty < \boldsymbol{\theta}_{2*} < \infty. \qquad (8.4.24)$$

From (8.4.13), the normalizing constant is

$$[c(m_2, k, v + m_1)]^{-1} |\mathbf{H}|^{m_2/2} |\mathbf{A}_{22\cdot 1}^{-1}|^{k/2}. \qquad (8.4.25)$$

Thus, given $\boldsymbol{\theta}_{1*}$, the $k \times m_2$ matrix of parameters $\boldsymbol{\theta}_{2*}$ is distributed as $t_{km_2}(\tilde{\boldsymbol{\theta}}_{2*}, \mathbf{H}^{-1}, \mathbf{A}_{22\cdot 1}, v + m_1)$. For the marginal distribution of $\boldsymbol{\theta}_{1*}$, since

$$p(\boldsymbol{\theta}_{1*} | \mathbf{y}) = \frac{p(\boldsymbol{\theta} | \mathbf{y})}{p(\boldsymbol{\theta}_{2*} | \boldsymbol{\theta}_{1*}, \mathbf{y})},$$

use of (8.4.23) through (8.4.25) yields

$$p(\boldsymbol{\theta}_{1*} | \mathbf{y}) \propto |\mathbf{H}|^{-m_2/2} |\mathbf{I}_k + (\mathbf{X}'\mathbf{X})(\boldsymbol{\theta}_{1*} - \hat{\boldsymbol{\theta}}_{1*})\mathbf{A}_{11}^{-1}(\boldsymbol{\theta}_{1*} - \hat{\boldsymbol{\theta}}_{1*})'|^{-\frac{1}{2}(v+k+m-1)}$$
$$\propto |\mathbf{I}_{m_1} + \mathbf{A}_{11}^{-1}(\boldsymbol{\theta}_{1*} - \hat{\boldsymbol{\theta}}_{1*})'\mathbf{X}'\mathbf{X}(\boldsymbol{\theta}_{1*} - \hat{\boldsymbol{\theta}}_{1*})|^{-\frac{1}{2}(v+k+m_1-1)},$$
$$-\infty < \boldsymbol{\theta}_{1*} < \infty. \qquad (8.4.26)$$

That is, the $k \times m_1$ matrix of parameters $\boldsymbol{\theta}_{1*}$ is distributed as

$$t_{km_1}[\hat{\boldsymbol{\theta}}_{1*}, (\mathbf{X}'\mathbf{X})^{-1}, \mathbf{A}_{11}, v].$$

Marginal Distribution of a Particular Column of $\boldsymbol{\theta}$

In particular, by setting $m_1 = 1$ in (8.4.21), the marginal distribution of $\boldsymbol{\theta}_{1*} = \boldsymbol{\theta}_1$ is the k-dimensional multivariate t distribution

$$p(\boldsymbol{\theta}_1 | \mathbf{y}) \propto [1 + a_{11}^{-1}(\boldsymbol{\theta}_1 - \hat{\boldsymbol{\theta}}_1)'\mathbf{X}'\mathbf{X}(\boldsymbol{\theta}_1 - \hat{\boldsymbol{\theta}}_1)]^{-\frac{1}{2}(v+k)}, \qquad -\infty < \boldsymbol{\theta}_1 < \infty, \qquad (8.4.27)$$

where $a_{11} = \mathbf{A}_{11}$ is now a scalar, that is, $\boldsymbol{\theta}_1 \sim t_k[\hat{\boldsymbol{\theta}}_1, v^{-1} a_{11} (\mathbf{X}'\mathbf{X})^{-1}, v]$.

By mere relabeling we may conclude that the marginal distribution of the ith column of $\boldsymbol{\theta}$ in (8.4.2) is

$$p(\boldsymbol{\theta}_i | \mathbf{y}) \propto [1 + a_{ii}^{-1}(\boldsymbol{\theta}_i - \hat{\boldsymbol{\theta}}_i)'\mathbf{X}'\mathbf{X}(\boldsymbol{\theta}_i - \hat{\boldsymbol{\theta}}_i)]^{-\frac{1}{2}(v+k)}, \qquad -\infty < \boldsymbol{\theta}_i < \infty. \qquad (8.4.28)$$

It will be noted that this distribution is identical to that obtained in (8.3.13) when only a single output was considered, except that in (8.3.13)

$\nu = n - k$, but in (8.4.28) $\nu = n - k - (m - 1)$. In a certain sense, the reduction of the degrees of freedom by $m - 1$ is not surprising. In adopting the multivariate framework, $m(m-1)/2$ additional parameters σ_{ij} $(i \neq j)$ are introduced. A part of the information from the sample is therefore utilized to estimate these parameters and $(m - 1)$ of them $(\sigma_{i1}, \ldots, \sigma_{i(i-1)}, \sigma_{i(i+1)}, \ldots, \sigma_{im})$ are connected with $\mathbf{y}_i$. We may say that 'one degree of freedom is lost' for each of the $(m - 1)$ additional parameters.

On the other hand, it is somewhat puzzling that if we ignored the multivariate structure of the problem and treated $\mathbf{y}_i$ as the output of a *univariate* response, then on the basis of the noninformative prior $p(\boldsymbol{\theta}_i, \sigma_{ii}) \propto \sigma_{ii}^{-1}$, we would obtain a posterior multivariate t distribution for $\boldsymbol{\theta}_i$ with $(m - 1)$ additional degrees of freedom. This would seem to imply that, by ignoring the information from the other $(m - 1)$ responses $\mathbf{y}_1, \ldots, \mathbf{y}_{i-1}, \mathbf{y}_{i+1}, \ldots, \mathbf{y}_m$, more precise inference about $\boldsymbol{\theta}_i$ could be made than when all the m responses were considered jointly. This phenomenon is related to the "paradox" pointed out by Dempster (1963) and the criticisms of the prior in (8.2.14) by Stone (1964).

The above implication is admittedly perplexing, and further research is needed to clarify the situation. We feel, however, that the multivariate results presented in this chapter are of considerable interest and certainly provide a sensible basis for inference in the common practical situation when $(n - k)$ is large relative to m.

Distribution of $\boldsymbol{\theta}$ *Expressed as a Product of Multivariate t Distributions*

We note that for the partition in (8.4.19), if we set $m_1 = m - 1$ and $m_2 = 1$, then from (8.4.20) the conditional distribution of $\boldsymbol{\theta}_{2*} = \boldsymbol{\theta}_m$, given $\boldsymbol{\theta}_{1*} = [\boldsymbol{\theta}_1, \ldots, \boldsymbol{\theta}_{m-1}]$, is

$$\boldsymbol{\theta}_m \sim t_k[\tilde{\boldsymbol{\theta}}_m, (\nu + m - 1)^{-1} a_{m\cdot 12\ldots(m-1)} \mathbf{H}^{-1}, \nu + m - 1], \qquad (8.4.29)$$

where

$$a_{m\cdot 12\ldots(m-1)} = \mathbf{A}_{22\cdot 1}.$$

From the marginal distribution of $\boldsymbol{\theta}_{1*} = [\boldsymbol{\theta}_1, \ldots, \boldsymbol{\theta}_{m-1}]$ in (8.4.21) if we partition $\boldsymbol{\theta}_{1*}$ into $[\boldsymbol{\theta}_{s*} \,\vdots\, \boldsymbol{\theta}_{m-1}]$ where $\boldsymbol{\theta}_{s*} = [\boldsymbol{\theta}_1, \ldots, \boldsymbol{\theta}_{m-2}]$, it is clear that the conditional distribution of $\boldsymbol{\theta}_{m-1}$, given $\boldsymbol{\theta}_{s*}$, is again a k-dimensional multivariate t distribution. It follows by repeating the process $m - 1$ times that, if we express $p(\boldsymbol{\theta} \mid \mathbf{y})$ as the product

$$p(\boldsymbol{\theta} \mid \mathbf{y}) = p(\boldsymbol{\theta}_1 \mid \mathbf{y}) p(\boldsymbol{\theta}_2 \mid \boldsymbol{\theta}_1, \mathbf{y}) \cdots p(\boldsymbol{\theta}_m \mid \boldsymbol{\theta}_1, \ldots, \boldsymbol{\theta}_{m-1}, \mathbf{y}), \qquad (8.4.30)$$

then each factor on the right-hand side is a k-dimensional multivariate t distribution.

Marginal and Conditional Distributions of Rows of $\boldsymbol{\theta}$

Results very similar to those given above for column decomposition of $\boldsymbol{\theta}$ can now be obtained for the rows of $\boldsymbol{\theta}$. Consider the partitions

$$\boldsymbol{\theta}' = \underset{\quad k_1 \qquad\quad k_2}{[\boldsymbol{\theta}_{(1)*} \,\vdots\, \boldsymbol{\theta}_{(2)*}]}\, m, \qquad \hat{\boldsymbol{\theta}}' = \underset{\quad k_1 \qquad\quad k_2}{[\hat{\boldsymbol{\theta}}_{(1)*} \,\vdots\, \hat{\boldsymbol{\theta}}_{(2)*}]}\, m \tag{8.4.31}$$

$$(\mathbf{X}'\mathbf{X})^{-1} = \mathbf{C} = \begin{bmatrix} \mathbf{C}_{11} & \mathbf{C}_{12} \\ \mathbf{C}_{21} & \mathbf{C}_{22} \end{bmatrix} \begin{matrix} k_1 \\ k_2 \end{matrix}, \qquad k_1 + k_2 = k,$$

where it is to be remembered that $\boldsymbol{\theta}'_{(1)*}$ are the first k_1 rows and $\boldsymbol{\theta}'_{(2)*}$ the remaining k_2 rows of $\boldsymbol{\theta}$. Since the $m \times k$ matrix $\boldsymbol{\theta}'$ is distributed as $t_{mk}[\hat{\boldsymbol{\theta}}', \mathbf{A}, (\mathbf{X}'\mathbf{X})^{-1}, \nu]$, it can be readily shown that

a) given $\boldsymbol{\theta}_{(1)*}$,

$$\boldsymbol{\theta}_{(2)*} \sim t_{mk_2}(\tilde{\boldsymbol{\theta}}_{(2)*}, \mathbf{G}, \mathbf{C}_{22\cdot1}, \nu + k_1), \tag{8.4.32}$$

where

$$\mathbf{C}_{22\cdot1} = \mathbf{C}_{22} - \mathbf{C}_{21}\,\mathbf{C}_{11}^{-1}\,\mathbf{C}_{12},$$

$$\tilde{\boldsymbol{\theta}}_{(2)*} = \hat{\boldsymbol{\theta}}_{(2)*} + (\boldsymbol{\theta}_{(1)*} - \hat{\boldsymbol{\theta}}_{(1)*})\,\mathbf{C}_{11}^{-1}\,\mathbf{C}_{12},$$

and

$$\mathbf{G} = \mathbf{A} + (\boldsymbol{\theta}_{(1)*} - \hat{\boldsymbol{\theta}}_{(1)*})\,\mathbf{C}_{11}^{-1}\,(\boldsymbol{\theta}_{(1)*} - \hat{\boldsymbol{\theta}}_{(1)*})'.$$

b) Marginally,

$$\boldsymbol{\theta}_{(1)*} \sim t_{mk_1}[\hat{\boldsymbol{\theta}}_{(1)*}, \mathbf{A}, \mathbf{C}_{11}, \nu] \tag{8.4.33}$$

or equivalently,

$$\boldsymbol{\theta}'_{(1)*} \sim t_{k_1 m}[\hat{\boldsymbol{\theta}}'_{(1)*}, \mathbf{C}_{11}, \mathbf{A}, \nu].$$

c) The gth row of $\boldsymbol{\theta}$, $\boldsymbol{\theta}'_{(g)}$, is distributed as

$$\boldsymbol{\theta}_{(g)} \sim t_m[\hat{\boldsymbol{\theta}}_{(g)}, \nu^{-1} c_{gg}\,\mathbf{A}, \nu] \tag{8.4.34}$$

where c_{gg} is the (gg)th element of $\mathbf{C}$.

d) The distribution of $\boldsymbol{\theta}$ can alternatively be expressed as the product

$$p(\boldsymbol{\theta} \mid \mathbf{y}) = p(\boldsymbol{\theta}_{(1)} \mid \mathbf{y})\, p(\boldsymbol{\theta}_{(2)} \mid \boldsymbol{\theta}_{(1)}, \mathbf{y}) \cdots p(\boldsymbol{\theta}_{(k)} \mid \boldsymbol{\theta}_{(1)}, \ldots, \boldsymbol{\theta}_{(k-1)}, \mathbf{y}), \tag{8.4.35}$$

where, parallel to (8.4.30), each factor on the right-hand side is an m-dimensional multivariate t distribution.

Comparing expression (8.4.34) with the result in (8.4.12), we see that, as was the case with a column vector of $\boldsymbol{\theta}$, the two distributions are of the same form except for the difference in the "degrees of freedom." They now differ by $(k - 1)$, simply because an additional $(k - 1)$ "input variables" are included in the model.

Marginal Distribution of a Block of Elements of $\boldsymbol{\theta}$

Finally consider now the partitions

$$\boldsymbol{\theta} = \overset{\qquad m_1 \qquad\quad m_2}{\begin{bmatrix} \boldsymbol{\theta}_{11} & \boldsymbol{\theta}_{12} \\ \boldsymbol{\theta}_{21} & \boldsymbol{\theta}_{22} \end{bmatrix}} \begin{matrix} k_1 \\ k_2 \end{matrix}, \qquad \hat{\boldsymbol{\theta}} = \overset{\qquad m_1 \qquad\quad m_2}{\begin{bmatrix} \hat{\boldsymbol{\theta}}_{11} & \hat{\boldsymbol{\theta}}_{12} \\ \hat{\boldsymbol{\theta}}_{21} & \hat{\boldsymbol{\theta}}_{22} \end{bmatrix}} \begin{matrix} k_1 \\ k_2 \end{matrix}. \tag{8.4.36}$$

It follows from (8.4.21) and (8.4.33) that the $k_1 \times m_1$ matrix of parameters $\boldsymbol{\theta}_{11}$ is distributed as

$$\boldsymbol{\theta}_{11} \sim t_{k_1 m_1}(\hat{\boldsymbol{\theta}}_{11}, \mathbf{C}_{11}, \mathbf{A}_{11}, \nu). \tag{8.4.37}$$

The marginal distributions of $\boldsymbol{\theta}_{12}$, $\boldsymbol{\theta}_{21}$, $\boldsymbol{\theta}_{22}$, and indeed that of any *block* of elements of $\boldsymbol{\theta}$ can be similarly obtained, and are left to the reader.

In the above we have shown that the marginal and the conditional distributions of certain subsets of $\boldsymbol{\theta}$ are of the matric-variate t form. This is, however, not true in general. For example, one can show that neither the marginal distribution of $(\boldsymbol{\theta}_{11}, \boldsymbol{\theta}_{22})$ nor the conditional distribution of $(\boldsymbol{\theta}_{12}, \boldsymbol{\theta}_{21})$ given $(\boldsymbol{\theta}_{11}, \boldsymbol{\theta}_{22})$ is a matric-variate t distribution. The problem of obtaining explicit expressions for the marginal and the conditional distributions in general is quite complex, and certain special cases have recently been considered by Dreze and Morales (1970) and Tiao, Tan, and Chang (1970).

Means and Covariance Matrix of $\boldsymbol{\theta}$

From (8.4.28), the matrix of means of the posterior distribution of $\boldsymbol{\theta}$ is

$$E(\boldsymbol{\theta}) = \hat{\boldsymbol{\theta}} \tag{8.4.38}$$

and the covariance matrix of $\boldsymbol{\theta}_i$ is

$$\text{Cov}(\boldsymbol{\theta}_i) = \frac{a_{ii}}{\nu - 2}(\mathbf{X}'\mathbf{X})^{-1}, \qquad i = 1, \ldots, m. \tag{8.4.39}$$

For the covariance matrix of $\boldsymbol{\theta}_i$ and $\boldsymbol{\theta}_j$, with no loss in generality we consider the case $i = 1$ and $j = 2$. Now

$$E(\boldsymbol{\theta}_1 - \hat{\boldsymbol{\theta}}_1)(\boldsymbol{\theta}_2 - \hat{\boldsymbol{\theta}}_2)' = \underset{\boldsymbol{\theta}_1}{E}(\boldsymbol{\theta}_1 - \hat{\boldsymbol{\theta}}_1) \underset{\boldsymbol{\theta}_2|\boldsymbol{\theta}_1}{E}(\boldsymbol{\theta}_2 - \hat{\boldsymbol{\theta}}_2)'.$$

If we set $m_1 = 2$ in (8.4.21) and perform a column decomposition of the $k \times 2$ matrix $\boldsymbol{\theta}_{1*} = [\boldsymbol{\theta}_1, \boldsymbol{\theta}_2]$, it is then clear from (8.4.20) that

$$\underset{\boldsymbol{\theta}_2|\boldsymbol{\theta}_1}{E}(\boldsymbol{\theta}_2 - \hat{\boldsymbol{\theta}}_2) = a_{11}^{-1} a_{12}(\boldsymbol{\theta}_1 - \hat{\boldsymbol{\theta}}_1)$$

so that, as might be expected,

$$E(\boldsymbol{\theta}_1 - \hat{\boldsymbol{\theta}}_1)(\boldsymbol{\theta}_2 - \hat{\boldsymbol{\theta}}_2)' = \frac{a_{12}}{\nu - 2}(\mathbf{X}'\mathbf{X})^{-1}. \tag{8.4.40}$$

Thus,

$$E\begin{bmatrix} \boldsymbol{\theta}_1 - \hat{\boldsymbol{\theta}}_1 \\ \boldsymbol{\theta}_2 - \hat{\boldsymbol{\theta}}_2 \end{bmatrix}[(\boldsymbol{\theta}_1 - \hat{\boldsymbol{\theta}}_1)', (\boldsymbol{\theta}_2 - \hat{\boldsymbol{\theta}}_2)'] = \frac{1}{\nu - 2}\begin{bmatrix} a_{11}(\mathbf{X}'\mathbf{X})^{-1} & a_{12}(\mathbf{X}'\mathbf{X})^{-1} \\ a_{12}(\mathbf{X}'\mathbf{X})^{-1} & a_{22}(\mathbf{X}'\mathbf{X})^{-1} \end{bmatrix}$$

$$= \frac{1}{\nu - 2}\begin{bmatrix} a_{11} & a_{12} \\ a_{12} & a_{22} \end{bmatrix} \otimes (\mathbf{X}'\mathbf{X})^{-1} \tag{8.4.41}$$

where $\otimes$ denotes the Kronecker product—see Appendix A8.4. In general, if we write the elements of $\boldsymbol{\theta}$ and $\hat{\boldsymbol{\theta}}$ as

$$\boldsymbol{\Theta}' = (\boldsymbol{\theta}_1', \ldots, \boldsymbol{\theta}_m'), \qquad \hat{\boldsymbol{\Theta}}' = (\hat{\boldsymbol{\theta}}_1', \ldots, \hat{\boldsymbol{\theta}}_m'), \tag{8.4.42a}$$

where $\boldsymbol{\Theta}$ and $\hat{\boldsymbol{\Theta}}$ are $km \times 1$ vectors, then

$$\mathrm{Cov}\,(\boldsymbol{\Theta}) = E(\boldsymbol{\Theta} - \hat{\boldsymbol{\Theta}})\,(\boldsymbol{\Theta} - \hat{\boldsymbol{\Theta}})' = \frac{1}{\nu - 2}\,\mathbf{A} \otimes (\mathbf{X}'\mathbf{X})^{-1}. \tag{8.4.42b}$$

By a similar argument, if we write

$$\boldsymbol{\Theta}_*' = (\boldsymbol{\theta}_{(1)}', \ldots, \boldsymbol{\theta}_{(k)}'), \qquad \hat{\boldsymbol{\Theta}}_* = (\hat{\boldsymbol{\theta}}_{(1)}', \ldots, \hat{\boldsymbol{\theta}}_{(k)}'), \tag{8.4.43a}$$

then

$$\mathrm{Cov}\,(\boldsymbol{\Theta}_*) = \frac{1}{\nu - 2}\,(\mathbf{X}'\mathbf{X})^{-1} \otimes \mathbf{A}. \tag{8.4.43b}$$

Linear Transformation of $\boldsymbol{\theta}$

Let $\mathbf{P}$ be a $k_1 \times k$ $(k_1 \leqslant k)$ matrix of rank k_1 and $\mathbf{Q}$ be a $m \times m_1$ $(m_1 \leqslant m)$ matrix of rank m_1. Suppose $\boldsymbol{\phi}$ is the $k_1 \times m_1$ matrix of random variables obtained from the linear transformation

$$\boldsymbol{\phi} = \mathbf{P}\,\boldsymbol{\theta}\,\mathbf{Q}. \tag{8.4.44}$$

Then $\boldsymbol{\phi}$ is distributed as $t_{k_1 m_1}[\mathbf{P}\,\hat{\boldsymbol{\theta}}\,\mathbf{Q}, \mathbf{P}(\mathbf{X}'\mathbf{X})^{-1}\mathbf{P}', \mathbf{Q}'\mathbf{A}\mathbf{Q}, \nu]$. The proof is left as an exercise for the reader.

Asymptotic Distribution of $\boldsymbol{\theta}$

When ν tends to infinity, the distribution of $\boldsymbol{\theta}$ approaches a km dimensional multivariate Normal distribution,

$$\lim_{\nu \to \infty} p(\boldsymbol{\theta} \mid \mathbf{y}) = (\sqrt{2\pi})^{-mk} |\hat{\boldsymbol{\Sigma}}^{-1}|^{k/2}\, |\mathbf{X}'\mathbf{X}|^{m/2}$$

$$\times \exp\left[-\tfrac{1}{2}\,\mathrm{tr}\,\hat{\boldsymbol{\Sigma}}^{-1}\,(\boldsymbol{\theta} - \hat{\boldsymbol{\theta}})'\,\mathbf{X}'\mathbf{X}\,(\boldsymbol{\theta} - \hat{\boldsymbol{\theta}})\right], \qquad -\infty < \boldsymbol{\theta} < \infty, \tag{8.4.45}$$

where

$$\hat{\boldsymbol{\Sigma}} = \nu^{-1}\mathbf{A},$$

and we shall say that, asymptotically, $\boldsymbol{\theta} \sim N_{mk}[\hat{\boldsymbol{\theta}}, \hat{\boldsymbol{\Sigma}} \otimes (\mathbf{X}'\mathbf{X})^{-1}]$.

To see this, in (8.4.15) let

$$\mathbf{Q} = \nu\,\mathbf{A}^{-1}(\boldsymbol{\theta} - \hat{\boldsymbol{\theta}})'\,\mathbf{X}'\mathbf{X}(\boldsymbol{\theta} - \hat{\boldsymbol{\theta}}).$$

$$= \hat{\boldsymbol{\Sigma}}^{-1}\,(\boldsymbol{\theta} - \hat{\boldsymbol{\theta}})'\,\mathbf{X}'\mathbf{X}(\boldsymbol{\theta} - \hat{\boldsymbol{\theta}}).$$

Then, we may write

$$|\mathbf{I}_m + \nu^{-1}\mathbf{Q}| = \prod_{i=1}^{m} (1 + \nu^{-1}\lambda_i) \tag{8.4.46}$$

where $(\lambda_1, ..., \lambda_m)$ are the latent roots of $\mathbf{Q}$. Thus, as $\nu \to \infty$

$$\lim_{\nu\to\infty} |\mathbf{I}_m + \nu^{-1}\mathbf{Q}|^{-\frac{1}{2}(\nu+k+m-1)} = \exp\left(-\frac{1}{2}\sum_{i=l}^{m} \lambda_i\right)$$

$$= \exp(-\tfrac{1}{2}\operatorname{tr}\mathbf{Q}).$$

Since

$$\operatorname{tr}\mathbf{Q} = (\mathbf{\Theta} - \hat{\mathbf{\Theta}})'\,\hat{\mathbf{\Sigma}}^{-1} \otimes (\mathbf{X}'\mathbf{X})(\mathbf{\Theta} - \hat{\mathbf{\Theta}}) \tag{8.4.47}$$

where $\mathbf{\Theta}$ and $\hat{\mathbf{\Theta}}$ are the $km \times 1$ vectors defined in (8.4.42a), and noting that

$$|\hat{\mathbf{\Sigma}}^{-1} \otimes (\mathbf{X}'\mathbf{X})| = |\hat{\mathbf{\Sigma}}^{-1}|^{k/2}\,|\mathbf{X}'\mathbf{X}|^{m/2},$$

the desired result follows at once.

It follows that

$$E(\mathbf{\Theta}) = \hat{\mathbf{\Theta}}, \qquad \operatorname{Cov}(\mathbf{\Theta}) = \hat{\mathbf{\Sigma}} \otimes (\mathbf{X}'\mathbf{X})^{-1} \tag{8.4.48a}$$

or, alternatively,

$$E(\mathbf{\Theta}_*) = \hat{\mathbf{\Theta}}_*, \qquad \operatorname{Cov}(\mathbf{\Theta}_*) = (\mathbf{X}'\mathbf{X})^{-1} \otimes \hat{\mathbf{\Sigma}} \tag{8.4.48b}$$

where $(\mathbf{\Theta}, \hat{\mathbf{\Theta}})$ and $(\mathbf{\Theta}_*, \hat{\mathbf{\Theta}}_*)$ are defined in (8.4.42a) and (8.4.43a), respectively.

8.4.4 H.P.D. Regions of $\boldsymbol{\theta}$

Expressions (8.4.28) and (8.4.34) allow us to make inferences about a specific column or row of $\boldsymbol{\theta}$. Using properties of the multivariate t distribution, H.P.D. regions of the elements of a row or a column can be easily determined.

We now discuss a procedure for the complete set of parameters $\boldsymbol{\theta}$, which makes it possible to decide whether a general point $\boldsymbol{\theta} = \boldsymbol{\theta}_0$ is or is not included in an H.P.D. region of approximate content $(1 - \alpha)$.

It is seen in expression (8.4.15) that the posterior distribution of $\boldsymbol{\theta}$ is a monotonic increasing function of the quantity $U(\boldsymbol{\theta})$, where

$$U(\boldsymbol{\theta}) = \frac{|\mathbf{A}|}{|\mathbf{A} + (\boldsymbol{\theta} - \hat{\boldsymbol{\theta}})'\,\mathbf{X}'\mathbf{X}(\boldsymbol{\theta} - \hat{\boldsymbol{\theta}})|} = |\mathbf{I}_m\,\mathbf{A} + ^{-1}(\boldsymbol{\theta} - \hat{\boldsymbol{\theta}})'\,\mathbf{X}'\mathbf{X}\,(\boldsymbol{\theta} - \hat{\boldsymbol{\theta}})|^{-1} \tag{8.4.49}$$

Consequently, the parameter point $\boldsymbol{\theta} = \boldsymbol{\theta}_0$ lies inside the $(1 - \alpha)$ H.P.D. region if and only if

$$\Pr\{U(\boldsymbol{\theta}) > U(\boldsymbol{\theta}_0) \mid \mathbf{y}\} \leqslant (1 - \alpha), \tag{8.4.50}$$

8.4.5 Distribution of $U(\boldsymbol{\theta})$

To obtain the posterior distribution of $U(\boldsymbol{\theta})$ so that (8.4.50) can be calculated, we first derive the moments of $U(\boldsymbol{\theta})$. Applying the integral identity (8.4.13) the

hth moment of $U(\boldsymbol{\theta})$ is found to be

$$E[U(\boldsymbol{\theta})^h \mid \mathbf{y}] = \frac{c(m, k, v + 2h)}{c(m, k, v)}$$

$$= \prod_{s=1}^{m} \frac{\Gamma[\frac{1}{2}(v - 1 + s) + h]\,\Gamma[\frac{1}{2}(v - 1 + k + s)]}{\Gamma[\frac{1}{2}(v - 1 + s)]\,\Gamma[\frac{1}{2}(v - 1 + k + s) + h]} \qquad (8.4.51)$$

From (8.4.46) and (8.4.49) it follows that $U = U(\boldsymbol{\theta})$ is a random variable defined on the interval (0, 1) so that the distribution of U is uniquely determined by its moments. Further, expression (8.4.51) shows that distribution of U is a function of (m, k, v). Adopting the symbol $U_{(m,k,v)}$ to mean a random variable whose probability distribution is that implied by the moments in (8.4.51), we have the following general result.

Theorem 8.4.1 Let $\hat{\boldsymbol{\theta}}$ be a $k \times m$ matrix of constants, $\mathbf{X}'\mathbf{X}$ and $\mathbf{A}$ be, respectively, a $k \times k$ and a $m \times m$ positive definite symmetric matrix of constants, and $v > 0$. If the $k \times m$ matrix of random variables

$$\boldsymbol{\theta} \sim t_{km}[\hat{\boldsymbol{\theta}}, (\mathbf{X}'\mathbf{X})^{-1}, \mathbf{A}, v],$$

then

$$U(\boldsymbol{\theta}) \sim U_{(m,k,v)}$$

where

$$U(\boldsymbol{\theta}) = |\mathbf{I}_m + \mathbf{A}^{-1}(\boldsymbol{\theta} - \hat{\boldsymbol{\theta}})'\,\mathbf{X}'\mathbf{X}\,(\boldsymbol{\theta} - \hat{\boldsymbol{\theta}})|^{-1}.$$

As noted by Geisser (1965a), expression (8.4.51) correspond exactly to that for the hth sampling moment of $U(\boldsymbol{\theta}_0)$ in the sampling theory framework when $\boldsymbol{\theta}$ are regarded as fixed and $\mathbf{y}$ random variables. Thus, the Bayesian probability $\Pr\{U(\boldsymbol{\theta}) > U(\boldsymbol{\theta}_0) \mid \mathbf{y}\}$ is numerically equivalent to the significance level associated with the null hypothesis $\boldsymbol{\theta} = \boldsymbol{\theta}_0$ against the alternative $\boldsymbol{\theta} \neq \boldsymbol{\theta}_0$.

Some Distributional Properties of $U_{(m,k,v)}$.

Following the development, for example, in Anderson (1958), we now discuss some properties of the distribution of $U_{(m,k,v)}$. It will be noted that the notation $U_{(m,k,v)}$ here is slightly different from the one used in Anderson. Specifically, in his notation, v is replaced by $v + m - 1$.

a) Since from (8.4.18) $c(k, m, v) = c(k, m, v)$, the hth moment in (8.4.51) can be alternatively expressed as

$$E[U(\boldsymbol{\theta})^h \mid \mathbf{y}] = \prod_{t=1}^{k} \frac{\Gamma[\frac{1}{2}(v - 1 + t) + h]\,\Gamma[\frac{1}{2}(v - 1 + m + t)]}{\Gamma[\frac{1}{2}(v - 1 + t)]\,\Gamma[\frac{1}{2}(v - 1 + m + t) + h]}\,. \qquad (8.4.52)$$

By comparing (8.4.51) with (8.4.52), we see that the roles played by m and k can be interchanged. That is, the distribution of $U_{(m,k,v)}$ is identical to that of $U_{(k,m,v)}$. In other words, the distribution of $U = U(\boldsymbol{\theta})$ arising from a multivariate model with m output variables and k regression coefficients for each output is identical to that from a multivariate model with k output variables and m regression coefficients for each output. With no loss in generality, we shall proceed with the m-output model, i.e., the $U_{(m,k,\nu)}$ distribution.

b) Now (8.4.51) can be written

$$E(U^h \mid \mathbf{y}) = \prod_{s=1}^{m} B\left(\frac{\nu - 1 + s}{2} + h, \frac{k}{2}\right) \Big/ B\left(\frac{\nu - 1 + s}{2}, \frac{k}{2}\right), \tag{8.4.53}$$

where $B(p, q)$ is the complete beta function. The right-hand side is the hth moment of the product of m independent variables $x_1, \ldots, x_m$ having beta distributions with parameters

$$\left(\frac{\nu - 1 + s}{2}, \frac{k}{2}\right), \qquad s = 1, \ldots, m.$$

It follows that U is distributed as the product $x_1 \ldots x_m$.

c) Suppose m is even. Then we can write (8.4.53) as

$$E(U^h \mid \mathbf{y}) = \prod_{t=1}^{m/2} \frac{B\left[\frac{\nu + 2(t-1)}{2} + h, \frac{k}{2}\right] B\left(\frac{\nu - 1 + 2t}{2} + h, \frac{k}{2}\right)}{B\left[\frac{\nu + 2(t-1)}{2}, \frac{k}{2}\right] B\left(\frac{\nu - 1 + 2t}{2}, \frac{k}{2}\right)}. \tag{8.4.54}$$

Using the duplication formula

$$\Gamma(p + \tfrac{1}{2})\,\Gamma(p) = \frac{\sqrt{\pi}\,\Gamma(2p)}{2^{2p-1}} \tag{8.4.55}$$

so that

$$B(p + \tfrac{1}{2}, q)\, B(p, q) = 2^{2q}\, B(2p, 2q)\, B(q, q), \tag{8.4.56}$$

we obtain

$$\begin{aligned} E(U^h \mid \mathbf{y}) &= \prod_{t=1}^{m/2} \frac{B[\nu + 2(t - 1 + h), k]}{B[\nu + 2(t-1), k]} \\ &= E(z_1^2 \cdots z_{m/2}^2)^h, \end{aligned} \tag{8.4.57}$$

where $z_1, \ldots, z_{m/2}$ are $m/2$ independent random variables having beta distributions with parameters $[\nu + 2(t-1), k]$, $t = 1, \ldots, m/2$, respectively. Thus, U is distributed as the product $z_1^2 \cdots z_{m/2}^2$.

The Case $m = 1$.

When $m = 1$ so that $\nu = n - k$, it follows from (8.4.53) that, U has the beta distribution with parameters $[(n-k)/2, k/2]$ so that the quantity $(n-k)\,(1-U)/(kU)$ is distributed as F with $(k, n-k)$ degrees of freedom. This result is of course to be expected, since for $m = 1$, we have $|\mathbf{A}| = (n - k)s^2$, where

$$s^2 = (n - k)^{-1} (\mathbf{y} - \hat{\mathbf{y}})' (\mathbf{y} - \hat{\mathbf{y}}),$$

so that

$$\left(\frac{1 - U}{U}\right)\left(\frac{n - k}{k}\right) = \frac{(\boldsymbol{\theta} - \hat{\boldsymbol{\theta}})' \mathbf{X}'\mathbf{X}(\boldsymbol{\theta} - \hat{\boldsymbol{\theta}})}{ks^2} \tag{8.4.58}$$

which, from (2.7.21), has the F distribution with $(k, n - k)$ degrees of freedom.

The Case $m = 2$.

When $m = 2, v = n - k - 1$ and from (8.4.57) $U^{1/2}$ is distributed as a beta variable with parameters $(n - k - 1, k)$. Thus, the quantity

$$\left(\frac{1 - U^{1/2}}{U^{1/2}}\right)\left(\frac{n - k - 1}{k}\right) = (|\mathbf{I}_2 + \mathbf{A}^{-1} (\boldsymbol{\theta} - \hat{\boldsymbol{\theta}})' \mathbf{X}'\mathbf{X}(\boldsymbol{\theta} - \hat{\boldsymbol{\theta}})|^{1/2}) - 1)\left(\frac{n - k - 1}{k}\right) \tag{8.4.59}$$

has the F distribution with $[2k, 2(n - k - 1)]$ degrees of freedom.

8.4.6 An Approximation to the Distribution of U for General m

For $m \geqslant 3$, the exact distribution of U is complicated, see e.g. Schatzoff (1966) and Pillai and Gupta (1969). We now give an approximation method following Bartlett (1938) and Box (1949). In expression (8.4.51), we make the substitutions

$$M = -\phi v \log U, \qquad t = -h/(\phi v),$$
$$x = \tfrac{1}{2}v, \qquad a_s = \tfrac{1}{2}(s + k - 1) \quad \text{and} \quad b_s = \tfrac{1}{2}(s - 1), \tag{8.4.60}$$

where ϕ is some arbitrary positive number, so that

$$E(U^h \mid \mathbf{y}) = E(e^{tM} \mid \mathbf{y})$$
$$= \prod_{s=1}^{m} \frac{\Gamma(x + a_s)}{\Gamma(x + b_s)} \frac{\Gamma[\phi x(1 - 2t) + x(1 - \phi) + b_s]}{\Gamma[\phi x(1 - 2t) + x(1 - \phi) + a_s]}. \tag{8.4.61}$$

In terms of the random variable M, (8.4.61) is then its moment-generating function. Taking logarithms and employing Stirling's series (see Appendix A2.2), we obtain the cumulant generating function of M as

$$\kappa_M(t) = -\frac{mk}{2}\log(1 - 2t) - \sum_{r=1}^{\infty} \omega_r[(1 - 2t)^{-r} - 1], \tag{8.4.62}$$

where

$$\omega_r = \frac{(-1)^r}{r(r + 1)(\phi x)^r} \sum_{s=1}^{m} \{B_{r+1}[x(1 - \phi) + b_s] - B_{r+1}[x(1 - \phi) + a_s]\}$$

and $B_r(z)$ is the Bernoulli polynomial of degree r and order one. The asymptotic expansion in (8.4.62) is valid provided ϕ is so chosen that $x(1-\phi)$ is bounded. In this case, ω_r is of order $O[(\phi x)^{-r}]$ in magnitude.

The series in (8.4.62) is of the same type as the one obtains in (2.12.15) for the comparison of the spread of k Normal populations. In particular, the distribution of $M = -\phi\nu \log U$ can be expressed as a weighted series of χ^2 densities, the leading term having mk degrees of freedom.

The Choice of ϕ

It follows that, if we take the leading term alone, then to order $O[(\phi x)^{-1}] = O[(\phi\frac{1}{2}\nu)^{-1}]$ the quantity

$$M = -\phi\nu \log U \tag{8.4.63}$$

is distributed as χ^2_{mk}, provided $\frac{1}{2}\nu(1-\phi)$ is bounded. In particular, if we take $\phi = 1$, we then have that $M = -\nu \log U$ is distributed approximately as χ^2_{mk}.

For moderate values of ν, the accuracy of the χ^2 approximation can be improved by suitably choosing ϕ so that $\omega_1 = 0$. This is because when $\omega_1 = 0$, the quantity M will be distributed as χ^2_{mk} to order $O[(\phi\frac{1}{2}\nu)^{-2}]$. Using the fact that

$$B_2(z) = z^2 - z + \tfrac{1}{6} \tag{8.4.64}$$

it is straight forward to verify that for $\omega_1 = 0$, we require

$$\phi = 1 + \frac{1}{2\nu}(m + k - 3). \tag{8.4.65}$$

This choice of ϕ gives very close approximations in practice. An example with $\nu = 9$, $m = k = 2$ will be given later in Section 8.4.8 to compare the approximation with the exact distribution.

It follows from the above discussion that to order $O[(\phi\frac{1}{2}\nu)^{-2}]$,

$$\Pr\{U(\boldsymbol{\theta}) > U(\boldsymbol{\theta}_0) \mid \mathbf{y}\} \doteq \Pr\{\chi^2_{mk} < -\phi\nu \log U(\boldsymbol{\theta}_0)\} \tag{8.4.66}$$

with

$$\phi = 1 + \frac{1}{2\nu}(m + k - 3),$$

$$\log U(\boldsymbol{\theta}_0) = -\log |\mathbf{I}_m + \mathbf{A}^{-1}(\boldsymbol{\theta}_0 - \hat{\boldsymbol{\theta}})'\mathbf{X}'\mathbf{X}(\boldsymbol{\theta}_0 - \hat{\boldsymbol{\theta}})|$$

which can be employed to decide whether the parameter point $\boldsymbol{\theta} = \boldsymbol{\theta}_0$ lies approximately inside or outside the $(1-\alpha)$ H.P.D. region.

8.4.7 Inferences about a General Parameter Point of a Block Submatrix of $\boldsymbol{\theta}$

In the above, we have discussed inference procedures for (a) a specific column of $\boldsymbol{\theta}$, (b) a specific row of $\boldsymbol{\theta}$, and (c) a parameter point $\boldsymbol{\theta}_0$ for the complete set of $\boldsymbol{\theta}$.

In some problems, we may be interested in making inferences about the parameters belonging to a certain block submatrix of $\boldsymbol{\theta}$. Without loss of generality we consider only the problem for the $k_1 \times m_1$ matrix $\boldsymbol{\theta}_{11}$ defined in (8.4.36). From (8.4.37) and Theorem 8.4.1 (on page 449), it follows that the quantity

$$U(\boldsymbol{\theta}_{11}) = |\mathbf{I}_{m_1} + \mathbf{A}_{11}^{-1}(\boldsymbol{\theta}_{11} - \hat{\boldsymbol{\theta}}_{11})' \mathbf{C}_{11}^{-1}(\boldsymbol{\theta}_{11} - \hat{\boldsymbol{\theta}}_{11})|^{-1} \qquad (8.4.67)$$

is distributed as $U_{(m_1,k_1,v)}$. This distribution would then allow us to decide whether a particular value of $\boldsymbol{\theta}_{11}$ lay inside or outside a desired H.P.D. region. In particular, for $m_1 > 2$ and $k_1 > 2$, we may then make use of the approximation

$$-\phi_1 v \log U(\boldsymbol{\theta}_{11}) \,\dot{\sim}\, \chi^2_{m_1 k_1}, \qquad (8.4.68)$$

where

$$\phi_1 = 1 + \frac{1}{2v}(m_1 + k_1 - 3),$$

so that the parameter point $\boldsymbol{\theta}_{11,0}$ lies inside the $(1-\alpha)$ H.P.D. region if and only if

$$\Pr\{U(\boldsymbol{\theta}_{11}) > U(\boldsymbol{\theta}_{11,0}) \mid \mathbf{y}\} \doteq \Pr\{\chi^2_{m_1 k_1} < -\phi_1 v \log U(\boldsymbol{\theta}_{11,0})\} < (1-\alpha). \quad (8.4.69)$$

8.4.8 An Illustrative Example

An experiment was conducted to study the effect of temperature on the yield of the product y_1 and the by-product y_2 of a chemical process. Twelve runs were made at different temperature settings ranging from 161.3°F to 195.7°F. The data are given in Table 8.4.1.

The average temperature employed is $\bar{T} = 177.86$. We suppose a model to be entertained whereby, over the range of temperature explored, the relationships between product yield and temperature and by-product yield and temperature were nearly linear so that to an adequate approximation

$$E(y_{u1}) = \theta_{11} + \theta_{21} x_u,$$

$$E(y_{u2}) = \theta_{12} + \theta_{22} x_u, \qquad u = 1, \ldots, 12, \qquad (8.4.70)$$

where $x_u = (T_u - \bar{T})/100$, the divisor 100 being introduced for convenience in calculation. The parameters θ_{11} and θ_{12} will thus determine the locations of the yield-temperature lines at the average temperature $\bar{T}$ while θ_{21} and θ_{22} will represent the slopes of these lines. The experimental runs were set up independently and we should therefore expect experimental errors to be independent from *run to run*. However, in any particular run, we should expect the error in y_1 to be correlated with that in y_2 since slight aberations in reaction

Table 8.4.1
Yield of product and by-product of a chemical process

Temp. °F	Product y_1	By-product y_2
161.3	63.7	20.3
164.0	59.5	24.2
165.7	67.9	18.0
170.1	68.8	20.5
173.9	66.1	20.1
176.2	70.4	17.5
177.6	70.0	18.2
181.7	73.7	15.4
185.6	74.1	17.8
189.0	79.6	13.3
193.5	77.1	16.7
195.7	82.8	14.8

conditions or in analytical procedures could simultaneously affect observations of both product and by-product yields. Finally, then, the tentative model was

$$\begin{aligned} y_{u1} &= \theta_{11} x_{u1} + \theta_{21} x_{u2} + \varepsilon_{u1} \\ y_{u2} &= \theta_{12} x_{u1} + \theta_{22} x_{u2} + \varepsilon_{u2}, \end{aligned} \tag{8.4.71}$$

where $x_{u2} = x_u$, and $x_{u1} = 1$ is a dummy variable introduced to "carry" the parameters θ_{11} and θ_{12}. It was supposed that $(\varepsilon_{u1}, \varepsilon_{u2})$ followed the bivariate Normal distribution $N_2(\mathbf{0}, \boldsymbol{\Sigma})$.

Given this setup, we apply the results arrived at earlier in this section to make inferences about the regression coefficients

$$\boldsymbol{\theta} = \begin{bmatrix} \theta_{11} & \theta_{12} \\ \theta_{21} & \theta_{22} \end{bmatrix} = [\boldsymbol{\theta}_1 \, , \, \boldsymbol{\theta}_2] = \begin{bmatrix} \boldsymbol{\theta}'_{(1)} \\ \boldsymbol{\theta}'_{(2)} \end{bmatrix} \tag{8.4.72}$$

against the background of a noninformative reference prior distribution for $\boldsymbol{\theta}$ and $\boldsymbol{\Sigma}$.

The relevant sample quantities are summarized below:

$$n = 12 \qquad m = k = 2$$

$$\mathbf{X}'\mathbf{X} = \begin{bmatrix} 12 & 0 \\ 0 & 0.14546 \end{bmatrix} \qquad \mathbf{C} = (\mathbf{X}'\mathbf{X})^{-1} = \begin{bmatrix} 0.0833 & 0 \\ 0 & 6.8750 \end{bmatrix} \tag{8.4.73}$$

$$\mathbf{A} = \begin{bmatrix} 61.4084 & -38.4823 \\ -38.4823 & 36.7369 \end{bmatrix} \qquad \mathbf{A}^{-1} = \begin{bmatrix} 0.0474 & 0.0496 \\ 0.0496 & 0.0792 \end{bmatrix}$$

and

$$\hat{\boldsymbol{\theta}} = \begin{bmatrix} 71.1417 & 18.0666 \\ 54.4355 & -20.0933 \end{bmatrix} = [\hat{\boldsymbol{\theta}}_1\ ,\ \hat{\boldsymbol{\theta}}_2] = \begin{bmatrix} \hat{\boldsymbol{\theta}}'_{(1)} \\ \hat{\boldsymbol{\theta}}'_{(2)} \end{bmatrix}.$$

The fitted lines

$$\hat{y}_1 = \hat{\theta}_{11} + \hat{\theta}_{21}(T - \bar{T}) \times 10^{-2}$$

$$\hat{y}_2 = \hat{\theta}_{12} + \hat{\theta}_{22}(T - \bar{T}) \times 10^{-2}$$

together with the data are shown in Fig. 8.4.1. As explained earlier, in a real data analysis, we should pause at this point to critically examine the conditional inference by study of residuals. We shall here proceed with further analysis supposing that such checks have proved satisfactory.

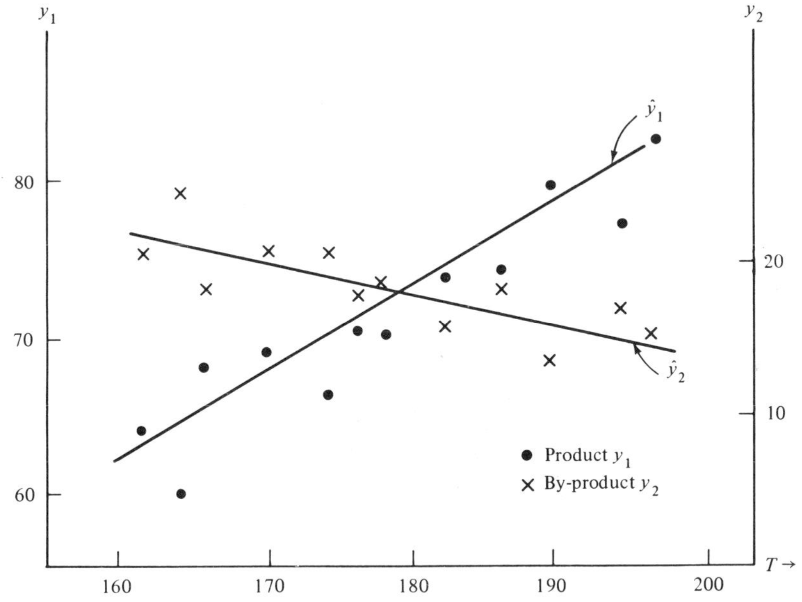

Fig. 8.4.1 Scatter diagram of the product y_1 and the by-product y_2 together with the best fitting lines.

Inferences about $\boldsymbol{\theta}_1 = (\theta_{11}, \theta_{21})'$

When interest centers primarily on the parameters $\boldsymbol{\theta}_1$ for the product y_1, we have from (8.4.28) that

$$p(\boldsymbol{\theta}_1 \mid \mathbf{y}) \propto [61.4084 + (\boldsymbol{\theta}_1 - \hat{\boldsymbol{\theta}}_1)' \mathbf{X}'\mathbf{X}(\boldsymbol{\theta}_1 - \hat{\boldsymbol{\theta}}_1)]^{-11/2} \qquad (8.4.74)$$

that is, a bivariate $t_2[\hat{\boldsymbol{\theta}}_1, 6.825(\mathbf{X}'\mathbf{X})^{-1}, 9]$ distribution. Since the matrix $\mathbf{X}'\mathbf{X}$ is diagonal, θ_{11} and θ_{21} are uncorrelated (but of course not independent).

Figure 8.4.2a shows contours of the 50, 75 and 95 per cent H.P.D. regions together with the mode $\hat{\boldsymbol{\theta}}_1$, from which overall conclusions about $\boldsymbol{\theta}_1$ may be drawn.

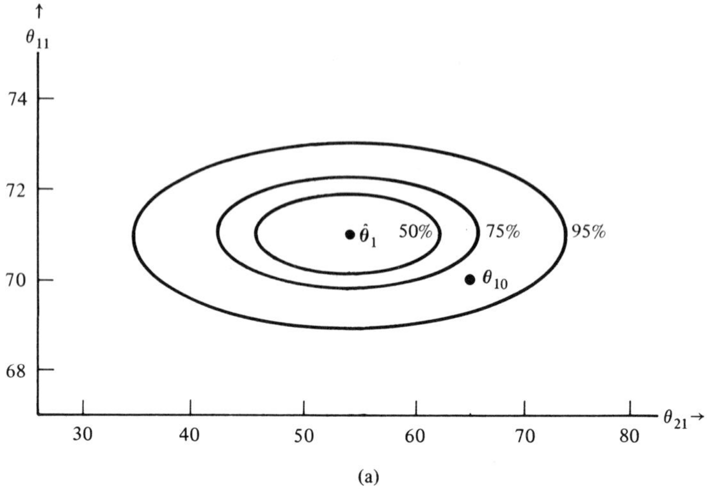

Fig. 8.4.2a Contours of the posterior distribution of $\boldsymbol{\theta}_1$, the parameters of the product straight line.

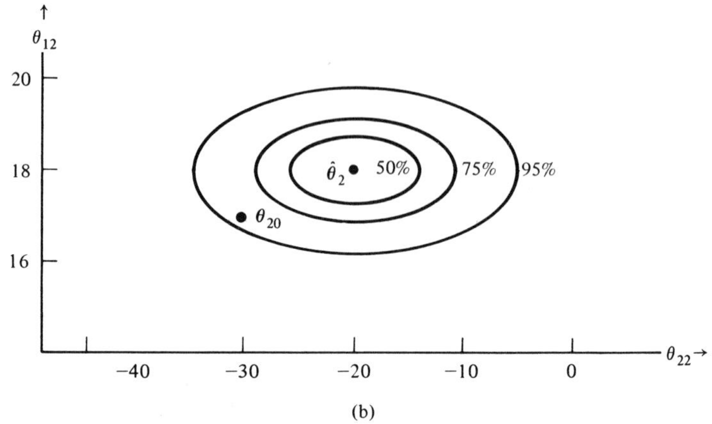

Fig. 8.4.2b Contours of the posterior distribution of $\boldsymbol{\theta}_2$, the parameters of the by-product straight line.

Inferences about $\boldsymbol{\theta}_2 = (\theta_{12}, \theta_{22})'$

Similarly, from (8.4.28), the posterior distribution of $\boldsymbol{\theta}_2$ for the by-product y_2 is

$$p(\boldsymbol{\theta}_2 \mid \mathbf{y}) \propto [36.7369 + (\boldsymbol{\theta}_2 - \hat{\boldsymbol{\theta}}_2)' \mathbf{X}'\mathbf{X}(\boldsymbol{\theta}_2 - \hat{\boldsymbol{\theta}}_2)]^{-11/2}, \tag{8.4.75}$$

which is a $t_2[\hat{\boldsymbol{\theta}}_2, 4.08(\mathbf{X}'\mathbf{X})^{-1}, 9]$ distribution. Again, the parameters θ_{12} and θ_{22} are uncorrelated. The 50, 75 and 95 per cent H.P.D. contours together with the mode $\hat{\boldsymbol{\theta}}_2$ for this distribution are shown in Fig. 8.4.2b. The contours have exactly the same shape and orientation as those in Fig. 8.4.2a because the same $\mathbf{X}'\mathbf{X}$ matrix is employed; the spread for $\boldsymbol{\theta}_2$ is however smaller than that for $\boldsymbol{\theta}_1$ since the sample variance from y_2 is less than that from y_1.

Inferences about $\boldsymbol{\theta}_{(2)} = (\theta_{21}, \theta_{22})'$

In problems of the kind considered, interest often centers on $\boldsymbol{\theta}'_{(2)} = (\theta_{21}, \theta_{22})$ which measure respectively the slopes of the yield/temperature lines for the product and by-product. From (8.4.34), the posterior distribution of $\boldsymbol{\theta}_{(2)}$ is

$$p(\boldsymbol{\theta}_{(2)} \mid \mathbf{y}) \propto [6.875 + (\boldsymbol{\theta}_{(2)} - \hat{\boldsymbol{\theta}}_{(2)})'\mathbf{A}^{-1}(\boldsymbol{\theta}_{(2)} - \hat{\boldsymbol{\theta}}_{(2)})]^{-11/2}, \tag{8.4.76}$$

that is, a $t_2[\hat{\boldsymbol{\theta}}_{(2)}, 0.764\mathbf{A}, 9]$ distribution. Figure 8.4.3 shows the 50, 75 and 95 per cent contours together with the mode $\hat{\boldsymbol{\theta}}_{(2)}$. Also shown in the same figure are the marginal distributions $t(\hat{\theta}_{21}, 46.90, 9)$ and $t(\hat{\theta}_{22}, 28.05, 9)$ and the

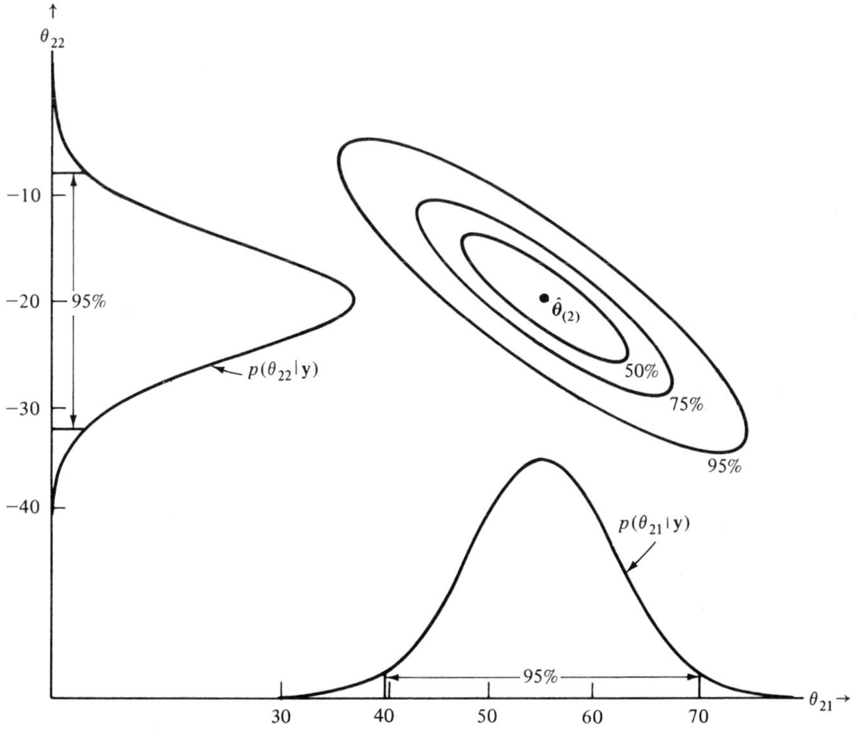

Fig. 8.4.3 Contours of the posterior distribution of $\boldsymbol{\theta}_{(2)}$ and the associated marginal distributions for the product and by-product data.

corresponding 95 per cent H.P.D. intervals for θ_{21} and θ_{22}. Figure 8.4.3 summarizes, then, the information about the slopes $(\theta_{21}, \theta_{22})$ coming from the data on the basis of a noninformative reference prior. The parameters are negatively correlated; the correlation between $(\theta_{21}, \theta_{22})$ is in fact the sample correlation between the errors $(\varepsilon_{u1}, \varepsilon_{u2})$

$$r_{12} = \frac{a_{12}}{(a_{11}\,a_{22})^{1/2}} = -\,0.81. \tag{8.4.77}$$

It is clear from the figure that care must be taken to distinguish between individual and joint inferences about $(\theta_{21}, \theta_{22})$. It could be exceedingly misleading to make inferences from the individual H.P.D. intervals about the parameters jointly (see the discussion in Section 2.7).

Joint Inferences about $\boldsymbol{\theta}$

To make overall inferences about the parameters $[\boldsymbol{\theta}_1, \boldsymbol{\theta}_2]$, we need to calculate the distribution of the quantity $U(\boldsymbol{\theta})$ defined in (8.4.49). For instance, suppose we wish to decide whether or not the parameter point

$$\boldsymbol{\theta}_0 = [\boldsymbol{\theta}_{10}\ ,\ \boldsymbol{\theta}_{20}] = \begin{bmatrix} 70 & 17 \\ 65 & -30 \end{bmatrix} \tag{8.4.78}$$

lies inside the 95 per cent H.P.D. region for $\boldsymbol{\theta}$. We have

$$(\boldsymbol{\theta}_0 - \hat{\boldsymbol{\theta}})'\mathbf{X}'\mathbf{X}(\boldsymbol{\theta}_0 - \hat{\boldsymbol{\theta}}) = \begin{bmatrix} 31.8761 & -0.6108 \\ -0.6108 & 27.9272 \end{bmatrix} \tag{8.4.79}$$

so that

$$U(\boldsymbol{\theta}_0) = \frac{|\mathbf{A}|}{|\mathbf{A} + (\boldsymbol{\theta}_0 - \hat{\boldsymbol{\theta}})'\mathbf{X}'\mathbf{X}(\boldsymbol{\theta}_0 - \hat{\boldsymbol{\theta}})|} = \frac{775.0668}{4503.8878} = 0.1720, \tag{8.4.80}$$

Since $m = 2$, we may use the result in (8.4.59) to calculate the exact probability that $U(\boldsymbol{\theta})$ exceeds $U(\boldsymbol{\theta}_0)$. We obtain

$$\Pr\{U(\boldsymbol{\theta}) > U(\boldsymbol{\theta}_0) \mid \mathbf{y}\} = 1 - I_{\sqrt{0.1720}}(9, 2) = 1 - 0.0022 = 0.9978. \tag{8.4.81}$$

From (8.4.50), we conclude that the point $\boldsymbol{\theta}_0$ lies outside the 95 per cent H.P.D region of $\boldsymbol{\theta}$. Note that while the point $\boldsymbol{\theta}_0 = (\boldsymbol{\theta}_{10}, \boldsymbol{\theta}_{20})$ is excluded from the 95 per cent region, Figs. 8.4.2a,b show that both the points $\boldsymbol{\theta}_{10}$ and $\boldsymbol{\theta}_{20}$ are included in the corresponding marginal 95 per cent H.P.D. regions. This serves to illustrate once more the distinction between joint inferences and marginal inferences.

Approximating Distribution of $U = U(\boldsymbol{\theta})$

It is informative to compare the exact distribution of $U(\boldsymbol{\theta})$ with the approximation in (8.4.63) using the present example ($\nu = 9$, $m = k = 2$). From (8.4.59), the exact distribution of U is found to be

$$p(U) = \frac{1}{2B(2, 9)} U^{3.5} (1 - U^{1/2}), \qquad 0 < U < 1. \tag{8.4.82}$$

Using the approximation given in (8.4.63) to (8.4.65), we find

$$\phi = \tfrac{19}{18}, \qquad \phi\nu = 9.5, \qquad M = -9.5 \log U \tag{8.4.83}$$

and

$$p(M) \doteq \tfrac{1}{4} M \exp(-\tfrac{1}{2}M), \qquad 0 < M < \infty.$$

This implies that the distribution of U is approximately

$$p(U) \doteq (22.5625)(-\log U)\, U^{3.75}, \qquad 0 < U < 1. \tag{8.4.84}$$

Table 8.4.2 gives a specimen of the exact and the approximate densities of U calculated from (8.4.82) and (8.4.84). Although the sample size is only 10, the agreement is very close.

Table 8.4.2

Comparison of the exact and the approximate distributions of U for $n = 12$ and $m = k = 2$

	$p(U)$	
U	Exact	Approximate
0.05	0.00098	0.00089
0.10	0.00973	0.00924
0.20	0.08900	0.08688
0.30	0.30098	0.29731
0.40	0.66947	0.66548
0.50	1.16498	1.16239
0.60	1.69708	1.69718
0.70	2.10935	2.11244
0.80	2.17561	2.18045
0.90	1.59706	1.60130
0.95	0.95220	0.95478

8.5 SOME ASPECTS OF THE DISTRIBUTION OF Σ FOR THE CASE OF A COMMON DERIVATIVE MATRIX X

We discuss in this section certain results pertaining to the posterior distribution of the elements of the covariance matrix $\boldsymbol{\Sigma} = \{\sigma_{ij}\}$.†

† For the important problem of making inferences about the latent roots and vectors of Σ which is not discussed in this book, see Geisser (1965a) and Tiao and Fienberg (1969).

8.5.1 Joint Distribution of $(\boldsymbol{\theta}, \boldsymbol{\Sigma})$

When the $\mathbf{X}_i$'s are common, the joint posterior distribution of $(\boldsymbol{\theta}, \boldsymbol{\Sigma})$ in (8.2.17) can be written

$$p(\boldsymbol{\theta}, \boldsymbol{\Sigma} \mid \mathbf{y}) \propto |\boldsymbol{\Sigma}|^{-\frac{1}{2}(v+k+2m)} \exp\{-\tfrac{1}{2}\operatorname{tr}\boldsymbol{\Sigma}^{-1}[\mathbf{A} + (\boldsymbol{\theta} - \hat{\boldsymbol{\theta}})'\mathbf{X}'\mathbf{X}(\boldsymbol{\theta} - \hat{\boldsymbol{\theta}})]\},$$
$$-\infty < \boldsymbol{\theta} < \infty, \quad \boldsymbol{\Sigma} > \mathbf{0}, \qquad (8.5.1)$$

where $v = n - (k + m) + 1$ and use is made of (8.4.4) and (8.4.5). The individual and joint inferences about $(\theta_{21}, \theta_{22})$. It could be exceedingly distribution can be written as the product $p(\boldsymbol{\theta}, \boldsymbol{\Sigma} \mid \mathbf{y}) = p(\boldsymbol{\theta} \mid \boldsymbol{\Sigma}, \mathbf{y})\, p(\boldsymbol{\Sigma} \mid \mathbf{y})$.

Conditional Distribution of $\boldsymbol{\theta}$ *given* $\boldsymbol{\Sigma}$

Given $\boldsymbol{\Sigma}$, we have that

$$p(\boldsymbol{\theta} \mid \boldsymbol{\Sigma}, \mathbf{y}) \propto \exp[-\tfrac{1}{2}\operatorname{tr}\boldsymbol{\Sigma}^{-1}(\boldsymbol{\theta} - \hat{\boldsymbol{\theta}})'\mathbf{X}'\mathbf{X}(\boldsymbol{\theta} - \hat{\boldsymbol{\theta}})], \qquad -\infty < \boldsymbol{\theta} < \infty, \qquad (8.5.2)$$

which by comparison with (8.4.45), is the km-dimensional Normal distribution

$$N_{mk}[\hat{\boldsymbol{\theta}}, \boldsymbol{\Sigma} \otimes (\mathbf{X}'\mathbf{X})^{-1}]. \qquad (8.5.3)$$

Marginal Distribution of $\boldsymbol{\Sigma}$

Thus, the marginal posterior distribution of $\boldsymbol{\Sigma}$ is

$$p(\boldsymbol{\Sigma} \mid \mathbf{y}) \propto |\boldsymbol{\Sigma}|^{-(\frac{1}{2}v+m)} \exp(-\tfrac{1}{2}\operatorname{tr}\boldsymbol{\Sigma}^{-1}\mathbf{A}), \qquad \boldsymbol{\Sigma} > \mathbf{0}, \qquad (8.5.4)$$

From the Jacobian in (8.2.13), the distribution of $\boldsymbol{\Sigma}^{-1}$ is

$$p(\boldsymbol{\Sigma}^{-1} \mid \mathbf{y}) \propto |\boldsymbol{\Sigma}^{-1}|^{\frac{1}{2}v-1} \exp(-\tfrac{1}{2}\operatorname{tr}\boldsymbol{\Sigma}^{-1}\mathbf{A}), \qquad \boldsymbol{\Sigma}^{-1} > \mathbf{0}, \qquad (8.5.5)$$

which, by comparing with (8.2.20), is recognized as the Wishart distribution $W_m(\mathbf{A}^{-1}, v)$ provided $v > 0$. The distribution of $\boldsymbol{\Sigma}$ in (8.5.4) may thus be called an m-dimensional "inverted" Wishart distribution with v degrees of freedom, and be denoted by $W_m^{-1}(\mathbf{A}, v)$. From (8.2.21) the normalizing constant for the distributions of $\boldsymbol{\Sigma}$ and $\boldsymbol{\Sigma}^{-1}$ is

$$|\mathbf{A}|^{\frac{1}{2}(v+m-1)}\, 2^{-\frac{1}{2}m(v+m-1)} \left[\Gamma_m\left(\frac{v+m-1}{2}\right)\right]^{-1}. \qquad (8.5.6)$$

Note that when $m = 1$ and $v = n - k$ the distribution in (8.5.4) reduces to an inverted χ^2 distribution,

$$p(\sigma_{11} \mid \mathbf{y}) \propto \sigma_{11}^{-\frac{1}{2}(n-k+2)} \exp\left(-\frac{a_{11}}{2\sigma_{11}}\right), \qquad \sigma_{11} > 0, \qquad (8.5.7)$$

which is the posterior distribution of $\sigma^2 = \sigma_{11}$ with data from a univariate multiple regression problem (see Section 2.7).

From the results in (8.5.1), (8.5.3), (8.5.4) and (8.4.15), we have the following useful theorem.

Theorem 8.5.1 Let $\hat{\boldsymbol{\theta}}$ be a $k \times m$ matrix of constants, $\mathbf{X}'\mathbf{X}$ and $\mathbf{A}$ be, respectively, a $k \times k$ and a $m \times m$ positive definite symmetric matrix of constants, and $v > 0$. If the joint distribution of the elements of the $k \times m$ matrix $\boldsymbol{\theta}$ and the $m \times m$ positive definite symmetric matrix $\boldsymbol{\Sigma}$ is

$$p(\boldsymbol{\theta}, \boldsymbol{\Sigma} \mid \mathbf{y}) \propto |\boldsymbol{\Sigma}|^{-\frac{1}{2}(v+k+2m)} \exp\{-\tfrac{1}{2} \operatorname{tr} \boldsymbol{\Sigma}^{-1}[\mathbf{A} + (\boldsymbol{\theta} - \hat{\boldsymbol{\theta}})' \mathbf{X}'\mathbf{X}(\boldsymbol{\theta} - \hat{\boldsymbol{\theta}})]\},$$

then, (a) given $\boldsymbol{\Sigma}$, $\boldsymbol{\theta} \sim N_{mk}[\hat{\boldsymbol{\theta}}, \boldsymbol{\Sigma} \otimes (\mathbf{X}'\mathbf{X})^{-1}]$, (b) $\boldsymbol{\Sigma} \sim W_m^{-1}(\mathbf{A}, v)$ and (c) $\boldsymbol{\theta} \sim t_{mk}[\hat{\boldsymbol{\theta}}, (\mathbf{X}'\mathbf{X})^{-1}, \mathbf{A}, v]$.

8.5.2 Some Properties of the Distribution of Σ

Consider the partition

$$\boldsymbol{\Sigma} = \begin{array}{c} \begin{matrix} m_1 & m_2 \end{matrix} \\ \left[\begin{array}{c|c} \boldsymbol{\Sigma}_{11} & \boldsymbol{\Sigma}_{12} \\ \hline \boldsymbol{\Sigma}_{21} & \boldsymbol{\Sigma}_{22} \end{array}\right] \begin{matrix} m_1 \\ m_2 \end{matrix} \end{array}, \qquad m_1 + m_2 = m. \tag{8.5.8}$$

We now proceed to obtain the posterior distribution of $(\boldsymbol{\Sigma}_{11}, \boldsymbol{\Omega}, \mathbf{T})$, where

$$\boldsymbol{\Omega} = \boldsymbol{\Sigma}_{22} - \boldsymbol{\Sigma}_{21}\boldsymbol{\Sigma}_{11}^{-1}\boldsymbol{\Sigma}_{12}$$

and

$$\mathbf{T} = \boldsymbol{\Sigma}_{11}^{-1}\boldsymbol{\Sigma}_{12}. \tag{8.5.9}$$

It is to be remembered that $\boldsymbol{\Sigma}_{11}$ is the covariance matrix of the m_1 errors $(\varepsilon_{u1}, \ldots, \varepsilon_{um_1})$, $\boldsymbol{\Omega}$ and $\mathbf{T}$ are, respectively, the covariance matrix and the $m_1 \times m_2$ matrix of "regression coefficients" for the conditional distribution of the remaining m_2 errors $(\varepsilon_{u(m_1+1)}, \ldots, \varepsilon_{um})$ given $(\varepsilon_{u1}, \ldots, \varepsilon_{um_1})$. Denoting

$$\mathbf{A} = \begin{array}{c} \begin{matrix} m_1 & m_2 \end{matrix} \\ \left[\begin{array}{c|c} \mathbf{A}_{11} & \mathbf{A}_{12} \\ \hline \mathbf{A}_{21} & \mathbf{A}_{22} \end{array}\right] \begin{matrix} m_2 \\ m_1 \end{matrix} \end{array}, \qquad \mathbf{A}_{22\cdot1} = \mathbf{A}_{22} - \mathbf{A}_{21}\,\mathbf{A}_{11}^{-1}\,\mathbf{A}_{12} \tag{8.5.10}$$

and

$$\hat{\mathbf{T}} = \mathbf{A}_{11}^{-1}\,\mathbf{A}_{12},$$

it can be readily shown that

a) $\boldsymbol{\Sigma}_{11}$ is distributed independently of $(\mathbf{T}, \boldsymbol{\Omega})$,

b) $$\boldsymbol{\Sigma}_{11} \sim W_{m_1}^{-1}(\mathbf{A}_{11}, v), \tag{8.5.11}$$

c) $$\boldsymbol{\Omega} \sim W_{m_2}^{-1}(\mathbf{A}_{22\cdot1}, v + m_1), \tag{8.5.12}$$

d) $$\mathbf{T} \sim t_{m_1 m_2}(\hat{\mathbf{T}}, \mathbf{A}_{11}^{-1}, \mathbf{A}_{22\cdot1}, v + m_1), \tag{8.5.13}$$

$$U(\mathbf{T}) = |\mathbf{I}_{m_2} + \mathbf{A}_{22\cdot1}^{-1}(\mathbf{T} - \hat{\mathbf{T}})'\mathbf{A}_{11}(\mathbf{T} - \hat{\mathbf{T}})|^{-1}$$

$$\sim U_{(m_2, m_1, v+m_1)}. \tag{8.5.14}$$

The above results may be verified as follows. Since $\boldsymbol{\Sigma}$ is positive definite, we can express the determinant and the inverse of $\boldsymbol{\Sigma}$ as

$$|\boldsymbol{\Sigma}| = |\boldsymbol{\Sigma}_{11}|\,|\boldsymbol{\Omega}|$$

$$\boldsymbol{\Sigma}^{-1} = \begin{bmatrix} \boldsymbol{\Sigma}_{11}^{-1} & \mathbf{0} \\ \mathbf{0} & \mathbf{0} \end{bmatrix} + \mathbf{M}, \tag{8.5.15}$$

where

$$\mathbf{M} = \begin{bmatrix} \mathbf{T}\boldsymbol{\Omega}^{-1}\mathbf{T}' & -\mathbf{T}\boldsymbol{\Omega}^{-1} \\ -\boldsymbol{\Omega}^{-1}\mathbf{T}' & \boldsymbol{\Omega}^{-1} \end{bmatrix}.$$

Expression (8.5.15) may be verified by showing $\boldsymbol{\Sigma}^{-1}\boldsymbol{\Sigma} = \mathbf{I}$. Thus, the distribution in (8.5.4) can be written

$$p(\boldsymbol{\Sigma} \mid \mathbf{y}) \propto [|\boldsymbol{\Sigma}_{11}|\,|\boldsymbol{\Omega}|]^{-(\frac{1}{2}v+m)} \exp\left(-\tfrac{1}{2}\operatorname{tr}\boldsymbol{\Sigma}_{11}^{-1}\mathbf{A}_{11} - \tfrac{1}{2}\operatorname{tr}\mathbf{MA}\right), \qquad \boldsymbol{\Sigma} > \mathbf{0}. \tag{8.5.16}$$

For fixed $\boldsymbol{\Sigma}_{11}$, it is readily seen by making use of (A8.1.1) in Appendix A8.1 that the Jacobian of the transformation from $(\boldsymbol{\Sigma}_{12}, \boldsymbol{\Sigma}_{22})$ to $(\mathbf{T}, \boldsymbol{\Omega})$ is

$$J = \left|\frac{\partial(\boldsymbol{\Sigma}_{12}, \boldsymbol{\Sigma}_{22})}{\partial(\mathbf{T}, \boldsymbol{\Omega})}\right| = |\boldsymbol{\Sigma}_{11}|^{m_2}. \tag{8.5.17}$$

Noting that $\mathbf{M}$ does not depend on $\boldsymbol{\Sigma}_{11}$, it follows from (8.5.16) and (8.5.17) that $\boldsymbol{\Sigma}_{11}$ is independent of $(\mathbf{T}, \boldsymbol{\Omega})$ so that

$$p(\boldsymbol{\Sigma}_{11}, \mathbf{T}, \boldsymbol{\Omega} \mid \mathbf{y}) = p(\mathbf{T}, \boldsymbol{\Omega} \mid \mathbf{y})\, p(\boldsymbol{\Sigma}_{11} \mid \mathbf{y}) \tag{8.5.18}$$

where

$$p(\mathbf{T}, \boldsymbol{\Omega} \mid \mathbf{y}) \propto |\boldsymbol{\Omega}|^{-(\frac{1}{2}v+m)} \exp\left(-\tfrac{1}{2}\operatorname{tr}\mathbf{MA}\right), \qquad \boldsymbol{\Omega} > \mathbf{0}, \quad -\infty < \mathbf{T} < \infty \tag{8.5.19}$$

and

$$P(\boldsymbol{\Sigma}_{11} \mid \mathbf{y}) \propto |\boldsymbol{\Sigma}_{11}|^{-(\frac{1}{2}v+m_1)} \exp\left(-\tfrac{1}{2}\operatorname{tr}\boldsymbol{\Sigma}_{11}^{-1}\mathbf{A}_{11}\right), \qquad \boldsymbol{\Sigma}_{11} > \mathbf{0}. \tag{8.5.20}$$

Thus, $\boldsymbol{\Sigma}_{11}$ is distributed as the inverted Wishart given in (8.5.11).

From (8.5.10) and (8.5.15), we may express the exponent of the distribution in (8.5.19) as

$$\begin{aligned} \operatorname{tr}\mathbf{MA} &= \operatorname{tr}(\mathbf{T}\boldsymbol{\Omega}^{-1}\mathbf{T}'\mathbf{A}_{11} - \mathbf{T}\boldsymbol{\Omega}^{-1}\mathbf{A}_{21} - \boldsymbol{\Omega}^{-1}\mathbf{T}'\mathbf{A}_{12} + \boldsymbol{\Omega}^{-1}\mathbf{A}_{22}) \\ &= \operatorname{tr}\boldsymbol{\Omega}^{-1}[\mathbf{A}_{22\cdot1} + (\mathbf{T} - \hat{\mathbf{T}})'\mathbf{A}_{11}(\mathbf{T} - \hat{\mathbf{T}})] \end{aligned} \tag{8.5.21}$$

In obtaining the extreme right of (8.5.21), repeated use is made of the fact that $\operatorname{tr}\mathbf{AB} = \operatorname{tr}\mathbf{BA}$.

Thus,

$$p(\mathbf{T}, \boldsymbol{\Omega} \mid \mathbf{y}) \propto |\boldsymbol{\Omega}|^{-\frac{1}{2}(v'+m_1+2m_2)} \exp\left\{-\tfrac{1}{2}\operatorname{tr}\boldsymbol{\Omega}^{-1}[\mathbf{A}_{22\cdot1} + (\mathbf{T} - \hat{\mathbf{T}})'\mathbf{A}_{11}(\mathbf{T} - \hat{\mathbf{T}})]\right\},$$
$$-\infty < \mathbf{T} < \infty, \quad \boldsymbol{\Omega} > \mathbf{0},$$

where $v' = v + m_1$, which is in exactly the same form as the joint distribution of $(\boldsymbol{\Sigma}, \boldsymbol{\theta})$ in (8.5.1). The results in (8.5.12) and (8.5.13) follow by application of Theorem 8.5.1 (on p. 461). Further, by using Theorem 8.4.1 (on p. 449), we obtain (8.5.14).

Distribution of σ_{11}

By setting $m_1 = 1$ in (8.5.11), the distribution of σ_{11} is

$$p(\sigma_{11} \mid \mathbf{y}) \propto \sigma_{11}^{-(\frac{1}{2}v+1)} \exp\left(-\frac{a_{11}}{2\sigma_{11}}\right), \qquad \sigma_{11} > 0, \tag{8.5.22}$$

a χ^{-2} distribution with $v = n - k - (m - 1)$ degrees of freedom. Comparing with the univariate case in (8.5.7), we see that the two distributions are identical except, as expected, that the degrees of freedom differ by $(m - 1)$. The difference is of minor importance when m is small relative to $n - k$ but can be appreciable otherwise—see the discussion about the posterior distribution of $\boldsymbol{\theta}_i$ in (8.4.28).

It is clear that the distribution of σ_{ii}, the ith diagonal element of $\boldsymbol{\Sigma}$, is given by simply replacing the subscripts (1,1) in (8.5.22) by (i, i).

The Two Regression Matrices $\boldsymbol{\Sigma}_{11}^{-1}\boldsymbol{\Sigma}_{12}$ *and* $\boldsymbol{\Sigma}_{22}^{-1}\boldsymbol{\Sigma}_{21}$

The $m_1 \times m_2$ matrix of "regression coefficients" $\mathbf{T} = \boldsymbol{\Sigma}_{11}^{-1}\boldsymbol{\Sigma}_{12}$ measures the dependence of the conditional expectation of the errors $(\varepsilon_{u(m_1+1)}, \ldots, \varepsilon_{um})$ on $(\varepsilon_{u1}, \ldots, \varepsilon_{um_1})$. From the posterior distribution of $\mathbf{T}$ in (8.5.13) the plausibility of different values of the measure $\mathbf{T}$ may be compared in terms of e.g. the H.P.D. regions. In deciding whether a parameter point $\mathbf{T} = \mathbf{T}_0$ lies inside or outside a desired H.P.D. region, from (8.5.14) and (8.4.66) one may then calculate

$$\Pr\{U(\mathbf{T}) > U(\mathbf{T}_0) \mid \mathbf{y}\} \doteq \Pr\{\chi^2_{m_1 m_2} < -\phi'(v + m_1)\log U(\mathbf{T}_0)\}, \tag{8.5.23}$$

where

$$\phi' = 1 + \frac{m_1 + m_2 - 3}{2(v + m_1)}.$$

In particular, if $\mathbf{T}_0 = \mathbf{0}$ which corresponds to $\boldsymbol{\Sigma}_{12} = \mathbf{0}$, that is, $(\varepsilon_{u1}, \ldots, \varepsilon_{um_1})$ are independent of $(\varepsilon_{u(m_1+1)}, \ldots, \varepsilon_{um})$, then

$$U(\mathbf{T}_0 = \mathbf{0}) = \frac{|\mathbf{A}|}{|\mathbf{A}_{11}|\,|\mathbf{A}_{22}|}. \tag{8.5.24}$$

Consider now the $m_2 \times m_1$ matrix of "regression coefficients"

$$\mathbf{Z} = \boldsymbol{\Sigma}_{22}^{-1}\boldsymbol{\Sigma}_{21} \tag{8.5.25}$$

It is clear from the development leading to (8.5.13) and (8.5.14) that by interchanging the roles of m_1 and m_2, we have

$$\mathbf{Z} \sim t_{m_2 m_1}(\hat{\mathbf{Z}}, \mathbf{A}_{22}^{-1}, \mathbf{A}_{11\cdot 2}, v + m_2) \tag{8.5.26}$$

where

$$\hat{\mathbf{Z}} = \mathbf{A}_{22}^{-1}\mathbf{A}_{21}, \qquad \mathbf{A}_{11\cdot 2} = \mathbf{A}_{11} - \mathbf{A}_{12}\mathbf{A}_{22}^{-1}\mathbf{A}_{21},$$

and

$$U(\mathbf{Z}) = |\mathbf{I}_{m_1} + \mathbf{A}_{11\cdot 2}^{-1}(\mathbf{Z} - \hat{\mathbf{Z}})'\mathbf{A}_{22}(\mathbf{Z} - \hat{\mathbf{Z}})|^{-1}$$
$$\sim U_{(m_1, m_2, v+m_2)}. \tag{8.5.27}$$

Thus, in deciding whether the parameter point $\mathbf{Z}_0$ is included in a desired H.P.D. region, we calculate

$$\Pr\{U(\mathbf{Z}) > U(\mathbf{Z}_0) \mid \mathbf{y}) \doteq \Pr\{\chi^2_{m_2 m_1} < -\phi''(v + m_2)\log U(\mathbf{Z}_0)\} \qquad (8.5.28)$$

where

$$\phi'' = 1 + \frac{m_1 + m_2 - 3}{2(v + m_2)}.$$

In particular, if $\mathbf{Z}_0 = \mathbf{0}$ which again corresponds to $\boldsymbol{\Sigma}_{12} = \mathbf{0}$, then

$$U(\mathbf{Z}_0 = \mathbf{0}) = \frac{|\mathbf{A}|}{|\mathbf{A}_{11}|\,|\mathbf{A}_{22}|}. \qquad (8.5.29)$$

Although $U(\mathbf{Z}_0 = \mathbf{0}) = U(\mathbf{T}_0 = \mathbf{0})$ the probabilities on the right-hand sides of (8.5.23) and (8.5.28) will be different whenever $m_1 \neq m_2$. This is not a surprising result. For, it can be seen from (8.5.18) that the distribution of $\mathbf{T}$ is in fact proportional to the conditional distribution of $\boldsymbol{\Sigma}_{12}$, given $\boldsymbol{\Sigma}_{11}$. Inferences about the parameter point $\boldsymbol{\Sigma}_{12} = \mathbf{0}$ in terms of the probability $\Pr\{\chi^2_{m_1 m_2} < -\phi'(v + m_1)\log U(\mathbf{T}_0 = \mathbf{0})\}$ can thus be interpreted as made conditionally for fixed $\boldsymbol{\Sigma}_{11}$. That is to say that we are comparing the plausibility of $\boldsymbol{\Sigma}_{12} = \mathbf{0}$ with other values of $\boldsymbol{\Sigma}_{12}$ in relation to a fixed $\boldsymbol{\Sigma}_{11}$. On the other hand, in terms of $\Pr\{\chi^2_{m_1 m_2} < -\phi''(v + m_2)\log U(\mathbf{Z}_0 = \mathbf{0})\}$, inferences about $\boldsymbol{\Sigma}_{12} = \mathbf{0}$ can be regarded as conditional on fixed $\boldsymbol{\Sigma}_{22}$. Thus, one would certainly not expect that, in general, the two types of conditional inferences about $\boldsymbol{\Sigma}_{12} = \mathbf{0}$ will be identical.

8.5.3 An Example

For illustration, consider again the product and by-product data in Table 8.4.1. The relevant sample quantities are given in (8.4.73).

When interest centers on the variance σ_{11} of the error ε_{u1} corresponding to the product y_1, we have from (8.5.22)

$$\sigma_{11} \sim 61.4084\,\chi_9^{-2} \qquad (8.5.30)$$

Using Table II (at the end of this book), limits of the 95 per cent H.P.D. interval of $\log\sigma_{11}$ in terms of σ_{11} are (3.02, 20.79).

Similarly, the posterior distribution of σ_{22} is such that

$$\sigma_{22} \sim 36.7369\,\chi_9^{-2} \qquad (8.5.31)$$

and the limits of the corresponding 95 per cent H.P.D. interval are (1.81, 12.44).

From (8.5.13), and since $m_1 = m_2 = 1$, the posterior distribution of $T = \sigma_{11}^{-1}\sigma_{12}$ is the univariate $t(\hat{T}, s_1^2, v_1)$ distribution where

$$\hat{T} = a_{11}^{-1}a_{12} = -0.627, \qquad v_1 = 10,$$

and

$$s_1^2 = \nu_1^{-1} a_{11}^{-1} a_{22\cdot 1} = \frac{a_{22} - a_{12}^2/a_{11}}{\nu_1 \times a_{11}} = 0.0206.$$

Thus,

$$\frac{T + 0.627}{0.143} \sim t_{10} \tag{8.5.32}$$

so that from Table IV at the end of this book, limits of the 95 per cent H.P.D. interval are (– 0.95, – 0.31). In particular, the parameter point $\sigma_{11}^{-1}\sigma_{12} = 0$ (corresponds to $\sigma_{12} = 0$) is excluded from the 95 per cent interval.

Finally, from (8.5.26) $Z = \sigma_{22}^{-1}\sigma_{12}$ is distributed as $t(\hat{Z}, s_2^2, \nu_2)$ where

$$\hat{Z} = a_{22}^{-1} a_{12} = -1.05, \qquad \nu_2 = 10$$

and

$$s_2^2 = \nu_2^{-1} a_{22}^{-1} a_{11\cdot 2} = \frac{a_{11} - a_{12}^2/a_{22}}{\nu_2 \times a_{22}} = 0.0574.$$

Thus,

$$\frac{Z + 1.05}{0.240} \sim t_{10}. \tag{8.5.33}$$

Limits of the 95 per cent H.P.D. interval are (– 1.58, – 0.52) and the point $\sigma_{22}^{-1}\sigma_{12} = 0$ is again excluded. Further, from (8.5.14), (8.5.27), (8.5.32) and (8.5.33)

$$\Pr\{U(T) > U(0) \mid \mathbf{y}\} = \Pr\{U(Z) > U(0) \mid \mathbf{y}\} = \Pr\{|t_{10}| > 4.37\} \tag{8.5.34}$$

so that inferences about $\sigma_{12} = 0$ in terms of either T or Z are identical. This is of course to be expected since, for this example, $m_1 = m_2 = 1$.

8.5.4 Distribution of the Correlation Coefficient ρ_{12}

The two regression matrices $\boldsymbol{\Sigma}_{11}^{-1}\boldsymbol{\Sigma}_{12}$ and $\boldsymbol{\Sigma}_{22}^{-1}\boldsymbol{\Sigma}_{21}$ are measures of the dependence (or association) between the two set of responses $(y_{u1}, \ldots, y_{um_1})$ and $(y_{u(m_1+1)}, \ldots, y_{um})$. When interest centers at the association between two specific responses y_{ui} and y_{uj}, the most natural measure of association is the correlation coefficient $\rho_{ij} = \sigma_{ij}/(\sigma_{ii}\sigma_{jj})^{1/2}$. Without loss of generality, we now consider how inferences may be made about ρ_{12}.

By setting $m_1 = 2$ in the distribution of $\boldsymbol{\Sigma}_{11}$ in (8.5.11), we can follow the development in Jeffreys (1961, p. 174) to obtain the posterior distribution of the correlation coefficient ρ_{12} as

$$p(\rho \mid \mathbf{y}) \propto (1 - \rho^2)^{\frac{1}{2}(\nu-2)} \int_0^\infty \omega^{-1} \left(\omega + \frac{1}{\omega} - 2\rho r\right)^{-(\nu+1)} d\omega, \qquad -1 < \rho < 1, \tag{8.5.35}$$

where $\rho = \rho_{12}$,

$$r = r_{12} = \frac{a_{12}}{(a_{11}\, a_{22})^{1/2}}$$

is the sample correlation coefficient, and the normalizing constant is

$$2(1 - r^2)^{(\nu+1)/2}\,\Gamma(\nu + 1)\Big/\left[\pi^{1/2}\,\Gamma\left(\frac{\nu}{2}\right)\Gamma\left(\frac{\nu+1}{2}\right)\right].$$

It is noted that this distribution depends only upon the sample correlation coefficient r.

To see this, for $m_1 = 2$, the posterior distribution of the elements $(\sigma_{11}, \sigma_{22}, \sigma_{12})$ of $\boldsymbol{\Sigma}_{11}$ in (8.5.11) is

$$p(\sigma_{11}, \sigma_{22}, \sigma_{12} \mid \mathbf{y}) \propto [\sigma_{11}\,\sigma_{22}\,(1 - \rho^2)]^{-(\frac{1}{2}\nu + 2)}$$
$$\times \exp\left\{-\frac{1}{2(1-\rho^2)}\left[\frac{a_{11}}{\sigma_{11}} + \frac{a_{22}}{\sigma_{22}} - \frac{2\rho\, a_{12}}{(\sigma_{11}\,\sigma_{22})^{1/2}}\right]\right\},$$
$$\sigma_{11} > 0, \quad \sigma_{22} > 0, \quad \sigma_{11}\sigma_{22} > \sigma_{12}^2, \qquad (8.5.36)$$

where from (8.5.6), the normalizing constant is,

$$(a_{11}a_{22} - a_{12}^2)^{(\nu+1)/2}\Big/\left\{2^{(\nu+1)}\,\pi^{1/2}\prod_{i=1}^{2}\Gamma[\tfrac{1}{2}(\nu + 2 - i)]\right\}$$

We now make the transformation, due to Fisher (1915),

$$x = \left(\frac{\sigma_{11}\,\sigma_{22}}{a_{11}\,a_{22}}\right)^{1/2}, \qquad \omega = \left(\frac{\sigma_{11}\,a_{22}}{\sigma_{22}\,a_{11}}\right)^{1/2}, \qquad \rho = \frac{\sigma_{12}}{(\sigma_{11}\sigma_{22})^{1/2}}. \qquad (8.5.37)$$

The Jacobian is

$$\left|\frac{\partial(\sigma_{11}\,\sigma_{22}\,\sigma_{12})}{\partial(x, \omega, \rho)}\right| = 2a_{11}\,\sigma_{22}(\sigma_{11}\,\sigma_{22})^{1/2} \qquad (8.5.38)$$

so that the distribution of (x, ω, ρ) is

$$p(x, \omega, \rho \mid \mathbf{y}) \propto (1 - \rho^2)^{\frac{1}{2}(\nu+4)}\,\omega^{-1}\,x^{-(\nu+2)} \exp\left[-\frac{1}{2(1-\rho^2)x}\left(\omega + \frac{1}{\omega} - 2\rho r\right)\right],$$
$$\omega > 0, \quad x > 0, \quad -1 < \rho < 1. \qquad (8.5.39)$$

Upon integrating out x,

$$p(\omega, \rho \mid \mathbf{y}) \propto (1 - \rho^2)^{\frac{1}{2}(\nu-2)}\,\omega^{-1}\left(\omega + \frac{1}{\omega} - 2\,\rho r\right)^{-(\nu+1)} \qquad \omega > 0, \quad -1 < \rho < 1,$$
$$(8.5.40)$$

from which we obtain the distribution of ρ given in (8.5.35).

The Special Case When $r = 0$

When $r = 0$, the distribution in (8.5.35) reduces to

$$p(\rho \mid r = 0) \propto (1 - \rho^2)^{\frac{1}{2}(\nu - 2)}, \qquad -1 < \rho < 1, \tag{8.5.41}$$

which is symmetric at $\rho = 0$, and is identical in form to the sampling distribution of r on the null hypothesis that $\rho = 0$. In this case, if we make the transformation

$$\rho = \frac{t}{(\nu + t^2)^{1/2}}, \tag{8.5.42}$$

then the distribution of t is

$$p(t) \propto (1 + t^2/\nu)^{-\frac{1}{2}(\nu + 1)}, \qquad -\infty < t < \infty,$$

so that the quantity t is distributed as $t(0, 1, \nu)$.

The General Case When $r \neq 0$

In general, the density function (8.5.35) cannot be expressed in terms of simple functions of r. With the availability of a computer, it can always be evaluated by numerical integration, however. To illustrate, consider again the bivariate product and by-product data introduced in Table 8.4.1. Figure 8.5.1 shows the posterior distribution of ρ calculated from (8.5.35). For this example $\nu = 9$ and $r = -0.81$. The distribution is skewed to the right and concentrated rather sharply about its mode at $\rho \doteq -0.87$; it practically rules out values of ρ exceeding -0.3.

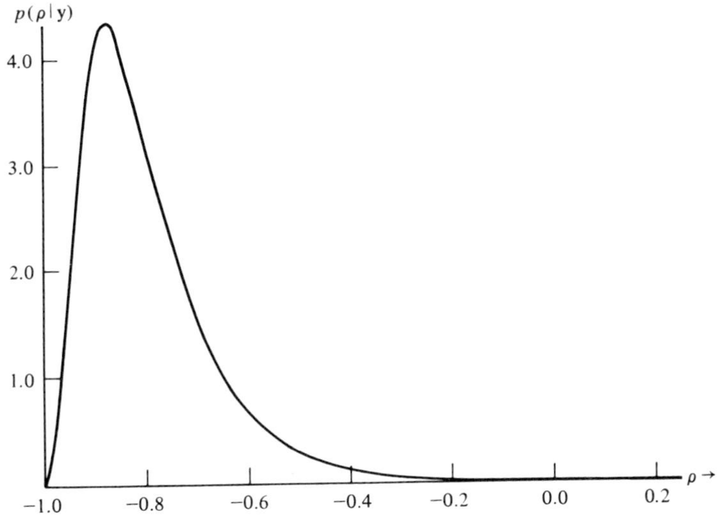

Fig. 8.5.1 Posterior distributions of ρ for the product and by-product data.

Series Expansions of $p(\rho \mid \mathbf{y})$

The distribution in (8.5.35) can be put in various different forms. In particular, it can be expressed as

$$p(\rho \mid \mathbf{y}) \propto \frac{(1-\rho^2)^{\frac{1}{2}(\nu-2)}}{(1-\rho r)^{\nu+\frac{1}{2}}} S_\nu(\rho, r), \qquad -1 < \rho < 1, \tag{8.5.43}$$

where

$$S_\nu(\rho, r) = 1 + \sum_{l=1}^{\infty} \frac{1}{l!} \left(\frac{1+\rho r}{8}\right)^l \prod_{s=1}^{l} \frac{(2s-1)^2}{(\nu + s + \frac{1}{2})}$$

is a hypergeometric series, and the normalizing constant is

$$(1-r^2)^{\frac{1}{2}(\nu+1)} [\Gamma(\nu+1)]^2 \Big/ \left[2^{(\nu-\frac{1}{2})} \Gamma\left(\frac{\nu}{2}\right) \Gamma\left(\frac{\nu+1}{2}\right) \Gamma(\nu + \tfrac{3}{2})\right].$$

To see this, the integral in (8.5.35) can be written

$$\int_0^\infty \omega^{-1} \left[\omega + \frac{1}{\omega} - 2\rho r\right]^{-(\nu+1)} d\omega = 2 \int_1^\infty \omega^{-1} \left[\omega + \frac{1}{\omega} - 2\rho r\right]^{-(\nu+1)} d\omega. \tag{8.5.44}$$

On the right-hand side of (8.5.44), we may make the substitution, again due to Fisher,

$$u = 1 - \frac{1-\rho r}{\frac{1}{2}[\omega + (1/\omega)] - \rho r} = \frac{\frac{1}{2}[\omega + (1/\omega)] - 1}{\frac{1}{2}[\omega + (1/\omega)] - \rho r}. \tag{8.5.45}$$

Noting that

$$\frac{1}{4}\left(\omega + \frac{1}{\omega}\right)^2 - \frac{1}{4}\left(\omega - \frac{1}{\omega}\right)^2 = 1,$$

we have

$$\frac{\partial u}{\partial \omega} = \omega^{-1}(1-u)(2u)^{1/2}[1 - \tfrac{1}{2}(1+\rho r)u]^{1/2}(1-\rho r)^{-1/2} \tag{8.5.46}$$

so that

$$p(\rho \mid \mathbf{y}) \propto \frac{(1-\rho^2)^{\frac{1}{2}(\nu-2)}}{(1-\rho r)^{\nu+\frac{1}{2}}} \int_0^1 \frac{(1-u)^\nu}{(2u)^{1/2}} [1 - \tfrac{1}{2}(1+\rho r)u]^{-1/2}\, du, \qquad -1 < \rho < 1. \tag{8.5.47}$$

Expanding the last term in the integrand in powers of u and integrating term by term, each term being a complete beta function, we obtain the result in (8.5.43).

We remark here that an alternative series representation of the distribution of ρ can be obtained as follows. In the integral of (8.5.35), since

$$\omega^{-1}\left(\omega + \frac{1}{\omega} - 2\rho r\right)^{-(\nu+1)} = \omega^{\nu}(\omega^2 - 2\rho r\omega + 1)^{-(\nu+1)}, \tag{8.5.48}$$

by completing the square in the second factor on the right-hand side of (8.5.48), we can write the distribution in (8.5.35) as

$$p(\rho \mid \mathbf{y}) \propto \frac{(1-\rho^2)^{\frac{1}{2}(\nu-2)}}{(1-\rho^2 r^2)^{(\nu+1)}} \int_1^{\infty} \omega^{\nu} \left[1 + \frac{(\omega - \rho r)^2}{1-\rho^2 r^2}\right]^{-(\nu+1)} d\omega \tag{8.5.49}$$

Upon repeated integration by parts, the above integral can be expressed as a finite series involving powers of $[(1-\rho r)/(1+\rho r)]^{1/2}$ and Student's t integrals. The density function of ρ can thus be calculated from a table of t distribution. This process becomes very tedious when ν is moderately large so its practical usefulness is limited.

The series $S_\nu(\rho, r)$ in (8.5.43) has its leading term equal to one, followed by terms of order $\nu^{-l}, l = 1, 2, \ldots$. When ν is moderately large, we may simply take the first term so that approximately

$$p(\rho \mid \mathbf{y}) \doteq c\, \frac{(1-\rho^2)^{\frac{1}{2}(\nu-2)}}{(1-\rho r)^{\nu+\frac{1}{2}}}, \qquad -1 < \rho < 1, \tag{8.5.50}$$

where c is the normalizing constant

$$c^{-1} = \int_{-1}^{1} \frac{(1-\rho^2)^{\frac{1}{2}(\nu-2)}}{(1-\rho r)^{\nu+\frac{1}{2}}}\, d\rho.$$

Although evaluation of c would still require the use of numerical methods, it is much simpler to calculate the distribution of ρ using (8.5.50) than to evaluate the integral in (8.5.35) for every value of ρ. Table 8.5.1 compares the exact distribution with the approximation using the data in Table 8.4.1. In spite of the fact that ν is only 9, the agreement is very close.

It is easily seen that the density function (8.5.50) is greatest when ρ is near r. However, except when $r = 0$, the distribution is asymmetrical.

The asymmetry can be reduced by making the transformation,

$$\zeta = \tanh^{-1}\rho = \tfrac{1}{2}\log\frac{1+\rho}{1-\rho}, \tag{8.5.51}$$

due to Fisher (1921). Following his argument, it is found that ζ is approximately Normal

$$N\left[\tanh^{-1} r - \frac{5r}{2(\nu+1)}, (\nu+1)^{-1}\right].$$

Setting $m = 2$, $k = 1$ so that $\nu = n - 2$, the distribution in (8.5.43) is identical to that given by Jeffreys for the case of sampling from a bivariate Normal population. Finally, we note that while we have obtained above the distribution of the specific correlation coefficient $\rho = \rho_{12}$, it is clear that the distribution of any correlation $\rho_{ij}, i \neq j$, is given simply by setting $r = r_{ij} = a_{ij}/(a_{ii}\,a_{jj})^{1/2}$ in (8.5.35) and its associated expressions.

Table 8.5.1

Comparison of the exact and the approximate distributions of ρ for $\nu = 9$ and $r = -0.81$

	$p(\rho \mid \mathbf{y})$	
ρ	Exact	Approximate
−0.98	0.2286	0.2283
−0.96	1.2164	1.2150
−0.94	2.4867	2.4844
−0.92	3.5159	3.5134
−0.90	4.1200	4.1180
−0.88	4.3281	4.3271
−0.86	4.2448	4.2448
−0.84	3.9782	3.9790
−0.82	3.6141	3.6157
−0.80	3.2127	3.2149
−0.70	1.5182	1.5210
−0.60	0.6584	0.6603
−0.50	0.2853	0.2865
−0.40	0.1260	0.1267
−0.30	0.0569	0.0572
−0.20	0.0261	0.0263

8.6 A SUMMARY OF FORMULAE AND CALCULATIONS FOR MAKING INFERENCES ABOUT $(\boldsymbol{\theta}, \boldsymbol{\Sigma})$

Using the product, by-product data in Table 8.4.1 for illustration, Table 8.6.1 below provides a short summary of the formulae and calculations required for making inferences about the elements of $(\boldsymbol{\theta}, \boldsymbol{\Sigma})$ for the linear model with common derivative matrix defined in (8.4.1). Specifically, the model is

$$\mathbf{y} = \mathbf{X}\boldsymbol{\theta} + \boldsymbol{\varepsilon} \tag{8.6.1}$$

where $\mathbf{y} = [\mathbf{y}_1, \ldots, \mathbf{y}_m]$ is a $n \times m$ matrix of observations, $\mathbf{X}$ is a $n \times k$ matrix of fixed elements with rank k, $\boldsymbol{\theta} = [\boldsymbol{\theta}_1, \ldots, \boldsymbol{\theta}_m]$ is a $k \times m$ matrix of parameters and $\boldsymbol{\varepsilon} = [\boldsymbol{\varepsilon}_{(1)}, \ldots, \boldsymbol{\varepsilon}_{(n)}]'$ is a $n \times m$ matrix of errors. It is assumed that $\boldsymbol{\varepsilon}_{(u)}$, $u = 1, \ldots, n$ are independently distributed as $N_m(\mathbf{0}, \boldsymbol{\Sigma})$.

Table 8.6.1

Summarized calculations for the linear model $\mathbf{y} = \mathbf{X}\boldsymbol{\theta} + \boldsymbol{\varepsilon}$

1. From (8.4.1), (8.4.4), (8.4.7) and (8.4.13), obtain

$$m=2, \qquad k=2, \qquad n=12, \qquad \nu=n-(m+k)+1=9$$

$$\mathbf{X}'\mathbf{X} = \begin{bmatrix} 12 & 0 \\ 0 & 0.15 \end{bmatrix}, \qquad \mathbf{C} = (\mathbf{X}'\mathbf{X})^{-1} = \begin{bmatrix} 0.08 & 0 \\ 0 & 6.88 \end{bmatrix}$$

$$\hat{\boldsymbol{\theta}} = (\mathbf{X}'\mathbf{X})^{-1}\mathbf{X}'\mathbf{y} = \begin{bmatrix} 71.14 & 18.07 \\ 54.44 & -20.09 \end{bmatrix}. \qquad \mathbf{A}=\{a_{ij}\} \quad i,j=1,\ldots,m$$

$$a_{ij}=(\mathbf{y}_i-\mathbf{X}\hat{\boldsymbol{\theta}}_i)'(\mathbf{y}_j-\mathbf{X}\hat{\boldsymbol{\theta}}_j), \qquad \text{and} \qquad \mathbf{A} = \begin{bmatrix} 61.41 & -38.48 \\ -38.48 & 36.74 \end{bmatrix}.$$

2. Inferences about a specific column or row of $\boldsymbol{\theta}$:
   Writing

$$\boldsymbol{\theta} = [\boldsymbol{\theta}_1, \ldots, \boldsymbol{\theta}_m] = \begin{bmatrix} \boldsymbol{\theta}'_{(1)} \\ \vdots \\ \boldsymbol{\theta}'_{(k)} \end{bmatrix}, \qquad \hat{\boldsymbol{\theta}} = [\hat{\boldsymbol{\theta}}_1, \ldots, \hat{\boldsymbol{\theta}}_m] = \begin{bmatrix} \hat{\boldsymbol{\theta}}'_{(1)} \\ \vdots \\ \hat{\boldsymbol{\theta}}'_{(k)} \end{bmatrix}$$

then from (8.4.28) and (8.4.34),

$$\boldsymbol{\theta}_i \sim t_k\left(\hat{\boldsymbol{\theta}}_i, \frac{a_{ii}}{\nu}\mathbf{C}, \nu\right), \qquad i=1,\ldots,m,$$

and

$$\boldsymbol{\theta}_{(g)} \sim t_m\left(\hat{\boldsymbol{\theta}}_{(g)}, \frac{c_{gg}}{\nu}\mathbf{A}, \nu\right), \qquad g=1,\ldots,k.$$

Thus, for $i=1$ and $g=2$

$$\boldsymbol{\theta}_1 \sim t_2\left\{\begin{pmatrix} 71.14 \\ 54.44 \end{pmatrix}, 6.83 \times \begin{bmatrix} 0.08 & 0 \\ 0 & 6.88 \end{bmatrix}, 9\right\}$$

$$\boldsymbol{\theta}_{(2)} \sim t_2\left\{\begin{pmatrix} 54.44 \\ 20.09 \end{pmatrix}, 0.76 \times \begin{bmatrix} 61.41 & -38.48 \\ -38.48 & 36.74 \end{bmatrix}, 9\right\}.$$

3. H.P.D. regions of $\boldsymbol{\theta}$: To decide if a general parameter point $\boldsymbol{\theta}_0$ lies inside or outside the $(1-\alpha)$ H.P.D. region, from (8.4.63) and (8.4.66), use the approximation

$$-\phi\nu \log U \,\dot{\sim}\, \chi^2_{mk},$$

where

$$U=U(\boldsymbol{\theta}) = |\mathbf{A}|\,|\mathbf{A}+(\boldsymbol{\theta}-\hat{\boldsymbol{\theta}})'\,\mathbf{X}'\mathbf{X}(\boldsymbol{\theta}-\hat{\boldsymbol{\theta}})|^{-1} \quad \text{and} \quad \phi=1+\frac{m+k-3}{2\nu},$$

so that $\boldsymbol{\theta}_0$ lies inside the $1-\alpha$ region if

$$-\phi\nu \log U(\boldsymbol{\theta}_0) < \chi^2(mk, \alpha).$$

Table 8.6.1 *Continued*

For the example, $\phi = 19/18$. Thus, if

$$\boldsymbol{\theta}_0 = \begin{bmatrix} 70 & 17 \\ 65 & -30 \end{bmatrix}, \qquad \text{then} \qquad U(\boldsymbol{\theta}_0) = 0.17$$

and the point lies inside the $1-\alpha$ region if

$$-9.5 \log 0.17 = 16.7 < \chi^2(4, \alpha).$$

4. H.P.D. regions for a block submatrix of $\boldsymbol{\theta}$: Let

$$\boldsymbol{\theta} = \overset{\quad m_1 \qquad m_2}{\begin{bmatrix} \boldsymbol{\theta}_{11} & \boldsymbol{\theta}_{12} \\ \boldsymbol{\theta}_{21} & \boldsymbol{\theta}_{22} \end{bmatrix}} \begin{matrix} k_1 \\ k_2 \end{matrix} \qquad \hat{\boldsymbol{\theta}} = \overset{\quad m_1 \qquad m_2}{\begin{bmatrix} \hat{\boldsymbol{\theta}}_{11} & \hat{\boldsymbol{\theta}}_{12} \\ \hat{\boldsymbol{\theta}}_{21} & \hat{\boldsymbol{\theta}}_{22} \end{bmatrix}} \begin{matrix} k_1 \\ k_2 \end{matrix}$$

$$\mathbf{A} = \overset{\quad m_1 \qquad m_2}{\begin{bmatrix} \mathbf{A}_{11} & \mathbf{A}_{12} \\ \mathbf{A}_{21} & \mathbf{A}_{22} \end{bmatrix}} \begin{matrix} m_1 \\ m_2 \end{matrix} \qquad \mathbf{C} = \overset{\quad k_1 \qquad k_2}{\begin{bmatrix} \mathbf{C}_{11} & \mathbf{C}_{12} \\ \mathbf{C}_{21} & \mathbf{C}_{22} \end{bmatrix}} \begin{matrix} k_1 \\ k_2 \end{matrix}$$

To decide, for example, if the parameter point $\boldsymbol{\theta}_{11,0}$ lies inside or outside the $(1-\alpha)$ H.P.D. region for $\boldsymbol{\theta}_{11}$, use the approximation in (8.4.68),

$$\phi_1 \nu \log U(\boldsymbol{\theta}_{11}) \mathrel{\dot\sim} \chi^2_{m_1 k_1},$$

where

$$U(\boldsymbol{\theta}_{11}) = |\mathbf{A}_{11}|\, |\mathbf{A}_{11} + (\boldsymbol{\theta}_{11} - \hat{\boldsymbol{\theta}}_{11})' \mathbf{C}_{11}^{-1} (\boldsymbol{\theta}_{11} - \hat{\boldsymbol{\theta}}_{11})|^{-1}$$

and $\phi_1 = 1 + (1/2\nu)(m_1 + k_1 - 3)$, so that $\boldsymbol{\theta}_{11,0}$ lies inside the $(1-\alpha)$ region if

$$-\phi_1 \nu \log U(\boldsymbol{\theta}_{11,0}) < \chi^2(m_1 k_1, \alpha),$$

Thus if $m_1 = k_1 = 1$, $\theta_{11,0} = 70$, then $\phi_1 = 17/18$ and

$$U(\theta_{11,0} = 70) = 61.41 \Bigg/ \left[61.41 + \frac{(70 - 71.14)^2}{0.08}\right] = 0.79,$$

so that $\theta_{11,0}$ lies inside the $(1-\alpha)$ region if

$$-8.5 \log 0.79 = 2.0 < \chi^2(1, \alpha).$$

Note that since $m_1 = k_1 = 1$, exact results could be obtained and the above is for illustration only.

5. Inferences about the diagonal elements of $\boldsymbol{\Sigma}$:
From (8.5.22)

$$\sigma_{ii} \sim a_{ii} \chi_\nu^{-2}, \qquad i = 1, \ldots, m.$$

Thus, for $i = 1$, $\sigma_{11} \sim 61.41 \chi_9^{-2}$

6. H.P.D. regions for the "regression matrix" $\mathbf{T}$: Let

$$\boldsymbol{\Sigma} = \overset{\quad m_1 \qquad m_2}{\begin{bmatrix} \boldsymbol{\Sigma}_{11} & \boldsymbol{\Sigma}_{12} \\ \boldsymbol{\Sigma}_{21} & \boldsymbol{\Sigma}_{22} \end{bmatrix}} \begin{matrix} m_1 \\ m_2 \end{matrix}, \qquad \mathbf{T} = \boldsymbol{\Sigma}_{11}^{-1} \boldsymbol{\Sigma}_{12}, \qquad \hat{\mathbf{T}} = \mathbf{A}_{11}^{-1} \mathbf{A}_{12},$$

$$\mathbf{A}_{22 \cdot 1} = \mathbf{A}_{22} - \mathbf{A}_{21} \mathbf{A}_{11}^{-1} \mathbf{A}_{12}$$

Table 8.6.1 *Continued*

To decide if a parameter point $\mathbf{T}_0$ lies inside or outside the $(1-\alpha)$ H.P.D. region, from (8.5.23), use the approximation

$$-\phi'(v+m_1)\log U(\mathbf{T}) \;\dot{\sim}\; \chi^2_{m_1m_2}$$

where

$$\phi' = 1 + \frac{m_1+m_2-3}{2(v+m_1)}$$

and

$$U(\mathbf{T}) = |\mathbf{A}_{22\cdot1}|\,|\mathbf{A}_{22\cdot1}+(\mathbf{T}-\hat{\mathbf{T}})'\mathbf{A}_{11}(\mathbf{T}-\hat{\mathbf{T}})|^{-1}$$

so that $\mathbf{T}_0$ lies inside the region if

$$-\phi'(v+m_1)\log U(\mathbf{T}_0) < \chi^2(m_1m_2,\alpha).$$

Thus, for $m_1=m_2=1$, $\phi'(v+m_1)=9.5$, $\mathbf{A}_{22\cdot1}=a_{22\cdot1}=12.63$ and $\hat{T}=0.63$. If $T_0=0$, then

$$U(T_0=0) = 12.63/[12.63+(0.63)^2\times 61.41]=0.34$$

and the point $T_0=0$ will lie inside the $(1-\alpha)$ region if

$$-9.5\log 0.34 = 10.3 < \chi^2(1,\alpha).$$

Note that (i) $U(\mathbf{0})=|\mathbf{A}|\{|\mathbf{A}_{11}|\,|\mathbf{A}_{22}|\}^{-1}$, and (ii) since $m_1=m_2=1$, exact results are, of course, available.

7. Inferences about the correlation coefficient ρ_{ij}: Use the approximating distribution in (8.5.50),

$$p(\rho\mid\mathbf{y}) \;\dot{\propto}\; \frac{(1-\rho^2)^{\frac{1}{2}(v-2)}}{(1-\rho r)^{v+\frac{1}{2}}}, \qquad -1<\rho<1,$$

where

$$\rho=\rho_{ij} \quad\text{and}\quad r=r_{ij}=a_{ij}/(a_{ii}a_{jj})^{1/2}, \qquad i,j=1,\ldots,m.$$

Thus,

$$p(\rho\mid\mathbf{y}) \;\dot{\propto}\; \frac{(1-\rho^2)^{3.5}}{(1+0.81\rho)^{9.5}}, \qquad -1<\rho<1.$$

The normalizing constant of the distribution can be obtained by numerical integration when desired.

APPENDIX A8.1

The Jacobians of Some Matrix Transformations

We here give the Jacobians of some matrix transformations useful in multivariate problems. In what follows the signs of the Jacobians are ignored.

a) Let $\mathbf{X}$ be a $k\times m$ matrix of km distinct random variables. Let $\mathbf{A}$ and $\mathbf{B}$ be, respectively, a $k\times k$ and $m\times m$ matrices of fixed elements. If

$$\mathbf{Z}_1 = \mathbf{A}\,\mathbf{X}, \qquad \mathbf{Z}_2 = \mathbf{X}\,\mathbf{B}, \qquad \mathbf{Z}_3 = \mathbf{A}\,\mathbf{X}\,\mathbf{B},$$

then the Jacobians are, respectively,

$$\left|\frac{\partial \mathbf{Z}_1}{\partial \mathbf{X}}\right| = |\mathbf{A}|^m, \qquad \left|\frac{\partial \mathbf{Z}_2}{\partial \mathbf{X}}\right| = |\mathbf{B}|^k, \qquad \left|\frac{\partial \mathbf{Z}_3}{\partial \mathbf{X}}\right| = |\mathbf{A}|^m\,|\mathbf{B}|^k. \quad \text{(A8.1.1)}$$

b) Let $\mathbf{X}$ be a $m \times m$ symmetric matrix consisting of $\frac{1}{2}m(m+1)$ distinct elements and let $\mathbf{C}$ be a $m \times m$ matrix of fixed elements. If

$$\mathbf{Y} = \mathbf{C}\,\mathbf{X}\,\mathbf{C}',$$

then

$$\left|\frac{\partial \mathbf{Y}}{\partial \mathbf{X}}\right| = |\mathbf{C}|^{m+1}. \quad \text{(A8.1.2)}$$

For proofs, see Deemer and Olkin (1951), based on results given by P. L. Hsu, also Anderson (1958, p. 162).

c) Let $\mathbf{X}$ be a $m \times m$ nonsingular matrix of random variables and $\mathbf{Y} = \mathbf{X}^{-1}$. Then,

$$\frac{\partial \mathbf{Y}}{\partial z} = -\,\mathbf{Y}\frac{\partial \mathbf{X}}{\partial z}\mathbf{Y}, \quad \text{(A8.1.3)}$$

where z is any function of the elements of $\mathbf{X}$.

Proof: Since $\mathbf{X}\,\mathbf{Y} = \mathbf{I}$, it follows that

$$\frac{\partial}{\partial z}(\mathbf{X}\,\mathbf{Y}) = \left(\frac{\partial \mathbf{X}}{\partial z}\right)\mathbf{Y} + \mathbf{X}\frac{\partial \mathbf{Y}}{\partial z} = \mathbf{0}.$$

Hence

$$\frac{\partial \mathbf{Y}}{\partial z} = -\,\mathbf{Y}\left(\frac{\partial \mathbf{X}}{\partial z}\right)\mathbf{Y}.$$

The Jacobians of two special cases of $\mathbf{X}$ are of particular interest.

a) If $\mathbf{X}$ has m^2 distinct random variables, then from (A8.1.1)

$$\left|\frac{\partial \mathbf{Y}}{\partial \mathbf{X}}\right| = |\mathbf{Y}|^{2m}. \quad \text{(A8.1.4)}$$

b) If $\mathbf{X}$ is symmetric, and consists of $\frac{1}{2}m(m+1)$ distinct random variables, then from (A8.1.2)

$$\left|\frac{\partial \mathbf{Y}}{\partial \mathbf{X}}\right| = |\mathbf{Y}|^{m+1}. \quad \text{(A8.1.5)}$$

APPENDIX A8.2

The Determinant of the Information Matrix of $\boldsymbol{\Sigma}^{-1}$

We now obtain the determinant of the information matrix of $\boldsymbol{\Sigma}^{-1}$ for the m-dimensional Normal distribution $N_m(\boldsymbol{\mu}, \boldsymbol{\Sigma})$. The density is

$$p(\mathbf{y} \mid \boldsymbol{\mu}, \boldsymbol{\Sigma}) = (2\pi)^{-m/2} |\boldsymbol{\Sigma}^{-1}|^{1/2} \exp\left[-\tfrac{1}{2} \operatorname{tr} \boldsymbol{\Sigma}^{-1} (\mathbf{y} - \boldsymbol{\mu})(\mathbf{y} - \boldsymbol{\mu})'\right],$$
$$-\infty < \mathbf{y} < \infty, \qquad \text{(A8.2.1)}$$

where $\mathbf{y} = (y_1, \ldots, y_m)'$, $\boldsymbol{\mu} = (\mu_1, \ldots, \mu_m)'$, $\boldsymbol{\Sigma} = \{\sigma_{ij}\}$ and $\boldsymbol{\Sigma}^{-1} = \{\sigma^{ij}\}$, $i, j = 1, \ldots, m$. We assume that $\boldsymbol{\Sigma}$ (and $\boldsymbol{\Sigma}^{-1}$) consists of $\frac{1}{2}m(m+1)$ distinct elements. Taking logarithms of the density function, we have

$$\log p = -\frac{m}{2} \log(2\pi) + \tfrac{1}{2} \log |\boldsymbol{\Sigma}^{-1}| - \tfrac{1}{2} \operatorname{tr} \boldsymbol{\Sigma}^{-1} (\mathbf{y} - \boldsymbol{\mu})(\mathbf{y} - \boldsymbol{\mu})'. \qquad \text{(A8.2.2)}$$

Differentiating $\log p$ with respect to σ^{ij}, $i, j = 1, \ldots, m$, $i \geqslant j$,

$$\frac{\partial \log p}{\partial \sigma^{ij}} = \frac{1}{2} \frac{1}{|\boldsymbol{\Sigma}^{-1}|} \frac{\partial |\boldsymbol{\Sigma}^{-1}|}{\partial \sigma^{ij}} - (y_i - \mu_i)(y_j - \mu_j). \qquad \text{(A8.2.3)}$$

Since $\partial |\boldsymbol{\Sigma}^{-1}| / \partial \sigma^{ij} = \alpha^{ij}$ where α^{ij} is the cofactor of σ^{ij}, it follows that the first term on the right-hand side of (A8.2.3) is simply $\frac{1}{2}\sigma_{ij}$. Thus, the second derivatives are

$$\frac{\partial^2 \log p}{\partial \sigma^{ij}\, \partial \sigma^{kl}} = \frac{1}{2} \frac{\partial \sigma_{ij}}{\partial \sigma^{kl}}, \qquad \begin{pmatrix} i, j = 1, \ldots, m; & i \geqslant j \\ k, l = 1, \ldots, m; & k \geqslant l \end{pmatrix}. \qquad \text{(A8.2.4)}$$

Consequently, the determinant of the information matrix is proportional to

$$|\mathscr{I}(\boldsymbol{\Sigma}^{-1})| = \left| -E\left\{\frac{\partial^2 \log P}{\partial \sigma^{ij}\, \partial \sigma^{kl}}\right\}\right| \propto \left|\frac{\partial \boldsymbol{\Sigma}}{\partial \boldsymbol{\Sigma}^{-1}}\right|. \qquad \text{(A8.2.5)}$$

From (A8.1.5), we have that

$$\left|\frac{\partial \boldsymbol{\Sigma}}{\partial \boldsymbol{\Sigma}^{-1}}\right| = |\boldsymbol{\Sigma}|^{m+1}. \qquad \text{(A8.2.6)}$$

APPENDIX A8.3

The Normalizing Constant of the $t_{km}[\hat{\boldsymbol{\theta}}, (\mathbf{X}'\mathbf{X})^{-1}, \mathbf{A}, \nu]$ Distribution

Let $\boldsymbol{\theta}$ be a $k \times m$ matrix of variables. We now show that

$$\int_{-\infty < \boldsymbol{\theta} < \infty} |\mathbf{I}_m + \mathbf{A}^{-1} (\boldsymbol{\theta} - \hat{\boldsymbol{\theta}})' \mathbf{X}'\mathbf{X} (\boldsymbol{\theta} - \hat{\boldsymbol{\theta}})|^{-\frac{1}{2}(\nu + k + m - 1)} d\boldsymbol{\theta} \qquad \text{(A8.3.1)}$$
$$= c(m, k, \nu)\, |\mathbf{X}'\mathbf{X}|^{-m/2}\, |\mathbf{A}^{-1}|^{-k/2},$$

where $\nu > 0$, $\hat{\boldsymbol{\theta}}$ is a $k \times m$ matrix, $\mathbf{X}'\mathbf{X}$ and $\mathbf{A}^{-1}$ are, respectively, a $k \times k$ and a $m \times m$ positive definite symmetric matrix,

$$c(m, k, \nu) = [\Gamma(\tfrac{1}{2})]^{mk} \frac{\Gamma_m[\frac{1}{2}(\nu + m - 1)]}{\Gamma_m[\frac{1}{2}(\nu + k + m - 1)]}$$

and $\Gamma_p(b)$ is the generalized Gamma function defined in (8.2.22).

Since $\mathbf{X}'\mathbf{X}$ and $\mathbf{A}$ are assumed definite, there exist a $k \times k$ and a $m \times m$ nonsingular matrices $\mathbf{G}$ and $\mathbf{H}$ such that

$$\mathbf{X}'\mathbf{X} = \mathbf{G}'\mathbf{G} \qquad \text{and} \qquad \mathbf{A}^{-1} = \mathbf{H}\mathbf{H}'. \tag{A8.3.2}$$

Let $\mathbf{T}$ be a $k \times m$ matrix such that

$$\mathbf{T} = \mathbf{G}(\boldsymbol{\theta} - \hat{\boldsymbol{\theta}})\mathbf{H}. \tag{A8.3.3}$$

Using the identity (8.4.11) we can write

$$|\mathbf{I}_m + \mathbf{A}^{-1}(\boldsymbol{\theta} - \hat{\boldsymbol{\theta}})'\mathbf{X}'\mathbf{X}(\boldsymbol{\theta} - \hat{\boldsymbol{\theta}})| = |\mathbf{I}_m + \mathbf{H}'(\boldsymbol{\theta} - \hat{\boldsymbol{\theta}})'\mathbf{G}'\mathbf{G}(\boldsymbol{\theta} - \hat{\boldsymbol{\theta}})\mathbf{H}| = |\mathbf{I}_m + \mathbf{T}'\mathbf{T}|$$
$$= |\mathbf{I}_k + \mathbf{T}\mathbf{T}'|. \tag{A8.3.4}$$

From (A8.1.1) the Jacobian of the transformation (A8.3.3) is

$$\left|\frac{\partial \mathbf{T}}{\partial \boldsymbol{\theta}}\right| = |\mathbf{G}|^m |\mathbf{H}|^k = |\mathbf{X}'\mathbf{X}|^{m/2} |\mathbf{A}^{-1}|^{k/2}. \tag{A8.3.5}$$

Thus, the integral on the left-hand side of (A8.3.1) is

$$|\mathbf{X}'\mathbf{X}|^{-m/2} |\mathbf{A}^{-1}|^{-k/2} Q_m, \tag{A8.3.6}$$

where

$$Q_m = \int_{-\infty < \mathbf{T} < \infty} |\mathbf{I}_k + \mathbf{T}\mathbf{T}'|^{-\frac{1}{2}(\nu + k + m - 1)} \, d\mathbf{T}. \tag{A8.3.7}$$

Let $\mathbf{T} = [\mathbf{t}_1, \ldots, \mathbf{t}_m] = [\mathbf{T}_1, \mathbf{t}_m]$ where $\mathbf{t}_i$ is a $k \times 1$ vector, $i = 1, \ldots, m$. Then,

$$\begin{aligned} |\mathbf{I}_k + \mathbf{T}\mathbf{T}'| &= |\mathbf{I}_k + \mathbf{T}_1\mathbf{T}_1' + \mathbf{t}_m\mathbf{t}_m'| \\ &= |\mathbf{I}_k + \mathbf{T}_1\mathbf{T}_1'| \, [1 + \mathbf{t}_m'(\mathbf{I}_k + \mathbf{T}_1\mathbf{T}_1')^{-1}\mathbf{t}_m]. \end{aligned} \tag{A8.3.8}$$

It follows that

$$Q_m = \int_{-\infty < \mathbf{T}_1 < \infty} |\mathbf{I}_k + \mathbf{T}_1\mathbf{T}_1'|^{-\frac{1}{2}(\nu + k + m - 1)} q_m \, d\mathbf{T}_1, \tag{A8.3.9}$$

where

$$q_m = \int_{-\infty < \mathbf{t}_m < \infty} [1 + \mathbf{t}_m'(\mathbf{I}_k + \mathbf{T}_1\mathbf{T}_1')^{-1}\mathbf{t}_m]^{-\frac{1}{2}(\nu + m - 1 + k)} d\mathbf{t}_m.$$

From (A2.1.12) in Appendix A2.1,

$$q_m = [\Gamma(\tfrac{1}{2})]^k \frac{\Gamma[\frac{1}{2}(\nu + m - 1)]}{\Gamma[\frac{1}{2}(\nu + k + m - 1)]} |\mathbf{I}_k + \mathbf{T}_1\mathbf{T}_1'|^{1/2}.$$

Thus,

$$Q_m = [\Gamma(\tfrac{1}{2})]^k \frac{\Gamma[\frac{1}{2}(\nu + m - 1)]}{\Gamma[\frac{1}{2}(\nu + k + m - 1)]} Q_{m-1},$$

where

$$Q_{m-1} = \int_{-\infty < \mathbf{T}_1 < \infty} |\mathbf{I}_k + \mathbf{T}_1\mathbf{T}_1|^{-\frac{1}{2}(\nu+m-2+k)}\, d\mathbf{T}_1.$$

The result in (A8.3.1) follows by repeating the process $m - 1$ times.

APPENDIX A8.4

The Kronecker Product of Two Matrices

We summarize in this appendix some properties of the Kronecker product of two matrices.

Definition: If $\mathbf{A}$ is a $m \times m$ matrix and $\mathbf{B}$ is a $n \times n$ matrix, then the Kronecker product of $\mathbf{A}$ and $\mathbf{B}$ in that order is the $(mn) \times (mn)$ matrix

$$\mathbf{A} \otimes \mathbf{B} = \begin{bmatrix} a_{11}\mathbf{B} & \cdots & a_{1m}\mathbf{B} \\ \vdots & & \\ a_{m1}\mathbf{B} & & a_{mm}\mathbf{B} \end{bmatrix}$$

Properties:

i) $(\mathbf{A} \otimes \mathbf{B})' = \mathbf{A}' \otimes \mathbf{B}'$

ii) If $\mathbf{A}$ and $\mathbf{B}$ are symmetric, then $\mathbf{A}\otimes\mathbf{B}$ is symmetric.

iii) When $\mathbf{A}$ and $\mathbf{B}$ are non-singular, then

$$(\mathbf{A} \otimes \mathbf{B})^{-1} = \mathbf{A}^{-1} \otimes \mathbf{B}^{-1}$$

iv) $\operatorname{tr} \mathbf{A} \otimes \mathbf{B} = \operatorname{tr} \mathbf{A} \operatorname{tr} \mathbf{B}$

v) $|\mathbf{A} \otimes \mathbf{B}| = |\mathbf{A}|^n\, |\mathbf{B}|^m$.

vi) If $\mathbf{C}$ is a $m \times m$ matrix and $\mathbf{D}$ is a $n \times n$ matrix, then

$$(\mathbf{A} + \mathbf{C}) \otimes \mathbf{B} = \mathbf{A} \otimes \mathbf{B} + \mathbf{C} \otimes \mathbf{B}$$

and

$$(\mathbf{A} \otimes \mathbf{B})\,(\mathbf{C} \otimes \mathbf{D}) = (\mathbf{AC}) \otimes (\mathbf{BD}).$$

CHAPTER 9

ESTIMATION OF COMMON REGRESSION COEFFICIENTS

9.1 INTRODUCTION: PRACTICAL IMPORTANCE OF COMBINING INFORMATION FROM DIFFERENT SOURCES

Problems occur, notably in engineering, economics, and business, where it is difficult to obtain sufficiently reliable estimates of critical parameters. One way to improve estimates is to seek more data of the same kind. But in engineering this may mean running further costly experiments while in economics and business further data from the same source may not be available. Another way in which estimates can sometimes be improved is by combining information from *different kinds* of data. Thus, in Section 8.2.6 we considered a chemical system producing three different products where the yield of each product depended on one or both of two chemical rate parameters. It was demonstrated that, by appropriately utilizing observations on all three products, much more precise inferences could be drawn than would have been possible with only one.

As a further illustration, in a study of the effect of interest rates on the demand for durables, an economist might wish to appropriately combine demand data from m different communities. Let $\mathbf{y}_i = (y_{1i}, \ldots, y_{ni})'$ be an $n \times 1$ vector, representing demand over n specific periods, for the ith community. Suppose that

$$\begin{aligned} \mathbf{y}_i &= E(\mathbf{y}_i) + \boldsymbol{\varepsilon}_i, \\ E(\mathbf{y}_i) &= \boldsymbol{\eta}_i(\boldsymbol{\xi}_i, \theta_i), \qquad i = 1, \ldots, m, \end{aligned} \tag{9.1.1}$$

where θ_i is a parameter for the ith community, measuring the effect on demand of a change in the interest rate, $\boldsymbol{\xi}_i$ is the corresponding $n \times 1$ vector of interest rates, and $\boldsymbol{\varepsilon}_i = (\varepsilon_{1i}, \ldots, \varepsilon_{ni})'$ is the associated $n \times 1$ vector of errors.

To complete the model we must make specific assumptions about (i) the relationship among $(\theta_1, \ldots, \theta_m)$, (ii) the nature of the expectation functions $\boldsymbol{\eta}_i(\boldsymbol{\xi}_i, \theta_i)$ and (iii) the form of the probability distribution of $\boldsymbol{\varepsilon}_i$. We discuss these assumptions below:

(i) If there is reason to believe that the parameters $(\theta_1, \ldots, \theta_m)$ for the m different communities resemble one another, then, irrespective of whether or not there are correlations among the m vectors of errors, $(\boldsymbol{\varepsilon}_1, \ldots, \boldsymbol{\varepsilon}_m)$, information about θ_i from the ith response vector $\mathbf{y}_i$ ought to contribute to knowledge about θ_j. A formulation which takes account of one kind of resemblence, supposes that $\theta_1, \ldots, \theta_m$

are m independent observations from some population distributed as $f(\theta)$. In particular, the random effect model discussed in Section 7.2 was of this kind. If the variance of $f(\theta)$ was expected to be small, the m vectors of expectation functions $\boldsymbol{\eta}_i$ might then be regarded as containing a *common* parameter $\theta_1 = \ldots = \theta_m = \theta_c$ and data from all m communities could be used to make inferences about θ_c. We have already considered a multivariate setup of this kind in Section 8.2.6. It was there possible to show that, even in the general situation where the expectation functions were nonlinear in the parameters, exact results could be obtained yielding the posterior distribution of the common parameters in a form which was mathematically simple.

ii) Mathematical simplicity does not invariably lead to tractible computation however, and we now consider the problem with the added simplification that the expectation functions are *linear* in the parameter θ_c. As usual, consideration of the more manageable linear case has more than one motivation. On the one hand, practical problems do occur where the appropriate expectation functions are indeed linear in the parameters. On the other hand, for nonlinear problems, we rarely need to concern ourselves with nonlinearity over the whole permissible region of variation of θ_c, but only in some smaller sub-region of the parameter space—for example, over a $(1 - \alpha)$ H.P.D. region for some small α. A function which is "globally" highly nonlinear in θ can be, in this sense, "locally" nearly linear. In such a case, as is indicated in Section 8.3.1, linear distribution theory can provide useful approximations. In the Bayesian formulation, we are concerned only with approximate linearity in relation to the one sample of observations which has in fact occurred and not with some hypothetical "repeated samples from the same population" which have not. It follows that we can, *for any given data sample*, check the adequacy of the assumption of local linearity by direct numerical computation as part of the statistical analysis.

iii) We shall suppose that the n vectors $\boldsymbol{\varepsilon}_{(u)} = (\varepsilon_{u1}, \ldots, \varepsilon_{um})$, $u = 1, \ldots, n$, are a sample of independent observations from the m-dimensional Normal population $N_m(\mathbf{0}, \boldsymbol{\Sigma})$, where $\boldsymbol{\Sigma} = \{\sigma_{ij}\}$.

In general, the relationships connecting demand with interest rate will contain, not one, but k common parameters $\boldsymbol{\theta}_c$. Thus, with the above assumptions we are led to a linear, Normal, common parameter model, which can be written

$$\mathbf{y}_i = \mathbf{X}_i \boldsymbol{\theta}_c + \boldsymbol{\varepsilon}_i, \qquad i = 1, \ldots, m \tag{9.1.2}$$

with $\mathbf{X}_i$ an $n \times k$ matrix appropriate to the ith community and $\boldsymbol{\theta}_c$ a $k \times 1$ vector of parameters common to all communities.

This model (9.1.2) will be recognized as a special case of the general linear model set out in (8.3.2)

$$\mathbf{y}_i = \mathbf{X}_i \boldsymbol{\theta}_i + \boldsymbol{\varepsilon}_i, \qquad i = 1, \ldots, m \tag{9.1.3}$$

appropriate to the circumstance that the regression coefficients $\boldsymbol{\theta}_1, \ldots, \boldsymbol{\theta}_m$ associated with the m vectors of responses are *common*, so that

$$\boldsymbol{\theta}_1 = \ldots = \boldsymbol{\theta}_m = \boldsymbol{\theta}_c. \tag{9.1.4}$$

In what follows we shall first discuss the situation, where the m response vectors $\mathbf{y}_1, \ldots, \mathbf{y}_m$ are supposed independent of one another so that $\boldsymbol{\Sigma}$ is diagonal. The posterior distribution of $\boldsymbol{\theta}_c$ is then a product of m multivariate t distributions. A useful and simple approximation procedure is developed for making inferences about subsets of $\boldsymbol{\theta}_c$ for the case of two responses. The more general situation is next considered where the m responses are correlated. Except when the derivative matrices $\mathbf{X}_1, \ldots, \mathbf{X}_m$ are identical, the resulting posterior distribution of $\boldsymbol{\theta}_c$ is rather complicated analytically. A useful approximation procedure will, however, be developed for $m = 2$.

As we have pointed out in Section 8.2.6, care must be exercised in the pooling of information to estimate common parameters. A natural question that arises in the investigator's mind will be whether the evidence from the data are *compatible* with the supposition that $\boldsymbol{\theta}_1 = \ldots = \boldsymbol{\theta}_m = \boldsymbol{\theta}_c$. In the following section, this question is considered for the important special case of estimating the assumed common mean of two Normal populations which have possibly unequal variances—a problem which has an interesting history.

9.2 THE CASE OF m INDEPENDENT RESPONSES

Suppose there are m independent responses related to the same parameters $\boldsymbol{\theta}_c = (\theta_1, \ldots, \theta_k)'$ by possibly different linear models. Suppose further that the number of observations for each response is not necessarily the same. Then the likelihood function is

$$l(\boldsymbol{\theta}_c, \sigma_{11}, \ldots, \sigma_{mm} \mid \mathbf{y}) \propto \left\{\prod_{i=1}^{m} \sigma_{ii}^{-\frac{1}{2}n_i}\right\} \exp\left\{-\frac{1}{2}\sum_{i=1}^{m} \sigma_{ii}^{-1}(\mathbf{y}_i - \mathbf{X}_i\boldsymbol{\theta}_c)'(\mathbf{y}_i - \mathbf{X}_i\boldsymbol{\theta}_c)\right\}, \tag{9.2.1}$$

where $\mathbf{y}_i$ is an $n_i \times 1$ vector of observations and $\mathbf{X}_i$ a $n_i \times k$ matrix of fixed elements. Relative to the noninformative reference prior

$$p(\boldsymbol{\theta}_c, \sigma_{11}, \ldots, \sigma_{mm}) \propto \prod_{i=1}^{m} \sigma_{ii}^{-1},$$

the posterior distribution is

$$p(\boldsymbol{\theta}_c, \sigma_{11}, \ldots, \sigma_{mm} \mid \mathbf{y}) \propto \left\{\prod_{i=1}^{m} \sigma_{ii}^{-(\frac{1}{2}n_i+1)}\right\}$$

$$\times \exp\left[-\frac{1}{2}\sum_{i=1}^{m} \frac{a_{ii} + (\boldsymbol{\theta}_c - \hat{\boldsymbol{\theta}}_i)'\mathbf{X}_i'\mathbf{X}_i(\boldsymbol{\theta}_c - \hat{\boldsymbol{\theta}}_i)}{\sigma_{ii}}\right], \tag{9.2.2}$$

where $\quad (\mathbf{X}_i'\mathbf{X}_i)\hat{\boldsymbol{\theta}}_i = \mathbf{X}_i'\mathbf{y}_i \quad$ and $\quad a_{ii} = (\mathbf{y}_i - \mathbf{X}_i\hat{\boldsymbol{\theta}}_i)'(\mathbf{y}_i - \mathbf{X}_i\hat{\boldsymbol{\theta}}_i).$

Integrating out $\sigma_{11}, \ldots, \sigma_{mm}$, the posterior distribution of the common parameters $\boldsymbol{\theta}_c$ is then proportional to the product of m multivariate t distributions

$$p(\boldsymbol{\theta}_c \mid \mathbf{y}) \propto \prod_{i=1}^{m} [a_{ii} + (\boldsymbol{\theta}_c - \hat{\boldsymbol{\theta}}_i)'\mathbf{X}_i'\mathbf{X}_i(\boldsymbol{\theta}_i - \hat{\boldsymbol{\theta}}_i)]^{-\frac{1}{2}n_i}, \qquad -\infty < \boldsymbol{\theta}_c < \infty. \tag{9.2.3}$$

Properties of this distribution have been considered by Dickey (1967a, 1968) and others.

9.2.1 The Weighted Mean Problem

The simplest problem of this kind has come to be called the problem of the *weighted mean*, Yates (1939). Suppose that $m = 2$, $k = 1$ and $(\mathbf{X}_1 = \mathbf{1}_{n_1}, \mathbf{X}_2 = \mathbf{1}_{n_2})$ are columns of ones. Typically we have two sets of Normally distributed observations $\mathbf{y}_1 = (y_{11}, \ldots, y_{n_1 1})'$ and $\mathbf{y}_2 = (y_{12}, \ldots, y_{n_2 2})'$ independent of one another and having different variances $\sigma_1^2 = \sigma_{11}$ and $\sigma_2^2 = \sigma_{22}$ but both assumed to have the same mean θ. The posterior distribution of θ is then proportional to the product of two t distributions

$$p(\theta \mid \mathbf{y}) \propto p(\theta \mid \mathbf{y}_1)p(\theta \mid \mathbf{y}_2) \propto \left[1 + \frac{n_1(\theta - \bar{y}_1)^2}{v_1 s_1^2}\right]^{-\frac{1}{2}(v_1+1)}$$

$$\times \left[1 + \frac{n_2(\theta - \bar{y}_2)^2}{v_2 s_2^2}\right]^{-\frac{1}{2}(v_2+1)}, \qquad -\infty < \theta < \infty, \tag{9.2.4}$$

where

$$v_i = n_i - 1, \qquad v_i s_i^2 = \Sigma(y_{ui} - \bar{y}_i)^2, \qquad i = 1, 2,$$

a distribution given by Jeffreys (1961). Simultaneously Fisher (1961a, b) obtained an identical distribution using a fiducial argument. Except for the normalizing constant, the density (9.2.4) can be easily calculated from a table of the density function of Student's t—e.g. Bracken and Schleifer (1964).

An Example

The following results were obtained in estimating a physical constant by two independent methods

	Method 1	*Method* 2
$\bar{y}$	107.9	109.5
s^2	12.1	39.4
n	11	13

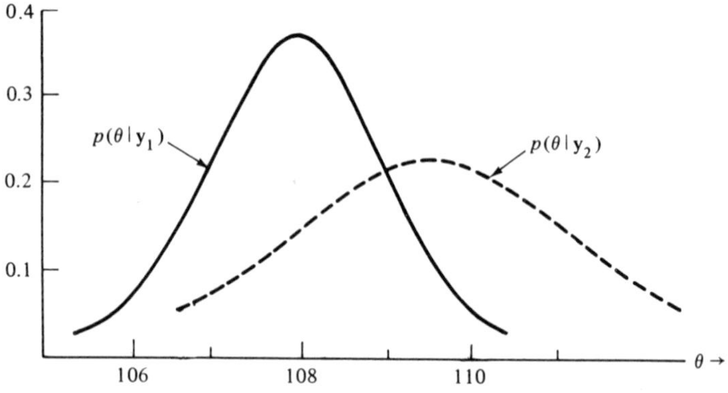

Fig. 9.2.1 Individual posterior distributions of θ: estimation of a common physical constant.

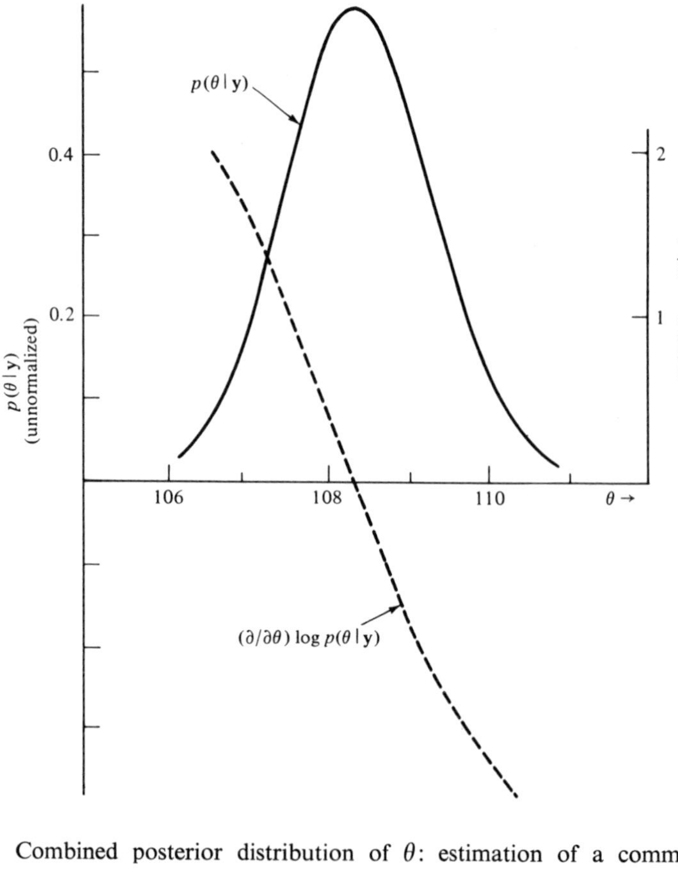

Fig. 9.2.2 Combined posterior distribution of θ: estimation of a common physical constant.

If θ_1 and θ_2 are the means for the two methods, then crucial to what follows is the assumption that $\theta_1 = \theta_2 = \theta$. We shall entertain this assumption for the moment and later discuss it further in Section 9.2.3. With this assumption, then, the two t distributions shown in Fig. 9.2.1 centered at 107.9 and 109.5 with scale factors $s_1/\sqrt{n_1} = 1.03$ and $s_2/\sqrt{n_2} = 1.74$ are the individual posterior distributions $p(\theta \mid \mathbf{y}_1)$ and $p(\theta \mid \mathbf{y}_2)$ obtained respectively from the results of methods A and B. The *unnormalized* posterior distribution of θ which combines information from both samples, is obtained by multiplying the t densities and is shown by the "t-like" distribution in Fig. 9.2.2. As might be expected, the combined distribution is "located" between the individual t distributions. For this example, $p(\theta \mid \mathbf{y}_1)$ has a narrower spread than $p(\theta \mid \mathbf{y}_2)$ and consequently exerts a stronger influence on the final posterior distribution.

9.2.2 Some Properties of $p(\theta \mid \mathbf{y})$

Although the distribution $p(\theta \mid \mathbf{y})$ in (9.2.4) is the product of two distributions which are symmetric, the distribution itself will not be symmetric unless either $\bar{y}_1 = \bar{y}_2$ or $n_1 = n_2$ and $s_1^2 = s_2^2$. We see for example in Fig. 9.2.2 that the distribution is slightly skewed to the right.

Upon differentiating the logarithm of $p(\theta \mid \mathbf{y})$ with respect to θ, it is easy to see that for $\bar{y}_1 < \bar{y}_2$

$$\frac{\partial}{\partial\theta} \log p(\theta \mid \mathbf{y}) \gtrless 0 \qquad \text{for} \qquad \begin{matrix} \theta < \bar{y}_1 \\ , \\ \theta > \bar{y}_2 \end{matrix} \tag{9.2.5}$$

where

$$\frac{\partial}{\partial\theta} \log p(\theta \mid \mathbf{y}) = -\frac{n_1^2(\theta - \bar{y}_1)}{\nu_1 s_1^2 + n_1(\theta - \bar{y}_1)^2} - \frac{n_2^2(\theta - \bar{y}_2)}{\nu_2 s_2^2 + n_2(\theta - \bar{y}_2)^2}, \tag{9.2.6}$$

so that the distribution is increasing in $(-\infty, \bar{y}_1)$ and decreasing in $(\bar{y}_2, \infty)$ and consequently the mode(s) of the distribution will lie between $(\bar{y}_1, \bar{y}_2)$.

Bimodality

When $\bar{y}_1$ and $\bar{y}_2$ are wide enough apart, the distribution becomes bimodal. By setting $(\partial/\partial\theta) \log p(\theta \mid \mathbf{y}) = 0$, it can be verified that when $n_1 = n_2 = n$ (so that $\nu_1 = \nu_2 = \nu$) a sufficient and necessary condition for bimodality is

$$\left[d - 2\nu\left(\frac{s_1^2}{n} + \frac{s_2^2}{n}\right)\right]^3 - 27d\left(\frac{\nu s_1^2}{n} - \frac{\nu s_2^2}{n}\right)^2 > 0, \tag{9.2.7}$$

where

$$d = (\bar{y}_1 - \bar{y}_2)^2.$$

To consider the implications of this condition, suppose for the moment that the sample variances are nearly equal, so that $s_1^2 \doteq s_2^2 = s^2$ and the second term on the left can be ignored, then (9.2.7) implies that the distribution will be bimodal only if

$$|\bar{y}_2 - \bar{y}_1| > 2\sqrt{\nu}\,(s/\sqrt{n}), \tag{9.2.8}$$

i.e., if

$$|\bar{y}_2 - \bar{y}_1| > 2\sqrt{\nu - 2}\,(\text{s.d.}),$$

where

$$\text{s.d.} = \frac{s}{\sqrt{n}}\sqrt{\frac{\nu}{\nu - 2}}$$

is the standard deviation of the individual posterior distributions $p(\theta \mid \mathbf{y}_1)$ and $p(\theta \mid \mathbf{y}_2)$. For example, when $\nu = 11$ this means that $\bar{y}_1$ and $\bar{y}_2$ must be at least 6 standard deviations apart before the distribution becomes bimodal.

The practical implication of this analysis is now considered. A question mentioned already and to be discussed in more detail in Section 9.2.3 is the appropriateness of the assumption of *compatibility* of the means. That is, the assumption that, to a sufficient approximation, $\theta_1 = \theta_2 = \theta$. One might at first think that bimodality in the posterior distribution of θ or its absence could provide a guide as to whether the compatibility assumption was reasonable. Expression (9.2.8) makes it clear that while bimodality would normally imply incompatibility, lack of bimodality certainly need not imply the converse.

To see why this is so, we have only to remember that for sufficiently large ν the component t distributions will approach the Normal form. Thus, for large sample sizes the right-hand side of (9.2.4) approaches the product of two Normal densities. But the product of two Normal densities, whatever their separation, is proportional to another Normal density, which is of course unimodal. It is not surprising then that, relative to their spread, the separation of the two component posterior density functions necessary to induce bimodality increases without limit as ν is increased.

Normal Approximation to $p(\theta \mid \mathbf{y})$

When the two sample means $\bar{y}_1$ and $\bar{y}_2$ are not widely discrepant, the combined posterior distribution (9.2.4) will be unimodal and located between the two individual t distributions. Although the unnormalized density function of θ can be easily determined, it does not seem possible to express the posterior probability integrals in a simple form. Thus, for example, exact evaluation of the probability content of a specific H.P.D. interval would require the use of numerical methods. However, we see that when both ν_1 and ν_2 are large the distribution in (9.2.4) is approximately

$$p(\theta \mid \mathbf{y}) \mathrel{\dot{\propto}} \exp\left[-\tfrac{1}{2}n'(\theta - \bar{\theta})^2\right], \tag{9.2.9}$$

where

$$n' = \frac{n_1}{s_1^2} + \frac{n_2}{s_2^2}, \qquad \bar{\theta} = \frac{1}{n'}\left(\frac{n_1}{s_1^2}\bar{y}_1 + \frac{n_2}{s_2^2}\bar{y}_2\right).$$

As shown by Fisher (1961b), the above expression can be made the leading term of an asymptotic expansion of the distribution in (9.2.4) in powers of ν_1^{-1} and ν_2^{-1}. The "correction" terms in the asymptotic series are polynomials in θ. In Appendix A9.1 the asymptotic procedure is given for the more general situation when $m = 2$ but $\boldsymbol{\theta}_c$ consists of more than one element. An alternative and simpler approximation to the distribution of $\boldsymbol{\theta}_c$ based upon multivariate t distribution will be developed in Section 9.3.

Returning to the distribution shown in Fig. 9.2.2 for the physical constant example, we see that the distribution is only slightly asymmetrical. Using the Normal approximation in (9.2.9), we have

$$\bar{\theta} = \left(\frac{11}{12.1} + \frac{13}{39.4}\right)^{-1} \left[\frac{11}{12.1}107.9 + \frac{13}{39.4}109.5\right] = 108.34,$$

which nearly coincides with the mode of the distribution in the figure. This itself, of course, does not imply that the distribution can be closely approximated by a Normal distribution without "correction" terms. A convenient device advocated by Barnard† for checking Normality is to graph the first derivative of the logarithm of the posterior density function against the parameter value. A unique characteristic of the Normal distribution is that this plot produces a straight line. A plot of the derivative in (9.2.6) in the range where the density is appreciable is shown by the broken line in Fig. 9.2.2. It is nearly linear, so that the posterior distribution $p(\theta \mid \mathbf{y})$ may be treated approximately as $N(108.34, 0.807)$.

9.2.3 Compatibility of the Means

Before information from m sets of measurements obtained by m different methods are combined, one should first consider whether the methods are measuring the same thing. In particular, it is helpful to compare individual posterior distributions, as was done for the numerical example in Fig. 9.2.1. More formally, questions of compatibility may be studied by considering the distribution of contrasts among the means derived from the joint "unconstrained" distribution of $\theta_1, \ldots, \theta_m$, when these parameters are *not* regarded as common. We illustrate with the case $m = 2$ and, as before, use an asterisk $*$ to denote unconstrained distributions, the constraint being that $\theta_1 = \theta_2$. As we have seen in Section 2.5, relative to the noninformative reference prior distribution

$$p^*(\theta_1, \theta_2, \sigma_1, \sigma_2) \propto (\sigma_1\sigma_2)^{-1}, \tag{9.2.10}$$

the joint posterior distribution of θ_1 and θ_2 is

$$p^*(\theta_1, \theta_2 \mid \mathbf{y}) \propto \left[1 + \frac{n_1(\theta_1 - \bar{y}_1)^2}{\nu_1 s_1^2}\right]^{-\frac{1}{2}(\nu_1+1)} \left[1 + \frac{n_2(\theta_2 - \bar{y}_2)^2}{\nu_2 s_2^2}\right]^{-\frac{1}{2}(\nu_2+1)},$$

$$-\infty < \theta_1 < \infty, \quad -\infty < \theta_2 < \infty. \tag{9.2.11}$$

† Personal communication.

Figure (9.2.3) shows the contours of (θ_1, θ_2) as well as various other aspects of the distribution.

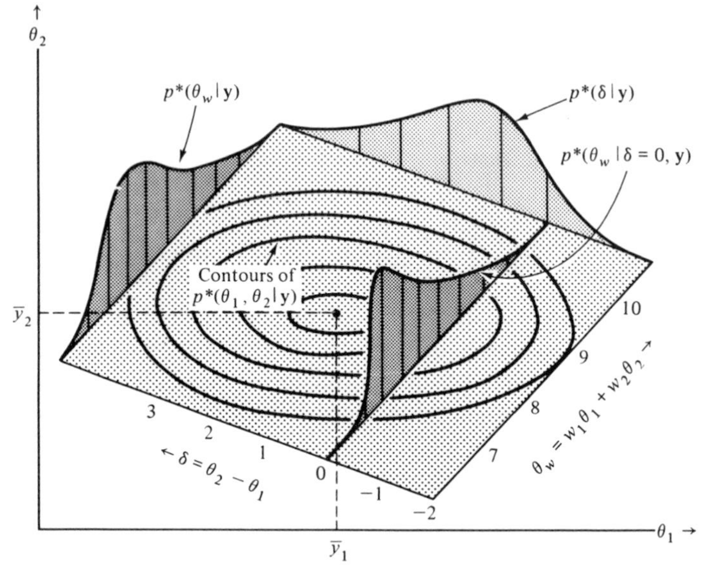

Fig. 9.2.3 Various aspects of the unconstrained posterior distribution of (θ_1, θ_2).

To obtain the distribution of the difference in means we write

$$\delta = \theta_2 - \theta_1 \qquad \text{and} \qquad \theta_w = w_1\theta_1 + w_2\theta_2, \tag{9.2.12}$$

where

$$0 \leqslant w_1 \leqslant 1 \qquad \text{and} \qquad w_1 + w_2 = 1.$$

Then the joint distribution of δ and θ_w is

$$p^*(\delta, \theta_w \mid \mathbf{y}) \propto \left[1 + \frac{n_1(\theta_w - w_2\delta - \bar{y}_1)^2}{\nu_1 s_1^2}\right]^{-\frac{1}{2}(\nu_1+1)} \left[1 + \frac{n_2(\theta_w + w_1\delta - \bar{y}_2)^2}{\nu_2 s_2^2}\right]^{-\frac{1}{2}(\nu_2+1)},$$

$$-\infty < \delta < \infty, \qquad -\infty < \theta_w < \infty. \tag{9.2.13}$$

The distribution can be written as the product

$$p^*(\delta, \theta_w \mid \mathbf{y}) = p^*(\delta \mid \mathbf{y})\, p^*(\theta_w \mid \delta, \mathbf{y}). \tag{9.2.14}$$

Whatever the choice of the weights w_1 and w_2, the density $p^*(\delta \mid \mathbf{y})$ will be proportional to the Behrens–Fisher distribution in (2.5.12) appropriate for making inferences about the difference $\delta = \theta_2 - \theta_1$. The distribution $p(\theta \mid \mathbf{y})$ of the common mean (9.2.4) is $p^*(\theta_w \mid \delta = 0, \mathbf{y})$, the conditional distribution of θ_w *given that it is known* that $\delta = \theta_1 - \theta_2 = 0$.

To shed light on the question of whether a supposition that δ was close to zero was, or was not, tenable in the light of the data one could inspect the distribution $p^*(\delta \mid \mathbf{y})$, the appropriately scaled Behrens–Fisher distribution for the difference in means. In a case like that illustrated in Fig. 9.2.4 the relevance of the distribution $p(\theta \mid \mathbf{y})$ in (9.2.4) would be thrown in doubt. Such evidence might suggest that, instead of attempting to pool incompatible information, the investigator ought to direct his efforts to explain possible bias in the methods.

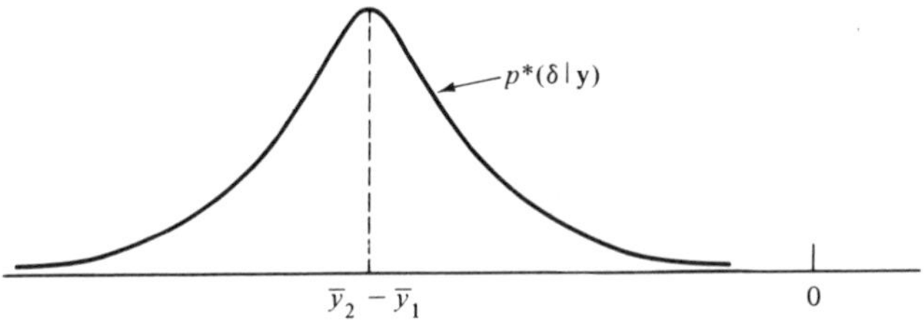

Fig. 9.2.4 Posterior distribution of the difference δ in relation to the value $\delta = 0$.

9.2.4 The Fiducial Distributions of θ Derived by Yates and by Fisher

Yates (1939), in his discussion of the problem of the weighted mean, considered the fiducial distribution of

$$\theta_w = w_1 \theta_1 + w_2 \theta_2$$

in the special case where

$$w_1 = \frac{n_1}{s_1^2} \Big/ \left(\frac{n_1}{s_1^2} + \frac{n_2}{s_2^2}\right), \qquad w_2 = \frac{n_2}{s_2^2} \Big/ \left(\frac{n_1}{s_1^2} + \frac{n_2}{s_2^2}\right) \tag{9.2.15}$$

and found that

$$\theta_w = \left(\frac{n_1}{s_1^2} + \frac{n_2}{s_2^2}\right)^{-1} \left(\frac{n_1}{s_1^2}\theta_1 + \frac{n_2}{s_2^2}\theta_2\right)$$

had a Behrens–Fisher distribution.

However, Yates' fiducial distribution is *not* the same as the fiducial distribution subsequently derived by Fisher. This, at first sight perplexing, situation is clarified by the Bayes analysis.

We have already discussed Fisher's result corresponding to (9.2.4). We now consider Yates' result from this same viewpoint. The marginal posterior distribution of

$\theta_w = w_1\theta_1 + w_2\theta_2$ is obtained in general by integrating out δ in (9.2.14),

$$p^*(\theta_w \mid \mathbf{y}) = \int_{-\infty}^{\infty} p^*(\theta_w \mid \delta, \mathbf{y})\, p^*(\delta \mid \mathbf{y})\, d\delta. \tag{9.2.16}$$

To see that this yields a density function numerically identical to Yates' fiducial distribution, in (9.2.11) write

$$t_1 = \frac{\theta_1 - \bar{y}_1}{s_1/\sqrt{n_1}}, \quad t_2 = \frac{\theta_2 - \bar{y}_2}{s_2/\sqrt{n_2}} \tag{9.2.17}$$

and

$$\tau = (\theta_w - \bar{y}_w) \Big/ \left(\frac{s_1^2}{n_1} w_1^2 + \frac{s_2^2}{n_2} w_2^2\right)^{1/2} = t_2 \cos\phi - t_1 \sin\phi$$
$$z = t_2 \sin\phi + t_1 \cos\phi \tag{9.2.18}$$

where

$$\bar{y}_w = w_1\bar{y}_1 + w_2\bar{y}_2 \quad \text{and} \quad \tan\phi = -\frac{(s_1/\sqrt{n_1})\, w_1}{(s_2/\sqrt{n_2})\, w_2},$$

Then, whatever weights w_1 and $w_2 = 1 - w_1$ are employed, the distribution of τ is

$$p(\tau \mid \nu_1, \nu_2, \phi) \propto \int_{-\infty}^{\infty} \left[1 + \frac{(\tau \sin\phi - z\cos\phi)^2}{\nu_1}\right]^{-\frac{1}{2}(\nu_1+1)} \left[1 + \frac{(z\sin\phi + \tau\cos\phi)^2}{\nu_2}\right]^{-\frac{1}{2}(\nu_2+1)} dz,$$
$$-\infty < \tau < \infty, \tag{9.2.19}$$

which is a Behrens–Fisher distribution. Yates result is obtained by using the particular weighting of (9.2.15).

The basic difference between the two results $p^*(\theta_w \mid \delta = 0, \mathbf{y})$ and $p^*(\theta_w \mid \mathbf{y})$ (see Fig. 9.2.3) is now readily seen. Fisher's result for the distribution of the common mean, which corresponds with Jeffreys' Bayesian solution, is the *conditional* distribution of θ_w (or θ_1 or θ_2), *given that* $\theta_1 = \theta_2$. That is to say, given that $\delta = \theta_1 - \theta_2 = 0$. On the other hand, Yates' result corresponds not to any conditional distribution but to a *marginal* distribution, that of $\theta_w = w_1\theta_1 + w_2\theta_2$.

The relevance of Yates' result, is clearly open to question. For, *if* it can be assumed that $\theta_1 = \theta_2$ and hence that $\delta = \theta_1 - \theta_2 = 0$, one should employ the conditional distribution (9.2.4). Conversely, if θ_1 and θ_2 cannot be assumed equal (implying that one or both contain unknown *systematic* error), there seems to be no basis for the weighting

$$\frac{w_1}{w_2} = \frac{n_1/s_1^2}{n_2/s_2^2}$$

which is based only on *sampling* variation.

There will be cases where it cannot be plausibly asserted that $\theta_1 = \theta_2$ exactly, but nevertheless we may feel fairly sure that the difference $\delta = \theta_1 - \theta_2$ is small. In this case the Fisher–Jeffreys solution may be used as an approximation. Specifically, if an *informative* prior distribution $p(\delta)$ is introduced which might for example be a Normal distribution centered at $\delta = 0$ with standard deviation σ_δ, then, on the same assumptions as before, the posterior distribution of θ_w will be

$$p^*(\theta_w \mid \mathbf{y}) \propto \int_{-\infty}^{\infty} p^*(\theta_w \mid \delta, \mathbf{y})\, p^*(\delta \mid \mathbf{y})\, p(\delta)\, d\delta. \tag{9.2.20}$$

Now (i) if we make σ_δ very large corresponding to the assertion that little is known about δ, we obtain the marginal distribution in (9.2.16), which however will depend on the *particular* weights w_1 and w_2 adopted; (ii) on the other hand, if we make σ_δ small so that $p(\delta)$ approaches a delta function at $\delta = 0$, then the distribution of θ_w tends to the Fisher–Jeffreys solution (9.2.4) *whatever the values of the weights* w_1 *and* w_2.

9.3 SOME PROPERTIES OF THE DISTRIBUTION OF THE COMMON PARAMETERS FOR TWO INDEPENDENT RESPONSES

Until further notice, we shall use the symbol $\boldsymbol{\theta}$ to denote the $k \times 1$ vector of regression coefficients common to all m responses, so that

$$\boldsymbol{\theta}' = \boldsymbol{\theta}_c' = (\theta_1, \ldots, \theta_k).$$

When there are two independednt responses, from (9.2.3) the posterior distribution of $\boldsymbol{\theta}$ is proportional to the product of two multivariate t distributions with different degrees of freedom,

$$p(\boldsymbol{\theta} \mid \mathbf{y}) \propto \prod_{i=1}^{2} \left[1 + \frac{(\boldsymbol{\theta} - \hat{\boldsymbol{\theta}}_i)' \mathbf{X}_i'\mathbf{X}_i(\boldsymbol{\theta} - \hat{\boldsymbol{\theta}}_i)}{\nu_i s_i^2}\right]^{-\frac{1}{2}(\nu_i + k)}, \qquad -\infty < \boldsymbol{\theta} < \infty, \tag{9.3.1}$$

where

$$\nu_i = n_i - k, \qquad s_i^2 = \frac{1}{\nu_i}(\mathbf{y}_i - \mathbf{X}_i\hat{\boldsymbol{\theta}}_i)'(\mathbf{y}_i - \mathbf{X}_i\hat{\boldsymbol{\theta}}_i),$$

and $\hat{\boldsymbol{\theta}}_i$ satisfies the normal equations $(\mathbf{X}_i'\mathbf{X}_i)\hat{\boldsymbol{\theta}}_i = \mathbf{X}_i'\mathbf{y}_i$. To obtain the marginal distribution of a subset of $\boldsymbol{\theta}$, say $\boldsymbol{\theta}_l' = (\theta_1, \ldots, \theta_1)$, directly it would be necessary to evaluate, for each value of $\boldsymbol{\theta}_l$, a $k - l$ dimensional integral. The difficulty may be avoided by writing the distribution in a different form which, in turn, leads to a simple approximation.

9.3.1 An Alternative Form for $p(\boldsymbol{\theta} \mid \mathbf{y})$

Writing $\sigma_1^2 = \sigma_{11}$ and $\sigma_2^2 = \sigma_{22}$ in (9.2.2) with $m = 2$ the posterior distribution of $(\sigma_1^2, \sigma_2^2, \boldsymbol{\theta})$ is

$$p(\sigma_1^2, \sigma_2^2, \boldsymbol{\theta} \mid \mathbf{y}) \propto (\sigma_1^2)^{-(\frac{1}{2}n_1 + 1)} (\sigma_2^2)^{-(\frac{1}{2}n_2 + 1)} \exp\left\{-\frac{1}{2}\left[\frac{\nu_1 s_1^2 + R_1(\boldsymbol{\theta})}{\sigma_1^2} + \frac{\nu_2 s_2^2 + R_2(\boldsymbol{\theta})}{\sigma_2^2}\right]\right\},$$

$$\sigma_1^2 > 0, \quad \sigma_2^2 > 0, \quad -\infty < \boldsymbol{\theta} < \infty, \tag{9.3.2}$$

where $$R_i(\boldsymbol{\theta}) = (\boldsymbol{\theta} - \hat{\boldsymbol{\theta}}_i)' \mathbf{X}_i'\mathbf{X}_i(\boldsymbol{\theta} - \hat{\boldsymbol{\theta}}_i), \qquad i = 1, 2.$$

If we integrate out (σ_1^2, σ_2^2) directly from (9.3.2), we obtain the form (9.3.1). Alternatively, we may make the transformation from $(\sigma_1^2, \sigma_2^2, \boldsymbol{\theta})$ to $(\sigma_1^2, w, \boldsymbol{\theta})$ where

$$w = \sigma_1^2/\sigma_2^2 \tag{9.3.3}$$

and integrate out σ_1^2 to obtain

$$p(\boldsymbol{\theta}, w \mid \mathbf{y}) \propto w^{\frac{1}{2}n_2 - 1} \{v_1 s_1^2 + R_1(\boldsymbol{\theta}) + w[v_2 s_2^2 + R_2(\boldsymbol{\theta})]\}^{-\frac{1}{2}(n_1+n_2)},$$
$$0 < w < \infty, \quad -\infty < \boldsymbol{\theta} < \infty. \tag{9.3.4}$$

This distribution can be written as the product

$$p(\boldsymbol{\theta}, w \mid \mathbf{y}) = p(\boldsymbol{\theta} \mid w, \mathbf{y})\, p(w \mid \mathbf{y}). \tag{9.3.5}$$

Assuming that at least one of the matrices $\mathbf{X}_1$ and $\mathbf{X}_2$ is of full rank, the identity (A7.1.1) in Appendix A7.1 may be used to combine the two quadratic forms $R_1(\boldsymbol{\theta})$ and $wR_2(\boldsymbol{\theta})$ in (9.3.4) into

$$R_1(\boldsymbol{\theta}) + wR_2(\boldsymbol{\theta}) = R(\boldsymbol{\theta} \mid w) + S_c(w) \tag{9.3.6}$$

where

$$R(\boldsymbol{\theta} \mid w) = [\boldsymbol{\theta} - \bar{\boldsymbol{\theta}}(w)]'(\mathbf{X}_1'\mathbf{X}_1 + w\mathbf{X}_2'\mathbf{X}_2)[\boldsymbol{\theta} - \bar{\boldsymbol{\theta}}(w)],$$

$$S_c(w) = w(\hat{\boldsymbol{\theta}}_1 - \hat{\boldsymbol{\theta}}_2)'\mathbf{X}_1'\mathbf{X}_1(\mathbf{X}_1'\mathbf{X}_1 + w\mathbf{X}_2'\mathbf{X}_2)^{-1}\mathbf{X}_2'\mathbf{X}_2(\hat{\boldsymbol{\theta}}_1 - \hat{\boldsymbol{\theta}}_2),$$

and

$$\bar{\boldsymbol{\theta}}(w) = (\mathbf{X}_1'\mathbf{X}_1 + w\mathbf{X}_2'\mathbf{X}_2)^{-1}(\mathbf{X}_1'\mathbf{X}_1\hat{\boldsymbol{\theta}}_1 + w\mathbf{X}_2'\mathbf{X}_2\hat{\boldsymbol{\theta}}_2).$$

It follows that the conditional distribution of $\boldsymbol{\theta}$ given w is

$$p(\boldsymbol{\theta} \mid w, \mathbf{y}) \propto [(n_1 + n_2 - k)s^2(w) + R(\boldsymbol{\theta} \mid w)]^{-\frac{1}{2}(n_1+n_2)}, \qquad -\infty < \boldsymbol{\theta} < \infty, \tag{9.3.7}$$

where

$$s^2(w) = \frac{1}{n_1 + n_2 - k}[v_1 s_1^2 + w v_2 s_2^2 + S_c(w)],$$

which is the k-dimensional multivariate t distribution

$$t_k[\bar{\boldsymbol{\theta}}(w),\ s^2(w)(\mathbf{X}_1'\mathbf{X}_1 + w\mathbf{X}_2'\mathbf{X}_2)^{-1},\ n_1 + n_2 - k].$$

To obtain the marginal distribution $p(w \mid \mathbf{y})$, we substitute (9.3.6) into (9.3.4) and integrate out $\boldsymbol{\theta}$ yielding

$$p(w \mid \mathbf{y}) = d\,|\mathbf{X}_1'\mathbf{X}_1 + w\mathbf{X}_2'\mathbf{X}_2|^{-1/2}\, w^{\frac{1}{2}n_2 - 1}\,[s^2(w)]^{-\frac{1}{2}(n_1+n_2-k)}, \qquad 0 < w < \infty, \tag{9.3.8}$$

where d is the appropriate normalizing constant such that

$$d^{-1} = \int_0^\infty |\mathbf{X}_1'\mathbf{X}_1 + w\mathbf{X}_2'\mathbf{X}_2|^{-1/2}\, w^{\frac{1}{2}n_2 - 1}\,[s^2(w)]^{-\frac{1}{2}(n_1+n_2-k)}\, dw.$$

An alternative form for the posterior distribution of θ is therefore

$$p(\theta \mid y) = \int_0^\infty p(\theta \mid w, y) \, p(w \mid y) \, dw, \qquad -\infty < \theta < \infty, \qquad (9.3.9)$$

where $p(\theta \mid w, y)$ and $p(w \mid y)$ are given in (9.3.7) and (9.3.8), respectively.

By comparing (9.3.9) with (9.3.1), it can be verified that the normalizing constant for the expression (9.3.1) is

$$\frac{\Gamma\left(\frac{n_1}{2}\right) \Gamma\left(\frac{n_2}{2}\right) (n_1 + n_2 - k)^{\frac{1}{2}(n_1 + n_2 - k)}}{[\Gamma(\frac{1}{2})]^k \, \Gamma\left(\frac{n_1 + n_2 - k}{2}\right) (v_1 s_1^2)^{\frac{1}{2} n_1} (v_2 s_2^2)^{\frac{1}{2} n_2}} \, d \qquad (9.3.10)$$

where d^{-1} is the integral given in (9.3.8).

Using properties of the multivariate t distribution, and partitioning

$$C(w) = (X_1' X_1 + w X_2' X_2)^{-1} = \begin{array}{c} {\scriptstyle l} \qquad {\scriptstyle p} \\ \left[\begin{array}{c|c} C_{ll}(w) & C_{lp}(w) \\ \hline C_{pl}(w) & C_{pp}(w) \end{array} \right] \begin{array}{c} l \\ p \end{array} \end{array} \qquad \bar{\theta}(w) = \left[\begin{array}{c} \bar{\theta}_l(w) \\ \bar{\theta}_p(w) \end{array} \right] \begin{array}{c} l \\ p \end{array},$$

$$p = k - l, \qquad (9.3.11)$$

the marginal distribution of the subset $\theta_l' = (\theta_1, \ldots, \theta_l)$, $(l \leqslant k)$, is

$$p(\theta_l \mid y) = \int_0^\infty p(\theta_l \mid w, y) \, p(w \mid y) \, dw, \qquad -\infty < \theta_l < \infty, \qquad (9.3.12)$$

where $p(\theta_l \mid w, y)$ is the $t_l[\bar{\theta}_l(w), s^2(w) C_{ll}(w), n_1 + n_2 - k]$ distribution. A similar expression can be obtained for the remaining elements $\theta_p' = (\theta_{l+1}, \ldots, \theta_k)$. Although it does not seem possible to express (9.3.12) in closed form, its evaluation only requires numerical integration of one-dimensional integrals instead of the $(k - l)$-dimensional integration implied by (9.3.1)—a very significant simplification.

9.3.2 Further Simplifications for the Distribution of θ

In evaluating the posterior distribution of θ and its associated marginals for subsets of θ, if we were to employ directly the expressions in (9.3.9) and (9.3.12), it would be necessary to calculate the determinant and the inverse of the matrix $(X_1' X_1 + w X_2' X_2)$ for every w. We now show how this problem can be made more manageable.

The special case where $\mathbf{X}_1'\mathbf{X}_1$ *and* $\mathbf{X}_2'\mathbf{X}_2$ *are proportional*

Consider first the special case where

$$\mathbf{X}_2'\mathbf{X}_2 = a\mathbf{X}_1'\mathbf{X}_1. \tag{9.3.13}$$

The expressions in (9.3.6) then become

$$R(\boldsymbol{\theta} \mid w) = (1 + aw)[\boldsymbol{\theta} - \bar{\boldsymbol{\theta}}(w)]' \, \mathbf{X}_1'\mathbf{X}_1 \, [\boldsymbol{\theta} - \bar{\boldsymbol{\theta}}(w)]$$

$$S_c(w) = \frac{aw}{1 + aw}(\hat{\boldsymbol{\theta}}_1 - \hat{\boldsymbol{\theta}}_2)' \, \mathbf{X}_1'\mathbf{X}_1 \, (\hat{\boldsymbol{\theta}}_1 - \hat{\boldsymbol{\theta}}_2) \tag{9.3.14}$$

$$\bar{\boldsymbol{\theta}}(w) = \frac{1}{1 + aw}(\hat{\boldsymbol{\theta}}_1 + aw\hat{\boldsymbol{\theta}}_2),$$

so that (a) the conditional distribution of $\boldsymbol{\theta}$ given w, $p(\boldsymbol{\theta} \mid w, \mathbf{y})$, is

$$p(\boldsymbol{\theta} \mid w, \mathbf{y}) \propto \{(n_1 + n_2 - k)s^2(w) + (1 + aw)\,[\boldsymbol{\theta} - \bar{\boldsymbol{\theta}}(w)]' \, \mathbf{X}_1'\mathbf{X}_1[\boldsymbol{\theta} - \bar{\boldsymbol{\theta}}(w)]\}^{-\frac{1}{2}(n_1+n_2)},$$

$$-\infty < \boldsymbol{\theta} < \infty, \tag{9.3.15}$$

where

$$(n_1 + n_2 - k)s^2(w) = \nu_1 s_1^2 + w\nu_2 s_2^2 + \frac{aw}{1 + aw}(\hat{\boldsymbol{\theta}}_1 - \hat{\boldsymbol{\theta}}_2)' \, \mathbf{X}_1'\mathbf{X}_1 \, (\hat{\boldsymbol{\theta}}_1 - \hat{\boldsymbol{\theta}}_2),$$

(b) the conditional distribution for the subset $\boldsymbol{\theta}_l$, $p(\boldsymbol{\theta}_l \mid w, \mathbf{y})$, is

$$p(\boldsymbol{\theta}_l \mid w, \mathbf{y})$$

$$\propto \{(n_1 + n_2 - k)s^2(w) + (1 + aw)[\boldsymbol{\theta}_l - \bar{\boldsymbol{\theta}}_l(w)]' \mathbf{D}_{ll}^{-1} \, [\boldsymbol{\theta}_l - \bar{\boldsymbol{\theta}}_l(w)]\}^{-\frac{1}{2}(n_1+n_2-k+l)},$$

$$-\infty < \boldsymbol{\theta}_l < \infty, \tag{9.3.16}$$

where

$$(\mathbf{X}_1'\mathbf{X}_1)^{-1} = \begin{bmatrix} \mathbf{D}_{ll} & \mathbf{D}_{lp} \\ \mathbf{D}_{pl} & \mathbf{D}_{pp} \end{bmatrix} \begin{matrix} p \\ l \end{matrix}, \quad \text{(columns } l, p\text{)}$$

and (c) the marginal distribution of w, $p(w \mid \mathbf{y})$, reduces to

$$p(w \mid \mathbf{y}) = d|\mathbf{X}_1'\mathbf{X}_1|^{-1/2}\,(1 + aw)^{-k/2}\, w^{\frac{1}{2}n_2 - 1}\,[s^2(w)]^{-\frac{1}{2}(n_1+n_2-k)}, \quad 0 < w < \infty. \tag{9.3.17}$$

Employing (9.3.15)–(9.3.17), the posterior distribution of $\boldsymbol{\theta}$ in (9.3.9) and its associated marginal distributions for subsets of $\boldsymbol{\theta}$ in (9.3.12) can now be conveniently evaluated on a computer with little numerical complication.

The posterior distribution of $\boldsymbol{\theta}$ thus obtained is of very much the same form as the posterior distribution of $\boldsymbol{\phi}$ in (7.4.33) and (7.4.41) for the BIBD model. In both cases we are combining information about regression coefficients from two sources where the variance ratio is not assumed known. The main distinction lies in the fact that for the BIBD model we have the added constraint $w = \sigma_{be}^2/\sigma_e^2 > 1$ on the variance ratio while no such restriction exists here.

The General Case when $\mathbf{X}_1'\mathbf{X}_1$ and $\mathbf{X}_2'\mathbf{X}_2$ are not Proportional

In obtaining (9.3.6), we have assumed that at least one of the two matrices, $\mathbf{X}_1'\mathbf{X}_1$ and $\mathbf{X}_2'\mathbf{X}_2$, is positive definite. To be specific, suppose that $\mathbf{X}_1'\mathbf{X}_1$ is positive definite. Then there exists a non-singular $k \times k$ matrix $\mathbf{P}$ such that

$$\mathbf{P}'\mathbf{X}_1'\mathbf{X}_1\mathbf{P} = \mathbf{I} \qquad \text{and} \qquad \mathbf{P}'\mathbf{X}_2'\mathbf{X}_2\mathbf{P} = \mathbf{\Lambda} \tag{9.3.18}$$

where $\mathbf{\Lambda} = \{\lambda_{jj}\}$ is a $k \times k$ diagonal matrix with non-negative diagonal elements. It can then be shown that, in the integral of (9.3.9) and (9.3.12),

a) the conditional distribution $p(\boldsymbol{\theta} \mid w, \mathbf{y})$ becomes

$$p(\boldsymbol{\theta} \mid w, \mathbf{y})$$
$$\propto \{(n_1 + n_2 - k)s^2(w) + [\boldsymbol{\theta} - \bar{\boldsymbol{\theta}}(w)]'(\mathbf{P}')^{-1}(\mathbf{I} + w\mathbf{\Lambda})\mathbf{P}^{-1}[\boldsymbol{\theta} - \bar{\boldsymbol{\theta}}(w)]\}^{-\frac{1}{2}(n_1+n_2)},$$
$$-\infty < \boldsymbol{\theta} < \infty, \tag{9.3.19}$$

with

$$\bar{\boldsymbol{\theta}}(w) = \mathbf{P}(\mathbf{I} + w\mathbf{\Lambda})^{-1}(\hat{\boldsymbol{\phi}}_1 + w\mathbf{\Lambda}\hat{\boldsymbol{\phi}}_2), \qquad \hat{\boldsymbol{\phi}}_i = \mathbf{P}^{-1}\hat{\boldsymbol{\theta}}_i, \quad i = 1, 2$$

and

$$(n_1 + n_2 - k)s^2(w) = v_1 s_1^2 + w v_2 s_2^2 + \sum_{j=1}^{k} \frac{w\lambda_{jj}}{1 + w\lambda_{jj}} (\hat{\phi}_{1j} - \hat{\phi}_{2j})^2,$$

b) the conditional distribution $p(\boldsymbol{\theta}_l \mid w, \mathbf{y})$ of the subset $\boldsymbol{\theta}_l$ is

$$p(\boldsymbol{\theta}_l \mid w, \mathbf{y})$$
$$\propto \{(n_1 + n_2 - k)s^2(w) + [\boldsymbol{\theta}_l - \bar{\boldsymbol{\theta}}_l(w)]'\mathbf{C}_{ll}^{-1}(w)[\boldsymbol{\theta}_l - \bar{\boldsymbol{\theta}}_l(w)]\}^{-\frac{1}{2}(n_1+n_2-k+l)},$$
$$-\infty < \boldsymbol{\theta}_l < \infty, \tag{9.3.20}$$

where

$$\mathbf{C}_{ll}(w) = \mathbf{P}_1(\mathbf{I} + w\mathbf{\Lambda})^{-1}\mathbf{P}_1'$$

and

$$\mathbf{P} = \begin{bmatrix} \mathbf{P}_1 \\ \cdots \\ \mathbf{P}_2 \end{bmatrix}\begin{matrix} l \\ p \end{matrix},$$

and (c) the marginal distribution $p(w \mid \mathbf{y})$ reduces to

$$p(w \mid \mathbf{y}) = d\,|\mathbf{X}_1'\mathbf{X}_1|^{-1/2} \left[\prod_{j=1}^{k} (1 + w\lambda_{jj})\right]^{-1/2} w^{\frac{1}{2}n_2 - 1}\,[s^2(w)]^{-\frac{1}{2}(n_1+n_2-k)},$$
$$0 < w < \infty. \tag{9.3.21}$$

To obtain these results, we may write

$$\mathbf{X}_1'\mathbf{X}_1 = (\mathbf{P}')^{-1}\mathbf{P}^{-1} \qquad \text{and} \qquad \mathbf{X}_2'\mathbf{X}_2 = (\mathbf{P}')^{-1}\mathbf{\Lambda}\mathbf{P}^{-1} \tag{9.3.22}$$

so that

$$(\mathbf{X}_1'\mathbf{X}_1 + w\mathbf{X}_2'\mathbf{X}_2)^{-1} = \mathbf{P}(\mathbf{I} + w\mathbf{\Lambda})^{-1}\mathbf{P}'. \tag{9.3.23}$$

It follows that for the expressions of $\bar{\boldsymbol\theta}(w)$ and $S_c(w)$ in (9.3.6) we now have

$$\bar{\boldsymbol\theta}(w) = \mathbf{P}(\mathbf{I} + w\boldsymbol\Lambda)^{-1}(\mathbf{P}^{-1}\hat{\boldsymbol\theta}_1 + w\boldsymbol\Lambda\mathbf{P}^{-1}\hat{\boldsymbol\theta}_2)$$
$$= \mathbf{P}(\mathbf{I} + w\boldsymbol\Lambda)^{-1}(\hat{\boldsymbol\phi}_1 + w\boldsymbol\Lambda\hat{\boldsymbol\phi}_2) \tag{9.3.24}$$

with

$$\hat{\boldsymbol\phi}_i = \mathbf{P}^{-1}\hat{\boldsymbol\theta}_i, \qquad i = 1, 2$$

and

$$S_c(w) = w(\hat{\boldsymbol\theta}_1 - \hat{\boldsymbol\theta}_2)'(\mathbf{P}')^{-1}\mathbf{P}^{-1}\mathbf{P}(\mathbf{I} + w\boldsymbol\Lambda)^{-1}\mathbf{P}'(\mathbf{P}')^{-1}\boldsymbol\Lambda\mathbf{P}^{-1}(\hat{\boldsymbol\theta}_1 - \hat{\boldsymbol\theta}_2)$$
$$= w(\hat{\boldsymbol\phi}_1 - \hat{\boldsymbol\phi}_2)'(\mathbf{I} + w\boldsymbol\Lambda)^{-1}\boldsymbol\Lambda(\hat{\boldsymbol\phi}_1 - \hat{\boldsymbol\phi}_2)$$
$$= \sum_{j=1}^{k}\left(\frac{w\lambda_{jj}}{1 + w\lambda_{jj}}\right)(\hat{\phi}_{1j} - \hat{\phi}_{2j})^2 \tag{9.3.25}$$

Further, the determinant $|\mathbf{X}_1'\mathbf{X}_1 + w\mathbf{X}_2'\mathbf{X}_2|$ is

$$|\mathbf{X}_1'\mathbf{X}_1 + w\mathbf{X}_2'\mathbf{X}_2| = |(\mathbf{P}')^{-1}\mathbf{P}^{-1} + w(\mathbf{P}')^{-1}\boldsymbol\Lambda\mathbf{P}^{-1}| = |\mathbf{P}'|^{-1}|\mathbf{P}^{-1}|\,|\mathbf{I} + w\boldsymbol\Lambda|$$
$$= |\mathbf{X}_1'\mathbf{X}_1|\prod_{j=1}^{k}(1 + w\lambda_{jj}) \tag{9.3.26}$$

Substituting (9.3.24)—(9.3.26) into (9.3.9) and (9.3.12), we obtain the desired results.

In computing the posterior distributions in (9.3.9) and (9.3.12), standard numerical methods can be used to obtain first the matrices $\mathbf{P}$ and $\boldsymbol\Lambda$. The distribution of $\boldsymbol\theta$ and its marginals of subsets of the elements of $\boldsymbol\theta$ can then be conveniently calculated using (9.3.19)–(9.3.21) with the appropriate normalizing constants inserted. All that is required is numerical evaluation of one-dimensional integrals involving simple functions. With the availability of a computer, this presents no problem.

9.3.3 Approximations to the Distribution of $\boldsymbol\theta$

We have shown in (7.4.50) that the distribution of $\boldsymbol\phi$ for the BIBD model can be closely approximated by setting

$$p(\boldsymbol\phi \mid \mathbf{y}) \doteq p(\boldsymbol\phi \mid \hat{w}, \mathbf{y})$$

where $\hat{w}$ is the mode of the posterior distribution of $\log w$. A similar argument can be applied to the distribution of $\boldsymbol\theta$ in (9.3.9). We can write

$$p(\boldsymbol\theta \mid \mathbf{y}) = \underset{\log w}{E}\; p(\boldsymbol\theta \mid w, \mathbf{y}) \doteq p(\boldsymbol\theta \mid \hat{w}, \mathbf{y}) \tag{9.3.27}$$

where $p(\boldsymbol\theta \mid \hat{w}, \mathbf{y})$ is obtained by inserting $\hat{w}$ for w in (9.3.19) and $\hat{w}$ is the mode of the density

$$p(\log w \mid \mathbf{y}) \propto w^{n_2/2}\left[\prod_{j=1}^{k}(1 + w\lambda_{jj})\right]^{-1/2}[s^2(w)]^{-\frac{1}{2}(n_1+n_2-k)},$$
$$-\infty < \log w < \infty, \tag{9.3.28}$$

which is derived from (9.3.21) by inserting the Jacobian $[(d/dw)\log w]^{-1} = w$. Upon taking logarithms of (9.3.28) and differentiating, it can be verified that $\hat{w}$ is the appropriate root of the equation

$$h(w) = 0, \tag{9.3.29}$$

$$h(w) = n_2 w^{-1} - \sum_{j=1}^{k} \lambda_{jj}(1 + w\lambda_{jj})^{-1}$$

$$- [s^2(w)]^{-1}\left[\nu_2 s_2^2 + \sum_{j=1}^{k} \lambda_{jj}(1 + \lambda_{jj} w)^{-2}(\hat{\phi}_{1j} - \hat{\phi}_{2j})^2\right],$$

which maximizes (9.3.28).

To obtain the root $\hat{w}$, it seems most convenient to employ the standard Newton–Raphson iteration procedure. That is, with an initial guessed value w_0, expand

$$h(w) \doteq h(w_0) + h'(w_0)(w_1 - w_0) = 0 \tag{9.3.30}$$

where

$$h'(w) = -n_2 w^{-2} + \sum_{j=1}^{k}\left[\frac{\lambda_{jj}}{1 + w\lambda_{jj}}\right]^2$$

$$+ 2[s^2(w)]^{-1}\left[\sum_{j=1}^{k} \lambda_{jj}^2(1 + w\lambda_{jj})^{-3}(\hat{\phi}_{1j} - \hat{\phi}_{2j})^2\right]$$

$$+ (n_1 + n_2 - k)[s^2(w)]^{-2}\left[\nu_2 s_2^2 + \sum_{j=1}^{k} \lambda_{jj}(1 + w\lambda_{jj})^{-2}(\hat{\phi}_{1j} - \hat{\phi}_{2j})^2\right]^2.$$

Thus, we find a new guessed value w_1 such that

$$w_1 = w_0 - \frac{h(w_0)}{h'(w_0)} \tag{9.3.31}$$

which can now be used as the next guessed value, and the process repeated until convergence occurs. A convenient choice for the initial value is $w_0 = s_1^2/s_2^2$.

To this degree of approximation, then, $\boldsymbol{\theta}$ is distributed as the k dimensional multivariate t distribution $t_k[\bar{\boldsymbol{\theta}}(\hat{w}), s^2(\hat{w})(\mathbf{X}_1'\mathbf{X}_1 + \hat{w}\mathbf{X}_2'\mathbf{X}_2)^{-1}, n_1 + n_2 - k]$ from which corresponding approximations to marginal distributions of subsets of $\boldsymbol{\theta}$ can also be obtained. The accuracy of this approximation will be illustrated by an example in the next section.

An alternative approximation employing an asymptotic expansion of the distribution in (9.3.1) in powers of (ν_1^{-1}, ν_2^{-1}) is given in Appendix A.9.1.

9.3.4 An Econometric Example

For illustration, we analyse a simple econometric investment model with annual time series data, 1935–54, taken from Boot and De Witt (1960), relating to two large corportaions, General Electric and Westinghouse. In this model, price deflated gross investment is assumed to be a linear function of expected profitability and beginning of year real capital stock. Following Grunfeld (1958), the value of outstanding shares at the beginning of the year is taken as a measure of a firm's expected profitability. The two investment relations are

$$\begin{aligned} y_{u1} &= \theta_{01} + \theta_1 x_{u11} + \theta_2 x_{u21} + \varepsilon_{u1}, \\ y_{u2} &= \theta_{02} + \theta_1 x_{u12} + \theta_2 x_{u22} + \varepsilon_{u2}, \end{aligned} \tag{9.3.32}$$

where u denotes the value of a variable in year u, $u = 1, 2, \ldots, 20$, and

Variable	General Electric	Westinghouse
Annual real gross investment	y_{u1}	y_{u2}
Value of shares at beginning of year	x_{u11}	x_{u12}
Real capital stock at beginning of year	x_{u21}	x_{u22}
Error term	ε_{u1}	ε_{u2}

The parameters θ_1 and θ_2 in (9.3.32) are taken to be the same for the two firms; however, θ_{01} and θ_{02} are assumed to be different to allow for certain possible differences in the investment behavior of the two firms. Further, ε_{u1} and ε_{u2} are assumed to be independently and Normally distributed for all u as $N(0, \sigma_1^2)$ and $N(0, \sigma_2^2)$, respectively.

We may write the model in (9.3.32) in the form

$$\mathbf{y}_i = \mathbf{T}_i \boldsymbol{\gamma}_i + \boldsymbol{\varepsilon}_i \tag{9.3.33}$$

where

$$\boldsymbol{\gamma}_i = \begin{bmatrix} \theta_{0i} \\ \cdots \\ \boldsymbol{\theta} \end{bmatrix}, \qquad \boldsymbol{\theta} = \begin{bmatrix} \theta_1 \\ \theta_2 \end{bmatrix},$$

$$\mathbf{T}_i = \begin{bmatrix} 1 & x_{11i} & x_{12i} \\ \vdots & \vdots & \vdots \\ 1 & x_{u1i} & x_{u2i} \\ \vdots & \vdots & \vdots \\ 1 & x_{n1i} & x_{n2i} \end{bmatrix}, \qquad \begin{aligned} \mathbf{y}_i' &= (y_{1i}, \ldots, y_{ui}, \ldots, y_{ni}) \\ \boldsymbol{\varepsilon}_i' &= (\varepsilon_{1i}, \ldots, \varepsilon_{ui}, \ldots, \varepsilon_{ni}), \\ i &= 1, 2 \quad \text{and} \quad n = 20. \end{aligned}$$

Thus, the two vectors of responses $(\mathbf{y}_1, \mathbf{y}_2)$ are independently Normally distributed, and contain *some* (but not all) common parameters $\boldsymbol{\theta}$. On the assumption that $(\theta_{01}, \theta_{02}, \boldsymbol{\theta})$ and $\log \sigma_1$ and $\log \sigma_2$ are locally uniform and independent *a priori*, the posterior distribution of the parameters is

$$p(\theta_{01}, \theta_{02}, \boldsymbol{\theta}, \sigma_1^2, \sigma_2^2 \mid \mathbf{y}) \propto \prod_{i=1}^{2} (\sigma_i^2)^{-(\frac{1}{2}n+1)} \exp\left\{ -\frac{1}{2\sigma_i^2} [\nu s_i^2 + (\boldsymbol{\gamma}_i - \hat{\boldsymbol{\gamma}}_i)' \mathbf{T}_i' \mathbf{T}_i (\boldsymbol{\gamma}_i - \hat{\boldsymbol{\gamma}}_i)] \right\},$$

$$\sigma_i^2 > 0, \quad -\infty < \theta_{0i} < \infty, \quad -\infty < \boldsymbol{\theta} < \infty, \tag{9.3.34}$$

where

$$\hat{\gamma}_i = \begin{bmatrix} \hat{\theta}_{0i} \\ \hat{\boldsymbol{\theta}}_i \end{bmatrix} = (\mathbf{T}_i'\mathbf{T}_i)^{-1}\mathbf{T}_i'\mathbf{y}_i, \qquad \nu = n - 3$$

and

$$\nu s_i^2 = (\mathbf{y}_i - \mathbf{T}_i\hat{\gamma}_i)'(\mathbf{y}_i - \mathbf{T}_i\hat{\gamma}_i)$$

Integrating out (θ_{01} and θ_{02}), we obtain

$$p(\boldsymbol{\theta}, \sigma_1^2, \sigma_2^2 \mid \mathbf{y}) \propto \prod_{i=1}^{2} (\sigma_i^2)^{-[\frac{1}{2}(n-1)+1]} \exp\left\{-\frac{1}{2\sigma_i^2}[\nu s_i^2 + (\boldsymbol{\theta} - \hat{\boldsymbol{\theta}}_i)'\mathbf{X}_i'\mathbf{X}_i(\boldsymbol{\theta} - \hat{\boldsymbol{\theta}}_i)]\right\},$$

$$\sigma_i^2 > 0, \quad -\infty < \boldsymbol{\theta} < \infty, \qquad (9.3.35)$$

where

$$\mathbf{X}_i = \begin{bmatrix} x_{11i} - \bar{x}_{1i} & x_{12i} - \bar{x}_{2i} \\ \vdots & \vdots \\ x_{u1i} - \bar{x}_{1i} & x_{u2i} - \bar{x}_{2i} \\ \vdots & \vdots \\ x_{n1i} - \bar{x}_{1i} & x_{n2i} - \bar{x}_{2i} \end{bmatrix}, \qquad \bar{x}_{ji} = \frac{1}{n}\sum_{u=1}^{n} x_{uji}, \qquad j = 1, 2.$$

This distribution is of exactly the same form as that in (9.3.2) with $n_i = n - 1$, $\nu_i = \nu$ and $k = 2$. Eliminating σ_1^2 and σ_2^2 by integration, the distribution of $\boldsymbol{\theta}$ then takes the form of the product of two bivariate t distributions as shown in (9.3.1) or alternatively can be expressed in the integral form (9.3.9). Numerical values for sample quantities needed in our subsequent analysis are given below.

General Electric	*Westinghouse*
$\mathbf{X}_1'\mathbf{X}_1 = 10^6 \begin{bmatrix} 3.254 & 0.233 \\ 0.233 & 1.193 \end{bmatrix}$	$\mathbf{X}_2'\mathbf{X}_2 = 10^6 \begin{bmatrix} 0.940 & 0.195 \\ 0.195 & 0.074 \end{bmatrix}$
$\hat{\boldsymbol{\theta}}_1 = \begin{pmatrix} \hat{\theta}_{11} \\ \hat{\theta}_{21} \end{pmatrix} = \begin{pmatrix} 0.02655 \\ 0.15170 \end{pmatrix}$	$\hat{\boldsymbol{\theta}}_2 = \begin{pmatrix} \hat{\theta}_{12} \\ \hat{\theta}_{22} \end{pmatrix} = \begin{pmatrix} 0.05289 \\ 0.09241 \end{pmatrix}$
$s_1^2 = 0.777 \times 10^3$	$s_2^2 = 0.104 \times 10^3$
$\nu_1 = 17$	$\nu_2 = 17$
$n_1 = 19$	$n_2 = 19$

A plot of the contours of the distribution of $\boldsymbol{\theta}$ is shown in Fig. 9.3.1. The three contours shown by the solid curves are drawn such that

$$p(\boldsymbol{\theta} \mid \mathbf{y}) = c\,p(\hat{\boldsymbol{\theta}} \mid \mathbf{y}) \qquad \text{for} \qquad c = 0.50, 0.25, 0.05$$

where $\hat{\boldsymbol{\theta}} \doteq (0.037, 0.145)$ is the mode of the distribution. In other words, they are boundaries of the H.P.D. regions containing approximately 50, 75 and 95 per cent of the probability mass, respectively. Also shown in the same figure are the 95 per cent contours (labelled by G.E. and W.H.) of the two factors of the distribution of $\boldsymbol{\theta}$ together with the corresponding centers $(\hat{\boldsymbol{\theta}}_1, \hat{\boldsymbol{\theta}}_2)$. As expected, the distribution of $\boldsymbol{\theta}$ is located "between" the two individual bivariate t distributions, and the spread of the combined distribution is smaller than either of its components. Further, the influence of the first factor—G.E.—is seen to be greater in determining the overall distribution because

$$s_1^4 \, |\mathbf{X}_1'\mathbf{X}_1|^{-1} < s_2^4 |\mathbf{X}_2'\mathbf{X}_2|^{-1}.$$

The solid contours in Fig. 9.3.1 are nearly elliptical suggesting that the distribution of $\boldsymbol{\theta}$ might well be approximated by the bivariate t distribution in (9.3.27). Since comparison of exact and approximate contours is rather difficult to make, the exact and approximate marginal distributions of θ_1 and θ_2 will be compared.

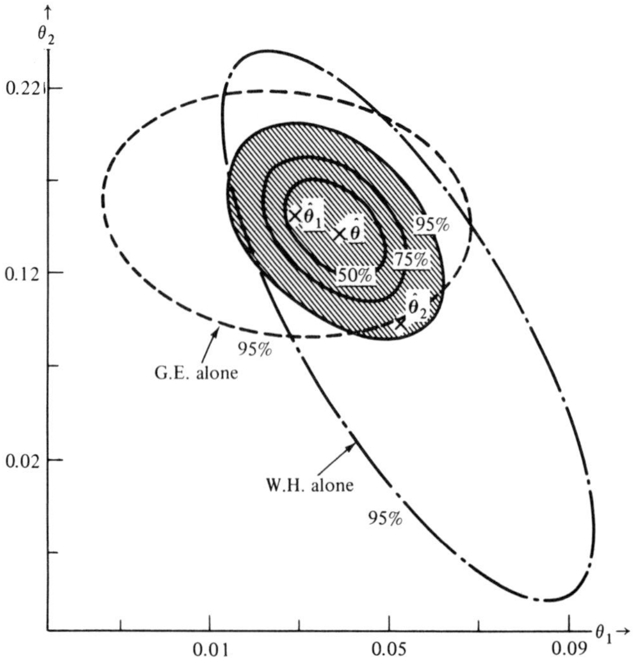

Fig. 9.3.1 Contours of the posterior distribution of $\boldsymbol{\theta}$ and its component factors: the investment data.

The solid curves in Fig. 9.3.2a and b show, respectively, the distributions of θ_1 and θ_2 calculated from (9.3.12). The two factors in the integrand were

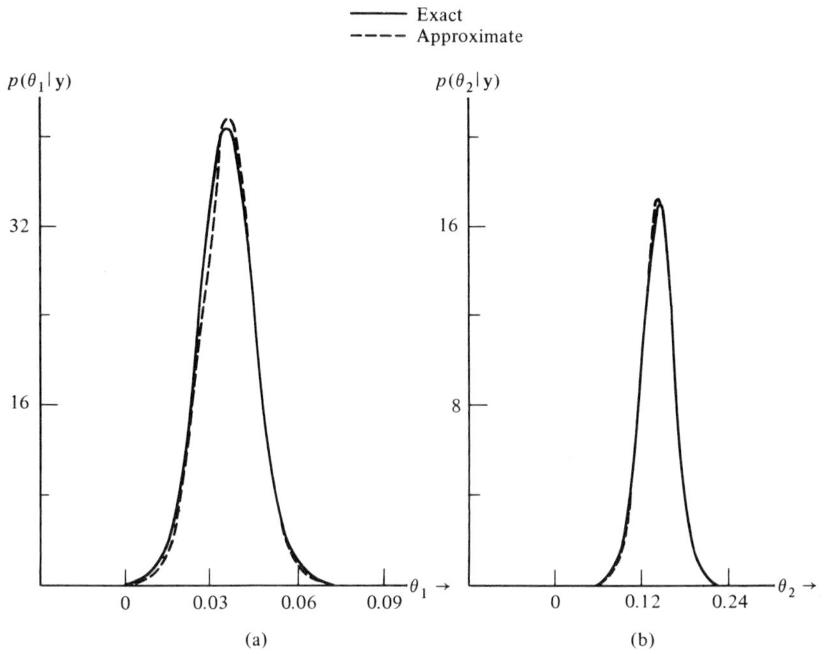

Fig. 9.3.2 Comparison of the exact and the approximate posterior distributions of θ_1 and θ_2: the investment data.

determined from the expressions in (9.3.20) and (9.3.21). The matrices **P** and **Λ** in (9.3.18) and the vectors $\hat{\boldsymbol{\phi}}_i$ in (9.3.19) were found to be

$$\mathbf{P} = 10^{-3}\begin{bmatrix} -0.19814 & 0.52196 \\ 0.89469 & 0.22306 \end{bmatrix} \qquad \boldsymbol{\Lambda} = \begin{bmatrix} 0.02703 & 0 \\ 0 & 0.30513 \end{bmatrix}$$

$$\hat{\boldsymbol{\phi}}_1 = \mathbf{P}^{-1}\hat{\boldsymbol{\theta}}_1 = 10^3\begin{bmatrix} 0.14331 \\ 0.10527 \end{bmatrix}, \qquad \hat{\boldsymbol{\phi}}_2 = \mathbf{P}^{-1}\hat{\boldsymbol{\theta}}_2 = 10^3\begin{bmatrix} 0.07128 \\ 0.12839 \end{bmatrix}.$$

The approximating distributions shown by the broken curves in Figs. 9.3.2 and 9.3.3 were determined using the result in (9.3.27). The root $\hat{w}$ of the equation $h(w) = 0$ in (9.3.29) was calculated by employing the iterative process in (9.3.30) and (9.3.31). Starting with the initial value $w_0 = s_1^2/s_2^2 \doteq 7.47$, we found after two iterations, $\hat{w} \doteq 7.349$. Thus, $\boldsymbol{\theta}$ is approximately distributed as $t_2[\bar{\boldsymbol{\theta}}(\hat{w}), s^2(\hat{w})\mathbf{P}(\mathbf{I} + \hat{w}\boldsymbol{\Lambda})^{-1}\mathbf{P}', 36]$ where

$$\bar{\boldsymbol{\theta}}(\hat{w}) = \begin{bmatrix} 0.03726 \\ 0.14460 \end{bmatrix} \quad \text{and} \quad s^2(\hat{w})\mathbf{P}(\mathbf{I} + \hat{w}\boldsymbol{\Lambda})^{-1}\mathbf{P}' = 10^{-4}\begin{bmatrix} 0.8898 & -0.8535 \\ -0.8535 & 5.2060 \end{bmatrix}.$$

It follows that to this degree of approximation, the quantities

$$\frac{\theta_1 - 0.03726}{0.00943} \quad \text{and} \quad \frac{\theta_2 - 0.1446}{0.0228}$$

are individually distributed as $t(0, 1, 36)$ from which the broken curves were drawn. For this example, the agreement between the exact distribution obtained from numerical integration, and the approximating distributions are sufficiently close for most practical purposes. This, together with the near elliptical shape of the contours in Fig. 9.3.1, suggests that the approximation in (9.3.27) will be useful for higher dimensional distributions.

9.4 INFERENCES ABOUT COMMON PARAMETERS FOR THE GENERAL LINEAR MODEL WITH A COMMON DERIVATIVE MATRIX

Returning to the linear multivariate model in (9.1.2), we now consider the case where not only the parameters but also the derivative matrices are assumed common, i.e., $\boldsymbol{\theta}_1 = \ldots = \boldsymbol{\theta}_m = \boldsymbol{\theta}_c$ and $\mathbf{X}_1 = \ldots = \mathbf{X}_m = \mathbf{X}$. The elements of the multivariate observation vectors $\boldsymbol{\varepsilon}_{(u)} = (\varepsilon_{u1}, \ldots, \varepsilon_{um})'$, $u = 1, \ldots, n$ are, however, not assumed independent.

For example, suppose observations were made by $m = 3$ different observers of a temperature which was linearly increasing with time. An appropriate model for the observations y_{u1}, y_{u2}, y_{u3} made at time t_u might be

$$\begin{aligned} y_{u1} &= \theta_1 + \theta_2 t_u + \varepsilon_{u1} \\ y_{u2} &= \theta_1 + \theta_2 t_u + \varepsilon_{u2} \\ y_{u3} &= \theta_1 + \theta_2 t_u + \varepsilon_{u3} \end{aligned} \tag{9.4.1}$$

If observations were made at n distinct times, we should have a model of the common parameter, common derivative form with

$$\boldsymbol{\theta}_1' = \boldsymbol{\theta}_2' = \boldsymbol{\theta}_3' = \boldsymbol{\theta}_c' = (\theta_1, \theta_2) \qquad \text{and} \qquad \mathbf{X}_1 = \mathbf{X}_2 = \mathbf{X}_3 = \mathbf{X},$$

where $\mathbf{X}$ is the $n \times 2$ matrix whose uth row is $(1, t_u)$, $u = 1, \ldots, n$.

In general, the model may be written in the form of equation (8.4.1) with the $k \times m$ matrix $\boldsymbol{\theta}$ of (8.4.2) having common parameters within every row so that

$$\boldsymbol{\theta} = \boldsymbol{\theta}_c \mathbf{1}_m', \qquad \boldsymbol{\theta}_c' = (\theta_1, \theta_2, \ldots, \theta_k). \tag{9.4.2}$$

From (8.4.8), it is readily shown that the posterior distribution of $\boldsymbol{\theta}_c$ is

$$p(\boldsymbol{\theta}_c \mid \mathbf{y}) \propto [1 + d(\boldsymbol{\theta}_c - \bar{\boldsymbol{\theta}}_c)'\mathbf{V}^{-1}(\boldsymbol{\theta} - \bar{\boldsymbol{\theta}}_c)]^{-\frac{1}{2}n}, \qquad -\infty < \boldsymbol{\theta}_c < \infty, \tag{9.4.3}$$

where

$$d = \mathbf{1}_m'\mathbf{A}^{-1}\mathbf{1}_m, \qquad \bar{\boldsymbol{\theta}}_c = \frac{1}{d}\hat{\boldsymbol{\theta}}\mathbf{A}^{-1}\mathbf{1}_m,$$

and

$$\mathbf{V} = (\mathbf{X}'\mathbf{X})^{-1} + \hat{\boldsymbol{\theta}}\left[\mathbf{A}^{-1} - \frac{1}{d}\mathbf{A}^{-1}\mathbf{1}_m\mathbf{1}_m'\mathbf{A}^{-1}\right]\hat{\boldsymbol{\theta}}',$$

which is the k-dimensional multivariate t distribution

$$t_k\left[\bar{\boldsymbol{\theta}}_c, \frac{1}{(n-k)d}\mathbf{V}, n-k\right].$$

To show this, from (8.4.8) the posterior distribution of $\boldsymbol{\theta}_c$ is

$$p(\boldsymbol{\theta}_c \mid \mathbf{y}) \propto |\mathbf{A} + (\boldsymbol{\theta}_c\mathbf{1}'_m - \hat{\boldsymbol{\theta}})'\,\mathbf{X}'\mathbf{X}(\boldsymbol{\theta}_c\mathbf{1}'_m - \hat{\boldsymbol{\theta}})|^{-\frac{1}{2}n}, \qquad -\infty < \boldsymbol{\theta}_c < \infty, \tag{9.4.4}$$

where it is to be remembered that $\hat{\boldsymbol{\theta}}$ is a $k \times m$ matrix. Now, using (8.4.11)

$$|\mathbf{A} + (\boldsymbol{\theta}_c\mathbf{1}'_m - \hat{\boldsymbol{\theta}})'\mathbf{X}'\mathbf{X}(\boldsymbol{\theta}_c\mathbf{1}'_m - \hat{\boldsymbol{\theta}})| = |\mathbf{A}|\,|\mathbf{X}'\mathbf{X}|\,|(\mathbf{X}'\mathbf{X})^{-1} + (\boldsymbol{\theta}_c\mathbf{1}'_m - \hat{\boldsymbol{\theta}})\mathbf{A}^{-1}\,(\boldsymbol{\theta}_c\mathbf{1}'_m - \hat{\boldsymbol{\theta}})'|. \tag{9.4.5}$$

We can write

$$(\boldsymbol{\theta}_c\mathbf{1}'_m - \hat{\boldsymbol{\theta}})\mathbf{A}^{-1}\,(\boldsymbol{\theta}_c\mathbf{1}'_m - \hat{\boldsymbol{\theta}})' = d(\boldsymbol{\theta}_c - \bar{\boldsymbol{\theta}}_c)\,(\boldsymbol{\theta}_c - \bar{\boldsymbol{\theta}}_c)' + \hat{\boldsymbol{\theta}}\mathbf{B}\hat{\boldsymbol{\theta}}', \tag{9.4.6}$$

where

$$\mathbf{B} = \mathbf{A}^{-1} - \frac{1}{d}\mathbf{A}^{-1}\mathbf{1}_m\mathbf{1}'_m\,\mathbf{A}^{-1}.$$

Making use of (9.4.6) and (8.4.11), we see that the third factor on the right-hand side of (9.4.5) is proportional to

$$[1 + d(\boldsymbol{\theta}_c - \bar{\boldsymbol{\theta}}_c)'\,\mathbf{V}^{-1}(\boldsymbol{\theta}_c - \bar{\boldsymbol{\theta}}_c)], \tag{9.4.7}$$

where

$$\mathbf{V} = (\mathbf{X}'\mathbf{X})^{-1} + \hat{\boldsymbol{\theta}}\,\mathbf{B}\hat{\boldsymbol{\theta}}',$$

and the desired result follows.

The special case for which $k = 1$ and $\mathbf{X}$ a $n \times 1$ column of ones was discussed by Geisser (1965b). The result for the situation when we have common parameters within rows of a block submatrix of $\boldsymbol{\theta}$ is similar to (9.4.3) and is left to the reader.

9.5 GENERAL LINEAR MODEL WITH COMMON PARAMETERS

We now discuss certain aspects of the problem of making inferences about the common parameters $\boldsymbol{\theta}_c$ in the general situation when the derivative matrices $\mathbf{X}_i$ are not common and the responses are correlated. For simplicity in notation, we shall again suppress the subscript c and denote $\boldsymbol{\theta} = \boldsymbol{\theta}_c$. By setting $\boldsymbol{\theta}_1 = \ldots = \boldsymbol{\theta}_m = \boldsymbol{\theta}$ in (8.2.23), the posterior distribution of the common parameters $\boldsymbol{\theta}$ is then

$$p(\boldsymbol{\theta} \mid \mathbf{y}) \propto |\mathbf{S}(\boldsymbol{\theta})|^{-\frac{1}{2}n}, \qquad -\infty < \boldsymbol{\theta} < \infty, \tag{9.5.1}$$

where

$$\left.\begin{aligned} \mathbf{S}(\boldsymbol{\theta}) &= \{S_{ij}(\boldsymbol{\theta})\} \\ S_{ij}(\boldsymbol{\theta}) &= (\mathbf{y}_i - \mathbf{X}_i\boldsymbol{\theta})'(\mathbf{y}_j - \mathbf{X}_j\boldsymbol{\theta}) \\ \boldsymbol{\theta}' &= (\theta_1, \ldots, \theta_k) \end{aligned}\right\} \qquad i = 1, \ldots, m, \quad j = 1, \ldots, m.$$

Since $\mathbf{X}_i \neq \mathbf{X}_j$, we cannot in general express $S_{ij}(\boldsymbol{\theta})$ in (9.5.1) in the form given in (8.4.4), and thus it is no longer possible to reduce the distribution of $\boldsymbol{\theta}$ to the much simpler form in (9.4.3). When the vector $\boldsymbol{\theta}$ consists of only one or two elements, whether or not the model is linear in the parameters the distribution can be plotted and the contributions from the individual vector of responses $\mathbf{y}_l$ assessed, as was done for example in Section 8.2.6.

As always, however, as the number of parameters increases, the situation becomes more and more complicated. Further, to obtain the marginal distribution of a subset, say $\boldsymbol{\theta}_l = (\theta_1, ..., \theta_l)'$, of $\boldsymbol{\theta}$, it is necessary to evaluate a $k - l$ dimensional integral for every value of $\boldsymbol{\theta}_l$. In what follows we shall discuss the important special case $m = 2$ for the linear model and develop an approximation procedure for the distribution of $\boldsymbol{\theta}$.

9.5.1 The Case $m = 2$

When $m = 2$, the distribution in (9.5.1) can be written

$$p(\boldsymbol{\theta} \mid \mathbf{y}) \propto [S_{11}(\boldsymbol{\theta})]^{-\frac{1}{2}n} \left[S_{22}(\boldsymbol{\theta}) - \frac{S_{12}^2(\boldsymbol{\theta})}{S_{11}(\boldsymbol{\theta})}\right]^{-\frac{1}{2}n}, \qquad -\infty < \boldsymbol{\theta} < \infty. \tag{9.5.2}$$

The first factor $[S_{11}(\boldsymbol{\theta})]^{-\frac{1}{2}n}$ is clearly in the form of the multivariate t distribution $t[\hat{\boldsymbol{\theta}}_1, s_1^2(\mathbf{X}_1'\mathbf{X}_1)^{-1}, \nu]$ where, assuming $\mathbf{X}_1$ is of full rank,

$$\begin{aligned} \hat{\boldsymbol{\theta}}_1 &= (\mathbf{X}_1'\mathbf{X}_1)^{-1}\mathbf{X}_1'\mathbf{y}_1, \\ s_1^2 &= (\mathbf{y}_1 - \mathbf{X}_1\hat{\boldsymbol{\theta}}_1)'(\mathbf{y}_1 - \mathbf{X}_1\hat{\boldsymbol{\theta}}_1)/\nu \end{aligned} \tag{9.5.3}$$

and $\nu = n - k$. This factor can be thought of as representing the contribution from the first response vector $\mathbf{y}_1$. The nature of the second factor, $[S_{22}(\boldsymbol{\theta}) - S_{12}^2(\boldsymbol{\theta})/S_{11}(\boldsymbol{\theta})]^{-\frac{1}{2}n}$, which provides the "extra information" from $\mathbf{y}_2$, is not easily recognizable. We now proceed to develop an alternative representation of the distribution of $\boldsymbol{\theta}$.

It will be recalled that if (y_1, y_2) are jointly Normal $N_2(\boldsymbol{\mu}, \boldsymbol{\Sigma})$, the distribution can be written

$$p(y_1, y_2 \mid \boldsymbol{\mu}, \boldsymbol{\Sigma}) = p(y_1 \mid \mu_1, \sigma_{11})\, p(y_2 \mid y_1, \boldsymbol{\mu}, \boldsymbol{\Sigma}) \tag{9.5.4}$$

where

$$p(y_1 \mid \mu_1, \sigma_{11}) = \frac{1}{\sqrt{2\pi\sigma_{11}}} \exp\left[-\frac{(y_1 - \mu_1)^2}{2\sigma_{11}}\right], \qquad -\infty < y_1 < \infty,$$

$$p(y_2 \mid y_1, \boldsymbol{\mu}, \boldsymbol{\Sigma}) = \frac{1}{\sqrt{2\pi\sigma_{22\cdot1}}} \exp\left[-\frac{(y_2 - \mu_{2\cdot1})^2}{2\sigma_{22\cdot1}}\right], \qquad -\infty < y_2 < \infty,$$

$$\sigma_{22\cdot1} = \sigma_{22} - \frac{\sigma_{12}^2}{\sigma_{11}}, \qquad \mu_{2\cdot1} = \mu_2 + \beta(y_1 - \mu_1), \qquad \beta = \frac{\sigma_{12}}{\sigma_{11}}.$$

For $m = 2$ and with common parameters $\boldsymbol{\theta}$, the likelihood function of $\boldsymbol{\theta}$ and $(\sigma_{11}, \sigma_{22\cdot1}, \beta)$ is

$$l(\boldsymbol{\theta}, \sigma_{11}, \sigma_{22\cdot1}, \beta \mid \mathbf{y}) \propto \sigma_{11}^{-\frac{1}{2}n} \sigma_{22\cdot1}^{-\frac{1}{2}n} \exp\left\{-\frac{1}{2\sigma_{11}}(\mathbf{y}_1 - \mathbf{X}_1\boldsymbol{\theta})'(\mathbf{y}_1 - \mathbf{X}_1\boldsymbol{\theta})\right.$$

$$\left. - \frac{1}{2\sigma_{22\cdot1}}[\mathbf{y}_2 - \mathbf{X}_2\boldsymbol{\theta} - \beta(\mathbf{y}_1 - \mathbf{X}_1\boldsymbol{\theta})]'[\mathbf{y}_2 - \mathbf{X}_2\boldsymbol{\theta} - \beta(\mathbf{y}_1 - \mathbf{X}_1\boldsymbol{\theta})]\right\}. \tag{9.5.5}$$

From the noninformative reference prior distribution of $\boldsymbol{\theta}$ and $\boldsymbol{\Sigma}$ in (8.2.8) and (8.2.14), we make the transformation from $(\sigma_{11}, \sigma_{22}, \sigma_{12})$ to $(\sigma_{11}, \sigma_{22\cdot1}, \beta)$ to obtain the prior

$$p(\boldsymbol{\theta}, \sigma_{11}, \sigma_{22\cdot1}, \beta) \propto \sigma_{11}^{-1/2} \sigma_{22\cdot1}^{-3/2}. \tag{9.5.6}$$

Consequently, the posterior distribution of $(\boldsymbol{\theta}, \sigma_{11}, \sigma_{22\cdot1}, \beta)$ is

$$p(\boldsymbol{\theta}, \sigma_{11}, \sigma_{22\cdot1}, \beta \mid \mathbf{y}) \propto \sigma_{11}^{-[\frac{1}{2}(n-1)+1]} \sigma_{22\cdot1}^{-[\frac{1}{2}(n+1)+1]} \exp\left[-\frac{S_{11}(\boldsymbol{\theta})}{2\sigma_{11}} - \frac{S_{22\cdot1}(\boldsymbol{\theta}, \beta)}{2\sigma_{22\cdot1}}\right],$$

$$\sigma_{11} > 0,\ \sigma_{22\cdot1} > 0, \quad -\infty < \beta < \infty, \quad -\infty < \boldsymbol{\theta} < \infty, \tag{9.5.7}$$

where

$$S_{11}(\boldsymbol{\theta}) = (\mathbf{y}_1 - \mathbf{X}_1\boldsymbol{\theta})'(\mathbf{y}_1 - \mathbf{X}_1\boldsymbol{\theta})$$

$$S_{22\cdot1}(\boldsymbol{\theta}, \beta) = [\mathbf{y}_2 - \mathbf{X}_2\boldsymbol{\theta} - \beta(\mathbf{y}_1 - \mathbf{X}_1\boldsymbol{\theta})]'[\mathbf{y}_2 - \mathbf{X}_2\boldsymbol{\theta} - \beta(\mathbf{y}_1 - \mathbf{X}_1\boldsymbol{\theta})].$$

If we were to integrate out $(\sigma_{11}, \sigma_{22\cdot1}, \beta)$ from (9.5.7), we would of course obtain the form given in (9.5.2). Alternatively, we may make the transformation from $(\sigma_{11}, \sigma_{22\cdot1})$ to (σ_{11}, w) where

$$w = \sigma_{11}/\sigma_{22\cdot1} \tag{9.5.8}$$

to obtain

$$p(\boldsymbol{\theta}, w, \beta \mid \mathbf{y}) \propto w^{\frac{1}{2}(n+1)-1}[S_{11}(\boldsymbol{\theta}) + wS_{22\cdot1}(\boldsymbol{\theta}, \beta)]^{-n},$$

$$0 < w < \infty, \quad -\infty < \beta < \infty, \quad -\infty < \boldsymbol{\theta} < \infty. \tag{9.5.9}$$

This distribution can be written as

$$p(\boldsymbol{\theta}, w, \beta \mid \mathbf{y}) = p(\boldsymbol{\theta} \mid w, \beta, \mathbf{y})p(w, \beta \mid \mathbf{y}). \tag{9.5.10}$$

Now, we may write

$$S_{11}(\boldsymbol{\theta}) + wS_{22\cdot1}(\boldsymbol{\theta}, \beta) = (2n - k)s^2(w, \beta) + R(\boldsymbol{\theta} \mid w, \beta), \tag{9.5.11}$$

where

$$R(\boldsymbol{\theta} \mid w, \beta) = [\boldsymbol{\theta} - \hat{\boldsymbol{\theta}}(w, \beta)]'[\mathbf{X}_1'\mathbf{X}_1 + w\mathbf{X}_{2\cdot1}'(\beta)\mathbf{X}_{2\cdot1}(\beta)][\boldsymbol{\theta} - \hat{\boldsymbol{\theta}}(w, \beta)],$$

$$(2n - k)s^2(w, \beta) = [\mathbf{y}_1 - \mathbf{X}_1\hat{\boldsymbol{\theta}}(w, \beta)]'[\mathbf{y}_1 - \mathbf{X}_1\hat{\boldsymbol{\theta}}(w, \beta)]$$

$$+ w[\mathbf{y}_{2\cdot1}(\beta) - \mathbf{X}_{2\cdot1}(\beta)\hat{\boldsymbol{\theta}}(w, \beta)]'[\mathbf{y}_{2\cdot1}(\beta) - \mathbf{X}_{2\cdot1}(\beta)\hat{\boldsymbol{\theta}}(w, \beta)],$$

$$\mathbf{X}_{2\cdot1}(\beta) = \mathbf{X}_2 - \beta\mathbf{X}_1, \qquad \mathbf{y}_{2\cdot1}(\beta) = \mathbf{y}_2 - \beta\mathbf{y}_1$$

and

$$\hat{\boldsymbol{\theta}}(w, \beta) = [\mathbf{X}_1'\mathbf{X}_1 + w\mathbf{X}_{2\cdot 1}'(\beta)\mathbf{X}_{2\cdot 1}(\beta)]^{-1}[\mathbf{X}_1'\mathbf{y}_1 + w\mathbf{X}_{2\cdot 1}'(\beta)\mathbf{y}_{2\cdot 1}(\beta)]$$

Making use of (9.5.11), we see that

a) the conditional distribution of $\boldsymbol{\theta}$, given (w, β), is

$$p(\boldsymbol{\theta} \mid w, \beta, \mathbf{y}) \propto [(2n - k)s^2(w, \beta) + R(\boldsymbol{\theta} \mid w, \beta)]^{-n}, \qquad -\infty < \boldsymbol{\theta} < \infty, \tag{9.5.12}$$

which is the

$$t_k\{\hat{\boldsymbol{\theta}}(w, \beta), s^2(w, \beta)[\mathbf{X}_1'\mathbf{X}_1 + w\mathbf{X}_{2\cdot 1}'(\beta)\mathbf{X}_{2\cdot 1}(\beta)]^{-1}, 2n - k\}$$

distribution.

b) partitioning

$$\boldsymbol{\theta} = \begin{bmatrix} \boldsymbol{\theta}_l \\ \boldsymbol{\theta}_p \end{bmatrix} \begin{matrix} l \\ p \end{matrix}, \qquad \hat{\boldsymbol{\theta}}(w, \beta) = \begin{bmatrix} \hat{\boldsymbol{\theta}}_l(w, \beta) \\ \hat{\boldsymbol{\theta}}_p(w, \beta) \end{bmatrix} \begin{matrix} l \\ p \end{matrix}$$

and (9.5.13)

$$[\mathbf{X}_1'\mathbf{X}_1 + w\mathbf{X}_{2\cdot 1}'(\beta)\mathbf{X}_{2\cdot 1}(\beta)]^{-1} = \overset{\qquad\quad l \qquad\qquad\quad p}{\begin{bmatrix} \mathbf{C}_{ll}(w, \beta) & \mathbf{C}_{lp}(w, \beta) \\ \mathbf{C}_{lp}'(w, \beta) & \mathbf{C}_{pp}(w, \beta) \end{bmatrix}} \begin{matrix} l \\ p \end{matrix},$$

the conditional distribution of the subset $\boldsymbol{\theta}_l$, given (w, β), is

$$p(\boldsymbol{\theta}_l \mid w, \beta, \mathbf{y})$$
$$\propto \{(2n - k)s^2(w, \beta) + [\boldsymbol{\theta}_l - \hat{\boldsymbol{\theta}}_l(w, \beta)]'\mathbf{C}_{ll}^{-1}(w, \beta)[\boldsymbol{\theta}_l - \hat{\boldsymbol{\theta}}_l(w, \beta)]\}^{-\frac{1}{2}(2n-k+l)},$$
$$-\infty < \boldsymbol{\theta}_l < \infty, \tag{9.5.14}$$

a $t_l[\hat{\boldsymbol{\theta}}_l(w, \beta), s^2(w, \beta)\mathbf{C}_{ll}(w, \beta), 2n - k]$ distribution, and

c) the posterior distribution of (w, β) is

$$p(w, \beta \mid \mathbf{y}) \propto |\mathbf{X}_1'\mathbf{X}_1 + w\mathbf{X}_{2\cdot 1}'(\beta)\mathbf{X}_{2\cdot 1}(\beta)|^{-1/2}\, w^{\frac{1}{2}(n+1)-1}\, [s^2(w, \beta)]^{-\frac{1}{2}(2n-k)},$$
$$0 < w < \infty, \quad -\infty < \beta < \infty. \tag{9.5.15}$$

Thus, we may write

$$p(\boldsymbol{\theta} \mid \mathbf{y}) = \int_0^\infty \int_{-\infty}^\infty p(\boldsymbol{\theta} \mid w, \beta, \mathbf{y})\, p(w, \beta \mid \mathbf{y})\, d\beta\, dw, \qquad -\infty < \boldsymbol{\theta} < \infty, \tag{9.5.16}$$

which is an alternative representation of the distribution in (9.5.2). The usefulness of this representation lies chiefly in the associated form for a subset $\boldsymbol{\theta}_l$,

$$p(\boldsymbol{\theta}_l \mid \mathbf{y}) = \int_0^\infty \int_{-\infty}^\infty p(\boldsymbol{\theta}_l \mid w, \beta, \mathbf{y})\, p(w, \beta \mid \mathbf{y})\, d\beta\, dw, \qquad -\infty < \boldsymbol{\theta}_l < \infty. \tag{9.5.17}$$

If exact evaluation of the marginal distribution of $\boldsymbol{\theta}_l$ is desired, it will only be necessary to calculate the double integral in (9.5.17), whatever the value of k, instead of a $k - l$ dimensional integral implied by the form in (9.5.2). This is particularly useful whenever $k - l$ is greater than two. In the situation where the responses were uncorrelated, (9.3.12) could be simplified according to (9.3.20) and (9.3.21). It does not seem possible in general here to find a matrix $\mathbf{P}$ to similarly diagonalize $[\mathbf{X}_1'\mathbf{X}_1 + w\mathbf{X}_{2\cdot 1}'(\beta)\mathbf{X}_{2\cdot 1}(\beta)]$ for all values of (β, w), and consequently it is necessary to evaluate its determinant and inverse as functions of β and w.

9.5.2 Approximations to the Distribution of $\boldsymbol{\theta}$ for $m = 2$

From (9.5.16), one might attempt to approximate the distribution of $\boldsymbol{\theta}$ by writing

$$p(\boldsymbol{\theta} \mid \mathbf{y}) = \underset{\mathbf{f}(w,\beta)}{E}\; p(\boldsymbol{\theta} \mid w, \beta, \mathbf{y}) \doteq p(\boldsymbol{\theta} \mid \hat{w}, \hat{\beta}, \mathbf{y}) \tag{9.5.18}$$

where $(\hat{w}, \hat{\beta})$ is the mode of the distribution of some functions $\mathbf{f}(w, \beta)$ of (w, β). However, this modal value $(\hat{w}, \hat{\beta})$ is not easy to determine since the distribution $p(w, \beta \mid \mathbf{y})$ in (9.5.15) is a rather complicated function of (w, β). We now develop an alternative approach.

The distribution in (9.5.9) may be written as

$$p(\boldsymbol{\theta}, \beta, w \mid \mathbf{y}) = p(\boldsymbol{\theta}, \beta \mid w, \mathbf{y}) p(w \mid \mathbf{y}) \tag{9.5.19}$$

where the first factor is

$$p(\boldsymbol{\theta}, \beta \mid w, \mathbf{y}) \propto [S_{11}(\boldsymbol{\theta}) + wS_{22.1}(\boldsymbol{\theta}, \beta)]^{-n}, \qquad -\infty < \beta < \infty, \qquad -\infty < \boldsymbol{\theta} < \infty. \tag{9.5.20}$$

For a given value of w, let $(\hat{\boldsymbol{\theta}}_w, \hat{\beta}_w)$ be the mode of the conditional distribution (9.5.20). Employing Taylor's theorem, we may expand $S_{11}(\boldsymbol{\theta}) + wS_{22\cdot 1}(\boldsymbol{\theta}, \beta)$ into

$$S_{11}(\boldsymbol{\theta}) + wS_{22\cdot 1}(\boldsymbol{\theta}, \beta) \doteq (2n - k - 1)s^2(w) + Q(\boldsymbol{\theta}, \beta \mid w) \tag{9.5.21}$$

where

$$Q(\boldsymbol{\theta}, \beta \mid w) = (\boldsymbol{\theta}' - \hat{\boldsymbol{\theta}}_w', \beta - \hat{\beta}_w)\mathbf{H}(w) \begin{pmatrix} \boldsymbol{\theta} - \hat{\boldsymbol{\theta}}_w \\ \beta - \hat{\beta}_w \end{pmatrix}$$

$$\mathbf{H}(w) = \begin{array}{c} \begin{matrix} \quad k & \quad\quad 1 \end{matrix} \\ \left[\begin{array}{c:c} \mathbf{H}_{11}(w) & \mathbf{h}_{12}(w) \\ \hdashline \mathbf{h}_{12}'(w) & h_{22}(w) \end{array}\right] \begin{matrix} k \\ 1 \end{matrix} \end{array},$$

$$\mathbf{H}_{11}(w) = \mathbf{X}_1'\mathbf{X}_1 + w\mathbf{X}_{2\cdot 1}'(\hat{\beta}_w)\mathbf{X}_{2\cdot 1}(\hat{\beta}_w)$$

$$h_{22}(w) = w(\mathbf{y}_1 - \mathbf{X}_1\hat{\boldsymbol{\theta}}_w)'(\mathbf{y}_1 - \mathbf{X}_1\hat{\boldsymbol{\theta}}_w)$$

$$\mathbf{h}_{12}(w) = w\{\mathbf{X}_{2\cdot 1}'(\hat{\beta}_w)(\mathbf{y}_1 - \mathbf{X}_1\hat{\boldsymbol{\theta}}_w) + \mathbf{X}_1'[\mathbf{y}_{2\cdot 1}(\hat{\beta}_w) - \mathbf{X}_{2\cdot 1}(\hat{\beta}_w)\hat{\boldsymbol{\theta}}_w]\}$$

and

$$(2n - k - 1)s^2(w) = S_{11}(\hat{\boldsymbol{\theta}}_w) + wS_{22\cdot 1}(\hat{\boldsymbol{\theta}}_w, \hat{\beta}_w).$$

To this degree of approximation, the conditional distribution of $(\boldsymbol{\theta}, \beta)$ given w is in the form of a $(k+1)$-dimensional t distribution. Integrating out β from (9.5.20) using the approximating form (9.5.21), we have that

$$p(\boldsymbol{\theta} \mid w, \mathbf{y}) \mathrel{\dot{\propto}} [(2n - k - 1)s^2(w) + Q(\boldsymbol{\theta} \mid w)]^{-\frac{1}{2}(2n-1)}, \qquad (9.5.22)$$

where

$$Q(\boldsymbol{\theta} \mid w) = (\boldsymbol{\theta} - \hat{\boldsymbol{\theta}}_w)' \, \mathbf{G}(w) \, (\boldsymbol{\theta} - \hat{\boldsymbol{\theta}}_w)$$

$$\mathbf{G}(w) = \mathbf{H}_{11}(w) - h_{22}^{-1}(w)\mathbf{h}_{12}(w)\mathbf{h}'_{12}(w),$$

which is the $t_k[\hat{\boldsymbol{\theta}}_w, s^2(w)\mathbf{G}^{-1}(w), 2n - k - 1]$ distribution. Further, substituting (9.5.21) into (9.5.9) and integrating out $(\boldsymbol{\theta}, \beta)$, the posterior distribution of w is, approximately,

$$p(w \mid \mathbf{y}) \mathrel{\dot{\propto}} |\mathbf{H}(w)|^{-1/2} \, w^{\frac{1}{2}(n+1)-1} \, [s^2(w)]^{-\frac{1}{2}(2n-k-1)}, \qquad 0 < w < \infty. \qquad (9.5.23)$$

We may now adopt an argument similar to that in (9.3.27) to approximate the distribution of $\boldsymbol{\theta}$ as

$$p(\boldsymbol{\theta} \mid \mathbf{y}) = \underset{\log w}{E} \; p(\boldsymbol{\theta} \mid w, \mathbf{y}) \doteq p(\boldsymbol{\theta} \mid \hat{w}, \mathbf{y}), \qquad (9.5.24)$$

where $p(\boldsymbol{\theta} \mid \hat{w}, \mathbf{y})$ is obtained by inserting $\hat{w}$ in (9.5.22) and $\hat{w}$ is the mode of the density

$$p(\log w \mid \mathbf{y}) \mathrel{\dot{\propto}} |\mathbf{H}(w)|^{-1/2} \, w^{\frac{1}{2}(n+1)} \, [s^2(w)]^{-\frac{1}{2}(2n-k-1)}, \qquad -\infty < \log w < \infty. \qquad (9.5.25)$$

Calculation of $(\hat{\boldsymbol{\theta}}_w, \hat{\beta}_w, \hat{w})$.

In obtaining the approximating form (9.5.22), the basic problem is to find, for a given value of w, the conditional mode $\hat{\boldsymbol{\theta}}_w, \hat{\beta}_w$ of (9.5.20). From (9.5.7), we see that the quantity $S_{11}(\boldsymbol{\theta}) + wS_{22\cdot1}(\boldsymbol{\theta}, \beta)$ is quadratic in either β or $\boldsymbol{\theta}$ given the other. Thus, conditional on $\boldsymbol{\theta}$, $S_{11}(\boldsymbol{\theta}) + wS_{22\cdot1}(\boldsymbol{\theta}, \beta)$ is minimized when

$$\hat{\beta}(\boldsymbol{\theta}) = \frac{(\mathbf{y}_1 - \mathbf{X}_1\boldsymbol{\theta})' \, (\mathbf{y}_2 - \mathbf{X}_2\boldsymbol{\theta})}{(\mathbf{y}_1 - \mathbf{X}_1\boldsymbol{\theta})' \, (\mathbf{y}_1 - \mathbf{X}_1\boldsymbol{\theta})}, \qquad (9.5.26)$$

and, conditional on β, it is minimized for

$$\hat{\boldsymbol{\theta}}(w, \beta) = [\mathbf{X}_1'\mathbf{X}_1 + w\mathbf{X}'_{2\cdot1}(\beta)\mathbf{X}_{2\cdot1}(\beta)]^{-1}[\mathbf{X}_1'\mathbf{y}_1 + w\mathbf{X}'_{2\cdot1}(\beta)\mathbf{y}_{2\cdot1}(\beta)] \qquad (9.5.27)$$

as given in (9.5.11) Thus, given an initial choice of $\boldsymbol{\theta}$, say $\boldsymbol{\theta} = \boldsymbol{\theta}_0$, expression (9.5.26) can be employed to obtain $\beta_0 = \hat{\beta}(\boldsymbol{\theta}_0)$ which in turn can be used to calculate $\hat{\boldsymbol{\theta}}(w, \beta_0)$ from (9.5.27) and so on. A convenient choice for $\boldsymbol{\theta}_0$ is the center of $S_{11}(\boldsymbol{\theta})$, namely $\boldsymbol{\theta}_0 = (\mathbf{X}_1'\mathbf{X}_1)^{-1}\mathbf{X}_1'\mathbf{y}_1$. This iterative procedure will converge under favorable circumstances.

Once the conditional mode $(\hat{\boldsymbol{\theta}}_w, \hat{\beta}_w)$ is determined for a given value of w, the corresponding density of $\log w$ in (9.5.25) for that value can be readily obtained. The mode $\hat{w}$ may thus be found from (9.5.25) by employing standard numerical search procedures on a computer. A convenient preliminary estimate of $\hat{w}$ is given by

$$w_0 = s_1^2/s_{2\cdot 1}^2, \tag{9.5.28}$$

where s_1^2 is given in (9.5.3), and

$$s_{2\cdot 1}^2 = s_2^2 - \frac{s_{12}^2}{s_1^2} \tag{9.5.29}$$

with

$$s_2^2 = \frac{1}{v}(\mathbf{y}_2 - \mathbf{X}_2\hat{\boldsymbol{\theta}}_2)'(\mathbf{y}_2 - \mathbf{X}_2\hat{\boldsymbol{\theta}}_2), \qquad s_{12} = \frac{1}{v}(\mathbf{y}_1 - \mathbf{X}_1\hat{\boldsymbol{\theta}}_1)'(\mathbf{y}_2 - \mathbf{X}_2\hat{\boldsymbol{\theta}}_2),$$

$$\hat{\boldsymbol{\theta}}_2 = (\mathbf{X}_2'\mathbf{X}_2)^{-1}\mathbf{X}_2'\mathbf{y}_2 \qquad \text{and} \qquad v = n - k.$$

The accuracy of the approximation will be illustrated by an example in the next section.

9.5.3 An Example

To illustrate the theory developed above, we consider the following example.

An experiment was conducted to assess the effect of change in pressure from 30 psi to 50 psi on a chemical process yield, known to be subject to large errors. Twelve batches of raw materials were randomly selected and, from each batch, a pair of samples were taken by two chemists A and B for separate experimentation. Each chemist ran half of his experiments at high pressure and half at low pressure and they arranged their pressure settings such that each level of pressure run by chemist A was paired with each level of pressure run by chemist B an equal number of times. The data are given in Table 9.5.1.

Table 9.5.1

Pressure effect independently assessed by two chemists

	Chemist A		Chemist B	
Batch no.	Pressure	Yield y_{1u}	Pressure	Yield y_{2u}
1	50	79.5	30	73.2
2	50	76.4	50	85.0
3	50	77.2	30	60.3
4	50	76.0	50	72.9
5	50	78.6	30	66.5
6	50	82.9	50	80.0
7	30	61.4	30	63.6
8	30	67.4	50	78.3
9	30	63.5	30	73.3
10	30	69.1	50	81.3
11	30	61.2	30	54.2
12	30	69.7	50	76.3

It is assumed that in the region of the experimentation, the following model is appropriate

$$\begin{aligned} y_{u1} &= \theta_{01} + \theta x_{u1} + \varepsilon_{u1} \\ y_{u2} &= \theta_{02} + \theta x_{u2} + \varepsilon_{u2} \end{aligned} \tag{9.5.30}$$

where

$$x_{ui} = \frac{\text{Pressure} - 40}{10}$$

so that the x_{ui} assume only two values $(1, -1)$. The pressure effect θ is assumed common for both chemists but the parameters θ_{01} and θ_{02} are taken to be different to allow for possible systematic differences in mean yield. The error terms $(\varepsilon_{u1}, \varepsilon_{u2})$ are assumed correlated because each pair of observations (y_{u1}, y_{u2}) is made from the same batch of material. In terms of the linear model in (8.3.2) we have

$$\mathbf{y}_i = \mathbf{X}_i\boldsymbol{\theta}_i + \boldsymbol{\varepsilon}_i, \qquad i = 1, 2, \tag{9.5.31}$$

where

$$\boldsymbol{\theta}_i = \begin{pmatrix} \theta_{0i} \\ \theta \end{pmatrix}, \qquad \mathbf{X}_i = [\mathbf{1} \,\vdots\, \mathbf{x}_i], \qquad \mathbf{x}_i' = (x_{1i}, \ldots, x_{ni}), \qquad n = 12$$

and $\mathbf{1}$ is a 12×1 vector of ones.

We are thus in a situation in which some of the parameters $\boldsymbol{\theta}_i$ and part of the derivative matrix $\mathbf{X}_i$ are common between the two responses.

Assuming that $(\varepsilon_{u1}, \varepsilon_{u2})$ has the bivariate Normal distribution $N_2(\mathbf{0}, \boldsymbol{\Sigma})$ and on the basis of the reference prior distribution in (9.5.6), the posterior distribution of $(\theta_{01}, \theta_{02}, \theta, \sigma_{11}, \sigma_{22\cdot 1}, \beta)$ is

$$p(\theta_{01}, \theta_{02}, \theta, \sigma_{11}, \sigma_{22\cdot 1}, \beta \mid \mathbf{y}) \propto g_1(\theta_{01}, \theta, \sigma_{11})\, g_2(\theta_{02}, \theta, \sigma_{22\cdot 1}, \beta),$$

$$-\infty < \theta_{01} < \infty, \quad -\infty < \theta_{02} < \infty, \quad -\infty < \theta < \infty, \; \sigma_{11} > 0, \; \sigma_{22\cdot 1} > 0,$$
$$-\infty < \beta < \infty, \tag{9.5.32}$$

where

$$g_1(\theta_{01}, \theta, \sigma_{11}) \propto \sigma_{11}^{-[(11/2)+1]} \exp\left[-\frac{1}{2\sigma_{11}}(\mathbf{y}_1 - \mathbf{X}_1\boldsymbol{\theta}_1)'(\mathbf{y}_1 - \mathbf{X}_1\boldsymbol{\theta}_1)\right]$$

$$g_2(\theta_{02}, \theta, \sigma_{22\cdot 1}, \beta) \propto \sigma_{22}^{-[(13/2)+1]}$$

$$\times \exp\left\{-\frac{1}{2\sigma_{22\cdot 1}}[\mathbf{y}_2 - \mathbf{X}_2\boldsymbol{\theta}_2 - \beta(\mathbf{y}_1 - \mathbf{X}_1\boldsymbol{\theta}_1)]'[\mathbf{y}_2 - \mathbf{X}_2\boldsymbol{\theta}_2 - \beta(\mathbf{y}_1 - \mathbf{X}_1\boldsymbol{\theta}_1)]\right\}$$

Since main interest centers on the pressure effect θ, the parameters $(\theta_{01}, \theta_{02})$ need first to be eliminated. Writing

$$\mathbf{y}_i - \mathbf{X}_i\boldsymbol{\theta}_i = \dot{\mathbf{y}}_i - \mathbf{x}_i\theta + (\theta_{0i} - \bar{y}_i)\mathbf{1}, \tag{9.5.33}$$

where

$$\dot{\mathbf{y}}_i = \mathbf{y}_i - \bar{y}_i\mathbf{1}, \;\; \bar{y}_i = \frac{1}{n}\sum_{u=1}^{n} y_{ui}, \qquad i = 1, 2,$$

and noting that $\mathbf{1}'\dot{\mathbf{y}}_i = \mathbf{1}'\mathbf{x}_i = 0$, we have that

$$(\mathbf{y}_1 - \mathbf{X}_1\boldsymbol{\theta}_1)'(\mathbf{y}_1 - \mathbf{X}_1\boldsymbol{\theta}_1) = (\dot{\mathbf{y}}_1 - \mathbf{x}_1\theta)'(\dot{\mathbf{y}}_1 - \mathbf{x}_1\theta) + 12(\theta_{01} - \bar{y}_1)^2 \qquad (9.5.34)$$

and

$$\begin{aligned}[]&[\mathbf{y}_2 - \mathbf{X}_2\boldsymbol{\theta}_2 - \beta(\mathbf{y}_1 - \mathbf{X}_1\boldsymbol{\theta}_1)]'[\mathbf{y}_2 - \mathbf{X}_2\boldsymbol{\theta}_2 - \beta(\mathbf{y}_1 - \mathbf{X}_1\boldsymbol{\theta}_1)] \\ &\quad = [\dot{\mathbf{y}}_2 - \mathbf{x}_2\theta - \beta(\dot{\mathbf{y}}_1 - \mathbf{x}_1\theta)]'[\dot{\mathbf{y}}_2 - \mathbf{x}_2\theta - \beta(\dot{\mathbf{y}}_1 - \mathbf{x}_1\theta)] \\ &\qquad + 12[\theta_{02} - \bar{y}_2 - \beta(\theta_{01} - \bar{y}_1)]^2.\end{aligned}$$

Substituting (9.5.34) into (9.5.32), $(\theta_{01}, \theta_{02})$ can be easily integrated out, yielding

$$p(\theta, \sigma_{11}, \sigma_{22\cdot 1}, \beta \mid \mathbf{y}) \propto g_1^*(\theta, \sigma_{11})\, g_2^*(\theta, \sigma_{22\cdot 1}, \beta),$$
$$-\infty < \theta < \infty, \quad -\infty < \beta < \infty, \quad \sigma_{11} > 0,\ \sigma_{22\cdot 1} > 0, \qquad (9.5.35)$$

where

$$g_1^*(\theta, \sigma_{11}) \propto \sigma_{11}^{-[(10/2)+1]} \exp\left[-\frac{1}{2\sigma_{11}}(\dot{\mathbf{y}}_1 - \mathbf{x}_1\theta)'(\dot{\mathbf{y}}_1 - \mathbf{x}_1\theta)\right]$$

and

$$g_2^*(\theta, \sigma_{22\cdot 1}, \beta) \propto \sigma_{22}^{-[(12/2)+1]}$$
$$\times \exp\left\{-\frac{1}{2\sigma_{22\cdot 1}}[\dot{\mathbf{y}}_2 - \mathbf{x}_2\theta - \beta(\dot{\mathbf{y}}_1 - \mathbf{x}_1\theta)]'[\dot{\mathbf{y}}_2 - \mathbf{x}_2\theta - \beta(\dot{\mathbf{y}}_1 - \mathbf{x}_1\theta)]\right\}$$

which is of the form in (9.5.7) with $n = 11$, and the results in Sections 9.5.1–2 are now applicable.

Figure 9.5.1 shows the posterior distribution of θ calculated from (9.5.2). The normalizing constant was computed by numerical integration. Also shown in the same figure are the normalized curves of the two component factors

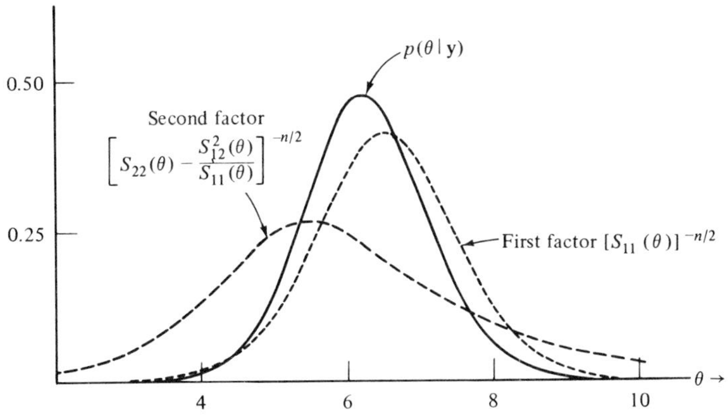

Fig. 9.5.1 Posterior distribution of θ and its component factors: the pressure data.

of the distribution. The first factor, which is a univariate t distribution centered at $\theta = 6.53$, represents information about θ coming from the results of chemist A, and the second factor, having mode at $\theta = 5.49$ and a long tail toward the right, represents the "extra information" provided by the results of chemist B. As expected, the overall distribution is located between the two component factors. It is a t-like distribution centered at $\theta \doteq 6.22$ and skewed slightly to the right.

Although for this example it will be scarcely necessary to approximate the distribution of θ, for illustrative purposes we have obtained the approximate distribution from (9.5.24). Starting with the preliminary estimate

$$w_0 = s_1^2/s_{2\cdot 1}^2 = 0.312$$

we found by trial and error that $\hat{w} \doteq 0.377$. The corresponding $\hat{\beta}_{\hat{w}}$ and $\hat{\theta}_{\hat{w}}$ are

$$\hat{\beta}_{\hat{w}} \doteq 0.6948, \qquad \hat{\theta}_{\hat{w}} \doteq 6.189$$

from which we obtained

$$p(\theta \mid \mathbf{y}) \;\dot{\propto}\; [227.9735 + 16.4171(\theta - 6.189)^2]^{-21/2}, \qquad -\infty < \theta < \infty.$$

That is, the quantity

$$\frac{\theta - 6.189}{0.833}$$

is approximately distributed as $t(0, 1, 20)$. Figure 9.5.2 compares the accuracy of the approximation for this example. The agreement between the exact and the approximate distributions is very close.

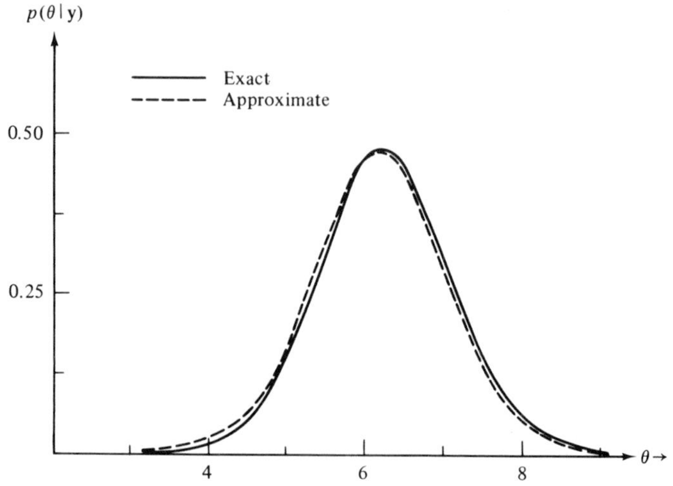

Fig. 9.5.2 Comparison of the exact and the approximate posterior distributions of θ: the pressure data.

9.6 A SUMMARY OF VARIOUS POSTERIOR DISTRIBUTIONS FOR COMMON PARAMETERS $\boldsymbol{\theta}$

In Table 9.6.1, we provide a summary of the various posterior distributions for making inferences about the common parameters $\boldsymbol{\theta}' = (\theta_1, \ldots, \theta_k)$ discussed in the preceding sections.

Table 9.6.1

A summary of various posterior distributions for the common parameters $\boldsymbol{\theta}$

Independent Responses:

1. The model is

$$\mathbf{y}_i = \mathbf{X}_i\boldsymbol{\theta} + \boldsymbol{\varepsilon}_i, \qquad i=1, \ldots, m,$$

where $\mathbf{y}_i$ is a $n_i \times 1$ vector of observations $\boldsymbol{\theta}$ is a $k \times 1$ vector of parameters, $\mathbf{X}_i$ is a $n_i \times k$ matrix of fixed elements, and $\boldsymbol{\varepsilon}_i$ is a $n_i \times 1$ vector of errors distributed as Normal $N_{n_i}(\mathbf{0}, \sigma_{ii}\mathbf{I})$. The $\boldsymbol{\varepsilon}_i$'s are assumed to be independent of one another.

2. The distribution of $\boldsymbol{\theta}$ is, from (9.2.3)

$$p(\boldsymbol{\theta} \mid \mathbf{y}) \propto \prod_{i=1}^{m} [a_{ii} + (\boldsymbol{\theta} - \hat{\boldsymbol{\theta}}_i)'\mathbf{X}_i'\mathbf{X}_i(\boldsymbol{\theta} - \hat{\boldsymbol{\theta}}_i)]^{-\frac{1}{2}n_i}, \qquad -\infty < \boldsymbol{\theta} < \infty,$$

where

$$a_{ii} = (\mathbf{y}_i - \mathbf{X}_i\hat{\boldsymbol{\theta}}_i)'(\mathbf{y}_i - \mathbf{X}_i\hat{\boldsymbol{\theta}}_i) \quad \text{and} \quad (\mathbf{X}_i'\mathbf{X}_i)\hat{\boldsymbol{\theta}}_i = \mathbf{X}_i'\mathbf{y}_i.$$

3. For $m=2$, $\mathbf{X}_1'\mathbf{X}_1$ positive definite, an alternative expression for $p(\boldsymbol{\theta} \mid \mathbf{y})$ is in (9.3.12)

$$p(\boldsymbol{\theta} \mid \mathbf{y}) = \int_0^\infty p(\boldsymbol{\theta} \mid w, \mathbf{y})p(w \mid \mathbf{y})\, dw$$

where from (9.3.8),

$$p(w \mid \mathbf{y}) \propto |\mathbf{X}_1'\mathbf{X}_1 + w\mathbf{X}_2'\mathbf{X}_2|^{-1/2}\, w^{\frac{1}{2}n_2 - 1}[s^2(w)]^{-\frac{1}{2}(n_1+n_2-k)}, \qquad 0 < w < \infty,$$

and, from (9.3.7), the conditional distribution $p(\boldsymbol{\theta} \mid w, \mathbf{y})$ is the multivariate t distribution

$$t_k[\bar{\boldsymbol{\theta}}(w),\ s^2(w)(\mathbf{X}_1'\mathbf{X}_1 + w\mathbf{X}_2'\mathbf{X}_2)^{-1},\ n_1 + n_2 - k]$$

with

$$\bar{\boldsymbol{\theta}}(w) = (\mathbf{X}_1'\mathbf{X}_1 + w\mathbf{X}_2'\mathbf{X}_2)^{-1}(\mathbf{X}_1'\mathbf{X}_1\hat{\boldsymbol{\theta}}_1 + w\mathbf{X}_2'\mathbf{X}_2\hat{\boldsymbol{\theta}}_2)$$

$$s^2(w) = \frac{1}{n_1 + n_2 - k}[\nu_1 s_1^2 + w\nu_2 s_2^2 + S_c(w)],$$

$$S_c(w) = w(\hat{\boldsymbol{\theta}}_1 - \hat{\boldsymbol{\theta}}_2)'\, \mathbf{X}_1'\mathbf{X}_1(\mathbf{X}_1'\mathbf{X}_1 + w\mathbf{X}_2'\mathbf{X}_2)^{-1}\, \mathbf{X}_2'\mathbf{X}_2(\hat{\boldsymbol{\theta}}_1 - \hat{\boldsymbol{\theta}}_2),$$

$$\nu_i = n_i - k, \qquad \nu_i s_i^2 = a_{ii}, \qquad i = 1, 2.$$

Table 9.6.1 *Continued*

4. In particular, partitioning

$$\boldsymbol{\theta} = \begin{bmatrix} \boldsymbol{\theta}_l \\ \boldsymbol{\theta}_p \end{bmatrix}\begin{matrix} l \\ p \end{matrix} \qquad \bar{\boldsymbol{\theta}}(w) = \begin{bmatrix} \bar{\boldsymbol{\theta}}_l(w) \\ \bar{\boldsymbol{\theta}}_p(w) \end{bmatrix}; \qquad (\mathbf{X}_1'\mathbf{X}_1 + w\mathbf{X}_2'\mathbf{X}_2)^{-1} = \begin{bmatrix} \mathbf{C}_{ll}(w) & \mathbf{C}_{lp}(w) \\ \mathbf{C}_{pl}(w) & \mathbf{C}_{pp}(w) \end{bmatrix}\begin{matrix} l \\ p \end{matrix}$$

the marginal distribution of $\boldsymbol{\theta}_l$ is

$$p(\boldsymbol{\theta}_l \mid \mathbf{y}) = \int_0^\infty p(\boldsymbol{\theta}_l \mid w, \mathbf{y})\, p(w \mid \mathbf{y})\, dw,$$

where $p(\boldsymbol{\theta}_l \mid w, \mathbf{y})$ is the multivariate t distribution

$$t_l[\bar{\boldsymbol{\theta}}_l(w),\ s^2(w)\mathbf{C}_{ll}(w),\ n_1+n_2-k].$$

5. From (9.3.18) to (9.2.21) computation of $p(\boldsymbol{\theta} \mid \mathbf{y})$ and $p(\boldsymbol{\theta}_l \mid \mathbf{y})$ can be simplified by first finding a $k \times k$ non-singular matrix $\mathbf{P}$ and a $k \times k$ diagonal matrix $\boldsymbol{\Lambda} = \{\lambda_{jj}\}$, $j = 1, \ldots, k$, such that

$$\mathbf{X}_1'\mathbf{X}_1 = (\mathbf{P}')^{-1}\mathbf{P}^{-1} \qquad \text{and} \qquad \mathbf{X}_2'\mathbf{X}_2 = (\mathbf{P}')^{-1}\boldsymbol{\Lambda}\mathbf{P}^{-1}$$

so that

$$\bar{\boldsymbol{\theta}}(w) = \mathbf{P}(\mathbf{I} + w\boldsymbol{\Lambda})^{-1}(\hat{\boldsymbol{\phi}}_1 + w\boldsymbol{\Lambda}\hat{\boldsymbol{\phi}}_2), \qquad \hat{\boldsymbol{\phi}}_i = \mathbf{P}^{-1}\hat{\boldsymbol{\theta}}_i, \qquad i = 1, 2$$

$$S_c(w) = \sum_{j=1}^{k} \left(\frac{w\lambda_{jj}}{1 + w\lambda_{jj}}\right)(\hat{\phi}_{1j} - \hat{\phi}_{2j})^2$$

$$|\mathbf{X}_1'\mathbf{X}_1 + w\mathbf{X}_2'\mathbf{X}_2| = |\mathbf{X}_1'\mathbf{X}_1| \prod_{j=1}^{k} (1 + w\lambda_{jj})$$

$$(\mathbf{X}_1'\mathbf{X}_1 + w\mathbf{X}_2'\mathbf{X}_2)^{-1} = \mathbf{P}(\mathbf{I} + w\boldsymbol{\Lambda})^{-1}\mathbf{P}'$$

and

$$\mathbf{C}_{ll}(w) = \mathbf{P}_1(\mathbf{I} + w\boldsymbol{\Lambda})^{-1}\mathbf{P}_1' \qquad \text{where} \qquad \mathbf{P} = \begin{bmatrix} \mathbf{P}_1 \\ \mathbf{P}_2 \end{bmatrix}\begin{matrix} l \\ p \end{matrix}$$

6. A simple approximation to $p(\boldsymbol{\theta} \mid \mathbf{y})$ is given by (9.3.27)

$$p(\boldsymbol{\theta} \mid \mathbf{y}) \doteq p(\boldsymbol{\theta} \mid \hat{w}, \mathbf{y}),$$

i.e.

$$\boldsymbol{\theta} \sim t_k[\bar{\boldsymbol{\theta}}(\hat{w}),\ s^2(\hat{w})(\mathbf{X}_1'\mathbf{X}_1 + \hat{w}\mathbf{X}_2'\mathbf{X}_2)^{-1},\ n_1+n_2-k]$$

where from (9.3.28) and (9.3.29), $\hat{w}$ is the root of the equation

$$n_2 w^{-1} - \sum_{j=1}^{k} \lambda_{jj}(1 + w\lambda_{jj})^{-1} - [s^2(w)]^{-1}\left[\nu_2 s_2^2 + \sum_{j=1}^{k} \lambda_{jj}(1 + w\lambda_{jj})^{-2}(\hat{\phi}_{1j} - \hat{\phi}_{2j})^2\right] = 0$$

maximizing $wp(w \mid \mathbf{y})$. An iterative procedure for finding w is described in (9.3.30) and (9.3.31).

Table 9.6.1 *Continued*

Correlated Responses with Common Derivative Matrix **X**:

7. The model is

$$\mathbf{y}_i = \mathbf{X}\boldsymbol{\theta}_c + \boldsymbol{\varepsilon}_i, \qquad i=1, \ldots, m,$$

where $\mathbf{y}_i$ is a $n \times 1$ vector of observations, $\boldsymbol{\theta}_c$ a $k \times 1$ vector of parameters, **X** a $n \times k$ matrix of fixed elements with rank k and $\boldsymbol{\varepsilon}_i = (\varepsilon_{1i}, \ldots, \varepsilon_{ni})'$ a $n \times 1$ vector of errors. It is assumed that $\boldsymbol{\varepsilon}_{(u)} = (\varepsilon_{u1}, \ldots, \varepsilon_{um})'$, $u=1, \ldots, m$, are independently distributed as $N_m(\mathbf{0}, \boldsymbol{\Sigma})$.

8. Let

$$\hat{\boldsymbol{\theta}} = [\hat{\boldsymbol{\theta}}_1, \ldots, \hat{\boldsymbol{\theta}}_m], \qquad \hat{\boldsymbol{\theta}}_i = (\mathbf{X}'\mathbf{X})^{-1}\mathbf{X}'\mathbf{y}_i, \qquad i=1, \ldots, m,$$

$$\mathbf{A} = \{a_{ij}\}, \qquad a_{ij} = (\mathbf{y}_i - \mathbf{X}\hat{\boldsymbol{\theta}}_i)'(\mathbf{y}_j - \mathbf{X}\hat{\boldsymbol{\theta}}_j), \qquad i,j=1, \ldots, m,$$

and $\mathbf{1}_m$ be a $m \times 1$ vector of ones. Then, from (9.4.3) the posterior distribution of $\boldsymbol{\theta}_c$ is the multivariate t distribution

$$\boldsymbol{\theta}_c \sim t_k\left[\bar{\boldsymbol{\theta}}_c, \frac{1}{(n-k)d}\mathbf{V}, n-k\right]$$

where

$$\bar{\boldsymbol{\theta}}_c = \frac{1}{d}\hat{\boldsymbol{\theta}}\mathbf{A}^{-1}\mathbf{1}_m, \qquad d = \mathbf{1}_m'\mathbf{A}^{-1}\mathbf{1}_m$$

and

$$\mathbf{V} = (\mathbf{X}'\mathbf{X})^{-1} + \hat{\boldsymbol{\theta}}\left[\mathbf{A}^{-1} - \frac{1}{d}\mathbf{A}^{-1}\mathbf{1}_m\mathbf{1}_m'\mathbf{A}^{-1}\right]\hat{\boldsymbol{\theta}}'$$

General Linear Model:

9. The model is

$$\mathbf{y}_i = \mathbf{X}_i\boldsymbol{\theta} + \boldsymbol{\varepsilon}_i, \qquad i=1, \ldots, m,$$

where $\mathbf{y}_i$ is a $n \times 1$ vector of observations, $\mathbf{X}_i$ is a $n \times k$ matrix of fixed elements, $\boldsymbol{\theta}$ a $k \times 1$ vector of parameters and $\boldsymbol{\varepsilon}_i$ a $n \times 1$ vector of errors. It is assumed that $\boldsymbol{\varepsilon}_{(u)} = (\varepsilon_{u1}, \ldots, \varepsilon_{um})'$, $u=1,\ldots, m$, are independently distributed as $N_m(\mathbf{0}, \boldsymbol{\Sigma})$.

10. The posterior distribution of $\boldsymbol{\theta}$ is, from (9.5.1),

$$p(\boldsymbol{\theta} \mid \mathbf{y}) \propto |\mathbf{S}(\boldsymbol{\theta})|^{-n/2}, \qquad -\infty < \boldsymbol{\theta} < \infty,$$

where

$$\mathbf{S}(\boldsymbol{\theta}) = \{S_{ij}(\boldsymbol{\theta})\} \qquad i,j=1, \ldots, m$$

$$S_{ij}(\boldsymbol{\theta}) = (\mathbf{y}_i - \mathbf{X}_i\boldsymbol{\theta})'(\mathbf{y}_j - \mathbf{X}_j\boldsymbol{\theta}).$$

Table 9.6.1 *Continued*

11. For $m=2$, an alternative expression of the posterior distribution of $\boldsymbol{\theta}$ is given in (9.5.16)

$$p(\boldsymbol{\theta} \mid \mathbf{y}) = \int_0^\infty \int_{-\infty}^\infty p(\boldsymbol{\theta} \mid w, \beta, \mathbf{y}) p(w, \beta \mid \mathbf{y})\, d\beta\, dw, \qquad -\infty < \boldsymbol{\theta} < \infty,$$

where, from (9.5.15),

$$p(w, \beta \mid \mathbf{y}) \propto |\mathbf{X}_1'\mathbf{X}_1 + w\mathbf{X}_{2\cdot 1}'(\beta)\mathbf{X}_{2\cdot 1}(\beta)|^{-1/2}\, w^{\frac{1}{2}(n+1)-1}[s^2(w)]^{-\frac{1}{2}(2n-k)},$$

$$0 < w < \infty, \qquad -\infty < \beta < \infty$$

and, from (9.5.12), the conditional posterior distribution $p(\boldsymbol{\theta} \mid w, \beta, \mathbf{y})$ is the multivariate t distribution

$$\boldsymbol{\theta} \sim t_k\{\hat{\boldsymbol{\theta}}(w, \beta),\, s^2(w, \beta)[\mathbf{X}_1'\mathbf{X}_1 + w\mathbf{X}_{2\cdot 1}'(\beta)\mathbf{X}_{2\cdot 1}(\beta)]^{-1},\ 2n-k\}$$

with

$$\hat{\boldsymbol{\theta}}(w, \beta) = [\mathbf{X}_1'\mathbf{X}_1 + w\mathbf{X}_{2\cdot 1}'(\beta)\mathbf{X}_{2\cdot 1}(\beta)]^{-1}[\mathbf{X}_1'\mathbf{y}_1 + w\mathbf{X}_{2\cdot 1}'(\beta)\mathbf{y}_{2.1}(\beta)]$$

$$\mathbf{X}_{2\cdot 1}(\beta) = \mathbf{X}_2 - \beta\mathbf{X}_1, \qquad \mathbf{y}_{2\cdot 1}(\beta) = \mathbf{y}_2 - \beta\mathbf{y}_1$$

and

$$(2n-k)s^2(w, \beta) = [\mathbf{y}_1 - \mathbf{X}_1\hat{\boldsymbol{\theta}}(w, \beta)]'[\mathbf{y}_1 - \mathbf{X}_1\hat{\boldsymbol{\theta}}(w, \beta)]$$
$$+ w[\mathbf{y}_{2\cdot 1}(\beta) - \mathbf{X}_{2\cdot 1}(\beta)\hat{\boldsymbol{\theta}}(w, \beta)]'[\mathbf{y}_{2\cdot 1}(\beta) - \mathbf{X}_{2\cdot 1}(\beta)\hat{\boldsymbol{\theta}}(w, \beta)].$$

12. By partitioning

$$\boldsymbol{\theta} = \begin{bmatrix} \boldsymbol{\theta}_l \\ \boldsymbol{\theta}_p \end{bmatrix} \begin{matrix} l \\ p \end{matrix}, \qquad \hat{\boldsymbol{\theta}}(w, \beta) = \begin{bmatrix} \hat{\boldsymbol{\theta}}_l(w, \beta) \\ \hat{\boldsymbol{\theta}}_p(w, \beta) \end{bmatrix} \begin{matrix} l \\ p \end{matrix}$$

and

$$[\mathbf{X}_1'\mathbf{X}_1 + w\mathbf{X}_{2\cdot 1}'(\beta)\mathbf{X}_{2\cdot 1}(\beta)]^{-1} = \begin{matrix} \quad l & \quad p \\ \begin{bmatrix} \mathbf{C}_{ll}(w, \beta) & \mathbf{C}_{lp}(w, \beta) \\ \mathbf{C}_{lp}'(w, \beta) & \mathbf{C}_{pp}(w, \beta) \end{bmatrix} \end{matrix} \begin{matrix} l \\ p \end{matrix},$$

the marginal posterior distribution of $\boldsymbol{\theta}_l$ is in (9.5.17),

$$p(\boldsymbol{\theta}_l \mid \mathbf{y}) = \int_0^\infty \int_{-\infty}^\infty p(\boldsymbol{\theta}_l \mid w, \beta, \mathbf{y}) p(w, \beta \mid \mathbf{y})\, d\beta\, dw, \qquad -\infty < \boldsymbol{\theta}_l < \infty,$$

where $p(\boldsymbol{\theta}_l \mid w, \beta, \mathbf{y})$ is the multivariate t distribution

$$\boldsymbol{\theta}_l \sim t_l\{\hat{\boldsymbol{\theta}}_l(w, \beta),\ s^2(w, \beta)\mathbf{C}_{ll}(w, \beta),\ 2n-k\}.$$

13. A procedure for approximating the distribution $p(\boldsymbol{\theta} \mid \mathbf{y})$ is developed in Section 9.5.2.

APPENDIX A9.1

Asymptotic Expansion of the Posterior Distribution of $\boldsymbol{\theta}$ for Two Independent Responses

We now develop an asymptotic expansion for the distribution of $\boldsymbol{\theta}$ in (9.3.1),

$$p(\boldsymbol{\theta} \mid \mathbf{y}) \propto \prod_{i=1}^{2} \left[1 + \frac{(\boldsymbol{\theta} - \hat{\boldsymbol{\theta}}_i)' \mathbf{X}_i'\mathbf{X}_i(\boldsymbol{\theta} - \hat{\boldsymbol{\theta}}_i)}{v_i s_i^2}\right]^{-\frac{1}{2}(v_i+k)}, \qquad -\infty < \boldsymbol{\theta} < \infty. \qquad \text{(A9.1.1)}$$

The results can be used to approximate the distribution as well as the associated marginal distributions of subsets of $\boldsymbol{\theta}$. The procedure is a generalization of Fisher's result (1961b) for the case of the weighted mean discussed in Section 9.2.1.

For simplicity in notation, we write

$$Q_i = (\boldsymbol{\theta}_i - \hat{\boldsymbol{\theta}}_i)' \mathbf{M}_i(\boldsymbol{\theta}_i - \hat{\boldsymbol{\theta}}_i), \qquad i = 1, 2, \qquad \text{(A9.1.2)}$$

where

$$\mathbf{M}_i = \frac{1}{s_i^2}\mathbf{X}_i'\mathbf{X}_i.$$

Expression (A9.1.1) then becomes

$$p(\boldsymbol{\theta} \mid \mathbf{y}) = c^{-1}\, g(Q_1, Q_2), \qquad -\infty < \boldsymbol{\theta} < \infty,$$

where

$$g(Q_1, Q_2) = \left(1 + \frac{Q_1}{v_1}\right)^{-\frac{1}{2}(v_1+k)} \left(1 + \frac{Q_2}{v_2}\right)^{-\frac{1}{2}(v_2+k)} \qquad \text{(A9.1.3)}$$

and

$$c = \int_{-\infty<\boldsymbol{\theta}<\infty} g(Q_1, Q_2)\, d\boldsymbol{\theta}$$

The expression $(1 + Q_1/v_1)^{-\frac{1}{2}(v_1+k)}$ can be written

$$\left(1 + \frac{Q_1}{v_1}\right)^{-\frac{1}{2}(v_1+k)} = \exp\left(-\tfrac{1}{2}Q_1\right) \exp\left[\tfrac{1}{2}Q_1 - \frac{v_1+k}{2}\log\left(1 + \frac{Q_1}{v_1}\right)\right].$$

Expanding the second factor on the right in powers of v_1^{-1}, we obtain

$$\left(1 + \frac{Q_1}{v_1}\right)^{-\frac{1}{2}(v_1+k)} = \exp\left(-\tfrac{1}{2}Q_1\right) \sum_{i=0}^{\infty} p_i v_1^{-i}, \qquad \text{(A9.1.4)}$$

where

$$p_0 = 1,$$
$$p_1 = \tfrac{1}{4}(Q_1^2 - 2kQ_1),$$
$$p_2 = \tfrac{1}{96}[3Q_1^4 - 4(3k+4)Q_1^3 + 12k(k+2)Q_1^2], \text{ etc.}$$

Similarly, we have that

$$\left(1 + \frac{Q_2}{v_2}\right)^{-\frac{1}{2}(v_2+k)} = \exp\left(-\tfrac{1}{2}Q_2\right) \sum_{i=0}^{\infty} q_i v_2^{-i}, \qquad \text{(A9.1.5)}$$

where
$$q_0 = 1,$$
$$q_1 = \tfrac{1}{4}(Q_2^2 - 2kQ_2),$$
$$q_2 = \tfrac{1}{96}[3Q_2^4 - 4(3k + 4)Q_2^3 + 12k(k + 2)Q_2^2], \text{ etc.}$$

Substituting (A9.1.4) and (A9.1.5) into (A9.1.3) and after a little reduction, we can express the posterior distribution as

$$p(\boldsymbol{\theta} \mid \mathbf{y}) = w^{-1} h(\boldsymbol{\theta}), \qquad -\infty < \boldsymbol{\theta} < \infty, \tag{A9.1.6}$$

where

$$h(\boldsymbol{\theta}) = \frac{|\mathbf{M}|^{1/2}}{(2\pi)^{\frac{1}{2}k}} \exp\left(-\tfrac{1}{2}Q\right) \sum_{i=0}^{\infty} \sum_{j=0}^{\infty} p_i q_j v_1^{-i} v_2^{-j},$$

$$Q = (\boldsymbol{\theta} - \tilde{\boldsymbol{\theta}})' \mathbf{M}(\boldsymbol{\theta} - \tilde{\boldsymbol{\theta}}),$$

$$\mathbf{M} = \mathbf{M}_1 + \mathbf{M}_2, \qquad \tilde{\boldsymbol{\theta}} = \mathbf{M}^{-1}(\mathbf{M}_1\hat{\boldsymbol{\theta}}_1 + \mathbf{M}_2\hat{\boldsymbol{\theta}}_2) \qquad \text{and} \qquad w = \int_{-\infty<\boldsymbol{\theta}<\infty} h(\boldsymbol{\theta})\, d\boldsymbol{\theta}. \tag{A9.1.7}$$

The integral w in (A9.1.7) can be evaluated term by term. From (A9.1.4) and (A9.1.5), we see that each term is a bivariate polynomial in the mixed moments of the quadratic forms Q_1 and Q_2 where the variables $\boldsymbol{\theta}$ have a multivariate Normal distribution $N_k(\tilde{\boldsymbol{\theta}}, \mathbf{M}^{-1})$. For this problem, it appears much simpler to obtain the mixed moments indirectly by first finding the mixed cumulants. It is straightforward to verify that the joint cumulant generating function of Q_1 and Q_2 is

$$\begin{aligned}\kappa(t_1, t_2) &= \log \int_{-\infty<\boldsymbol{\theta}<\infty} \frac{|\mathbf{M}|^{1/2}}{(2\pi)^{\frac{1}{2}k}} \exp\left(t_1 Q_1 + t_2 Q_2 - \tfrac{1}{2}Q\right) d\boldsymbol{\theta} \\ &= -\tfrac{1}{2} \log |\mathbf{I} - 2\mathbf{M}^{-1}(t_1\mathbf{M}_1 + t_2\mathbf{M}_2)| + t_1\boldsymbol{\eta}_1'\mathbf{M}_1\boldsymbol{\eta}_1 + t_2\boldsymbol{\eta}_2'\mathbf{M}_2\boldsymbol{\eta}_2 \\ &\quad + 2(t_1\mathbf{M}_1\boldsymbol{\eta}_1 + t_2\mathbf{M}_2\boldsymbol{\eta}_2)'(\mathbf{M} - 2t_1\mathbf{M}_1 - 2t_2\mathbf{M}_2)^{-1}(t_1\mathbf{M}_1\boldsymbol{\eta}_1 + t_2\mathbf{M}_2\boldsymbol{\eta}_2),\end{aligned} \tag{A9.1.8}$$

where

$$\boldsymbol{\eta}_1 = \tilde{\boldsymbol{\theta}} - \hat{\boldsymbol{\theta}}_1 \qquad \text{and} \qquad \boldsymbol{\eta}_2 = \tilde{\boldsymbol{\theta}} - \hat{\boldsymbol{\theta}}_2.$$

Upon differentiating (A9.1.8) and after some algebraic reduction, we find (see Appendix A9.2)

$$\left.\begin{aligned}\kappa_{10} &= \operatorname{tr} \mathbf{M}^{-1}\mathbf{M}_1 + \boldsymbol{\eta}_1'\mathbf{M}_1\boldsymbol{\eta}_1, \\ \kappa_{01} &= \operatorname{tr} \mathbf{M}^{-1}\mathbf{M}_2 + \boldsymbol{\eta}_2'\mathbf{M}_2\boldsymbol{\eta}_2, \\ \kappa_{rs} &= 2^{r+s-1}(r + s - 2)![(r + s - 1) \operatorname{tr} \mathbf{M}^{-1}\mathbf{G}^{rs} \\ &+ (r\boldsymbol{\eta}_1 + s\boldsymbol{\eta}_2)' \mathbf{G}^{rs}(r\boldsymbol{\eta}_1 + s\boldsymbol{\eta}_2) - r\boldsymbol{\eta}_1'\mathbf{G}^{rs}\boldsymbol{\eta}_1 - s\boldsymbol{\eta}_2'\mathbf{G}^{rs}\boldsymbol{\eta}_2], \qquad r + s \geqslant 2,\end{aligned}\right\} \tag{A9.1.9}$$

where

$$\mathbf{G}^{rs} = \mathbf{M}(\mathbf{M}^{-1}\mathbf{M}_1)^r (\mathbf{M}^{-1}\mathbf{M}_2)^s.$$

Employing the bivariate moment–cumulant inversion formulae as given by Cook (1951), the integral w in (A9.1.7) can be written as

$$w = \sum_{i=0}^{\infty} \sum_{j=0}^{\infty} b_{ij} \nu_1^{-i} \nu_2^{-j}, \qquad \text{(A9.1.10)}$$

where

$$\begin{aligned}
b_{00} &= 1,\\
b_{10} &= \tfrac{1}{4}(\kappa_{20} + \kappa_{10}^2 - 2k\kappa_{10}),\\
b_{01} &= \tfrac{1}{4}(\kappa_{02} + \kappa_{01}^2 - 2k\kappa_{01}),\\
b_{11} &= \tfrac{1}{16}[\kappa_{22} + \kappa_{20}\kappa_{02} + 2\kappa_{11}^2 + 4\kappa_{11}\kappa_{01}\kappa_{10} + \kappa_{10}^2\kappa_{01}^2 + 2\kappa_{21}\kappa_{01} + 2\kappa_{12}\kappa_{10}\\
&\quad + \kappa_{20}\kappa_{01}^2 + \kappa_{02}\kappa_{10}^2 - 2k(\kappa_{12} + \kappa_{21} + \kappa_{02}\kappa_{10} + \kappa_{20}\kappa_{01} + 2\kappa_{11}\kappa_{10}\\
&\quad + 2\kappa_{11}\kappa_{01} + \kappa_{10}\kappa_{01}^2 + \kappa_{01}\kappa_{10}^2) + 4k^2(\kappa_{11} - \kappa_{01}\kappa_{10})],\\
b_{20} &= \tfrac{1}{96}[3(\kappa_{40} + 3\kappa_{20}^2 + 4\kappa_{30}\kappa_{10} + 6\kappa_{20}\kappa_{10}^2 + \kappa_{10}^4)\\
&\quad - 4(3k+4)(\kappa_{30} + 3\kappa_{20}\kappa_{10} + \kappa_{10}^3) + 12k(k+2)(\kappa_{20} + \kappa_{10}^2)],\\
b_{02} &= \tfrac{1}{96}[3(\kappa_{04} + 3\kappa_{02}^2 + 4\kappa_{03}\kappa_{01} + 6\kappa_{02}\kappa_{01}^2 + \kappa_{01}^4) - 4(3k+4)\\
&\quad - 4(3k+4)(\kappa_{03} + 3\kappa_{02}\kappa_{01} + \kappa_{01}^3) + 12k(k+2)(\kappa_{02} + \kappa_{01}^2)], \text{ etc.}
\end{aligned}$$

Substituting the results in (A9.1.10) into (A9.1.6), we obtain the following asymptotic expression for the posterior distribution of $\boldsymbol{\theta}$,

$$p(\boldsymbol{\theta} \mid \mathbf{y}_2) = \frac{|\mathbf{M}|^{1/2}}{(2\pi)^{\frac{1}{2}k}} \exp[-\tfrac{1}{2}(\boldsymbol{\theta} - \tilde{\boldsymbol{\theta}})' \mathbf{M}(\boldsymbol{\theta} - \tilde{\boldsymbol{\theta}})] \sum_{i=0}^{\infty} \sum_{j=0}^{\infty} d_{ij} \nu_1^{-i} \nu_2^{-j},$$

$$-\infty < \boldsymbol{\theta} < \infty, \qquad \text{(A9.1.11)}$$

where

$$\begin{aligned}
&d_{00} = 1, \qquad d_{10} = p_1 - b_{10}, \qquad d_{01} = q_1 - b_{01},\\
&d_{11} = (p_1 - b_{10})(q_1 - b_{01}) + b_{10}b_{01} - b_{11},\\
&d_{20} = p_2 - b_{20} - p_1 b_{10} + b_{10}^2, \qquad d_{02} = q_2 - b_{02} - q_1 b_{01} + b_{01}^2,
\end{aligned}$$

and $\tilde{\boldsymbol{\theta}}$ and $\mathbf{M}$ are given in (A9.1.7).

Expressions for additional terms d_{12}, d_{21}, d_{22}, etc., can similarly be found if desired.

The posterior distribution is thus expressed in the form of a multivariate Normal distribution multiplied by a power series in ν_1^{-1} and ν_2^{-1}. When both ν_1 and ν_2 tend to infinity, all terms of the power series except the leading one vanish so that, in the limit, the posterior distribution is multivariate Normal $N_k(\tilde{\boldsymbol{\theta}}, \mathbf{M}^{-1})$. For finite values of ν_1 and ν_2, the terms in the power series can be regarded as "corrections" in a Normal approximation to the distribution in (A9.1.1). From

(A9.1.4), (A9.1.5) and (A9.1.9), we see that numerical evaluation of the coefficients in the power series involves merely matrix inversions and multiplications, operations which are easily performed on an electronic computer.

We note that when the posterior distribution is a univariate distribution as in (9.2.4), the results in (A9.1.11) are in exact agreement with those obtained by Fisher (1961b). In Fisher's derivation, each term of the integral w in (A9.1.7) was expressed in terms of the moments of a univariate Normal distribution. It can therefore be evaluated directly without making use of the mixed-cumulant formulae given in (A9.1.9) which seem more convenient for the multivariate case considered here.

For the univariate case, posterior probabilities can be calculated using the formulae given in Fisher's paper. When $k > 1$, the corresponding formulae for the evaluation of joint probabilities become exceedingly cumbersome and are not given here.

The Marginal Posterior Distribution

When interest centers on a subset of the elements of $\boldsymbol{\theta}$, say $\boldsymbol{\theta}_l' = (\theta_1, ..., \theta_l)$, an asymptotic expression for the corresponding marginal posterior distribution can be obtained by integrating out the remaining elements, $\boldsymbol{\theta}_r' = (\theta_{l+1}, ..., \theta_k)$ from the joint distribution in (A9.1.11). We have that

$$p(\boldsymbol{\theta}_l \mid \mathbf{y}) = \frac{|\mathbf{M}|^{1/2}}{(2\pi)^{\frac{1}{2}k}} \int_{-\infty<\boldsymbol{\theta}_r<\infty} \exp\left(-\tfrac{1}{2}Q\right) \sum_{i=0}^{\infty}\sum_{j=0}^{\infty} d_{ij} v_1^{-i} v_2^{-j} d\boldsymbol{\theta}_r. \qquad \text{(A9.1.12)}$$

Partitioning $\tilde{\boldsymbol{\theta}}$ into $\tilde{\boldsymbol{\theta}}' = (\tilde{\boldsymbol{\theta}}_l' \,\vdots\, \tilde{\boldsymbol{\theta}}_r')$ and the matrices $\mathbf{M}$ and $\mathbf{M}^{-1}$ into

$$\mathbf{M} = \begin{array}{c} \quad l \qquad r \\ \left[\begin{array}{c:c} \mathbf{M}_{ll} & \mathbf{M}_{lr} \\ \hdashline \mathbf{M}_{rl} & \mathbf{M}_{rr} \end{array}\right]\begin{array}{c} l \\ r \end{array} \end{array} \qquad \mathbf{M}^{-1} = \begin{array}{c} \quad l \qquad r \\ \left[\begin{array}{c:c} \mathbf{V}_{ll} & \mathbf{V}_{lr} \\ \hdashline \mathbf{V}_{rl} & \mathbf{V}_{rr} \end{array}\right]\begin{array}{c} l \\ r \end{array} \end{array}$$

we can write the marginal posterior distribution as

$$p(\boldsymbol{\theta}_l \mid \mathbf{y}) = \frac{|\mathbf{V}_{ll}^{-1}|^{1/2}}{(2\pi)^{\frac{1}{2}l}} \exp\left[-\tfrac{1}{2}(\boldsymbol{\theta}_l - \tilde{\boldsymbol{\theta}}_l)'\mathbf{V}_{ll}^{-1}(\boldsymbol{\theta}_l - \tilde{\boldsymbol{\theta}}_l)\right] f(\boldsymbol{\theta}_l), \qquad -\infty < \boldsymbol{\theta}_l < \infty, \qquad \text{(A9.1.13)}$$

where

$$f(\boldsymbol{\theta}_l) = \frac{|\mathbf{M}_{rr}|^{1/2}}{(2\pi)^{\frac{1}{2}r}} \int_{-\infty<\boldsymbol{\theta}_r<\infty} \exp\left[-\tfrac{1}{2}(\boldsymbol{\theta}_r - \bar{\boldsymbol{\theta}}_r)'\mathbf{M}_{rr}(\boldsymbol{\theta}_r - \bar{\boldsymbol{\theta}}_r)\right] \sum_{i=0}^{\infty}\sum_{j=0}^{\infty} d_{ij}\, v_1^{-i} v_2^{-j}\, d\boldsymbol{\theta}_r$$

with

$$\bar{\boldsymbol{\theta}}_r = \tilde{\boldsymbol{\theta}}_r - \mathbf{M}_{rr}^{-1}\mathbf{M}_{rl}\,(\boldsymbol{\theta}_l - \tilde{\boldsymbol{\theta}}_l).$$

From the expression for d_{ij} given in (A9.1.11), we see that each term in the integral $f(\boldsymbol{\theta}_l)$ is a bivariate polynomial in the quadratic form Q_1 and Q_2 where $\boldsymbol{\theta}_l$ is now considered fixed and $\boldsymbol{\theta}_r$ has a multivariate Normal distribution

$N_r(\bar{\boldsymbol{\theta}}_r, \mathbf{M}_{rr}^{-1})$. Adopting the same procedure as that described in the preceding section and by setting

$$\hat{\boldsymbol{\theta}}_1' = (\hat{\boldsymbol{\theta}}_{1l}' \,\vdots\, \hat{\boldsymbol{\theta}}_{1r}'), \qquad \hat{\boldsymbol{\theta}}_2' = (\hat{\boldsymbol{\theta}}_{2l}' \,\vdots\, \hat{\boldsymbol{\theta}}_{2r}') \tag{A9.1.14}$$

$$\mathbf{M}_1 = \begin{array}{@{}c@{}l@{}} \begin{array}{cc} \scriptstyle l & \scriptstyle r \end{array} & \\ \left[\begin{array}{c:c} \mathbf{B}_{ll} & \mathbf{B}_{lr} \\ \hdashline \mathbf{B}_{rl} & \mathbf{B}_{rr} \end{array}\right] & \begin{array}{l} \scriptstyle l \\ \scriptstyle r \end{array} \end{array}, \qquad \mathbf{M}_1^{-1} = \begin{array}{@{}c@{}l@{}} \begin{array}{cc} \scriptstyle l & \scriptstyle r \end{array} & \\ \left[\begin{array}{c:c} \mathbf{E}_{ll} & \mathbf{E}_{lr} \\ \hdashline \mathbf{E}_{rl} & \mathbf{E}_{rr} \end{array}\right] & \begin{array}{l} \scriptstyle l \\ \scriptstyle r \end{array} \end{array},$$

$$\mathbf{M}_2 = \begin{array}{@{}c@{}l@{}} \begin{array}{cc} \scriptstyle l & \scriptstyle r \end{array} & \\ \left[\begin{array}{c:c} \mathbf{C}_{ll} & \mathbf{C}_{lr} \\ \hdashline \mathbf{C}_{rl} & \mathbf{C}_{rr} \end{array}\right] & \begin{array}{l} \scriptstyle l \\ \scriptstyle r \end{array} \end{array}, \qquad \mathbf{M}_2^{-1} = \begin{array}{@{}c@{}l@{}} \begin{array}{cc} \scriptstyle l & \scriptstyle r \end{array} & \\ \left[\begin{array}{c:c} \mathbf{F}_{ll} & \mathbf{F}_{lr} \\ \hdashline \mathbf{F}_{rl} & \mathbf{F}_{rr} \end{array}\right] & \begin{array}{l} \scriptstyle l \\ \scriptstyle r \end{array} \end{array},$$

we obtain the mixed cumulants of Q_1 and Q_2

$$\begin{aligned} \omega_{10} &= \operatorname{tr} \mathbf{M}_{rr}^{-1} \mathbf{B}_{rr} + \boldsymbol{\gamma}_1' \mathbf{B}_{rr} \boldsymbol{\gamma}_1 + (\boldsymbol{\theta}_l - \hat{\boldsymbol{\theta}}_{1l})' \mathbf{E}_{ll}^{-1} (\boldsymbol{\theta}_l - \hat{\boldsymbol{\theta}}_{1l}) \\ \omega_{01} &= \operatorname{tr} \mathbf{M}_{rr}^{-1} \mathbf{C}_{rr} + \boldsymbol{\gamma}_2' \mathbf{C}_{rr} \boldsymbol{\gamma}_2 + (\boldsymbol{\theta}_l - \hat{\boldsymbol{\theta}}_{2l})' \mathbf{F}_{ll}^{-1} (\boldsymbol{\theta}_l - \hat{\boldsymbol{\theta}}_{2l}) \\ \omega_{rs} &= 2^{r+s-1}(r+s-2)!\,[(r+s-1) \operatorname{tr} \mathbf{M}_{rr}^{-1} \mathbf{H}^{rs} \\ &\quad + (r\boldsymbol{\gamma}_1 + s\boldsymbol{\gamma}_2)' \mathbf{H}^{rs} (r\boldsymbol{\gamma}_1 + s\boldsymbol{\gamma}_2) - r\boldsymbol{\gamma}_1' \mathbf{H}^{rs} \boldsymbol{\gamma}_1 - s\boldsymbol{\gamma}_2' \mathbf{H}^{rs} \boldsymbol{\gamma}_2], \qquad r + s \geqslant 2, \end{aligned} \tag{A9.1.15}$$

where

$$\begin{aligned} \mathbf{H}^{rs} &= \mathbf{M}_{rr} (\mathbf{M}_{rr}^{-1} \mathbf{B}_{rr})^r (\mathbf{M}_{rr}^{-1} \mathbf{C}_{rr})^s \\ \boldsymbol{\gamma}_1 &= (\tilde{\boldsymbol{\theta}}_r - \hat{\boldsymbol{\theta}}_{1r}) + (\mathbf{B}_{rr}^{-1} \mathbf{B}_{rl} - \mathbf{M}_{rr}^{-1} \mathbf{M}_{rl})(\boldsymbol{\theta}_l - \hat{\boldsymbol{\theta}}_{1l}) \\ \boldsymbol{\gamma}_2 &= (\tilde{\boldsymbol{\theta}}_r - \hat{\boldsymbol{\theta}}_{2r}) + (\mathbf{C}_{rr}^{-1} \mathbf{C}_{rl} - \mathbf{M}_{rr}^{-1} \mathbf{M}_{rl})(\boldsymbol{\theta}_l - \hat{\boldsymbol{\theta}}_{2l}). \end{aligned}$$

Using the results in (A9.1.15), we can express the marginal posterior distribution of $\boldsymbol{\theta}_l$ as

$$p(\boldsymbol{\theta}_l \mid \mathbf{y}) = \frac{|\mathbf{V}_{ll}|^{-1/2}}{(2\pi)^{\frac{1}{2}l}} \exp\left[-\tfrac{1}{2}(\boldsymbol{\theta}_l - \tilde{\boldsymbol{\theta}}_l)' \mathbf{V}_{ll}^{-1} (\boldsymbol{\theta}_l - \tilde{\boldsymbol{\theta}}_l)\right] \sum_{i=0}^{\infty} \sum_{j=0}^{\infty} \delta_{ij} v_1^{-i} v_2^{-j},$$

$$-\infty < \boldsymbol{\theta}_l < \infty, \tag{A9.1.16}$$

with

$$\begin{aligned} &\delta_{00} = 1, \qquad \delta_{10} = g_{10} - b_{10}, \qquad \delta_{01} = g_{01} - b_{01}, \\ &\delta_{11} = g_{11} - b_{11} - g_{10} b_{01} - g_{01} b_{10} + 2b_{01} b_{10}, \\ &\delta_{20} = g_{20} - b_{20} - g_{10} b_{10} + b_{10}^2, \qquad \delta_{02} = g_{02} - b_{02} - g_{01} b_{01} + b_{01}^2, \text{ etc.,} \end{aligned}$$

and the quantities g_{ij} are functions of the mixed cumulants ω_{ij} with the functional relationships exactly the same as those between b_{ij} and κ_{ij} shown in (A9.1.10).

It is seen that the leading term in the expansion is a multivariate Normal distribution $N_l(\tilde{\boldsymbol{\theta}}_l, \mathbf{V}_{ll})$. When $\boldsymbol{\theta}_l$ consists of only one variable, the quantities

δ_{ij} in (A9.1.16) are simply polynomials in that variable. Employing the well-known expression for the moments of a Normal variable, one can easily derive an asymptotic expression for the moments. In addition, probability integrals can also be approximated using methods given in Fisher's previously cited paper (1961b).

APPENDIX A9.2

Mixed Cumulants of the Two Quadratic Forms Q_1 and Q_2

From the joint cumulant generating function of the quadratic forms Q_1 and Q_2 given in (A9.1.8), we now derive the expressions for the mixed cumulants shown in (A9.1.9). In our development, we shall make use of the following lemma.

Lemma Let $\mathbf{P}_1$ be a $n \times n$ positive definite symmetric matrix and $\mathbf{P}_2$ be a $n \times n$ nonnegative definite symmetric matrix. Then, for sufficiently small i, we have

$$\log |\mathbf{I} - i\mathbf{P}_1\mathbf{P}_2| = -\sum_{r=1}^{\infty} \frac{i^r}{r} \operatorname{tr} (\mathbf{P}_1\mathbf{P}_2)^r. \tag{A9.2.1}$$

Employing the above lemma and for sufficiently small values of t_1 and t_2, we can expand the first term on the right of (A9.1.8) into

$$-\tfrac{1}{2}\log |\mathbf{I} - 2\mathbf{M}^{-1}(t_1\mathbf{M}_1 + t_2\mathbf{M}_2)| = \sum_{r=1}^{\infty} \frac{2^{r-1}}{r} \operatorname{tr} (t_1\mathbf{M}^{-1}\mathbf{M}_1 + t_2\mathbf{M}^{-1}\mathbf{M}_2)^r. \tag{A9.2.2}$$

In (A9.1.8), the quadratic form $t_1\boldsymbol{\eta}_1'\mathbf{M}_1\boldsymbol{\eta}_1$ can be written

$$\begin{aligned} t_1\boldsymbol{\eta}_1'\mathbf{M}_1\boldsymbol{\eta}_1 &= t_1\boldsymbol{\eta}_1'\mathbf{M}_1(\mathbf{M} - 2t_1\mathbf{M}_1 - 2t_2\mathbf{M}_2)^{-1}(\mathbf{M} - 2t_1\mathbf{M}_1 - 2t_2\mathbf{M}_2)\boldsymbol{\eta}_1 \\ &= t_1\boldsymbol{\eta}_1'\mathbf{M}_1(\mathbf{M} - 2t_1\mathbf{M}_1 - 2t_2\mathbf{M}_2)^{-1}\mathbf{M}\boldsymbol{\eta}_1 - 2t_1^2\boldsymbol{\eta}_1'\mathbf{M}_1(\mathbf{M} - 2t_1\mathbf{M}_1 \\ &\quad -2t_2\mathbf{M}_2)^{-1}\mathbf{M}_1\boldsymbol{\eta}_1 - 2t_1t_2\boldsymbol{\eta}_1'\mathbf{M}_1(\mathbf{M} - 2t_1\mathbf{M}_1 - 2t_2\mathbf{M}_2)^{-1}\mathbf{M}_2\boldsymbol{\eta}_1. \end{aligned} \tag{A9.2.3}$$

Similarly,

$$\begin{aligned} t_2\boldsymbol{\eta}_2'\mathbf{M}_2\boldsymbol{\eta}_2 &= t_2\boldsymbol{\eta}_2'\mathbf{M}_2(\mathbf{M} - 2t_1\mathbf{M}_1 - 2t_2\mathbf{M}_2)^{-1}\mathbf{M}\boldsymbol{\eta}_2 - 2t_2^2\boldsymbol{\eta}_2'\mathbf{M}_2(\mathbf{M} - 2t_1\mathbf{M}_1 \\ &\quad - 2t_2\mathbf{M}_2)^{-1}\mathbf{M}_2\boldsymbol{\eta}_2 - 2t_1t_2\boldsymbol{\eta}_2'\mathbf{M}_2(\mathbf{M} - 2t_1\mathbf{M}_1 - 2t_2\mathbf{M}_2)^{-1}\mathbf{M}_1\boldsymbol{\eta}_2. \end{aligned} \tag{A9.2.4}$$

Thus, the expression in (A9.1.8) becomes

$$\begin{aligned} \kappa(t_1, t_2) &= \sum_{r=1}^{\infty} \frac{2^{r-1}}{r} \operatorname{tr} (t_1\mathbf{M}^{-1}\mathbf{M}_1 + t_2\mathbf{M}^{-1}\mathbf{M}_2)^r + t_1\boldsymbol{\eta}_1'\mathbf{M}_1(\mathbf{I} - 2t_1\mathbf{M}^{-1}\mathbf{M}_1 \\ &\quad -2t_2\mathbf{M}^{-1}\mathbf{M}_2)^{-1}\boldsymbol{\eta}_1 + t_2\boldsymbol{\eta}_2'\mathbf{M}_2(\mathbf{I} - 2t_1\mathbf{M}^{-1}\mathbf{M}_1 - 2t_2\mathbf{M}^{-1}\mathbf{M}_2)^{-1}\boldsymbol{\eta}_2 \\ &\quad - 2t_1t_2(\boldsymbol{\eta}_1 - \boldsymbol{\eta}_2)'\mathbf{M}_1(\mathbf{I} - 2t_1\mathbf{M}^{-1}\mathbf{M}_1 - 2t_2\mathbf{M}^{-1}\mathbf{M}_2)^{-1}\mathbf{M}^{-1}\mathbf{M}_2(\boldsymbol{\eta}_1 - \boldsymbol{\eta}_2). \end{aligned} \tag{A9.2.5}$$

Since $\mathbf{M} = \mathbf{M}_1 + \mathbf{M}_2$, it is easy to see that the matrix $\mathbf{M}_1\mathbf{M}^{-1}\mathbf{M}_2$ is symmetric. By virtue of this property, we have

$$(t_1\mathbf{M}^{-1}\mathbf{M}_1 + t_2\mathbf{M}^{-1}\mathbf{M}_2)^r = \sum_{i=0}^{r} \binom{r}{i} t_1^i t_2^{r-i}(\mathbf{M}^{-1}\mathbf{M}_1)^i(\mathbf{M}^{-1}\mathbf{M}_2)^{r-i} \tag{A9.2.6}$$

and, for sufficiently small values of t_1 and t_2,

$$(\mathbf{I} - 2t_1\mathbf{M}^{-1}\mathbf{M}_1 - 2t_2\mathbf{M}^{-1}\mathbf{M}_2)^{-1} = \sum_{i=0}^{\infty}\sum_{j=0}^{\infty} 2^{i+j}t_1^i t_2^j\binom{i+j}{i}(\mathbf{M}^{-1}\mathbf{M}_1)^i(\mathbf{M}^{-1}\mathbf{M}_2)^j. \tag{A9.2.7}$$

Substituting (A9.2.6–7) into (A9.2.5) and after a little rearrangement, we find

$$\begin{aligned}\kappa(t_1, t_2) = 1 &+ \sum_{r=1}^{\infty} 2^{r-1}t_1^r\left[\frac{1}{r}\operatorname{tr}(\mathbf{M}^{-1}\mathbf{M}_1)^r + \boldsymbol{\eta}_1'\mathbf{M}(\mathbf{M}^{-1}\mathbf{M}_1)^r\boldsymbol{\eta}_1\right]\\ &+ \sum_{r=1}^{\infty} 2^{r-1}t_2^r\left[\frac{1}{r}\operatorname{tr}(\mathbf{M}^{-1}\mathbf{M}_2)^r + \boldsymbol{\eta}_2'\mathbf{M}(\mathbf{M}^{-1}\mathbf{M}_2)^r\boldsymbol{\eta}_2\right]\\ &+ \sum_{r=1}^{\infty}\sum_{s=1}^{\infty} 2^{r+s-1}t_1^r t_2^s\frac{(r+s-2)!}{r!s!}[(r+s-1)\operatorname{tr}\mathbf{M}^{-1}\mathbf{G}^{rs}\\ &+ (r\boldsymbol{\eta}_1 + s\boldsymbol{\eta}_2)'\mathbf{G}^{rs}(r\boldsymbol{\eta}_1 + s\boldsymbol{\eta}_2) - r\boldsymbol{\eta}_1'\mathbf{G}^{rs}\boldsymbol{\eta}_1 - s\boldsymbol{\eta}_2'\mathbf{G}^{rs}\boldsymbol{\eta}_2],\end{aligned} \tag{A9.2.8}$$

where

$$\mathbf{G}^{rs} = \mathbf{M}(\mathbf{M}^{-1}\mathbf{M}_1)^r(\mathbf{M}^{-1}\mathbf{M}_2)^s.$$

Upon differentiating (A9.2.8), we obtain

$$\kappa_{r0} = 2^{r-1}(r-1)![\operatorname{tr}(\mathbf{M}^{-1}\mathbf{M}_1)^r + r\boldsymbol{\eta}_1'\mathbf{M}(\mathbf{M}^{-1}\mathbf{M}_1)^r\boldsymbol{\eta}_1], \tag{A9.2.9}$$

$$\kappa_{0s} = 2^{s-1}(s-1)![\operatorname{tr}(\mathbf{M}^{-1}\mathbf{M}_2)^s + s\boldsymbol{\eta}_2'\mathbf{M}(\mathbf{M}^{-1}\mathbf{M}_2)^s\boldsymbol{\eta}_2], \tag{A9.2.10}$$

$$\begin{aligned}\kappa_{rs} = 2^{r+s-1}(r+s-2)![(r+s-1)\operatorname{tr}\mathbf{M}^{-1}\mathbf{G}^{rs} + (r\boldsymbol{\eta}_1 + s\boldsymbol{\eta}_2)'\,\mathbf{G}^{rs}\,(r\boldsymbol{\eta}_1 + s\boldsymbol{\eta}_2)\\ - r\boldsymbol{\eta}_1'\mathbf{G}^{rs}\boldsymbol{\eta}_1 - s\boldsymbol{\eta}_2'\mathbf{G}^{rs}\boldsymbol{\eta}_2], \qquad r, s \geqslant 1\end{aligned} \tag{A9.2.11}$$

which can then be combined into the expressions given in (A9.1.9).

CHAPTER 10

TRANSFORMATION OF DATA

10.1 INTRODUCTION

In this chapter we discuss the problem of data transformation in relation to the linear model

$$\mathbf{y} = E(\mathbf{y}) + \boldsymbol{\varepsilon}, \qquad E(\mathbf{y}) = \mathbf{X}\boldsymbol{\theta}, \tag{10.1.1}$$

where $\mathbf{y}$ is an $n \times 1$ vector of observations, $\mathbf{X}$ a $n \times k$ matrix of fixed elements, $\boldsymbol{\theta}$ a $k \times 1$ vector of regression coefficients, and $\boldsymbol{\varepsilon}$ a $n \times 1$ vector of errors.

In Section 2.7, we discussed in detail the linear model (10.1.1) under the assumption that $\boldsymbol{\varepsilon}$ is distributed as $N_n(\mathbf{0}, \sigma^2\mathbf{I})$. This Normal theory linear model is of great practical importance and has been used in a wide variety of applications. These include, for example, regression analysis and the analysis of designed experiments such as k-way classification designs, factorials, randomized blocks, latin squares, incomplete blocks, and response surface designs. Normal theory analysis, whether from a Bayesian or sampling point of view, is attractive because of its simplicity and transparency. The possibility of data transformation greatly widens its realm of adequate validity.

In using the Normal linear model, we make assumptions not only about (i) the adequacy of the expectation function $E(\mathbf{y}) = \mathbf{X}\boldsymbol{\theta}$ but also about the suitability of the spherical Normal error function to represent the probability distribution of $\boldsymbol{\varepsilon}$. Specifically, we assume (ii) constancy of error variance from one observation to another (iii), Normality of the distributions of the observations and (iv) independence of these distributions.† The last assumption (iv) is perhaps the most important of all. Its violation leads to dramatic consequences. Its relaxation leads to the consideration of a whole new range of important time series and dynamic models, which, however, are not treated here.

Nevertheless, at least for data generated by designed experiments, the independence assumption is one whose applicability can to some extent be assured by the physical conduct of the experiment, and we shall here suppose it to be

† We are, of course, also assuming that there are no "aberrant" observations. In practice, due to the possible anomalies in the experimental setup, some of the observations might have been generated, not from the postulated model, but from an alternative model which has a large bias or a much larger variance. For a Bayesian analysis of this problem, see Box and Tiao (1968b).

valid. Assumptions (i), (ii), and (iii) are usually not under the experimenter's control, but it happens rather frequently that, although these assumptions are not true for the observations in the original metric, they are reasonably well satisfied when the observations are suitably transformed. We now consider why this is so.

10.1.1 A Factorial Experiment on Grinding

Consider an experiment on the grinding of particles which are approximately spherical. Suppose that observations on the final size of the particles were taken after standard material had been fed to three different grinding machines for four different grinding periods, using a 3×4 factorial design, and that the primary purpose of the experiment was to determine how different machines and different grinding periods affected the final size of the particles.

Now suppose it happened to be true that *if "size" were measured by particle radius* y, then, to an adequate approximation,† the Normal linear model (10.1.1) could represent the data and the effects of machines and of periods of grinding would behave additively. Specifically, suppose that, if "particle size" was measured in terms of *radius*, then, to a sufficient approximation,

a) there would be no interaction between machines and grinding periods,

b) the error variances would be constant, and

c) the error distribution would be Normal.

Now it might be inconvenient to measure particle size in terms of the radius y and it might not occur to the investigator to do so. Instead, he might measure the area of the circular section of the particle seen under the microscope (proportional to y^2) or he might weigh the particle and so effectively work with the volume (proportional to y^3). In either case he would probably report his results in terms of what was actually measured.

But if, in terms of the y's, the effects were additive and the errors spherically Normal, this would certainly not be true, for example, of the effects and errors in terms of the y^2's. Specifically, suppose that

$$y_{ij} = E(y_{ij}) + \varepsilon_{ij}, \qquad i = 1, 2, 3; \quad j = 1, \ldots, 4, \tag{10.1.2}$$

where $E(y_{ij}) = \theta_i + \phi_j$ is the average response for the ith machine run and the jth period of time, and ε_{ij} is distributed as $N(0, \sigma^2)$. Then, for the particle size measured in terms of area, we should have the model

$$y_{ij}^2 = E(y_{ij}^2) + e_{ij} \tag{10.1.3}$$

where

$$E(y_{ij}^2) = \theta_i^2 + \phi_j^2 + 2\theta_i\phi_j + \sigma^2 \tag{10.1.4}$$

---

† Since the particular radius could not be negative, the Normal assumption could not be true exactly.

and

$$e_{ij} = 2(\theta_i + \phi_j)\,\varepsilon_{ij} + \varepsilon_{ij}^2 - \sigma^2 \tag{10.1.5}$$

with

$$E(e_{ij}) = 0, \text{Var}\,(e_{ij}) = 4(\theta_i + \phi_j)^2\sigma^2 + 2\sigma^4. \tag{10.1.6}$$

We note that for the response y_{ij}^2

a) a nonadditive (interaction) term $\theta_i \phi_j$ now appears in the expectation function $E(y_{ij}^2)$ in (10.1.4),

b) the error e_{ij} in (10.1.5) has a variance which is not constant but depends directly on $E(y_{ij}) = \theta_i + \phi_j$, and

c) since ε_{ij} has a Normal distribution the distribution of e_{ij} must certainly be non-Normal.

Now in spite of this it is nevertheless true that, if the investigator made an appropriate transformation (in this case the square root transformation) of his "particle area" data, the simple Normal linear theory analysis would apply.

In general, there is often surprisingly little to be said in favour of the original metric in which data happen to be taken. For example, temperature can be measured in terms of °C when it is related to the expansion of a thread of mercury in a thermometer. However it could equally well be measured in terms of molecular velocity $V \propto (T + 273)^{1/2}$ which would be of more immediate relevance in some contexts.

Bearing these arguments in mind, it is perhaps not surprising that numerous examples occur where the assumptions underlying the Normal theory linear model, although not true for observations in the original metric, provide an adequate representation *after suitable transformation.* In some instances appropriate transformations have been arrived at from theoretical considerations alone, in some cases by analysis of the data alone, and in some instances by a mixture of both. We shall here consider estimation of transformations from the data.

It is not of course suggested that interaction, heterogeneity of variance and non-Normality can *always* be eliminated by suitable transformation of the data. In fact, cases of interaction, inhomogeneity of variance and non-Normality can each be divided into two classes: *transformable* and *non-transformable.* In the first class the phenomena of interaction, of variance inequality or of non-Normality are anomalies arising only because of unsuitable choice of metric. In the second class the phenomenon is of more fundamental character which transformation cannot eliminate.

Finally it will not necessarily be true that the same transformation of the data will *simultaneously* eliminate all discrepancies from assumption. Cases can arise where, for example, approximate additivity can be achieved by a particular transformation but not simultaneously with variance homogeneity.

In summary, we should like to employ the model (10.1.1) with

a) the expectation function $E(\mathbf{y}) = \mathbf{X}\boldsymbol{\theta}$ having the simplest possible form,
b) the individual errors ε_i, $i = 1, \ldots, n$, having the same variance σ^2, and
c) the individual errors ε_i independently and Normally distributed.

If a measurement of y does not possess these properties, a suitable nonlinear transformation of y may improve the situation. It is often the case that the logarithm, the reciprocal, the square root, or some other transformations of y will make possible an analysis which strains assumptions less and in terms of which a simpler representation is possible.

10.2 ANALYSIS OF THE BIOLOGICAL AND THE TEXTILE DATA

We now introduce two sets of data and consider in some detail the question of choice of appropriate models, following joint work with D.R. Cox.

10.2.1 Some Biological Data

Table 10.2.1 shows the survival times of $n = 48$ animals exposed to three different poisons and subject to four different treatments. The experiment was set out in a 3×4 factorial design with a fourfold replication (four animals per group). We shall analyze the data using the linear expectation function $E(\mathbf{y}) = \mathbf{X}\boldsymbol{\theta}$ and we now discuss precisely what this involves and the justification for doing so.

Table 10.2.1

Survival time (in 10-hr units) of animals in a 3×4 factorial experiment

	Treatment			
Poison	*A*	*B*	*C*	*D*
I	0.31	0.82	0.43	0.45
	0.45	1.10	0.45	0.71
	0.46	0.88	0.63	0.66
	0.43	0.72	0.76	0.62
II	0.36	0.92	0.44	0.56
	0.29	0.61	0.35	1.02
	0.40	0.49	0.31	0.71
	0.23	1.24	0.40	0.38
III	0.22	0.30	0.23	0.30
	0.21	0.37	0.25	0.36
	0.18	0.38	0.24	0.31
	0.23	0.29	0.22	0.33

The use of the linear expectation function $E(\mathbf{y}) = \mathbf{X}\boldsymbol{\theta}$ for this biological data and for most other examples is an admitted approximation to the truth. In most cases there undoubtedly exists some true underlying functional relationship

$$E(\mathbf{y}) = f(\boldsymbol{\xi}, \boldsymbol{\phi}), \tag{10.2.1}$$

which is probably nonlinear in the parameters $\boldsymbol{\phi}$ and which describes *in mechanistic terms* how $E(\mathbf{y})$ is affected by some set of basic input variables $\boldsymbol{\xi}$, involving some set of parameters $\boldsymbol{\phi}$. We have tacitly decided to replace it with the empirical linear model (10.1.1). This may be either because, in the present state of the art, this "true" relationship is unknown and would be too difficult to find out, or because the known mechanism is too complicated to use.

For this biological example, it might be possible to write differential equations which represented the absorption of poison into the blood stream of the animals and to represent mechanistically by other equations the effect of the treatments. If this could be done, one might run appropriate experiments to test the model and to obtain estimates of the basic constants it contained. In the absence of such fundamental knowledge, we would have to set our sights lower. We could, for example, simply set out to estimate the "effects" of the treatments in terms of the increase in survival time they produced. This could be done in terms of a purely empirical linear model.

Specifically, we might postulate that, associated with the tth treatment ($t = 1, 2, 3, 4$) and the pth poison ($p = 1, 2, 3$), there was a mean survival time θ_{tp}. If we were principally interested in differences associated with the various poison-treatment combinations, we could write the model in the form

$$E(y_i) = \theta_0 x_{0i} + (\theta_{11} - \theta_0) x_{11i} + (\theta_{12} - \theta_0) x_{12i} + \cdots + (\theta_{43} - \theta_0) x_{43i} + \varepsilon,$$
$$i = 1, \ldots, 48, \tag{10.2.2}$$

where $x_{0i} = 1$ and the indicator variable x_{tpi} is 1 if the ith animal received the pth poison and the tth treatment, and zero otherwise. One way of dealing with the problem that the resulting 48×13 derivative matrix would not be of full rank, would be to omit, say, the final term $(\theta_{43} - \theta_0) x_{43i}$. The resulting empirical model which would be of the form of (10.1.1) would contain 12 functionally independent parameters.

In some circumstances it would be reasonable to expect that a simpler empirical model containing fewer parameters could be found. In particular, it might be true, to an adequate approximation, that the effects of poisons and treatments were additive. Specifically, if $\theta_{t.} - \theta_0$ was the mean change in survival time produced by the tth treatment and $\theta_{.p} - \theta_0$ the mean change produced by the pth poison, then the model could be written

$$E(y_i) = \theta_0 x_{0i} + (\theta_{1.} - \theta_0) u_{1i} + \cdots + (\theta_{4.} - \theta_0) u_{4i} + (\theta_{.1} - \theta_0) w_{1i}$$
$$+ \cdots + (\theta_{.3} - \theta_0) w_{3i} + \varepsilon_i, \tag{10.2.3}$$

where u_{ti} is 1 or 0 depending on whether the ith animal had or did not have the tth treatment and w_{pi} is 1 or 0 depending on whether the ith animal had the pth poison. Again we could omit, say, $(\theta_{4.} - \theta_0)\, u_{4i}$ and $(\theta_{.3} - \theta_0)\, w_{3i}$ to retain an indicator matrix $\mathbf{X}$ of full rank.

We shall call the general model in (10.2.2) the *interaction model* and the more specialized model in (10.2.3) the *additive model.* It is often convenient to write the more general interaction model in the form of the additive model with additional terms specifically carrying parameter combinations which measure interaction. For the present example, we could write (10.2.2) alternatively as

$$E(y_i) = \theta_0\, x_{0i} + \sum_{t=1}^{3} (\theta_{t.} - \theta_0)\, u_{ti} + \sum_{p=1}^{2} (\theta_{.p} - \theta_0)\, w_{pi} + \sum_{t=1}^{3} \sum_{p=1}^{2} (\theta_{tp} - \theta_{t.} - \theta_{.p} + \theta_0)\, x_{tpi} \,. \tag{10.2.4}$$

By omitting the last summation containing the six independent interaction parameters, we would have the nonsingular form of the additive model of (10.2.3).

Whereas the interaction model contains twelve functionally independent parameters, the additive model contains only six. In the interest of simplicity we naturally would wish to use the additive model if this were adequate. While the additive model might not be appropriate in the original metric, it might become so if a suitable data transformation were employed.

Representation in terms of the smallest possible number of parameters we call "parsimonious parametrization." Clearly if by, say, a power transformation from y to y^λ we could validate a simple additive model, we would have served the interest of parsimony. That is, by including one extra parameter λ we would have made it possible to eliminate six interaction parameters.

We now consider Normality and constancy of variance. Inspection of the sample variances in the 12 cells of Table 10.2.1 shows that they differ very markedly and that they tend to increase as the cell mean increases. It might be that a transformation could make it more plausible that population variances were equal. Again, while one would not expect on this limited amount of data that there would be a great deal of information about its Normality or otherwise, it might very well be that some transformation of the data could improve the Normality of the distributions of the errors.

10.2.2 The Textile Data

As a further example, we consider the data of Table 10.2.2. These data came originally from an unpublished report to the Technical Committee, International Wool Textile Organization, in which Drs. A. Barella and A. Sust described some experiments on the behaviour of worsted yarn under cycles of repeated loading.

Table 10.2.2 gives the numbers of cycles to failure, y, obtained in a 3^3 experiment in which the factors were

ξ_1: length of test specimen (250, 300, 350 mm),

ξ_2: amplitude of loading cycle (8, 9, 10 mm),

ξ_3: load (40, 45, 50 g).

Table 10.2.2

Cycles to failure of worsted yarn in a 3^3 factorial experiment

Factor levels			Cycles to failure
x_1	x_2	x_3	y
−1	−1	−1	674
−1	−1	0	370
−1	−1	+1	292
−1	0	−1	338
−1	0	0	266
−1	0	+1	210
−1	+1	−1	170
−1	+1	0	118
−1	+1	+1	90
0	−1	−1	1,414
0	−1	0	1,198
0	−1	+1	634
0	0	−1	1,022
0	0	0	620
0	0	+1	438
0	+1	−1	442
0	+1	0	332
0	+1	+1	220
+1	−1	−1	3,636
+1	−1	0	3,184
+1	−1	+1	2,000
+1	0	−1	1,568
+1	0	0	1,070
+1	0	+1	566
+1	+1	−1	1,140
+1	+1	0	884
+1	+1	+1	360

$x_1 = (\xi_1 - 300)/50$, $x_2 = \xi_2 - 9$ and $x_3 = (\xi_3 - 45)/5$.

Although the form $E(y) = f(\boldsymbol{\xi}, \boldsymbol{\phi})$ was unknown, one might hope to locally represent this function by expanding it about the average levels $\boldsymbol{\xi}_0$ in a Taylor series. We would then have

$$E(y) = f_0 + \sum_{t=1}^{3} f_t(\xi_t - \xi_{t0}) + \sum_{t=1}^{3}\sum_{j=1}^{3} f_{tj}(\xi_t - \xi_{t0})(\xi_j - \xi_{j0}) + \text{higher-order terms}, \qquad (10.2.5)$$

where

$$f_0 = f(\boldsymbol{\xi}_0, \boldsymbol{\phi}), \qquad f_t = \left.\frac{\partial f}{\partial \xi_t}\right|_{\boldsymbol{\xi}=\boldsymbol{\xi}_0}, \qquad f_{tj} = \left.\frac{\partial^2 f}{\partial \xi_t\, \partial \xi_j}\right|_{\boldsymbol{\xi}=\boldsymbol{\xi}_0},$$

and

$$\boldsymbol{\xi}_0 = (\xi_{10}, \xi_{20}, \xi_{30})' = (300, 9, 45)'.$$

If it was assumed, for example, that third- and higher-order terms could be neglected, then we could employ for the expectation function the second-degree polynomial in ξ_1, ξ_2, ξ_3 which could be written in the form

$$E(y) = \theta_0 x_0 + \sum_{t=1}^{3} \theta_t x_t + \sum_{t=1}^{3}\sum_{j=t}^{3} \theta_{tj} x_{tj} \qquad (10.2.6)$$

where

$$x_0 = 1, \qquad x_t = (\xi_t - \xi_{t0})/C_t, \qquad x_{tj} = x_t x_j,$$

$$C_1 = 50, \qquad C_2 = 1, \qquad C_3 = 5,$$

and

$$\theta_0 = f_0, \qquad \theta_t = C_t f_t, \qquad \theta_{tj} = C_t C_j f_{tj}.$$

The expectation of the ith observation y_i would then be linear in the parameters $(\theta_0, \theta_1, \theta_2, \theta_3, \theta_{11}, \theta_{12}, \theta_{13}, \theta_{22}, \theta_{23}, \theta_{33})$ and depend on the values of (x_1, x_2, x_3) as set out in Table 10.2.2.

It seems highly probable that the full second-degree polynomial model containing ten terms could represent the data fairly well. However, inspection of the data shows that the response appears to be monotonic in x_1, x_2 and x_3. It is possible, therefore, that after some suitable data transformation a simpler model linear in $\boldsymbol{\xi} = (\xi_1, \xi_2, \xi_3)$ omitting the six second-degree terms $(\theta_{11}, \theta_{22}, \theta_{33}, \theta_{12}, \theta_{13}, \theta_{23})$ might be suitable, and improvement in variance homogeneity and in Normality of the error distribution might be achieved also.

In summary, the task of the data analyst is to make explicit the model which underlies a given body of data. In attempting to relate data to any kind of model the analyst runs two kinds of risks. Obviously, he may force the data to a model which is inadequate. Alternatively, he may so encumber the model with unnecessary parameters that the corresponding analysis is too complex to be worthy of the name. A complex analysis is justifiable when the phenomenon itself causes the complexity. Often complexity is introduced unnecessarily.

For example, we shall show that the textile data are better represented by a transformed model linear in the experimental variables $\boldsymbol{\xi}$ and containing only six parameters (θ_0, θ_1, θ_2, θ_3, σ, λ,) than it is by a model in which an untransformed y is represented as a quadratic function of $\boldsymbol{\lambda}$, containing eleven parameters ($\theta_0, \theta_1, \theta_2, \theta_3, \theta_{11}, \theta_{22}, \theta_{33}, \theta_{12}, \theta_{13}, \theta_{23}, \sigma$). In this situation we can say that the complexity has not arisen from the physical situation but rather from failure to work with the appropriate metric. Again, in the biological example, it will be shown that the use of a simple transformation avoids the necessity for interaction parameters and also for postulating different variances in the groups.

The analyst's problem is how to allow diversity without producing chaos. The principle of parsimony, aptly christened by Tukey (1961), says that parameters should be introduced sparingly and in such a way that the maximum amount of resolution is achieved for each parameter introduced. Parsimonious parametrization is frequently made possible by suitable data transformation.

10.3 ESTIMATION OF THE TRANSFORMATION

We shall use the notation $y^{(\boldsymbol{\lambda})}$ to define a parametric family of nonlinear transformations in which $\boldsymbol{\lambda}$ may be a scalar or a vector containing τ elements $\lambda_1, \ldots, \lambda_\tau$. We suppose that $y^{(\boldsymbol{\lambda})}$ is a monotonic function of y over the admissible range. Two transformations of particular usefulness are $y^{(\lambda)} = y^\lambda$ and $y^{(\boldsymbol{\lambda})} = (y + \lambda_2)^{\lambda_1}$.

We use these transformations in the forms

$$y^{(\lambda)} = \begin{cases} \dfrac{y^\lambda - 1}{\lambda} & (\lambda \neq 0), \\ \log y & (\lambda = 0), \end{cases} \qquad y > 0 \tag{10.3.1}$$

$$y^{(\boldsymbol{\lambda})} = \begin{cases} \dfrac{(y + \lambda_2)^{\lambda_1} - 1}{\lambda_1} & (\lambda_1 \neq 0), \\ \log (y + \lambda_2) & (\lambda_1 = 0), \end{cases} \qquad y > - \lambda_2. \tag{10.3.2}$$

because they are then continuous at $\lambda = 0$ and $\lambda_1 = 0$, respectively. This class of transformations includes as special cases the logarithmic transformation, the reciprocal transformation, and the square root transformation.

Now suppose that a suitable family of transformations $y^{(\boldsymbol{\lambda})}$ has been selected and assume that, for *some chosen* $\boldsymbol{\lambda}$, to a sufficient approximation,

$$\mathbf{y}^{(\boldsymbol{\lambda})} = \mathbf{X}\boldsymbol{\theta} + \boldsymbol{\varepsilon}, \tag{10.3.3}$$

where $\mathbf{y}^{(\boldsymbol{\lambda})} = (y_1^{(\boldsymbol{\lambda})}, \ldots, y_n^{(\boldsymbol{\lambda})})'$, $\mathbf{y} = (y_1, \ldots, y_n)'$ is a $n \times 1$ vector of observations, $\mathbf{X}$ is a $n \times k$ matrix of fixed elements having rank k the first column of which consists of n ones, $\boldsymbol{\theta}$ is a $k \times 1$ vector of parameters, and $\boldsymbol{\varepsilon}$ is a $n \times 1$ vector of errors having spherical Normal distribution $N_n(\mathbf{0}, \sigma^2 \mathbf{I})$.

10.3.1 Prior and Posterior Distributions of λ

The probability density function of the *original untransformed* observations $\mathbf{y}$ is

$$p(\mathbf{y} \mid \boldsymbol{\lambda}, \boldsymbol{\theta}, \sigma^2) = \frac{1}{(2\pi)^{\frac{1}{2}n}\,\sigma^n} \exp\left[-\frac{(\mathbf{y}^{(\lambda)} - \mathbf{X}\boldsymbol{\theta})'\,(\mathbf{y}^{(\lambda)} - \mathbf{X}\boldsymbol{\theta})}{2\sigma^2}\right] J(\boldsymbol{\lambda}; \mathbf{y}),$$

$$-\infty < \mathbf{y}^{(\lambda)} < \infty, \qquad (10.3.4)$$

where the Jacobian $J(\boldsymbol{\lambda}; \mathbf{y})$ is

$$J(\boldsymbol{\lambda}; \mathbf{y}) = \prod_{i=1}^{n} \left|\frac{dy_i^{(\lambda)}}{dy_i}\right|. \qquad (10.3.5)$$

Writing

$$S_\lambda = (\mathbf{y}^{(\lambda)} - \mathbf{X}\,\hat{\boldsymbol{\theta}}_\lambda)'\,(\mathbf{y}^{(\lambda)} - \mathbf{X}\,\hat{\boldsymbol{\theta}}_\lambda) \qquad (10.3.6)$$

with

$$\hat{\boldsymbol{\theta}}_\lambda = (\mathbf{X}'\mathbf{X})^{-1}\,\mathbf{X}'\,\mathbf{y}^{(\lambda)},$$

the joint posterior distribution is

$$p(\boldsymbol{\theta}, \log\sigma, \boldsymbol{\lambda} \mid \mathbf{y}) \propto \sigma^{-n} \exp\left[-\frac{S_\lambda + (\boldsymbol{\theta} - \hat{\boldsymbol{\theta}}_\lambda)'\,\mathbf{X}'\mathbf{X}(\boldsymbol{\theta} - \hat{\boldsymbol{\theta}}_\lambda)}{2\sigma^2}\right] J(\boldsymbol{\lambda}; \mathbf{y})p(\boldsymbol{\theta},\ \log\sigma, \boldsymbol{\lambda}),$$

$$-\infty < \log\sigma < \infty, \quad -\infty < \boldsymbol{\theta} < \infty, \quad -\infty < \boldsymbol{\lambda} < \infty, \qquad (10.3.7)$$

where $p(\boldsymbol{\theta}, \log\sigma, \boldsymbol{\lambda})$ is the prior distribution.

Choice of the Prior Distribution $p(\boldsymbol{\theta}, \log\sigma, \boldsymbol{\lambda})$

We now consider what to choose for the prior $p(\boldsymbol{\theta}, \log\sigma, \boldsymbol{\lambda})$. Writing

$$p(\boldsymbol{\theta}, \log\sigma, \boldsymbol{\lambda}) = p(\boldsymbol{\lambda})\,p(\boldsymbol{\theta}, \log\sigma \mid \boldsymbol{\lambda}), \qquad (10.3.8)$$

we shall suppose that, *for any specific* $\boldsymbol{\lambda}$, $p(\boldsymbol{\theta}, \log\sigma \mid \boldsymbol{\lambda})$ is locally uniform. On the other hand, $p(\boldsymbol{\theta}, \log\sigma \mid \boldsymbol{\lambda})$ must clearly be dependent on $\boldsymbol{\lambda}$ for a change in $\boldsymbol{\lambda}$ will magnify or diminish all the data and will hence change the value of this locally uniform density. We conclude then that

$$p(\boldsymbol{\theta}, \log\sigma \mid \boldsymbol{\lambda}) \quad \propto \quad g(\boldsymbol{\lambda}). \qquad (10.3.9)$$

To decide the form of the function $g(\boldsymbol{\lambda})$, we employ the following argument. Let us assume that over the range of the data we can write approximately

$$y_i^{(\lambda)} \doteq a_\lambda + l_\lambda\,y_i, \qquad i = 1, 2, \ldots, n \qquad (10.3.10)$$

where l_λ is some representative value of the gradient $(dy^{(\lambda)}/dy)$.

Taking expectations, we have

$$E(y_i^{(\lambda)}) \doteq a_\lambda + l_\lambda\,E(y_i). \qquad (10.3.11)$$

Since

$$\boldsymbol{\theta} = (\mathbf{X}'\mathbf{X})^{-1}\,\mathbf{X}'\,E\,[\mathbf{y}^{(\lambda)}], \qquad (10.3.12)$$

it follows that

$$\boldsymbol{\theta} \doteq a_\lambda (\mathbf{X}'\mathbf{X})^{-1} \mathbf{X}' \mathbf{1} + l_\lambda \boldsymbol{\phi}, \tag{10.3.13}$$

where $\boldsymbol{\phi} = (\mathbf{X}'\mathbf{X})^{-1} \mathbf{X}' E(\mathbf{y})$ is a set of linear functions of the expectations of the observations $\mathbf{y}$. Also, let σ_y be the standard deviation of y_i. Then from (10.3.10),

$$\sigma \doteq |l_\lambda| \, \sigma_y, \tag{10.3.14}$$

so that

$$\log \sigma \doteq \log |l_\lambda| + \log \sigma_y. \tag{10.3.15}$$

If, as it seems reasonable, we take $\boldsymbol{\phi}$ and $\log \sigma_y$ to be locally uniform and independent *a priori*, then (10.3.13) and (10.3.15) imply that

$$p(\boldsymbol{\theta}, \log \sigma \mid \lambda) \propto |l_\lambda|^{-k}. \tag{10.3.16}$$

The value we shall use for $|l_\lambda|$ is the geometric mean of the absolute values of the derivatives $dy^{(\lambda)}/dy$ at the actual data points, that is,

$$|l_\lambda| = \left[\prod_{i=1}^{n} \left| \frac{dy_i^{(\lambda)}}{dy_i} \right| \right]^{1/n} = J(\lambda; \mathbf{y})^{1/n} = J_\lambda^{1/n}. \tag{10.3.17}$$

Thus,

$$p(\boldsymbol{\theta}, \log \sigma \mid \lambda) \propto g(\lambda) \propto J_\lambda^{-k/n} \tag{10.3.18}$$

and finally

$$p(\boldsymbol{\theta}, \log \sigma, \lambda) \propto J_\lambda^{-k/n} \, p(\lambda). \tag{10.3.19}$$

Posterior Distribution of λ

With this prior distribution, the joint posterior distribution for $\boldsymbol{\theta}, \log \sigma$, and λ is

$$p(\boldsymbol{\theta}, \log \sigma, \lambda \mid \mathbf{y}) \propto \sigma^{-n} \exp \left[- \frac{S_\lambda + (\boldsymbol{\theta} - \hat{\boldsymbol{\theta}}_\lambda)' \mathbf{X}'\mathbf{X} (\boldsymbol{\theta} - \hat{\boldsymbol{\theta}}_\lambda)}{2\sigma^2} \right] J_\lambda^{(n-k)/n} \, p(\lambda),$$

$$-\infty < \log \sigma < \infty, \quad -\infty < \boldsymbol{\theta} < \infty, \quad -\infty < \lambda < \infty, \tag{10.3.20}$$

where it should be noted that J_λ involves the observations but not the parameters $\boldsymbol{\theta}$ and σ. Integrating out $\boldsymbol{\theta}$ and $\log \sigma$, we obtain

$$p(\lambda \mid \mathbf{y}) \propto p_u(\lambda \mid \mathbf{y}) \, p(\lambda), \qquad -\infty < \lambda < \infty, \tag{10.3.21}$$

where

$$p_u(\lambda \mid \mathbf{y}) \propto (S_\lambda / J_\lambda^{2/n})^{-\frac{1}{2}(n-k)}$$

is the posterior distribution for a uniform reference prior for λ. Alternatively, if we work with the "normalized" data

$$z^{(\lambda)} = J_\lambda^{-1/n} \, y^{(\lambda)}, \tag{10.3.22}$$

then, writing $\hat{\mathbf{z}}^{(\lambda)} = \mathbf{X}(\mathbf{X}'\mathbf{X})^{-1}\mathbf{X}'\mathbf{z}^{(\lambda)}$ where $\mathbf{z}^{(\lambda)} = (z_1^{(\lambda)}, \ldots, z_n^{(\lambda)})'$, we have

$$p_u(\lambda \mid \mathbf{y}) \propto [S(\lambda, \mathbf{z})]^{-\frac{1}{2}(n-k)}, \qquad -\infty < \lambda < \infty, \tag{10.3.23}$$

with

$$S(\lambda, \mathbf{z}) = (\mathbf{z}^{(\lambda)} - \hat{\mathbf{z}}^{(\lambda)})' (\mathbf{z}^{(\lambda)} - \hat{\mathbf{z}}^{(\lambda)}).$$

In practice, in choosing a transformation that may hopefully result in parsimonious parametrization, constancy of variance, and Normality, we can compute the residual sum of squares $S(\boldsymbol{\lambda}, \mathbf{z})$ for a suitable range of values of $\boldsymbol{\lambda}$, and hence, $[S(\boldsymbol{\lambda}, \mathbf{z})]^{-\frac{1}{2}(n-k)}$. Normalization of $[S(\boldsymbol{\lambda}, \mathbf{z})]^{-\frac{1}{2}(n-k)}$, will then produce $p_u(\boldsymbol{\lambda} \mid \mathbf{y})$. If desired, $p_u(\boldsymbol{\lambda} \mid \mathbf{y})$ may then be combined with any prior weight function $p(\boldsymbol{\lambda})$.

It should be noted that by regarding $J_{\boldsymbol{\lambda}}$ as a constant, (10.3.22) implies that the logarithm of the standard deviation of $z^{(\boldsymbol{\lambda})}$ can be written

$$\log \sigma(z^{(\boldsymbol{\lambda})}) \doteq -\frac{1}{n}\log J_{\boldsymbol{\lambda}} + \log \sigma. \tag{10.3.24}$$

Using the approximation (10.3.15) with $|l_{\boldsymbol{\lambda}}| = J_{\boldsymbol{\lambda}}^{1/n}$, we have

$$\log \sigma \doteq \frac{1}{n}\log J_{\boldsymbol{\lambda}} + \log \sigma_y. \tag{10.3.25}$$

It follows that

$$\log \sigma(z^{(\boldsymbol{\lambda})}) \doteq \log \sigma_y \tag{10.3.26}$$

so that, to the degree of approximation (10.3.15), the standard deviation of $z^{(\boldsymbol{\lambda})}$ is the same for any $\boldsymbol{\lambda}$.

10.3.2 The Simple Power Transformation

In practice, the power transformation in (10.3.1) is of particular interest. In the z form we have $\boldsymbol{\lambda} = \lambda$,

$$z^{(\lambda)} = \begin{cases} (y^{\lambda} - 1)/(\lambda\, \dot{y}^{\lambda-1}), & (\lambda \neq 0) \\ \dot{y}\log y, & (\lambda = 0) \end{cases} \tag{10.3.27}$$

where $\dot{y} = (\prod_{i=1}^{n} y_i)^{1/n}$ is the geometric mean of the data.

In obtaining the posterior distribution of λ, the chief labor is in calculating the residual sum of squares $S(\lambda, \mathbf{z})$ for a range of values of λ. Computer programs are now rather generally available for analyzing linear models and in particular in producing, via an analysis of variance table, the residual sum of squares.

Writing $w = y/\dot{y}$, we have for the power transformations

$$z^{(\lambda)} = \begin{cases} \dot{y}\,\lambda^{-1}\, w^{\lambda} + c_{\lambda}, & (\lambda \neq 0) \\ \dot{y}\log w + c_0, & (\lambda = 0) \end{cases} \tag{10.3.28}$$

where c_{λ} is a constant depending on λ. Now the size of the constant c_{λ} will have no effect on any element in the analysis of variance table except the "correction for the mean." Consequently, when the simple power transformations are being considered, $S(\lambda, \mathbf{z})$ may be obtained simply by computing $w_i = y_i/\dot{y}, u = 1, 2, \ldots, n$ and performing an analysis of variance on the variates

$$\begin{cases} \dot{y}\,\lambda^{-1}\,w^{\lambda}, & (\lambda \neq 0) \\ \dot{y}\log w, & (\lambda = 0) \end{cases} \tag{10.3.29}$$

(note that $\dot{y}\,\lambda^{-1}\,w^{\lambda} = y$ when $\lambda = 1$).

So far as the calculation of $p_u(\lambda \mid \mathbf{y})$ is concerned, we can work equally well with the variates $\lambda^{-1}\,w^{\lambda}$ and $\log w$, since the fixed multiplication constant $\dot{y}$ will have no effect on $p_u(\lambda \mid \mathbf{y})$, which is in any case normalized so that $\int p_u(\lambda \mid \mathbf{y})d\lambda = 1$.

10.3.3 The Biological Example

For the biological data in Table 10.2.1 we would like, if possible, to find a transformation for which:

a) the "six parameter" additive model rather than the "twelve parameter" interaction model is applicable.

b) the cell variances are equal and

c) the errors are Normal.

These properties seem unlikely to apply for the original data. In particular simple plotting demonstrates a strong tendency for the within-cell variances to increase as the within-cell means increase. Furthermore, the analysis of variance in Table 10.3.1 for the untransformed data suggests the possibility of interaction. Interpreting this table from the Bayes point of view discussed in Section 2.7, we can say that in the space of the six interaction parameters the point $(0, 0, 0, 0, 0, 0)$ barely lies within the 90 per cent H.P.D. region. (The appropriate mean square ratio is $41.7/22.2 = 1.88$ while the 10 per cent point for F with 6 and 36 degrees of freedom is 1.95.)

Table 10.3.1

Analyses of variance of the biological data

	Mean squares × 1000			
	Degrees of freedom	Untransformed	Degrees of freedom	Reciprocal transformation (z form)
Poisons	2	516.3	2	568.7
Treatments	3	307.1	3	221.9
$P \times T$	6	41.7	6	8.5
Within groups	36	22.2	36(35)	7.8(8.0)

Since it is hoped that after transformation the *additive* model will be adequate, the appropriate residual sum of squares to use in calculating $p_u(\lambda \mid \mathbf{y})$ is that obtained after fitting the additive model and is based on 42 degrees of freedom. We denote it by $S_{42}(\lambda, \mathbf{z})$.

Table 10.3.2

Values of $S_{42}(\lambda, \mathbf{z})$ and of $p_u(\lambda \mid \mathbf{y})$ over a range of λ where the density is appreciable: the biological data

λ	$S_{42}(\lambda, \mathbf{z})$	λ	$p_u(\lambda \mid \mathbf{y})$
1.0	1.0509	0.0	0.01
0.5	0.6345	−0.1	0.02
0.0	0.4239	−0.2	0.08
−0.2	0.3752	−0.3	0.26
−0.4	0.3431	−0.4	0.49
−0.6	0.3258	−0.5	0.94
−0.8	0.3225	−0.6	1.46
−1.0	0.3331	−0.7	1.82
−1.2	0.3586	−0.8	1.82
−1.4	0.4007	−0.9	1.42
−1.6	0.4625	−1.0	0.92
−2.0	0.6639	−1.1	0.47
−2.5	1.1331	−1.2	0.19
−3.0	2.0489	−1.3	0.07
		−1.5	0.01

Table 10.3.2 shows values of $S_{42}(\lambda, \mathbf{z})$, and of the resulting $p_u(\lambda \mid \mathbf{y})$, over a range of λ in which the density is appreciable. Upon normalizing $[S_{42}(\lambda, \mathbf{z})]^{-21}$ by numerical integration, we find

$$p_u(\lambda \mid \mathbf{y}) = 0.866 \times 10^{-10} [S_{42}(\lambda, \mathbf{z})]^{-21}, \qquad -\infty < \lambda < \infty. \qquad (10.3.30)$$

The posterior distribution $p_u(\lambda \mid \mathbf{y})$ shown in Fig. 10.3.1 is approximately Normal with mean -0.75 and standard deviation 0.22. The 95 per cent H.P.D. interval extends from about -1.18 to about -0.32. We notice in particular that $\lambda = 1$ (no transformation), $\lambda = 0$ (log transformation), are untenable on this data. On the other hand, the reciprocal y^{-1}, which has a natural appeal for the analysis of mortality time data because it can be interpreted as the "rate of dying", possesses most of the advantages obtainable from transformation. A plot of within-cell sample variances against within-cell means for the transformed data now shows no dependence. Furthermore there is now no suspicion of lack of additivity.

The analysis of variance table for the untransformed data and for the reciprocal transformation (in the z form) is shown in Table 10.3.1. The

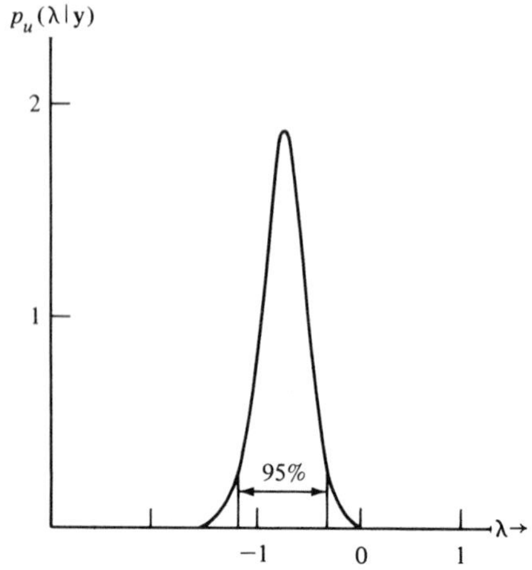

Fig. 10.3.1 The posterior distribution $p_u(\lambda \mid \mathbf{y})$ for the biological data. (Arrows show approximate 95 per cent H.P.D. interval)

Normal assumptions are, of course, much more appropriate for the transformed data and this results in a much greater sensitivity of the analysis. The within-groups mean square is reduced to about a third of its previous value relative to the poison and treatment mean squares which remain about the same size. This implies, for example, that the spread of the individual marginal posterior distributions of the effects will be reduced by a factor of about $\sqrt{3}$ when transformed back to the original metric.

10.3.4 The Textile Example

Consider now the textile data of Table 10.2.2. In the original analysis, a second-degree polynomial in x_1, x_2 and x_3 was employed to represent $E(y)$. However, since inspection of the data suggests that y is monotonic in x_1, x_2 and x_3, there is a possibility that a first-degree polynomial might provide adequate representation for $E(y^\lambda)$, with λ suitably chosen.

We therefore work with the transformed variate (10.3.27), hoping that after such transformation:

a) the expected value of the transformed response can be adequately represented by a model *linear* in the x's,
b) the error variance will be constant, and
c) the observations will be Normally distributed.

The appropriate residual sum of squares to be considered is that after fitting a linear function to the x's. The sum of squares has $(27 - 4) = 23$ degrees

Table 10.3.3

Values of $S_{23}(\lambda, \mathbf{z})$ and of $p_u(\lambda \mid \mathbf{y})$ over a range of λ where the density is appreciable: the textile data

λ	$S_{23}(\lambda, \mathbf{z})$	λ	$p_u(\lambda \mid \mathbf{y})$
1.00	5.4810	0.20	0.02
0.80	2.9978	0.15	0.09
0.60	1.5968	0.10	0.42
0.40	0.8178	0.05	1.58
0.20	0.4115	0.00	4.18
0.00	0.2519	−0.05	5.64
−0.20	0.2920	−0.10	4.66
−0.40	0.5378	−0.15	2.36
−0.60	1.1035	−0.20	0.77
−0.80	2.1396	−0.25	0.19
−1.00	3.9955	−0.30	0.04
		−0.35	0.01

of freedom. Table 10.3.3 shows the value of $S_{23}(\lambda, \mathbf{z})$ and of the resulting $p_u(\lambda \mid \mathbf{y})$ over a range of λ in which the density is appreciable. Upon normalizing $[S_{23}(\lambda, \mathbf{z})]^{-11.5}$ by numerical integration, we find

$$p_u(\lambda \mid \mathbf{y}) = 0.540 \times 10^{-6} [S_{23}(\lambda, \mathbf{z})]^{-11.5}, \qquad -\infty < \lambda < \infty. \qquad (10.3.31)$$

The distribution $p_u(\lambda \mid \mathbf{y})$ is plotted in Fig. 10.3.2. It has its mean at -0.06 and the 95 per cent H.P.D. region extends from -0.20 to 0.08, determining the appropriate transformation very closely. The log transformation ($\lambda = 0$) has considerable practical advantages for this example and is strongly supported by the data. The analysis of variance for the untransformed data and for the logarithmically transformed data taken in the z form is shown in Table 10.3.4.

Table 10.3.4

Analyses of variance of the textile data

		Mean squares × 1000		
	Degrees of freedom	Untransformed	Degrees of freedom	Logarithmic transformation (z form)
Linear	3	4,916.2	3	2,374.4
Quadratic	6	704.1	6	8.1
Residual	17	73.9	17(16)	11.9(12.6)

It is seen that transformation eliminates the need for second-order terms in the equation, while the use of the more appropriate model greatly increases the sensitivity of the analysis.

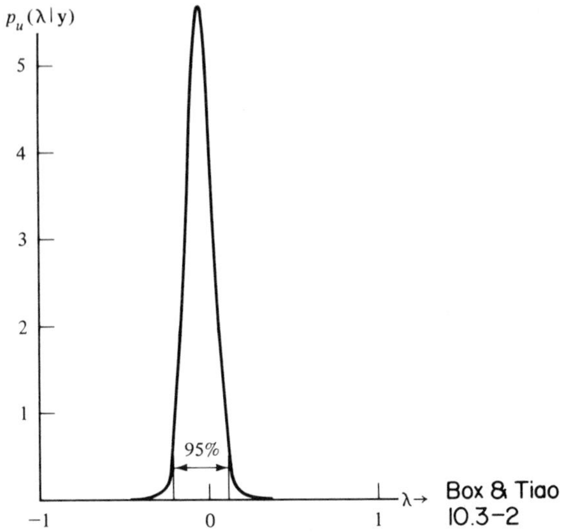

Fig. 10.3.2 The posterior distribution $p_u(\lambda \mid \mathbf{y})$ for the textile data. (Arrows show approximate 95 per cent H.P.D. interval)

10.3.5 Two Parameter Transformation

To illustrate the estimation of a two-parameter transformation, we apply to the textile data the transformation (10.3.2), which, writing gm for "geometric mean", can be written in the z form (10.3.22) as

$$z^{(\lambda)} = \begin{cases} \dfrac{(y+\lambda_2)^{\lambda_1} - 1}{\lambda_1 [\mathrm{gm}\,(y+\lambda_2)]^{\lambda_1 - 1}}, & (\lambda_1 \neq 0) \\ \mathrm{gm}\,(y+\lambda_2)\log(y+\lambda_2), & (\lambda_1 = 0) \end{cases} \tag{10.3.32}$$

where

$$\mathrm{gm}\,(y+\lambda_2) = \left[\prod_{i=1}^{n} (y_i + \lambda_2)\right]^{1/n}.$$

Figure 10.3.3 shows the contours of the joint posterior distribution $p_u(\lambda_1, \lambda_2 \mid \mathbf{y})$ obtained from a 7×11 grid of values of

$$\log p_u(\lambda_1, \lambda_2 \mid \mathbf{y}) = \text{const} - (11.5) \log S(\lambda_1, \lambda_2, \mathbf{z}). \tag{10.3.33}$$

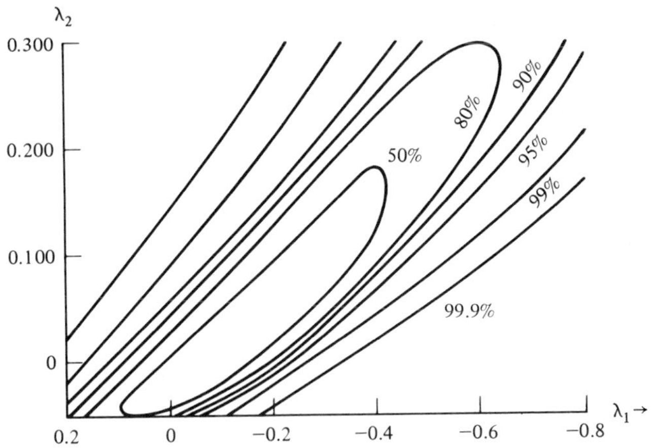

Fig. 10.3.3 Contours of the posterior distribution $p_u(\lambda_1, \lambda_2 \mid \mathbf{y})$ for the transformation $(y + \lambda_2)^{\lambda_1}$: the textile data. (Contours labeled enclose approximate H.P.D. regions)

As before, if the k-dimensional posterior distribution $p(\boldsymbol{\lambda} \mid \mathbf{y})$ were multivariate Normal, then a $(1 - \alpha)$ H.P.D. region for parameters $\boldsymbol{\lambda}$ would be defined by

$$\log p(\boldsymbol{\lambda} \mid \mathbf{y}) - \log p(\boldsymbol{\lambda} \mid \mathbf{y}) < \tfrac{1}{2}\chi^2(k, \alpha), \tag{10.3.34}$$

Although, for this example, the contours show that the Normal approximation is likely to be a poor one, we have nevertheless as a rough guide labeled H.P.D. regions with per cent probabilities using (10.3.34). It is evident that, for this example, there is likely to be no particular advantage in including a nonzero value for λ_2.

10.4 ANALYSIS OF THE EFFECTS θ AFTER TRANSFORMATION

So far we have considered only the marginal posterior distribution $p(\boldsymbol{\lambda} \mid \mathbf{y})$. Although there will be some circumstances in which inferences about $\boldsymbol{\lambda}$ will be the major objective, in many cases the chief interest of the analysis would be in making inferences about $\boldsymbol{\theta}$. The status of $\boldsymbol{\lambda}$ will then be that of a (very valuable) vector of nuisance parameters.

As usual, we can eliminate $\boldsymbol{\lambda}$ by integrating $p(\boldsymbol{\theta} \mid \boldsymbol{\lambda}, \mathbf{y})$ with weight function $p(\boldsymbol{\lambda} \mid \mathbf{y})$ in accordance with

$$p(\boldsymbol{\theta} \mid \mathbf{y}) = \int p(\boldsymbol{\theta}, \boldsymbol{\lambda} \mid \mathbf{y}) d\boldsymbol{\lambda} = \int p(\boldsymbol{\theta} \mid \boldsymbol{\lambda}, \mathbf{y}) \, p(\boldsymbol{\lambda} \mid \mathbf{y}) d\boldsymbol{\lambda}. \tag{10.4.1}$$

Again in the spirit of Section 1.6, we should be cautious in practical problems of relying solely on this integration, for if $p(\boldsymbol{\theta} \mid \boldsymbol{\lambda}, \mathbf{y})$ were changing drastically over the range in which $p(\boldsymbol{\lambda} \mid \mathbf{y})$ was appreciable, it would be important to know about it. Thus, in principle, we should always make a study of $p(\boldsymbol{\theta} \mid \boldsymbol{\lambda}, \mathbf{y})$ for a number of values of $\boldsymbol{\lambda}$ which are of interest.

From the distribution in (10.3.20) with $p(\boldsymbol{\lambda})$ assumed uniform, the joint distribution $p(\boldsymbol{\theta}, \boldsymbol{\lambda} \mid \mathbf{y})$ in $z^{(\boldsymbol{\lambda})}$ form is

$$p(\boldsymbol{\theta}, \boldsymbol{\lambda} \mid \mathbf{y}) \propto J_{\boldsymbol{\lambda}}^{-k/n} [S(\boldsymbol{\lambda}, \mathbf{z}) + Q(\boldsymbol{\theta}, \boldsymbol{\lambda})]^{-n/2}, \qquad -\infty < \boldsymbol{\theta} < \infty, \quad -\infty < \boldsymbol{\lambda} < \infty, \tag{10.4.2}$$

where

$$Q(\boldsymbol{\theta}, \boldsymbol{\lambda}) = J_{\boldsymbol{\lambda}}^{-2/n} (\boldsymbol{\theta} - \hat{\boldsymbol{\theta}}_{\boldsymbol{\lambda}})' \mathbf{X}'\mathbf{X}(\boldsymbol{\theta} - \hat{\boldsymbol{\theta}}_{\boldsymbol{\lambda}}).$$

Thus, the conditional distribution $p(\boldsymbol{\theta} \mid \boldsymbol{\lambda}, \mathbf{y})$ is

$$p(\boldsymbol{\theta} \mid \boldsymbol{\lambda}, \mathbf{y}) \propto \left[1 + \frac{Q(\boldsymbol{\theta}, \boldsymbol{\lambda})}{s_\nu^2(\boldsymbol{\lambda})\, \nu}\right]^{-\frac{1}{2}(\nu + k)}, \qquad -\infty < \boldsymbol{\theta} < \infty, \tag{10.4.3}$$

where

$$\nu = n - k, \qquad \nu s_\nu^2(\boldsymbol{\lambda}) = S(\boldsymbol{\lambda}, \mathbf{z}),$$

which is the k-dimensional multivariate t distribution

$$t_k [\hat{\boldsymbol{\theta}}_{\boldsymbol{\lambda}}, s_\nu^2(\boldsymbol{\lambda}) J_{\boldsymbol{\lambda}}^{2/n} (\mathbf{X}'\mathbf{X})^{-1}, \nu],$$

and the marginal distribution of $\boldsymbol{\theta}$ is

$$p(\boldsymbol{\theta} \mid \mathbf{y}) \propto \int_{-\infty < \boldsymbol{\lambda} < \infty} J_{\boldsymbol{\lambda}}^{-k/n} [S(\boldsymbol{\lambda}, \mathbf{z}) + Q(\boldsymbol{\theta}, \boldsymbol{\lambda})]^{-n/2}\, d\boldsymbol{\lambda}, \qquad -\infty < \boldsymbol{\theta} < \infty. \tag{10.4.4}$$

From (10.3.23), the mode $\hat{\boldsymbol{\lambda}}$ of the marginal distribution of $\boldsymbol{\lambda}$ is the value minimizing $S(\boldsymbol{\lambda}, \mathbf{z})$. We may employ Taylor's theorem to write

$$S(\boldsymbol{\lambda}, \mathbf{z}) + Q(\boldsymbol{\theta}, \boldsymbol{\lambda}) \doteq S(\hat{\boldsymbol{\lambda}}, \mathbf{z}) + (\boldsymbol{\lambda} - \hat{\boldsymbol{\lambda}})'\mathbf{G}(\boldsymbol{\lambda} - \hat{\boldsymbol{\lambda}}) + Q(\boldsymbol{\theta}, \hat{\boldsymbol{\lambda}}) + g(\boldsymbol{\theta}, \boldsymbol{\lambda}), \tag{10.4.5}$$

where $2\mathbf{G}$ is the matrix of the second derivatives of $S(\boldsymbol{\lambda}, \mathbf{z})$ with respect to $\boldsymbol{\lambda}$ evaluated at $\hat{\boldsymbol{\lambda}}$ and

$$g(\boldsymbol{\theta}, \boldsymbol{\lambda}) \doteq \sum_{i=1}^{\tau} \left.\frac{\partial Q}{\partial \lambda_i}\right|_{\hat{\boldsymbol{\lambda}}} (\lambda_i - \hat{\lambda}_i) + \frac{1}{2} \sum_{i=1}^{\tau} \sum_{j=1}^{\tau} \left.\frac{\partial^2 Q}{\partial \lambda_i\, \partial \lambda_j}\right|_{\hat{\boldsymbol{\lambda}}} (\lambda_i - \hat{\lambda}_i)(\lambda_j - \hat{\lambda}_j). \tag{10.4.6}$$

For moderate n, the influence of $g(\boldsymbol{\theta}, \boldsymbol{\lambda})$ will be small and can be ignored. Expression (10.4.4) can thus be approximated to yield

$$p(\boldsymbol{\theta} \mid \mathbf{y}) \mathrel{\dot{\propto}} [S(\boldsymbol{\lambda}, \mathbf{z}) + Q(\boldsymbol{\theta}, \boldsymbol{\lambda})]^{-\frac{1}{2}(\nu - \tau + k)} \underset{\boldsymbol{\lambda}}{E}(J_{\boldsymbol{\lambda}}^{-k/n}), \tag{10.4.7}$$

where the expectation is taken over the distribution

$$p_1(\boldsymbol{\lambda}) \propto \left[1 + \frac{(\boldsymbol{\lambda} - \hat{\boldsymbol{\lambda}})'\mathbf{G}(\boldsymbol{\lambda} - \hat{\boldsymbol{\lambda}})}{S(\hat{\boldsymbol{\lambda}}, \mathbf{z}) + Q(\boldsymbol{\theta}, \hat{\boldsymbol{\lambda}})}\right]^{-\frac{1}{2}n}, \qquad -\infty < \boldsymbol{\lambda} < \infty. \tag{10.4.8}$$

Now for moderate n,

$$\underset{\boldsymbol{\lambda}}{E}(J_{\boldsymbol{\lambda}}^{-k/n}) \doteq J_{\hat{\boldsymbol{\lambda}}}^{-k/n} \tag{10.4.9}$$

which does not involve $\boldsymbol{\theta}$, so that finally we can express the marginal distribution of $\boldsymbol{\theta}$ approximately as

$$p(\boldsymbol{\theta} \mid \mathbf{y}) \mathrel{\dot{\propto}} \left[1 + \frac{Q(\boldsymbol{\theta}, \hat{\boldsymbol{\lambda}})}{s^2_{\nu-\tau}(\hat{\boldsymbol{\lambda}})(\nu - \tau)}\right]^{-\frac{1}{2}(\nu-\tau+k)}, \qquad -\infty < \boldsymbol{\theta} < \infty, \tag{10.4.10}$$

where

$$(\nu - \tau)\, s^2(\hat{\boldsymbol{\lambda}}) = S(\hat{\boldsymbol{\lambda}}, \mathbf{z}),$$

which is the k-dimensional multivariate t distribution

$$t_k\,[\boldsymbol{\theta}_{\hat{\boldsymbol{\lambda}}}, s^2_{\nu-\tau}(\hat{\boldsymbol{\lambda}})\, J_{\hat{\boldsymbol{\lambda}}}^{2/n}\,(\mathbf{X}'\mathbf{X})^{-1}, \nu - \tau].$$

Thus, approximately the marginal posterior distribution of $\boldsymbol{\theta}$ is obtained by analysing $z^{(\hat{\boldsymbol{\lambda}})}$ as if $\hat{\boldsymbol{\lambda}}$ were a known fixed parameter and reducing the residual degrees of freedom by τ, the number of transformation parameters.

Having obtained $\hat{\boldsymbol{\lambda}}$, we can thus approximately justify performing a standard analysis with the transformed variate $z^{(\hat{\boldsymbol{\lambda}})}$ as if the transformation had been given in advance. The only modification will be the reduction of the residual degrees of freedom.

In the two examples, we have suggested that detailed analysis for $\boldsymbol{\theta}$ might be carried through not for $\hat{\lambda}$ (which was -0.75 for the biological sample and -0.06 for the textile sample), but we might use the reciprocal transformation ($\lambda = -1$) in the first case and the logarithmic transformation ($\lambda = 0$) in the second. Two arguments can be used to justify this.

a) In the neighborhood of $\hat{\lambda}, p(\boldsymbol{\theta} \mid \lambda, \mathbf{y})$ changes only slowly in λ so that there is little practical difference between using $p(\boldsymbol{\theta} \mid \hat{\lambda}, \mathbf{y})$ which, after appropriate modification in the degrees of freedom, approximates (10.4.1) and in using $p(\boldsymbol{\theta} \mid \lambda_0, \mathbf{y})$, where λ_0 is -1 and 0, respectively, for the two examples.

b) For the biological example, the reciprocal transformation, which says that the rate of dying of the animals is additive so far as the effects of poisons and treatments are concerned, is one which makes some biological sense.

Also, in many fields of technology, relationships often occur of the form $y = \text{const.}\ \xi_1^{\beta_1} \xi_2^{\beta_2}, \ldots, \xi_k^{\beta_k}\, \varepsilon$, suggesting a logarithmic transformation of both the dependent and the independent variables. Now, in the textile example, the independent variables $\xi_1, \xi_2, \ldots, \xi_k$ have been changed over such narrow ranges relative to their means that there the log transformations would be approximately linear and could be omitted. On this argument, a logarithmic transformation of the response y alone is a natural one to employ for the textile example.

Thus, the values $\lambda = -1$ for the first example and $\lambda = 0$ for the second were choices in which there was in fact reasonably stronger prior belief. Now a reasonably strong prior distribution centered at -1, for the first example, and 0 for the second would cause the posterior distribution $p(\lambda \mid \mathbf{y})$ to be even more sharply peaked and when substituted in (10.4.1) would justify approximately an analysis in terms of reciprocal and logarithmic transformations respectively.

Finally, then, in the actual analysis of these examples, an appropriate transformation λ_0 close to $\hat{\lambda}$ was made and a standard analysis performed with the transformed variate y^{λ_0}, but with the number of residual degrees of freedom reduced by one. From a Bayesian viewpoint, this analysis rests on the approximate use of the multivariate t distribution $p(\boldsymbol{\theta} \mid \lambda_0, \mathbf{y})$ with reduced degrees of freedom to represent $p(\boldsymbol{\theta} \mid \mathbf{y})$.

Bayesian justification of the analysis of variance table and calculation of the posterior distributions of particular subsets, differences, and other linear functions of the elements of $\boldsymbol{\theta}$ only make use of particular aspects of this basic multivariate t distribution.

The bracketed items in Tables 10.3.1 and 10.3.4 show the modifications to the degrees of freedom and mean squares which would be appropriate in the analysis of variance table on the above argument.

10.5 FURTHER ANALYSIS OF THE BIOLOGICAL DATA

It is an inherent assumption in the foregoing analysis that a transformation exists which simultaneously achieves simplicity of the model, homogeneity of variance, and Normality. This is clearly a much *less* restricted assumption than the more usual one of supposing that these requirements are all met *without* transformation. However, it is of interest to perform further analysis which in some way separates the issues of

a) simplicity of the linear model,

b) homogeneity of variance,

c) Normality

and which makes it possible to see to what extent these may be achieved with the same transformation.

Consider again the biological data. We have in our previous analysis tacitly supposed that there existed a transformed variable $y^{(\lambda)}$ in (10.3.1) for which *simultaneously*

a) $y_i^{(\lambda)}$ had a Normal distribution $N\{E[y_i^{(\lambda)}], \sigma_i^2\}$,

b) $\sigma_i^2 = \sigma^2$, and

c) $E[y_i^{(\lambda)}]$ was adequately represented by *additive* row and column parameters, $i = 1, \ldots, n$.

Further light is shed on the situation by supposing first that λ was chosen to satisfy (a) only, then considering how the situation is changed by adding requirement (b), and finally considering the effect of adding requirement (c).

If we merely suppose that a transformation $y^{(\lambda)}$ of the form (10.3.1) exists which simultaneously induces Normality in all cells but not necessarily constancy of cell variances additivity, then, for each cell, we have

$$\mathbf{y}_j^{(\lambda)} = \theta_j \mathbf{1} + \boldsymbol{\varepsilon}_j, \qquad j = 1, \ldots, k, \tag{10.5.1}$$

where for the jth cell corresponding to the tth treatment and pth poison, $\boldsymbol{\varepsilon}_j$ is spherical Normal $N_{n_j}(\mathbf{0}, \sigma_j^2 \mathbf{I})$ (for this example, $k = 12, n_j = 4$). Following the argument in Section 10.3.1, we have the prior distribution

$$p(\boldsymbol{\theta}, \log \boldsymbol{\sigma}, \lambda) \propto J_\lambda^{-k/n} p(\lambda), \tag{10.5.2}$$

where $n = \Sigma n_j$, $\boldsymbol{\theta} = (\theta_1, \ldots, \theta_k)$, $\log \boldsymbol{\sigma} = (\log \sigma_1, \ldots, \log \sigma_k)'$ and, as before,

$$J_\lambda = \prod_{i=1}^{n} \left| \frac{dy_i^{(\lambda)}}{dy_i} \right|.$$

Writing $p_u(\lambda \mid N)$ to indicate the posterior distribution under Normality (N) only and with $p(\lambda)$ assumed locally uniform, it can be readily be shown that

$$p_u(\lambda \mid N) \propto \prod_{j=1}^{k} [S_{\nu_j}(\lambda, \mathbf{z})]^{-\frac{1}{2}\nu_j}, \qquad -\infty < \lambda < \infty, \tag{10.5.3}$$

where $S_{\nu_j}(\lambda, \mathbf{z})$ is the sum of squares of deviations from the cell mean in terms of $z^{(\lambda)}$ for the jth cell having $\nu_j = n_j - 1$ degrees of freedom ($\nu_j = 3$ for the present example). The ordinates of $p_u(\lambda \mid N)$ are shown in the fourth column of Table 10.5.1 and the distribution is plotted in Fig. 10.5.1.

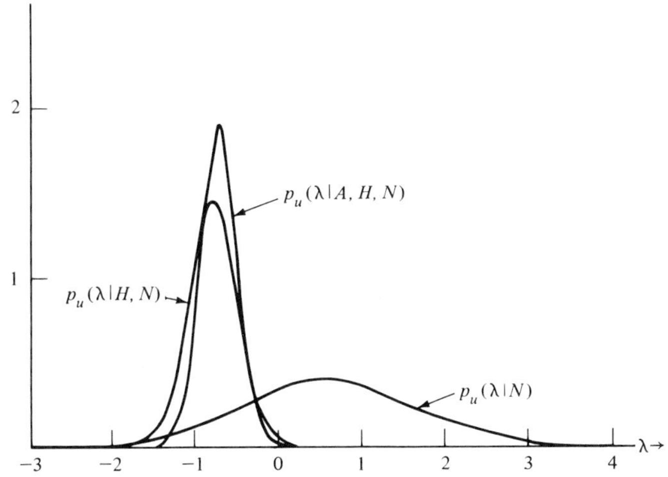

Fig. 10.5.1 Posterior distributions of λ for different models: the biological data.

It is well known that small samples are not able to tell us much about the Normality or otherwise of a distribution so that it is scarcely surprising that $p_u(\lambda \mid N)$ is found to cover an extremely wide range of λ.

Table 10.5.1

Ordinates of posterior distribution of λ for different models: the biological data

λ	$p_u(\lambda \mid A, H, N)$	$p_u(\lambda \mid H, N)$	$p_u(\lambda \mid N)$
1.0			0.335
0.5		0.006	0.398
0.0	0.006	0.021	0.342
−0.1	0.023	0.055	0.324
−0.2	0.076	0.127	0.304
−0.3	0.257	0.261	0.283
−0.4	0.492	0.471	0.261
−0.5	0.942	0.754	0.240
−0.6	1.462	1.059	0.218
−0.7	1.823	1.320	0.196
−0.8	1.823	1.430	0.173
−0.9	1.419	1.360	0.153
−1.0	0.923	1.136	0.134
−1.1	0.468	0.850	0.116
−1.2	0.194	0.558	0.099
−1.3	0.067	0.329	0.083
−1.4	0.019	0.170	0.069
−1.5	0.005	0.078	0.058
−1.6	0.001	0.032	0.050
−1.7		0.009	

We now consider the effect of assuming that a transformation $y^{(\lambda)}$ exists in terms of which not only Normality but also constant variance is obtained (but not necessarily additivity). Using the model in (10.5.1) but now with $\sigma_1^2 = \sigma_2^2 = \cdots = \sigma_k^2$, the within-cells sum of square $S_w(\lambda, \mathbf{z}) = \sum_{j=1}^{k} S_{\nu_j}(\lambda, \mathbf{z})$ is pooled from the $k = 12$ groups of four animals and has $\nu_w = \sum_{j=1}^{k} \nu_j = 36$ degrees of freedom. The resulting posterior distribution of λ,

$$p_u(\lambda \mid H, N) \propto [S_w(\lambda, \mathbf{z})]^{-\frac{1}{2}\nu_w}, \qquad -\infty < \lambda < \infty, \tag{10.5.4}$$

shown in Fig. 10.5.1, is now very much sharper.

Finally, if we assume that additivity can also be achieved by transformation, we have the probability distribution $p_u(\lambda \mid A, H, N)$ already given in Fig. 10.3.1 and also shown in Fig. 10.5.1 which is even sharper than $p_u(\lambda \mid H, N)$.

If we denote by $S_I(\lambda, \mathbf{z})$ the sum of squares for interaction which has $\nu_I = 6$ degrees of freedom, then the residual sum of squares $S(\lambda, \mathbf{z})$ originally considered is, in our new notation $S(\lambda, \mathbf{z}) = S_w(\lambda, \mathbf{z}) + S_I(\lambda, \mathbf{z})$ and

$$p_u(\lambda \mid A, H, N) \propto [S_w(\lambda, \mathbf{z}) + S_I(\lambda, \mathbf{z})]^{-\frac{1}{2}(\nu_w + \nu_I)}, \qquad -\infty < \lambda < \infty. \tag{10.5.5}$$

Since in each case we obtain the posterior distribution of λ *conditional* on the truth of a model containing given restrictions, we need some practical answer to the question of whether a constraint is justified or not.

10.5.1 Successive Constraints

Let C denote a particular constraint. Then, as we have seen earlier in (1.5.5), we can write

$$p(\lambda \mid C) = p(\lambda) \times \frac{p(C \mid \lambda)}{p(C)}, \tag{10.5.6}$$

where $p(C) = \underset{\lambda}{E}\,[p(C \mid \lambda)]$ is a constant independent of λ.

We apply this to the present example with $p_u(\lambda \mid N)$ the "unconstrained" density for which it is only assumed that Normality can be achieved for *some* λ, and with $p_u(\lambda \mid H, N)$ the "constrained" density for which it is assumed that homogeneity of variance can also be achieved. With the dependence on Normality (N) understood, we can write

$$p_u(\lambda \mid H) = p_u(\lambda)\frac{p_u(H \mid \lambda)}{p_u(H)} \propto p_u(\lambda)\, p_u(H \mid \lambda). \tag{10.5.7}$$

Thus, the distribution $p_u(\lambda \mid H)$ is decomposed into two factors both of which are functions of λ: the unconstrained "prior" distribution $p_u(\lambda)$ and the "likelihood" factor $p_u(H \mid \lambda)$ representing the effect of the constraint H. Now, we can obtain $p_u(H \mid \lambda)$ from the relationship

$$p_u(H \mid \lambda) \propto \frac{p_u(\lambda \mid H)}{p_u(\lambda)} \propto [W_H(\lambda, \mathbf{z})]^{1/2}, \tag{10.5.8}$$

where from (10.5.3) and (10.5.4),

$$W_H(\lambda, \mathbf{z}) = \prod_{j=1}^{k} \frac{[S_{\nu_j}(\lambda, \mathbf{z})/\nu_j]^{\nu_j}}{[S_w(\lambda, \mathbf{z})/\nu_w]^{\nu_w}}. \tag{10.5.9}$$

In other words,

$$p_u(\lambda \mid \sigma_1^2 = \ldots = \sigma_k^2) \propto p_u(\lambda)\, p_u(\sigma_1^2 = \ldots = \sigma_k^2 \mid \lambda) \tag{10.5.10}$$

where

$$p_u(\sigma_1^2 = \ldots = \sigma_k^2 \mid \lambda) \propto [W_H(\lambda, \mathbf{z})]^{1/2}. \tag{10.5.11}$$

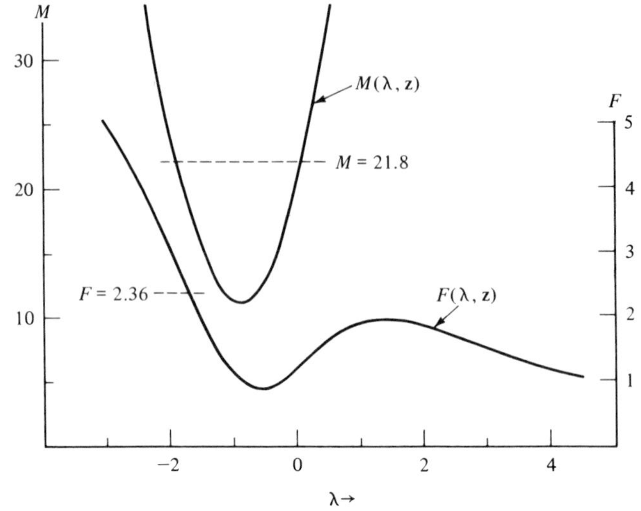

Fig. 10.5.2 Values of $M(\lambda, \mathbf{z})$ and $F(\lambda, \mathbf{z})$ as functions of λ: the biological data.

Now recalling our previous discussion of the comparison of variances in Section 2.12, if we consider the $k - 1$ linear contrasts in $\log \sigma_i$, $\phi_i = \log \sigma_i - \log \sigma_k$, the second factor on the right of (10.5.10) is proportional to the probability density for $\boldsymbol{\phi}_0 = \mathbf{0}$ conditional on the choice of a particular value for λ. Light may be shed, therefore, on the question of whether *for a particular value of* λ the constraint H is acceptable or not by considering whether the point $\boldsymbol{\phi}_0 = \mathbf{0}$ falls in or outside an H.P.D. region which includes some reasonably large proportion $1 - \alpha$ of the posterior distribution. As we have already seen in (2.12.13) and (2.12.21), this can be done by referring

$$M(\lambda, \mathbf{z}) = -2 \log W_H(\lambda, \mathbf{z})$$

to

$$\left\{1 + \frac{1}{3(k-1)}\left[\sum_{j=1}^{k} \frac{1}{\nu_j} - \frac{1}{\nu_w}\right]\right\} \chi^2(k-1, \alpha).$$

Thus, in the biological problem, for a particular λ, the point $\boldsymbol{\phi}_0 = \mathbf{0}$ would lie outside a 95 per cent H.P.D. region if $M(\lambda, \mathbf{z}) > 21.8$. The values for $M(\lambda, \mathbf{z})$ for various values of λ are plotted in Fig. 10.5.2. From this we see that the constraint $\sigma_1^2 = \sigma_2^2 = \ldots = \sigma_k^2$ is, in fact, compatible with values of λ in a range which includes, for example, the reciprocal transformation.

We can now proceed in a similar way to determine the appropriateness of the further constraint of additivity. With dependence on Normality (N) understood, we can write

$$p_u(\lambda \mid A, H) = p_u(\lambda \mid H)\frac{p_u(A \mid H, \lambda)}{p_u(A \mid H)} \propto p_u(\lambda \mid H)\, p_u(A \mid H, \lambda). \qquad (10.5.12)$$

The "likelihood" factor $p_u(A \mid H, \lambda)$ can be obtained from

$$p_u(A \mid H, \lambda) \propto \frac{p_u(\lambda \mid A, H)}{p_u(\lambda \mid H)} \propto [W_A(\lambda, \mathbf{z})]^{1/2}, \tag{10.5.13}$$

where, from (10.5.4) and (10.5.5),

$$W_A(\lambda, \mathbf{z}) = \frac{[S_w(\lambda, \mathbf{z})]^{\nu_w}}{[S_w(\lambda, \mathbf{z}) + S_I(\lambda, \mathbf{z})]^{\nu_w + \nu_I}}. \tag{10.5.14}$$

Writing $\boldsymbol{\theta}_I$ to denote the I interaction parameters, we thus have

$$p_u(\lambda \mid \boldsymbol{\theta}_I = \mathbf{0}, \sigma_1^2 = \ldots = \sigma_k^2) \propto p_u(\lambda \mid \sigma_1^2 = \ldots = \sigma_k^2) p_u(\boldsymbol{\theta}_I = \mathbf{0} \mid \sigma_1^2 = \ldots = \sigma_k^2, \lambda) \tag{10.5.15}$$

where the second factor on the right

$$p_u(\boldsymbol{\theta}_I = \mathbf{0} \mid \sigma_1^2 = \ldots = \sigma_k^2, \lambda) \propto [W_A(\lambda, \mathbf{z})]^{1/2} \tag{10.5.16}$$

is the density of $\boldsymbol{\theta}_I$ at $\boldsymbol{\theta}_I = \mathbf{0}$, conditional on the choice of λ and given that the cell variances are constant.

From (10.4.3), it is readily seen that $\boldsymbol{\theta}_I$ is distributed *a posteriori* as $t_I[\hat{\boldsymbol{\theta}}_I(\lambda), s_w^2(\lambda) J_\lambda^{2/n} \mathbf{C}, \nu_w]$ where $\hat{\boldsymbol{\theta}}_I(\lambda)$ is the $I \times 1$ sub-vector of $\hat{\boldsymbol{\theta}}_\lambda$ corresponding to $\boldsymbol{\theta}_I$, $\nu_w s_w^2(\lambda) = S_w(\lambda, \mathbf{z})$ and $\mathbf{C}$ is proportional to the covariance matrix of $\boldsymbol{\theta}_I$. The interaction sum of square $S_I(\lambda, \mathbf{z})$ is in fact

$$S_I = S_I(\lambda, \mathbf{z}) = J_\lambda^{-2/n} \hat{\boldsymbol{\theta}}_I(\lambda)' \mathbf{C}^{-1} \hat{\boldsymbol{\theta}}_I(\lambda) \tag{10.5.17}$$

and, with the dependence on λ and $\mathbf{z}$ understood, expression (10.5.16) can be written

$$p_u(\boldsymbol{\theta}_I = \mathbf{0} \mid \sigma_1^2 = \ldots = \sigma_k^2, \lambda) \propto (s_w^2)^{-\frac{1}{2}\nu_I} \left(1 + \frac{J_\lambda^{-2/n} \hat{\boldsymbol{\theta}}_I' \mathbf{C}^{-1} \hat{\boldsymbol{\theta}}_I}{\nu_w s_w^2}\right)^{-\frac{1}{2}(\nu_w + \nu_I)}. \tag{10.5.18}$$

The question of whether *for a particular value of* λ the constraint A is acceptable or not may be resolved by considering whether the parameter point $\boldsymbol{\theta}_I = \mathbf{0}$ falls in or outside an appropriate H.P.D. region. Since $\boldsymbol{\theta}_I$ follows a multivariate t distribution, the question is answered by referring the quantity

$$F(\lambda, \mathbf{z}) = (J^{-2/n} \hat{\boldsymbol{\theta}}_I' \mathbf{C}^{-1} \hat{\boldsymbol{\theta}}_I)/(\nu_I s_w^2)$$
$$= \frac{S_I/\nu_I}{S_w/\nu_w} \tag{10.5.19}$$

to an F table with ν_I and ν_w degrees of freedom. For the present example, F has 6 and 36 degrees of freedom. Thus, the point $\boldsymbol{\theta}_I = \mathbf{0}$ will lie outside a 95 per cent H.P.D. region if $F(\lambda, \mathbf{z}) > 2.36$. The plot of $F(\lambda, \mathbf{z})$ as a function of λ given in Fig. 10.5.2 shows that the further constraint is acceptable over a fairly wide range which includes the interesting region close to the reciprocal transformation.

10.5.2 Summary of Analysis

Finally, then, the overall posterior distribution $p_u(\lambda \mid A, H, N)$ shown in Fig. 10.3.1 can be decomposed into three functions,

$$p_u(\lambda \mid A, H, N) \propto p_u(\lambda \mid N)\, p_u(H \mid N, \lambda)\, p_u(A \mid H, N, \lambda). \qquad (10.5.20)$$

The first factor is the posterior distribution of λ, supposing only that the transformation $y^{(\lambda)}$ exists which induces Normality simultaneously in all the cells. Its product with a second factor, measuring the probability density associated with variance homogeneity for each value of λ, is proportional to $p_u(\lambda \mid H, N)$, the posterior distribution of λ, supposing a transformation $y^{(\lambda)}$ exists which induces both Normality *and* homogeneity of variance. Finally, the product with a third factor, measuring the probability density associated with zero interaction for each value of λ, is proportional to $p_u(\lambda \mid A, H, N)$, the posterior distribution supposing that a transformation $y^{(\lambda)}$ exists which induces Normality, homogeneity of variance, and additivity. The plausibility of the constraints of homogeneity of variance and of additivity can be assessed for every contemplated λ by considering whether the constraint defines a parameter point inside or outside a suitable H.P.D. region.

For the present data the constraints appear justified. However, situations commonly occur where, for example, nonadditivity is not reduceable by a given family of transformations. One would be warned of this possibility if the F plot of Fig. 10.5.2 indicated that for no value of λ was the point $\boldsymbol{\theta}_I = \mathbf{0}$ included in, say, a 95 per cent H.P.D. region. Similarly, situations are common in which variance homogeneity cannot be induced by transformation. This possibility would be assessed by the M plot of Fig. 10.5.2.

10.6 FURTHER ANALYSIS OF THE TEXTILE DATA

A similar decomposition is possible for the textile data analysis. In our original analysis, it was assumed that a transformation $y^{(\lambda)}$ was possible which induced linearity in the response surface *as well as* spherical Normality. It could be true, of course, that no such transformation was possible, in which case we could fall back on the possibility of inducing spherical Normality with the original quadratic model. The introduction of additivity which we have previously discussed is a special case of the induction of parsimony in the parametrization of the linear model. In general we have

$$\mathbf{y}^{(\lambda)} = \mathbf{X}_1 \boldsymbol{\theta}_1 + \mathbf{X}_2 \boldsymbol{\theta}_2 + \boldsymbol{\varepsilon}, \qquad (10.6.1)$$

in which $\boldsymbol{\theta}_2$ are the parameters which hopefully may not be needed.

In the particular example of the textile data, the elements of $\boldsymbol{\theta}_2$ are the coefficients of the second-degree terms. We have

$$p_u(\lambda \mid \boldsymbol{\theta}_2 = \mathbf{0}, H, N) \propto p_u(\lambda \mid H, N)\, p(\boldsymbol{\theta}_2 = \mathbf{0} \mid H, N, \lambda). \qquad (10.6.2)$$

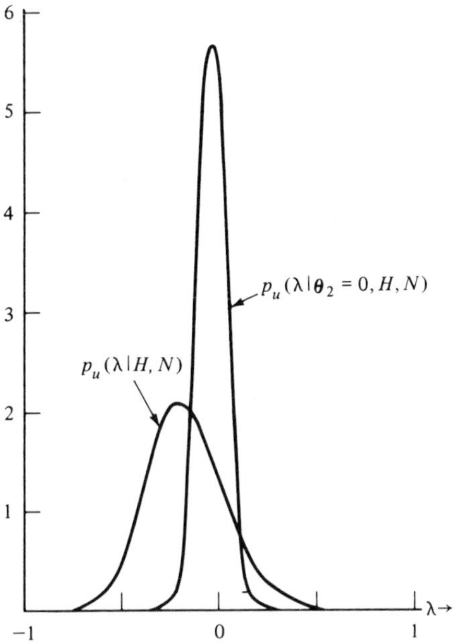

Fig. 10.6.1 Posterior distributions of λ for different models: the textile data.

The "constrained" and "unconstrained" probability density of λ for the textile data are shown in Fig. 10.6.1. Writing $(S_1, S_2, S_R, \nu_1, \nu_2, \nu_R)$ for the sums of squares and degrees of freedom associated with linear terms, quadratic terms, and residuals, we have

$$(S_R + S_2)^{-(\nu_R+\nu_2)/2} = S_R^{-\nu_R/2} \frac{S_R^{\nu_R/2}}{(S_R + S_2)^{(\nu_R+\nu_2)/2}}. \tag{10.6.3}$$

We find by the previous argument that the second factor on the right is proportional to the ordinate at $\boldsymbol{\theta}_2 = \mathbf{0}$ of the multivariate t distribution for $\boldsymbol{\theta}_2$. The appropriateness of the constraint $\boldsymbol{\theta}_2 = \mathbf{0}$ for any value of λ may thus be judged by considering whether $\boldsymbol{\theta}_2 = \mathbf{0}$ is included in an appropriate $(1 - \alpha)$ H.P.D. region, which in turn is equivalent to referring

$$F(\lambda, \mathbf{z}) = \frac{S_2/\nu_2}{S_R/\nu_R}$$

to the α significance point of the F-distribution with ν_2 and ν_R degrees of freedom. A plot of $F(\lambda, \mathbf{z})$ in Fig. 10.6.2 shows that, in this case, for values of λ in the region around $\lambda = 0$, there is no reason to question the applicability of the constraint that the surface in the transformed response could be planar.

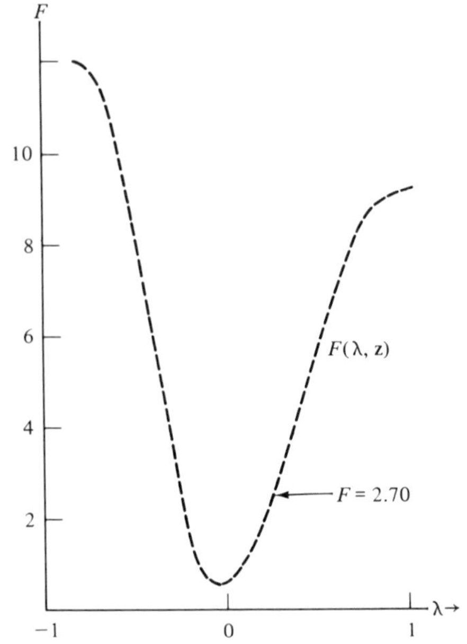

Fig. 10.6.2 Values of $F(\lambda, \mathbf{z})$ as a function of λ: the textile data.

10.7 A SUMMARY OF FORMULAE FOR VARIOUS PRIOR AND POSTERIOR DISTRIBUTIONS

In this chapter the frequently made assumption that the Normal linear model is adequate in the original metric is relaxed, and instead it is assumed only that there exists some *transformation* $y^{(\lambda)}$ of the observation y for which the linear model is appropriate. Specifically, the model is given in (10.3.3) as

$$\mathbf{y}^{(\boldsymbol{\lambda})} = \mathbf{X}\boldsymbol{\theta} + \boldsymbol{\varepsilon},$$

where $\mathbf{y}^{(\boldsymbol{\lambda})} = (y_1^{(\boldsymbol{\lambda})}, \ldots, y_n^{(\boldsymbol{\lambda})})'$, $\mathbf{y} = (y_1, \ldots, y_n)'$ is a $n \times 1$ vector of observations, $\mathbf{X}$ is a $n \times k$ matrix of fixed elements, $\boldsymbol{\theta}$ is a $k \times 1$ vector of parameters and $\boldsymbol{\varepsilon}$ is a $n \times 1$ vector of errors distributed as $N_n(\mathbf{0}, \sigma^2\mathbf{I})$. Table 10.7.1 provides a short summary of the formulae for the prior and posterior distributions of $(\boldsymbol{\theta}, \sigma^2, \boldsymbol{\lambda})$.

Table 10.7.1

A summary of various prior and posterior distributions

1. The prior is, from (10.3.8) to (10.3.19),

$$p(\boldsymbol{\theta}, \log\sigma, \boldsymbol{\lambda}) = p(\boldsymbol{\lambda})p(\boldsymbol{\theta}, \log\sigma \mid \boldsymbol{\lambda}),$$

where $p(\boldsymbol{\lambda})$ is some prior for $\boldsymbol{\lambda}$ and

$$p(\boldsymbol{\theta}, \log\sigma \mid \boldsymbol{\lambda}) \mathrel{\dot{\propto}} J_{\lambda}^{-k/n}, \qquad J_{\lambda} = \prod_{i=1}^{n} \left| \frac{dy_i^{(\boldsymbol{\lambda})}}{dy_i} \right|.$$

Table 10.7.1 *Continued*

2. Conditional on $\boldsymbol{\lambda}$, the posterior distribution of $(\boldsymbol{\theta}, \log\sigma)$ is, from (10.3.20),

$$p(\boldsymbol{\theta}, \log\sigma \mid \boldsymbol{\lambda}, \mathbf{y}) \propto \sigma^{-n} \exp\left[-\frac{S_{\lambda}+(\boldsymbol{\theta}-\hat{\boldsymbol{\theta}}_{\lambda})'\mathbf{X}'\mathbf{X}(\boldsymbol{\theta}-\hat{\boldsymbol{\theta}}_{\lambda})}{2\sigma^2}\right],$$

$$-\infty < \boldsymbol{\theta} < \infty, \quad \sigma > 0,$$

where

$$\hat{\boldsymbol{\theta}}_{\lambda} = (\mathbf{X}'\mathbf{X})^{-1}\mathbf{X}'\mathbf{y}^{(\lambda)} \quad \text{and} \quad S_{\lambda} = (\mathbf{y}^{(\lambda)}-\mathbf{X}\hat{\boldsymbol{\theta}}_{\lambda})'(\mathbf{y}^{(\lambda)}-\mathbf{X}\hat{\boldsymbol{\theta}}_{\lambda}).$$

3. Working with the normalized data

$$z^{(\lambda)} = J_{\lambda}^{-1/n} y^{(\lambda)},$$

the posterior distribution of $\boldsymbol{\lambda}$ with $p(\boldsymbol{\lambda})$ a uniform reference prior is in (10.3.23),

$$p_u(\boldsymbol{\lambda} \mid \mathbf{y}) \propto [S(\boldsymbol{\lambda}, \mathbf{z})]^{-\frac{1}{2}(n-k)}, \qquad -\infty < \boldsymbol{\lambda} < \infty,$$

where

$$S(\boldsymbol{\lambda}, \mathbf{z}) = (\mathbf{z}^{(\lambda)}-\hat{\mathbf{z}}^{(\lambda)})'(\mathbf{z}^{(\lambda)}-\hat{\mathbf{z}}^{(\lambda)})$$

and

$$\hat{\mathbf{z}}^{(\lambda)} = \mathbf{X}(\mathbf{X}'\mathbf{X})^{-1}\mathbf{X}'\mathbf{z}^{(\lambda)}.$$

4. In particular, if $\boldsymbol{\lambda}=\lambda$ and for the simple power transformation in (10.3.1),

$$y^{(\lambda)} = \begin{cases} (y^{\lambda}-1)/\lambda & (\lambda \neq 0) \\ \log y & (\lambda = 0) \end{cases} \qquad y>0,$$

then, from (10.3.27)

$$z^{(\lambda)} = \begin{cases} (y^{\lambda}-1)/(\lambda\dot{y}^{\lambda-1}) & (\lambda \neq 0) \\ \dot{y}\log y & (\lambda = 0) \end{cases}$$

where

$$\dot{y} = \left(\prod_{i=1}^{n} y_i\right)^{1/n}.$$

For the two-parameter transformation in (10.3.2),

$$y^{(\boldsymbol{\lambda})} = \begin{cases} [(y+\lambda_2)^{\lambda_1}-1]/\lambda_1 & (\lambda_1 \neq 0) \\ \log(y+\lambda_2) & (\lambda_1 = 0) \end{cases} \qquad y > -\lambda_2,$$

then, from (10.3.32)

$$z^{(\lambda)} = \begin{cases} [(y+\lambda_2)^{\lambda_1}-1]/\{\lambda_1[\text{gm}(y+\lambda_2)]^{\lambda_1-1}\} & (\lambda_1 \neq 0) \\ \text{gm}(y+\lambda_2)\log(y+\lambda_2) & (\lambda_1 = 0) \end{cases}$$

where

$$\text{gm}(y+\lambda_2) = \left[\prod_{i=1}^{n}(y_i+\lambda_2)\right]^{1/n}.$$

Table 10.7.1 *Continued*

5. The posterior distribution of $\boldsymbol{\theta}$ with $p(\boldsymbol{\lambda})$ a uniform reference prior is in (10.4.4),

$$p(\boldsymbol{\theta} \mid \mathbf{y}) \propto \int_{-\infty < \boldsymbol{\lambda} < \infty} J_{\boldsymbol{\lambda}}^{-k/n} [S(\boldsymbol{\lambda}, \mathbf{z}) + Q(\boldsymbol{\theta}, \boldsymbol{\lambda})]^{-n/2} \, d\boldsymbol{\lambda}, \qquad -\infty < \boldsymbol{\theta} < \infty,$$

where

$$Q(\boldsymbol{\theta}, \boldsymbol{\lambda}) = J_{\boldsymbol{\lambda}}^{-2/n} (\boldsymbol{\theta} - \hat{\boldsymbol{\theta}}_{\boldsymbol{\lambda}})' \, \mathbf{X}'\mathbf{X} (\boldsymbol{\theta} - \hat{\boldsymbol{\theta}}_{\boldsymbol{\lambda}}).$$

This distribution can be approximated by (10.4.10),

$$p(\boldsymbol{\theta} \mid \mathbf{y}) \mathrel{\dot{\propto}} \left[1 + \frac{Q(\boldsymbol{\theta}, \hat{\boldsymbol{\lambda}})}{s_{\nu-\tau}^2(\hat{\boldsymbol{\lambda}})(\nu - \tau)} \right]^{-\frac{1}{2}(\nu-\tau+k)}, \qquad -\infty < \boldsymbol{\theta} < \infty,$$

that is, a k-dimensional multivariate t distribution

$$t_k[\hat{\boldsymbol{\theta}}_{\hat{\boldsymbol{\lambda}}}, \, s_{\nu-\tau}^2(\hat{\boldsymbol{\lambda}}) J_{\hat{\boldsymbol{\lambda}}}^{2/n} (\mathbf{X}'\mathbf{X})^{-1}, \, \nu - \tau],$$

where $\nu = n - k$, τ is the number of elements in $\boldsymbol{\lambda}$, $\hat{\boldsymbol{\lambda}}$ is the value minimizing $S(\boldsymbol{\lambda}, \mathbf{z})$, and $(\nu - \tau) s_{\nu-\tau}^2(\hat{\boldsymbol{\lambda}}) = S(\hat{\boldsymbol{\lambda}}, \mathbf{z})$.

TABLES

Table I

Area under the Normal curve

$$\alpha = \int_{u_\alpha}^{\infty} \phi(t)\,dt \quad \text{where} \quad \phi(t) = \frac{1}{\sqrt{2\pi}} e^{-\frac{1}{2}t^2}$$

u_α	**0.00**	**0.01**	**0.02**	**0.03**	**0.04**	**0.05**	**0.06**	**0.07**	**0.08**	**0.09**
0.0	0.5000	0.4960	0.4920	0.4880	0.4840	0.4801	0.4761	0.4721	0.4681	0.4641
0.1	0.4602	0.4562	0.4522	0.4483	0.4443	0.4404	0.4364	0.4325	0.4286	0.4247
0.2	0.4207	0.4168	0.4129	0.4090	0.4052	0.4013	0.3974	0.3936	0.3897	0.3859
0.3	0.3821	0.3783	0.3745	0.3707	0.3669	0.3632	0.3594	0.3557	0.3520	0.3483
0.4	0.3446	0.3409	0.3372	0.3336	0.3300	0.3264	0.3228	0.3192	0.3156	0.3121
0.5	0.3085	0.3050	0.3015	0.2981	0.2946	0.2912	0.2877	0.2843	0.2810	0.2776
0.6	0.2743	0.2709	0.2676	0.2643	0.2611	0.2578	0.2546	0.2514	0.2483	0.2451
0.7	0.2420	0.2389	0.2358	0.2327	0.2296	0.2266	0.2236	0.2206	0.2177	0.2148
0.8	0.2119	0.2090	0.2061	0.2033	0.2005	0.1977	0.1949	0.1922	0.1894	0.1867
0.9	0.1841	0.1814	0.1788	0.1762	0.1736	0.1711	0.1685	0.1660	0.1635	0.1611
1.0	0.1587	0.1562	0.1539	0.1515	0.1492	0.1469	0.1446	0.1423	0.1401	0.1379
1.1	0.1357	0.1335	0.1314	0.1292	0.1271	0.1251	0.1230	0.1210	0.1190	0.1170
1.2	0.1151	0.1131	0.1112	0.1093	0.1075	0.1056	0.1038	0.1020	0.1003	0.0985
1.3	0.0968	0.0951	0.0934	0.0918	0.0901	0.0885	0.0869	0.0853	0.0838	0.0823
1.4	0.0808	0.0793	0.0778	0.0764	0.0749	0.0735	0.0721	0.0708	0.0694	0.0681
1.5	0.0668	0.0655	0.0643	0.0630	0.0618	0.0606	0.0594	0.0582	0.0571	0.0559
1.6	0.0548	0.0537	0.0526	0.0516	0.0505	0.0495	0.0485	0.0475	0.0465	0.0455
1.7	0.0446	0.0436	0.0427	0.0418	0.0409	0.0401	0.0392	0.0384	0.0375	0.0367
1.8	0.0359	0.0351	0.0344	0.0366	0.0329	0.0322	0.0314	0.0307	0.0301	0.0294
1.9	0.0287	0.0281	0.0274	0.0268	0.0262	0.0256	0.0250	0.0244	0.0239	0.0233
2.0	0.0228	0.0222	0.0217	0.0212	0.0207	0.0202	0.0197	0.0192	0.0188	0.0183
2.1	0.0179	0.0174	0.0170	0.0166	0.0162	0.0158	0.0154	0.0150	0.0146	0.0143
2.2	0.0139	0.0136	0.0132	0.0129	0.0125	0.0122	0.0119	0.0116	0.0113	0.0110
2.3	0.0107	0.0104	0.0102	0.0099	0.0096	0.0094	0.0091	0.0089	0.0087	0.0084
2.4	0.0082	0.0080	0.0078	0.0075	0.0073	0.0071	0.0069	0.0068	0.0066	0.0064
2.5	0.0062	0.0060	0.0059	0.0057	0.0055	0.0054	0.0052	0.0051	0.0049	0.0048
2.6	0.0047	0.0045	0.0044	0.0043	0.0041	0.0040	0.0039	0.0038	0.0037	0.0036
2.7	0.0035	0.0034	0.0033	0.0032	0.0031	0.0030	0.0029	0.0028	0.0027	0.0026
2.8	0.0026	0.0025	0.0024	0.0023	0.0023	0.0022	0.0021	0.0021	0.0020	0.0019
2.9	0.0019	0.0018	0.0018	0.0017	0.0016	0.0016	0.0015	0.0015	0.0014	0.0014
3.0	0.0013	0.0013	0.0013	0.0012	0.0012	0.0011	0.0011	0.0011	0.0010	0.0010
3.1	0.0010	0.0009	0.0009	0.0009	0.0008	0.0008	0.0008	0.0008	0.0007	0.0007
3.2	0.0007	0.0007	0.0006	0.0006	0.0006	0.0006	0.0006	0.0005	0.0005	0.0005
3.3	0.0005	0.0005	0.0005	0.0004	0.0004	0.0004	0.0004	0.0004	0.0004	0.0003
3.4	0.0003	0.0003	0.0003	0.0003	0.0003	0.0003	0.0003	0.0003	0.0003	0.0002
3.5	0.0002	0.0002	0.0002	0.0002	0.0002	0.0002	0.0002	0.0002	0.0002	0.0002
3.6	0.0002	0.0002	0.0001	0.0001	0.0001	0.0001	0.0001	0.0001	0.0001	0.0001
3.7	0.0001	0.0001	0.0001	0.0001	0.0001	0.0001	0.0001	0.0001	0.0001	0.0001
3.8	0.0001	0.0001	0.0001	0.0001	0.0001	0.0001	0.0001	0.0001	0.0001	0.0001
3.9	0.0000	0.0000	0.0000	0.0000	0.0000	0.0000	0.0000	0.0000	0.0000	0.0000
u_α	**0.00**	**0.01**	**0.02**	**0.03**	**0.04**	**0.05**	**0.06**	**0.07**	**0.08**	**0.09**

Table II

Values of $\underline{\chi}^2$, $\bar{\chi}^2$

$\underline{\chi}^2$ and $\bar{\chi}^2$ are determined by the conditions

(1) $(\underline{\chi}^2)^{n/2} e^{-\frac{1}{2}\underline{\chi}^2} = (\bar{\chi}^2)^{n/2} e^{-\frac{1}{2}\bar{\chi}^2}$

(2) shaded area $\alpha = \int_0^{\underline{\chi}^2} f(t)\,dt + \int_{\bar{\chi}^2}^{\infty} f(t)\,dt$

where

$$f(t) = \frac{1}{\Gamma(n/2)2^{n/2}} t^{(n/2)-1} e^{-\frac{1}{2}t}$$

	$\alpha = 0.05$		$\alpha = 0.01$	
n	$\underline{\chi}^2$	$\bar{\chi}^2$	$\underline{\chi}^2$	$\bar{\chi}^2$
1	$0.0^2 31593$	7.8168	$0.0^3 13422$	11.345
2	0.084727	9.5303	0.017469	13.285
3	0.29624	11.191	0.101048	15.127
4	0.60700	12.802	0.26396	16.901
5	0.98923	14.369	0.49623	18.621
6	1.4250	15.897	0.78565	20.296
7	1.9026	17.392	1.1221	21.931
8	2.4139	18.860	1.4978	23.533
9	2.9532	20.305	1.9069	25.106
10	3.5162	21.729	2.3444	26.653
11	4.0995	23.135	2.8069	28.178
12	4.7005	24.525	3.2912	29.683
13	5.3171	25.900	3.7949	31.170
14	5.9477	27.263	4.3161	32.641
15	6.5908	28.614	4.8530	34.097
16	7.2453	29.955	5.4041	35.540
17	7.9100	31.285	5.9683	36.971
18	8.5842	32.607	6.5444	38.390
19	9.2670	33.921	7.1316	39.798
20	9.9579	35.227	7.7289	41.197
21	10.656	36.525	8.3358	42.586
22	11.361	37.818	8.9515	43.967
23	12.073	39.103	9.5755	45.340
24	12.791	40.383	10.2073	46.706
25	13.514	41.658	10.846	48.064
26	14.243	42.927	11.492	49.416
27	14.977	44.192	12.145	50.761
28	15.716	45.451	12.803	52.100
29	16.459	46.707	13.468	53.434
30	17.206	47.958	14.138	54.762

Taken, with permission, from D. V. Lindley, D. A. East, and P. A. Hamilton, "Tables for Making Inferences About the Variance of a Normal Distribution," *Biometrika* **47,** 1960, pages 433-8.

Table III

Percentage points of the χ^2 distribution

$$\alpha = \int_{\chi^2}^{\infty} f(t)\, dt$$

where

$$f(t) = \frac{1}{\Gamma(n/2)\, 2^{n/2}} t^{(n/2)-1}\, e^{-\frac{1}{2}t}$$

n	α				
	0.01	0.05	0.10	0.25	0.50
1	6.63490	3.84146	2.70554	1.32330	0.454937
2	9.21034	5.99147	4.60517	2.77259	1.38629
3	11.3449	7.81473	6.25139	4.10835	2.36597
4	13.2767	9.48773	7.77944	5.38527	3.35670
5	15.0863	11.0705	9.23635	6.62568	4.35146
6	16.8119	12.5916	10.6446	7.84080	5.34812
7	18.4753	14.0671	12.0170	9.03715	6.34581
8	20.0902	15.5073	13.3616	10.2188	7.34412
9	21.6660	16.9190	14.6837	11.3887	8.34283
10	23.2093	18.3070	15.9871	12.5489	9.34182
11	24.7250	19.6751	17.2750	13.7007	10.3410
12	26.2170	21.0261	18.5494	14.8454	11.3403
13	27.6883	22.3621	19.8119	15.9839	12.3398
14	29.1413	23.6848	21.0642	17.1170	13.3393
15	30.5779	24.9958	22.3072	18.2451	14.3389
16	31.9999	26.2962	23.5418	19.3688	15.3385
17	33.4087	27.5871	24.7690	20.4887	16.3381
18	34.8053	28.8693	25.9894	21.6049	17.3379
19	36.1908	30.1435	27.2036	22.7178	18.3376
20	37.5662	31.4104	28.4120	23.8277	19.3374
21	38.9321	32.6705	29.6151	24.9348	20.3372
22	40.2894	33.9244	30.8133	26.0393	21.3370
23	41.6384	35.1725	32.0069	27.1413	22.3369
24	42.9798	36.4151	33.1963	28.2412	23.3367
25	44.3141	37.6525	34.3816	29.3389	24.3366
26	45.6417	38.8852	35.5631	30.4345	25.3364
27	46.9630	40.1133	36.7412	31.5284	26.3363
28	48.2782	41.3372	37.9159	32.6205	27.3363
29	49.5879	42.5569	39.0875	33.7109	28.3362
30	50.8922	43.7729	40.2560	34.7998	29.3360

Taken, with permission, from E. S. Pearson and H. O. Hartley (editors), *Biometrika Tables for Statisticians,* Volume 1, 1958, Cambridge University Press, Cambridge, England.

Table IV

Percentage points of Student's t distribution

$$\alpha/2 = \int_{t_{\alpha/2}}^{\infty} p(t)\,dt$$

where

$$p(t) = \frac{1}{B(\frac{1}{2}, \frac{1}{2}n)\sqrt{n}}(1 + t^2/n)^{-\frac{1}{2}(n+1)}$$

n	α				
	0.01	0.05	0.10	0.25	0.50
1	63.657	12.706	6.314	2.414	1.000
2	9.925	4.303	2.920	1.604	0.816
3	5.841	3.182	2.353	1.423	0.765
4	4.604	2.776	2.132	1.344	0.741
5	4.032	2.571	2.015	1.301	0.727
6	3.707	2.447	1.943	1.273	0.718
7	3.499	2.365	1.895	1.254	0.711
8	3.355	2.306	1.860	1.240	0.706
9	3.250	2.262	1.833	1.230	0.703
10	3.169	2.228	1.812	1.221	0.700
11	3.106	2.201	1.796	1.215	0.697
12	3.055	2.179	1.782	1.209	0.695
13	3.012	2.160	1.771	1.204	0.694
14	2.977	2.145	1.761	1.200	0.692
15	2.947	2.131	1.753	1.197	0.691
16	2.921	2.120	1.746	1.194	0.690
17	2.898	2.110	1.740	1.191	0.689
18	2.878	2.101	1.734	1.189	0.688
19	2.861	2.093	1.729	1.187	0.688
20	2.845	2.086	1.725	1.185	0.687
21	2.831	2.080	1.721	1.183	0.686
22	2.819	2.074	1.717	1.182	0.686
23	2.807	2.069	1.714	1.180	0.685
24	2.797	2.064	1.711	1.179	0.685
25	2.787	2.060	1.708	1.178	0.684
26	2.779	2.056	1.706	1.177	0.684
27	2.771	2.052	1.703	1.176	0.684
28	2.763	2.048	1.701	1.175	0.683
29	2.756	2.045	1.699	1.174	0.683
30	2.750	2.042	1.697	1.173	0.683
∞	2.576	1.960	1.645	1.150	0.674

Taken, with permission, from M. Merrington, "Table of Percentage Points of the t-distribution," *Biometrika,* Volume 32, 1941–42, p. 300.

Table V

Values of $\underline{F}$, $\overline{F}$

$\underline{F}$ and $\overline{F}$ are determined by the conditions

$$(1)\quad \underline{F}^{\nu_1/2}\left(1+\frac{\nu_1}{\nu_2}\underline{F}\right)^{-(\nu_1+\nu_2)/2} = \overline{F}^{\nu_1/2}\left(1+\frac{\nu_1}{\nu_2}\overline{F}\right)^{-(\nu_1+\nu_2)/2}$$

$$(2)\quad \alpha = \int_0^{\underline{F}} p(F)\,dF + \int_{\overline{F}}^{\infty} p(F)\,dF = \text{shaded area.}$$

where

$$p(F) = \frac{1}{B(\nu_1/2, \nu_2/2)}(\nu_1/\nu_2)^{\nu_1/2} F^{(\nu_1/2)-1}\left(1+\frac{\nu_1}{\nu_2}F\right)^{-(\nu_1+\nu_2)/2}$$

ν_1	ν_2	$\alpha = 0.01$ $\underline{F}$	$\alpha = 0.01$ $\overline{F}$	$\alpha = 0.05$ $\underline{F}$	$\alpha = 0.05$ $\overline{F}$	$\alpha = 0.10$ $\underline{F}$	$\alpha = 0.10$ $\overline{F}$	$\alpha = 0.25$ $\underline{F}$	$\alpha = 0.25$ $\overline{F}$
5	5	0.067	14.939	0.140	7.146	0.198	5.050	0.331	3.022
6	5	0.085	14.094	0.163	6.787	0.222	4.821	0.355	2.914
	6	0.090	11.072	0.172	5.820	0.233	4.284	0.369	2.712
	5	0.100	13.497	0.181	6.536	0.241	4.659	0.373	2.835
7	6	0.107	10.570	0.192	5.590	0.254	4.131	0.388	2.635
	7	0.113	8.883	0.200	4.995	0.264	3.787	0.400	2.498
	5	0.114	13.051	0.196	6.347	0.257	4.537	0.388	2.777
8	6	0.122	10.193	0.208	5.416	0.271	4.014	0.404	2.577
	7	0.128	8.549	0.218	4.832	0.282	3.676	0.417	2.441
	8	0.133	7.496	0.226	4.433	0.291	3.438	0.427	2.342
	5	0.125	12.713	0.209	6.203	0.270	4.442	0.399	2.731
	6	0.134	9.908	0.222	5.281	0.285	3.925	0.417	2.532
9	7	0.142	8.290	0.233	4.705	0.297	3.589	0.430	2.396
	8	0.148	7.257	0.241	4.312	0.307	3.353	0.441	2.297
	9	0.153	6.541	0.248	4.026	0.315	3.179	0.450	2.221
	5	0.135	12.436	0.220	6.086	0.280	4.368	0.409	2.694
	6	0.145	9.673	0.234	5.175	0.296	3.854	0.427	2.495
	7	0.154	8.083	0.245	4.603	0.309	3.519	0.441	2.359
10	8	0.160	7.066	0.255	4.213	0.320	3.285	0.453	2.261
	9	0.166	6.361	0.262	3.931	0.329	3.111	0.462	2.186
	10	0.171	5.847	0.269	3.716	0.336	2.978	0.470	2.126
	5	0.144	12.220	0.229	5.989	0.290	4.305	0.417	2.664
	6	0.155	9.487	0.244	5.087	0.306	3.795	0.436	2.465
	7	0.164	7.917	0.256	4.520	0.320	3.462	0.451	2.330
11	8	0.172	6.908	0.266	4.132	0.331	3.229	0.463	2.230
	9	0.178	6.212	0.275	3.852	0.340	3.057	0.473	2.155
	10	0.183	5.705	0.282	3.640	0.348	2.923	0.481	2.096
	11	0.188	5.320	0.288	3.474	0.355	2.818	0.488	2.048
	5	0.152	12.039	0.237	5.913	0.297	4.254	0.424	2.639
	6	0.164	9.330	0.253	5.014	0.315	3.746	0.444	2.440
12	7	0.173	7.775	0.266	4.451	0.329	3.415	0.459	2.304
	8	0.181	6.778	0.276	4.066	0.341	3.182	0.472	2.206

Table V—(*continued*)

ν_1	ν_2	$\alpha = 0.01$ $\underline{F}$	$\alpha = 0.01$ $\bar{F}$	$\alpha = 0.05$ $\underline{F}$	$\alpha = 0.05$ $\bar{F}$	$\alpha = 0.10$ $\underline{F}$	$\alpha = 0.10$ $\bar{F}$	$\alpha = 0.25$ $\underline{F}$	$\alpha = 0.25$ $\bar{F}$
12	9	0.188	6.088	0.285	3.786	0.351	3.010	0.482	2.130
	10	0.194	5.587	0.293	3.575	0.359	2.877	0.490	2.071
	11	0.199	5.206	0.299	3.410	0.366	2.772	0.498	2.023
	12	0.204	4.906	0.305	3.277	0.372	2.686	0.505	1.982
	5	0.159	11.884	0.244	5.845	0.304	4.210	0.430	2.618
	6	0.171	9.194	0.261	4.952	0.323	3.704	0.450	2.419
	7	0.182	7.658	0.274	4.390	0.337	3.374	0.466	2.283
	8	0.190	6.670	0.285	4.009	0.349	3.143	0.479	2.184
13	9	0.198	5.986	0.295	3.731	0.360	2.971	0.490	2.108
	10	0.204	5.488	0.303	3.520	0.368	2.839	0.499	2.049
	11	0.209	5.109	0.310	3.357	0.376	2.733	0.506	2.001
	12	0.214	4.812	0.316	3.224	0.382	2.647	0.513	1.961
	13	0.219	4.573	0.321	3.115	0.388	2.577	0.519	1.927
	5	0.165	11.745	0.250	5.789	0.310	4.174	0.435	2.600
	6	0.178	9.085	0.268	4.899	0.329	3.668	0.456	2.400
	7	0.189	7.555	0.282	4.341	0.344	3.340	0.472	2.265
	8	0.198	6.574	0.293	3.961	0.357	3.108	0.486	2.166
	9	0.206	5.895	0.303	3.684	0.368	2.936	0.496	2.090
14	10	0.213	5.402	0.311	3.475	0.377	2.805	0.506	2.030
	11	0.219	5.026	0.319	3.310	0.385	2.700	0.514	1.982
	12	0.224	4.731	0.325	3.178	0.391	2.614	0.521	1.942
	13	0.228	4.494	0.331	3.069	0.397	2.544	0.527	1.908
	14	0.233	4.299	0.336	2.978	0.403	2.484	0.532	1.879
	5	0.170	11.630	0.256	5.740	0.315	4.141	0.440	2.584
	6	0.184	8.989	0.274	4.853	0.335	3.636	0.461	2.385
	7	0.196	7.465	0.288	4.299	0.351	3.310	0.478	2.249
	8	0.205	6.490	0.300	3.917	0.364	3.079	0.491	2.150
	9	0.214	5.820	0.310	3.643	0.375	2.908	0.503	2.074
15	10	0.221	5.328	0.319	3.434	0.384	2.776	0.512	2.014
	11	0.227	4.953	0.327	3.270	0.392	2.671	0.520	1.965
	12	0.232	4.660	0.334	3.138	0.399	2.585	0.527	1.925
	13	0.237	4.424	0.339	3.030	0.406	2.514	0.534	1.891
	14	0.242	4.232	0.345	2.939	0.411	2.455	0.539	1.862
	15	0.246	4.070	0.349	2.862	0.416	2.403	0.544	1.838
	5	0.175	11.531	0.261	5.696	0.320	4.113	0.444	2.569
	6	0.190	8.901	0.279	4.813	0.340	3.610	0.466	2.372
	7	0.202	7.390	0.294	4.259	0.356	3.282	0.483	2.235
	8	0.212	6.421	0.307	3.882	0.370	3.052	0.496	2.136
	9	0.220	5.750	0.317	3.606	0.381	2.882	0.508	2.060
16	10	0.228	5.262	0.326	3.398	0.391	2.749	0.518	2.000
	11	0.235	4.891	0.334	3.234	0.399	2.645	0.526	1.951
	12	0.240	4.600	0.341	3.103	0.407	2.560	0.533	1.911
	13	0.245	4.364	0.347	2.995	0.413	2.489	0.540	1.877
	14	0.250	4.172	0.353	2.905	0.419	2.429	0.546	1.847
	15	0.254	4.011	0.358	2.827	0.424	2.377	0.551	1.823
	16	0.258	3.875	0.362	2.761	0.429	2.333	0.555	1.801
	5	0.180	11.448	0.265	5.656	0.324	4.089	0.448	2.558
17	6	0.195	8.824	0.284	4.776	0.345	3.586	0.470	2.359
	7	0.207	7.323	0.300	4.224	0.361	3.260	0.487	2.223

Table V—(*continued*)

ν_1	ν_2	$\alpha = 0.01$ $\underline{F}$	$\alpha = 0.01$ $\bar{F}$	$\alpha = 0.05$ $\underline{F}$	$\alpha = 0.05$ $\bar{F}$	$\alpha = 0.10$ $\underline{F}$	$\alpha = 0.10$ $\bar{F}$	$\alpha = 0.25$ $\underline{F}$	$\alpha = 0.25$ $\bar{F}$
	8	0.218	6.357	0.312	3.847	0.375	3.030	0.501	2.123
	9	0.227	5.689	0.323	3.573	0.387	2.858	0.513	2.047
	10	0.234	5.205	0.333	3.366	0.397	2.727	0.523	1.987
17	11	0.241	4.834	0.341	3.203	0.405	2.622	0.531	1.938
	12	0.247	4.545	0.348	3.072	0.413	2.537	0.539	1.898
	13	0.253	4.311	0.354	2.964	0.420	2.466	0.545	1.864
	14	0.258	4.119	0.360	2.873	0.426	2.406	0.551	1.834
	15	0.262	3.959	0.365	2.797	0.431	2.355	0.557	1.809
	16	0.266	3.823	0.370	2.731	0.436	2.311	0.561	1.787
	17	0.270	3.707	0.374	2.673	0.440	2.272	0.566	1.768
	5	0.184	11.370	0.269	5.625	0.328	4.067	0.451	2.547
	6	0.199	8.755	0.288	4.747	0.349	3.565	0.473	2.348
	7	0.212	7.262	0.304	4.195	0.366	3.238	0.491	2.212
	8	0.233	6.300	0.317	3.819	0.380	3.008	0.505	2.112
	9	0.232	5.638	0.329	3.546	0.392	2.838	0.517	2.036
	10	0.241	5.153	0.338	3.338	0.402	2.707	0.527	1.976
18	11	0.248	4.784	0.347	3.176	0.411	2.602	0.536	1.927
	12	0.254	4.496	0.354	3.044	0.419	2.516	0.544	1.886
	13	0.260	4.264	0.361	2.936	0.426	2 446	0.551	1.852
	14	0.265	4.073	0.367	2.846	0.432	2.386	0.557	1.823
	15	0.269	3.913	0.372	2.770	0.437	2.335	0.562	1.797
	16	0.274	3.778	0.377	2.703	0.442	2.290	0.567	1.776
	17	0.277	3.661	0.381	2.646	0.447	2.251	0.571	1.756
	18	0.281	3.560	0.385	2.595	0.451	2.217	0.575	1.739
	5	0.187	11.301	0.273	5.594	0.332	4.048	0.454	2.536
	6	0.203	8.701	0.293	4.718	0.353	3.547	0.477	2.338
	7	0.217	7.205	0.309	4.168	0.370	3.220	0.494	2.202
	8	0.228	6.251	0.322	3.792	0.385	2.991	0.509	2.102
	9	0.238	5.589	0.334	3.519	0.397	2.821	0.521	2.026
	10	0.246	5.107	0.344	3.312	0.407	2.689	0.532	1.966
	11	0.254	4.740	0.352	3.150	0.416	2.584	0.541	1.917
19	12	0.260	4.452	0.360	3.020	0.424	2.499	0.548	1.876
	13	0.266	4.221	0.367	2.912	0.431	2.428	0.555	1.841
	14	0.271	4.030	0.373	2.821	0.438	2.368	0.561	1.812
	15	0.276	3.871	0.378	2.744	0.443	2.316	0.567	1.787
	16	0.280	3.736	0.383	2.678	0.448	2.272	0.572	1.765
	17	0.284	3.621	0.388	2.621	0.453	2.233	0.576	1.745
	18	0.288	3.519	0.392	2.571	0.457	2.199	0.580	1.728
	19	0.291	3.432	0.396	2.526	0.461	2.168	0.584	1.712
	5	0.191	11.239	0.276	5.566	0.335	4.029	0.457	2.528
	6	0.207	8.646	0.296	4.693	0.356	3.529	0.479	2.330
	7	0.221	7.157	0.313	4.143	0.374	3.203	0.498	2 193
	8	0.232	6.207	0.327	3.769	0.389	2.973	0.512	2.093
	9	0.242	5.546	0.338	3.497	0.401	2.803	0.525	2.017
	10	0.251	5.066	0.348	3.290	0.412	2.672	0.535	1.956
	11	0.259	4.700	0.357	3.127	0.421	2.567	0.544	1.907
20	12	0.266	4.413	0.365	2.996	0.429	2.482	0.552	1.867
	13	0.272	4.182	0.372	2.889	0.436	2.411	0.559	1.832
	14	0.277	3.993	0.379	2.799	0.443	2.351	0.566	1.803
	15	0.282	3.834	0.384	2.722	0.449	2.300	0.571	1.777

Table V—(*continued*)

ν_1	ν_2	$\alpha = 0.01$ $\underline{F}$	$\bar{F}$	$\alpha = 0.05$ $\underline{F}$	$\bar{F}$	$\alpha = 0.10$ $\underline{F}$	$\bar{F}$	$\alpha = 0.25$ $\underline{F}$	$\bar{F}$
20	16	0.287	3.699	0.389	2.657	0.454	2.256	0.576	1.755
	17	0.291	3.584	0.394	2.599	0.459	2.216	0.581	1.735
	18	0.295	3.483	0.398	2.549	0.463	2.182	0.585	1.718
	19	0.298	3.395	0.402	2.504	0.467	2.152	0.589	1.702
	20	0.301	3.318	0.406	2.464	0.471	2.124	0.592	1.688
	5	0.194	11.182	0.279	5.543	0.338	4.014	0.459	2.520
	6	0.211	8.598	0.300	4.670	0.360	3.514	0.482	2.322
	7	0.225	7.116	0.316	4.123	0.378	3.188	0.500	2.185
	8	0.237	6.163	0.331	3.748	0.392	2.959	0.516	2.085
	9	0.247	5.508	0.343	3.476	0.405	2.789	0.528	2.008
	10	0.256	5.027	0.353	3.269	0.416	2.657	0.539	1.948
	11	0.264	4.665	0.362	3.108	0.425	2.553	0.548	1.899
	12	0.271	4.377	0.370	2.977	0.434	2.467	0.556	1.858
21	13	0.277	4.147	0.377	2.869	0.441	2.396	0.563	1.824
	14	0.283	3.958	0.384	2.779	0.448	2.336	0.570	1.794
	15	0.288	3.800	0.390	2.702	0.454	2.285	0.575	1.769
	16	0.293	3.666	0.395	2.636	0.459	2.240	0.581	1.746
	17	0.297	3.550	0.400	2.579	0.464	2.201	0.585	1.727
	18	0.301	3.450	0.404	2.528	0.468	2.167	0.589	1.709
	19	0.305	3.362	0.408	2.484	0.473	2.136	0.593	1.693
	20	0.308	3.285	0.412	2.444	0.476	2.109	0.597	1.679
	21	0.311	3.216	0.415	2.409	0.480	2.084	0.600	1.667
	5	0.197	11.131	0.282	5.518	0.340	3.999	0.461	2.513
	6	0.214	8.555	0.303	4.650	0.363	3.499	0.485	2.314
	7	0.228	7.075	0.320	4.102	0.381	3.175	0.503	2.177
	8	0.241	6.126	0.334	3.728	0.396	2.944	0.518	2.077
	9	0.251	5.471	0.347	3.456	0.409	2.774	0.531	2.001
	10	0.260	4.994	0.357	3.251	0.420	2.643	0.542	1.940
	11	0.268	4.630	0.366	3.088	0.429	2.539	0.551	1.891
	12	0.276	4.346	0.375	2.958	0.438	2.454	0.560	1.850
	13	0.282	4.115	0.382	2.851	0.445	2.383	0.567	1.816
22	14	0.288	3.927	0.389	2.760	0.452	2.322	0.573	1.786
	15	0.293	3.769	0.394	2.684	0.458	2.272	0.579	1.760
	16	0.298	3.634	0.400	2.618	0.464	2.226	0.584	1.738
	17	0.302	3.520	0.405	2.561	0.469	2.188	0.589	1.718
	18	0.307	3.420	0.409	2.510	0.473	2.153	0.593	1.701
	19	0.310	3.332	0.413	2.465	0.478	2.122	0.597	1.685
	20	0.314	3.255	0.417	2.425	0.481	2.095	0.601	1.671
	21	0.317	3.186	0.421	2.390	0.485	2.070	0.604	1.658
	22	0.320	3.125	0.424	2.358	0.488	2.047	0.607	1.646
	5	0.200	11.079	0.285	5.498	0.343	3.985	0.464	2.506
	6	0.217	8.511	0.306	4.631	0.365	3.485	0.487	2.307
	7	0.232	7.040	0.323	4.082	0.384	3.161	0.506	2.171
	8	0.244	6.092	0.338	3.709	0.399	2.933	0.521	2.071
	9	0.255	5.439	0.350	3.439	0.412	2.763	0.534	1.994
23	10	0.264	4.964	0.361	3.232	0.423	2.631	0.545	1.933
	11	0.273	4.600	0.370	3.071	0.433	2.526	0.554	1.884
	12	0.280	4.316	0.379	2.941	0.442	2.441	0.563	1.843
	13	0.287	4.087	0.386	2.833	0.449	2.370	0.570	1.809
	14	0.293	3.898	0.393	2.744	0.456	2.310	0.577	1.779

Table V—(*continued*)

ν_1	ν_2	$\alpha = 0.01$		$\alpha = 0.05$		$\alpha = 0.10$		$\alpha = 0.25$	
		$\underline{F}$	$\bar{F}$	$\underline{F}$	$\bar{F}$	$\underline{F}$	$\bar{F}$	$\underline{F}$	$\bar{F}$
23	15	0.298	3.740	0.399	2.667	0.463	2.258	0.583	1.753
	16	0.303	3.607	0.405	2.601	0.468	2.214	0.588	1.731
	17	0.308	3.492	0.410	2.543	0.473	2.175	0.593	1.710
	18	0.312	3.392	0.414	2.493	0.478	2.140	0.597	1.693
	19	0.316	3.304	0.418	2.449	0.482	2.110	0.601	1.678
	20	0.319	3.228	0.422	2.409	0.486	2.082	0.605	1.663
	21	0.323	3.159	0.426	2.373	0.490	2.057	0.608	1.650
	22	0.326	3.097	0.429	2.341	0.493	2.035	0.611	1.638
	23	0.329	3.042	0.433	2.312	0.496	2.014	0.614	1.628
	5	0.202	11.037	0.287	5.481	0.345	3.972	0.465	2.500
	6	0.220	8.474	0.308	4.613	0.368	3.475	0.489	2.301
	7	0.235	7.003	0.326	4.067	0.386	3.150	0.508	2.165
	8	0.247	6.063	0.341	3.694	0.402	2.921	0.524	2.064
	9	0.258	5.409	0.353	3.423	0.415	2.751	0.537	1.988
	10	0.268	4.933	0.364	3.217	0.427	2.620	0.548	1.927
	11	0.277	4.573	0.374	3.056	0.437	2.515	0.557	1.877
	12	0.284	4.288	0.383	2.925	0.445	2.430	0.566	1.837
	13	0.291	4.060	0.390	2.818	0.453	2.358	0.573	1.802
24	14	0.297	3.872	0.397	2.728	0.460	2.299	0.580	1.772
	15	0.303	3.714	0.403	2.652	0.466	2.247	0.586	1.747
	16	0.308	3.581	0.409	2.585	0.472	2.202	0.591	1.724
	17	0.313	3.466	0.414	2.528	0.477	2.163	0.596	1.704
	18	0.317	3.367	0.419	2.478	0.482	2.129	0.600	1.686
	19	0.321	3.279	0.423	2.433	0.487	2.098	0.605	1.670
	20	0.325	3.202	0.427	2.393	0.491	2.070	0.608	1.656
	21	0.328	3.134	0.431	2.358	0.494	2.045	0.612	1.643
	22	0.331	3.072	0.434	2.325	0.498	2.023	0.615	1.632
	23	0.334	3.017	0.438	2.296	0.501	2.002	0.618	1.621
	24	0.337	2.967	0.441	2.269	0.504	1.984	0.621	1.611
	5	0.204	10.997	0.289	5.464	0.347	3.962	0.467	2.494
	6	0.222	8.441	0.311	4.594	0.370	3.464	0.491	2.296
	7	0.238	6.973	0.329	4.051	0.389	3.139	0.510	2.158
	8	0.251	6.035	0.344	3.680	0.405	2.910	0.526	2.059
	9	0.262	5.383	0.357	3.408	0.418	2.741	0.539	1.982
	10	0.272	4.907	0.368	3.203	0.430	2.610	0.550	1.921
	11	0.280	4.546	0.378	3.041	0.440	2.504	0.560	1.872
	12	0.288	4.263	0.386	2.911	0.449	2.419	0.568	1.831
	13	0.295	4.036	0.394	2.803	0.457	2.348	0.576	1.796
25	14	0.301	3.847	0.401	2.713	0.464	2.287	0.583	1.766
	15	0.307	3.690	0.407	2.637	0.470	2.236	0.589	1.740
	16	0.312	3.557	0.413	2.571	0.476	2.191	0.594	1.718
	17	0.317	3.443	0.418	2.514	0.481	2.152	0.599	1.697
	18	0.322	3.343	0.423	2.463	0.486	2.118	0.604	1.680
	19	0.326	3.256	0.428	2.419	0.491	2.087	0.608	1.664
	20	0.330	3.179	0.432	2.379	0.495	2.059	0.612	1.650
	21	0.333	3.111	0.436	2.343	0.499	2.034	0.615	1.637
	22	0.336	3.049	0.439	2.311	0.502	2.012	0.619	1.625
	23	0.339	2.994	0.442	2.281	0.505	1.991	0.622	1.614
	24	0.342	2.944	0.446	2.254	0.509	1.973	0.624	1.604
	25	0.345	2.898	0.448	2.230	0.511	1.955	0.627	1.595

Table V—(*continued*)

ν_1	ν_2	$\alpha = 0.01$ $\underline{F}$	$\alpha = 0.01$ $\bar{F}$	$\alpha = 0.05$ $\underline{F}$	$\alpha = 0.05$ $\bar{F}$	$\alpha = 0.10$ $\underline{F}$	$\alpha = 0.10$ $\bar{F}$	$\alpha = 0.25$ $\underline{F}$	$\alpha = 0.25$ $\bar{F}$
	5	0.207	10.961	0.291	5.448	0.349	3.950	0.469	2.489
	6	0.225	8.409	0.313	4.581	0.372	3.454	0.493	2.291
	7	0.240	6.943	0.331	4.038	0.391	3.128	0.512	2.154
	8	0.254	6.005	0.346	3.664	0.407	2.901	0.528	2.054
	9	0.265	5.358	0.359	3.395	0.421	2.732	0.541	1.976
	10	0.275	4.884	0.371	3.189	0.433	2.599	0.553	1.916
	11	0.284	4.523	0.381	3.029	0.443	2.495	0.562	1.866
	12	0.292	4.240	0.390	2.898	0.452	2.409	0.571	1.825
	13	0.299	4.013	0.397	2.790	0.460	2.338	0.579	1.790
	14	0.305	3.824	0.405	2.700	0.467	2.278	0.585	1.760
	15	0.311	3.668	0.411	2.624	0.474	2.226	0.592	1.734
26	16	0.317	3.535	0.417	2.558	0.479	2.182	0.597	1.712
	17	0.322	3.421	0.422	2.500	0.485	2.142	0.602	1.692
	18	0.326	3.321	0.427	2.450	0.490	2.108	0.607	1.674
	19	0.330	3.234	0.432	2.405	0.494	2.077	0.611	1.658
	20	0.334	3.157	0.436	2.365	0.499	2.049	0.615	1.643
	21	0.338	3.089	0.440	2.330	0.503	2.024	0.618	1.630
	22	0.341	3.028	0.443	2.297	0.506	2.002	0.622	1.618
	23	0.344	2.972	0.447	2.268	0.510	1.981	0.625	1.607
	24	0.347	2.922	0.450	2.241	0.513	1.962	0.628	1.597
	25	0.350	2.877	0.453	2.217	0.516	1.945	0.630	1.588
	26	0.353	2.835	0.456	2.194	0.518	1.929	0.633	1.580
	5	0.209	10.928	0.293	5.433	0.351	3.941	0.470	2.484
	6	0.227	8.379	0.315	4.569	0.374	3.445	0.494	2.286
	7	0.243	6.918	0.333	4.025	0.394	3.120	0.514	2.149
	8	0.256	5.981	0.349	3.652	0.410	2.892	0.530	2.048
	9	0.268	5.333	0.362	3.382	0.423	2.721	0.543	1.971
	10	0.278	4.861	0.374	3.176	0.435	2.590	0.555	1.911
	11	0.287	4.501	0.384	3.015	0.446	2.486	0.565	1.861
	12	0.295	4.219	0.393	2.885	0.455	2.400	0.573	1.819
	13	0.302	3.991	0.401	2.777	0.463	2.329	0.581	1.785
	14	0.309	3.804	0.408	2.688	0.470	2.268	0.588	1.755
27	15	0.315	3.648	0.415	2.611	0.477	2.217	0.594	1.729
	16	0.321	3.514	0.421	2.545	0.483	2.172	0.600	1.706
	17	0.326	3.400	0.426	2.488	0.488	2.133	0.605	1.686
	18	0.330	3.301	0.431	2.437	0.493	2.099	0.609	1.668
	19	0.335	3.214	0.436	2.393	0.498	2.067	0.614	1.652
	20	0.339	3.137	0.440	2.353	0.502	2.040	0.618	1.638
	21	0.342	3.069	0.444	2.318	0.506	2.015	0.621	1.625
	22	0.346	3.008	0.448	2.285	0.510	1.992	0.625	1.613
	23	0.349	2.953	0.451	2.255	0.513	1.972	0.628	1.602
	24	0.352	2.902	0.454	2.229	0.517	1.953	0.631	1.592
	25	0.355	2.857	0.457	2.204	0.520	1.935	0.634	1.583
	26	0.358	2.815	0.460	2.182	0.522	1.920	0.636	1.574
	27	0.360	2.777	0.463	2.161	0.525	1.905	0.638	1.566
28	5	0.210	10.898	0.295	5.420	0.353	3.932	0.472	2.480
	6	0.229	8.352	0.317	4.555	0.376	3.435	0.496	2.281
	7	0.245	6.896	0.336	4.011	0.396	3.112	0.516	2.145
	8	0.259	5.958	0.351	3.641	0.412	2.884	0.532	2.044
	9	0.271	5.311	0.365	3.370	0.426	2.713	0.545	1.967

Table V—(*continued*)

ν_1	ν_2	$\alpha = 0.01$ $\underline{F}$	$\alpha = 0.01$ $\bar{F}$	$\alpha = 0.05$ $\underline{F}$	$\alpha = 0.05$ $\bar{F}$	$\alpha = 0.10$ $\underline{F}$	$\alpha = 0.10$ $\bar{F}$	$\alpha = 0.25$ $\underline{F}$	$\alpha = 0.25$ $\bar{F}$
28	10	0.281	4.839	0.376	3.166	0.438	2.582	0.557	1.906
	11	0.290	4.482	0.387	3.004	0.448	2.477	0.567	1.856
	12	0.298	4.199	0.396	2.873	0.457	2.392	0.576	1.815
	13	0.306	3.971	0.404	2.767	0.466	2.320	0.583	1.780
	14	0.313	3.784	0.411	2.676	0.473	2.260	0.590	1.750
	15	0.319	3.628	0.418	2.600	0.480	2.209	0.597	1.724
	16	0.324	3.496	0.424	2.534	0.486	2.164	0.602	1.701
	17	0.329	3.382	0.429	2.476	0.492	2.124	0.607	1.681
	18	0.334	3.282	0.435	2.426	0.497	2.089	0.612	1.663
	19	0.339	3.196	0.439	2.382	0.501	2.059	0.616	1.647
	20	0.343	3.119	0.444	2.341	0.506	2.031	0.620	1.633
	21	0.346	3.050	0.448	2.306	0.510	2.006	0.624	1.619
	22	0.350	2.989	0.451	2.273	0.513	1.983	0.628	1.607
	23	0.353	2.934	0.455	2.244	0.517	1.963	0.631	1.596
	24	0.356	2.884	0.458	2.217	0.520	1.944	0.634	1.586
	25	0.359	2.838	0.461	2.192	0.523	1.927	0.637	1.577
	26	0.362	2.797	0.464	2.170	0.526	1.911	0.639	1.569
	27	0.365	2.758	0.467	2.149	0.529	1.896	0.642	1.561
	28	0.367	2.724	0.470	2.130	0.531	1.882	0.644	1.553
29	5	0.212	10.867	0.296	5.408	0.354	3.925	0.473	2.476
	6	0.231	8.330	0.319	4.543	0.378	3.428	0.498	2.277
	7	0.247	6.873	0.338	4.000	0.398	3.104	0.517	2.140
	8	0.261	5.937	0.353	3.630	0.414	2.875	0.533	2.040
	9	0.273	5.291	0.367	3.359	0.428	2.706	0.547	1.962
	10	0.284	4.820	0.379	3.154	0.440	2.574	0.559	1.901
	11	0.293	4.461	0.389	2.993	0.451	2.470	0.569	1.851
	12	0.301	4.181	0.399	2.863	0.460	2.383	0.578	1.810
	13	0.309	3.953	0.407	2.756	0.468	2.312	0.586	1.775
	14	0.316	3.766	0.414	2.666	0.476	2.252	0.593	1.745
	15	0.322	3.610	0.421	2.589	0.483	2.201	0.599	1.719
	16	0.328	3.478	0.427	2.523	0.489	2.156	0.605	1.697
	17	0.333	3.364	0.433	2.466	0.495	2.116	0.610	1.676
	18	0.338	3.264	0.438	2.415	0.500	2.081	0.615	1.658
	19	0.342	3.178	0.443	2.370	0.504	2.051	0.619	1.642
	20	0.347	3.101	0.447	2.331	0.509	2.023	0.623	1.627
	21	0.350	3.033	0.451	2.295	0.513	1.998	0.627	1.615
	22	0.354	2.972	0.455	2.263	0.517	1.975	0.630	1.603
	23	0.357	2.917	0.459	2.233	0.520	1.954	0.633	1.591
	24	0.361	2.867	0.462	2.206	0.524	1.935	0.636	1.582
	25	0.364	2.821	0.465	2.181	0.527	1.918	0.639	1.572
	26	0.366	2.779	0.468	2.159	0.530	1.902	0.642	1.564
	27	0.369	2.741	0.471	2.138	0.532	1.887	0.644	1.555
	28	0.372	2.706	0.473	2.119	0.535	1.874	0.647	1.548
	29	0.374	2.674	0.476	2.101	0.537	1.861	0.649	1.541
30	5	0.214	10.841	0.298	5.393	0.356	3.916	0.474	2.472
	6	0.233	8.308	0.321	4.532	0.380	3.421	0.499	2.273
	7	0.249	6.852	0.340	3.991	0.399	3.097	0.519	2.136
	8	0.263	5.918	0.356	3.618	0.416	2.868	0.535	2.036
	9	0.276	5.272	0.369	3.349	0.430	2.698	0.549	1.958
	10	0.286	4.801	0.381	3.144	0.442	2.567	0.561	1.897

Table V—(*continued*)

		α = 0.01		α = 0.05		α = 0.10		α = 0.25	
ν_1	ν_2	$\underline{F}$	$\bar{F}$	$\underline{F}$	$\bar{F}$	$\underline{F}$	$\bar{F}$	$\underline{F}$	$\bar{F}$
	11	0.296	4.444	0.392	2.983	0.453	2.462	0.571	1.847
	12	0.304	4.162	0.401	2.854	0.462	2.377	0.580	1.806
	13	0.312	3.935	0.410	2.746	0.471	2.305	0.588	1.771
	14	0.319	3.750	0.417	2.656	0.478	2.245	0.595	1.741
	15	0.325	3.594	0.424	2.579	0.485	2.193	0.601	1.715
	16	0.331	3.461	0.430	2.513	0.492	2.148	0.607	1.692
30	17	0.336	3.347	0.436	2.456	0.497	2.109	0.612	1.672
	18	0.341	3.248	0.441	2.405	0.503	2.074	0.617	1.654
	19	0.346	3.162	0.446	2.361	0.507	2.043	0.621	1.637
	20	0.350	3.085	0.450	2.321	0.512	2.015	0.625	1.623
	21	0.354	3.017	0.455	2.284	0.516	1.990	0.629	1.610
	22	0.358	2.955	0.459	2.252	0.520	1.968	0.633	1.598
	23	0.361	2.900	0.462	2.223	0.524	1.946	0.636	1.587
	24	0.365	2.851	0.466	2.196	0.527	1.928	0.639	1.577
	25	0.368	2.805	0.469	2.171	0.530	1.910	0.642	1.567
	26	0.370	2.764	0.472	2.149	0.533	1.894	0.645	1.559
	27	0.373	2.725	0.475	2.128	0.536	1.879	0.647	1.551
	28	0.376	2.690	0.477	2.109	0.538	1.866	0.649	1.543
	29	0.378	2.658	0.480	2.091	0.541	1.853	0.652	1.536
	30	0.381	2.628	0.482	2.074	0.543	1.841	0.654	1.529

Reproduced from G. C. Tiao and R. H. Lochner, "Tables for the Comparison of the Spread of Two Normal Distributions." Technical Report 88. Department of Statistics, the University of Wisconsin (1966).

Table VI

Percentage points of the F distribution

$$\alpha = \int_F^\infty p(F)\,dF \quad \text{where} \quad p(F) = \frac{1}{B(\nu_1/2, \nu_2/2)} (\nu_1/\nu_2)^{\nu_1/2} F^{(\nu_1/2)-1} \left(1 + \frac{\nu_1}{\nu_2} F\right)^{-(\nu_1+\nu_2)/2}$$

$\alpha = 0.5$

$\nu_2 \backslash \nu_1$	1	2	3	4	5	6	8	12	15	20	30	60	∞
1	1.00	1.50	1.71	1.82	1.89	1.94	2.00	2.07	2.09	2.12	2.15	2.17	2.20
2	0.667	1.00	1.13	1.21	1.25	1.28	1.32	1.36	1.38	1.39	1.41	1.43	1.44
3	0.585	0.881	1.00	1.06	1.10	1.13	1.16	1.20	1.21	1.23	1.24	1.25	1.27
4	0.549	0.828	0.941	1.00	1.04	1.06	1.09	1.13	1.14	1.15	1.16	1.18	1.19
5	0.528	0.799	0.907	0.965	1.00	1.02	1.05	1.09	1.10	1.11	1.12	1.14	1.15
6	0.515	0.780	0.886	0.942	0.977	1.00	1.03	1.06	1.07	1.08	1.10	1.11	1.12
7	0.506	0.767	0.871	0.926	0.960	0.983	1.01	1.04	1.05	1.07	1.08	1.09	1.10
8	0.499	0.757	0.860	0.915	0.948	0.971	1.00	1.03	1.04	1.05	1.07	1.08	1.09
9	0.494	0.749	0.852	0.906	0.939	0.962	0.990	1.02	1.03	1.04	1.05	1.07	1.08
10	0.490	0.743	0.845	0.899	0.932	0.954	0.983	1.01	1.02	1.03	1.05	1.06	1.07
11	0.486	0.739	0.840	0.893	0.926	0.948	0.977	1.01	1.02	1.03	1.04	1.05	1.06
12	0.484	0.735	0.835	0.888	0.921	0.943	0.972	1.00	1.01	1.02	1.03	1.05	1.06
13	0.481	0.731	0.832	0.885	0.917	0.939	0.967	0.996	1.01	1.02	1.03	1.04	1.05
14	0.479	0.729	0.828	0.881	0.914	0.936	0.964	0.992	1.00	1.01	1.03	1.04	1.05
15	0.478	0.726	0.826	0.878	0.911	0.933	0.960	0.989	1.00	1.01	1.02	1.03	1.05
16	0.476	0.724	0.823	0.876	0.908	0.930	0.958	0.986	0.997	1.01	1.02	1.03	1.04
17	0.475	0.722	0.821	0.874	0.906	0.928	0.955	0.983	0.995	1.01	1.02	1.03	1.04
18	0.474	0.721	0.819	0.872	0.904	0.926	0.953	0.981	0.992	1.00	1.02	1.03	1.04
19	0.473	0.719	0.818	0.870	0.902	0.924	0.951	0.979	0.990	1.00	1.01	1.02	1.04
20	0.472	0.718	0.816	0.868	0.900	0.922	0.950	0.977	0.989	1.00	1.01	1.02	1.03
21	0.471	0.716	0.815	0.867	0.899	0.921	0.948	0.976	0.987	0.998	1.01	1.02	1.03
22	0.470	0.715	0.814	0.866	0.898	0.919	0.947	0.974	0.986	0.997	1.01	1.02	1.03
23	0.470	0.714	0.813	0.864	0.896	0.918	0.945	0.973	0.984	0.996	1.01	1.02	1.03
24	0.469	0.714	0.812	0.863	0.895	0.917	0.944	0.972	0.983	0.994	1.01	1.02	1.03
25	0.468	0.713	0.811	0.862	0.894	0.916	0.943	0.971	0.982	0.993	1.00	1.02	1.03
26	0.468	0.712	0.810	0.861	0.893	0.915	0.942	0.970	0.981	0.992	1.00	1.01	1.03
27	0.467	0.711	0.809	0.861	0.892	0.914	0.941	0.969	0.980	0.991	1.00	1.01	1.03
28	0.467	0.711	0.808	0.860	0.892	0.913	0.940	0.968	0.979	0.990	1.00	1.01	1.02
29	0.466	0.710	0.808	0.859	0.891	0.912	0.940	0.967	0.978	0.990	1.00	1.01	1.02
30	0.466	0.709	0.807	0.858	0.890	0.912	0.939	0.966	0.978	0.989	1.00	1.01	1.02
40	0.463	0.705	0.802	0.854	0.885	0.907	0.934	0.961	0.972	0.983	0.994	1.01	1.02
60	0.461	0.701	0.798	0.849	0.880	0.901	0.928	0.956	0.967	0.978	0.989	1.00	1.01
120	0.458	0.697	0.793	0.844	0.875	0.896	0.923	0.950	0.961	0.972	0.983	0.994	1.01
	[illegible]	0.693	0.789	0.839	0.870	0.891	0.918	0.945	0.956	0.967	0.978	0.989	1.00

Table VI (continued)

$\alpha = 0.25$

$\nu_2 \backslash \nu_1$	1	2	3	4	5	6	8	12	15	20	30	60	∞
1	5.83	7.50	8.20	8.58	8.82	8.98	9.19	9.41	9.49	9.58	9.67	9.76	9.85
2	2.57	3.00	3.15	3.23	3.28	3.31	3.35	3.39	3.41	3.43	3.44	3.46	3.48
3	2.02	2.28	2.36	2.39	2.41	2.42	2.44	2.45	2.46	2.46	2.47	2.47	2.47
4	1.81	2.00	2.05	2.06	2.07	2.08	2.08	2.08	2.08	2.08	2.08	2.08	2.08
5	1.69	1.85	1.88	1.89	1.89	1.89	1.89	1.89	1.89	1.88	1.88	1.87	1.87
6	1.62	1.76	1.78	1.79	1.79	1.78	1.78	1.77	1.76	1.76	1.75	1.74	1.74
7	1.57	1.70	1.72	1.72	1.71	1.71	1.70	1.68	1.68	1.67	1.66	1.65	1.65
8	1.54	1.66	1.67	1.66	1.66	1.65	1.64	1.62	1.62	1.61	1.60	1.59	1.58
9	1.51	1.62	1.63	1.63	1.62	1.61	1.60	1.58	1.57	1.56	1.55	1.54	1.53
10	1.49	1.60	1.60	1.59	1.59	1.58	1.56	1.54	1.53	1.52	1.51	1.50	1.48
11	1.47	1.58	1.58	1.57	1.56	1.55	1.53	1.51	1.50	1.49	1.48	1.47	1.45
12	1.46	1.56	1.56	1.55	1.54	1.53	1.51	1.49	1.48	1.47	1.45	1.44	1.42
13	1.45	1.55	1.55	1.53	1.52	1.51	1.49	1.47	1.46	1.45	1.43	1.42	1.40
14	1.44	1.53	1.53	1.52	1.51	1.50	1.48	1.45	1.44	1.43	1.41	1.40	1.38
15	1.43	1.52	1.52	1.51	1.49	1.48	1.46	1.44	1.43	1.41	1.40	1.38	1.36
16	1.42	1.51	1.51	1.50	1.48	1,47	1.45	1.43	1.41	1.40	1.38	1.36	1.34
17	1.42	1.51	1.50	1.49	1.47	1.46	1.44	1.41	1.40	1.39	1.37	1.35	1.33
18	1.41	1.50	1.49	1.48	1.46	1.45	1.43	1.40	1.39	1.38	1.36	1.34	1.32
19	1.41	1.49	1.49	1.47	1.46	1.44	1.42	1.40	1.38	1.37	1.35	1.33	1.30
20	1.40	1.49	1.48	1.47	1.45	1.44	1.42	1.39	1.37	1.36	1.34	1.32	1.29
21	1.40	1.48	1.48	1.46	1.44	1.43	1.41	1.38	1.37	1.35	1.33	1.31	1.28
22	1.40	1.48	1.47	1.45	1.44	1.42	1.40	1.37	1.36	1.34	1.32	1.30	1.28
23	1.39	1.47	1.47	1.45	1.43	1.42	1.40	1.37	1.35	1.34	1.32	1.30	1.27
24	1.39	1.47	1.46	1.44	1.43	1.41	1.39	1.36	1.35	1.33	1.31	1.29	1.26
25	1.39	1.47	1.46	1.44	1.42	1.41	1.39	1.36	1.34	1.33	1.31	1.28	1.25
26	1.38	1.46	1.45	1.44	1.42	1.41	1.38	1.35	1.34	1.32	1.30	1.28	1.25
27	1.38	1.46	1.45	1.43	1.42	1.40	1.38	1.35	1.33	1.32	1.30	1.27	1.24
28	1.38	1.46	1.45	1.43	1.41	1.40	1.38	1.34	1.33	1.31	1.29	1.27	1.24
29	1.38	1.45	1.45	1.43	1.41	1.40	1.37	1.34	1.32	1.31	1.29	1.26	1.23
30	1.38	1.45	1.44	1.42	1.41	1.39	1.37	1.34	1.32	1.30	1.28	1.26	1.23
40	1.36	1.44	1.42	1.40	1.39	1.37	1.35	1.31	1.30	1.28	1.25	1.22	1.19
60	1.35	1.42	1.41	1.38	1.37	1.35	1.32	1.29	1.27	1.25	1.22	1.19	1.15
120	1.34	1.40	1.39	1.37	1.35	1.33	1.30	1.26	1.24	1.22	1.19	1.16	1.10
∞	1.32	1.39	1.37	1.35	1.33	1.31	1.28	1.24	1.22	1.19	1.16	1.12	1.00

Table VI—(*continued*)

$\alpha = 0.1$

$\nu_2 \backslash \nu_1$	1	2	3	4	5	6	8	12	15	20	30	60	∞
1	39.86	49.50	53.59	55.83	57.24	58.20	59.44	60.71	61.22	61.74	62.26	62.79	63.33
2	8.53	9.00	9.16	9.24	9.29	9.33	9.37	9.41	9.42	9.44	9.46	9.47	9.49
3	5.54	5.46	5.39	5.34	5.31	5.28	5.25	5.22	5.20	5.18	5.17	5.15	5.13
4	4.54	4.32	4.19	4.11	4.05	4.01	3.95	3.90	3.87	3.84	3.82	3.79	3.76
5	4.06	3.78	3.62	3.52	3.45	3.40	3.34	3.27	3.24	3.21	3.17	3.14	3.10
6	3.78	3.46	3.29	3.18	3.11	3.05	2.98	2.90	2.87	2.84	2.80	2.76	2.72
7	3.59	3.26	3.07	2.96	2.88	2.83	2.75	2.67	2.63	2.59	2.56	2.51	2.47
8	3.46	3.11	2.92	2.81	2.73	2.67	2.59	2.50	2.46	2.42	2.38	2.34	2.29
9	3.36	3.01	2.81	2.69	2.61	2.55	2.47	2.38	2.34	2.30	2.25	2.21	2.16
10	3.29	2.92	2.73	2.61	2.52	2.46	2.38	2.28	2.24	2.20	2.16	2.11	2.06
11	3.23	2.86	2.66	2.54	2.45	2.39	2.30	2.21	2.17	2.12	2.08	2.03	1.97
12	3.18	2.81	2.61	2.48	2.39	2.33	2.24	2.15	2.10	2.06	2.01	1.96	1.90
13	3.14	2.76	2.56	2.43	2.35	2.28	2.20	2.10	2.05	2.01	1.96	1.90	1.85
14	3.10	2.73	2.52	2.39	2.31	2.24	2.15	2.05	2.01	1.96	1.91	1.86	1.80
15	3.07	2.70	2.49	2.36	2.27	2.21	2.12	2.02	1.97	1.92	1.87	1.82	1.76
16	3.05	2.67	2.46	2.33	2.24	2.18	2.09	1.99	1.94	1.89	1.84	1.78	1.72
17	3.03	2.64	2.44	2.31	2.22	2.15	2.06	1.96	1.91	1.86	1.81	1.75	1.69
18	3.01	2.62	2.42	2.29	2.20	2.13	2.04	1.93	1.89	1.84	1.78	1.72	1.66
19	2.99	2.61	2.40	2.27	2.18	2.11	2.02	1.91	1.86	1.81	1.76	1.70	1.63
20	2.97	2.59	2.38	2.25	2.16	2.09	2.00	1.89	1.84	1.79	1.74	1.68	1.61
21	2.96	2.57	2.36	2.23	2.14	2.08	1.98	1.87	1.83	1.78	1.72	1.66	1.59
22	2.95	2.56	2.35	2.22	2.13	2.06	1.97	1.86	1.81	1.76	1.70	1.64	1.57
23	2.94	2.55	2.34	2.21	2.11	2.05	1.95	1.84	1.80	1.74	1.69	1.62	1.55
24	2.93	2.54	2.33	2.19	2.10	2.04	1.94	1.83	1.78	1.73	1.67	1.61	1.53
25	2.92	2.53	2.32	2.18	2.09	2.02	1.93	1.82	1.77	1.72	1.66	1.59	1.52
26	2.91	2.52	2.31	2.17	2.08	2.01	1.92	1.81	1.76	1.71	1.65	1.58	1.50
27	2.90	2.51	2.30	2.17	2.07	2.00	1.91	1.80	1.75	1.70	1.64	1.57	1.49
28	2.89	2.50	2.29	2.16	2.06	2.00	1.90	1.79	1.74	1.69	1.63	1.56	1.48
29	2.89	2.50	2.28	2.15	2.06	1.99	1.89	1.78	1.73	1.68	1.62	1.55	1.47
30	2.88	2.49	2.28	2.14	2.05	1.98	1.88	1.77	1.72	1.67	1.61	1.54	1.46
40	2.84	2.44	2.23	2.09	2.00	1.93	1.83	1.71	1.66	1.61	1.54	1.47	1.38
60	2.79	2.39	2.18	2.04	1.95	1.87	1.77	1.66	1.60	1.54	1.48	1.40	1.29
120	2.75	2.35	2.13	1.99	1.90	1.82	1.72	1.60	1.55	1.48	1.41	1.32	1.19
∞	2.71	2.30	2.08	1.94	1.85	1.77	1.67	1.55	1.49	1.42	1.34	1.24	1.00

Table VI—(continued)

$\alpha = 0.05$

$\nu_2 \backslash \nu_1$	1	2	3	4	5	6	8	12	15	20	30	60	∞
1	161.4	199.5	215.7	224.6	230.2	234.0	238.9	243.9	245.9	248.0	250.1	252.2	254.3
2	18.51	19.00	19.16	19.25	19.30	19.33	19.37	19.41	19.43	19.45	19.46	19.48	19.50
3	10.13	9.55	9.28	9.12	9.01	8.94	8.85	8.74	8.70	8.66	8.62	8.57	8.53
4	7.71	6.94	6.59	6.39	6.26	6.16	6.04	5.91	5.86	5.80	5.75	5.69	5.63
5	6.61	5.79	5.41	5.19	5.05	4.95	4.82	4.68	4.62	4.56	4.50	4.43	4.36
6	5.99	5.14	4.76	4.53	4.39	4.28	4.15	4.00	3.94	3.87	3.81	3.74	3.67
7	5.59	4.74	4.35	4.12	3.97	3.87	3.73	3.57	3.51	3.44	3.38	3.30	3.23
8	5.32	4.46	4.07	3.84	3.69	3.58	3.44	3.28	3.22	3.15	3.08	3.01	2.93
9	5.12	4.26	3.86	3.63	3.48	3.37	3.23	3.07	3.01	2.94	2.86	2.79	2.71
10	4.96	4.10	3.71	3.48	3.33	3.22	3.07	2.91	2.85	2.77	2.70	2.62	2.54
11	4.84	3.98	3.59	3.36	3.20	3.09	2.95	2.79	2.72	2.65	2.57	2.49	2.40
12	4.75	3.89	3.49	3.26	3.11	3.00	2.85	2.69	2.62	2.54	2.47	2.38	2.30
13	4.67	3.81	3.41	3.18	3.03	2.92	2.77	2.60	2.53	2.46	2.38	2.30	2.21
14	4.60	3.74	3.34	3.11	2.96	2.85	2.70	2.53	2.46	2.39	2.31	2.22	2.13
15	4.54	3.68	3.29	3.06	2.90	2.79	2.64	2.48	2.40	2.33	2.25	2.16	2.07
16	4.49	3.63	3.24	3.01	2.85	2.74	2.59	2.42	2.35	2.28	2.19	2.11	2.01
17	4.45	3.59	3.20	2.96	2.81	2.70	2.55	2.38	2.31	2.23	2.15	2.06	1.96
18	4.41	3.55	3.16	2.93	2.77	2.66	2.51	2.34	2.27	2.19	2.11	2.02	1.92
19	4.38	3.52	3.13	2.90	2.74	2.63	2.48	2.31	2.23	2.16	2.07	1.98	1.88
20	4.35	3.49	3.10	2.87	2.71	2.60	2.45	2.28	2.20	2.12	2.04	1.95	1.84
21	4.32	3.47	3.07	2.84	2.68	2.57	2.42	2.25	2.18	2.10	2.01	1.92	1.81
22	4.30	3.44	3.05	2.82	2.66	2.55	2.40	2.23	2.15	2.07	1.98	1.89	1.78
23	4.28	3.42	3.03	2.80	2.64	2.53	2.37	2.20	2.13	2.05	1.96	1.86	1.76
24	4.26	3.40	3.01	2.78	2.62	2.51	2.36	2.18	2.11	2.03	1.94	1.84	1.73
25	4.24	3.39	2.99	2.76	2.60	2.49	2.34	2.16	2.09	2.01	1.92	1.82	1.71
26	4.23	3.37	2.98	2.74	2.59	2.47	2.32	2.15	2.07	1.99	1.90	1.80	1.69
27	4.21	3.35	2.96	2.73	2.57	2.46	2.31	2.13	2.06	1.97	1.88	1.79	1.67
28	4.20	3.34	2.95	2.71	2.56	2.45	2.29	2.12	2.04	1.96	1.87	1.77	1.65
29	4.18	3.33	2.93	2.70	2.55	2.43	2.28	2.10	2.03	1.94	1.85	1.75	1.64
30	4.17	3.32	2.92	2.69	2.53	2.42	2.27	2.09	2.01	1.93	1.84	1.74	1.62
40	4.08	3.23	2.84	2.61	2.45	2.34	2.18	2.00	1.92	1.84	1.74	1.64	1.51
60	4.00	3.15	2.76	2.53	2.37	2.25	2.10	1.92	1.84	1.75	1.65	1.53	1.39
120	3.92	3.07	2.68	2.45	2.29	2.17	2.02	1.83	1.75	1.66	1.55	1.43	1.25
∞	3.84	3.00	2.60	2.37	2.21	2.10	1.94	1.75	1.67	1.57	1.46	1.32	1.00

Table VI—(*continued*)

$\alpha = 0.01$

$\nu_2 \backslash \nu_1$	1	2	3	4	5	6	8	12	15	20	30	60	∞
1	4052	4999.5	5403	5625	5764	5859	5982	6106	6157	6209	6261	6313	6366
2	98.50	99.00	99.17	99.25	99.30	99.33	99.37	99.42	99.43	99.45	99.47	99.48	99.50
3	34.12	30.82	29.46	28.71	28.24	27.91	27.49	27.05	26.87	26.69	26.50	26.32	26.13
4	21.20	18.00	16.69	15.98	15.52	15.21	14.80	14.37	14.20	14.02	13.84	13.65	13.46
5	16.26	13.27	12.06	11.39	10.97	10.67	10.29	9.89	9.72	9.55	9.38	9.20	9.02
6	13.75	10.92	9.78	9.15	8.75	8.47	8.10	7.72	7.56	7.40	7.23	7.06	6.88
7	12.25	9.55	8.45	7.85	7.46	7.19	6.84	6.47	6.31	6.16	5.99	5.82	5.65
8	11.26	8.65	7.59	7.01	6.63	6.37	6.03	5.67	5.52	5.36	5.20	5.03	4.86
9	10.56	8.02	6.99	6.42	6.06	5.80	5.47	5.11	4.96	4.81	4.65	4.48	4.31
10	10.04	7.56	6.55	5.99	5.64	5.39	5.06	4.71	4.56	4.41	4.25	4.08	3.91
11	9.65	7.21	6.22	5.67	5.32	5.07	4.74	4.40	4.25	4.10	3.94	3.78	3.60
12	9.33	6.93	5.95	5.41	5.06	4.82	4.50	4.16	4.01	3.86	3.70	3.54	3.36
13	9.07	6.70	5.74	5.21	4.86	4.62	4.30	3.96	3.82	3.66	3.51	3.34	3.17
14	8.86	6.51	5.56	5.04	4.69	4.46	4.14	3.80	3.66	3.51	3.35	3.18	3.00
15	8.68	6.36	5.42	4.89	4.56	4.32	4.00	3.67	3.52	3.37	3.21	3.05	2.87
16	8.53	6.23	5.29	4.77	4.44	4.20	3.89	3.55	3.41	3.26	3.10	2.93	2.75
17	8.40	6.11	5.18	4.67	4.34	4.10	3.79	3.46	3.31	3.16	3.00	2.83	2.65
18	8.29	6.01	5.09	4.58	4.25	4.01	3.71	3.37	3.23	3.08	2.92	2.75	2.57
19	8.18	5.93	5.01	4.50	4.17	3.94	3.63	3.30	3.15	3.00	2.84	2.67	2.49
20	8.10	5.85	4.94	4.43	4.10	3.87	3.56	3.23	3.09	2.94	2.78	2.61	2.42
21	8.02	5.78	4.87	4.37	4.04	3.81	3.51	3.17	3.03	2.88	2.72	2.55	2.36
22	7.95	5.72	4.82	4.31	3.99	3.76	3.45	3.12	2.98	2.83	2.67	2.50	2.31
23	7.88	5.66	4.76	4.26	3.94	3.71	3.41	3.07	2.93	2.78	2.62	2.45	2.26
24	7.82	5.61	4.72	4.22	3.90	3.67	3.36	3.03	2.89	2.74	2.58	2.40	2.21
25	7.77	5.57	4.68	4.18	3.85	3.63	3.32	2.99	2.85	2.70	2.54	2.36	2.17
26	7.72	5.53	4.64	4.14	3.82	3.59	3.29	2.96	2.81	2.66	2.50	2.33	2.13
27	7.68	5.49	4.60	4.11	3.78	3.56	3.26	2.93	2.78	2.63	2.47	2.29	2.10
28	7.64	5.45	4.57	4.07	3.75	3.53	3.23	2.90	2.75	2.60	2.44	2.26	2.06
29	7.60	5.42	4.54	4.04	3.73	3.50	3.20	2.87	2.73	2.57	2.41	2.23	2.03
30	7.56	5.39	4.51	4.02	3.70	3.47	3.17	2.84	2.70	2.55	2.39	2.21	2.01
40	7.31	5.18	4.31	3.83	3.51	3.29	2.99	2.66	2.52	2.37	2.20	2.02	1.80
60	7.08	4.98	4.13	3.65	3.34	3.12	2.82	2.50	2.35	2.20	2.03	1.84	1.60
120	6.85	4.79	3.95	3.48	3.17	2.96	2.66	2.34	2.19	2.03	1.86	1.66	1.38
∞	6.63	4.61	3.78	3.32	3.02	2.80	2.51	2.18	2.04	1.88	1.70	1.47	1.00

... E. S. Pearson and H. O. Hartley (editors), *Biometrika Tables for Statisticians*, Vol. 1, Cambridge University Press, Cambridge

REFERENCES

PRINCIPAL SOURCE REFERENCES

The following papers formed the original basis for portions of the chapters indicated, and are not specifically referenced in the text.

Chapter 2

Box, G. E. P., and Tiao, G. C. (1965), "Multiparameter Problems from a Bayesian Viewpoint," *Ann. Math. Statist.* **36,** 1468

Chapter 3

Box, G. E. P., and Tiao, G. C. (1962), "A further Look at Robustness via Bayes's Theorem," *Biometrika* **49,** 419

Chapter 4

Box, G. E. P., and Tiao, G. C. (1964a), "A Bayesian Approach to the Importance of Assumptions Applied to the Comparison of Variances," *Biometrika* **51,** 153

Box, G. E. P., and Tiao, G. C. (1964b), "A Note on Criterion Robustness and Inference Robustness," *Biometrika* **51,** 169

Chapter 5

Tiao, G. C., and Tan, W. Y. (1965), "Bayesian Analysis of Random-Effect Models in the Analysis of Variance. I. Posterior Distribution of Variance Components," *Biometrika* **52,** 37

Tiao, G. C., and Box, G. E. P. (1967), "Bayesian Analysis of a Three-Component Hierarchical Design Model," *Biometrika* **54,** 109

Chapter 6

Tiao, G. C. (1966), "Bayesian Comparison of Means of a Mixed Model with Application to Regression Analysis," *Biometrika* **53,** 11

Chapter 7

Box, G. E. P., and Tiao, G. C. (1968a), "Bayesian Estimation of Means for the Random-Effect Model," *J. Amer. Statist. Assoc.* **63,** 174

Tiao, G. C., and Draper, N. R. (1968), "Bayesian Analysis of Linear Models with Two Random Components, with Special Reference to the Balanced Incomplete Block Design," *Biometrika* **55,** 101

Chapter 8

Tiao, G. C., and Zellner, A. (1964a), "On the Bayesian Estimation of Multivariate Regression," *J. Roy. Statist. Soc., Series B* **26,** 277

Box, G. E. P., and Draper, N. R. (1965), "The Bayesian Estimation of Common Parameters from Several Responses," *Biometrika* **52,** 355

Chapter 9

Tiao, G. C., and Zellner, A. (1964b), "Bayes's Theorem and the Use of Prior Knowledge in Regression Analysis," *Biometrika* **51,** 219

Chapter 10

Box, G. E. P., and Cox, D. R. (1964), "An Analysis of Transformations," *J. Roy. Statist. Soc., Series B* **26,** 211

GENERAL REFERENCES

Afonja, B. (1970), "Some Bayesian Considerations of the Analysis and Choice of a Class of Designs," Ph.D. thesis, the University of Wisconsin, Madison

Ali, M. M. (1969), "Some Aspects of the One-Way Random-Effects Model and the Linear Regression Model with Two Random Components," Ph.D. thesis, the University of Wisconsin, Madison

Anderson, T. W. (1958), *An Introduction to Multivariate Statistical Analysis,* New York: Wiley

Ando, A. and Kaufman, G. M. (1965), "Bayesian Analysis of the Independent Multinormal Process—Neither Mean nor Precision Known," *J. Amer. Statist. Assoc.* **60,** 347

Anscombe, F. J. (1948a), "The Transformation of Poisson, Binomial and Negative-Binomial Data," *Biometrika* **35,** 246

Anscombe, F. J. (1948b), "Contributions to the Discussion on D. G. Champernowne's Sampling Theory Applied to Autoregressive Sequences," *J. Roy. Statist. Soc., Series B* **10,** 239

Anscombe, F. J. (1961), "Examination of Residuals," *Proceedings of* 4*th Berkeley Symp. Math. Statist. Proc.* **1,** 1

Anscombe, F. J. (1963), "Bayesian Inference Concerning Many Parameters with Reference to Supersaturated Designs," *Bulletin Int. Stat. Inst.* **40–42,** 721

Anscombe, F. J. and Tukey, J. W. (1963), "The Examination and Analysis of Residuals," *Technometrics* **5,** 141

Barnard, G. A. (1947), "Significance Tests for 2×2 Tables," *Biometrika* **34,** 123

Barnard, G. A. (1949), "Statistical Inference," *J. Roy. Statist. Soc., Series B* **11,** 115

Barnard, G. A. (1954), "Sampling Inspection and Statistical Decisions," *J. Roy. Statist. Soc., Series B* **16,** 151

Barnard, G. A., Jenkins, G. M., and Winsten, C. B. (1962), "Likelihood Inference and Time Series," *J. Roy. Statist. Soc., Series A* **125,** 321

Bartlett, M. S. (1936), "The Square Root Transformation in Analysis of Variance," *Suppl. J. Roy. Statist. Soc.,* **8,** 27, 85

Bartlett, M. S. (1937), "Properties of Sufficiency and Statistical Test," *Proc. Roy. Soc., Series A* **160,** 268

Bartlett, M. S. (1938), "Further Aspects of the Theory of Multiple Regression," *Proc. Camb. Phil. Soc.* **34,** 33.

Bayes, T. R. (1763), "An Essay Towards Solving a Problem in the Doctrine of Chances," *Phil. Trans. Roy. Soc. London* **53,** 370 (reprinted in *Biometrika* (1958), **45,** 293)

Beale, E. M. L. (1960), "Confidence Regions in Nonlinear Estimation," *J. Roy. Statist. Soc., Series B* **22,** 41

Behrens, W. V. (1929), ((Ein Beitrag zur Fehlerberechnung bei Weniger Beobachtungen," *Landw. Jb.* **68,** 807

Birnbaum, A. (1962), "On the Foundation of Statistical Inference," *J. Amer. Statist. Assoc.* **57,** 269

Boot, J. C. G., and De Witt, G. M. (1960), "Investment Demand: An Empirical Contribution to the Aggregation Problem," *Intern. Econ. Review* **1,** 3

Box, G. E. P. (1949), "A General Distribution Theory for a class of Likelihood Criteria," *Biometrika* **36,** 317

Box, G. E. P. (1953a), "Non-normality and Tests on Variances," *Biometrika* **40,** 318

Box, G. E. P. (1953b), "A Note on Regions for Test of Kurtosis," *Biometrika* **40,** 465

Box, G. E. P. (1954), "Some Theorems on Quadratic Forms Applied in the Study of Analysis of Variance Problems: II. Effects of Inequality of Variance and of Correlation Between Errors in the Two-Way Classification," *Ann. Math. Statist.* **25,** 484

Box, G. E. P. (1957), "Use of Statistical Methods in the Elucidation of Basic Mechanisms," *Bull. Int. Stat. Inst.* **36,** 215

Box, G. E. P. (1960), "Fitting Empirical Data," *Ann. New York Academy of Sciences* **86,** 792

Box, G. E. P., and Andersen, S. L. (1955), "Permutation Theory in the Derivation of Robust Criteria and the Study of Departures from Assumptions," *J. Roy. Statist. Soc., Series B* **17,** 1

Box, G. E. P., and Jenkins, G. M. (1970), *Time Series Analysis, Forecasting and Control*, San Francisco: Holden–Day

Box, G. E. P., Erjavec, J., Hunter, W. G. and MacGregor, J. F. (1972), "Some Problems Associated with the Analysis of Multiresponse Data" (to appear in *Technometrics*)

Box, G. E. P., and Tiao, G. C. (1968b) "A Bayesian Approach to Some Outlier Problems," *Biometrika* **55,** 119

Box, G. E. P., and Watson, G. S. (1962), "Robustness to Non-Normality of Regression Tests," *Biometrika* **49,** 93

Bracken, J., and Schleifer, A. (1964), *Tables for Normal Sampling with Unknown Variance,* Cambridge, Harvard University Press

Brookner, R. J., and Wald, A. (1941), "On the Distribution of Wilks' Statistic for Testing the Independence of Several Groups of Variates," *Ann. Math. Statist.* **12,** 137

Bulmer, M. G. (1957), "Approximate Confidence Limits for Components of Variance," *Biometrika* **44,** 159

Bush, N., and Anderson, R. L. (1963), "A Comparison of Three Different Procedures for Estimating Variance Components," *Technometrics* **5,** 421

Carlton, G. A. (1946), "Estimating the Parameters of a Rectangular Distribution," *Ann. Math. Statist.* **17,** 355

Cochran, W. G. (1934), "The Distribution of Quadratic Forms in a Normal System, with Applications to the Analysis of Covariance," *Proc. Camb. Phil. Soc.* **30,** 178

Cochran, W. G., and Cox, G. M. (1950), *Experimental Designs,* second edition, New York: Wiley

Cook, M. B. (1951), "Bivariate κ-Statistics and Cumulants of their Joint Sampling Distribution," *Biometrika* **38,** 179

Cornish, E. A. (1954), "The Multivariate *t*-Distribution Associated with a Set of Normal Sample Deviates," *Aust. J. Phys.* **7,** 531

Crump, S. L. (1946), "Estimation of Variance Components in the Analysis of Variance," *Biometrics* **2,** 7

Daniel, C. (1959), "Use of Half Normal Plots in Interpreting Factorial Experiments," *Technometrics* **1,** 311

Daniels, H. E. (1939), "The Estimation of Components of Variance," *J. Roy. Statist. Soc., Supplement* **6,** 186

Davies, O. L. (editor), (1949), *Statistical Methods in Research and Production*, second edition, London: Oliver and Boyd

Davies, O. L. (editor), (1967), *Statistical Methods in Research and Production*, third edition, London: Oliver and Boyd

De Bruijn, N. G. (1961), *Asymptotic Methods in Analysis,* Amsterdam: North-Holland

Deemer, W. L., and Olkin, I. (1951), "The Jacobians of Certain Matrix Transformations Useful in Multivariate Analysis, Based on Lectures by P. L. Hsu," *Biometrika* **38,** 345

De Finetti, B. (1937), "La Prévision: Ses Lois Logiques, ses Sources Subjectives," *Ann. Inst. H. Poincaré* **7,** 1. English translation in *Studies in Subjective Probability,* H. E. Kyburg, Jr. and H. G. Smokler (editors), 1964, New York: Wiley

De Groot, M. H. (1970), *Optimal Statistical Decisions,* New York: McGraw–Hill

Dempster, A. P. (1963), "On a Paradox Concerning Inference about a Covariance Matrix," *Ann. Math. Statist.* **34,** 1414

Diananda, P. H. (1949), "Note on Some Properties of Maximum Likelihood Estimates," *Proc. Camb. Phil. Soc.* **45,** 536

Dickey, J. M. (1967a), "Expansions of *t* Densities and Related Complete Integrals," *Ann. Math. Statist.* **38,** 503

Dickey, J. M. (1967b), "Matric-Variate Generalizations of the Multivariate *t* Distribution and the Inverted Multivariate *t* Distribution," *Ann. Math. Statist.* **38,** 511

Dickey, J. M. (1968), "Three Multidimensional-Integral Identities with Bayesian Applications," *Ann. Math. Statist.* **39,** 1615

Dreze, J. H., and Morales, J. A. (1970), "Bayesian Full Information Analysis of the Simultaneous Equation Model," *Center for Operations Research and Econometrics*, Discussion Paper No. 7031, Université Catholique de Louvain

Dunnett, C. W., and Sobel, M. (1954), "A Bivariate Generalization of Student's *t*-Distribution With Tables for Certain Special Cases," *Biometrika* **41,** 153

Edwards, W., Lindman, H., and Savage, L. J. (1963), "Bayesian Statistical Inference for Psychological Research," *Psychological Rev.* **70,** 193

Eisenhart, C. (1947), "The Assumptions Underlying the Analysis of Variance," *Biometrics* **3,** 1

Federer, W. T. (1955), *Experimental Design*, New York: Macmillan

Fisher, R. A. (1915), "Frequency Distribution of the Value of the Correlation Coefficient in Samples from an Indefinitely Large Population," *Biometrika* **10,** 507

Fisher, R. A. (1921), "On the 'Probable Error' of a Coefficient of Correlation Deduced from a Small Sample," *Metron* **1,** 3

Fisher, R. A. (1922), "On the Mathematical Foundations of Theoretical Statistics," *Phil. Trans. Roy. Soc., Series A* **222,** 309

Fisher, R. A. (1924), "On a Distribution Yielding the Error Function of Several Well-Known Statistics," *Proc. Int. Math. Congress, Toronto*, 805

Fisher, R. A. (1925), "Theory of Statistical Estimation," *Proc. Camb. Phil. Soc.* **22,** 700

Fisher, R. A. (1930), "Inverse Probability," *Proc. Camb. Phil. Soc.* **26,** 528

Fisher, R. A. (1935), "The Fiducial Argument in Statistical Inference," *Ann. Eugen.* **6,** 391

Fisher, R. A. (1939), "The Comparison of Samples with Possibly Unequal Variances," *Ann. Eugen.* **9,** 174

Fisher, R. A. (1959), *Statistical Methods and Scientific Inference*, second edition, London: Oliver and Boyd

Fisher, R. A. (1960), *The Design of Experiments*, seventh edition, New York: Hafner

Fisher, R. A. (1961a), "Sampling the Reference Set," *Sankhyá, Series A* **23,** 3

Fisher, R. A. (1961b), "Weighted Mean of Two Samples With Unknown Variance Ratio," *Sankhyá, Series A* **23,** 103

Fraser, D. A. S. (1968), *The Structure of Inference,* New York: Wiley

Fraser, D. A. S., and Haq, M. S. (1969), "Structural Probability and Prediction for the Multivariate Model," *J. Roy. Statist. Soc., Series B* **31,** 317

Gates, C. E., and Shine, C. J. (1962), "The Analysis of Variance of the *S*-Stage Hierarchical Classification," *Biometrics* **18,** 529

Gayen, A. K. (1949), "The Distribution of 'Student's' *t* in Random Sample of any Size Drawn from Non-Normal Universe," *Biometrika* **36,** 353

Gayen, A. K. (1950), "The Distribution of the Variance Ratio in Random Samples of any Size Drawn from a Non-Normal Universe," *Biometrika* **37,** 236

Geary, R. C. (1936), "The Distribution of 'Student's' Ratio for Non-Normal Samples," *J. Roy. Statist. Soc., Supplement* **3,** 178

Geary, R. C. (1947), "Testing for Normality," *Biometrika* **34,** 209

Geisser, S. (1965a), "Bayesian Estimation in Multivariate Analysis," *Ann. Math. Statist.* **30,** 150

Geisser, S. (1965b), "A Bayes Approach for Combining Correlated Estimates," *J. Amer. Statist. Assoc.* **60,** 602

Geisser, S., and Cornfield, J. (1963), "Posterior Distributions for Multivariate Normal Parameters," *J. Roy. Statist. Soc., Series B* **25,** 368

Gossett, W. S. ["Student"] (1908), "The Probable Error of a Mean," *Biometrika* **6,** 1

Gower, J. C. (1962), "Variance Component Estimation for Unbalanced Hierarchical Classification," *Biometrics* **18,** 537

Graybill, F. A., and Weeks, D. L. (1959), "Combining Inter-Block and Intra-Block Information in Balanced Incomplete Blocks," *Ann. Math. Statist.* **30,** 799

Graybill, F. A., and Wortham, A. W. (1956), "A Note on Uniformly Best Unbiased Estimators for Variance Components," *J. Amer. Statist. Assoc.* **51,** 266

Grunfeld, Y. (1958), "The Determinants of Corporate Investment," Ph.D. thesis, University of Chicago

Guttman, I., and Meeter, D. A. (1965), "On Beale's Measure of Nonlinearity," *Technometrics* **7,** 623

Hartigan, J. A. (1964), "Invariant Prior Distributions," *Ann. Math. Statist.* **35,** 836

Hartigan, J. A. (1965), "The Asymptotic Unbiased Prior Distribution," *Ann. Math. Statist.* **36,** 1137

Hartley, H. O. (1940), "Testing the Homogeneity of a Set of Variances," *Biometrika* **31,** 249

Hartley, H. O. (1961), "The Modified Gauss–Newton Method for the Fitting of Nonlinear Regression Functions by Least Squares," *Technometrics* **3,** 269

Henderson, C. R. (1953), "Estimation of Variance and Covariance Components," *Biometrics* **9,** 226

Herbach, L. H. (1959), "Properties of Model II-type Analysis of Variance Tests," *Ann. Math. Statist.* **30,** 939

Hildreth, C. (1963), "Bayesian Statisticians and Remote Clients," *Econometrika* **32,** 422

Hill, B. M. (1965), "Inference About Variance Components in the One-Way Model," *J. Amer. Statist. Assoc.* **60,** 806

Hill, B. M. (1967), "Correlated Errors in the Random Model," *J. Amer. Statist. Assoc.* **62,** 1387

Hogg, R. V., and Craig, A. T. (1970), *Introduction to Mathematical Statistics,* second edition, New York: Macmillan

Huzurbazar, V. S. (1955), "Exact Forms of Some Invariants for Distributions Admitting Sufficient Statistics," *Biometrika* **43,** 533

Jackson, D. (1921), "Note on the Median of a Set of Numbers," *Bull. Amer. Math. Soc.* **27,** 160

James, W., and Stein, C. M. (1961), "Estimation with Quadratic Loss Function," *Proc. 4th Berkeley Symp. Math. Statist. Proc.* **1,** 361

Jaynes, E. T. (1968), "Prior Probabilities," *IEEE Trans. Systems Science and Cybernetics* **SSC–4,** 227

Jeffreys, H. (1961), *Theory of Probability*, third edition, Oxford: Clarendon Press

Jeffreys, H., and Swirlee, B. (1956), *Methods of Mathematical Physics*, Cambridge: Cambridge University Press

Johnson, R. A. (1967), "An Asymptotic Expansion for Posterior Distributions," *Ann. Math. Statist.* **38,** 1899

Johnson, R. A. (1970), "Asymptotic Expansions Associated with Posterior Distributions," *Ann. Math. Statist.* **41,** 851

Kahirsagar, A. M. (1960), "Some Extensions of the Multivariate *t*-Distribution and the Multivariate Generalization of the Distribution of the Regression Coefficients," *Proc. Camb. Phil. Soc.* **57,** 80

Kempthorne, O. (1952), *The Design and Analysis of Experiments,* New York: Wiley

Kendall, M. G., and Stuart, A. (1961), *The Advanced Theory of Statistics, Volume* 2, New York: Hafner

Klotz, J. H., Milton, R. C., and Zacks, S. (1969), "Mean Square Efficiency of Estimators of Variance Component," *J. Amer. Statist. Assoc.* **64,** 1383

Kolmogoroff, A. (1941), "Confidence Limits for an Unknown Distribution Function," *Ann. Math. Statist.* **12,** 461

Lehman, E. L. (1959), *Testing Statistical Hypotheses,* New York: Wiley

Lindley, D. V. (1965), *Introduction to Probability and Statistics from a Bayesian Viewpoint, Part* 2, *Inference*, Cambridge: Cambridge University Press

Lindley, D. V. (1971), "Bayesian Statistics, a Review," *Regional Conference Series in Applied Mathematics, S.I.A.M.*

Lund, D. R. (1967), "Parameter Estimation in a Class of Power Distributions," Ph.D. thesis, the University of Wisconsin, Madison

Marquardt, D. W. (1963), "An Algorithm For Least Squares Estimation of Nonlinear Parameters," *J. Soc. Ind. Appl. Math.* **11,** 431

Milne-Thomson, L. M. (1960), *The Calculus of Finite Differences,* New York: Macmillan

Mood, A. M., and Graybill, F. A. (1963), *Introduction to the Theory of Statistics*, New York: McGraw–Hill

Moriguti, S. (1954), "Confidence Limits for a Variance Component," *Rep. Stat. Appl. Res. Juse.* **3,** 29

Mosteller, F., and Wallace, D. L. (1964), *Inference and Disputed Authorship: The Federalist*, Reading, Mass.: Addison-Wesley

Nelder, J. A. (1954), "The Interpretation of Negative Component of Variance," *Biometrika* **41,** 544

Neyman, J., and Pearson, E. S. (1928), "On the Use and Interpretation of Certain Test Criteria for Purposes of Statistical Inference, Part I," *Biometrika* **20A,** 175

Novick, M. R. (1969), "Multiparameter Bayesian Indifference Procedures," *J. Roy. Statist. Soc. Series B* **31,** 29

Novick, M. R., and Hall, W. J. (1965), "A Bayesian Indifference Procedure," *J. Amer. Statist. Assoc.* **60,** 1104

Patil, V. H. (1964), "The Behrens–Fisher Problem and its Bayesian Solution," *J. Indian Statist. Assoc.* **2,** 21

Patnaik, P. B. (1949), "The Non-Central χ^2- and F-Distributions and Their Applications," *Biometrika* **36,** 202

Pearson, K. (1934), *Tables of the Incomplete Beta-Function*, Cambridge: Cambridge University Press

Perks, F. J. A. (1947), "Some Observations on Inverse Probability, Including a New Indifference Rule," *J. Inst. Actuaries* **73,** 285

Pillai, K. C. S., and Gupta, A. K. (1969), "On the Exact Distribution of Wilks' Criterion," *Biometrika* **56,** 109

Pitman, E. J. G. (1936), "Sufficient Statistics and Intrinsic Accuracy," *Proc. Camb. Phil. Soc.* **32,** 567

Portnoy, S. (1971), "Formal Bayes Estimation with Application to a Random Effect Model," *Ann. Math. Statist.* **42,** 1379

Pratt, J. W., Raiffa, H., and Schlaifer, R. (1965), *Introduction to Statistical Decision Theory.* New York: McGraw–Hill

Raiffa, H., and Schlaifer, R. (1961), *Applied Statistical Decision Theory*, Cambridge: Harvard University Press

Ramsey, F. P. (1931), *The Foundation of Mathematics and Other Logical Essays*, London: Routledge and Kegan Paul

Robbins, H. (1955), "An Empirical Bayes Approach to Statistics," *Proc. 3rd Berkeley Symp. Math. Statist. Proc.* 157

Robbins, H. (1964), "The Empirical Bayes Approach to Statistical Problems," *Ann. Math. Statist.* **35,** 1

Savage, L. J. (1954), *The Foundation of Statistics*, New York: Wiley

Savage, L. J. (1961a), "The Subject Basis of Statistical Practice," unpublished manuscript, the University of Michigan

Savage, L. J. (1961b), "The Foundation of Statistics Reconsidered," *Proc. 4th Berkeley Symp.* **1,** 575

Savage, L. J., *et al.* (1962), *The Foundation of Statistical Inference*, London: Methuen

Schatzoff, M. (1966), "Exact Distribution of Wilk's Likelihood Ratio Criterion," *Biometrika* **53,** 347

Scheffé, H. (1959), *The Analysis of Variance*, New York: Wiley

Schlaifer, R. (1959), *Probability and Statistics for Business Decision*, New York: McGraw-Hill

Searle, S. R. (1958), "Sampling Variances of Estimates of Components of Variance," *Ann. Math. Statist.* **29,** 167

Seshadri, V. (1966), "Comparison of Combined Estimators in Balanced Incomplete Blocks," *Ann. Math. Statist.* **37,** 1832

Siegel, C. L. (1935), "Ueber die Analytische Theorie der Quadratischen Formen," *Ann. Math.* **36,** 527

Stein, C. M. (1962), "Confidence Sets for the Mean of a Multivariate Normal Distribution," *J. Roy. Statist. Soc., Series B* **24,** 265

Stone, M. (1964), "Comments on a Posterior Distribution of Geisser and Cornfield," *J. Roy. Statist. Soc., Series B* **26,** 274

Stone, M., and Springer, B. G. F. (1965), "A Paradox Involving Quasi Prior Distributions," *Biometrika* **52,** 623

Sukhatmé, P. V. (1938), "On Fisher and Behrens' Test of Significance for the Difference in Means of Two Normal Samples," *Sankhyá* **4,** 39

Tan, W. Y. (1964), "Bayesian Analysis of Random Effect Models," Ph.D. thesis, the University of Wisconsin, Madison

Thompson, W. A., Jr. (1962), "The Problem of Negative Estimates of Variance Components," *Ann. Math. Statist.* **33,** 273

Thompson, W. A., Jr. (1963), "Non-Negative Estimates of Variance Components," *Technometrics* **5,** 441

Tiao, G. C., and Ali, M. M. (1971a), "Effect of Non-Normality on Inferences About Variance Components," *Technometrics* **13,** 635

Tiao, G. C., and Ali, M. M. (1971b), "Analysis of Correlated Random Effect: Linear Model with Two Random Components," *Biometrika* **58,** 37

Tiao, G. C., and Fienberg, S. (1969), "Bayesian Estimation of Latent Roots and Vectors with Special Reference to the Bivariate Normal Distribution," *Biometrika* **56,** 97

Tiao, G. C., and Guttman, I. (1965), "The Inverted Dirichlet Distribution with Applications," *J. Amer. Statist. Assoc.* **60,** 793

Tiao, G. C., and Lund, D. R. (1970), "The Use of OLUMV Estimators in Inference Robustness Studies of the Location Parameter of a Class of Symmetric Distributions," *J. Amer. Statist. Assoc.* **65,** 370

Tiao, G. C., and Tan, W. Y. (1966), "Bayesian Analysis of Random-Effect Models in the Analysis of Variance. II. Effect of Autocorrelated Errors," *Biometrika* **53,** 477

Tiao, G. C., Tan, W. Y., and Chang, Y. C. (1970), "A Bayesian Approach to Multivariate Regression Subject to Linear Constraints," paper presented to the *Second World Congress of the Econometric Society*, Cambridge, England

Tukey, J. W. (1956), "Variances of Variance Components: I. Balanced Designs," *Ann. Math. Statist.* **27,** 722

Tukey, J. W. (1961), "Discussion Emphasizing the Connection Between Analysis of Variance and Spectrum Analysis," *Technometrics* **3,** 191

Turner, M. C. (1960), "On Heuristic Estimation Methods," *Biometrics* **16,** 299

Wang, Y. Y. (1967), "A Comparison of Several Variance Component Estimators," *Biometrika* **54,** 301

Welch, B. L. (1938), "The Significance of the Difference Between Two Means When the Population Variances are Unequal," *Biometrika* **29,** 350

Welch, B. L. (1947), "The Generalization of 'Student's' Problem when Several Different Population Variances are Involved," *Biometrika* **34,** 28

Welch, B. L., and Peers, H. W. (1963), "On Formulae for Confidence Points Based on Integrals of Weighted Likelihood," *J. Roy. Statist. Soc. Series, B* **25,** 318

Wilks, S. S. (1962), *Mathematical Statistics*, New York: Wiley

Williams, J. S. (1962), "A Confidence Interval for Variance Components," *Biometrika* **49,** 278

Wishart, J. (1928), "The Generalized Product Moment Distribution in Samples from a Normal Multivariate Population," *Biometrika* **20A,** 32

Yates, F. (1939), "An Apparent Inconsistency Arising from Tests of Significance Based on Fiducial Distribution of Unknown Parameters," *Proc. Camb. Phil. Soc.* **35,** 579

Yates, F. (1940), "The Recovery of Inter-Block Information in Balanced Incomplete Block Designs," *Ann. Eugen.* **10,** 317

Zacks, S. (1967), "More Efficient Estimators of Variance Components," Technical Report No. 4, Department of Statistics, Kansas State University, Manhattan, Kansas

Zellner, A. (1962), "An Efficient Method of Estimating Seemingly Unrelated Regression and Tests for Aggregation Bias," *J. Amer. Statist. Assoc.* **57,** 348

Zellner, A. (1963), "Estimators for Seemingly Unrelated Regression Equations: Some Finite Sample Results," *J. Amer. Statist. Assoc.* **58,** 977

Zellner, A., and Tiao, G. C. (1964), "Bayesian Analysis of the Regression Model with Autocorrelated Errors," *J. Amer. Statist. Assoc.* **59,** 763

INDEXES

AUTHOR INDEX

SUBJECT INDEX

ABCDEFGH798765432